Coherent Flow Structures in Open Channels

Coherent Flow Structures in Open Channels

Edited by

PHILIP J. ASHWORTH
School of Geography, University of Leeds

SEAN J. BENNETT
Department of Earth Sciences, University of Leeds
Current address: USDA-ARS, National Sedimentation Laboratory

JAMES L. BEST
Department of Earth Sciences, University of Leeds

STUART J. McLELLAND
School of Geography and Department of Earth Sciences, University of Leeds

JOHN WILEY & SONS

Chichester • New York • Brisbane • Toronto • Singapore

National 01243 779777
International (+44) 1243 779777

Other Wiley Editorial Offices

John Wiley & Sons, Inc., 605 Third Avenue,
New York, NY 10158-0012, USA

Jacaranda Wiley Ltd, 33 Park Road, Milton,
Queensland 4064, Australia

John Wiley & Sons (Canada) Ltd, 22 Worcester Road,
Rexdale, Ontario M9W 1L1, Canada

John Wiley & Sons (Asia) Pte Ltd, 2 Clementi Loop #02-01,
Jin Xing Distripark, Singapore 0512

British Library Cataloguing in Publication Data

A catalogue record for this book is available from the British Library

ISBN 0-471-95723-2

Produced from camera-ready copy supplied by the editors
Printed and bound in Great Britain by Bookcraft (Bath) Ltd
This book is printed on acid-free paper responsibly manufactured from sustainable forestation, for which at least two trees are planted for each one used for paper production.

Contents

List of Contributors

P. S. Addison
Department of Civil and Transportation Engineering, Napier University, 10 Colinton Road, Merchiston Campus, Edinburgh, EH10 5DT, UK

Andrei Alabyan
Geographical Faculty, Moscow State University, GSP-3, Moscow, 119899, Russia

Philip J. Ashworth
School of Geography, University of Leeds, Leeds, LS2 9JT, UK

C. S. Babakaiff
Department of Geography, Simon Fraser University, Burnaby, British Columbia, V5A 1S6, Canada

Sean J. Bennett
Department of Earth Sciences, University of Leeds, Leeds, LS2 9JT, UK
Current address: USDA-ARS, National Sedimentation Lab., PO Box 1157, Oxford, MS, 38655, USA

James L. Best
Department of Earth Sciences, University of Leeds, Leeds, LS2 9JT, UK

Thomas Buffin-Bélanger
Département de géographie, Université de Montréal, C.P. 6128, Succursale Centre-Ville, Montréal, Québec, H3C 3J7, Canada

Nicholas J. Clifford
Department of Geography and Jackson Environment Institute, University College London, 26 Bedford Way, London, WC1H 0AP, UK

W. Brian Dade
Institute of Theoretical Geophysics, Department of Applied Mathematics & Theoretical Physics, Silver Street, Cambridge, CB3 9EW, UK

Carmen de Jong
Institut für Geographische, Wissenschaften Freie Universität Berlin, Grunewaldstraße 35, Berlin 12165, Germany

Andrea Defina
Institute of Hydraulics "Giovanni Poleni", Faculty of Engineering, University of Padova, via Loredan, 20-I 35131 Padova, Italy

Sylvain Deland
Département de géographie, Université de Montréal, C.P. 6128, Succursale Centre-Ville, Montréal, Québec, H3C 3J7, Canada

Andreas Dittrich
Institute of Hydraulic Structures and Agricultural Engineering, University of Karlsruhe, Kaiserstraße 12, D-76128 Karlsruhe, Germany

Peter Ergenzinger
Institut für Geographische, Wissenschaften Freie Universität Berlin, Grunewaldstraße 35, Berlin 12165, Germany

R. I. Ferguson
Department of Geography, University of Sheffield, Sheffield, S10 2TN, UK

M. W. Gallagher
Department of Engineering, University of Aberdeen, Fraser Noble Building, Aberdeen, AB9 2UE, UK

Marcelo García
Department of Civil Engineering, University of Illinois at Urbana-Champaign, 250 N. Mathews, Urbana, IL 61801, USA

A. J. Grass
Department of Civil and Environmental Engineering, University College London, Gower Street, London, WC1E 6BT, UK

Albert Gyr
Institute of Hydromechanics and Water Resources Management, Federal Institute of Technology, CH-8093 Zürich, Switzerland

Peter J. Hardcastle
Proudman Oceanographic Laboratory, Bidston Observatory, Birkenhead, Merseyside, L43 7RA, UK

Jack Hardisty
School of Geography and Earth Resources, University of Hull, Hull, HU6 7RX, UK

E. J. Hickin
Department of Geography, Simon Fraser University, Burnaby, British Columbia, V5A 1S6, Canada

Andrew J. Hogg
Institute of Theoretical Geophysics, Department of Applied Mathematics & Theoretical Physics, Silver Street, Cambridge, CB3 9EW, UK

Nick E. Hughes
School of Geography and Earth Resources, University of Hull, Hull, HU6 7RX, UK

Herbert E. Huppert
Institute of Theoretical Geophysics, Department of Applied Mathematics & Theoretical Physics, Silver Street, Cambridge, CB3 9EW, UK

D. P. Keogh
Department of Civil and Transportation Engineering, Napier University, 10 Colinton Road, Merchiston Campus, Edinburgh, EH10 5DT, UK

A. D. Kirkbride
Department of Geography, University of Lancaster, Lancaster, LA1 4YB, UK

Tadanori Kitamura
School of Global Environmental Engineering, Kyoto University, Sakyo-ku, Kyoto, 606, Japan

Ray Kostaschuk
Department of Geography, University of Guelph, Guelph, Ontario, N1G 2W1, Canada

Fabián López
Department of Civil Engineering, University of Illinois at Urbana-Champaign, 250 N. Mathews, Urbana, IL 61801, USA

Saleem Mahmood
FAP-24, House 96, Road 23, Banani, Dhaka, Bangladesh

M. Mansour-Tehrani
Department of Civil and Environmental Engineering, University College London, Gower Street, London, WC1E 6BT, UK

Stephen R. McLean
Mechanical and Environmental Engineering Department, University of California, Santa Barbara, CA 93106, USA

Stuart J. McLelland
School of Geography and Department of Earth Sciences, University of Leeds, Leeds, LS2 9JT, UK

Andreas Müller
Institute of Hydromechanics and Water Resources Management, Federal Institute of Technology, CH-8093 Zürich, Switzerland

C. Naish
Department of Civil Engineering, University of Bristol, Queen's Building, University Walk, Bristol, BS8 1TR, UK

Jonathon M. Nelson
National Research Program, Water Resources Division, U.S. Geological Survey, Box 25046, MS 413, Lakewood, CO 80225, USA

Franz Nestmann
Institute of Hydraulic Structures and Agricultural Engineering, University of Karlsruhe, Kaiserstraße 12, D-76128 Karlsruhe, Germany

Yarko Niño
Department of Civil Engineering, University of Illinois at Urbana-Champaign, 250 N. Mathews, Urbana, IL 61801, USA

Gary Parker
St. Anthony Falls Laboratory, University of Minnesota, Mississippi River at 3rd Avenue, Minneapolis, MN 55414, USA

Chris Paola
Department of Geology & Geophysics and St. Anthony Falls Laboratory, University of Minnesota, Minneapolis, MN 55455, USA

A. Pollard
Department of Mechanical Engineering, Queen's University, Kingston, Ontario, K7L 3N6, Canada

P. Rameshwaran
Department of Engineering, King's College, University of Aberdeen, Aberdeen, AB9 2UE, UK

Bruce L. Rhoads
Department of Geography, University of Illinois at Urbana-Champaign, Urbana, IL 61801, USA

W. R. Roy Richardson
Department of Geography, The University of Nottingham, University Park, Nottingham, NG7 2RD, UK

Helen L. Rouse
School of Geography and Earth Resources, University of Hull, Hull, HU6 7RX, UK

André G. Roy
Département de géographie, Université de Montréal, C.P. 6128, Succursale Centre-Ville, Montréal, Québec, H3C 3J7, Canada

R. H. J. Sellin
Department of Civil Engineering, University of Bristol, Queen's Building, University Walk, Bristol, BS8 1TR, UK

J. Shaw
Department of Geography, University of Alberta, Edmonton, T6G 2H4, Canada

Ronald L. Shreve
Department of Earth and Space Sciences and Institute of Geophysics and Planetary Physics, University of California, Los Angeles, CA 90024, USA

A. Sidorchuk
Laboratory of Soil Erosion and Fluvial Processes, Geographical Faculty, Moscow State University, GSP-3, Moscow 119899, Russia

C. R. Smith
Department of Mechanical Engineering and Mechanics, 19 Memorial Drive West, Lehigh University, Bethlehem, Pennsylvania 18015, USA

Richard L. Soulsby
HR Wallingford Ltd., Howbery Park, Wallingford, Oxfordshire, OX10 8BA, UK

S. J. Tait
Department of Civil and Structural Engineering, University of Sheffield, Sir Frederick Mappin Building, Mappin Street, Sheffield, S1 4DU, UK

Colin R. Thorne
Department of Geography, The University of Nottingham, University Park, Nottingham, NG7 2RD, UK

Peter D. Thorne
Proudman Oceanographic Laboratory, Bidston Observatory, Birkenhead, Merseyside, L43 7RA, UK

Tetsuro Tsujimoto
Department of Civil Engineering, Kanazawa University, 2-40-20, Kodatsuno, Kanazawa, 920, Japan

Paul Villard
Department of Geography, University of Guelph, Guelph, Ontario, N1G 2W1, Canada

N. WaKarani
Department of Mechanical Engineering, Queen's University, Kingston, Ontario, K7L 3N6, Canada

B. B. Willetts
Department of Engineering, University of Aberdeen, Fraser Noble Building, Aberdeen, AB9 2UE, UK

Peter R. Wormleaton
Department of Civil Engineering, Queen Mary and Westfield College, University of London, Mile End Road, London, E1 4NS, UK

Acknowledgements

This book arose from a three-day international conference held at the University of Leeds, UK 10-12 April 1995, which attracted a group of truly interdisciplinary participants from around the world. The contents represent both summaries of work on coherent flow structures over a range of scales and some of the very latest research within a range of disciplines. We hope the conference and this book will continue to foster the ongoing links between various branches of science on the nature of coherent flow structures, their role in shaping the form of open channels and the use of this understanding in applied earth sciences and engineering. We would like to thank all the participants and speakers at the meeting for contributing to such a stimulating and productive event. We are especially grateful to the keynote speakers (Gerrit Klaassen, Steve McLean, Gary Parker and Chuck Smith) for devoting their time and energy to giving presentations that acted as focal points for the different sessions.

The meeting and this book would not have been possible without financial assistance from the University of Leeds, the School of Geography Research Fund, the British Sedimentological Research Group and DANTEC plc., for which we are very grateful. A number of colleagues at Leeds provided their time and skills to ensure the meeting went smoothly: Tim Hadwin, Neil Woodhouse, Dave Staveneau and several other colleagues are thanked for their help with the conference logistics; Alison Manson, Alistair French and Lois Wright of the School of Geography Graphics Unit produced much of the conference literature and design work. We are especially grateful to Alison for her many hours of advice, discussion and patience! We would also like to thank the many reviewers of papers contained in these pages for their assistance under often tight deadlines. The River Surevy Project, FAP-24, Bangladesh, kindly provided access to the cover SPOT Image. Finally, three of the editors and the contributors to this volume are indebted to Stuart McLelland who has masterminded the final production of the proof copy, spending many long days and nights ensuring this book is both professionally produced and published close to one year after the conference – a schedule better than many journals. Abi Hudlass and Claire Walker at John Wiley and Sons provided invaluable technical support and encouragement throughout the production of this book for which we are extremely grateful.

Phil Ashworth
Sean Bennett
Jim Best
Stuart McLelland

Leeds, February 1996

1 Coherent Flow Structures in Smooth-wall Turbulent Boundary Layers: Facts, Mechanisms and Speculation

C.R. SMITH
Department of Mechanical Engineering and Mechanics, 19 Memorial Drive West, Lehigh University, Bethlehem, Pennsylvania 18015, USA

ABSTRACT

Prior to the late 1950s, turbulent boundary layers were studied as if they were stochastic entities, consisting of rather randomized property fluctuations riding on an otherwise mean flow. We now know that vortex deformation and evolution play a dominant role in creating and sustaining 'turbulence', and that turbulent boundary layers display a deterministic 'structure' of irregular, but repetitive, spatio-temporal flow patterns such as low-speed wall 'streaks', wall region 'bursts', 'ejections', and 'sweeps'. These flow elements comprise some of what is perceived as the 'coherent structure' of a turbulent boundary layer, whereby 'coherent' implies a repetitive and correlatable 'structure' to the observed and detected flow behaviour. Over the years, a number of excellent experimental and computational studies have provided detailed correlation, assessment, and categorization of the myriad three-dimensional patterns, or flow structure of turbulent boundary layers.

However, the ultimate goal of science is to move from descriptive/empirical results to a synthesis of cause-effect relationships implied by the results. The present paper presents such a synthesis by piecing together the intricacies of the coherent structure of turbulent boundary layers, attempting to extract a model of the physical processes and the complex behaviour that give rise to the rich symbiotic mix of vortices and flow interactions which we call turbulence. The particular importance of vortices, and their role in generating and maintaining turbulence will be examined, including vortex:

- *three-dimensional deformation*
- *migration*
- *intensification*
- *amalgamation and growth in scale*
- *surface interactions, including localized, pressure-induced surface-fluid eruptions, and spanwise and streamwise vortex regeneration*

Conceptual models hypothesizing how the various coherent structures interact and contribute to sustaining turbulent boundary layers are offered, as well as illustrations of how this conceptual model is modified and applies to flows with large roughness and/or movable sediment beds.

Coherent Flow Structures in Open Channels.
Edited by P.J. Ashworth, S.J. Bennett, J.L. Best and S.J. McLelland.

INTRODUCTION

Starting in 1956 and through the early 1960s, a series of experiments carried out at Stanford University definitively demonstrated that there is far greater structure or order in turbulent boundary-layer flows than had been previously believed. Using hydrogen bubble and dye visualization near the wall, Kline *et al.* (1967) observed that hydrogen bubbles would collect into long, spatially-separated streaks which are essentially aligned in the streamwise direction. They observed that the visualization medium migrated laterally from regions of relatively high-speed flow into flanking low-speed regions, thus concentrating into what appeared as low-speed 'streaks', as shown in Figure 1.1. Subsequent measurements of the instantaneous velocity field and streamwise observations of these spanwise-discrete streaks (e.g. Smith & Schwartz, 1983; Wallace & Balint, 1990) strongly suggested that streaks project in concentrated plumes well above the wall, as shown in Figure 1.2; these patterns suggest that the low-speed fluid concentrations are caused by the convergence of the flow in the cross-flow plane, accompanied by an outflow away from the wall. In contrast, the middle of the high-speed regions are characterized by flow toward the wall and laterally outward. The instantaneous streamwise velocity within a streak region is retarded (relative to the mean) and, hence the terminology 'low-speed streaks' is commonly applied. In contrast, in the zones between the streaks, the streamwise velocity is in excess of the mean velocity and are therefore termed 'high-speed' regions or streaks.

Wall-layer streaks may be observed readily in any turbulent flow adjacent to a wall using a variety of other visualization media, such as dye in water or smoke in

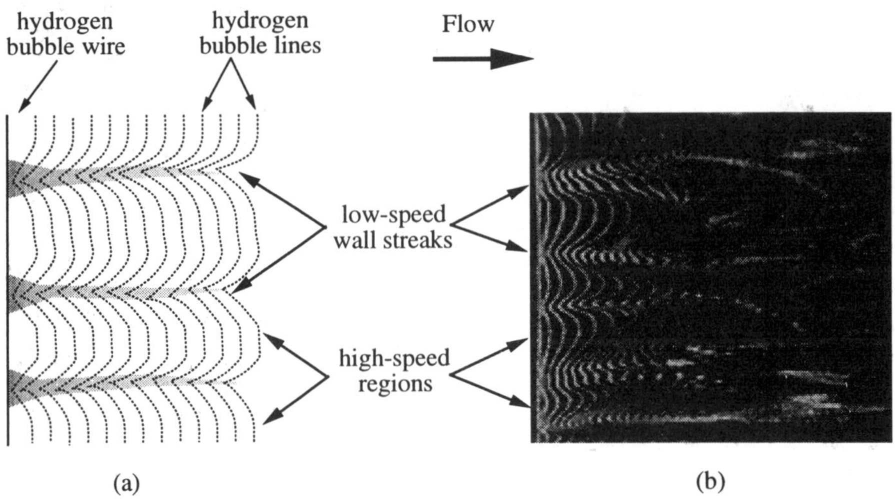

Figure 1.1. (a) Plan-view schematic of low-speed streak flow structure that occurs in the turbulent wall layer. (b) Actual hydrogen bubble photograph, $Re_\theta = 4300$

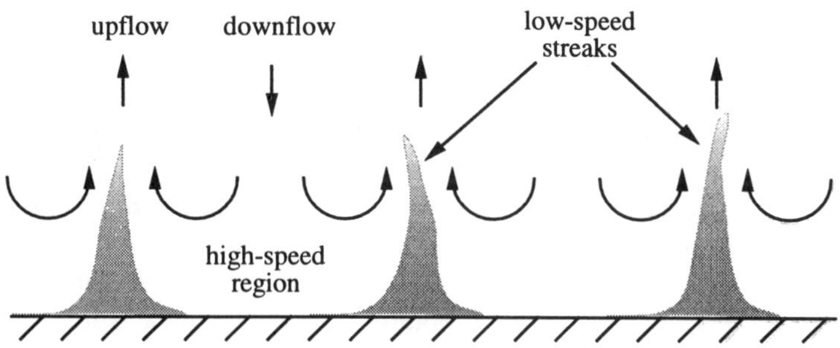

Figure 1.2. End-view schematic of instantaneous velocity field (looking upstream) in the cross-flow plane

air, and are now known to be a persistent and dominant feature of the near-wall flow for both smooth and rough boundaries (Grass, 1991; Defina, this volume; Grass & Mansour-Tehrani, this volume), and even within large atmospheric boundary layers (Klewicki *et al.*, 1995). Although streaks are unsteady phenomena and not uniformly distributed, it is possible to define a mean spanwise streak spacing λ. If streak spacing is non-dimensionalized on the turbulent wall-layer length scale, defined by ν/u_τ, the non-dimensional streak spacing $\lambda^+ = u_t\lambda/\nu$ ($u_\tau = (\tau_w/\rho)^{0.5}$, where ν is the kinematic viscosity of the fluid and τ_w is the wall shear stress; thus u_τ represents an equivalent 'shear' velocity) is found to have a mean value of 100 for smooth-wall flows over a wide range of Reynolds numbers (Kline *et al.*, 1967; Smith & Metzler, 1983; Klewicki *et al.*, 1995), with individual λ^+ values distributed between 50 and 300. Consequently, the mean streak spacing is comparable to the vertical extent of the wall layer, at least in the mean. However, even away from the wall, streak-like patterns (i.e. laterally spaced lower-speed regions) can be detected which appear to scale proportionally to distance from the wall (Smith & Metzler, 1983), although these regions appear more chaotic and less organized than the wall-layer streaks. The streamwise extent of the wall-layer streaks is typically of the order of 600 to 1000 in units of ν/u_τ.

The cause of the wall-layer streaks has been a matter of some debate, but the above observations led some authors (e.g. Blackwelder & Eckelmann, 1979; Blackwelder, 1983) to suggest that the streaks are due to the presence of relatively long counter-rotating pairs of vortices aligned in the streamwise direction. It was subsequently demonstrated (Robinson, 1991a) that although streamwise vortices are quite prevalent and associated with the development of low-speed streaks, they are quite short in downstream extent ($\Delta x^+ \approx 100 - 200$) and of unclear origin.

The discovery of this 'streaky' structure as a dominant wall-layer feature provoked considerable interest since this suggested that there are repeatable, deterministic structures in the near-wall flow. The terminology 'coherent structure' was subsequently coined to imply that the near-wall region contains certain basic flow modules or structures that give rise to the apparently ordered development observed in the wall layer, which was previously believed to be essentially random. However, the low-speed streak structure does not persist throughout the boundary layer; if one moves a hydrogen bubble wire progressively farther from the wall, as pointed out above, the streaky structure will become progressively more broadly spaced, and less organized; eventually (say at $y^+ > 50 - 60$) the local patterns appear to reflect a complex tangle of vortices with the scales of the vortices increasing with distance from the wall. Figure 1.3a is a schematic representation of the type of observed amalgam of advected vortex structures when a hydrogen bubble wire is oriented normal to the wall; Figure 1.3b is a typical vertical hydrogen bubble visualization reflecting the presence of both transverse and streamwise vortices in a low Reynolds number boundary layer as they are advected from upstream past the wire.

The sharp transition from low-speed streaks to various-scale vortices suggests that the time-dependent boundary-layer flow is roughly comprised of: (i) an outer region dominated by vortex motions of varying strengths, sizes and orientations; and (ii) an inner wall-layer exhibiting a persistent streaky structure. For most of the time the boundary layer maintains this double structure, with the wall-layer flow being driven primarily by the vortical structures which are advecting past in the outer region just above the wall layer. Kline *et al.* (1967) have referred to such time intervals as 'quiescent periods' which implies that the wall layer is locally intact with observable streaks present. There are, however, brief intervals when the two regions of the boundary layer interact strongly. Such strong interactions constitute a dominant and particularly important characteristic of turbulent wall-layer behaviour.

During these strong interactions, the quiescent period terminates in highly-localized breakdowns of the well-ordered, wall-layer flow. Such events were originally referred to (Kline *et al.*, 1967) as 'bursting' and are characterized by an abrupt eruption of a low-speed streak, culminating with an ejection of near-wall fluid away from the wall, often well into the outer region of the boundary layer. Viewed from the side, the latter stages of this event appear as illustrated in Figure 1.4, where a relatively narrow plume of fluid erupts from the wall and subsequently rolls up into a new vortex structure. Bursts always occur adjacent to a wall-layer streak with the streak first observed to lift from the wall, then rapidly breakdown and eject into the outer flow. In some of the original observations of bursting, Kim *et al.* (1971) reported that oscillations develop in the streamwise and spanwise direction, increasing in amplitude and scale until a breakdown occurs in the form of a wall-layer eruption. Originally, Kline attributed the bursting to some type of instability in the wall-layer flow. Following the ejection of wall-layer fluid, an event known as a 'sweep' occurs immediately (Corino & Brodkey, 1969) in which high-speed outer-layer fluid appears to move toward and penetrate the wall layer, sweeping away the chaotic motion associated with the remnants of the 'burst'. Upon the completion of a burst, the wall-layer streaks re-establish in the near-wall flow through development of new streaks and redevelopment of some of the

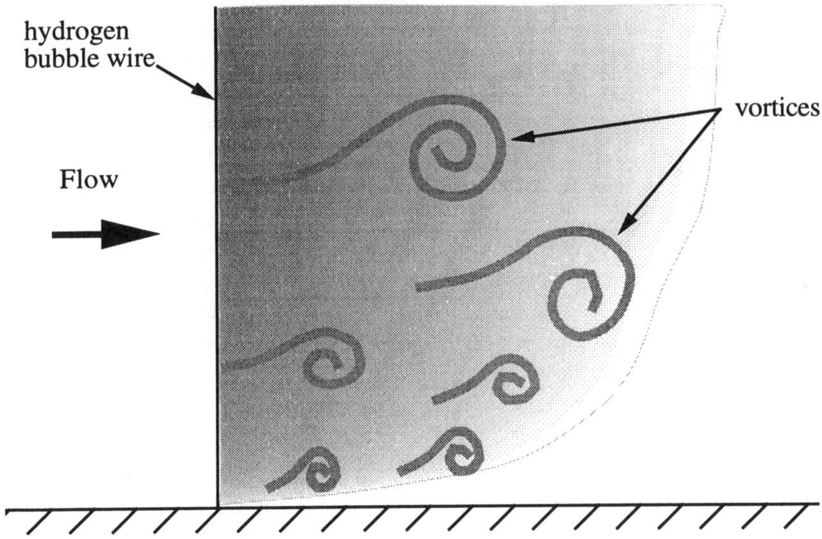

(a)

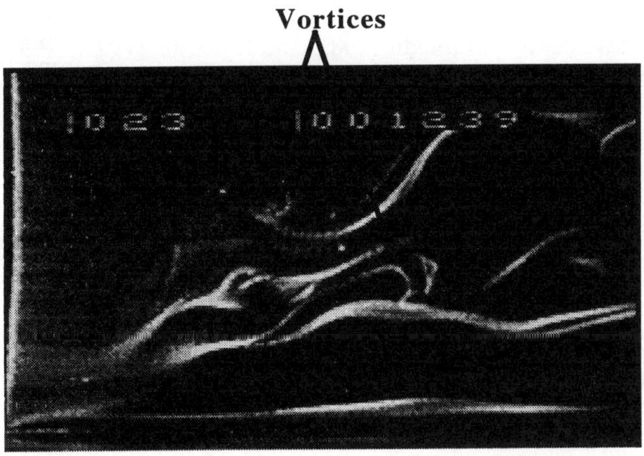

(b)

Figure 1.3. Side-view of turbulent boundary layer. (a) Schematic of idealized varying-scale vortices. (b) Actual hydrogen bubble wire visualization; wire oriented normal to the surface

existing streaks (by a process of reinforcement, Smith, 1984). Thus, the wall layer undergoes a cyclic (but not periodic) process of local, slow streak development, followed by a rapid, strong streak interaction with the external portion of the

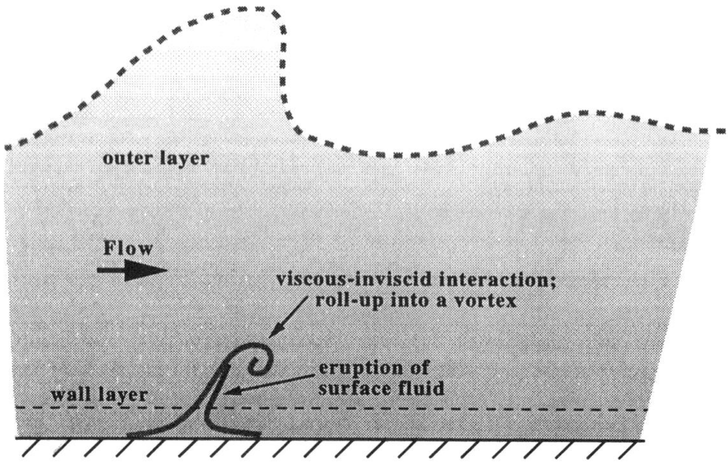

Figure 1.4. Schematic of turbulent bursting process in side view (not to scale)

boundary layer. However, over a large number of cycles, the *spatial distribution* of the streaks and bursting is found to behave randomly. Further details of the observed kinematics of bursting are discussed by Offen & Kline (1975), Blackwelder & Haritonidis (1983) and Robinson (1991a, 1991b).

The causes of both wall-layer streaks and bursting were unclear for many years, and are still subject to some debate. However, it is now generally accepted that both these features of the wall-layer flow are associated with vortex-surface interaction, as will be described subsequently in this paper. Note that bursting occurs over a brief period when the wall layer and the outer region of the boundary layer interact strongly in what may be classified as a strong, unsteady, viscous-inviscid interaction (Smith *et al.*, 1991). The duration of the burst relative to the quiescent period is quite short (Willmarth, 1975), however, the majority of the production of Reynolds stress occurs during these bursting events (Kim *et al.*, 1971; Willmarth, 1975; Cantwell, 1981). The bursting process also constitutes the fundamental process for production of new vorticity. This vorticity generation occurs at the wall due to localized pressure gradients imposed by passing vortices; during bursting, new vorticity is strongly concentrated at the wall followed by an abrupt ejection into the outer regions of the turbulent boundary layer. This regenerative process, in which the outer region is intermittently fed with concentrated new vorticity, is central to understanding the processes of Reynolds stress production near a wall and will be discussed later.

As pointed out by Townsend (1976), flow structures in a turbulent boundary layer can be roughly divided between 'active' and 'inactive' motions, where the active motions are ones which are active in extracting energy from the mean flow via stretching by velocity gradients (i.e. turbulence production) and the inactive motions are ones where dissipative processes dominate. In general, active motions predominate near the wall, with inactive motions dominating as the edge of the boundary layer is approached. Additionally, as shown for example by Falco (1977) and Smith *et al.* (1991), large outer region structures appear to span the majority of the turbulent boundary layer and appear to dominate the flow when viewed on a time-average basis (e.g. Kovasznay *et al.*, 1970). However, while significant energy is contained in these large mean motions, they are basically inactive, dissipative and extract little energy from the mean flow. It appears that these larger-scale structures are created by the coalescence and agglomeration of numerous smaller-scale, active vortical structures (Head & Bandyopadhyay, 1981; Smith *et al.*, 1991, Nagib *et al.*, 1993); this systematic growth in scales has been conceptually idealized by Perry & Chong (1982) as a 'hierarchy' of hairpin vortices. Thus, the most important, coherent, and active structures appear to be the small-scale structure which develops near the wall; it is a discussion of these smaller-scale structures and their process of evolution which will be the major focus of the present paper.

Note that the fluid dynamics and flow structure in the near-wall region are principally associated with the interaction of vortices with the wall; these vortices are three-dimensional in shape and generally referred to as hairpin vortices or horseshoe vortices, depending on the scale and shape of the vortex. Most of the phenomena observed in the near-wall flow can be interpreted in terms of the motion of such vortices, their mutual interactions and their influence on the viscous flow near the wall. In recent years, detailed research has elucidated much of the fundamental physics associated with hairpin-type vortices and their relationship to other observed flow patterns, such as low-speed streaks. Thus, the remainder of this paper will focus on the characteristics of hairpin-type vortices and their processes of: (i) generation; (ii) surface and mutual interaction; (iii) regeneration, evolution and amalgamation; and (iv) sustaining turbulent boundary layers.

VORTICES: THE MOVERS AND SHAKERS OF TURBULENCE

Because of the complexities and three-dimensionality of the near-wall flow, progress in understanding the basic dynamic processes has been quite protracted. Following an early period in which turbulence was considered a process of random, stochastic fluctuations, there was a realization that vortices must play a key role in developing and sustaining turbulence (e.g. Theodorsen, 1952; Hama, 1962; Black, 1968). Theodorsen (1952), at a time well in advance of the first published observations of coherent structure, proposed that vortices of the shape shown in Figure 1.5 are an important flow structure in wall turbulence. This type of vortex is referred to as a *hairpin vortex* since it is shaped like a metal hairpin, with the closed end downstream and the open end trailing upstream. The symmetric structure shown in Figure 1.5a is of the type originally proposed by Theodorsen (1952), and is the simplest conceptual model which can explain the salient features of the flow

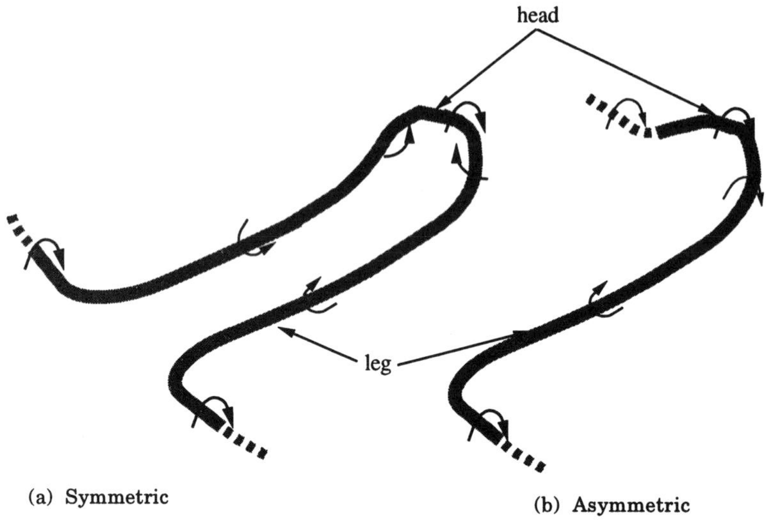

Figure 1.5. Schematic diagram of typical hairpin vortex configurations. Sense of vorticity indicated by arrows

behaviour near the wall. However, turbulent flow is a complex, unsteady environment of competing vortices and disturbances embedded in a strongly-sheared background flow; as a consequence, most vortices in a turbulent flow are expected to exhibit various degrees of asymmetry as reflected by Figure 1.5b. Recent evaluations (Kline & Robinson, 1989; Robinson, 1991a, 1991b) of the results of direct numerical simulations (DNS) for both a low-Reynolds number boundary-layer and a channel flow illustrate the dominance of these asymmetric hairpin-like vortices. Robinson (1991a, 1991b), from his evaluation of DNS results, identified arch (or horseshoe) vortices and 'quasi-streamwise' vortices (implying that the majority of the vorticity associated with the vortex is aligned in the streamwise direction), shown in Figure 1.6, as the two most commonly detected flow structures. In Figure 1.7, Robinson (1991a) indicates the relative distribution of these structures within the regions of the boundary layer. Note that the accompanying mean velocity profile in Figure 1.7 is plotted on a logarithmic scale, with the regions labelled according to the typical characterization of the velocity scaling that applies in the respective regions (see Hinze, 1978). It should be emphasized that the logarithmic scale is potentially misleading since the outer region actually comprises the vast majority of the boundary layer.

If one compares Figures 1.5b and 1.6, it is apparent that the asymmetric hairpin vortex shown in Figure 1.5 is essentially a composite of the two types of vortex structures shown in Figure 1.6. To understand how an essentially contiguous vortex can be detected as two separate structures, one must understand how vortex structures in general, and those sketched in Figures 1.6 and 1.7 in particular, are

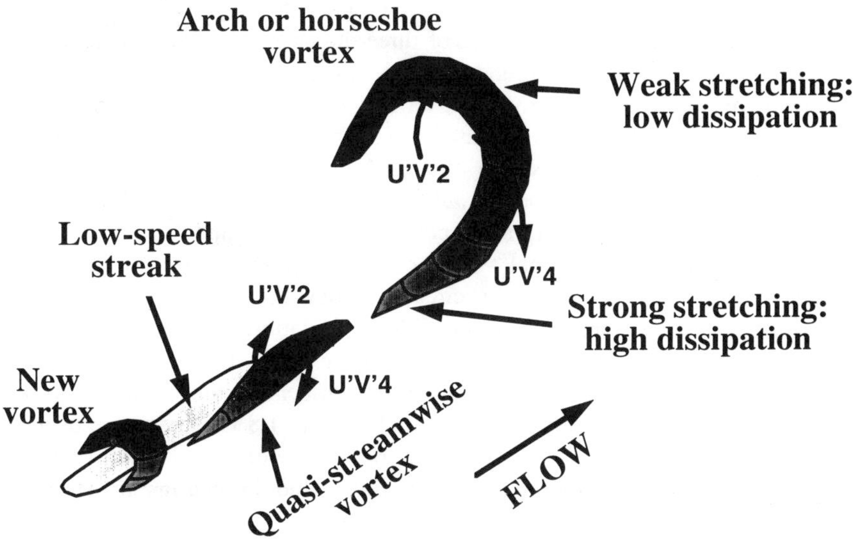

Figure 1.6. Schematic of vortex topologies and vortex regeneration in a low Reynolds number boundary layer (after Robinson, 1991b)

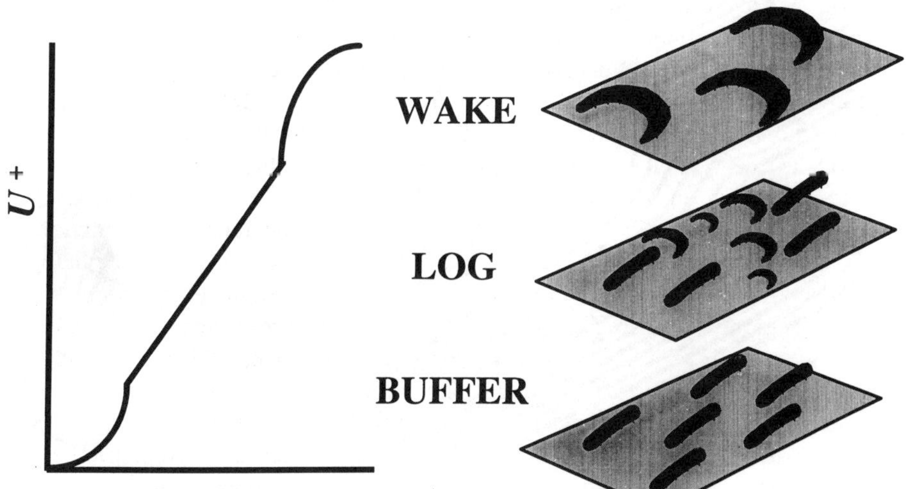

Figure 1.7. Schematic of vortical-structure populations in different regions of a low Reynolds number turbulent boundary layer (after Robinson, 1991b)

determined from DNS results. A direct numerical simulation of boundary-layer turbulence produces a vast data set in three-dimensional space and time (Moin & Kim, 1985, 1986; Spalart, 1988). The flow-field in this type of simulation at any instant in time can consist of a tangle of three-dimensional vorticity embedded in a region of high shear or background vorticity, especially near the wall. Extracting individual vortex structures from such data sets is by no means trivial. Originally, Kline & Robinson (1989) even raised questions about how a vortex should be defined in an environment where the background flow is already rotational. Historically, a vortex is considered as a region in the flow where a number of elementary vortex lines concentrate to comprise a vortex tube. The embodiment of this vortex tube is a rotational vortex core, which gives rise to instantaneous particle paths which are roughly circular about the vortex core. By tracing instantaneous vortex lines within both DNS simulations and recent experimental measurements (Moin & Kim, 1985; Grass *et al.*, 1991), numerous three-dimensional vortices are observed, with the associated vortex lines displaying patterns similar to those shown in Figure 1.5. Figure 1.8 is an example of the type of vortex line pattern that develops for an apparent symmetric hairpin-type vortex; the vortex is revealed by tracing a series of vortex lines originating from $y^+ \approx 7$ in

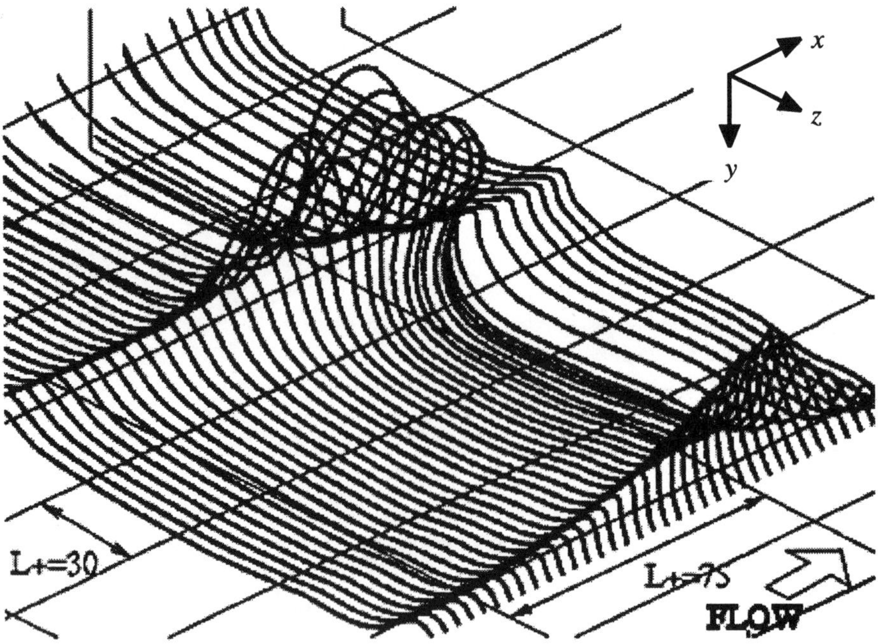

Figure 1.8. Vortex lines illustrating the formation of a symmetric hairpin-type vortex in the near-wall region of a smooth-wall channel flow (courtesy of C.H. Crawford & G.E. Karniadakis)

the near-wall region of a smooth-wall channel flow (Crawford & Karniadakis, pers. comm.). In general, at their extremities the concentrated vortex lines tend to diverge from one another or 'unroll' into the vorticity of the background flow so that it is difficult to precisely define the beginning and the end of the vortex; thus, the broken line terminating the vortex cores in Figure 1.5 is meant to indicate this dispersion of the vortex lines into the background vorticity, as illustrated clearly in Figure 1.8.

In contrast to the tracking of vortex lines, Robinson (1991b) identified a vortex structure as an advected low-pressure region, for which the instantaneous streamlines form a circular or spiral pattern around a vortex core (the rotation about the vortex core results in a reduced interior pressure within the vortex). Vortex cores in Robinson's (1991b) work were identified as three-dimensional, constant-pressure surfaces which were arbitrarily selected to optimally reveal the combined extent and strength of the rotational region; the sketches shown in Figures 1.6 and 1.7, reproduced from Robinson (1991b), were established from this approach. In contrast, the vortex configurations shown in Figure 1.5 are based on a model where the vortex lines in the fluid congregate to form a core, similar to that shown in Figure 1.8. Since the environment within which the vortex evolves will cause it to stretch and relax differentially along its extent, it is clear that employing closed contours of pressure for vortex detection will reveal differing regions of a vortex, depending on the level of pressure chosen for the isopressure contour. Thus, it seems apparent that the structures shown in Figures 1.5 and 1.6 must be closely related, if not identical, to one another. Indeed, Robinson noted that under the right conditions an arch vortex would be clearly coupled to a quasi-streamwise leg, reflecting essentially the structure shown in Figure 1.5b.

Returning to the configurations shown in Figure 1.5, the portion of the vortex farthest from the wall forms an arch which will subsequently be referred to as the *vortex head.* The portions below the head are usually inclined at a shallow angle (15° to 30°) to the wall and will be referred to as the vortex legs. The trailing part of a leg normally makes the closest approach to the wall. Comparing the arch and quasi-streamwise structures shown in Figure 1.6, it is evident that both may be considered to reflect at least part of an axisymmetric hairpin, as shown in Figure 1.5b, depending on the extent of the vortex head and/or leg associated with the total vortex structure. This point will be discussed in some detail subsequently with regard to the process of how hairpin legs and heads form.

The following sections discuss how the necessary dynamics for perpetuation of a near-wall turbulent boundary layer are provided by interaction of hairpin vortices with: (i) their environment; (ii) each other; and (iii) the viscous flow near the wall. These sections describe a composite physical picture based upon a large number of experimental, analytical and numerical investigations at various institutions of fully turbulent boundary-layer flows, as well as a number of 'kernel' studies (Smith *et al.*, 1991; Smith, 1993). Here the term 'kernel' is used to denote the study of a fundamental flow structure carried out to establish the fluid mechanic processes of the structure which may be relevant to the processes of a turbulent boundary layer. One property of turbulent near-wall flows, which has long inhibited understanding, is that they are a kaleidoscope of many competing events occurring almost simultaneously. A 'kernel' study attempts to isolate a basic structure considered to be relevant to turbulence, such that it can be studied in isolation. A particularly relevant kernel study that bears upon understanding the processes in turbulent boundary layers is the examination of the behaviour of hairpin vortices (Acarlar & Smith, 1987a, 1987b; Smith *et al.*, 1991; Haidari & Smith, 1994); once the basic features and behaviour of hairpin-like vortices are understood, it becomes possible to interpret developments in the more complicated environment of a turbulent wall-bounded flow.

THE EVOLUTION OF HAIRPIN VORTICES IN A SHEAR FLOW

A turbulent boundary layer sustains its 'turbulent' nature by: (i) the *generation* of subsequent vortices; and (ii) the *interaction* of the collective mix of vortices in the flow with both the surrounding shear flow and with the adjacent solid boundary. As will be described later, the generation/regeneration of hairpin-like vortices occurs due to an interaction of an existing vortex with wall-layer fluid, precipitating a viscous-inviscid interaction: a process in which fluid/vorticity at the wall is abruptly focused and ejected out of the wall layer. However, the present section describes how interactions of existing vortices with the background shear flow can also lead to the evolution of new vortex structures. This second process may be thought of as an essentially inviscid type reorganization and local concentration of the *existing* vorticity field. Note that in a turbulent boundary layer, both the regeneration of new vorticity and the redistribution of existing vorticity are critically important. Any vortex which initially has its vorticity concentrated in a small core will weaken as the vorticity in the core diffuses radially outward. In the absence of interactions, vorticity will eventually redistribute until the vortex is no longer identifiable. Since a turbulent boundary layer contains a large number of vortices at various stages of evolution, it is necessary to understand how new vortex structures continually develop.

Three-dimensional vortex interactions such as those characteristic of turbulence depend on the process of vortex stretching. As indicated by Lighthill (1963), a viscous, rotational flow consists of many elementary vortex tubes, all of which must form closed loops in all real fluids. Kelvin's Theorem indicates that for a vortex tube in an incompressible flow, the product of the cross-sectional area times the local vorticity magnitude must remain a constant. Thus, when a vortex tube is compressed along its length, the cross-sectional area expands and the local vorticity diminishes. Conversely, the vorticity will intensify when the tube is stretched. The importance of these processes to turbulence was first suggested by Lighthill (1963), who argued that a process of vorticity stretching near the wall and vortex relaxation in the outer part of the boundary layer is the only rational means for maintaining the high gradient of mean vorticity in the wall-region of a turbulent boundary layer.

An examination of three-dimensional stretching/compression in a turbulent-type shear flow was examined by Hon & Walker (1991), who considered the evolution of a three-dimensional distortion of an otherwise two-dimensional, spanwise vortex located above a plane wall. Their results indicate that if the fluid above the wall is either stagnant or moves with uniform speed, any distortion in a line vortex will gradually spread along the length of the vortex, but will not amplify. However, if the line vortex is located in a shear flow, any distortion of the line vortex (no matter how small) will spread quite rapidly as the vortex advects in the shear flow (Smith *et al.*, 1991). Figure 1.9 shows a Biot-Savart simulation of the rapid deformation and expansion experienced by an inviscid two-dimensional line vortex with a small initial three-dimensional distortion when placed in a region of uniform shear. Here the vortex is advected from left to right, with the sense of the vorticity indicated by the arrows. As shown in Figure 1.9b, a vortex head quickly forms and rises from the wall, subsequently bending back in the shear flow; simultaneously, vortex legs develop and move down toward the wall. As the process continues, the vortex head moves farther from the wall as the legs continue to approach the wall continually accentuating the streamwise extent of the distortion. Note that this process of vortex distortion is essentially independent of the initial amplitude of the distortion shown in Figure 1.9a. The original disturbance also spreads in the spanwise direction, as shown in Figure 1.9c, producing similar hairpin-type vortex structures, which Smith *et al.* (1991) term subsidiary vortices. Note that the characteristic spacing of the spanwise legs, indicated by λ in Figure 1.9c, is mainly dependent on the strength of the shear, with either higher background shear or a reduced vortex strength (relative to the shear) producing smaller λ spacings.

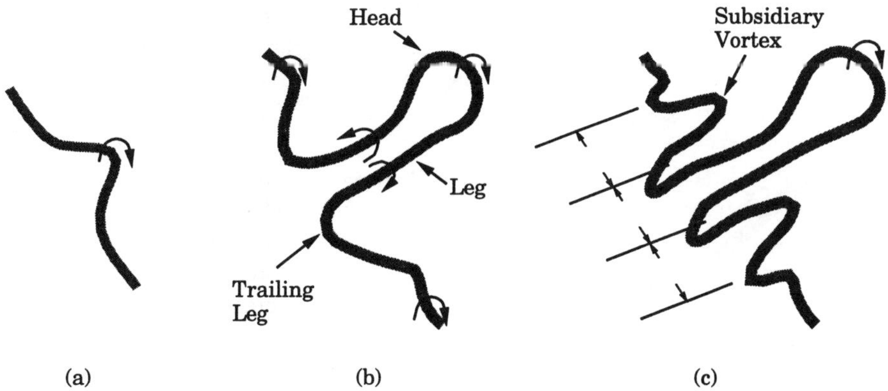

Figure 1.9. Evolution of a symmetrical hairpin vortex in a shear flow: (a) initial distribution; (b) development of vortex legs and head; and (c) evolution of subsidiary vortices and penetration of the legs toward the surface

Because most distortions of the vorticity field of a turbulent boundary layer are expected to be asymmetric, the effect of uniform background shear on asymmetric disturbances was also examined by Smith *et al.* (1991); a schematic diagram summarizing the development is shown in Figure 1.10. Here the initial vortex configuration is comprised of a pair of displaced two-dimensional vortex lines which are joined together through a section of high curvature. In this case, a 'one-leg' dominant or asymmetric hairpin vortex evolves from the initial distortion, with initially one dominant leg developing and moving toward the wall (Figure 1.10b). As the process continues, subsidiary hairpins again form, as shown in Figure 1.10c, with the development of the subsidiary vortices evolving in a more symmetric fashion; again the spanwise spacing of the subsidiary hairpins λ is strongly dependent on the level of background shear.

Note that for a turbulent-type boundary layer shear profile, the shear is not uniform, but is high at the wall and gradually diminishes toward the outer edge of the wall layer. Smith *et al.* (1991) show that a hairpin-like vortex developing within this turbulent-type shear profile (Figure 1.11) will develop legs that squeeze sharply together as they penetrate downward through regions of increasing shear, while the head will expand broadly as it moves away from the wall. This expansion of the head is similar to the observations of Robinson (1991b) that vortex 'arches' (referred to here as heads) in simulated turbulence were seen to 'send' legs down toward the wall and that the 'arches' develop a greater spanwise extent with increasing distance from the wall (*cf.* Figure 1.7).

The simulated behaviour of an initially-distorted inviscid vortex in a shear flow can be summarized as follows. Vortices advected in a shear flow develop

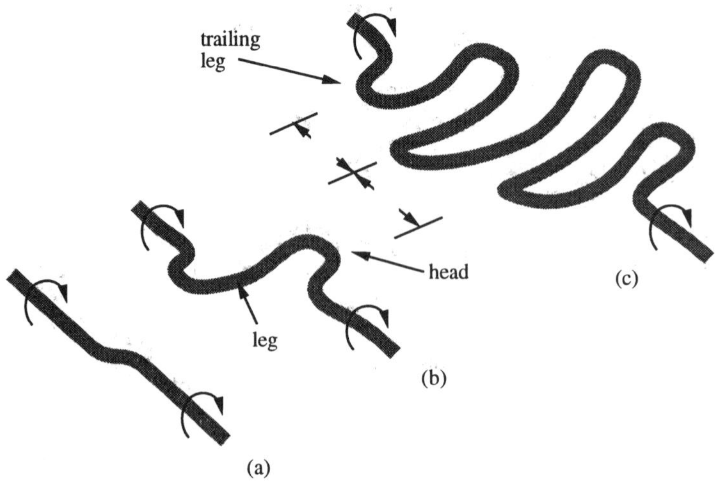

Figure 1.10. Evolution of asymmetric hairpin-like vortex structures in a uniform shear flow: (a) initial asymmetric distribution; (b) development of vortex legs and head; and (c) evolution of subsidiary vortices and penetration of the legs toward the surface

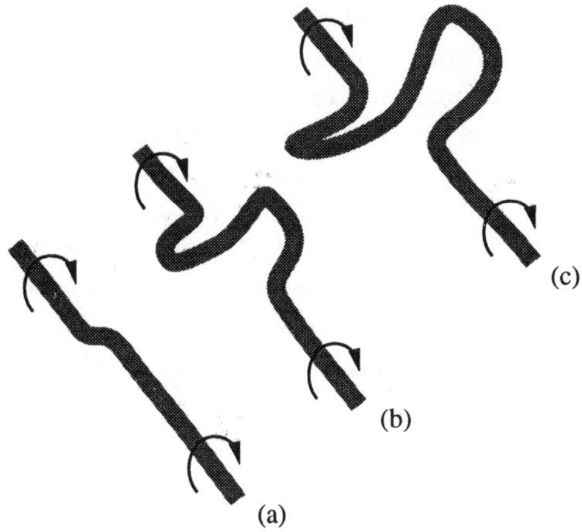

Figure 1.11. Evolution of asymmetric hairpin-like vortex structures in a turbulent shear flow: (a) initial distortion; (b) development of vortex legs and head; and (c) narrowing of leg spacing and expansion of head

characteristic hairpin-like vortex shapes from the portions of the vortex where the curvature of the vortex is relatively high, with most of these hairpin-like shapes developing as asymmetric or one-legged hairpins. The legs of the hairpin-like vortices always move downward toward the wall which has important consequences for vortex-surface interactions, as discussed below. As three-dimensional hairpin-like vortices develop, large portions of the vortex are transformed from spanwise to streamwise vorticity with the deformation of the vortex line progressively expanding both (i) over a larger streamwise extent, and (ii) in the spanwise direction via the evolution of subsidiary vortices. The relative spacing between subsidiary vortices decreases with increasing background shear and decreasing vortex strength.

The basic physics of hairpin-type vortices, as outlined above, is also provided by a number of kernel studies carried out both experimentally by Acarlar & Smith (1987a, 1987b), Haidari & Smith (1994) and computationally by Singer & Joslin (1994) on selectively-generated hairpin vortices in a laminar boundary layer. These studies confirm the non-linear (and primarily inviscid) evolution of three-dimensional vortices in a background shear flow. Of course, it should be appreciated that the evolution of a large number of subsidiary hairpin vortices from an initial distorted line vortex, as depicted in Figures 1.9c and 1.10c, will be neither ordered nor predictable in a turbulent boundary layer where a cluttered environment of competing vortices act to non-linearly influence the evolution of any single vortex. On the other hand, it is clear that a process of both streamwise and spanwise growth of three-dimensional vortices is supported and amplified by their presence in a sheared environment.

With an understanding of the behaviour of individual hairpin vortices, a physical model of some of the processes occurring in a turbulent boundary layer can be hypothesized. The principally inviscid vortex-interaction processes that are of importance are: (i) spanwise vortex compression and vortex stretching in zones of increasing shear; (ii) spanwise vortex expansion and vortex relaxation in regions of decreasing shear; (iii) vortex coalescence yielding growth of larger vortices; and (iv) vortex reconnection to form closed loop vortices. Although each process will be discussed separately, it is important to appreciate that most of the important processes in the boundary layer occur in two or more parallel modes. This applies to the effectively inviscid processes described in this section, as well as the viscous processes described subsequently. This 'parallelism' is unusual in physics and makes it very hard to give a balanced description of turbulent boundary-layer dynamics since the major events may develop concurrently. To help facilitate the discussion, the terms 'outer region' (where inviscid dynamics dominate) and 'wall layer' (viscous effects dominate) will be used. Unlike the mean flow, where a boundary between these regions may be clearly defined, the spatial-temporal interface between the two regions can continually vary, with the outer region penetrating close to the wall in some locations, but not in others.

Near the wall, but within the outer region, the flow field contains myriad advected hairpin vortices, the majority of which will be of the asymmetric type shown in Figure 1.5b. These vortices are of relatively small spanwise extent near the wall (typically on the order of $\Delta z^+ \approx 100$) and of varying strengths; these vortices compete with adjacent vortices for position, are strongly influenced by the induced velocity fields of their immediate neighbours and can coalesce or amalgamate with adjacent vortices. As each vortex develops, it sends vortex legs downward toward the wall into regions of increasing shear. This exposure to increased shear promotes a process of lateral vortex stretching (Lighthill, 1963) as well as longitudinal stretching, as suggested in Figures 1.10 and 1.11, causing intensification of the vorticity of the legs. The number of vortices in proximity to the wall and the penetration of the legs of the deformed vortices (as suggested schematically in Figures 1.9, 1.10 and 1.11) into the zones of increased shear near the wall, gives the impression of an increased number of streamwise vortices near the wall, as noted in Robinson's (1991b) synthesis of DNS results (*cf.* Figure 1.7). As suggested by Figure 1.11, one leg by itself can appear to be a new 'quasi-streamwise' vortex near the wall. It may be expected that if vortex legs are relatively weak, they will either amalgamate with adjacent vortices or dissipate as they penetrate the viscous flow near the wall. In contrast, the legs of stronger hairpin-type vortices will penetrate close to the wall and, via an interaction process described below, provoke the generation of new hairpin vortices.

The upper portions of the hairpin-like vortices should behave quite differently from the legs. As suggested by Figures 1.9, 1.10 and 1.11, a vortex head will rise through the shear flow and at the same time expand in scale as it enters regions of decreased shear, leading to a diminution in the local vorticity. The evolution of hairpin vortices into much larger, but similar, vortices in the outer part of the boundary layer has been observed in a number of experimental studies (Head &

Bandyopadhyay, 1981; Perry & Chong, 1982; Smith & Lu, 1990; Grass *et al.*, 1991; Gretta & Smith, 1993). In addition, a process involving vortex intertwining and reinforcement is hypothesized to occur as the upward-migrating vortices approach one another closely. This process of coalescence of smaller vortices into structures of larger scale has been noted in simulations of the interaction of multiple hairpin vortices in a shear flow by Smith *et al.* (1991) and is illustrated schematically in Figure 1.12. Here the amalgamation of three symmetric hairpin vortices and an asymmetric vortex are simulated for a uniform shear flow. As suggested by Figure 1.12, the initial vortices intertwine and amalgamate to produce an asymmetric hairpin vortex structure of larger spanwise scale. Note that an introduction of only a slight degree of initial asymmetry stimulates the evolution of a large-scale, asymmetric hairpin vortex. The calculations described by Smith *et al.* (1991) suggest that interactions such as Figure 1.12 are the rule, not the exception, and ultimately result in complicated amalgamations of vortex heads far from the wall.

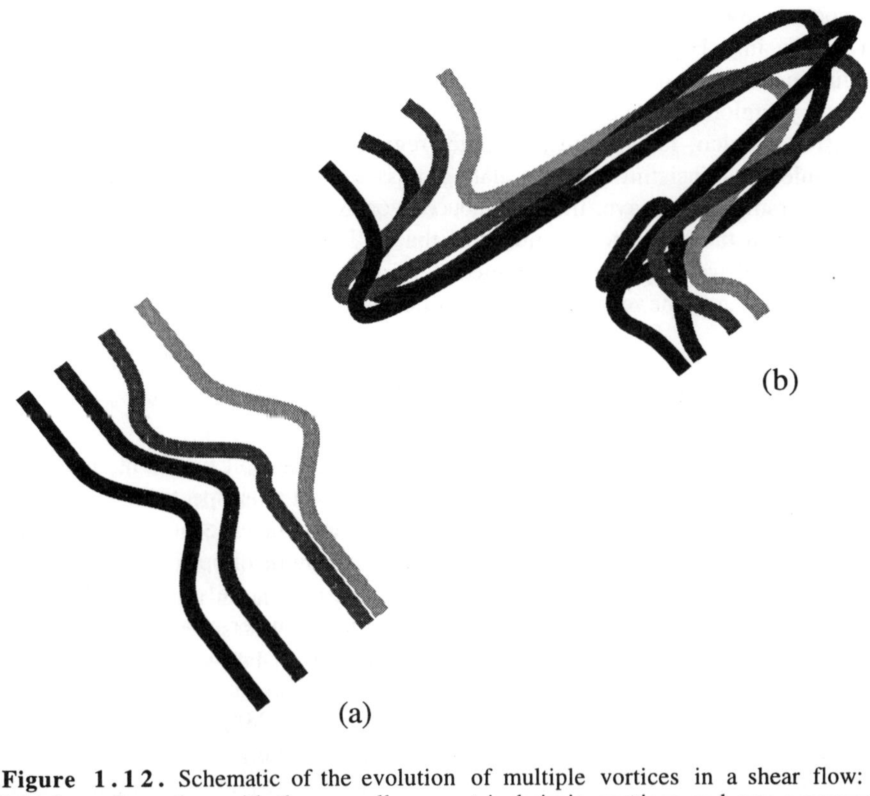

Figure 1.12. Schematic of the evolution of multiple vortices in a shear flow: (a) initial configuration with three small symmetric hairpin vortices and one asymmetric vortex; (b) evolved topology of a larger amalgamated asymmetric hairpin vortex

A limiting aspect of Biot-Savart simulations of vortex behaviour is that when two vortex cores move into close proximity such that the cores almost touch, the technique fails. Calculations carried out by Liu *et al.* (1986), as well as various experimental observations (e.g. Dhanak & De Bernardinis, 1981) suggest that when this situation occurs, the vortex cores often break and reconnect. Simulation of such behaviour requires a full Navier-Stokes solution – a process which has yet to be performed for complex turbulent-type flows. This failure of the Biot-Savart simulations in the final stages of multiple vortex head intertwinement was noted by Smith *et al.* (1991) and is suggestive that some form of vortex break/reconnection event must take place during actual amalgamation processes. One possible result of such events could be a vortex breaking and reconnection resulting in the release of closed vortex loops or rings, which have been proposed by Falco (1991) as an additionally important element (termed 'typical eddies') of boundary-layer turbulence. Although the physical origin and existence of typical eddies has been controversial, the calculations of Smith *et al.* (1991) suggest a possible process for this type of loop-vortex formation.

All of the manifold processes described above lead to both migration of vortices and vorticity away from the wall and a mechanism by which smaller initial vortices can evolve into larger vortex structures. These processes suggest that the large overturning motions observed in the outer region of the boundary layer are probably created by agglomeration of elementary vortices that originate near the wall (Smith *et al.*, 1991; Falco, 1991). Thus, the outer region may be regarded as a 'graveyard' for turbulence, consisting of recirculating masses of diffusing vorticity in a weak velocity gradient that have, through induced vortex motions, migrated away from their origin in the high-shear region near the wall. Attention will now be focused on the 'active' wall-layer region immediately adjacent to the wall, and how this region gives rise to the manifold initial vortices.

LOW-SPEED STREAKS: ORIGIN AND IMPORTANCE

Because low-speed streaks are a dominant feature of near-wall turbulent flows, and wall-layer bursting initiates near a streak, it was natural to suspect that the streaks are an important dynamic feature of the turbulent wall layer. Since the wall-layer streaks are commonly observed to have streamwise lengths of up to ten times their average spanwise spacing of $\lambda^+ \approx 100$ (in wall-layer units), the simplest conceptual model has been that the streaks are caused by long counter-rotating vortex pairs which are aligned in the streamwise direction. This was a common theme in the literature through the late 1980s (e.g. Blackwelder & Eckelmann, 1979; Ersoy & Walker, 1985; Aubrey *et al.*, 1988; Walker & Herzog, 1988). Although this type of model does represent some of the observed features, the origin of such counter-rotating pairs was unclear, and neither experiments nor low Reynolds-number DNS simulations (Moin & Kim, 1985, 1986; Spalart, 1988; Robinson, 1991a, 1991b) support the concept of long streamwise vortices.

It was speculated that the hypothesis of hairpin or horseshoe vortices originally introduced by Theodorsen might provide an active mechanism for creation of the streaks; this hypothesis has now been supported by a number of 'kernel' experiments (e.g. Acarlar & Smith, 1987a, 1987b; Haidari & Smith, 1994), DNS simulations of hairpin vortices (Singer & Joslin, 1994), as well as Robinson's (1991a, 1991b) synthesis of DNS results. In the experimental studies, controlled hairpin vortices were created in a laminar boundary layer using fluid injection through the wall. Employing dye and hydrogen bubble visualization, the motion induced by the passage of the vortex could then be observed and tracked. The results of these observations, as well as others discussed by Smith *et al.* (1991) and Robinson (1991a, 1991b), are generalized in Figure 1.13. As shown, the effect of the vortex motion near the wall, for both a symmetric and an asymmetric hairpin vortex, is to create a trail of upwelling fluid adjacent to each leg of a vortex as it passes over the plate. Figures 1.13a and 1.13b illustrate possible interactions for a symmetric hairpin vortex, where either two distinct streaks are produced or where two initial streaks merge, giving the appearance of a single streak (Haidari & Smith, 1994; Smith *et al.*, 1991). The basic streak-creating effect is not dependent on the degree of asymmetry in the hairpin vortex; the situation for a strongly asymmetric hairpin with only one dominant leg is shown in Figure 1.13c, wherein only one streak is created. Note that Figure 1.13c is similar to the model inferred by Robinson (1991b), as shown in Figure 1.6.

Note that not every hairpin vortex within the boundary layer will produce a low-speed streak; the observed streaks in the wall layer generally are a product of those vortices that are nearest the wall, with these nearest vortices also acting to shield the near-wall fluid from the influence of the majority of the farther-removed vortices. A streak will only initiate when a vortex is advected close to the wall or a vortex leg penetrates through the amalgam of vortices above and comes into close proximity to the wall. Conversely, when the causative vortex moves away from the immediate vicinity of the wall, a streak will dissipate due to viscous effects. Lateral oscillations or side-to-side meandering of low-speed streaks are also observed commonly in the wall layer (Smith, 1984; Smith *et al.*, 1991) and this is believed to be the result of weak, pressure-induced jostling action associated with upstream vortices overrunning existing streaks.

The above model suggests that a wall-layer streak is basically a trail of passive fluid induced by the passage of a hairpin vortex. The spanwise spacing of $\lambda^+ \approx 100$ most likely reflects the population and spacing of hairpin vortex legs closest to the wall which, in turn, is controlled by the level of background shear (see above). Because the streamwise velocity is much greater than typical spanwise velocities, the streaks (or vortex trails) are observed to be long with respect to their spanwise spacing. As a hairpin vortex advects downstream, the leading edge of the streak will move with a hairpin leg, originating from that side of the trailing vortex leg where

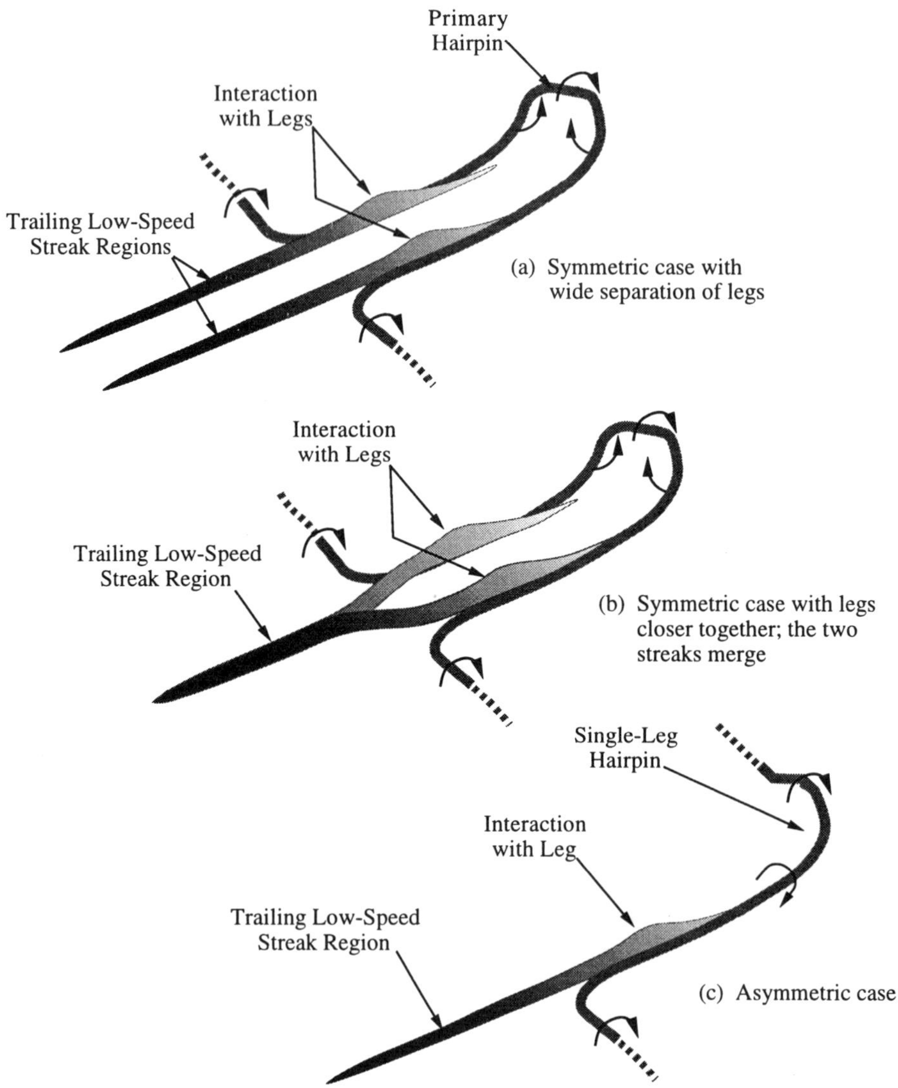

Figure 1.13. Schematic illustrating how advecting hairpin-like vortices generate low-speed streaks

the flow is away from the wall. However, in a high Reynolds number flow this situation cannot continue indefinitely since the trailing leg also imposes a zone of local adverse pressure gradient on the near-wall flow below and to the side of the leg where the streak forms. Eventually the presence of this local pressure gradient will cause a wall-layer eruption as described in the next section. Consequently, all streaks are a transient flow structure and must eventually terminate because either the causative hairpin (i) dissipates due to viscous effects, (ii) recirculates away from the wall (by mutual induction with neighbouring vortices) or (iii) provokes a local wall-layer eruption.

The literature on wall-layer streaks is extensive and is permeated with arguments as to whether the low-speed streaks are active or inactive. As discussed by Kline & Robinson (1989, 1990) and Robinson (1991b), as well as here, low-speed streaks which remain near the wall are inactive, representing hairpin vortex trails. On the other hand, streaks that lift away from the wall become active; such events are an indicator that an eruptive event is about to occur either directly above the streak or in its immediate vicinity, as described in the following section.

BURSTING/REGENERATION: THE BREEDING OF VORTICES

We now turn attention to the bursting process, during which the wall layer breaks down and erupts into the outer region, giving rise to the production of turbulent energy and the sustaining momentum exchange necessary to perpetuate turbulence; in this process, vorticity that has previously diffused outward from the wall is concentrated into a narrow longitudinal region within the wall layer and subsequently ejected. Of the characteristics of the turbulent wall layer discussed thus far, this is perhaps the most important. These eruptive wall-layer events intermittently feed regions of highly-concentrated vorticity into the outer region, creating shear layers which subsequently roll up into new hairpin-like vortices; these eruptions are thus the source of new turbulence. A second important aspect of this process is that the eruption is quickly followed by a sweeping motion of high-speed fluid from upstream that penetrates close to the wall. This penetration of high-speed fluid is then followed by a long period of relaxing flow (Blackwelder & Kaplan, 1976; Walker *et al.*, 1989), but on average the (mean) velocity near the wall is relatively high as compared to a laminar boundary layer (see, for example, the conditionally-averaged velocity profiles presented by Grass, 1971). Consequently, the 'bursts' or wall-layer breakdowns initiate the strong cross-stream momentum exchange and elevated shear stress characteristic of turbulent boundary layers.

It is important to note that since their inception, the terms 'burst', 'ejection' and 'sweep events' have been used inconsistently by researchers. For example, the quadrant analysis originally introduced by Wallace *et al.* (1972), and subsequently used by Willmarth & Lu (1972), categorizes local flow behaviour into 'quadrants', depending on the sign of the fluctuating streamwise and normal velocities u' and v'. Researchers employing this technique have termed fourth quadrant motions ($u' > 0$, $v' < 0$) 'sweeps' while second quadrant motions ($u' < 0$, $v' > 0$) are called 'ejections' (and equated by some with bursts). Because the quadrant method provides an unambiguous criterion, it is often used for analysis of pointwise experimental measurements and DNS data. Other pointwise 'burst' detection schemes are typified by the *VITA* technique, originally developed by Blackwelder & Kaplan (1976), which basically detects streamwise velocity excursions. However, such techniques using fixed point measurements are quite questionable for detecting the flow structures described in this paper (i.e. effects due to an advecting and deforming vortex). For example, quadrant analysis of DNS data easily associates second and fourth quadrant motion with induced flows adjacent to an advecting hairpin vortex, as shown in Figure 1.6 (from Robinson, 1991b), however, many regions of second and fourth quadrant motion show no clear relationship to a vortex (Lu & Smith, 1991). In addition, although such individual vortex motions do generate Reynolds stress $-u'v'$, such contributions arise principally from inviscid motions induced by the vortex and are self-cancelling with other associated inviscid motions in a long time-average (see Smith *et al.*, 1991). In this article the terms 'burst' and 'sweep' are used in the classical context (Corino & Brodkey, 1969; Blackwelder & Haritonidis, 1983) to imply much more dramatic and intermittent events associated with local breakdown of the wall-layer flow.

The breakdown of streaks and the wall layer can be explained by the viscous response of the fluid near a wall to the passage of wall-region vortices. This type of response has been demonstrated in a number of fundamental experimental and theoretical 'kernel' studies including: the motion induced by a two-dimensional vortex in an otherwise stagnant flow above a wall (Walker, 1978; Harvey & Perry, 1971; Peridier *et al.*, 1991a, 1991b), advected two-dimensional vortices in a uniform flow above a wall (Doligalski & Walker, 1984), vortex pairs near a wall (Ersoy & Walker, 1985) and vortex rings near a wall (Walker *et al.*, 1987; Falco, 1982). These studies showed that a sequence of events occur which initiate with a discrete eruption of wall-layer fluid which can precipitate a strong viscous-inviscid interaction between the essentially inviscid vortex and the viscous, erupting near-wall fluid, culminating in the generation and ejection of a new vortex. The experimental examination of the key stages of the process is very challenging, because the eruptive event initiates very close to the wall and quite abruptly; thus, identification of the key events was initially done via detailed numerical studies of the unsteady boundary-layer equations which for large Reynolds number describe the developing flow in an initially thin layer near the wall. However, calculations cannot, as yet, be carried through to the final ejection phase of the process, although this phase has been clearly observed experimentally. Employing collective experimental/theoretical results, the sequence of events due to a strong vortex interaction with a wall has been established (Smith *et al.*, 1991) and is summarized below. Note that while the initial vortex configuration affects the initial stages of the interaction, the final eruptive event is a generic boundary-layer response; a variety of other physical situations where such eruptive flows can occur are described by Doligalski *et al.* (1994).

Figure 1.14 summarizes the basic surface-eruption processes created by a two-dimensional vortex advecting in a uniform flow above a wall. When viewed by an

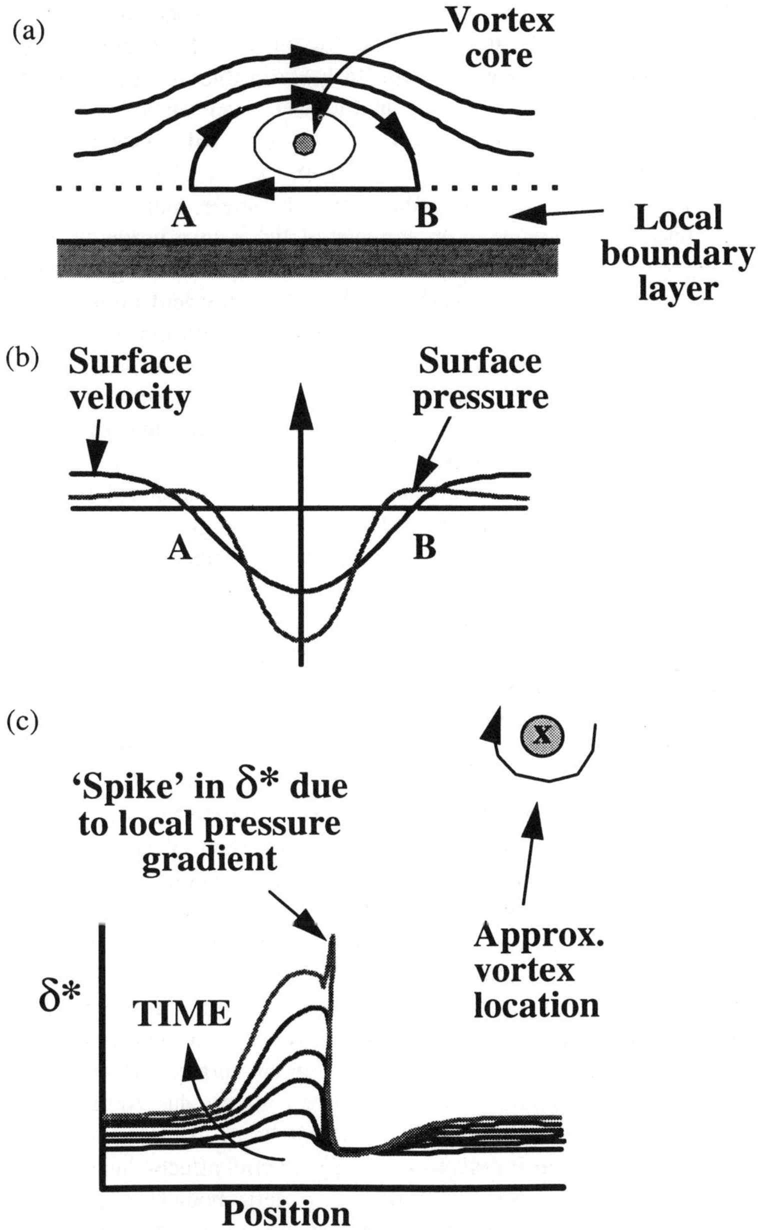

Figure 1.14. Schematic of induced viscous wall-layer behaviour due to advecting inviscid vortex near the wall. (a) inviscid streamline pattern; (b) surface velocity and pressure due to vortex; advected reference frame; and (c) temporal wall-layer displacement thickness

observer moving with the vortex, the instantaneous streamlines associated with the inviscid flow near the wall will appear as sketched in Figure 1.14a; note that the details of the vortex core geometry and the motion inside the core are not important. As the vortex moves, a thin unsteady boundary layer must develop in order to satisfy the no slip condition at the wall. Typical velocity and pressure distributions (as viewed advecting with the vortex) induced near the wall by the vortex are qualitatively sketched in Figure 1.14b; basically, it is the imposition of the adverse portion of this translating pressure gradient which stimulates the subsequent interaction process. The presence of the local adverse pressure gradient causes the formation of a moving zone of separation within the boundary layer which grows rapidly normal to the wall. For a vortex-induced boundary layer, the wall appears to be moving in a reference frame advecting with the vortex; the boundary-layer fluid beneath the vortex focuses into a narrow streamwise band and appears to leave the wall as a sharply-focused fluid spike (Peridier *et al.*, 1991a; Doligalski *et al.*, 1994). As shown in Figure 1.14c, a gradual thickening of the wall layer first occurs, followed by the evolution of a sharp 'spike' in the displacement surface as the boundary layer starts to separate locally from the wall. The work of Van Dommelen & Shen (1980) and Elliott *et al.* (1983) show that this type of eruptive boundary-layer response is expected in most two-dimensional, high Reynolds number flows, whenever a localized region of adverse streamwise pressure gradient is present. This phenomenon is known as 'unsteady separation' and implies a process in which an initially thin boundary layer grows rapidly outward and interacts strongly with the external flow. In physical terms, what is happening is that fluid particles just above the wall are compressed dramatically in the streamwise direction by the opposing inertia and pressure gradient forces; since the volume of each fluid particle is conserved (if incompressible), this strong streamwise compression results in a rapid and substantial vertical elongation.

An important effect of this compression process is a local concentration of the vorticity in the boundary layer into a narrow band. Under normal circumstances, vorticity diffuses slowly outward from the wall in response to a pressure distribution imposed at the outer edge of the boundary layer, such that at any instant the vorticity field is relatively smooth. However, once an event of the type illustrated in Figure 1.14 occurs, the vorticity field is concentrated locally into a double-sided shear layer which moves rapidly away from the wall; remarkably, once the eruptive phase develops, the local flow field 'forgets' the initial causative pressure field, thus this kind of generic eruptive response can result from a variety of adverse pressure-inducing situations. Note that the generation of eruptive events requires a strong interaction process that will generally occur only when a strong vortex is brought into close proximity of a wall; relatively weak vortices or vortices which are far from the wall may cause boundary-layer growth behind the vortex, but such a growth will be much more gradual and no 'spiky' response will occur (Doligalski *et al.*, 1994).

The three-dimensional response of a surface layer due to an advecting three-dimensional vortex, such as a hairpin shown in Figure 1.5, is rather more complicated than two-dimensional situations. Van Dommelen & Cowley (1990) have shown that an eruption stimulated by a three-dimensional vortex tends to develop as a ridge in the surface fluid, the shape of which depends on the portion of the vortex in proximity to the wall. A rapid outward penetration of fluid will initiate from such a ridge with the furthest penetration occurring at the point of closest approach by the vortex. Currently, the calculation of the subsequent eruption and viscous-inviscid interaction of such fluid ridges is currently beyond the scope of modern computational methods. However, several experiments have been able to resolve issues of subsequent behaviour, and are discussed by Smith *et al.* (1991) and Doligalski *et al.* (1994); these experiments show that generally these eruptive tongues of fluid rapidly roll-over into vortices, starting from the point of highest penetration and rolling progressively outboard. Of the experiments described by Smith *et al.* (1991), the most relevant to the dynamics of the turbulent boundary layer is a demonstration of the development and growth of a single symmetric hairpin vortex generated by discrete fluid injection through the wall beneath a laminar boundary layer (Haidari & Smith, 1994). Through the careful use of hydrogen bubble visualization, the following process of regeneration and production of new hairpin vortices was clearly observed.

Consider the symmetric hairpin vortex shown in Figure 1.15a. In the region immediately behind the vortex head, a region of adverse streamwise pressure gradient occurs which is very similar to that produced by a transverse vortex (see Figure 1.14). Farther behind the vortex head and in the region between the vortex legs, where the induced flow is away from the wall, the local spanwise pressure gradient is also locally adverse. In general, the streamwise adverse pressure gradient is generally weak, since the vortex head is relatively far from the wall and continues to move outward in the shear flow. In contrast, the vortex legs move progressively closer to the wall, which intensifies the spanwise adverse pressure gradient caused by the legs; in addition, because the legs move very little in the spanwise direction, the vortex remains in proximity of the wall, which increases the likelihood of an eruptive wall event (Doligalski & Walker, 1984).

As sketched in Figure 1.15, a symmetric hairpin vortex can generate surface-layer separations at three locations: behind the head and immediately inboard of *each* leg. For convenience, the hypothesized process in proximity to only one of the legs is shown in Figure 1.15 for a frame of reference moving with the vortex. As indicated in Figure 1.15a, separation initiates along U-shaped fronts (shown as broken lines), with the tip of the ejected tongue originating somewhere near the base of the U and moving outward, as indicated in Figure 1.15b. As this ejected tongue penetrates regions of increased streamwise velocity, the tongue rolls up into a new hairpin-like vortex as shown in Figure 1.15c through 1.15e. In the final stage of this viscous-inviscid interaction, the erupting ridge completely detaches from the surface layer below the parent vortex and a new *secondary* hairpin vortex is formed. The

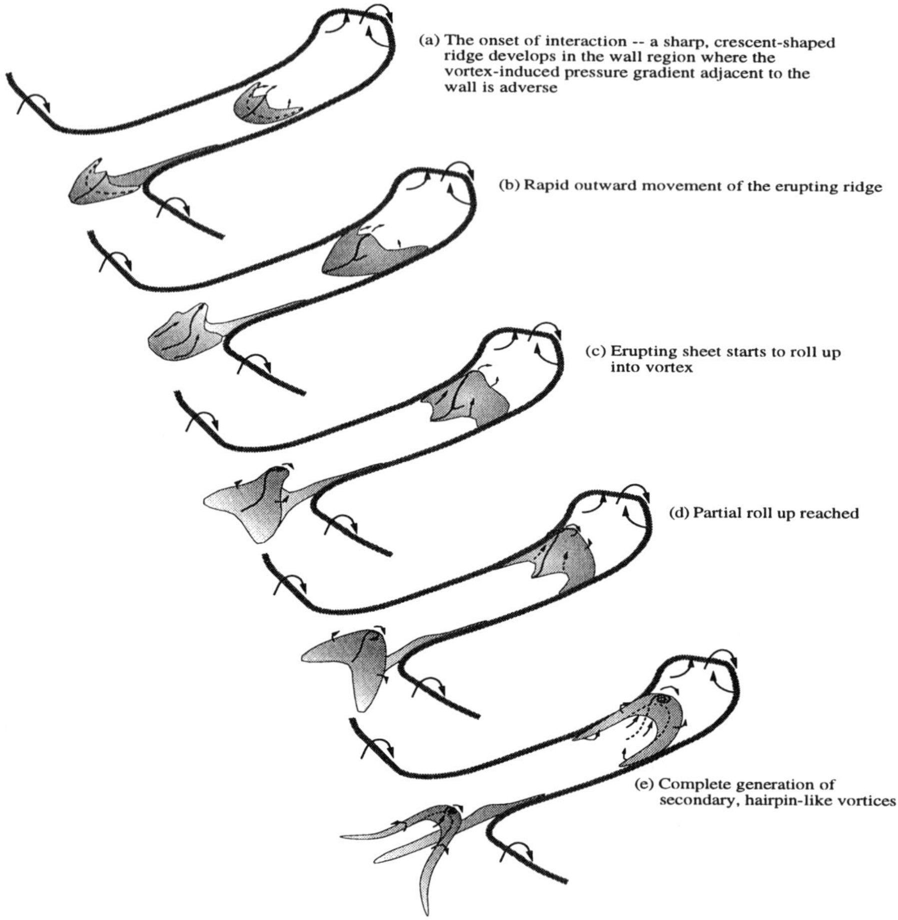

Figure 1.15. The generation of secondary vortices via surface interaction for a symmetric hairpin vortex

displacement of fluid outward by this eruptive/vortex formation process between stages 1.15d and 1.15e is countered by an inflow of faster moving fluid from immediately upstream due to continuity considerations; this appears as a 'sweep' of fluid bringing higher-momentum fluid to the wall.

Note that the events shown in Figure 1.15 occur intermittently and within a very short time-scale relative to that of the overall motion of the vortex. In addition, it is anticipated that the eruptive events shown in Figure 1.15 will not occur simultaneously near the head and the legs. For 'one-legged' hairpin vortices (or quasi-streamwise vortices), the process shown in Figure 1.15 is essentially the same except that the eruptive activity associated with the vortex head is expected to be less significant, with the bulk of the eruptions associated with the trailing legs.

The process outlined here is consistent with the description of Robinson (1991a, 1991b) of the evolution of 'new vortical arches' (i.e. new hairpin vortices) near 'quasi-streamwise' vortices from his synthesis of DNS results (*cf.* Figure 1.6). Kline & Robinson (1989) and Robinson (1991b) have observed from DNS results that the generation of new hairpin vortices near the wall occur in much the same way as shown in Figure 1.15; Robinson's schematics based on the DNS results are shown in Figure 1.16. Similar to Figure 1.15, Robinson identifies two types of regeneration which he referred to as (i) symmetric, and (ii) one-sided regeneration, depending on whether the process occurs respectively behind the vortex head or adjacent to the leg, with the one-sided regeneration by far the more common process.

The regenerative process outlined in Figure 1.15 was clearly observed by Haidari & Smith (1994), who observed that the production of new hairpins quickly leads to a streamwise and spanwise spread of the original disturbance in an otherwise laminar boundary layer. This occurs through complex processes of vortex generation coupled with vortex intertwining (as discussed above) followed by subsequent wall-layer eruption/hairpin generation cycles promoting continued streamwise and spanwise growth. Eventually, a large structure similar to a turbulent spot evolves from a single hairpin vortex. Recently, Singer & Joslin (1994) performed a DNS simulation of the Haidari & Smith experiment, and have shown essentially identical regenerative and interactive behaviour.

In a turbulent boundary layer, newly-created vortices may intertwine with the parent vortex or neighbouring vortices as discussed above. Alternatively, new vortices can act to induce further eruptions downstream, thereby perpetuating the generation of additional hairpin-like vortices. Note that only those vortices strongest and closest to the wall will induce eruptions; Robinson (1991b) detected that only a fraction (perhaps less than 50%) of lifted streaks roll up to form a new vortex, with the rest appearing to dissipate and disappear. Visualization of turbulent boundary layers suggests that the narrow-band eruptive process associated with the roll up of the lifted, active streaks is the dominant mode of eruption in the turbulent wall-layer. When viewed using a light sheet perpendicular to both the wall and flow direction, hydrogen bubbles or smoke released on the wall and immediately upstream of the light sheet are observed to concentrate into intermittent, thin eruptive spires which penetrate to distances on the order of $y^+ \approx 100$ (Smith *et al.*, 1991; Wallace & Balint, 1990). These spires, which appear as shown in Figure 1.2, are believed to be instantaneous slices through the thin eruptive cowls shown in Figure 1.15, with each spire representing a single realization of the local eruption of the wall layer. Consequently, these narrow eruptive plumes or cowls of concentrated vorticity represent the fundamental way in which fluid and vorticity from the wall region is exchanged with the outer part of the flow and turbulence is sustained.

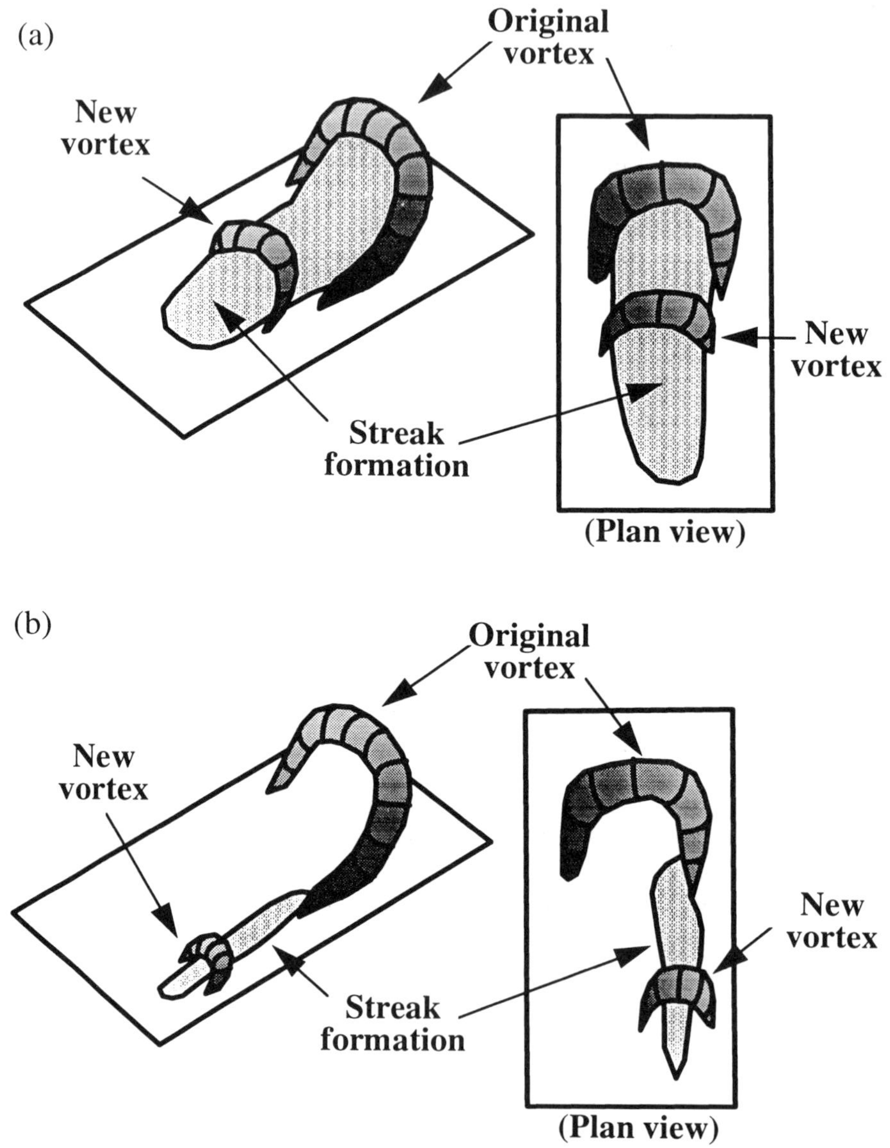

Figure 1.16. Schematic of regeneration of hairpin-like vortices in the near-wall region showing (a) symmetrical regeneration and (b) one-sided (asymmetric) regeneration (modified from Robinson, 1991b)

THE ROLE OF COHERENT STRUCTURE IN TURBULENT BOUNDARY LAYERS: A CONCEPTUAL MODEL

Based on the cumulative background discussed in the previous sections, a conceptual model can be formulated, hypothesizing how coherent structure is involved in the development, evolution, and maintenance of a turbulent boundary layer. The key element in this model is the understanding of the role of deformed vortices of a hairpin-like shape, although the various kernel studies and DNS evaluations suggest that the shape of the vortices is not as important as the proximity of the vortices to the wall.

It is clear from the studies cited that once hairpin-like vortices are present in a shear flow, they are able to: (i) interact with other three-dimensional vortices to yield larger-scale flow structures; and (ii) regenerate *new* vortices through an interaction with the viscous wall layer. A detailed model of near-wall turbulence based on hairpin-type vortices is described elsewhere (Smith *et al.*, 1991); the following description, in conjunction with Figures 1.17 and 1.18, summarize the key aspects of a conceptual model for both the near-wall region, and for the development of outer-region flow structures.

- Low-speed streaks are generated by the interaction of a passing streamwise or hairpin-like vortex with wall-region fluid, and comprise a narrow spire of low-speed fluid lifted from the wall. If the original vortex is strong this can precipitate a burst event wherein a streak(s) in proximity to the vortex will penetrate the outer flow (i.e. erupt), destabilize and roll over into a secondary hairpin-like vortex (vortices) via a viscous/inviscid interaction, resulting in the ejection of a portion of the streak into the outer region. If the vortex is weak, a streak(s) may form, but the vortex action may be insufficient to create a local breakdown and formation of a secondary hairpin vortex. In this latter case, a streak will either diffuse or be acted upon by subsequent impinging vortices. Action by subsequent streamwise or hairpin-like vortices (generated independently upstream) can either refocus or buffet the original streak. A refocusing can cause the original streak to grow further (possibly combining or amalgamating with other adjacent streaks) and subsequently breakdown; buffeting of a streak can occur when a vortex from upstream passes in proximity to the original streak, but not closely enough to precipitate a refocusing. In either case, the impingement of subsequent vortices will create the appearance of a waviness and swaying, a commonly observed behaviour for visualized low-speed streaks.
- A 'burst' is conceptualized as the local breakdown and ejection into the outer region of wall-layer fluid in proximity to a low-speed streak; this breakdown is a form of localized unsteady separation, precipitated by a local adverse pressure gradient created by an advecting wall-region vortex. The ejection associated with this burst can result in the formation of one or more secondary hairpin-like structures in the immediate wake of the initial vortex

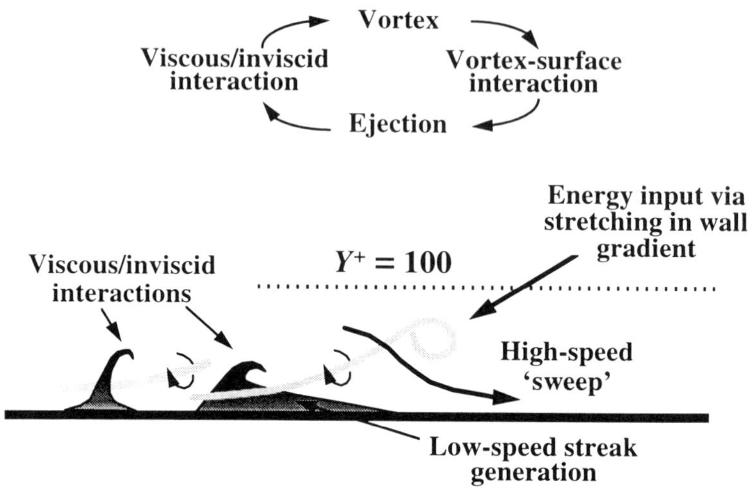

Figure 1.17. Generalized process of near-wall turbulence regeneration

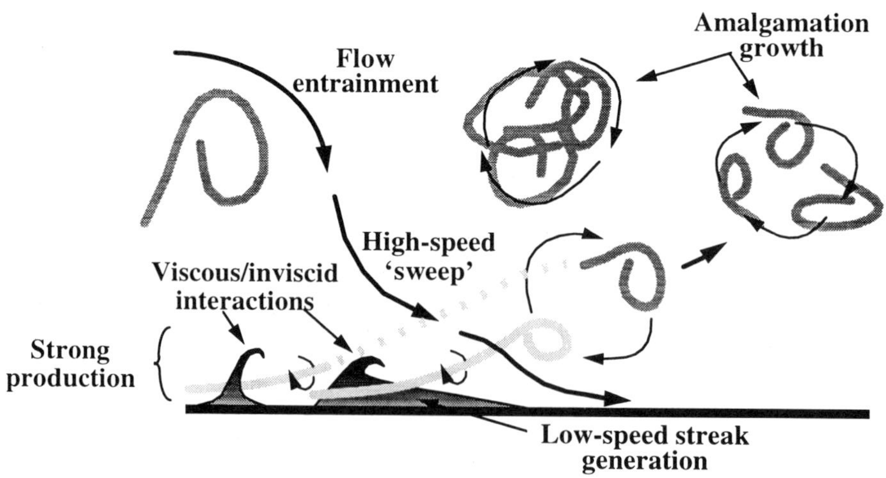

Figure 1.18. Conceptual model of turbulence regeneration, amalgamation and evolution of outer-region structures

via a viscous/inviscid interaction; a burst may also occur well behind the initial vortex due to the stimulation of a breakdown by the passage of a subsequent vortex (Haidari & Smith, 1994). In either case, the result is the rapid ejection of fluid from a streak into the outer flow, creating other hairpin-type flow structures.

- A 'sweep' is the three-dimensional inflow of high-speed fluid from the outer region due to both the formation and presence of hairpin-like vortices. This process takes conceptually two forms: (i) during the formation stage, when low-speed fluid is ejected outward from a streak, an inflow of higher-speed fluid will occur near the plane of symmetry of the streak, resulting in the local recovery of the mean velocity profile (Lu & Smith, 1991), which will appear and be detected as a local acceleration of the flow; (ii) alternatively, an advecting streamwise or hairpin-like vortex will induce higher-speed fluid toward the wall on the wallward-rotating portion of either the streamwise vortex or the legs of the hairpin vortex; this can result in the observation (visual studies) or detection (fixed probe studies) of what appears as an accelerated 'sweep' type behaviour (Robinson, 1991a, 1991b).
- As illustrated in Figure 1.17, the processes occurring in the near-wall are cyclical, although not periodic; wall-region vortices interact with and precipitate ejections of wall-region fluid which subsequently rolls up to form new vortices through viscous/inviscid interactions with the higher-speed, outer-region fluid. This process defines a continuing cycle which perpetuates both the elements which sustain turbulence (i.e. three-dimensional vortices) and the process for their generation (i.e. viscous-inviscid interactions).
- The engine that powers the turbulence regeneration process and is necessary to sustain the energy transfer from the free-stream to the near-wall is three-dimensional vortex stretching in the local velocity gradient. As suggested originally by Theodorsen (1952), the formation of hairpin-like vortices provides the logical mechanism for achieving this energy transfer process in the near-wall region with the energy input to the hairpin vortices supplied by the free-stream work done during vortex stretching in the local velocity gradient. During rapid stretching by the local velocity gradient, the angular momentum (proportional to ωr^2, where ω is the vorticity in the direction of the vortex centreline and r is distance normal to the vortex centreline) in a tilted streamwise vortex or leg of a hairpin-like vortex will be essentially conserved, while the energy (proportional to $\omega^2 r^2$) will increase significantly as the vortex tube narrows. Note also that this narrowing of a vortex tube sharply increases the velocity gradients within the vortex tube which strongly elevates viscous dissipation (proportional to the square of the velocity gradient in the tube); as is characteristic of turbulence, this dissipation will thus be highest in regions of high shear, such as in the vicinity of the wall.
- As Figure 1.18 shows, the process of growth to a fully-turbulent boundary layer can be explained by the proximity of multiple hairpin-like vortices in different phases of development, which creates a condition conducive to coalescence, or three-dimensional vortex amalgamation. This process of amalgamation, demonstrated in both simulations (Smith *et al.*, 1991; Figure 1.12) and experiment (Gretta & Smith, 1993), suggests that local

collections of hairpin-like vortices can intertwine and interact to yield essentially a hairpin-like structure of somewhat larger scale. Recent studies (Haidari & Smith, 1994; Singer & Joslin, 1994) have observed the development of just such an amalgamation process, tracking the controlled evolution of a single hairpin vortex into a multi-hairpin, turbulent spot-like structure. This clearly suggests that the outer region of a turbulent boundary layer can evolve from hairpin-like vortex structures and that the large, arch-type vortices (Robinson, 1991a) and rollers (Falco, 1977) observed in the outer region of turbulent boundary layers are essentially large amalgamations of initially smaller, deformed vortices. Since these larger outer-region structures directly interface with the free-stream flow, the overturning of the structures plays an instrumental part in inducing the flow of higher-speed fluid toward the wall (i.e. intermittent engulfment) which eventually finds its way to the wall along a tortuous gauntlet of vortex-induced motions, terminating in a 'sweep' motion at the wall, as shown in Figure 1.18. However, despite the overall size and strength of these outer structures, the stretching mechanism for energy transfer to these larger-scale vortical structures is strongly diminished because of the weak mean velocity gradient in the outer region. Following an initially strong energy input to the initial vortex scales near the wall, these outer structures will basically 'evolve' to larger and larger scales, but with no significant additional energy input, eventually succumbing to slow, viscous dissipation. The outer part of the boundary layer may thus be regarded as a 'graveyard' for vorticity, where the cumulative remnants of deformed wall-region vortices pass through a complicated process of dissipation, diffusion and mutual cancellation, similar to the hierarchy model of Perry & Chong (1982).

As a cautionary note, one should recognize that the above hypotheses on the mechanisms of energy exchange and growth of a turbulent boundary layer are based on studies which have been conducted primarily at low Reynolds number and predominantly on smooth walls. Thus, although one presumes that the basic processes outlined above should maintain in general, one should still exercise caution when extrapolating these processes to the very high Reynolds number behaviour encountered for flows over aircraft or large ships, or in environmental circumstances where roughness is particularly important, such as flows in rivers or in the atmospheric boundary layer. Indeed, the earlier cited work of Klewicki *et al.* (1995) in the near-wall of a true atmospheric boundary layer suggested consistency of low-speed streak spacing to high Reynolds number, but also noted several variations in the local statistics of the flow from accepted low Reynolds number flows, such as reduced levels of local velocity fluctuation. It is unclear whether such differences are indicative that a growth in boundary layer scale results in a small modification of the energy exchange process outlined above or a substantive change in the physical processes hypothesized from the low Reynolds number studies. The implications of these variations and whether they indicate significant changes in the turbulent energy exchanges process at high Reynolds number is a particularly important area for future research.

IMPLICATIONS FOR SEDIMENT TRANSPORT

The conceptual model outlined above has numerous implications for the impact/interaction of turbulence with sediment and bedforms. As has been illustrated by several researchers (Grass, 1982; Weedman & Slingerland, 1985; Best, 1992), the wall-region vortex interaction can give rise to the formation of sediment 'streaks' along a flat wall or on a flat, sandy bed. It is clear that these sediment streaks are a manifestation of the same vortex interaction processes that gives rise to the low-speed fluid streaks, although the sediment streaks appear more stable and are subject to varying spacing depending on the sediment grain size (i.e. smaller grain sizes form more closely-spaced streaks). This is speculated to be the result of the sediment streaks forming streamwise 'ridges' which tend to 'focus' the fluid behaviour in proximity to these ridges (Best, 1992), similar to the focusing behaviour observed for streamwise cylindrical rods of small diameter (Johansen & Smith, 1986).

However, at flow rates of practical importance these *flow-dominated* sediment flows generally undergo a metamorphosis to a *sediment-dominated* flow, wherein the sediment collects into larger ridges of transverse extent (Williams & Kemp, 1971) or the sediment is of significantly large dimension (Grass *et al.*, 1991; Kirkbride, 1993) such that the vertical turbulence scale of these sediment irregularities exceeds $k^+ = u_\tau k / \nu > 50 - 90$ (where k is the sediment size), which classifies the wall as *fully-rough*. For large roughness flows comprised of large sediment (clasts) of fully-rough scale, the low-speed streaks of a smooth wall essentially disappear and are replaced by wakes shed by the clasts, as pointed out by Kirkbride (1993). However, Grass *et al.* (1991) and Grass & Mansour-Tehrani (this volume) show that the region above fixed roughness elements displays a somewhat organized spanwise pattern akin to streaks, but with a spanwise spacing scaling on the roughness dimension (roughly three times k). This suggests that the spanwise spacing of the flow patterns scales with the roughness (i.e. clast) size. Additionally, it is well known (Ligrani, 1989) that the flow above a fully-rough wall still displays a mean velocity profile with a characteristic logarithmic scaling and the turbulence Reynolds stress distribution in the outer region undergoes only minor variations. These factors suggest that for a fully-rough wall, as shown in Figure 1.19, the scale of the initial hairpin-like vortices will increase proportionally with the roughness scale; however, the subsequent evolution of large-scale structures via outer-region interaction and amalgamation of the generated vortices should otherwise proceed in basically the same manner as for a smooth-wall flow. Clearly, an increase in wall roughness will elevate the wall drag and overall momentum exchange process, however, the basic outer-region process of vortex evolution

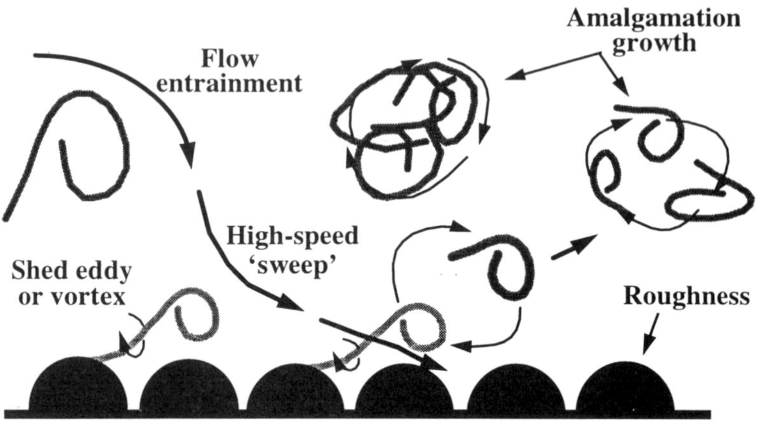

Figure 1.19. Conceptual model of turbulence generation in the presence of a fully-rough wall

should not be effectively modified. Indeed, Grass (1971) illustrates that the presence of roughness accentuates the amplitude of bursts and ejections in proximity to the rough surface, which clearly reflect an increased level of momentum exchange. What clearly will change is the elimination of the necessity for low-speed streaks and the subsequent viscous/inviscid interactions requisite for the formation of hairpin-like vortices on a smooth surface; note that to achieve a fully-rough wall, the roughness will be of the order of $k^+ \approx 50 - 100$, and essentially supplant the near-wall region of the boundary layer within which the streaks are formed. For the fully-rough wall, the separated wakes of the roughness elements supply the necessary spatial shear layers which spawn the generation of the requisite three-dimensional vortex structures by bluff-body shedding, these in-turn feed energy to the wall from the outer region by three-dimensional stretching of the shed vortices.

The behaviour of a fully-rough surface created by sediment ridges or ripples, modelled in Figure 1.20, is a little less clear cut, but this type of surface should behave similarly to the single clast-type roughness. The difficulty is that the formation of transverse ripples will generate separated shear layers, which will roll-up to form some type of three-dimensional vortex, but with much less discreteness than larger roughness elements. In addition, since the ripples will be composed of smaller sediment, they will more subject to transport by the higher-speed sweeps of the turbulence (Best, 1992). However, both of these effects may not be terribly important, since regardless of the source, once three-dimensional vortices are generated at the rough surface, the evolutionary process to larger scales should be quite similar to that for a smooth wall, although the shape and contours of the sediment surface (and potentially the initial vortex scales) will be in a continual state of change.

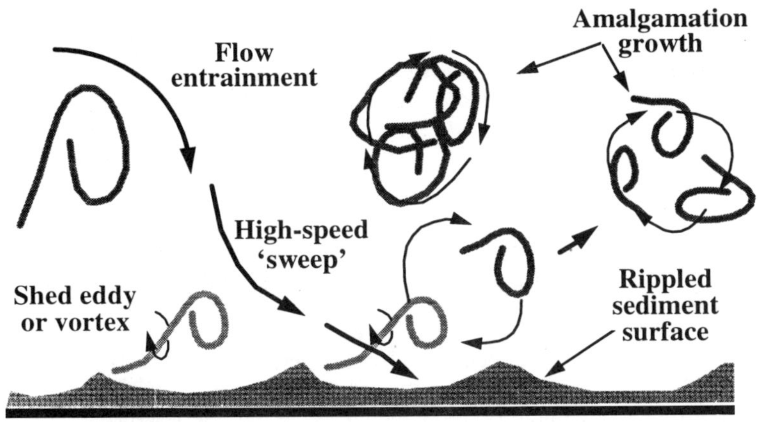

Figure 1.20. Conceptual model of turbulence/sediment interaction for a sediment-dominated flow

In general, it is expected that for all bedforms only the streak formation process will be significantly modified from the conceptual flow structure model for a smooth-wall turbulent boundary layer outlined above. In general, it is anticipated that once three-dimensional vortices form (by whatever mechanism) adjacent to a bounding surface, they will provide the mechanism for energy transfer from the outer to the inner region via vortex stretching in the local velocity gradient. The process of amalgamation and evolution of these vortices will then proceed in a fashion similar to the outer region of a smooth-wall flow, via a hierarchical process akin to that speculated originally by Perry & Chong (1982). As long as the boundary layer thickness or flow depth is large compared to the roughness scale, the above hypothesized process of roughness-generated vortices perpetuating and enhancing the growth of a turbulent boundary layer should apply. However, when bedform scales become comparable to, or exceed the boundary layer thickness, one wonders whether the concept of a 'turbulent' boundary layer is stretching the point. Clearly, the behaviour in these very large-scale roughness regimes is a fascinating and pragmatic area of research which warrants detailed study.

ACKNOWLEDGEMENTS

I would like to thank continuing support by the Air Force Office of Scientific Research, which has allowed me a protracted period to study both turbulent boundary layers and the vortex phenomena associated with such irksome flows. I would also like to thank my manifold past and present research students, whose collective efforts were critical to the development of my understanding of turbulent structure. I am also appreciative of the detailed review of the manuscript by Prof. Tony Grass, particularly for his very helpful comments and clarifications on the behaviour of rough-wall flows. Additionally, I want to express my thanks and admiration to my long-time collaborator and colleague, Dr. David Walker, whose marvellous analytical and computational skills have helped me to understand and appreciate the intricacies of vortices and vortex-surface interactions. And finally, I would like to thank Dr. Jim Best, who thought enough of my work and its implications to invite me to keynote a conference on sedimentology.

REFERENCES

Acarlar, M.S. & Smith, C.R. 1987a. 'A study of hairpin vortices in a laminar boundary layer. Part I. Hairpin vortices generated by hemisphere protuberances', *J. Fluid Mech.*, **175**, 1–41.

Acarlar, M.S. & Smith, C.R. 1987b. 'A study of hairpin vortices in a laminar boundary layer. Part 2. Hairpin vortices generated by fluid injection', *J. Fluid Mech.*, **175**, 43–83.

Aubrey, N., Holmes, P., Lumley, J.L. & Stone, E. 1988. 'The dynamics of coherent structures in the wall region of a turbulent boundary layer', *J. Fluid Mech.*, **192**, 115–173.

Best, J. 1992. 'On the entrainment of sediment and initiation of bed defects: insights from recent developments within turbulent boundary layer research', *Sedimentology*, **39**, 797–811.

Black, T. 1968. 'An analytical study of the measured wall pressure field under supersonic turbulent boundary layers', NASA-CR **888**, 103 pp.

Blackwelder, R.F. 1983. 'Analogies between transitional and turbulent boundary layers', *Phys. Fluids*, **26**, 2807–2815.

Blackwelder, R.F. & Eckelmann, H. 1979. 'Streamwise vortices associated with the bursting phenomenon', *J. Fluid Mech.*, **94**, 577–594.

Blackwelder, R.F. & Haritonidis, J.H. 1983. 'Scaling of the bursting frequency in turbulent boundary layers', *J. Fluid Mech.*, **132**, 87–103.

Blackwelder, R.F. & Kaplan, R.E. 1976. 'On the wall structure of the turbulent boundary layer', *J. Fluid Mech.*, **76**, 89–112.

Blackwelder, R.F. & Kovasznay, L.S.G. 1972. 'Time scales and correlations in a turbulent boundary layer', *Phys. Fluids*, **15**, 1545–1554.

Cantwell, B.J. 1981. 'Organized motion in turbulent flow', *Ann. Rev. Fluid Mech.*, **13**, 457–515.

Corino, E.R. & Brodkey, R.S. 1969. 'A visual investigation of the wall region in turbulent flow', *J. Fluid Mech.*, **37**, 1–30.

Dhanak, M.R. & De Bernardinis, B. 1981. 'The evolution of an elliptic vortex ring', *J. Fluid Mech.*, **109**, 189–216.

Doligalski, T.L. & Walker, J.D.A. 1984. 'Boundary layer induced by a convected two-dimensional vortex', *J. Fluid Mech.*, **139**, 1–28.

Doligalski, T.L., Smith, C.R. & Walker, J.D.A. 1994. 'Vortex interactions with walls', *Ann. Rev. Fluid Mech.*, **26**, 573–616.

Elliott, J.W., Cowley, S.J. & Smith, F.T. 1983. 'Breakdown of boundary layers, (i) on moving surfaces; (ii) in self-similar unsteady flow; (iii) in fully unsteady flow', *Geophys. Astrophys. Fluid Dynamics*, **25**, 77–138.

Ersoy, S. & Walker, J.D.A. 1985. 'Viscous flow induced by counterrotating vortices', *Phys. Fluids*, **28**, 2687–2698.

Falco, R.E. 1977. 'Coherent motions in the outer region of turbulent boundary layers', *Phys. Fluids*, **20**, S124–S132.

Falco, R.E. 1982. 'A synthesis and model of turbulence structure in the wall region', in Zartic, Z. (Ed) *Structure of Turbulence in Heat and Mass Transfer*, Hemisphere, 43–57.

Falco, R.E. 1991. 'A coherent structure model of the turbulent boundary layer and its ability to predict Reynolds number dependence', *Phil. Trans. Roy. Soc. London*, **A336**, 103–129.

Grass, A.J. 1971. 'Structural features of turbulent flow over smooth and rough boundaries', *J. Fluid Mech.*, **50**, 233–255.

Grass, A.J. 1983. 'The influence of boundary layer turbulence on the mechanics of sediment transport', in Sumer, B.M. & Müller, A. (Ed) *Mechanics of Sediment Transport*, Proc. of Euromech **156**, Balkema, Rotterdam, 3–18.

Grass, A.J., Stuart, R.J. & Mansour-Tehrani, M. 1991. 'Vortical structures and coherent motion in turbulent flow over rough boundaries', *Phil. Trans. R. Soc. London*, **A336**, 35–65.

Gretta, W.J. & Smith, C.R. 1993. 'The flow structure and statistics of a passive mixing tab', *ASME J. Fluid Engrg.*, **115**, 225–263.

Haidari, A.H. & Smith, C.R. 1994. 'The generation and regeneration of single hairpin vortices', *J. Fluid Mech.*, **277**, 135–162.

Hama, F.R. 1962. 'Progressive deformation of a curved vortex filament by its own induction', *Phys. Fluids*, **5**, 1156–1162.

Harvey, J.K. & Perry, F.J. 1971. 'Flowfield produced by trailing vortices in the vicinity of the ground', *AIAA J.*, **9**, 1659–1660.

Head, M.R. & Bandyopadhyay, P. 1981. 'New aspects of turbulent boundary layer structure', *J. Fluid Mech.*, **107**, 297–338.

Hinze, J.O. 1978. *Turbulence*, McGraw-Hill, New York, 790 pp.

Hon, T.L. & Walker, J.D.A. 1991. 'Evolution of hairpin vortices in a shear flow', *Computers and Fluids*, **20**, 343–358.

Johansen, J.B. & Smith, C.R. 1986.' The effects of cylindrical surface modifications on turbulent boundary layers', *AIAA J.*, **24**, 1081–1088.

Kim, H.T., Kline, S.J. & Reynolds, W.C. 1971. 'The production of turbulence near a smooth wall', *J. Fluid Mech.*, **50**, 133–160.

Kirkbride, A. 1993. 'Observations of the influence of bed roughness on turbulence structure in depth limited flows over gravel beds', in Clifford, N.J., French, J.R. & Hardisty, J. (Eds) *Turbulence, Perspectives on Flow and Sediment Transport*, John Wiley & Sons Ltd., Chichester, 185–196.

Klewicki, J.C., Metzger, M.M., Kelner, E. & Thurlow, E.M. 1995. 'Viscous sublayer flow visualization at $Re_\theta \approx 1{,}500{,}000$', *Phys. Fluids*, **7**, 857–865.

Kline, S.J. & Robinson, S.K. 1989. 'Quasi-coherent structures in the turbulent boundary layer. Part I, Status report on a community-wide summary of the data', in Kline, S.J. & Afgan, N.H. (Eds) *Near Wall Turbulence*, Hemisphere, 200–217.

Kline, S.J. & Robinson, S.K. 1990. 'Turbulent boundary layer structure, progress, status and challenges', in Gyr, A. (Ed) *Structure of Turbulence and Drag Reduction*, Springer-Verlag, 3–22.

Kline, S.J., Reynolds, W.C., Schraub, F.A. & Runstadler, P.W. 1967. 'The structure of turbulent boundary layers', *J. Fluid Mech.*, **95**, 741–773.

Kovasznay, L.S.G., Kibens, V. & Blackwelder, R.F. 1970. 'Large-scale motion in the intermittent region of a turbulent boundary layer', *J. Fluid Mech.*, **41**, 283–325.

Lighthill, M.J. 1963. 'Introduction. Boundary layer theory', in Rosenhead, L. (Ed) *Laminar Boundary Layers*, Oxford University Press, 46–109.

Ligrani, P.M. 1989. 'Structure of turbulent boundary layers', in Cheremisinoff, N.P. (Ed) *Encyclopedia of Fluid Mechanics, Vol. 8 Aerodynamics and Compressible Flow*, Gulf Publishing Company, Houston, 165–170.

Liu, C.H., Tavantzis, J. & Ting, L. 1986. 'Numerical studies of motion and decay of vortex filaments', *AIAA J.*, **24**, 1290–1297.

Lu, L.J. & Smith, C.R. 1991. 'Use of flow visualization data to examine spatial-temporal velocity and burst-type characteristics in a turbulent boundary layer', *J. Fluid Mech.*, **232**, 303–340.

Moin, P. & Kim, J. 1985. 'The structure of the vorticity field in turbulent channel flow. Part I, Analysis of instantaneous fields and statistical correlations', *J. Fluid Mech.*, **155**, 441–464.

Moin, P. & Kim, J. 1986. 'The structure of the vorticity field in turbulent channel flow. Part 2, Study of ensemble-averaged fields', *J. Fluid Mech.*, **162**, 339–363.

Morkovin, M.V. 1993. 'Bypass-transition research, issues and philosophy', in Ashpis, D.E., Gatski, T.B. & Hirsch, C. (Eds) *Instabilities and Turbulence in Engineering Flows*, Kluwer Academic Publishers, 3–30.

Nagib, H.M., Hathway, D.W., Naguib, A.M. & Wark, C.E. 1993. 'Characterization of dynamically active events in turbulent boundary layers', in So, R.M.C., Speziale, C.G. & Launder, B.E. (Eds) *Near-Wall Turbulent Flows*, Elsevier Science Publishers B.V., 949–962.

Offen, G. R. & Kline, S. J. 1975. 'A proposed model of the bursting process in turbulent boundary layers', *J. Fluid Mech.*, **70**, 209–228.

Peridier, V.J., Smith, F.T. & Walker, J.D.A. 1991a. 'Vortex-induced boundary-layer separation. Part 1. The limit problem $Re \to \infty$', *J. Fluid Mech.*, **232**, 99–131.

Peridier, V.J., Smith, F.T. & Walker, J.D.A. 1991b. 'Vortex-induced boundary-layer separation. Part 2. Unsteady interacting boundary-layer theory', *J. Fluid Mech.*, **232**, 133–165.

Perry, A.E. & Chong, M.S. 1982. 'On the mechanisms of wall turbulence', *J. Fluid Mech.*, **119**, 173–217.

Robinson, S.K. 1991a. 'Coherent motions in the turbulent boundary layer', *Ann. Rev. Fluid Mech.*, **23**, 601–639

Robinson, S.K. 1991b. 'The kinematics of turbulent boundary layer structure', *NASA Tech. Memo.*, ***103859***, 490 pp.

Singer, B.A. & Joslin, R.D. 1994. 'Metamorphosis of a hairpin vortex into a young turbulent spot', *Phys. Fluids.*, **6**, 3724–3730.

Smith, C.R. 1984. 'A synthesized model of the near-wall behaviour in turbulent boundary layers', in Zakin, J.L. & Patterson, G. (Eds) *Proc. of 8th Biennial Symp. on Turbulence*, Univ. of Missouri-Rolla, 299–327.

Smith, C.R. 1993. 'Use of 'kernel' experiments for modelling near-wall turbulence', in So, R.M.C., Speziale, C.G. & Launder, B.E. (Eds) *Near-Wall Turbulent Flows,* Elsevier Science Publishers B.V., 33–42.

Smith, C.R. & Lu, L.J. 1990. 'The use of a template matching technique to identify hairpin vortex flow structures in turbulent boundary layers', in Kline, S.J. & Afgan, N.H. (Eds) *Near Wall Turbulence,* Hemisphere, New York, 248–267.

Smith, C.R. & Metzler, S.P. 1983. 'The characteristics of low-speed streaks in the near-wall region of a turbulent boundary layer', *J. Fluid Mech.*, **129**, 27–54.

Smith, C.R., Walker, J.D.A., Haidari, A.H. & Sobrun, U. 1991. 'On the dynamics of near-wall turbulence', *Phil. Trans. Roy. Soc. London*, **A336**, 131–175.

Spalart, P.R. 1988. 'Direct simulation of a turbulent boundary layer up to $Re_\theta \approx 1410$', *J. Fluid Mech.*, **187**, 61–98.

Theodorsen, T. 1952. 'Mechanism of turbulence', in *Proceedings Second Midwestern Conference on Fluid Mechanics*, Bull. No. **149**, Ohio State Univ., Columbus, Ohio, 1–18.

Townsend, A.A. 1976. *The Structure of Turbulent Shear Flow*, 2nd Edition, Cambridge University Press, 150–158.

Van Dommelen, L.L. & Cowley, S.J. 1990. 'On the Lagrangian description of unsteady boundary-layer separation. Part 2. General theory', *J. Fluid Mech.*, **210**, 593–626.
Van Dommelen, L.L. & Shen, S.F. 1980. 'The spontaneous generation of the singularity in a separating boundary layer', *J. Comp. Phys.*, **38**, 125–140.
Walker, J.D.A. 1978. 'The Boundary Layer Due to Rectilinear Vortex', *Proc. Roy. Soc. London*, **A359**, 167–188.
Walker, J.D.A., Abbott, D.E., Scharnhorst, R.K. & Weigand, G.G. 1989. 'Wall-layer model for the velocity profile in turbulent flows', *AIAA J.*, **27**, 140–149.
Walker, J.D.A. & Herzog 1988. 'Eruption mechanisms for turbulent flow near walls', in Hirata, M. & Kasagi, N. (Eds) *Transport Phenomena in Turbulent Flows,* Hemisphere, 145–156.
Walker, J.D.A., Smith, C.R., Doligalski, T.L. & Cerra, A.W. 1987. 'Impact of a vortex ring on a wall', *J. Fluid Mech.*, **181**, 99–140.
Wallace, J.M. & Balint, J.L. 1990. ' Flow visualization study of the effects of trip type on the structure of the turbulent boundary layer', *Video Tape*, Turbulence Laboratory, Univ. of Maryland.
Wallace, J.M., Eckelman, H. & Brodkey, R.S. 1972. 'The wall region of turbulent shear flow', *J. Fluid Mech.*, **54**, 39–48.
Weedman, S.D. & Slingerland, R. 1985. 'Experimental study of sand streaks formed in turbulent boundary layers', *Sedimentology*, **32**, 133–145.
Willmarth, W.W. 1975. 'Structure of turbulence in boundary layers', *Adv. Appl. Mech.*, **15**, 159–254.
Willmarth, W.W. & Lu, S.S. 1972. 'Structure of the Reynolds stress near the wall', *J. Fluid Mech.*, **55**, 65–92.

NOTATION

k	vertical dimension of a roughness element
r	radius from centre of a vortex
Re_θ	Reynolds number based on momentum thickness
u_τ	wall 'shear' velocity, $(\tau_w/\rho)^{0.5}$
u v w	respectively streamwise, normal, and lateral flow velocities
u' v' w'	respectively streamwise, normal, and lateral velocity fluctuations
x y z	respectively streamwise, normal, and lateral Cartesian dimensions
Δ	indicates the change in a spatial dimension
λ	lateral spacing of low-speed streaks; may also indicate lateral spacing of hairpin vortex legs
ν	fluid kinematic viscosity
ρ	fluid density
τ_w	fluid shear stress at a wall boundary
ω	vorticity

2 Generalized Scaling of Coherent Bursting Structures in the Near-wall Region of Turbulent Flow over Smooth and Rough Boundaries

A.J. GRASS & M. MANSOUR-TEHRANI
Department of Civil and Environmental Engineering, University College London, Gower Street, London, WC1E 6BT, UK

ABSTRACT

Recent studies have shown that the distinctive pattern of low-velocity streaks observed and extensively studied in turbulent flow over smooth boundaries, is also a characteristic feature of the near-wall region of rough-boundary flows, more typical in the geophysical environment. It has been further demonstrated that these streaks are generated by, and reflect the scale of vortical bursting structures, which have been found to dominate the dynamics of turbulent boundary layers.

Experimental measurements of the spanwise spacing of low-velocity wall streaks, formed in a water channel flow over smooth and rough beds are presented and shown to scale with the viscous and roughness length-scales for hydrodynamically smooth and rough walls respectively. This finding is in accord with the predictions of dimensional analysis. Using a wall length-scale derived from the shear stress and mean strain rate (mean velocity gradient) immediately adjacent to the rough boundary, which define an effective wall 'eddy' viscosity, the low-speed streak spacing, at the wall, is shown to follow an apparently general universal relationship for hydrodynamically smooth, transitional and fully rough conditions. The resulting non-dimensional spacing, is approximately 100 effective viscosity length-scales, in remarkable agreement with the widely quoted result for smooth-wall flow.

INTRODUCTION

Turbulence transport processes dominate the majority of boundary shear flows of geophysical and engineering interest. These flows range from large-scale fluid motion in the atmosphere, rivers and oceans to flows in smaller-scale, man-made systems. The extraordinary complexity of turbulence phenomena, and the resulting theoretical intractability, has meant that most of our current understanding and knowledge stems largely from systematic physical experimentation over the past century. Progress in our understanding has accelerated in recent years, primarily due to advances in measurement instrumentation and increasingly important, the

Coherent Flow Structures in Open Channels.
Edited by P.J. Ashworth, S.J. Bennett, J.L. Best and S.J. McLelland.

availability of digital computers of ever growing speed and capacity.

It is apparent from casual observation, that turbulent motion is not simply a formless, randomly chaotic state. Strongly coherent patterns, indicative of dominant, deterministic physical processes, can be clearly identified in the visualized flow field. Research over the past three decades has concentrated attention on this conspicuous coherence in the turbulent structure.

Kline *et al.* (1967) gave a detailed description of the coherent, streaky nature of the flow structure in the viscous sublayer on a smooth wall with its alternating narrow, elongated zones of high and low fluid velocity. This streaky structure was linked to the bursting phenomenon, a random cyclical process of recurrent instability and turbulence generation in the wall region of the boundary layer, producing energetic fluid ejection and inrush events (see Smith, this volume, for a full review). Grass (1971) further clearly demonstrated the existence of bursting and both violent ejection and inrush events in flow over rough walls. It was concluded that these were common features of boundary layer turbulence irrespective of wall roughness condition.

The common presence of these dominant structural features suggests that a shared mechanism could be responsible for their generation. Theodorsen (1952) proposed a simple vortex model, with the idealized horseshoe configuration shown in Figure 2.1, as the central element of the turbulence generation process in shear flows. This remarkable prediction has been subsequently confirmed for a wide range of such flows (Grass *et al.,* 1991). The most convincing smooth wall evidence stems largely from the results of direct numerical simulation (DNS) studies reviewed, for example, by Kline & Robinson (1989), Robinson (1991) and Smith (this volume). The consensus view is that the violent dynamics of the bursting process over smooth walls is directly linked to the presence of powerful vortical structures, of distorted and fragmented horseshoe geometry, embedded in the wall flow field. Findings from parallel physical experiments reported by Grass *et al.* (1991), have demonstrated that the coherent vortex structures are also present in turbulent boundary layers over rough walls and that they are similarly associated with bursting events. These vortices appear to represent the central dynamic element in a recurring, highly non-linear cycle of turbulence and burst generating instability.

The three-dimensional vortices essentially act as pumps, entraining, lifting and transporting mass, momentum, vorticity and contaminants such as heat and bed sediment out from the wall region into the body of the flow. Footprints of these coherent flow structures take the form of elongated, streaky patterns of high- and low-velocity, and hence instantaneous shear stress, over the bed surface. Vortical bursting structures thus play a dominant role, not only in boundary layer dynamics, but also in the key process physics of sediment transport and bed morphology as discussed previously, for example, by Grass (1983), Leeder (1983), Drake *et al.* (1988), Williams *et al.* (1989) and Best (1992). They extract large amounts of kinetic energy from the mean flow and hence have a major retarding action on the surrounding fluid, by the simple mechanism of shear layer vortex stretching. This

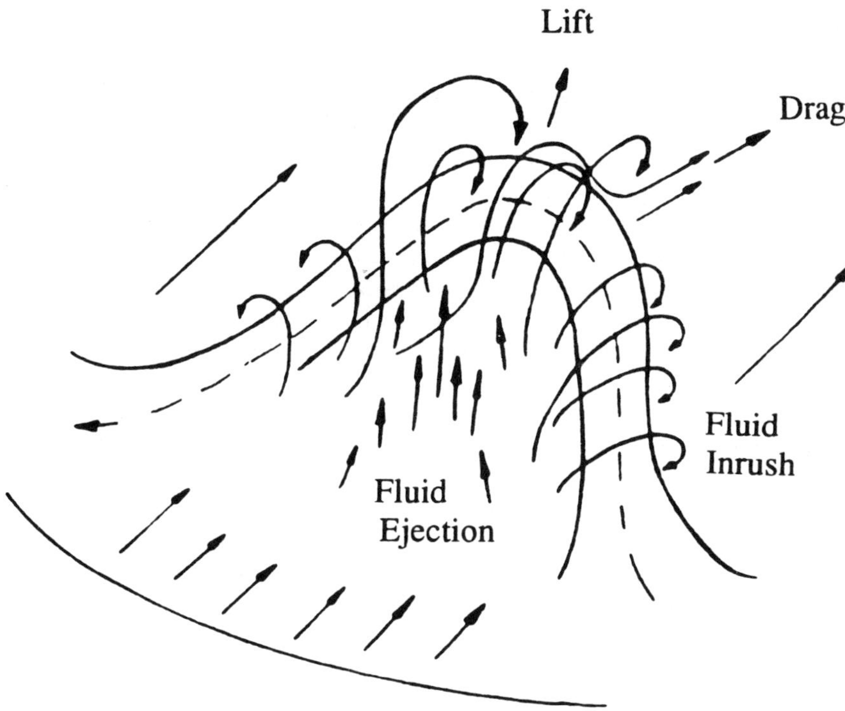

Figure 2.1. Horseshoe vortex model first proposed by Theodorsen (1952)

now firmly established key role, was anticipated by Kuchemann (1965), who most aptly described vortices as "the sinews and muscles of fluid motion".

The mean spacing, λ, between the low-speed streaks, is of central importance as it coincides with the dominant spanwise scale or wavelength of the vortical instability structures adjacent to the boundary. It therefore represents a key characteristic of the local, highly non-linear dynamics of the wall shear layer. Measurements from a wide range of different types of smooth-wall turbulent flows indicate very weak dependence of the non-dimensional mean streak spacing, $\lambda^+ = \lambda U_\tau / \nu$, on the overall flow Reynolds number, where U_τ is the bed shear velocity, ν is the fluid kinematic viscosity, and ν / U_τ is the relevant smooth-wall viscous length-scale. This considerable body of experimental evidence yields a λ^+ value close to 100, a number which has proved remarkably robust and invariant, and which is even reproduced in transitional turbulent spots. The reason why this particular dominant wavelength should prevail for the spanwise structure remains theoretically obscure. A number of investigators have explored the mechanics of low-speed streak formation on smooth boundaries and the link to horseshoe or hairpin type vortex structures in the near-wall flow. This work, including the findings from their own extensive theoretical and experimental studies, has been reviewed by Smith & Metzler (1983), Smith *et al.* (1991), and Smith (this volume).

Grass *et al.* (1991) reviewed and suggested new evidence in support of the 'attached-eddy' hypothesis first proposed by Townsend (1976) and extended by Perry & Chong (1982) and Perry *et al.* (1986, 1987), in which high Reynolds number boundary layers, such as occur in large-scale geophysical flows, are perceived to comprise a scale hierarchy of geometrically self-similar turbulence structures. The basic horseshoe type geometry of the structures found in wall shear flow (Figure 2.1), is one of a restricted range of coherent vortical forms, including also the vortex ring, which are recurring features in a large number of different turbulent flows. In a review of recent theoretical work, Hunt *et al.* (1994) argue that these vortex structures can be viewed as apparently robust eigensolutions of the Navier-Stokes equations. Barenblatt (1993) suggests a partially self-similar, fractal pattern for these solutions, implicit in his formulations of a Reynolds number dependent version of Prandtl's power-law relationship for the mean velocity profile in the wall layer of pipe flow.

The near-wall layer structures represent the smallest size elements in the scale hierarchy which ranges up to geometrically similar vortical structures of size commensurate with the thickness of the boundary layer. Grass *et al.* (1991) describe how these largest scale structures will impose a low-frequency modulation on the short term mean velocity close to the wall. They suggest the likelihood that this will be accompanied by similar modulation and time variation in the streak spacing. Evidence that such modulation does indeed occur has been recently provided by Brereton & Hwang (1994) who carried out tests in a pipe flow with a superimposed sinusoidal oscillation on the mean velocity. The measured, phase averaged, streak spacing in these experiments showed a corresponding sinusoidal variation with, as might be expected, a phase lag relative to the velocity oscillations. However, the long term average streak spacing, taken over many oscillation cycles, showed little significant variation about a mean λ^+ of approximately 100. This finding is consistent with the near Reynolds number independence of the mean streak spacing observed in turbulent boundary layers.

Kline *et al.* (1967) demonstrated that in steady flow, pressure gradients do have a clear influence on the spacing of low-speed streaks in the near-wall layer. Favourable and adverse pressure gradients were found to respectively increase and decrease the λ^+ values relative to the zero pressure gradient magnitude of 100. These pressure gradient induced effects on the λ^+ scaling have been confirmed in more recent experiments reported by Simpson (1991). This study was concerned with the structure of the near-wall region of two-dimensional, turbulent, separated flow where large pressure gradients were shown to have a major influence on the streak spacing.

In addition to establishing that fragmented horseshoe type vortices also represent the dominant dynamic flow structures over rough boundaries, in common with smooth boundary flow, the wider investigation reported by Grass *et al.* (1991) also confirmed that a streaky wall structure exists over rough boundaries. Grass *et al.* (1991) found that the vortical structures embedded in the near-wall flow field exhibited a well defined, preferred spanwise wavelength, which, for fully rough boundary conditions, was shown to scale with the wall roughness length-scale in accord with the predictions of dimensional analysis. In the case of spherical wall roughness-elements with maximum hexagonal packing density, for example, the spanwise streak spacing, λ, scaled by the sphere diameter, k, yielded a λ / k ratio of approximately 3.2. This compares extremely well with the measurements reported by Defina (1992) in which a λ / k ratio of 3.4 was recorded, again for flow over spherical roughness elements.

The early measurements of streak spacing reported by Grass *et al.* (1991) are summarized here for the reader's convenience. New data is presented concerned with the effects of wall distance and the influence of natural pebble roughness. Attention is focused on the question of appropriate scaling of streak spacing for smooth, transitional and fully rough conditions. Starting with the premise that the dominant spanwise wavelength of the instability process generating the wall structure is functionally related to the characteristics of the highly sheared thin layer of fluid immediately in contact with the wall, the possibility of deriving a general scaling relationship is examined.

EXPERIMENTAL ARRANGEMENTS AND PROCEDURES

Measurements were made in a fully developed turbulent boundary layer formed along the upper surface of a flat, horizontal glass plate, supported in an open channel water flow. The glass sided channel was 0.5 m wide and 6.2 m long and a uniform flow depth of approximately 50 mm above the plate was kept constant throughout the entire test programme. Great care was taken with the design and construction of the channel inlet system to eliminate disturbances and provide a closely uniform, low turbulence approach flow at the channel entrance which coincided with the leading edge of the bed plate. The measurement station was located 2 m downstream at a position where the flow was fully developed but remained suitably two-dimensional. Water was recirculated through the channel via a constant head tank which maintained very steady flow conditions. A laser Doppler anemometer was used for the general control measurements and to obtain the mean velocity and Reynolds stress profiles from which the linear total shear stress profile was derived and extrapolated to determine the bed shear stress.

The bed roughness condition was varied from hydrodynamically smooth to fully rough. This was achieved using a range of different diameter glass spheres (1.15, 6 and 12 mm), and also natural rounded pebbles (nominal 9 mm mean sieve diameter), packed in a single layer resting on the upper surface of the glass bed plate. The average height of the tops of the natural pebbles above the glass surface, which was taken as the appropriate physical roughness height, k, for this material, was 7.48 mm, reflecting the fact that the pebbles tended to lie with their smallest dimension vertical.

As described by Grass *et al.* (1991), the spacing characteristics of the low-speed streaks and hence the transverse scaling of the vortical structures in the wall layer were measured using the hydrogen bubble tracer technique. In the present experiments, the hydrogen bubbles were generated from a 25 μm diameter platinum-iridium alloy wire stretched perpendicular to the mean flow direction, and located at variable height, close to and parallel to the horizontal bed.

A pulsed negative voltage was supplied to this wire electrode, releasing thin lines of bubble tracers into the illuminated near-bed flow field. These tracer lines were photographed through a viewing plate just touching the free water-surface, using a 35 mm still camera. Sample photographs are shown in Figure 2.2 for the different bed roughness conditions. As can be seen, the newly released tracer lines retain coherence along their entire length. Photographic images were projected onto the surface of a tablet co-ordinate analyser to define the shape of these coherent tracer lines in digital format for spatial spectrum analysis. Ensemble averages based on 35 individual spectra were calculated to establish the existence or otherwise of any preferred spanwise wavelength and to measure such wavelengths from the spectral peaks. The mean spacing of the low-speed streaks was also determined by direct measurement using a marking technique similar to that adopted by Smith & Metzler (1983). It was found that this method gave closely similar results, confirming that the two approaches are basically focusing on the same physical flow feature. Whilst somewhat more subjective due to the human factor involved in identifying individual streaks, the direct measurement approach has the additional advantage of allowing determination of the distribution characteristics and statistics of the streak spacing.

The general test flow conditions, bed roughness characteristics and values of derived experiment parameters are set out in Table 2.1. Four constant flow rates, Q1, 2, 3, 4 were used throughout the entire test programme. These flow rates were fixed by means of different diameter orifice plates inserted into the supply line between the constant head tank and the channel. With the supply line control valve kept fully open, this system provided excellent day to day reproducibility for a particular test flow. As indicated in Table 2.1, a total of 12 experiments were carried out, three with a hydrodynamically smooth bed (S Q1, 2, 3), three each with the 1.15 and 6 mm diameter spherical roughness elements (K1 Q1, 2, 3 and K6 Q1, 2, 3 respectively), two with the 12 mm diameter spheres (K12 Q1, 2), and one test using the 9 mm rounded pebble roughness (K9 Q4).

Figure 2.2. Plan view of near-wall flow structure visualized by hydrogen bubble tracers generated on a transverse wire parallel to the bed. Flow direction ↑. Test number, bed roughness conditions and wire distance from top of roughness (y_1) as follows: (a) SQ3, smooth, y_1 = 1 mm; (b) K1 Q3, k = 1.15 mm, y_1 = 1 mm; (c) K6 Q3, k = 6 mm, y_1 = 3 mm; (d) K12 Q3, k = 12 mm, y_1 = 3.5 mm; and (e) K9 Q4, k = 7.48 mm, y_1 = 5.5 mm. *Shown on facing page*

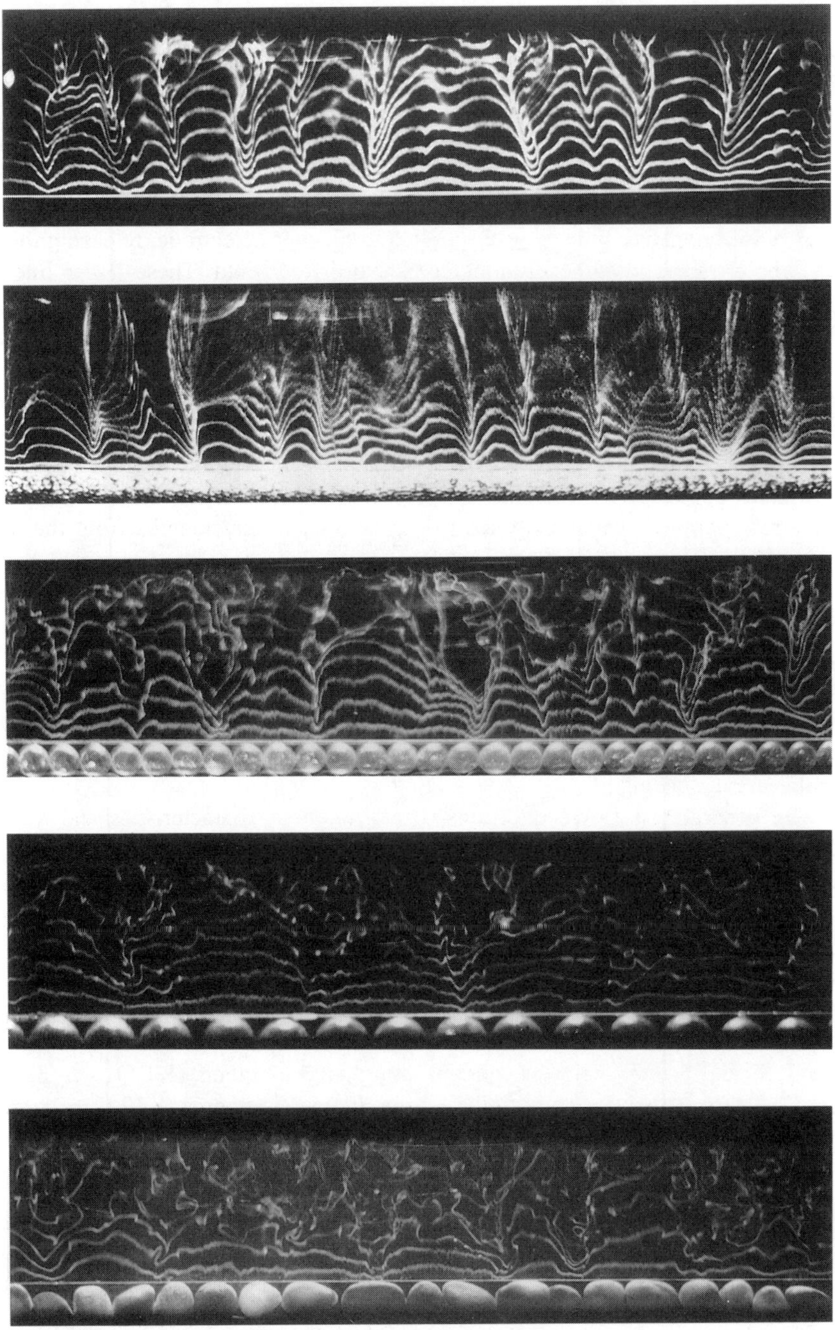

a)

b)

c)

d)

e)

Table 2.1. Flow and bed roughness conditions and derived variables

Bed roughness condition			Smooth					Spherical roughness				Pebble
Experiment		SQ1	SQ2	SQ3	k1Q1	k1Q2	k1Q3	k6Q1	k6Q2	k6Q3	k12Q2	k12Q3	k9Q4
Flow depth†	d mm	50.0	49.9	50.3	49.8	49.7	49.9	49.0	49.1	48.9	48.8	46.9	49.7
Depth average velocity	U mm s^{-1}	222.0	163.3	105.0	227.8	164.8	106.0	219.4	154.6	111.0	155.9	106.0	159.5
Kinematic viscosity	ν mm^2 s^{-1}	1.07	0.94	0.94	1.06	0.94	0.94	1.05	1.00	1.05	1.08	1.03	1.05
Effective bed viscosity	ν' mm^2 s^{-1}	1.07	0.94	0.94	1.39	1.11	1.09	3.03	2.50	1.49	5.21	3.21	3.05
Reynolds number	dU/ν	10335	8669	5613	10702	8713	5611	10239	75.91	5160	7044	5114	7550
Bed shear velocity	U_τ mm s^{-1}	11.02	8.47	6.07	12.48	9.18	6.46	15.91	13.15	8.46	14.82	9.21	13.72
Von Kármán constant	κ	0.394	0.393	0.395	0.392	0.398	0.387	0.37	0.395	0.384	0.420	0.382	0.387
Roughness height	k mm				1.15	1.15	1.15	6.0	6.0	6.0	12.0	12.0	7.48
Roughness Reynolds number	kU_τ/ν				13.6	11.2	7.9	90.9	78.9	48.5	164.7	107.3	97.8
Position of velocity origin‡	y_t mm				0.35	0.45	0.45	1.30	1.30	1.20	2.40	2.40	1.18
y_t / k					0.30	0.39	0.39	0.22	0.22	0.20	0.20	0.20	0.16
Roughness length scale	y_0 mm				0.0201	0.0207	0.0318	0.167	0.174	0.169	0.320	0.327	0.221
Equivalent sand roughness	k_s mm				0.96	0.96	0.96	5.01	5.22	5.07	9.60	9.82	6.63
Roughness Reynolds number	$k_s U_\tau/\nu$				11.3	9.4	6.6	75.9	68.6	40.9	131.7	79.2	86.6
k_s / k					0.83	0.83	0.83	0.84	0.87	0.85	0.80	0.82	0.88

† Measured from the velocity origin, ‡ Measured downwards from the top of the roughness elements

RESULTS AND DISCUSSION

The positions, y_t, of the effective velocity origins below the mean level of the tops of the roughness elements are given in Table 2.1. These origin locations were obtained using a combination of the progressive origin shift method, first suggested by Clauser (1956) and adapted by Perry & Joubert (1963), and linear regression, to establish the best-fit straight line in the logarithmic zone of the mean velocity profile. When scaled by the roughness height, *k*, the resulting y_t / k values indicate that for the fully rough spherical elements, the velocity origin is located close to one fifth of a diameter below the sphere tops. This observation is in very good agreement with previously published results (e.g. Einstein & El Samni, 1949).

The roughness Reynolds numbers, kU_τ / ν (Table 2.1), indicate that fully rough boundary conditions prevailed for tests K6 Q1, 2, K12 Q2, 3 and K9 Q4, whilst tests K1 Q1, 2, 3 and K6 Q3 were respectively in the lower and upper transitional roughness regimes. Under fully rough conditions, the wall roughness can be defined in terms of the Nikuradse equivalent sand/gravel diameter, k_s, defined by $k_s = 30y_0$, where y_0 is the wall roughness length-scale. The latter length-scale is derived from the zero velocity intercept between the logarithmic function fitted to the mean velocity profile in the wall zone and the vertical y axis. In the case of the fully rough experiments, k_s / k has a mean value of 0.83. This reflects the smooth geometry of both the spherical and pebble roughness elements used in the tests. The k_s value of 0.96 mm quoted in Table 2.1 for the 1.15 mm spherical roughness elements, is derived from $k_s = 0.83k$, assuming geometrical similarity with the larger spheres.

The extensive streamwise coherence of the low-speed streaks on the smooth wall, discussed in detail by Smith (this volume), is clearly illustrated in Figure 2.2a. Similar streak coherence is also seen to persist for the early transitional roughness conditions (Figure 2.2b) reflecting the continued dominance of viscous shear transfer and relatively weak molecular type diffusion at the wall. These low-velocity ejection zones are still clearly visible in the near-wall layer flow over the 6 mm diameter spheres (Figure 2.2c). However, their general characteristics and appearance have now changed significantly. The distance between the low-speed streaks has increased as has their width, whereas the extent of their streamwise coherence as visualized by the bubble tracer lines close to the bed, is apparently reduced and disrupted by the strong diffusive action of the intense, small-scale turbulence generated by the roughness elements. These trends are further advanced in the case of the fully rough 12 mm diameter spheres (Figure 2.2d) and the 9 mm pebble roughness (Figure 2.2e).

Alternative visualization techniques using dye and small particles as tracers, employed both in the present investigation and by Defina (1992; this volume), strongly confirm the basic streaky nature of the near-wall flow field over rough boundaries. Geophysical scale evidence is also provided by the elongated streaky patterns commonly observed in the near-ground air flow by wind blown dust, sand

and powder snow, and on water surfaces by wind generated waves (Grass *et al.*, 1991).

The clear presence of low-speed streak ejection zones over the pebble boundary with its randomly shaped and distributed roughness elements, provides convincing evidence that these dynamic structural features of near-wall turbulence are not simply a product of the regular packing geometry assumed by spheres. Nor is it the case that the spanwise position of the streaks are linked to particular individual roughness elements immediately upstream of the bubble tracer wire, since the streaks were observed to randomly shift their spanwise position with time.

As discussed and illustrated by Grass *et al.* (1991), a remarkable feature of the rough-wall flow is the apparent ability to order itself very rapidly over a small vertical distance above the roughness elements and become decoupled from the local influence of individual elements. This was demonstrated by contrasting the near-bed flow field characteristics visualized with the hydrogen bubble wire successively located at heights of 0.5 and 3 mm above the tops of the 12 mm spheres. Very close to the tops, at a height of 0.5 mm, the bubble tracers respond to the three-dimensional velocity field and pick up the intense, small-scale disturbances in the separating flow round the individual roughness elements. However, the visualization traces show very little evidence of this highly disturbed localized flow state with the wire raised to just 3 mm above the tops of the 12 mm spheres as shown in Figure 2.2d.

Ensemble-averaged spectra for the spanwise velocity fluctuations, corresponding to the mean flow and bed roughness conditions in Figures 2.2a–e exhibit a well defined single peak (Figures 2.3a–e) that indicates the presence of a dominant spanwise wavelength, λ, in the near-wall structure. The ability of the spectral peak method to pick out a preferred wavelength in the velocity perturbations was confirmed by the measured value of 12 mm with the hydrogen bubble wire 0.5 mm above the roughness tops which coincides exactly with the sphere diameter as expected (Grass *et al.*, 1991). In the case of the smooth-wall flow $\lambda^+ = \lambda U_\tau / \nu = 101$, in good agreement with previous measurements (Kline *et al.*, 1967; Smith & Metzler, 1983).

For the roughness Reynolds numbers under the present test conditions (Table 2.1), λ should be largely independent of Reynolds number for the 6 and 12 mm diameter roughness elements. In these circumstances, following the suggestion of Grass (1971), λ should scale with the roughness dimension, k. In other words, $\lambda / k =$ constant for the geometrically similar roughness elements and packing used in these tests. This expectation is closely confirmed with measured λ / k values of 3.25 and 3.18 for the 6 and 12 mm spheres, respectively, based on the peak λ values derived from Figures 2.3c and 2.3d (Grass *et al.*, 1991). The dominant spanwise wavelength for the pebble roughness (Figure 2.3e) produced a corresponding λ / k ratio of 3.11 in close agreement with the corresponding values for the spherical roughness elements. Defina (1992) also measured the mean spacing between the low-speed streaks formed over a bed comprised of spherical

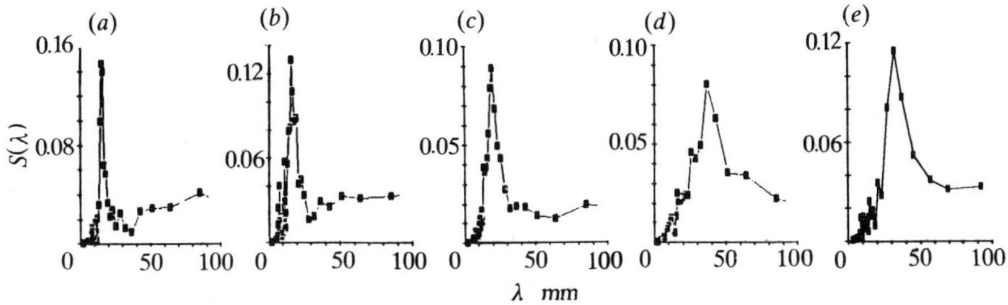

Figure 2.3. Ensemble averages of wavelength spectra of spanwise fluctuations in streamwise velocity derived from hydrogen bubble tracers as illustrated in Figure 2.2. Flow conditions and y_1 values of (a)–(e) correspond with Figure 2.2 (a)–(e) respectively

roughness elements with an average diameter of 10 mm, and found $\lambda / k = 3.4$ for fully rough conditions in remarkably good agreement with the λ / k values obtained in the present investigation.

The spacing between individual low-speed streaks was also measured directly by marking their visually identified positions. It was found that a log-normal probability function fitted all the resulting distribution histograms extremely closely. The coefficients of variation (standard deviation/mean) for the distributions were very narrowly grouped about a mean value of approximately 0.4 showing little systematic departure with either changing bed roughness condition or wall distance. All the measured distributions were therefore well represented by a single universal probability function of log-normal form when plotted with the streak spacing normalized by the mean value. These findings are in close accord with the results previously presented by Smith & Metzler (1983) relating exclusively to distributions of streak spacing in smooth-wall flow.

In the case of the fully rough-wall conditions involving relatively large roughness elements, a difficulty arises in objectively defining the height above the roughness tops at which the initial near-wall streak spacing should be measured. As previously discussed, with the bubble tracer wire located very close to the tops of the roughness elements, the observed sharp peak in the spacing spectra simply registers the transverse spacing of the adjacent element tops. It was found that as the wire height was progressively raised through a thin zone approximately $k/3$ thick above the roughness tops, the spectral peak associated with the direct influence of the local roughness geometry rapidly attenuated, leaving a broad band spectra with no dominant peak. At a height of approximately $k/3$ above the roughness tops, the spectra once again exhibit a very clear dominant peak but now coincident with the mean low-speed streak spacing and decoupled from the local topography of the bed roughness. A height of $k/3$ was therefore used for determination of the initial streak spacing at the wall. This height, not surprisingly, also coincided approximately with the level at which the near-bed mean velocity profiles measured above the tops and over gaps between the roughness elements, merged together on the vertical y-axis.

These observed features of the wall contact fluid dynamics strongly suggest that the thin layer of fluid flowing just below and above the tops of the roughness elements, and subjected to high shear and the violent chaotic interaction of small-scale vortical eddies generated in the separation zone downstream of individual elements, can be viewed as a reconstituted fluid with enhanced effective viscosity. This fluid might then be expected to be subject to the same complex mechanisms and patterns of instability as occur in the viscous sublayer on smooth walls, producing similar low-speed streaks. The generalized scaling concept described below, stems directly from this simple conceptual model of the strongly sheared layer of fluid immediately in contact with the bed.

On the basis of the evidence discussed previously, it has been established that low-speed streaks and the vortical bursting structures responsible for their generation, are a universal feature of wall layer shear flow *irrespective* of boundary roughness condition. If attention is concentrated specifically on the mean streak spacing, λ, at the wall, then dimensional analysis indicates that λ will scale with the appropriate wall layer length-scale. This length scale takes the form, ν / U_τ, for hydrodynamically smooth boundaries and, $k_s = 30y_0$, for fully rough boundaries. In the transitional roughness regime λ will scale with a functional combination of these viscous and physical roughness scales.

The measured values of λ at the wall, respectively scaled by ν / U_τ and k_s are shown plotted as a function of roughness Reynolds number, $k_s^+ = k_s U_\tau / \nu$, in Figures 2.4 and 2.5. Figure 2.4 indicates that $\lambda^+ = \lambda U_\tau / \nu$ tends to a limiting value close to 100 as k_s^+ tends to zero for smooth-wall conditions. Similarly, at the other extreme, Figure 2.5 demonstrates that for fully rough conditions λ / k_s becomes independent of viscosity and assumes a second apparently universal and constant limiting value of approximately 3.8.

The lower range transitional data produced by the 1.5 mm spherical roughness indicates progressive separation from the smooth wall $\lambda^+ = 100$ curve as k_s^+ increases (Figure 2.5). Also shown in Figure 2.5 is the roughness function derived from Nikuradse's (1933) calibration of the universal logarithmic relationship for the mean velocity profile in the inner region of circular pipe flow. The transitional zone data is seen to merge with the smooth-wall logarithmic relationship for low k_s^+, and with the Reynolds number independent rough wall version of the logarithmic profile for k_s^+ values in excess of 70. In the transition zone, the velocity profile data has a somewhat complex pattern as indicated in Figure 2.5. It is unfortunately not possible to identify any corresponding pattern in the λ / k_s variation due to the sparsity of data in the transition zone from the present tests. The tests reported by Defina (1992) provide additional data in the transition zone and give some indication that the curve fitted to the λ / k_s data in this region may drop slightly below the $\lambda / k_s = 3.8$ line plotted in Figure 2.5, prior to rising and merging with this fully rough-wall relationship.

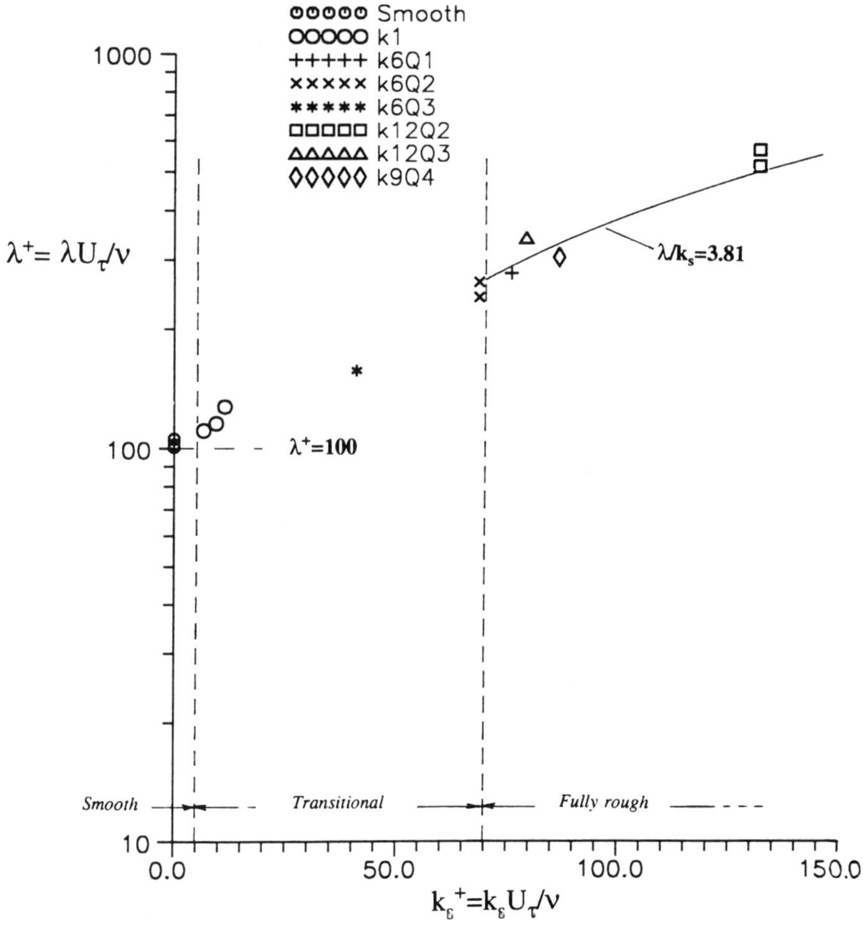

Figure 2.4. Mean wall-streak spacing $\lambda U_\tau/\nu$ as a function of roughness Reynolds number k_s^+

On either side of the transition zone marked in Figures 2.4 and 2.5, the relevantly non-dimensionalized mean streak spacing, λ, thus takes up two apparently universal constant values. The fact that the mechanics of streak generation have been shown to be closely similar in both these extreme cases of wall roughness condition, suggests the possibility of a generalized approach to scaling the near-wall flow structure. If it is assumed that the vortical bursting structures and associated low-speed streaks result from violent non-linear instability in the highly sheared vorticity generating layer in contact with the wall, then it can be plausibly speculated that the mean spanwise wavelength, λ, of such an

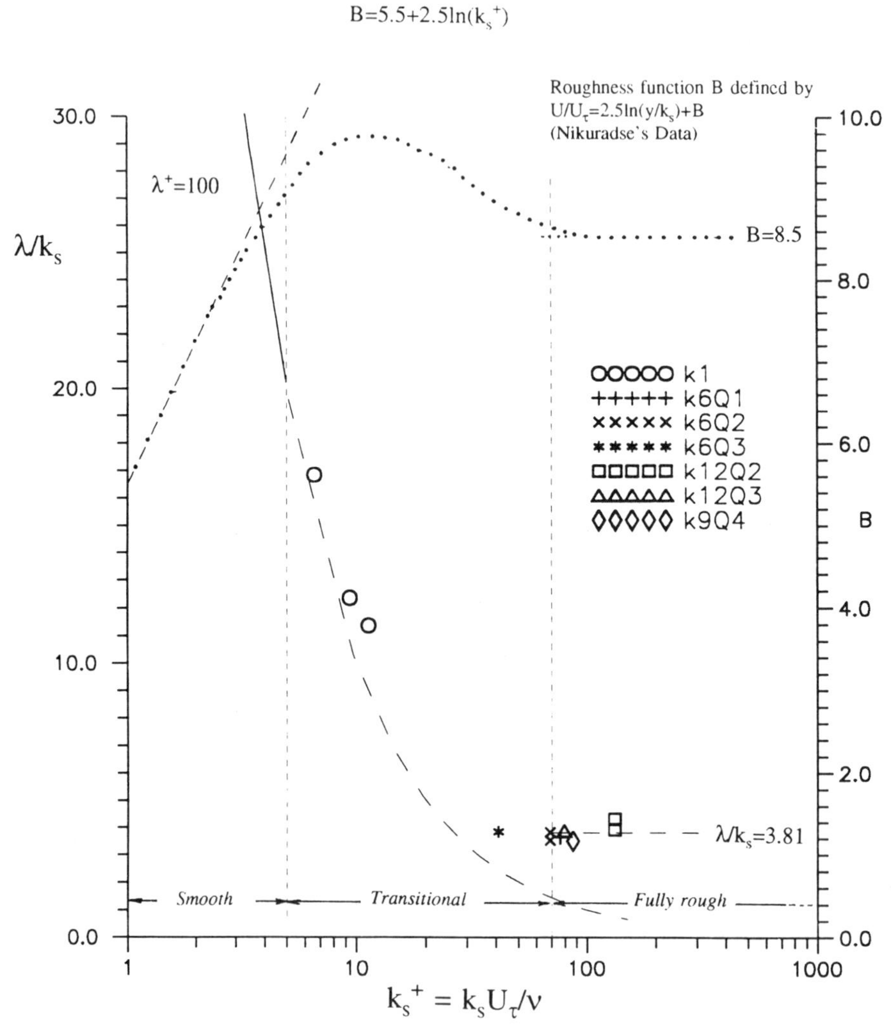

Figure 2.5. Mean wall-streak spacing λ/k_s as a function of roughness Reynolds number k_s^+

instability process is likely to be directly linked to and scaled by the defining physical characteristics of this shear layer.

Relevant defining characteristics for these wall contact shear layers are the fluid density, the shear stress transmitted by the layer which closely approximates to the wall shear stress, and the rate of strain or mean streamwise velocity gradient across the layer. Applied to the thin viscous sublayer on a smooth boundary, where the velocity profile is closely linear and the velocity gradient is constant, dimensional analysis gives ν / U_τ as the characteristic shear layer length-scale. This length scale has been shown to correlate the measured values of mean streak spacing on smooth walls extremely well. Returning to the previously proposed conceptual model of the thin shear layer in contact with rough walls, the fluid contained within this layer is considered to assume a reconstituted state with an enhanced effective viscosity, ν'. Dimensional analysis now yields what can be viewed as a generalized wall length-scale ν' / U_τ.

In order to test this hypothesis by applying the proposed generalized length-scale to the measured streak spacing data for the various wall roughness conditions, it was first necessary to estimate values of effective wall viscosity ν'. This in turn required estimation of the near-wall strain rate or velocity gradient. Estimating the mean velocity gradient close to a rough bed is unfortunately fraught with difficulties and uncertainties. The mean velocity profile in the layer containing the tops of the roughness elements is clearly sensitive to the location of the vertical measuring line relative to the individual elements, with the local mean velocity profile over the roughness element crests differing substantially from that over the gaps between the elements. To obtain a suitable overall mean velocity profile it was therefore necessary to take an ensemble average of four local velocity profiles (for each test) measured at a representative sample of locations over the rough bed surface. A curve was then fitted to this averaged profile to derive the velocity gradients close to the tops of the roughness elements. Another difficulty arises for rough-wall flows since, unlike the smooth boundary case, the velocity gradient and hence the effective wall viscosity ν', at the bed are not constant, but vary with height above and below the tops of the surface roughness. Faced with this inherent lack of uniqueness, in order to obtain a velocity gradient representative of the mean strain rate at the wall, it was decided on a physically plausible, albeit partially arbitrary basis, to calculate the velocity gradient, and hence ν', at the point midway between the roughness tops and the velocity origin ($y = y_t / 2$).

The λ data for the near-wall region is shown replotted in Figure 2.6 when scaled by the derived effective viscosity length-scale, ν' / U_τ. As can be seen, the resulting non-dimensionalized λ values are now largely independent of wall roughness Reynolds number k_s^+. This important observation appears to confirm ν' / U_τ as a generalized length-scale for the low-speed streak spacing and the associated near-wall turbulence structure irrespective of boundary roughness condition. A further new finding emerges from Figure 2.6. Namely, that by employing the eddy viscosity length-scale, the resulting non-dimensionalized streak spacing adjacent to the bed takes up an apparently universal value close to 100 irrespective of wall roughness condition, and identical to the smooth wall value. This evidence would appear to strongly support the proposed conceptual model of the thin shear layer enveloping the roughness tops, as a reconstituted fluid of enhanced effective viscosity and subject to similar modes of spanwise instability as occur in the viscous sublayer on smooth walls.

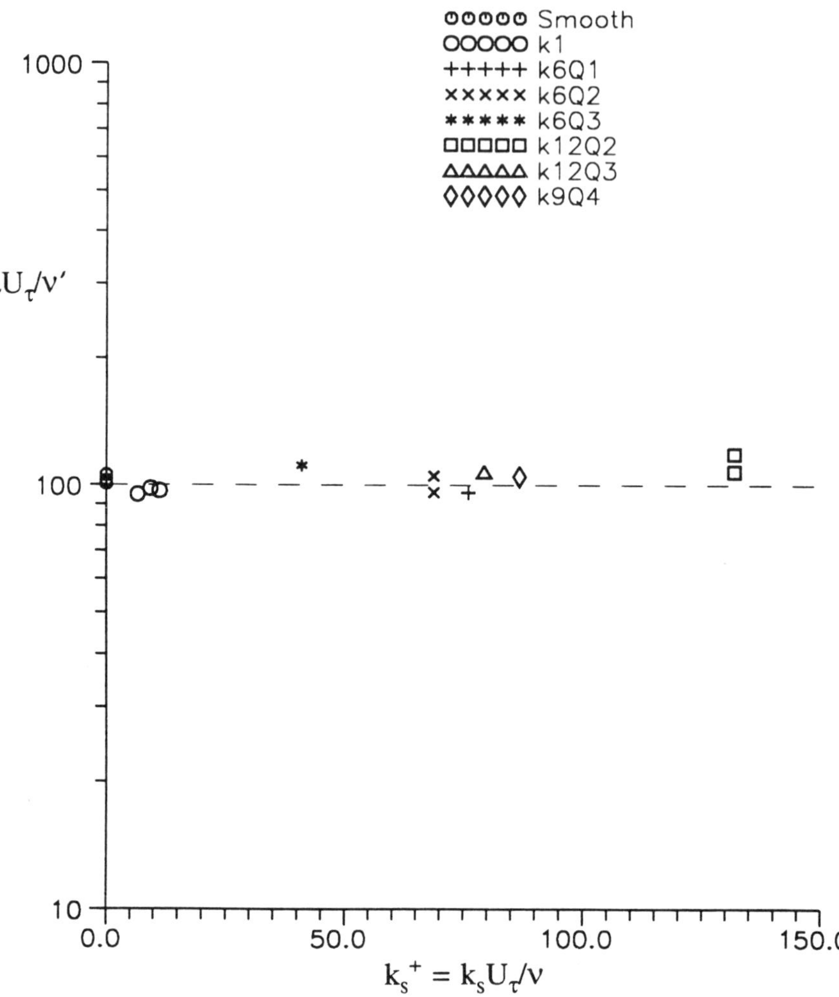

Figure 2.6. Mean wall-streak spacing $\lambda U_\tau/\nu$ (scaled by effective viscosity wall length-scale ν'/U_τ) as a function of roughness Reynolds number k_s^+

As discussed previously, the magnitude of ν' and therefore $\lambda U_\tau/\nu'$, is affected by the choice of vertical position, relative to the roughness tops, at which the velocity gradient is estimated. Since the curve fitted to the near-wall velocity profile data is semi-logarithmic in form, the local velocity gradients are inversely proportional to the origin distance, y. If, therefore, the wall velocity gradients and corresponding ν' values are calculated at the height of the roughness tops and at a position $2y_t/3$ below the roughness tops ($y = y_t/3$), the wall values of $\lambda U_\tau/\nu'$ become 50 and 150 respectively. The latter value can only be taken as a reasonable estimate however, since reliable velocity measurements could only be obtained down to the level of the roughness tops. Estimation of velocity gradients below this level thus becomes increasingly uncertain with increasing extrapolation distance. This highlights the future need for a full spatial mapping of the complex flow field above and below roughness element tops prior to drawing definitive conclusions. It should be noted in this context that the widely quoted value of $\lambda^+ = 100$ for smooth-wall streak spacing, is in fact obtained by effectively assuming that the relevant flow instability is scaled by the rate of strain at the bed. Since the exact location and nature of the instability process is far from fully understood , this is also an arbitrary choice which affects the λ^+ value. In summary therefore, whilst there remain real concerns over the quantitative determination of appropriate wall values of ν', it is felt that these considerations do not materially detract from the wider significance of the present findings.

Figure 2.7 shows a plot of low-speed streak spacing as a function of wall distance. It is once again evident that applying the effective viscosity length-scale produces an excellent correlation of all the present data for both smooth and rough walls. In common with the previously observed smooth-wall behaviour, outside the thin high shear layer at the wall ($yU_\tau/\nu' > 20 - 30$), the rough-wall streak spacing also increases relatively rapidly with distance from the boundary.

The apparent existence of generalized wall scaling adds considerable support to the hypothesis that the same hydrodynamic instability mechanism is responsible for generating the streaky flow pattern observed over both smooth and rough walls. Such patterns represent the imprint of the bursting events induced by and linked to the instability generated vortical structures discussed by Grass *et al.* (1991) and Smith (this volume). The relevant spanwise length-scale of this instability appears to be determined solely by the relevant layer shear stress and the strain rate, irrespective of whether or not the shear is carried by predominantly molecular or turbulence momentum transport. In principle therefore, as suggested by Grass *et al.* (1991), there appears to be no reason why this same instability mechanism should not manifest itself throughout the entire boundary layer thickness to produce the scale hierarchy of vortical structures proposed by Perry & Chong (1982). These arguments are further reinforced by the important findings of the numerical simulation study reported by Rogers & Moin (1987), which demonstrated that horseshoe vortices form in a sheared field of homogeneous turbulence.

CONCLUSIONS

Recent experimental and numerical simulation studies have shown that the dynamics of boundary layer turbulence over smooth and rough walls is dominated by powerful, three-dimensional vortical structures embedded in the shear flow.

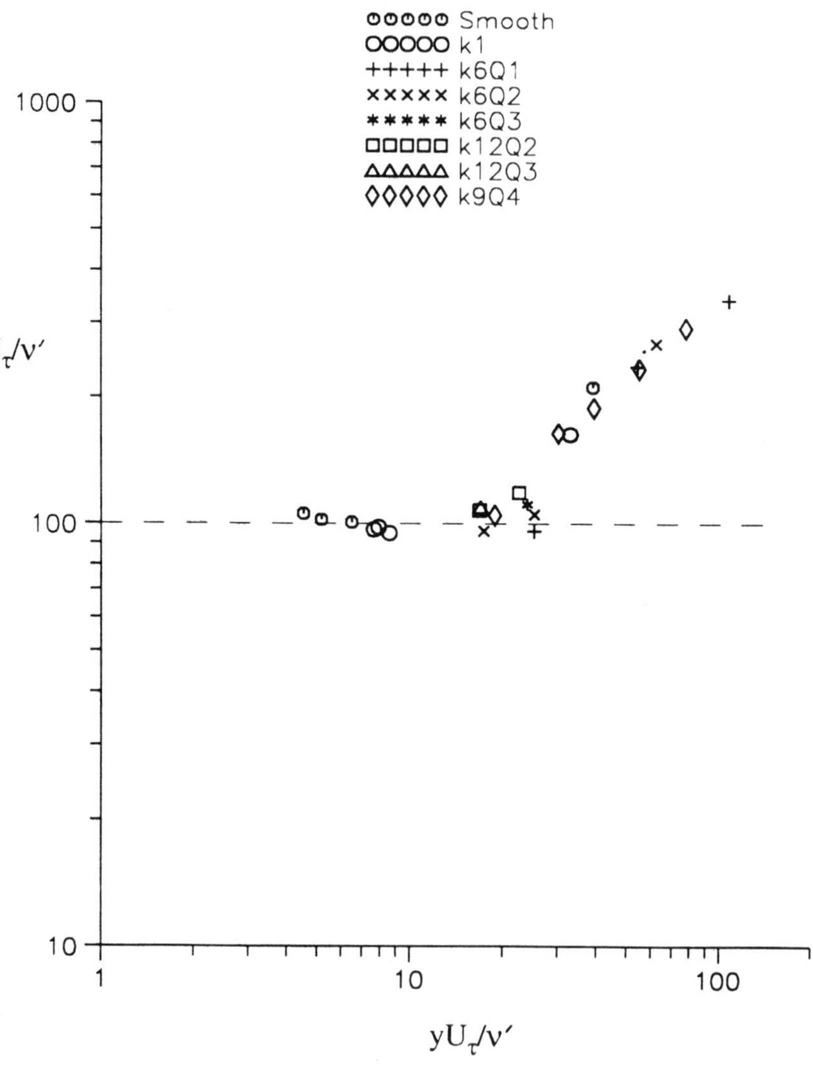

Figure 2.7. Mean streak spacing $\lambda U_\tau/\nu$ as a function of wall distance yU_τ/ν' (scaled by effective viscosity wall length-scale ν'/U_τ)

These flow structures are responsible for the bursting events in which low momentum fluid is violently ejected from the wall region, generating low-speed streaks.

The study reported in the present paper has focused attention on the spanwise spacing between these low-speed streaks which essentially coincides with the scale of the wall layer flow structures. A conceptual model is proposed in which the thin, highly sheared layer enveloping the tops of the roughness elements, is viewed as being comprised of a reconstituted fluid of enhanced effective viscosity. The defining dynamic characteristics of this layer, namely the transmitted shear stress and rate of strain, can then be used to establish a generalized wall length-scale.

This effective viscosity length-scale has been shown to provide a direct universal link between the smooth and rough-wall streak spacing, strongly supporting the basic validity of the underlying concepts in the shear layer model hypothesis. The findings are seen to have wider implications regarding the hierarchy of vortical structure scales thought to be present in large-scale, high Reynolds number boundary layers such as occur in geophysical flows. Most geophysical flows take place over fully rough boundaries where the spacing of low-speed streak, fluid ejection zones and their generating vortical structures adjacent to the wall, have been shown in this investigation to scale directly with the roughness element size, independent of Reynolds number. Using the present small-scale test results, it can therefore be directly predicted, for example, that the cross-wind spacing of fluid ejection zones (Figure 2.1), lifting and dispersing polluted street-level air in city centres will be of the order of 100 roughness length-scales ($100y_0$). This will typically translate as 5–20 building heights depending on their plan distribution and also the street orientation. In this context, the generating vortices can thus be perceived as performing the indispensable role of huge, geophysical scale vacuum cleaners. Similar spacing predictions can also be made for the zones of violent ejection and suspension of wind blown sand over desert dunes, and over corresponding dunes formed on sandy beds under large-scale, benthic boundary layer flows.

REFERENCES

Barenblatt, G.I. 1993. 'Scaling laws for fully developed turbulent shear flows. Part 1. Basic hypotheses and analysis', *J. Fluid Mech.*, **248**, 513–520.

Best, J. 1992. 'On the entrainment of sediment and initiation of bed defects : insights from recent developments within turbulent boundary layer research', *Sedimentology*, **39**, 797–811.

Brereton, G.J. & Hwang, J.L. 1994. 'The spacing of streaks in unsteady turbulent wall-bounded flow', *Phys. Fluids*, **6**, 2446–2454.

Clauser, F.H. 1956. 'Turbulent boundary layer', *Adv. Appl. Mech.*, **4**, 1–51.

Defina, A. 1992. 'Interasse traversale delle explosioni turbolente in regime di parete scabra', *Atti dell'Instituto Veneto di SS. LL. AA.*, **CL**, 1–13.

Drake, T.G., Shreve, R.L., Dietrich, W.E., Whiting, P.J. & Leopold L.B. 1988. 'Bedload transport of fine gravel observed by motion picture photography', *J. Fluid Mech.*, **192**, 193–217.

Einstein, H.A. & El Samni, E.A. 1949. 'Hydrodynamic forces on a rough wall', *Rev. Mod. Phys.*, **21**, 520–524.

Grass, A.J. 1971. 'Structural features of turbulent flow over smooth and rough boundaries', *J. Fluid Mech.* **50**, 233–255.

Grass, A.J. 1983. 'The influence of boundary layer turbulence on the mechanics of sediment transport', in Sumer, B.M. & Müller, A. (Eds) *Proc. Euromech 156: Mechanics of Sediment Transport*, A.A. Balkema, Rotterdam, 3–17.

Grass, A.J., Stuart, R.J. & Mansour-Tehrani, M. 1991. 'Vortical structures and coherent motion in turbulent flow over smooth and rough boundaries', *Phil. Trans. Roy. Soc. London,* **A336**, 35–65.

Hunt, J.C.R., Vassilicos, J.C. & Kevlahan, N.K.R. 1994. 'Turbulence: A state of nature or a collection of phenomena?', in Branover, H. & Unger, Y. (Eds) *Progress in Turbulence Research,* 7th Beer Sheva Int. Sem. on MHD flows and turbulence, Beer Sheva, Israel, Feb. 1993, Progress in Astronautics and Aeronautics Series, AIAA Washington, **162**, 1–18.

Kline, S.J., Reynolds, W.C., Schraub, F.A. & Runstadler, P.W. 1967. 'The structure of turbulence in boundary layers', *J. Fluid Mech.*, **30**, 741–773.

Kline, S.J., and Robinson, S.K. 1989. 'Turbulent boundary layer structure: Progress, status and challenges', in Gyr, A. (Ed) *Structure of Turbulence and Drag Reduction*, Hemisphere, New York, 3–22.

Kuchemann, D. 1965. 'Report on the IUTAM symposium on concentrated vortex motions in fluids', *J. Fluid Mech.*, **21**, 1–20

Leeder, M.R. 1983. 'On the dynamics of sediment suspension by residual Reynolds stresses -confirmation of Bagnold's theory', *Sedimentology,* **30**, 485–491.

Nikuradse, J. 1933. *Strömungsgesetze in Rauhen Rohren*, Kaiser Wilhelm-Institut für Stromungsforschung, Göttingen, Forschungsheft **361**, 22 pp. (Translation: *Nat. Advisory Comm. Aero.*, Tech. Mem. **1292**, 1950).

Perry, A.E. & Chong M.S. 1982 'On the mechanism of wall turbulence', *J. Fluid Mech.,* **119**, 173–217.

Perry, A.E. & Joubert, P.N. 1963 'Rough wall boundary layers in adverse pressure gradients', *J. Fluid Mech.,* **17**, 193–211.

Perry, A.E., Henblest, S. & Chong M.S. 1986. 'A theoretical and experimental study of wall turbulence', *J. Fluid Mech.,* **165**, 163–199.

Perry A.E., Lim K.L. & Henblest S.M. 1987. 'An experimental study of the turbulence structure in smooth- and rough-wall boundary layers', *J. Fluid Mech.*, **177**, 437–466.

Robinson, S.K. 1991. 'Coherent motions in the turbulent boundary layer', *Ann. Rev. Fluid Mech.*, **23**, 601–639.

Rogers, M.M. & Moin, P. 1987. 'The structure of the vorticity field in homogeneous turbulent flows', *J. Fluid Mech.,* **176**, 33–66.

Simpson, R.L. 1991. 'The structure of the near-wall region of two-dimensional turbulent separated flow', *Phil. Trans. Roy. Soc. London*, **A336**, 5–17.

Smith, C.R. & Metzler, S.P. 1983. 'The characteristics of low-speed streaks in the near-wall region of a turbulent boundary layer', *J. Fluid Mech.*, **129**, 27–54.

Smith, C.R., Walker, J.D.A., Haidari, A.H. & Sobrun, U. 1991. 'On the dynamics of near-wall turbulence', *Phil. Trans. Roy. Soc. London*, **A336**, 131–195.

Theodorsen, T. 1952. 'Mechanism of turbulence', *Proc. of the 2nd Midwestern Conf. on Fluid Mechanics.*, Ohio State University, Columbus, 1–18.

Townsend, A.A. 1976. *The Structure of Turbulent Shear Flow*, Cambridge University Press, 429 pp.

Williams, J.J., Thorne, P.D. & Heathershaw, A.D. 1989. 'Measurements of turbulence in the benthic boundary layer over a gravel bed', *Sedimentology,* **36**, 959–971.

NOTATION

d flow depth

k	roughness height
k_s	equivalent sand roughness diameter
k_s^+	$k_s U_\tau / \nu$, roughness Reynolds number
U	depth-average velocity
U_τ	bed shear-velocity
y	vertical height above velocity origin
y_1	height of bubble tracer wire above top of roughness elements
y_o	roughness length-scale
y_t	depth of velocity origin below roughness tops
κ	von Kármán constant
λ	mean transverse spacing between low-speed streaks
λ^+	dimensionless transverse spacing between low-speed streaks $= \lambda U_\tau / \nu$
ν	kinematic viscosity
ν'	effective (eddy) bed viscosity

3 Laboratory Observations of Particle Entrainment into Suspension by Turbulent Bursting

MARCELO GARCÍA, YARKO NIÑO & FABIÁN LÓPEZ
Department of Civil Engineering, University of Illinois at Urbana-Champaign, 250 N. Mathews, Urbana, IL 61801, USA

ABSTRACT

Experimental observations on the phenomenon of sediment particle entrainment into suspension in the wall region of a turbulent open-channel flow are reported over both smooth and transitionally rough beds. A high-speed video system, capable of recording up to 1000 frames per second, was used in the experiments to register flow visualizations and particle motion. In some of the experiments, flow visualizations were recorded synchronously with hot-film measurements of streamwise flow velocity, which yielded information on the velocity field associated with the coherent flow structures. Shear layers typically observed in the near-wall region of the open-channel flow interact with sediment particles lying on the channel bottom and eventually result in the particles being entrained into suspension. The results obtained here show that this process is effective in entraining particles even when they are totally immersed in the viscous sublayer, contrary to previous work. The experiments in flows with a rough bed indicate that hiding effects between grains tend to preclude the entrainment of particles with sizes finer than that of the roughness elements. The presence of roughness elements does not affect the main characteristics of flow ejections (quadrant-2 events), which are ultimately responsible for particle entrainment into suspension.

INTRODUCTION

The streamwise velocity field in the viscous sublayer and buffer region of a smooth-wall turbulent boundary layer is organized into alternating narrow streaks of high- and low-speed fluid, having a mean spacing of about 100 wall units near the solid boundary, which are quite persistent in time (Robinson, 1991). The majority of the turbulence production in the entire flat-bed boundary layer occurs in the near-wall region during intermittent, quasi-periodic events consisting of violent outward ejections of low-speed fluid and inrushes of high-speed fluid towards the wall. This near-wall turbulence production process is usually referred to as the 'burst-sweep' cycle. There also exists sufficient evidence that this process occurs, with similar characteristics, in the case of rough-wall boundary layers (Grass, 1971; Grass *et al.*,

Coherent Flow Structures in Open Channels.
Edited by P.J. Ashworth, S.J. Bennett, J.L. Best and S.J. M^c^Lelland.

1991; Grass & Mansour-Tehrani, this volume). The intermittent nature of the wall region is associated with the presence of coherent flow structures. For instance, flow visualizations have shown the existence of diverse vortical structures, with inclined 'horseshoes' or 'hairpins', although more recently a funnel shape has also been proposed (Kaftori *et al.*, 1994), whilst numerical data bases (e.g. Jimenez *et al.*, 1988; Guezennec *et al.*, 1989) as well as particle image velocimetry (PIV) (e.g. Liu *et al.*, 1991; Urushihara *et al.*, 1993) reveal the existence of inclined, thin shear layers of concentrated spanwise vorticity. All these more-or-less organized features have been called *coherent structures*, meaning 'a three-dimensional region of the flow over which at least one fundamental flow variable exhibits significant correlation with itself or with another variable over a range of space and/or time that is significantly larger than the smallest local scales of the flow'(Robinson, 1991).

Sutherland (1967), motivated by the observations of Vanoni (1964), speculated that the mechanism for entraining sediment grains into suspension would involve interactions between turbulent vortices and particles in the near-bed region of the flow. Although the details of such interactions as conceptualized by Sutherland in his experiments, and recently revisited by Wells (1992), are not totally correct in the light of present knowledge of turbulence structure in the wall region of boundary layer flows, his hypothesis delineates the basic mechanics of such phenomena which has been verified more recently by a number of experimental studies.

Several studies have investigated the implications of the bursting phenomenon on the mechanics of particle transport in the near-wall region of boundary layers. Indeed, different experimental techniques have been applied to visualize the flow field in this region with the aims of both inferring the mechanism by which particles are entrained into suspension and also, in some cases, analysing the modulation of turbulence due to the presence of particles (e.g. Grass, 1974; Sumer & Oguz, 1978; Sumer & Deigaard, 1981; Ashida & Fujita, 1986; Yung *et al.*, 1988; Rashidi *et al.*, 1990; Kaftori *et al.* 1995a, 1995b). In the most recent of these studies, Kaftori *et al.* proposed a funnel type of vortex as the dominant coherent structure near the wall of a turbulent boundary layer. They claim to have found sufficient evidence to conclusively assert that coherent structures are the main factor affecting particle motion near a solid boundary in turbulent flows, as well as deposition and entrainment. Nevertheless, Kaftori *et al.*'s results correspond only to polystyrene particles with a specific density of 1.05 and sizes in the range from 100 to 900 μm. Likewise, most of the other studies mentioned above have used almost neutrally buoyant particles of rather large sizes, which may exhibit a behaviour different to that of much heavier, albeit smaller, natural sand particles. From the studies cited above, only that of Grass (1974) involved natural sediment, however his rather brief report does not provide enough details about the experiments nor sufficient or conclusive data. In summary, to date few precise descriptions exist concerning the particle-turbulence interactions responsible for the transport of sediment in the wall region of open-channel flows. This may be due in part to a lack of sufficient data that provide both the information and spatial-temporal resolution needed to analyse kinematic and dynamic characteristics of sediment particle motion.

In this paper, some experimental results regarding particle-turbulence interactions in the near-wall region of a turbulent open-channel flow which result in particle entrainment into suspension are reported. Coherent flow structures associated with flow ejection events are characterized and the motion of particles moving over a smooth boundary as well as over a transitionally rough bed are investigated using both particle-flow visualization and particle tracking techniques. These experimental results differ from others in that heavy, although small, natural sand particles were observed to interact with the turbulence of the flow, as opposed to the light (almost neutrally buoyant), larger particles that have been used in previous studies. Also, the present investigation concentrates solely on the processes occurring in the wall layer of open-channel flows, as opposed to other studies that analyse almost the whole flow field, thus providing a better spatial resolution of particle motion during the initial stages of particle entrainment.

EXPERIMENTS

EXPERIMENTAL SETUP

The experiments were conducted in a rectangular open channel 18.6 m long, 0.30 m wide, and 0.28 m high, with the bed slope set to a value of about 9×10^{-4}. The 0.90 m long test section was located 12 m downstream from the entrance and had the right wall made of Plexiglas to allow particle tracking and flow visualization. A solution of white clay in water was used as the flow tracer with the injection point located approximately 25 mm upstream from the centre of the field of view of the camera. A high-speed video system, the Kodak Ektapro TR Motion Analyser, which has the capability to record up to 1000 frames s^{-1}, was used to record particle motion as well as flow structures developing near the bed. A strobe light with a flash-duration of 20 μs was synchronized with the high-speed video system to provide whole-field illumination. The video images acquired were downloaded from special high-speed Kodak tapes onto standard VHS tapes, and then digitized into a personal computer using a frame grabber. The images were then analysed using the National Institute of Health's *IMAGE* public domain software. A hot-film, ruggedized hemispheric velocity probe (TSI model 1239W), was used to measure the streamwise component of flow velocity at different distances from the bed. The sensor was located within the field of view of the camera approximately 40 mm downstream from the tracer-injection point. The sensing element is a quartz coated platinum film band located at the end of the hemisphere which is 0.12 mm wide and has a diameter of 0.75 mm. For the range of Reynolds numbers reported here, the dimensions of the probe correspond to approximately 12 to 24 viscous units. Using this kind of probe in water has the advantage that due to the high heat transfer rate between sensor and fluid, the heat conduction loss to the quartz rod becomes comparatively negligible, whilst keeping the frequency response flat and uniform up to very high frequencies (typically 30 kHz). Data acquisition was made with an analogue-to-digital card connected to a personal computer and appropriate software. For each flow depth, the mean bed shear stress was determined by a best fit to the logarithmic law using five mean-data points. The video system outputs a synchronization signal which was employed to trigger the acquisition of data with the hot-film sensor simultaneously with the video recording. A schematic view of the experimental set-up is shown in Figure 3.1.

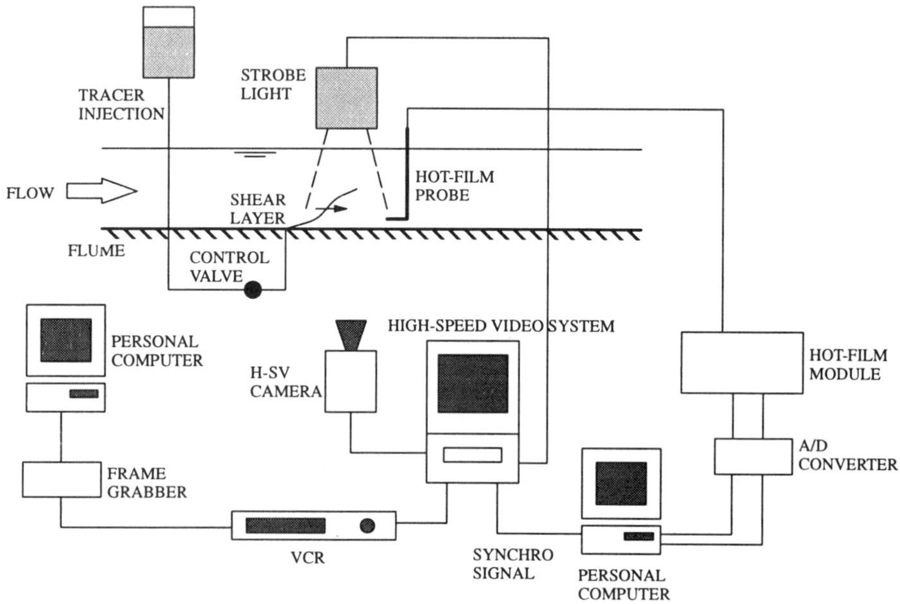

Figure 3.1. Schematic view of experimental set-up

EXPERIMENTAL METHOD

Two different sets of experiments were conducted: (i) Series F, which was aimed at characterizing near-bed coherent flow structures; and (ii) Series P, which was oriented towards characterization of the interactions between particles and flow structures leading to particle entrainment into suspension. The experiments of Series F involved synchronized measurement of flow velocity using the hot-film probe and recording of flow visualization images in the near-wall region using the high-speed video system, while the experiments of Series P consisted of visualization of particle motion during entrainment events, in some cases complemented with flow visualizations of coherent structures.

In the following discussion x, y and z denote co-ordinates in the streamwise, vertical and transverse directions, respectively. Also, wall units denote characteristic scales formed using the flow shear velocity, u_*, and the kinematic viscosity of water, ν, such that for example, ν/u_* represents a length scale, u_* represents a velocity scale, and ν / u_*^2 represents a time-scale. Variables with the superscript $^+$ have been made dimensionless with wall units.

The Series F experiments were conducted over a smooth surface only. The clay solution tracer was injected through an orifice in the channel bottom, to act as a marker for flow structures developing at the wall. The tracer injection velocity was not matched precisely with that of the flow. However, the injection rate was controlled to minimize disturbances to the flow and was adjusted to allow the plume of dye to remain attached to the bed for some distance from the injection point before being ejected away from the bottom. This technique allows visualization of flow ejections fairly well, but does not provide visual information concerning flow structure in the region above the ejection, nor is it efficient in marking sweep events. For the velocity measurements, the hot-film sensor was located at different distances from the bed. Results corresponding to measurements taken at four different dimensionless vertical locations above the bed are reported in this paper, namely $y^+ = 38$ ($Re = 5300$), $y^+ = 52$ ($Re = 12\,400$), $y^+ = 64$ ($Re = 18\,100$) and $y^+ = 80$ ($Re = 5300$), (or equivalently, $y/h = 0.10$, 0.07, 0.06 and 0.22, respectively), where $Re = Uh/\nu$ denotes the Reynolds number of the flow, U is the cross-sectional mean flow velocity and h is the flow depth. Hot-film data were acquired at 500 Hz, thus satisfying the Nyquist criteria with the present maximum frequencies of approximately 200 Hz of the velocity signal. Video images were recorded at 500 frames s^{-1}, hence keeping a direct correspondence between hot-film data points and video frames.

The Series P experiments included two different sub-series of experiments, one in which the threshold conditions for particle entrainment into suspension were examined and one in which the characteristics of the process of particle entrainment into suspension were analysed in detail. On the other hand, the Series P experiments considered two bed roughness conditions, which corresponded to smooth flows (herein termed Series P-S) and transitionally rough flows (termed Series P-T). In Series P-T, a k-type bed roughness (Perry *et al.*, 1969) was created by gluing sand particles with a D_{50} grain size of approximately 0.53 mm to the originally smooth surface of the channel bottom. In all the experiments in Series P, particles were fed externally into the flow at a distance sufficiently far upstream from the observation window to ensure particle motion reached independence from initial conditions before going into the field of view of the camera. Particle motion was recorded from the side using the high-speed video system set at a recording rate of 250 frames s^{-1}. Typical dimensions of the flow field registered by the camera were about 0.01 to 0.02 m in the streamwise direction and 0.008 to 0.015 m in the vertical direction, corresponding to flow fields of approximately 200 by 150 wall units to 550 by 400 wall units. These dimensions were found adequate to resolve spatially particle trajectories during entrainment. A number of experiments were conducted in which particle motion was recorded with simultaneous flow visualization using the tracer injection technique described previously.

EXPERIMENTAL CONDITIONS

In the Series F experiments, data corresponding to three different Reynolds numbers are reported. The experimental conditions are given in Table 3.1, where $Fr = U/\sqrt{gh}$ denotes the flow Froude number, with g denoting gravitational acceleration, and $Re_* = u_* h/\nu$ denotes the Reynolds number of the flow based on shear velocity.

In Series P five different particles were used, namely, glass beads with mean diameter, d_p, of 38 and 94 μm, and natural sand particles with d_p values of 112, 224, and 530 μm, respectively. All the particles had a submerged specific density, $R = (\rho_s - \rho)/\rho$, of about 1.65 where ρ_s and ρ denote the particle and fluid densities, respectively. A summary of the particle characteristics is presented in Table 3.2, including the dimensionless particle diameter $R_p = (gRd_p^3)^{0.5}/\nu$, and the particle settling velocity, v_s, which was estimated theoretically assuming the particles were spheres with a diameter equivalent to d_p. In the experiments of Series P-T the size of the particles transported by the flow, d_p, was different from the size of the particles forming the bed roughness elements, d_b. Values of the ratio d_p/d_b in the experiments of Series P-T are also shown in Table 3.2. The latter values indicate that in some of the Series P-S experiments the particles had sizes smaller than the corresponding thickness of the viscous sublayer. In the experiments of Series P-T, where the size of the particles transported by the flow, d_p, was different from the size of the particles forming the bed roughness elements, d_b, the ratio d_p/d_b had values in the range from 0.07 to 1.0. In Series P-S, values of the particle Reynolds number, $Re_{p*} = u_* d_p/\nu$, were in the range from 0.70 to about 20, which indicates that in some of these experiments particles had sizes smaller than the corresponding thickness of the viscous sublayer.

Table 3.1. Experimental conditions for Series F experiments

h (m)	U (m s^{-1})	u_* (m s^{-1})	Re	Re_*	Fr
0.023	0.231	0.016	5 300	370	0.485
0.036	0.344	0.020	12 400	720	0.580
0.044	0.412	0.022	18 100	970	0.626

Table 3.2. Particle properties

d_p (μm)	R_p	v_s (mm s^{-1})	d_p/d_b
38	0.9	1.3	0.072
94	3.7	7.0	0.177
112	4.8	9.6	0.211
224	13.5	30.0	0.423
530	49.1	89.1	1.000

Experimental conditions corresponding to those Series P experiments which analysed characteristics of particle entrainment into suspension are presented in Table 3.3, including values of the dimensionless Shields shear stress defined as $\tau_* = u_*^2/(gRd_p)$, where d_p is 224 μm.

Table 3.3. Experimental conditions for Series P experiments

Series	h (m)	u_* (m s^{-1})	Re	Re_{p*}	Re_*	τ_*
P-S	0.033	0.019	10760	627	4.3	0.100
	0.039	0.021	14470	803	4.6	0.117
	0.045	0.022	18650	995	5.0	0.135
	0.050	0.023	22500	1165	5.2	0.150
P-T	0.047	0.030	18840	1436	6.8	0.255
	0.051	0.032	21480	1634	7.2	0.284
	0.056	0.035	25240	1924	7.7	0.329

METHOD OF ANALYSIS

Analysis of the video images for the Series F experiments provided information concerning the shape, geometry, convective velocity and general behaviour of the flow structures observed. These structures consisted mainly of near-wall shear layers, as described by Robinson (1991), which originate in low-speed fluid ejections away from the wall. Video images of the flow visualizations were analysed to detect, by visual inspection, developing flow structures approaching the hot-film sensors. The field of view of the video enabled images of structures in the region from about 400–500 wall units upstream of the velocity sensor to be recorded. Sub-series of the hot-film data, each corresponding to a range of frames with a developing flow structure near the bed, were extracted from the original series and later used to perform conditional ensemble averages of the velocity signal. Taylor's frozen turbulence hypothesis, together with measured mean convective velocities of shear layers from the video images, were used to transform temporal data into spatial information for each event and allowed estimation of the spatial flow velocity field associated with the occurrence of the observed shear layers.

Video recordings of Series P experiments were analysed to define the regime of sediment transport prevalent in each particular experiment and determine the limiting conditions for particle entrainment into suspension. For conditions of generalized entrainment, images of particle motion and particle-flow interactions were analysed in order to elucidate the physical mechanisms involved in such phenomena. Selected frames of the video recordings were digitized into a personal computer and these images were then analysed to obtain the position of particles in successive frames. This procedure was performed manually in most cases, although an automatic particle tracking algorithm (Dill, 1994; Hassan *et al.*, 1992) was also used. From the digitized particle trajectories, kinematic and dynamic characteristics of particle motion during entrainment into suspension were obtained. This information complemented the visualizations of flow-particle interactions and helped develop a conceptual model of the mechanics of particle entrainment into suspension.

CHARACTERIZATION OF NEAR-BED COHERENT STRUCTURES

Flow visualization in Series F and P shows that low-speed streaks tend to be lifted-up as a consequence of quasi-periodic ejections of low-momentum fluid away from the channel bed. In such instances, shear layers developed which correspond to inclined, thin filaments of concentrated spanwise vorticity, similar to those revealed by the analysis of data bases generated through direct numerical simulations (e.g. Jimenez *et al.*, 1988; Guezennec *et al.*, 1989) and those observed experimentally using PIV (e.g. Liu *et al.*, 1991; Urushihara *et al.*, 1993). Observations showed these shear layers lifted from the bottom surface and were then stretched out as they rose away from the bed into the outer regions of the wall layer. In some cases (less than 10% of the detected shear layers), the tips of the structures were observed to roll over in the form of quasi-transverse vortices, with diameters of the order of Taylor's micro-length scale corresponding to this distance to the bed, which may be related to the cross-section of hairpin vortices. Vertical profiles of Taylor's micro-length scale, defined as $u_{rms} / \overline{((\partial u' / \partial x)^2)}^{0.5}$, were estimated from $\sqrt{15 \nu u_{rms}^2 / \varepsilon}$ (where ε is the turbulent dissipation rate) for both smooth and transitionally rough conditions and found to follow very closely the expression proposed by Nezu & Nakagawa (1993). In other cases the streamwise coalescence of lifted low-speed streaks was apparent, as suggested in the conceptual model of Smith (1984), although these phenomena had a rather low frequency of occurrence.

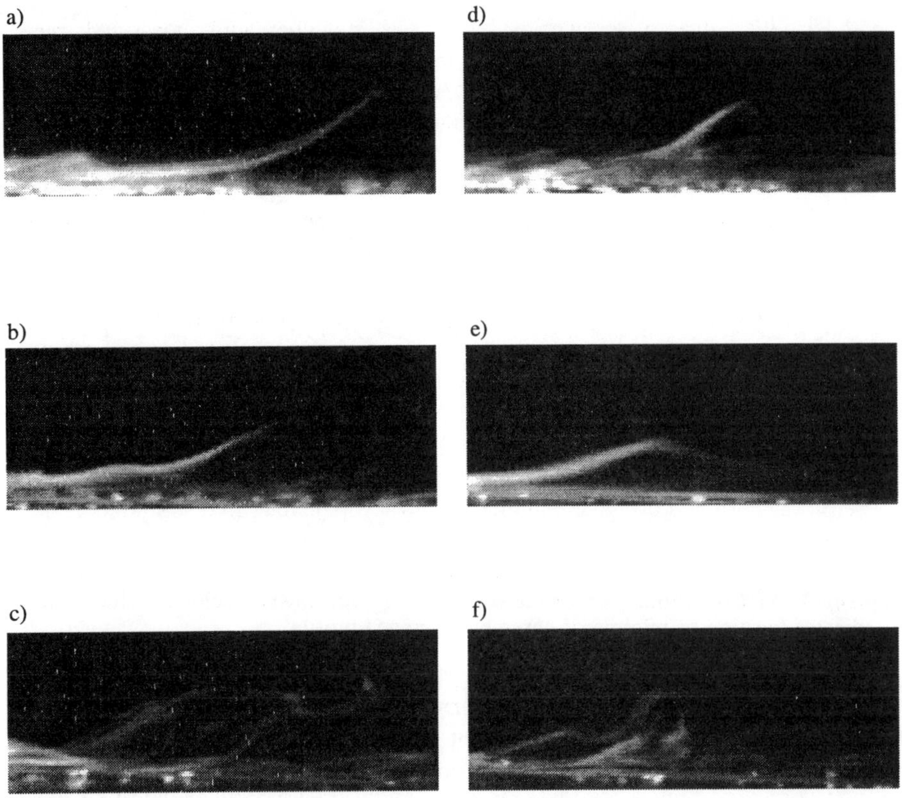

Figure 3.2. Shear layers at different stages of development. Experiments of Series P-S, *Re* = 15 750. Field of view of each image is approximately 328 by 125 wall units

Examples of the type of structures observed in these flow visualization experiments for the case of smooth flows are presented in Figure 3.2, where typical shear layers are shown at different stages of development. In some cases a quasi-transverse vortex can be observed at the tip of the structure (Figure 3.2d). The present results show no noticeable difference in the geometry and general characteristics of the shear layers observed in either the smooth and transitionally rough flows, implying that the same mechanism of formation and development of these structures may be operating in both cases. The shape of the shear layers, when expressed in wall units, appears to be invariant with both flow Reynolds number and bed roughness. The average shear layer, depicted as the solid line in Figure 3.3, extends a distance of approximately 100 wall units in the vertical and about 500 wall units in the streamwise direction. The angle to the bed of this mean shape varies with the distance from the wall, with an average value of approximately 14°.

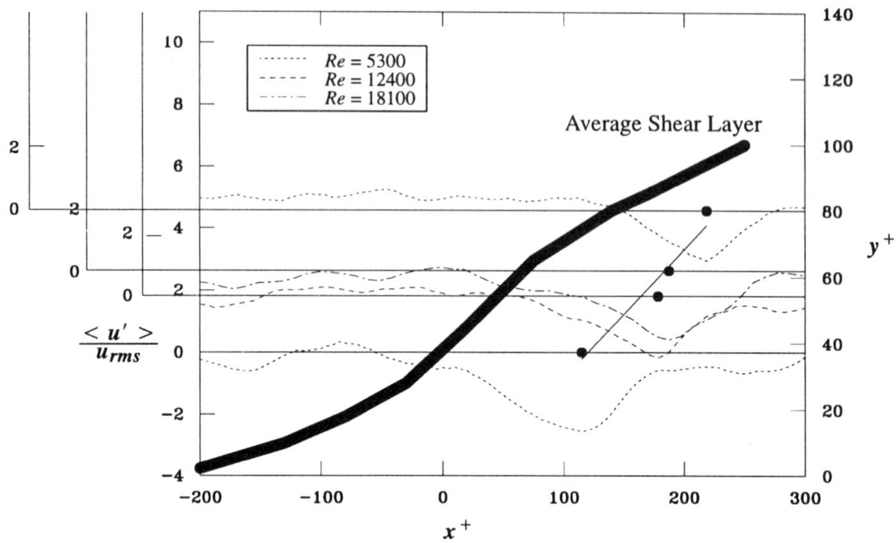

Figure 3.3. Conditional ensemble averages of streamwise velocity fluctuations. • indicates location of minima of ensemble-averaged profile

The observed shear layers may be related to the funnel vortices described by Kaftori *et al.* (1994), since they present themselves as inclined structures with angles to the bed similar to those reported for the funnel vortices. Both the shear layers and funnel vortices are related to flow ejections taking place near the bed and extending upwards within the limits of the wall layer. In fact, some flow visualization results showed shear layers evolving into some kind of helical motion (Niño, 1995), very similar to that described by Kaftori *et al.* (1994) for the funnel vortices. Nevertheless, the longitudinal extension of the shear layers observed herein (about 300–500 wall units, see Figure 3.3) seems to be slightly larger than the wavelength of the funnel vortices described by Kaftori *et al.* (about 150–300 wall units), the latter involving a complete revolution of the helical motion in the funnel vortices and thus implying a complete ejection-sweep cycle. Because of this, it appears that although in some cases shear layers and funnel vortices may be related, in general they seem to be a different kind of coherent flow structure emerging in the wall layer of a turbulent boundary layer.

For each flow Reynolds number, several shear layers were identified in the video recordings and their evolution followed through successive frames whilst keeping track of the associated streamwise velocity at the measuring point. This information was used to derive conditional ensemble-averaged values of dimensionless temporal velocity fluctuations during the passage of the structures at each Reynolds number and then converted to spatial information using Taylor's frozen turbulence approach. Figure 3.3 shows the results obtained for all three

Reynolds numbers of Series F computed from measurements taken at four distances from the bed of $y^+ = 38$, 52, 64 and 80, plotted together with the average dimensionless shear layer shape discussed above. Each profile is plotted so that its ordinate zero is located at the corresponding normalized distance to the bed shown at right. More details concerning these measurements, including associated bed shear stresses, are reported in García *et al.* (1995).

The conditionally averaged velocity distributions of Figure 3.3 clearly show the existence of an organized pattern, with a minimum in the streamwise velocity downstream of the shear layer at varying distances from the bed. The minima of the velocity distributions shown in Figure 3.3 define a line with a similar inclination to the bed as that of the observed coherent structures. The characteristics of the associated dimensionless flow field also indicate an invariance of the velocity distribution with Reynolds number, at least in the range of values investigated here, when made dimensionless using the local value of the corresponding standard deviation u_{rms}. The existence of the detected minima implies an organized contribution to the Reynolds stress in the second quadrant, since vertical flow ejections correlate with streamwise velocity deficits. In fact, it has been found experimentally that almost every contribution (90%) to the total Reynolds stress in the second quadrant of magnitude $k\overline{|u'v'|}$ with $k \geq 2$ where $\overline{|u'v'|}$ denotes the total mean Reynolds stress, is caused by ejections of low-speed streaks (Talmon *et al.*, 1986). PIV results also show the neighbouring regions of the near-wall shear layers as strong contributors to the Reynolds stresses (Liu *et al.*, 1991).

LIMITING CONDITIONS FOR PARTICLE ENTRAINMENT INTO SUSPENSION

The video recordings of particle motion in Series P were analysed to define the regime of sediment transport prevalent in each particular experiment. In general, the onset of particle entrainment into suspension was more or less abrupt. For low values of bed shear stress the particles moved in contact with the bed and in the case of smooth flows, the particles were displaced by rolling and sliding only. In the case of transitionally rough flows, the particles moved mainly by saltation, except those with small d_p/d_b ratios which moved within the interstices of the much larger bed roughness elements. As the shear stress was increased, progressively more frequent events occurred during which particles were suddenly entrained into the outer regions of the wall layer and remained in suspension for distances generally greater than 100 particle diameters. During these events, the particles clearly responded to large velocity fluctuations in the flow. Therefore, there appears to be a threshold level of the bed shear stress below which particles do not seem to respond to turbulent ejections in the near-bed region. The results obtained from Series P-S

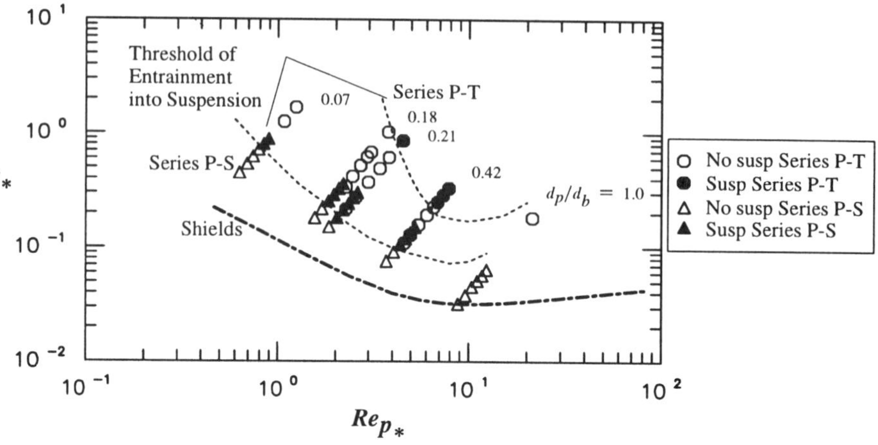

Figure 3.4. Threshold conditions for particle entrainment into suspension. Experimental results of Series P-S and P-T. Open symbols correspond to conditions of no suspension, and closed symbols represent conditions where suspension was observed. Values refer to ratios d_p/d_b

and P-T are plotted in Figure 3.4 in the parameter space (Re_{p*}, τ_*), where Shields' curve for initiation of motion is also plotted for comparison.

It is interesting to note that in Figure 3.4 those symbols corresponding to Re_{p*} values lower than about 5 are associated with particle sizes smaller than the thickness of the viscous sublayer. It is, therefore, apparent that particles totally immersed in the viscous sublayer are entrained into suspension which contradicts previous observations of Sumer & Oguz (1978) and Yung *et al.* (1988). It is also apparent from Figure 3.4 that, for the range of sizes studied here, as particle size decreases, progressively higher values of the dimensionless shear stress are required for entrainment. This is rather obvious since the flow turbulence intensities responsible for lifting particles into suspension become relatively less strong as the bottom wall is approached within the viscous sublayer, which is thicker at lower shear velocities and therefore lower Re_{p*} (Nezu & Nakagawa, 1993). This effect would explain the conclusions of Sumer & Oguz (1978) and Yung *et al.* (1988), in the sense that the values of shear stresses in their flows were probably not large enough to induce the entrainment of the particles totally immersed in the viscous sublayer.

It is clear from Figure 3.4 that the threshold curve defining conditions for the initiation of suspension in Series P-T are displaced toward higher values of τ_* as compared to Series P-S. This implies that as the bed roughness increases (from smooth to transitionally rough), higher values of dimensionless shear stress are required to entrain particles of the same size. This appears to be related to a hiding effect that tends to preclude particle entrainment and may be related to the modification of the turbulence structure surrounding the entrained particle by the shielding effect of the coarser particles of the bed. This effect tends to be enhanced as the ratio d_p/d_b decreases.

PARTICLE-TURBULENCE INTERACTIONS

Visualizations of particle motion in the near-bed region of the smooth flows of Series P-S showed that particles tend to be picked up from low-speed streaks, lifted away from the wall by some kind of ejection mechanism, and deposited back to the bed along the high-speed streaks. Particles then tend to be displaced laterally toward the low-speed streaks by the action of cross flows related to longitudinal vorticity. In the case of the transitionally rough flows of Series P-T, the particles did not tend to sort along wall streaks although they were observed to be entrained by a similar ejection mechanism as in the case of smooth flows.

Both the present work and much past research, based primarily on experimental evidence, suggests a consensus on the mechanism that causes ejection of particles away from the bed. Such particle ejections are related to interactions between particles and intermittent events associated with turbulent bursting during which low-momentum fluid is ejected toward the outer regions of the wall layer (Grass, 1974; Sumer & Oguz, 1978; Sumer & Deigaard, 1981; Ashida & Fujita, 1986; Yung *et al.*, 1989; Rashidi *et al.*, 1990; Kaftori *et al.*, 1995a, 1995b). The same conclusion was reached by Pedinotti *et al.* (1992) whose direct numerical simulations of particle-turbulence interactions in the near-wall region of a channel flow showed that particles are lifted from the bed by the action of upflows caused by quasi-streamwise vortices which detach low-speed fluid from the wall.

The present results seem to indicate that the shear layers described in previous sections interact with particles lying on the bed, such that the flow ejection occurring downstream of the shear layer (Figure 3.3) would induce the lift-up of particles away from the bed, and this mechanism would be the same regardless of the bed roughness condition, smooth or transitionally rough. This is illustrated in Figure 3.5 which shows a sequence of images of an inclined shear layer being convected and stretched in the streamwise direction, together with a sand particle which is being lifted-up from the bed at a distance of about 115 wall units downstream of the structure, apparently responding to an ejection event taking place in this region. Note that this distance corresponds well with the distance of velocity minima shown in Figure 3.3.

Figure 3.6 shows a conceptualization of the relative motion of an ejected particle with respect to the shear layer in a frame of reference moving at the convection velocity of the flow structure. In Figure 3.6 the relative flow velocity field around the shear layer is also sketched, based on the experimental measurements using PIV by Urushihara *et al.* (1993). According to those observations, there appears to be a relative stagnation point located over the shear layer which corresponds to a saddle

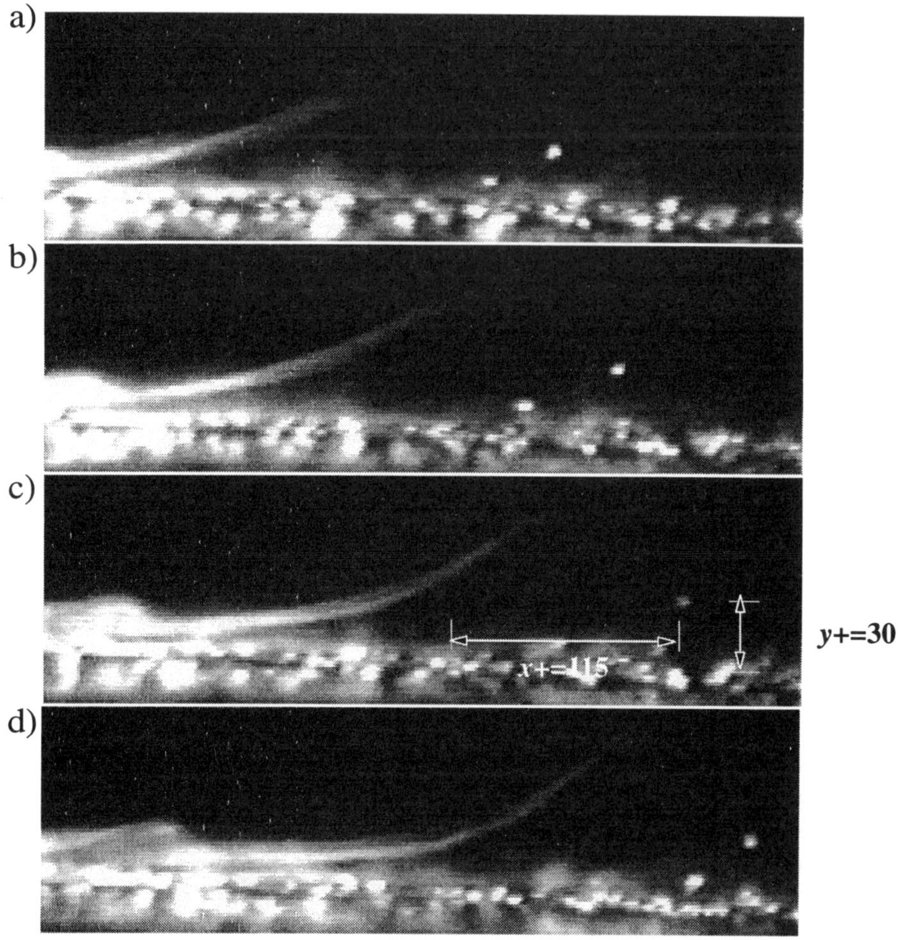

Figure 3.5. Sequence of images of particle-shear layer interaction. Experiments of Series P-S, Re = 15 000, d_p = 224 μm, Re_{p*} = 4.5. Field of view of each image is about 375 by 114 wall units, and the time interval between images is 0.008 s

point of the relative velocity vector field. The zone downstream from the structure corresponds to a low-velocity region where the relative flow field is in the upstream direction toward the shear layer. The zone upstream from the structure corresponds to a high-velocity region where the relative flow is in the downstream direction, also toward the shear layer. Along the shear layer, the relative velocity vectors are in the direction of the inclined structure and have a larger magnitude upstream from the structure than downstream from it, so the shear layer represents a zone of concentrated transverse vorticity. The initial ejection of the particle would take

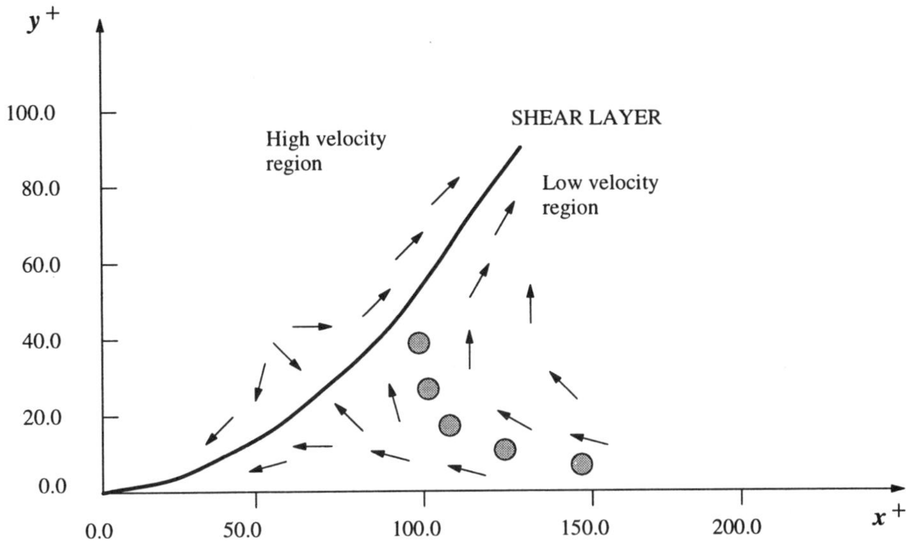

Figure 3.6. Schematic view of particle motion as seen from a frame of reference moving with the shear layer. Circles represent different relative positions of the particle as it is ejected away from the bed. Arrows sketch the relative flow velocity field surrounding the shear layer as measured by Urushihara *et al.* (1993)

place at a distance of about 100 to 200 wall units downstream from the structure and would then have a relative motion in the upstream direction toward the shear layer. As shear layers and ejected particles get closer together, they tend to have a more direct interaction, such that particles eventually get trapped within the flow field associated with the structures. In some cases it was observed that particles lined-up along the shear layer and followed its path as the structure was stretched toward the outer regions of the wall layer. However other particles lost correlation with the shear layer and were observed to lag behind it until they were either deposited back on the bed or were taken-up by a new, developing flow ejection. In these cases, particles fall from the structures in a phenomenon termed the 'crossing-trajectories' effect (Wells & Stock, 1983; Zhuang *et al.*, 1989) where the particle and flow structure lose correlation before the latter loses coherence.

PARTICLE MOTION DURING ENTRAINMENT

Analysis of particle trajectories during entrainment showed that typical particle ejection angles were in the range from 10 to 20°, similar to the angles typical of the shear layers observed in the present experiments. An example that is representative of the results obtained here for both smooth and transitionally rough

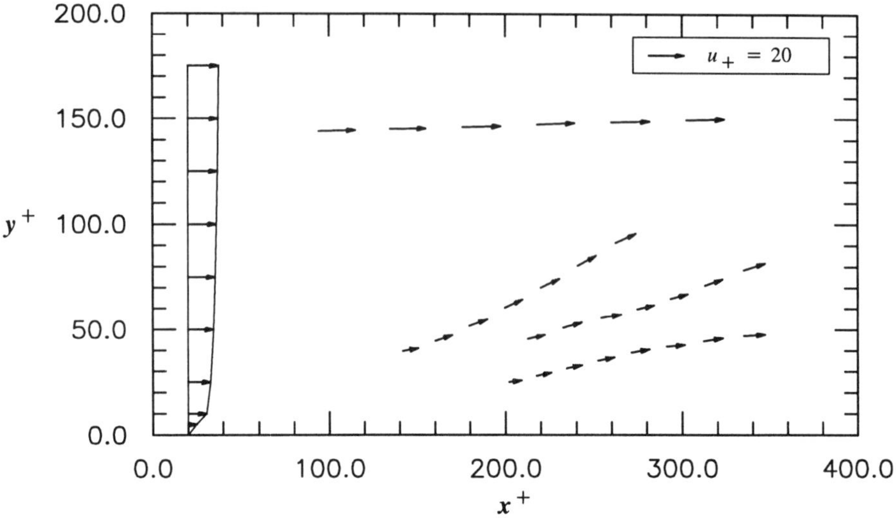

Figure 3.7. Velocity map of particles suspended in one ejection event. Each sequence of arrows corresponds to a different particle. The time interval between arrows of the same sequence corresponds to about 2 wall time units (0.004 s). The velocity profile on the left corresponds to the logarithmic mean velocity profile for smooth flows for $y^+ > 10$ and to the linear law for $y^+ > 10$. Experiments of Series P-S, $d_p = 112$ μm, $Re = 20\ 145$, $Re_{p*} = 2.6$

conditions, is presented in Figure 3.7, in which a velocity map of particles being entrained in a particular ejection event in one of the Series P-S experiments is shown. The sequence of arrows shows the particle trajectories and the temporal evolution of their velocity vectors. Figure 3.7 shows the path of two of the three near-bed particles being entrained toward the outer regions of the wall layer (values of y⁺Ê~ 100) at angles of approximately 20°, whilst the third particle will not reach heights greater than about 50 wall units before falling back to the bed (the cross-trajectories effect). Clearly, the horizontal components of the measured particle velocities during entrainment are considerably lower than the local mean flow velocity predicted by the logarithmic law for smooth flows and indicates that the streamwise component of the flow velocity fluctuations, u', in the ejection event responsible for the entrainment of the particles have a negative sign. Together with the obvious fact that the vertical component of such velocity fluctuations, v', is positive during entrainment, this shows that the flow ejection event responsible for the entrainment of particles in Figure 3.7 is an event of quadrant-2 ($u' < 0$, $v' > 0$) and thus an important contributor to the Reynolds stress $\overline{u'v'}$. It is also apparent from Figure 3.7 that, as a consequence of the momentum transfer from the flow to the particle, the magnitude of the particle velocity tends to increase as the particle is lifted-up away from the channel bottom and gets accelerated as it is being dragged by fluid of increasing momentum.

From the upward trajectories of ejected particles observed in the present experiments, vertical distributions of normalized ensemble-averaged values of the streamwise and vertical components of particle ejection velocity, u_p^+ and v_p^+ respectively, were computed together with their corresponding standard deviations. This was done by dividing the y^+ co-ordinate in intervals of dimension Δy^+ of 20, and ensemble averaging all the velocity data points (for all the experimental conditions available) contained in each of those intervals. A value of $\Delta y^+ = 20$ was used to be consistent with the analogous analysis of Sumer & Oguz (1978). The results obtained from Series P-S for v_p^+ and their corresponding standard deviations are plotted in Figure 3.8 as a function of y^+ in $0 < y^+ < 150$, together with the values of the dimensionless standard deviation of the vertical component of flow velocity fluctuations, v_{rms}^+, given by the model of Nakagawa & Nezu (1981), and the experimental data of Grass (1974), Sumer & Oguz (1978) and Sumer & Deigaard (1981).

As seen in Figure 3.8, the values of v_p^+ computed from the present data for values of y^+ in the range from 0 to 150 tend to be almost constant and equal to about 1.0–1.3, with values of the corresponding standard deviation ranging from 0.5 to 0.9. The values of v_p^+ tend to be much larger than the theoretical flow values of v_{rms}^+ close to the bed, but display a better correspondence in the outer regions of the wall layer. This again supports the argument that particle ejections

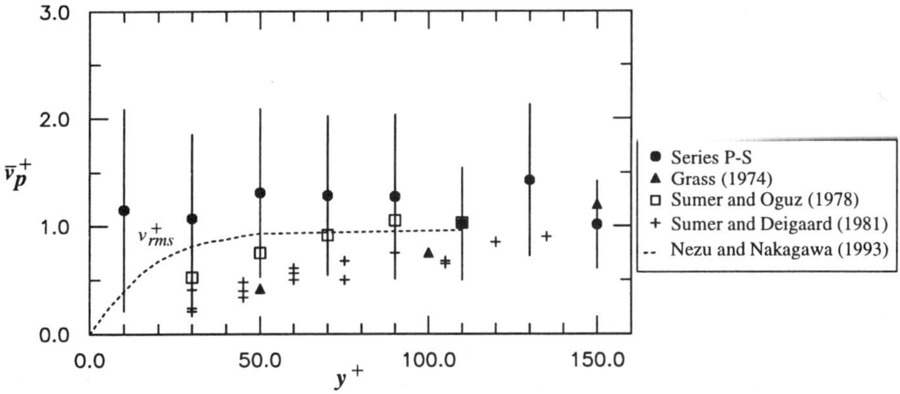

Figure 3.8. Dimensionless ensemble-averaged values of the vertical component of particle ejection velocity. Experiments of Series P-S, d_p = 224 μm, *Re* in the range from 10 760 to 22 500, Re_{p*} in the range from 4.3 to 5.2. Symbols denote mean values and vertical lines represent a total length of two standard deviations. Dashed line represents v_{rms}^+ values

are caused by rather strong quadrant-2 flow events, with typical vertical velocity fluctuations much larger than the corresponding long-term standard deviation. It is also apparent from Figure 3.8 that the values of v_p^+ computed from the present data tend to be much larger than the other experimental data presented, especially close to the bed, although such differences decrease in the outer regions of the wall layer.

Of the experimental data plotted in Figure 3.8, only those of Grass (1974) correspond closely to the present experimental conditions. However, Grass did not provide much detail of his experiments nor how the average particle ejection velocities were computed. It is possible that in his experiments, ejection events could have entrained into suspension a larger number of particles than in the present experiments, such that particle-particle interactions could have influenced ejection velocities (Grass, pers. comm.). These interactions may include wake interference effects which may become important as the concentration gets higher and even direct collisions between particles in the case of very high particle concentrations. Additionally, the experimental data of Sumer & Oguz (1978) and Sumer & Deigaard (1981) concern particles much larger, but less dense than those used in the present study. Although the ratio between the particle settling velocity, v_s, and the shear velocity in their experiments was in the range 0.4 to 2.0, very similar to the present range of 1.3 to 1.6, the dimensionless time constant, $t_p^+ = (\rho_s d_p^2 u_*^2)/(18\nu^2\rho)$, of their particles was of the order of 100, whilst that of the present particles ranges from 2.7 to 4.0. Since t_p^+ is a measure of the inertia of the particles, it is clear that Sumer & Oguz's and Sumer & Deigaard's particles would respond much slower to sudden accelerations of the flow than the present particles. In terms of the response to flow ejection events, it can be expected that particles with high values of t_p^+ will have ejection velocities of lower magnitude than that associated with the flow ejections, and to accelerate slowly to eventually match the flow velocity as they are lifted away from the bed. On the other hand particles with low values of t_p^+ would have ejection velocities of magnitude similar to those associated with the flow ejections and would tend to follow the turbulence of the flow rather closely. This would explain the generally low values of the ejection velocities exhibited by the particles in Sumer & Oguz's and Sumer & Deigaard's experiments in the wall region when compared both to the v_{rms}^+ values and the present experimental results.

In Figure 3.9, the results obtained from the analysis of the experiments of Series P-T for v_p^+ and their corresponding standard deviations are plotted as a function of y^+, together with the distribution of v_{rms}^+ for smooth flows as in Figure 3.8, the data of Grass (1971) and the data of Sumer & Deigaard (1981) corresponding to rough flows. The results obtained in Series P-S (Figure 3.8) are also presented for comparison. The values of v_p^+ computed from the data of Series P-T show some tendency to increase with y^+, from values close to about 1.0 near the wall to values

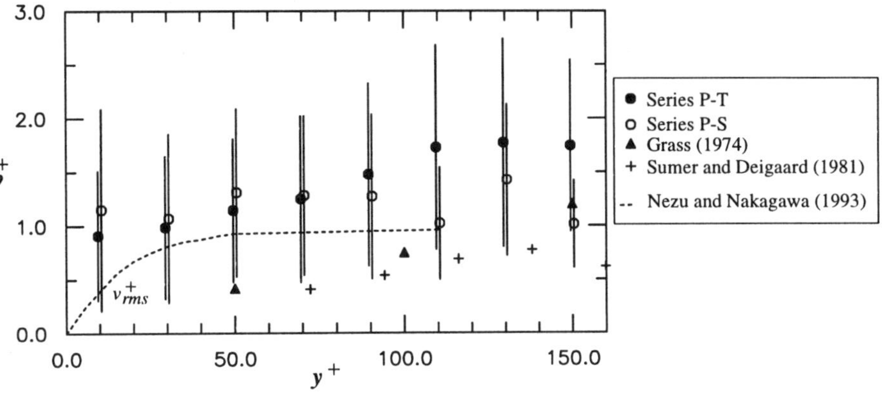

Figure 3.9. Dimensionless ensemble-averaged values of the vertical component of particle ejection velocity. Experiments of Series P-T, d_p = 224 μm, Re in the range from 18 840 to 25 240, Re_{p*} in the range from 6.8 to 7.7. Symbols denote mean values and vertical lines represent a total length of two standard deviations. Dashed line represents v_{rms}^+ values

close to 2.0 in the regions outside of the wall layer, with corresponding values of the standard deviation of 0.5 to 0.9. As in the smooth flow case, the values of v_p^+ are always larger than the theoretically estimated values of v_{rms}^+ in the range $0 < y^+ < 150$ examined here. Figure 3.9 shows that the v_p^+ values for Series P-T are very similar to those estimated in the smooth flow except for values of y^+ greater than approximately 100, where the values of v_p^+ for Series P-T are larger than those of Series P-S. This behaviour suggests that flow ejections over a rough bed are more intense than those in a smooth flow, in agreement with the observations of Sumer & Deigaard (1981) who also reported larger vertical particle ejection velocities in the case of rough flows than smooth flows. This is also supported by other experimental work, such as that of Grass (1971) who concluded that although the flow ejection mechanism over a rough bed would be similar to the case of a smooth surface, flow entrainment and ejections tend to be much more violent over rough beds. Nevertheless, as pointed out by Sumer & Deigaard (1981), this enhanced vertical flow intensity does not appear to influence significantly the behaviour of particles entrained from the rough bed, which does not differ appreciably from the smooth bed case.

Figure 3.9 also shows that the values of v_p^+ computed from Series P-T tend to be much larger than the experimental values reported by Grass (1974) and Sumer & Deigaard (1981), the latter corresponding to their experiments over a rough bed. It is also apparent that their estimated values of v_p^+ are smaller than the values of v_{rms}^+, which contradicts the notion of particularly strong quadrant-2 events driving particle ejections from the bed. As indicated previously, there is insufficient information available concerning Grass's experiments in order to explain the apparent differences with the present observations. However, the differences between the present experimental results for Series P-T and those of Sumer & Deigaard may be again explained by the different values of the dimensionless particle time constant, t_p^+, in both cases. The values of t_p^+ in Sumer & Deigaard's rough bed experiments were approximately 240, much larger than those in the present experiments of Series P-T which were only about 7 to 9.

CONCLUSIONS

A series of events may be envisaged to generate particle entrainment into suspension. Low-momentum fluid is lifted-up from the bed of open-channel flows as a consequence of quasi-periodic ejection events and subsequently evolves into different coherent structures. This mechanism appears to be common to flow over both smooth and rough surfaces. The most frequently observed coherent structures correspond to shear layers of concentrated spanwise vorticity, which have a typical inclination angle to the bed of about 14°. These structures extend a distance of approximately 100 and 500 wall units in the vertical and streamwise directions, respectively. An intense flow ejection (event of quadrant-2) occurs at a distance of about 100 to 200 wall units downstream from the shear layer.

The present observations indicate that particles totally immersed within the viscous sublayer can be entrained into suspension by the flow, contrary to some previous experimental results. Nevertheless, as particle size decreases progressively higher values of bed shear stress are needed for entrainment which seems to be a consequence of the rather abrupt drop of turbulent intensities within the viscous sublayer as the bed is approached. In the experiments of Series P-T the roughness elements appear to induce a hiding effect that tends to preclude particle entrainment by locally affecting the turbulence structure of the flow. This hiding effect seems to be dependent on the ratio d_p/d_b, such that as this ratio becomes smaller, progressively higher values of τ_* are needed to entrain the particles as compared to those needed to entrain the same particles from a smooth bed.

Particles are picked up from the bed by the flow ejection events occurring downstream of the shear layers. Particle ejection angles range from 10 to 20° and are similar to the angle of inclination of the shear layers. After the particles are lifted from the bed their relative motion is toward the shear layer. Different types of particle-shear layer interactions may occur. When particles become trapped within the flow field associated with the shear layers they are raised toward the outer regions of the wall layer as the flow structure becomes stretched into such regions.

A second type of interaction occurs when the particle lags the flow ejection such that it eventually falls from the structure back to the bed, without reaching elevations higher than about 50 wall units, defining the so-called crossing-trajectories effect in which particles lose correlation with the flow structure before the structure itself loses its coherence.

Dimensionless averaged values of the vertical component of particle velocity, v_p^+, during ejections are approximately constant in the range $0 < y^+ < 150$, for both smooth and transitionally rough conditions, with values of about 1.0 to 2.0. However, for transitionally rough flows there is a tendency for the values of v_p^+ to be somewhat larger than those corresponding to smooth conditions in the range $y^+ < 100$. The values of v_p^+ are much larger than estimates of fluid v_{rms}^+ close to the bed, but the two become more similar in the outer region of the wall layer. These vertical particle ejection velocities are also much larger than previous experimental determinations. However, these differences may be explained in terms of the dimensionless particle time constant, t_p^+ which is much smaller in the present experiments than in previous studies, thus indicating a much faster response of the present particles to sudden accelerations of the flow. In the case of transitionally rough flows, the values of the dimensionless ensemble-averaged vertical component of the particle velocity during ejections appear to be very similar to those observed in the smooth flows for $y^+ < 100$, but are larger in the region $0 < y^+ < 150$, reaching values of about 2.0 wall units in this region. This suggests that flow ejections in transitionally rough flows are more intense than those in the smooth flows and is in agreement with past observations. Nevertheless, these enhanced ejections do not seem to affect substantially the characteristics of the particle ejections as compared to those observed in the smooth-wall flows, which is also in agreement with previous experimental results.

Future work requires further elucidation of particle-particle interactions and experiments with higher particle concentrations, i.e. higher particle loadings. One problem that may arise is that high concentrations greatly complicate image analysis and this may require improvement of the video camera resolution whilst maintaining a high frame rate such as that used in the present study. An interesting experimental challenge would be to use high-speed video for particle tracking and particle-image-velocimetry to characterize the instantaneous flow field associated with particle entrainment and motion. This should be possible in the near future as better technology becomes available.

ACKNOWLEDGEMENTS

We thank the National Science Foundation (grant CTS-9210211), the Office of Naval Research (grant N00014-93-1-0044), and the donors of the Petroleum Research Fund, administered by the American Chemical Society (grant PRF 24328-G2) for supporting this research. Thanks are also due to Jim Best, University of Leeds, for his dedicated editorial work and valuable suggestions and comments.

REFERENCES

Ashida, K. & Fujita, M. 1986. 'Stochastic model for particle suspension in open channels', *J. Hydroscience and Hydr. Engrg.*, **4**, 21–46.

Dill, A.J. 1994. *Video-based Particle Tracking Velocimetry Technique for Measuring Flow Velocity in Porous Media*, Unpub. Masters Thesis, Univ. Illinois at Urbana-Champaign, Illinois, 89 pp.

García, M., López, F. & Niño, Y. 1995 'Characterization of near-bed coherent structures in turbulent open channel flow using synchronized high-speed video and hot-film measurements', *Exper. Fluids*, **19**, 16–28.

Grass, A.J. 1971. 'Structural features of turbulent flow over smooth and rough boundaries', *J. Fluid Mech.*, **50**, 233–255.

Grass, A.J. 1974. 'Transport of fine sand on a flat bed: turbulence and suspension mechanics', *Proc. Euromech 48, Inst. Hydrodynamic and Hydraulic Engrg., Tech. Univ. Denmark*, 33–34.

Grass, A.J., Stuart, R.J. & Mansour-Tehrani, M. 1991. 'Vortical structures and coherent motion in turbulent flow over smooth and rough boundaries', *Phil. Trans. R. Soc. London*, **A336**, 35–65.

Guezennec, Y.G., Piomelli, U. & Kim, J. 1989. 'On the shape and dynamics of wall structures in turbulent channel flow', *Phys. Fluids A*, **1**, 764–766.

Hassan, Y.A., Blanchat, T.K., Seeley, C.H. & Canaan, R.E. 1992. 'Simultaneous velocity measurements of both components of a two-phase flow using Particle Image Velocimetry', *Int. J. Multiphase Flow*, **18**, 371–395.

Head, M.R. & Bandyopadhyay, P. 1981. 'New aspects of turbulent boundary-layer structure', *J. Fluid Mech.*, **107**, 297–338.

Jimenez, J., Moin, P., Moser, R. & Keefe, L. 1988. 'Ejection mechanisms in the sublayer of a turbulent channel', *Phys. Fluids*, **31**, 1311–1313.

Kaftori D., Hetsroni, G. & Banerjee, S. 1994. 'Funnel-shaped vortical structures in wall turbulence', *Phys. Fluids*, **6**, 3035–3050.

Kaftori D., Hetsroni, G. & Banerjee, S. 1995a. 'Particle behaviour in the turbulent boundary layer. I. Motion, deposition, and entrainment', *Phys. Fluids*, **7**, 1095–1106.

Kaftori D., Hetsroni, G. & Banerjee 1995b. 'Particle behaviour in the turbulent boundary layer. II. Velocity and distribution profiles'. *Phys. Fluids*, **7**, 1107–1121.

Liu, Z., Landreth, C.C., Adrian, R.J. & Hanratty, T.J. 1991. 'Measurements in turbulent channel flow by high resolution Particle Image Velocimetry', *Exper. Fluids*, **10**, 301–312.

Nakagawa, H. & Nezu, I. 1981. 'Structure of space-time correlations of bursting phenomena in an open channel flow', *J. Fluid Mech.*, **104**, 1–43.

Nezu, I. & Nakagawa, H. 1993. *Turbulence in Open-channel Flows*, IAHR Monograph, A.A. Balkema, Rotterdam, 281 pp.

Niño, Y. 1995. *Particle Motion in the Near Bed Region of a Turbulent Open Channel Flow: Implications for Bedload Transport by Saltation and Sediment Entrainment into Suspension*, Unpub. PhD Thesis, Univ. Illinois at Urbana-Champaign, 370 pp.

Pedinotti, S., Mariotti, G. & Banerjee, S. 1992. 'Direct numerical simulation of particle behaviour in the wall region of turbulent flows in horizontal channels', *Int. J. Multiphase Flow*, **18**, 927–941.
Perry, A.E., Schofield, W.H. & Joubert, P.N. 1969. 'Rough wall turbulent boundary layers', *J. Fluid Mech.*, **37**, 383–413.
Rashidi, M., Hetsroni, G. & Banerjee, S. 1990. 'Particle-turbulence interaction in a boundary layer', *Int. J. Multiphase Flow*, **16**, 935–949.
Robinson, S.K. 1991. 'Coherent motions in the turbulent boundary layer', *Ann. Rev. Fluid Mech.*, **23**, 601–639.
Smith, C.R. 1984. 'A synthesized model of the near-wall behaviour in turbulent boundary layers', in Zakin, J.L. & Patterson, G. (Eds) *Proc. 8th Biennial Symp. Turbulence*, Missouri-Rolla, Missouri, 299–327.
Sumer, B.M. & Deigaard, R. 1981. 'Particle motions near the bottom in turbulent flow in an open channel. Part 2', *J. Fluid Mech.*, **109**, 311–337.
Sumer, B.M. & Oguz, B. 1978. 'Particle motions near the bottom in turbulent flow in an open channel', *J. Fluid Mech.*, **86**, 109–127.
Sutherland, A.J. 1967. 'Proposed mechanism for sediment entrainment by turbulent flows', *J. Geophys. Res.*, **72**, 191–198.
Talmon, A.M., Kunen, J.M.G. & Ooms, G. 1986. 'Simultaneous flow visualization and Reynolds-stress measurement in a turbulent boundary layer', *J. Fluid Mech.*, **163**, 459–478.
Urushihara, T., Meinhart, C.D. & Adrian, R.J. 1993. 'Investigation of the logarithmic layer in pipe flow using Particle Image Velocimetry', in So, R.M.C., Speziale, C.G. & Launder, B.E. (Eds) *Near-Wall Turbulent Flows*, Elsevier Science Publishers B.V. 433–446.
Vanoni, V.A. 1964. 'Measurements of critical shear stress for entraining fine sediments in a boundary layer', *Report No.* ***KH-R-7***, California Institute of Technology, Pasadena, 47 pp.
Wells, J. 1992. *Expériences sur les Impacts Liquide-solide et le Soulèvement de Particules par un Tourbillon Annulaire: Applications à L'Erosion des Sols et au Transport des Sédiments*, Unpub. PhD Thesis, University Joseph Fourier, Grenoble I, 289 pp.
Wells, M.R. & Stock, D.E. 1983. 'The effects of crossing trajectories on the dispersion of particles in a turbulent flow', *J. Fluid Mech.*, **136**, 31–62.
Yung, B.P.K., Merry, H. & Bott, T.R. 1988. 'The role of turbulent bursts in particle re-entrainment in aqueous systems', *Chemical Engrg. Science*, **44**, 873–882.
Zhuang, Y., Wilson, J.D. & Lozowski, E.P. 1989. 'A trajectory-simulation model for heavy particle motion in turbulent flow', *J. Fluids Engrg.*, **111**, 492–494.

NOTATION

d_p mean particle diameter
d_b mean diameter of particles forming the bed-roughness elements
Fr Froude number of the flow
h mean flow depth
R submerged specific density
Re flow Reynolds number
Re_* flow Reynolds number based on shear velocity

Re_{p*}	particle Reynolds number based on shear velocity
R_p	dimensionless particle diameter
t_p^+	particle dimensionless time constant
U	mean cross-sectional velocity
u'	streamwise component of flow velocity fluctuation
u_+	local mean velocity made dimensionless with shear velocity
u_p^+	ensemble-averaged value of streamwise component of particle ejection velocity made dimensionless with shear velocity
u_{rms}	standard deviation of streamwise flow velocity
$\overline{u'v'}$	Reynolds stress
u_*	flow shear velocity
v'	vertical component of flow velocity fluctuation
v_p^+	ensemble-averaged value of vertical component of particle ejection velocity made dimensionless with shear velocity
v_{rms}	standard deviation of vertical flow velocity
$x\ y\ z$	Cartesian co-ordinates in the streamwise, bed-normal and transverse directions, respectively
y^+	bed-normal co-ordinate made dimensionless with the viscous length-scale
ε	turbulent dissipation rate
ν	fluid kinematic viscosity
ρ	fluid density
ρ_s	particle density
τ_*	dimensionless Shields bed shear stress

4 Transverse Spacing of Low-speed Streaks in a Channel Flow over a Rough Bed

ANDREA DEFINA
Institute of Hydraulics "Giovanni Poleni", Faculty of Engineering, University of Padova, via Loredan, 20-I 35131 Padova, Italy

ABSTRACT

Long, low-speed streaks in a channelized flow over a rough bed were visualized using a passive tracer and the mean transverse spacing between these longitudinal streaks was measured at different distances from the wall. In the near-wall region, streak spacing was found to be approximately four times the equivalent sand roughness height and maintained this value irrespective of water depth and channel slope. This result confirms the experimental findings of Grass (1971, 1992) and Defina (1987, 1992). As distance from the wall increases, the spanwise spacing between longitudinal turbulent structures also grows, the rate of growth following a simple linear law. These results suggest that the vortical structures expand in scale with increasing wall distance to form the scale hierarchy suggested in recent models (e.g. Perry & Chong, 1982; Perry et al., 1986, 1991). It was also found that the largest hierarchy of vortices, which scale with the flow depth, correspond to the instantaneous secondary currents in a straight channel flow.

INTRODUCTION

The existence of turbulent coherent structures over a rough wall similar to those present over a smooth wall was first suggested and confirmed by Grass (1971). Grass (1971) used H_2 bubble flow visualization to identify and describe the near-wall structure of flow in transitionally rough and fully rough wall regimes. He found that both ejection and inrush events bear a striking resemblance to equivalent features observed over a smooth wall. Grass concluded that these were common features of boundary layer turbulence and argued that a shared mechanism could be responsible for the generation of turbulent structures irrespective of the type of wall roughness.

Defina (1987) subsequently investigated channel flow over a rough wall with the aim of visualizing a streaky pattern (if any) similar to the one detectable over a smooth wall, even in the absence of an equivalent viscous sublayer over a rough bed. Using two different visualization techniques, Defina clearly identified low-

Coherent Flow Structures in Open Channels.
Edited by P.J. Ashworth, S.J. Bennett, J.L. Best and S.J. McLelland.

speed streaks in transitionally rough and fully rough channel flows. Defina (1987) used a rough bed comprised of spherical pebbles approximately 10 mm in diameter arranged in a hexagonal pattern with an equivalent sand roughness height, k_s, of 7.5 mm. For roughness Reynolds numbers, $k_s^+ = k_s u_* / \nu$, (where u_* is the bed shear velocity and ν is the kinematic viscosity) in the range between 20 and 80, low-speed streaks were visualized using the hydrogen bubble technique. For higher k_s^+ values ($k_s^+ \approx 200$) Defina utilized white PVC particle tracers to visualize the flow structure. These visualization techniques, although somewhat subjective, allowed quantitative measurements of transverse streak spacing for the lower transitional roughness regime which were in good agreement with the previous work of Smith & Metzler (1983) for flow over a smooth wall. Under fully rough wall conditions Defina found that the mean streak spacing, λ, scaled with the size of the wall roughness elements, as previously suggested by Grass, such that $\lambda / k_s \approx 4.5$. These results were later confirmed (see Figure 4.1) in similar experiments by Grass *et al.* (1991), Grass (1992) and Defina (1992) who used short cotton tufts to visualize the flow.

A possible explanation of the formation of a streaky pattern over a rough wall comes from the numerical experiments of Lee *et al.* (1990) who studied the structural features produced in initially homogeneous turbulence that was subjected to a high shear rate. They found that under a shear rate equivalent to that found in the viscous sublayer of a wall-bounded shear flow, a streaky pattern formed with a

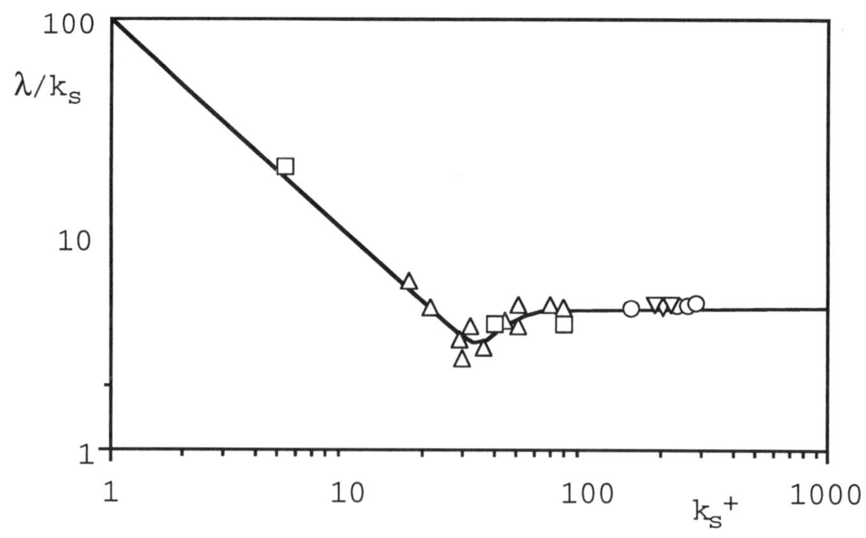

Figure 4.1. Transverse spacing of near-wall low-speed streaks scaled by equivalent sand roughness height, k_s, versus roughness Reynolds number $k_s^+ = k_s u_* / \nu$. Data of: Defina (1987): △, hydrogen bubbles; ▽, PVC particles; Defina (1992): ○, tufts; Grass *et al.* (1991): □, H_2 bubbles

mean transverse spacing comparable to that usually found in turbulent boundary layers. According to the authors, a strong shear rate is capable of organizing turbulence into coherent structures, whatever the type of flow.

Based on these results, Grass *et al.* (1991) suggested that a similar mechanism could be responsible for the streaky pattern produced under rough wall conditions: the 'debris' of turbulence released behind roughness elements is organized under a high shear rate into coherent elongated structures. Grass (1992) and Grass & Mansour-Tehrani (this volume) propose a universal scaling length as the ratio between a local 'effective eddy viscosity', ν_t and bed shear velocity, u_* such that $\lambda \approx 100\,\nu_t / u_*$ irrespective of the wall roughness condition.

If such a mechanism holds near the wall, then there is no apparent reason preventing the existence of shear-induced larger structures away from the bed in a channel flow high into the outer layer. Indeed, strong shear layers induced by near-wall features may promote the organization of new structures, building up hierarchies of vortices as assumed by Perry & Chong (1982). In other words, these turbulent structures self-organize and expand in scale with increasing wall distance. Nakagawa *et al.* (1981), using both visual techniques and hot-film correlation techniques, measured streak spacing in the near-wall and in the outer region of a channel flow over a smooth wall. They found a gradual increase in λ for $y^+(= yu_* / \nu) > 30$ and an asymptotic increase to $\lambda \approx 2y$ for $y^+ > 100$. Smith & Metzler (1983) found a slightly different result: they measured the transverse streak spacing at $y^+ = 1, 5, 10, 15, 20, 30$ and found a much more rapid increase of λ with distance from the wall.

Another question which arises from this previous work is whether the largest hierarchy of vortices correspond to secondary currents in straight channel flow. The term 'secondary current' is often used to designate both the instantaneous (or short-term averaged) and time-averaged currents which occur in a plane normal to the mean flow direction. The former, which are addressed in this paper, appear as pairs of counter-rotating vortices, scale with the flow depth and are very unstable in space and time (Gulliver & Halverson, 1987; Tamburrino, 1990). The latter secondary flows, hereafter termed 'Eulerian secondary currents', are instead stable but weak (Nezu & Rodi, 1985; Tominaga *et al.*, 1989). Eulerian secondary currents may result from the temporal mean of the transverse velocity field produced by quasi-instantaneous longitudinal vortices. Thus, where the quasi-instantaneous secondary currents are spatially stable, the Eulerian secondary currents can be detected. An example is given by considering the flow in a wide rectangular channel: sidewalls locally stabilize secondary currents and Eulerian measurements recognize persistent cellular currents near the corners (e.g. Tominaga *et al.*, 1989). Additionally, transverse bed undulations such as sand ridges and troughs (Nezu & Rodi, 1985) or spanwise variations of roughness (Müller & Studerus, 1979) may act to stabilize secondary currents. It is also interesting to note that secondary currents have been reported only for channel flow but are absent in turbulent boundary layer studies. This is likely because the free surface promotes vertical stability and enhances secondary currents.

The evidence for the existence of a near-wall streaky pattern, irrespective of wall condition, generated by horseshoe-type vortical structures is now very convincing. The existence of similar large-scale turbulent features in a channel flow is equally well documented. However, many uncertainties remain concerning the possible links between near-wall and outer coherent flow structures. Moreover, the existence of intermediate-scale structures is at present purely conjectural (Grass *et al.*, 1991) as there is still a lack of experimental evidence.

The objective of the present investigation is to explore the possibility that hierarchies of vortices exist in a channel flow which gradually expand in scale from roughness height up to the flow depth. In this experimental work, based upon flow visualization, attention was focused on the most apparent and detectable feature of turbulent coherent structures: the transverse spacing between longitudinal streaks.

EXPERIMENTAL APPARATUS AND PROCEDURE

The experiments were performed in a 0.38 m wide, 20 m long tilting flume with Plexiglas walls. The bottom slope can be adjusted to a maximum of 5% and the flume uses a downstream weir to achieve quasi-uniform flow conditions along the test section.

At the test section, which is 8 m long and located about 4 m downstream from the flume inlet, the bottom was artificially roughed by pasting spherical pebbles of ball-clay with a diameter ranging between $D = 9.6$ mm and $D = 10.5$ mm in the densest possible packing array. The pebbles were then painted white to improve dye visualization.

The tracer was fed into the flow through a number of holes located about 4.5 m downstream from the leading edge of the test section. In order to attenuate disturbances produced by the tracer injection, the openings were drilled at the interstices between grains along a line normal to the flow direction; transverse spacing between holes was approximately 5 mm (see Figure 4.2).

A small tank containing the tracer was attached below the test section in direct connection with the holes (Figure 4.2) in order to assure a uniform feed. The tracer used to visualize the flow was dyed milk. The small amount of fat (1.5–2.0%) in the milk was very effective in improving flow visualization (examples are given in Figure 4.3) and the results reported in the following were all obtained using this tracer. More details of the tracer and the experimental facilities are reported by Susin (1994).

Different mean velocity profiles were measured at the centre line of the flume with a very small Pitot tube, with a diameter of about 1.5 mm, which was connected to a pressure transducer. Velocity profiles were fitted to the logarithmic law-of-the-wall (Figure 4.4) in order to derive the equivalent sand roughness height, k_s. It was found that $k_s = 6.7$ mm or $k_s / D = 0.67$.

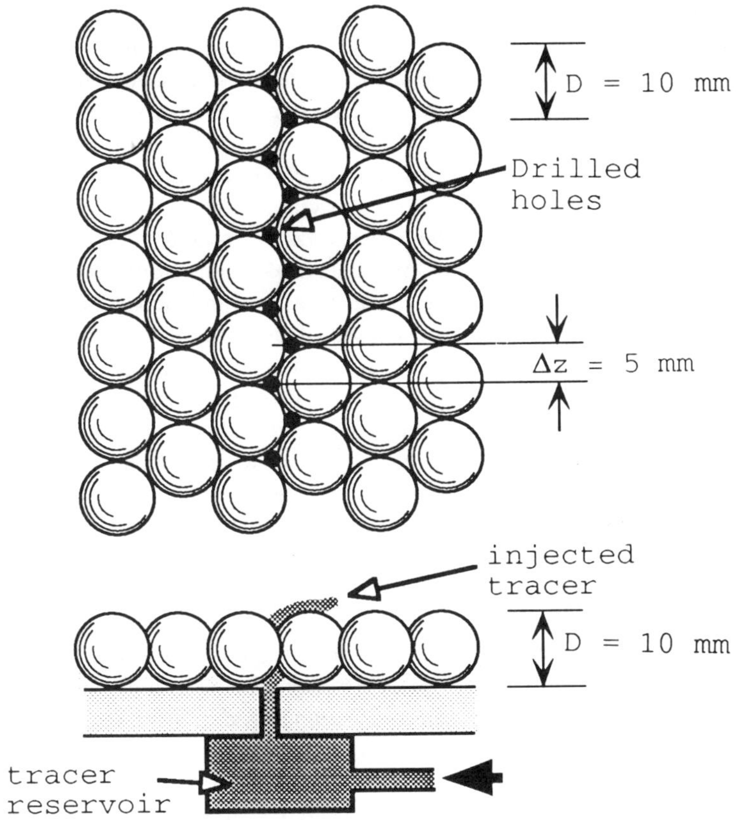

Figure 4.2. Schematic diagram of the tracer inlet (flow is from left to right). The tracer is contained in a small reservoir located below the flume and enters the flow through a number of small holes drilled at the interstices between grains. Transverse spacing between injection points is approximately 5 mm

Experiments were then performed by varying the bed channel slope and water depth. Experimental conditions are summarized in Table 4.1 and for each case listed in the table (i.e. bed channel slope), about 15 runs were performed with different water depths.

For each flow condition a number of photographs were taken of both side and plan view. A time lag of about 2 s was allowed between two consecutive plan view pictures. Side view photographs, taken during runs with $d > 3$ cm, were used to evaluate the downstream tracer profile: $y_t = y_t(x_t)$ where y_t is the thickness of the marked layer measured at the distance x_t downstream the tracer inlet. The thickness y_t was then taken to be a local measure of the 'distance from the wall' characterizing the marked turbulent structures. Linking y_t to the distance from the wall, y, adopted in Eulerian measurement techniques (e.g. H_2 bubbles and velocity correlation techniques) is not straightforward. It can be assumed, however, that quasi-longitudinal, counter-rotating vortices extending up to y_t, concentrate into streaks mainly H_2 bubbles which are generated at a distance from the wall, y, corresponding approximately to the core region of such vortical structures, such that $y \approx y_t / 2$.

About 20 plan view photographs were taken during each run and successively used to evaluate the mean transverse spacing between streaks. The counting procedure adopted was very similar to that used by Smith & Metzler (1983).

When a number of 'long enough' parallel streaks could be clearly identified in the photographs, their average spacing, λ, was measured and recorded together with the distance from the tracer inlet to the centroid of the streaks. Streaks were considered to be long enough when their length was not smaller than about three to four times the transverse spacing. Often, however, streak length was much larger than this threshold value. In order to reduce the general uncertainties and subjectivity inherent in this streak-counting procedure, measurements were performed by two different individuals and repeated after a lag of about three weeks with minor differences in the results. Errors in the average transverse spacing between streaks were within ±6%. Therefore, the scatter of data plotted below (Figure 4.7) must be ascribed to the small number of photographs that were processed.

The recorded distances from the tracer inlet were then transformed into corresponding wall distances with the aid of the tracer profiles as previously discussed. It was thus possible to analyse streak spacing as it changes with distance from the wall. This analysis procedure is summarized in the sketch of Figure 4.5.

EXPERIMENTAL RESULTS AND DISCUSSION

The measured tracer profiles are plotted in Figure 4.6 and show that, with a minor scatter, a single curve could be drawn for each channel slope to interpolate measurements when both the downstream distance from the tracer inlet, x_t, and the thickness of the marked layer, y_t, are scaled with the flow depth d. A slight difference between the curves due to the changes in channel bed slope was observed.

Plan view photographs were then used to evaluate the transverse spacing between longitudinal streaks. When the flow depth was smaller than approximately 15 mm (i.e. $d / k_s < 2.2$), a clear streaky pattern could not be recognized: ribbons of marked fluid were seen only randomly distributed in time and space. As this could be due to the relatively small Reynolds number ($\mathrm{Re} = \overline{U}d / \nu = 1200 - 2700$), some additional runs were performed increasing the bed channel slope to 0.007 and 0.01. Again when the flow depth did not exceed twice the equivalent sand roughness height the flow, viewed from above, did not reveal any regular streaky structure.

An explanation for this feature may be that at such small flow depths, the average shear rate, and hence vortex stretching mechanism, is not capable of

(a)

(b)

(c)

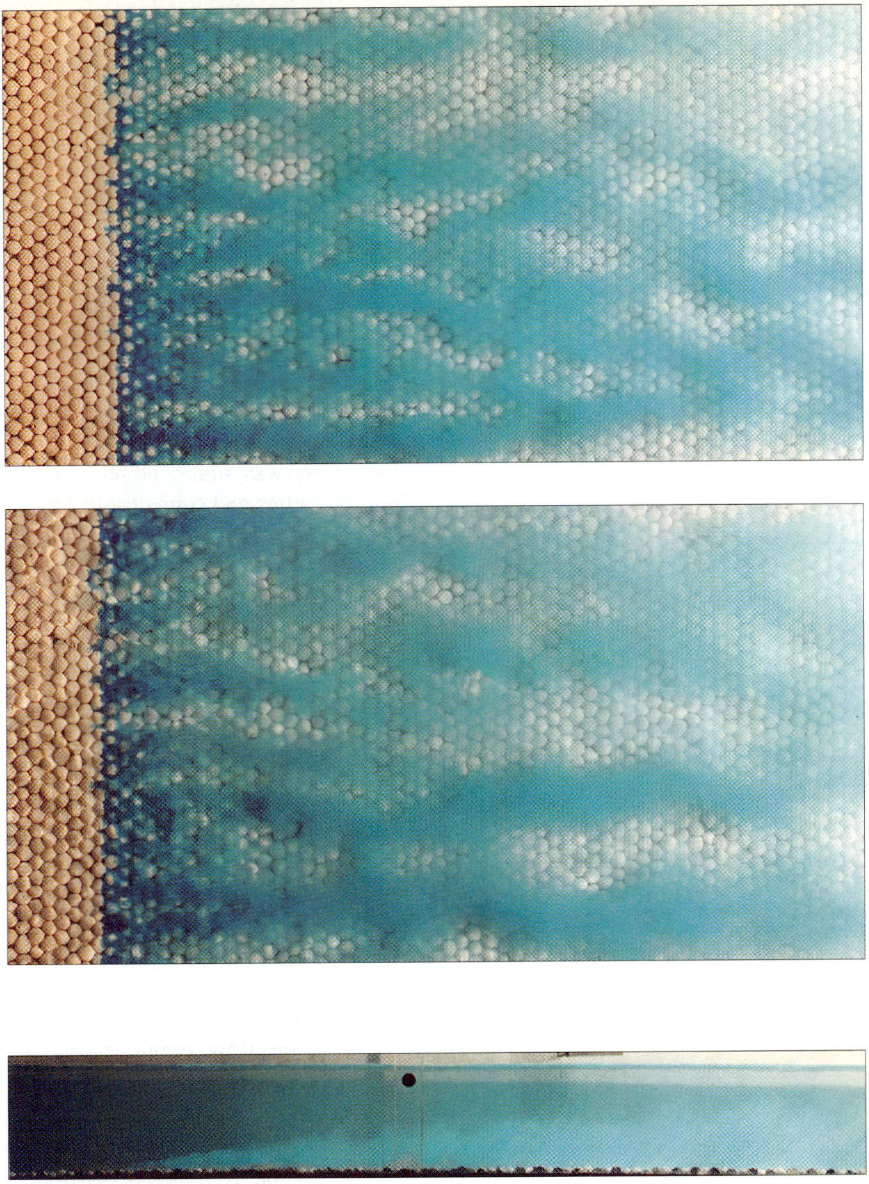

Figure 4.3. Examples of plan and side view photographs taken during the experiments: (a) and (b) streaky pattern visualized by dyed milk (Case S2); (a) $d = 49$ mm; $U = 0.39$ m s^{-1}; (b) $d = 98$ mm; $U = 0.60$ m s^{-1}; and (c) straight vortical features inclined at 45° to the wall

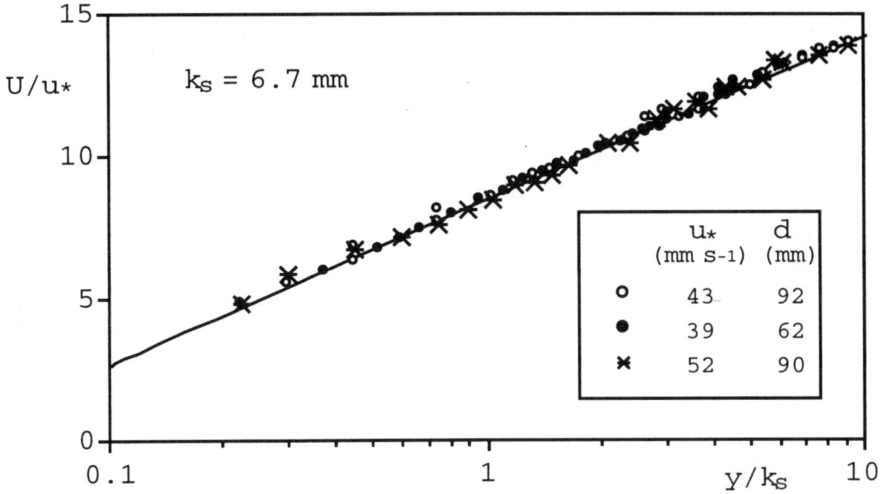

Figure 4.4. Measured velocity profiles, scaled by shear velocity, u_*, and roughness height, k_s. The straight line represents the logarithmic law of the wall: $U / u_* = \ln(30y / k_s) / \kappa$, where κ is the von Kármán constant $\cong$ 0.4

Table 4.1. Experimental conditions

case	slope S	flow depth d (mm)	mean velocity $\overline{U}$ (m s^{-1})	shear velocity $u_* = \sqrt{gdS}$ (mm s^{-1})	Reynolds number $Re = \overline{U}\,d / \nu$ $\times 10^{-3}$	roughness Reynolds number $k_s^+ = k_s u_* / \nu$
S2	0.002	23–98	0.17–0.60	21–44	3.9–58.8	140–295
S3	0.003	11–50	0.11–0.42	18–38	1.2–21.0	120–255
S5	0.005	10–54	0.14–0.60	22–51	1.4–32.4	150–340

organizing the vortical debris produced behind roughness elements. Moreover, since this threshold water depth is approximately half the spanwise spacing of near-wall turbulent structures, then the lack of coherence can be interpreted as additional indirect proof of the existence of a well-defined first hierarchy of vortical structures just above the roughness elements.

At higher flow depths, a streaky pattern could be visualized easily. The measured transverse spacing between longitudinal streaks, λ, is plotted versus wall distance, y_t, in Figure 4.7. Here both λ and y_t are scaled with the roughness height k_s. Near the wall, the transverse spacing maintains a constant value irrespective of water

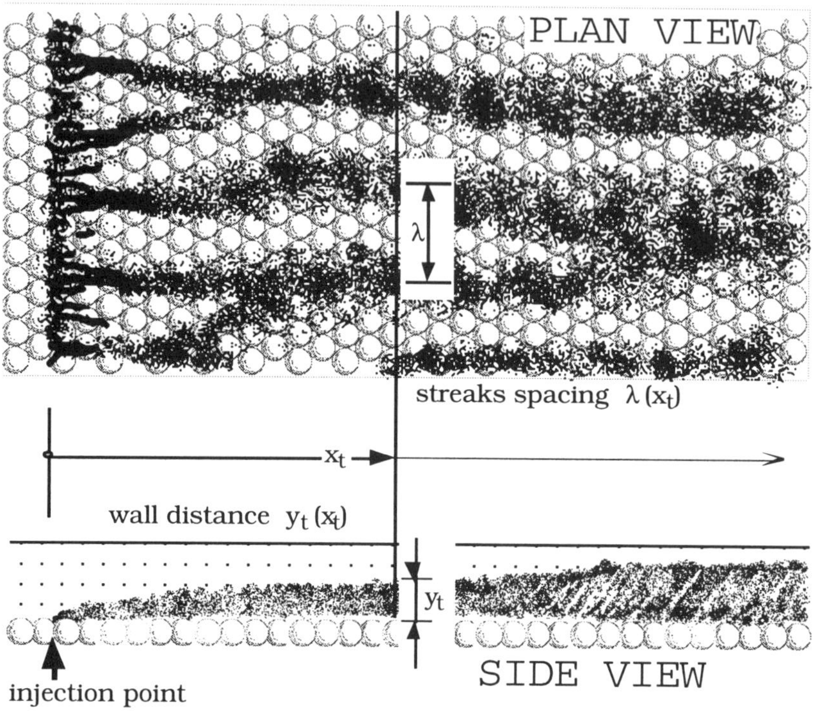

Figure 4.5. Schematic diagram of the measurement technique (flow is from left to right). When a number of parallel streaks are identified in the plan view photographs their average spacing, λ, and the distance from the tracer inlet to the centroid of the streaks, x_t, are measured. Distance from the tracer inlet, x_t, is then transformed into corresponding wall distance, y_t, with the aid of the tracer profile

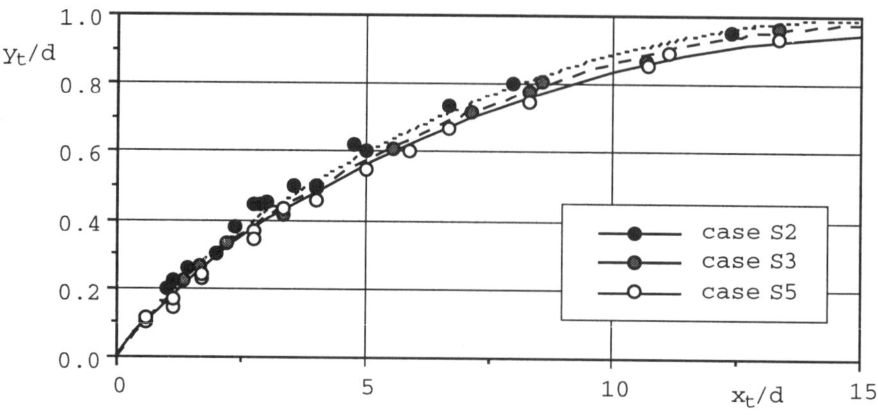

Figure 4.6. Measured tracer profiles scaled by water depth, d, for three different flume slopes. For each slope, experimental points (dots) are fitted by a single curve

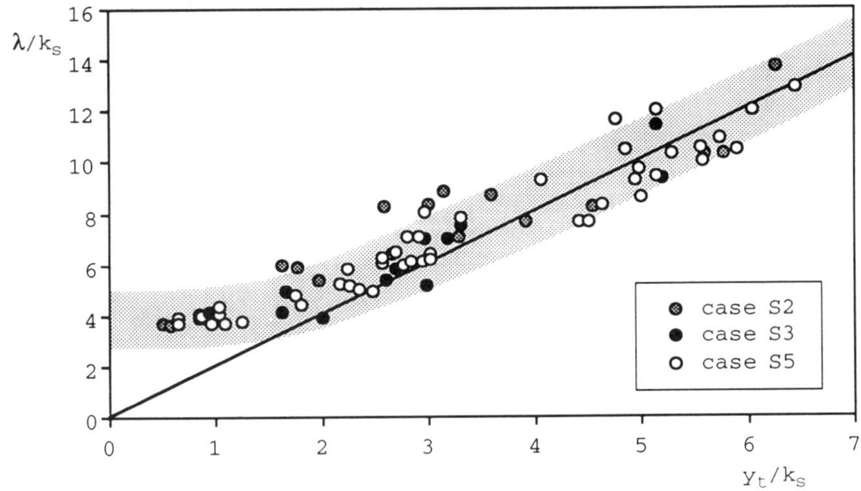

Figure 4.7. Transverse spacing of streaks scaled by equivalent sand roughness height, k_s, versus wall distance y_t / k_s. The straight line is $\lambda = 2y_t$

depth and channel slope, being about four times the equivalent sand roughness height as previously found by Defina (1987, 1992) and Grass *et al.* (1991).

The transverse spacing between streaks found in the near-wall layer retains a constant value for a small vertical extent because it is the same vortical hierarchy which collects and concentrates the dyed milk injected at the wall into longitudinal streaks, whilst pumping it away from the channel bottom (Figure 4.8). Further, away from the channel bottom, that is for $y_t / k_s > 2$, transverse spacing grows roughly according to the linear law: $\lambda = 2y_t$ (Figure 4.7).

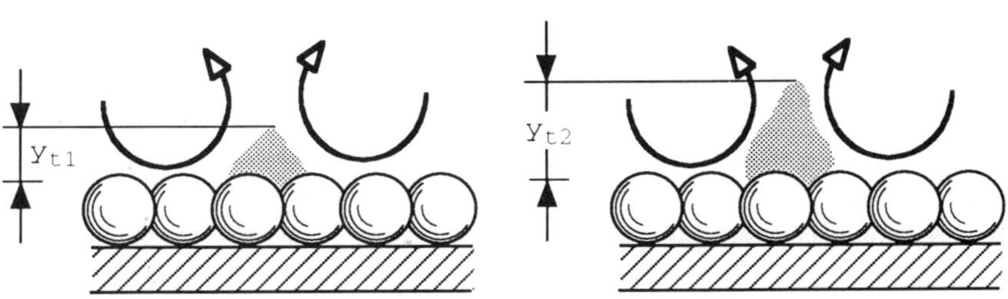

Figure 4.8. Schematic diagram of tracer pick-up by near-wall coherent structures. Near the wall it is the first hierarchy of vortices that picks up and concentrates the tracer. Therefore the same streak transverse spacing, λ, corresponds to different distances from the wall, y_t

The results presented here are in good agreement with the experimental findings of Grass (1992; Figure 6) who plotted transverse spacing of low-speed streaks λ, scaled by effective viscosity wall length-scale ν_t / u_* as a function of wall distance $y u_* / \nu_t$. Direct comparison between Grass' data and the present results is, however, possible only under the two following assumptions. First, it is assumed that the near-wall streak spacing, $\lambda \approx 100 \nu_t / u_*$, measured by Grass corresponds to the present $\lambda \approx 4 k_s$. This simply means that effective eddy viscosity ν_t, under rough wall flow conditions, should be evaluated at a wall distance commensurate with the roughness height. Secondly, it is assumed that the above suggested relationship $y_t \approx 2y$ is applicable. The data of Grass (1992; Figure 6) and the present study are both plotted in Figure 4.9, each with their own scaling length. For comparison, the data of Smith & Metzler (1983), who measured the variation in streak spacing with distance from the wall under smooth wall flow conditions, are also plotted in Figure 4.9. The agreement between all three data sets is fairly good and supports the suggestion that Eulerian measurements taken at a wall distance y relates to coherent flow structures characterized by a vertical size, y_t, which is about twice y. It is interesting to note that in Grass' experiments (Grass, 1992) different roughness heights were used while water depth and Reynolds number were kept roughly constant. In the present experiments, however, a constant roughness height was used while water depth and Reynolds number were allowed to vary over a logarithmic cycle (Table 4.1).

During the experiments it was also observed that far downstream from the tracer inlet only larger vortical structures were marked by the dyed fluid. Small-scale near-wall vortices and intermediate-scale vortical structures could not be detected at all (Figure 4.3). This may suggest that hierarchies of vortices exist which are dynamically related to each other by a well-defined feedback mechanism involving vorticity stretching. Strong internal shear layers, produced just above the near-wall low-speed streaks, may promote the downstream formation of new and larger, geometrically similar, flow structures. These structures, in turn, induce new shear layers further away from the wall and the process continues to the largest scale. The process described by Grass *et al.* (1991) in which these vortical structures "*extract energy from the mean shear flow by the vortex stretching mechanism and act as pumps transporting mass, momentum, vorticity and contaminants ... across the boundary layer*" therefore appears as a well-structured, dynamic process. This also supports the suggestion that Eulerian measurements taken away from the wall are not the result of an averaging process over different scale coherent structures, but are strongly related to a particular hierarchy of vortices.

It was also observed that when the dyed structures extended over the whole flow depth, and hence their transverse spacing was about twice the water depth, then they were characterized by a noticeable spatial persistence. At higher flow depths ($d > 40 - 50\,\text{mm}$) they could be clearly identified until they reached the downstream end of the test section. This may suggest that the largest hierarchy of longitudinal

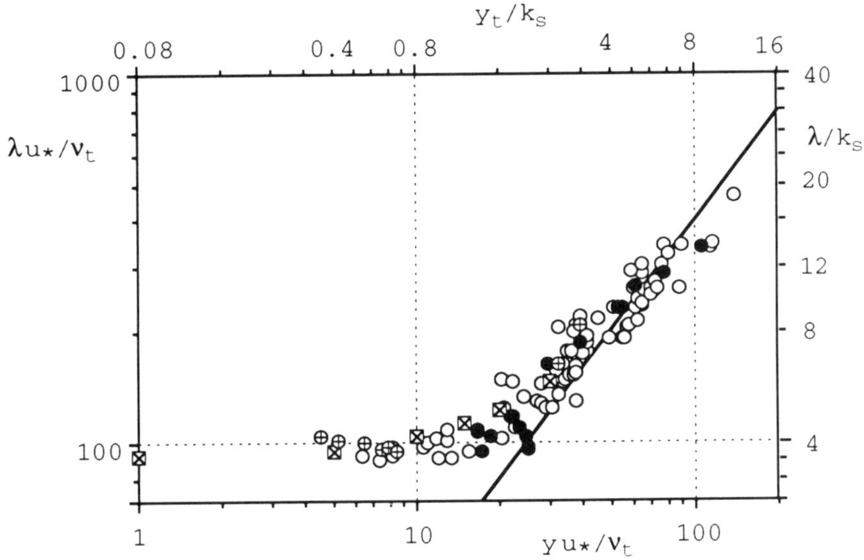

Figure 4.9. Variation of mean streak spacing with distance from the wall. Present study (λ/k_s versus y_t/k_s) : ○; Grass (1992; Figure 6) ($\lambda u_*/\nu_t$ versus yu_*/ν_t): ⊕, smooth wall; ●, rough wall; Smith & Metzler (1983) ($\lambda u_*/\nu$ versus yu_*/ν): ⊠, smooth wall. The straight line represents $\lambda = 2y_t$ or $\lambda = 4y$

vortices, responsible for the streak appearance, correspond to the instantaneous secondary currents found in a straight channel flow.

When the flow was viewed from the side, then the dyed layer appeared as a random agglomeration of very narrow and substantially straight features inclined at approximately 45° to the bottom. The overall picture (see Figure 4.3) is very similar to that described by Head & Bandyopadhyay (1981; Figure 5) but at this stage the possibility that these features are elongated hairpin vortices is largely conjectural. Moreover, comparison between the plan and side views was somewhat surprising: while transverse spacing and spanwise size of longitudinal streaks clearly increased with downstream distance from the tracer inlet, no apparent changes could be envisaged for these inclined features except for their vertical extension. It may be speculated that these features are something like a 'signature' of the vortex stretching mechanism, producing smaller scale secondary structures aligned along the direction of the maximum rate of strain.

CONCLUSIONS

It is now well established that horseshoe-type vortical structures are a common feature of both smooth and rough wall bounded shear flows. Moreover, they have been detected both near the wall and in the outer flow region, suggesting that a similar mechanism could be responsible for their formation and growth.

The results of the experimental investigation presented here, based on a flow visualization technique, confirm the existence of hierarchies of vortices with a characteristic length-scale which increases with distance from the wall.

When the flow depth is smaller than about twice the roughness height, then no streaky pattern is formed in a channel flow. At greater flow depths, low-speed streaks are clearly detectable with a near-wall transverse spacing which is about four times the equivalent sand roughness height as previously found by Defina (1987, 1992), Grass *et al.* (1991) and Grass (1992). Both these results suggest that a well defined first hierarchy of vortices exist that are formed just above the roughness elements.

As distance from the wall increases, then the transverse spacing between coherent flow structures gradually grows according to the simple linear law: $\lambda = 2y_t$. This result provides consistent support for the eddy hierarchy model proposed by Perry & Chong (1982) and later refined by Perry *et al.* (1986, 1991). Finally it is apparent from the present visualization study that instantaneous secondary currents in a straight channel flow simply coincide with the largest hierarchy of longitudinal vortices.

REFERENCES

Defina, A. 1987. *Indagine Sperimentale e Teorica Sull'interazione tra Particelle Solide e Strutture Fluide Coerenti Note col Termine di Esplosioni Turbolente*, Unpub. PhD Diss., Università di Firenze, Genova e Padova, 116 pp.

Defina, A. 1992. 'Interasse trasversale delle esplosioni turbolente in regime di parete scabra', *Atti dell'Istituto Veneto di SS. LL. AA*, **CL**, 1–13.

Grass, A.J. 1971. 'Structural features of turbulent flow over smooth and rough boundaries', *J. Fluid Mech.*, **50**, 233–255.

Grass, A.J. 1992. 'Experiments in turbulent motion', *Proc. 23rd Convegno di Idraulica e Costruzioni Idrauliche, Firenze, Italy*, **5**, 41–53.

Grass, A.J., Stuart, R.J. & Mansour-Tehrani, M. 1991. 'Vortical structures and coherent motion in turbulent flow over smooth and rough boundaries', *Phil. Trans. R. Soc. London*, **A336**, 35–65.

Gulliver, J.S. & Halverson, M.J. 1987. 'Measurements of large streamwise vortices in an open-channel flow', *Water Resour. Res.*, **23**, 115–123.

Head, M.R. & Bandyopadhyay, P. 1981. 'New aspects of turbulent boundary layer structure', *J. Fluid Mech.*, **107**, 297–338.

Lee, M.J., Kim, J. & Moin, P. 1990. 'Structure of turbulence at high shear rate', *J. Fluid Mech.*, **216**, 561–583.

Müller, A. & Studerus, X. 1979. 'Secondary flow in an open channel', *Proc. 18th IAHR Congr.*, **3**, 19–24.

Nakagawa, H., Nezu, I. & Tominaga, A. 1981. 'Spanwise streaky structure and macroturbulence in open-channel flows', *Mem. Fac. Eng. Kyoto Univ.*, **43-1**, 34–67.

Nezu, I. & Rodi, W. 1985. 'Experimental study on secondary currents in open channel flow', *Proc. 21st IAHR Congr.*, **2**, 115–119.

Perry, A.E. & Chong, M.S. 1982. 'On the mechanism of wall turbulence', *J. Fluid Mech.*, **119**, 173–217.

Perry, A.E., Henblest, S.M. & Chong, M.S. 1986. 'A theoretical and experimental study of wall turbulence', *J. Fluid Mech.*, **165**, 163–199.

Perry, A.E., Li, J.D. & Marusic, I. 1991. 'Towards a closure scheme for turbulent boundary layers using the attached eddy hypothesis', *Phil. Trans. R. Soc. London*, **A336**, 67–79.

Smith, C.R. & Metzler, S.P. 1983. 'The characteristics of low-speed streaks in the near-wall region of a turbulent boundary layer', *J. Fluid Mech.*, **129**, 27–54.

Susin, F.M. 1994. *Indagine Sperimentale sulle Strutture Vorticali Coerenti in una Corrente a Superficie Libera in Moto Turbolento di Parete Scabra*, Unpub. Master Thesis, Università di Firenze, Genova e Padova, 77 pp.

Tamburrino, A. 1990. 'Flow structure of the outer region in open channel flow', *Ann. Rev. Fluid Mech.*, **25**, 81–87.

Tominaga, A., Nezu, I., Ezaki, K. & Nakagawa, H. 1989. 'Three-dimensional turbulent structure in straight open channel flows', *J. Hydraul. Res.*, **27**, 149–173.

NOTATION

D	grain diameter
Re	Reynolds number ($= Ud/\nu$)
U $\overline{U}$	mean and depth-averaged downstream velocity
d	flow depth
k_s	equivalent sand roughness height
k_s^+	roughness Reynolds number ($= k_s u_* / \nu$)
u_*	bed shear velocity
x_t	distance from the tracer inlet
y	distance from the wall
y_t	thickness of the marked layer
κ	von Kármán's constant
λ	transverse streak spacing
ν	kinematic viscosity
ν_t	effective 'eddy' viscosity

5 A Model of an Impinging Jet on a Granular Bed, with Application to Turbulent, Event-driven Bedload Transport

ANDREW J. HOGG, W. BRIAN DADE, HERBERT E. HUPPERT
Institute of Theoretical Geophysics, Department of Applied Mathematics & Theoretical Physics, Silver Street, Cambridge, CB3 9EW, UK

RICHARD L. SOULSBY
HR Wallingford Ltd., Howbery Park, Wallingford, Oxfordshire, OX10 8BA, UK

ABSTRACT

A model of the impingement of a jet onto a granular bed is developed to calculate the volume of mobilized particles, as a function of characteristics of the jet and the grains. This model comprises a phenomenological description of bedload transport, which aims to form a new, quantitative link between coherent turbulent motions within turbulent boundary layers and the movement of sediment. The turbulent bursting phenomenon has for some time been identified as a dynamically important means of transporting momentum across the boundary layer and generating turbulent kinetic energy. It comprises two predominant types of motion, ejections and sweeps, both of which yield positive contributions to the Reynolds shear stress. While the abrupt ejection of fluid away from the boundary is a significant means for the suspension of particles, this paper considers an idealization of the impact of downward and fast moving sweep events on an erodible granular bed in terms of an impinging jet. A framework of analysis is proposed to model the way in which the impingement of the flow leads to the mobilization of particles from the boundary, the volume of entrained particles being specified as a function of the characteristics of the hydrodynamic event and the particles. By invoking a balance between the volume of entrained particles and the volume deposited between events, the bedload transport under equilibrium conditions is obtained. The predictions of this model are in accord with experimental observations as well as other derivations which adopt an entirely different approach. Furthermore, the quantitative calculations elucidated here lead to a mathematical relationship between the rate of bedload transport and the hydrodynamic characteristics of the flow which agree with previous empirical formulations.

Coherent Flow Structures in Open Channels.
Edited by P.J. Ashworth, S.J. Bennett, J.L. Best and S.J. McLelland.

INTRODUCTION

The experimental studies of Kline *et al.* (1967) and Grass (1971) were among the first to identify unsteady, yet coherent, three-dimensional motions within turbulent boundary layers. They discovered that various flow patterns which exhibit irregular, but repetitive structures are found within turbulent boundary layers and that these patterns are coherent in both a temporal and spatial sense. These experimental studies identified predominant features of the flow: an abrupt ejection of fluid away from the boundary with a compensating inrush (sweep) towards the boundary and low-speed streaks within the viscous sublayer (see review by Smith, this volume). Collectively these are known as the turbulent bursting phenomenon. Observations of such features are common to all turbulent boundary layers, although the low-speed streaks are less pronounced for flows over rough boundaries for which the roughness elements inhibit the establishment of a viscous sublayer. However, Grass *et al.* (1991), Grass & Mansour-Tehrani (this volume) and Defina (this volume) find low-speed streaks above the grain tops. In addition to these laboratory studies, the bursting phenomenon has been noted in direct numerical simulations of bounded turbulent flow (Moin & Kim, 1985) and in field studies, where the boundary layer is of geophysical dimensions (Gordon, 1974; Heathershaw, 1974).

It is postulated that the origin of these coherent turbulent motions is due to vortex deformation within the flow. While a precise dynamical description of this evolution and interaction has not yet been formulated, some recent studies (Robinson, 1991; Smith *et al.*, 1991; Smith, this volume) have sought to draw upon the conclusions of various illustrative studies of vortex motion to suggest a plausible mechanism for the generation and maintenance of turbulence. Central to these mechanisms is the bursting phenomenon because it is responsible for the production of turbulent kinetic energy. Furthermore, the ejection and inrush events (which will henceforth be referred to as ejections and sweeps) make substantial contributions to the vertical rate of transfer of horizontal momentum by turbulent motions, known as the Reynolds stress, and are hence essential momentum transport mechanisms across the boundary layer.

The role of turbulent motions in the process of transporting sedimentary particles has been the focus of considerable discussion. Studies have sought to form a link between the coherent turbulent motions within the turbulent boundary layer and the transport of sediment (Jackson, 1976; Sumer & Deigaard, 1981). At a most basic level, it seems apparent that the interactions of vortex-like structures within the flow must have a dominant influence on the dynamics of sediment transport. The ejection events transport particles away from the boundary, while sweep events mobilize particles which are in repose on the boundary. Whereas recent papers have examined the way in which ejection events may be involved in the process of sediment suspension (Hogg *et al.*, 1994), the current work examines the role of sweeps in the mobilization of particles at the bed.

Grass (1971) noted that not only do sweep events play an important role in the general mechanism of the vertical transport of momentum, but they were capable of mobilizing sand particles in repose on the smooth lower boundary of a laboratory flume and transporting them over considerable distances downstream. He found that the longitudinal scales of these events were much greater than the lateral scale. Drake *et al.* (1988) concluded that when a sweep impinged on a bed, particles were entrained at a rate many times the average and were transported downstream faster than the average flow velocity. They found that the sweep-transport events accounted for 70% of the total transport of sediment, even though they occurred for only 9% of the time. This reaffirms the considerable care which is needed when assessing sediment transport strictly from mean stresses rather than from instantaneous stresses, at least at low rates of transport. A similar conclusion may be drawn from the field study of Heathershaw & Thorne (1985). They used a novel acoustic measurement to link turbulent velocity fluctuations with the movement of sediment and found that the sweep events led to appreciable sediment transport. They also noted that the rarely occurring outward interaction events, which move faster than the mean flow and away from the boundary, mobilized more sediment per event than any other type of turbulent motion. Outward interactions occur relatively infrequently, however, rendering the sweep event as the most significant means for moving bedload particles. Heathershaw & Thorne (1985) also demonstrated that sediment movement exhibits a better correlation with streamwise velocity than with kinematic Reynolds stress. This observation suggests that bedload transport is not driven solely by bed shear stress, but additionally by forces arising from pressure fluctuations within the bed. In the studies of Drake *et al.* (1988) and Heathershaw & Thorne (1985), the beds were comprised of fine gravels, which were sufficiently large that they were unlikely to be entrained into suspension under most flow conditions, but rather were transported in the bedload, predominantly driven by the intermittent sweep events.

In this paper a framework for modelling the impingement of a jet on a granular bed is presented and possible applications to bedload transport are discussed. It is proposed that incoming sweep events may be treated as impinging jets and the model is used to study their interaction with an erodible bed of grains. The volume of particles mobilized by a single event is quantified as a function of the characteristics of the jet and the particles. The next section of this paper discusses experimental measurements of the rate of bedload transport and reviews the concept of bulk erosion, first introduced by Fernandez-Luque (1974), which suggests a new approach to modelling the bedload transport. The mathematical model of the impinging jet is described, separating it into essentially three distinct stages. This model permits evaluation of the rate of transport in equilibrium conditions.

BEDLOAD TRANSPORT

The motion of particles within the bedload may be distinguished from those in suspension by the following simple mechanistic definition. The weight of the particles within the bedload is never fully supported by the fluid motion, whereas within the suspended load this balance is possible. Hence, within the bedload, particles roll, slide and saltate along the underlying bed. With sufficiently energetic flows, which mobilize highly concentrated layers of particles close to the bed, it is possible for the weight of the particles to be supported by inter-granular interactions. This regime is termed 'sheet-flow', because the particles move over each other in highly concentrated layers.

There have been a number of experimental studies of steady-state bedload transport for a range of flow conditions and types of sediment. These experiments, conducted in laboratory flumes, have measured the flux of bedload transport, q_{bl}, which is defined by

$$q_{bl} = \int_0^D C(y)u(y)\mathrm{d}y \tag{5.1}$$

where $C(y)$ is the concentration of particles in the bedload, $u(y)$ is their horizontal velocity, y is the distance normal to the bed and D is the flow depth. This flux can be non-dimensionalized with respect to $\sqrt{(s-1)gd^3}$, where s, g, d are the ratio of the solid to fluid densities, gravitational acceleration and the diameter of the particles, respectively. The bedload transport flux is typically measured as a function of the Shields parameter (θ) based on the tractive stress exerted by the flow on the bed (skin friction). This parameter expresses the ratio of the drag exerted by the flow on a particle to the weight of the particle. For a turbulent flow in which the magnitude of the skin friction (σ_b) may be characterized by a friction velocity ($\sigma_b = \rho u_*^2$), the Shields parameter is given by

$$\theta = \frac{{u_*}^2}{(s-1)gd} \tag{5.2}$$

Shields (1936) studied the initiation of particle motion and elucidated the critical value of this ratio (θ_c) which must be exceeded before particle motion is initiated from a bed of non-cohesive grains. Shields demonstrated how this parameter varied with particle Reynolds number based on the particle diameter and found that for water flows and particle Reynolds numbers in excess of unity, $\theta_c \approx 0.05$. When the Shields parameter just exceeds the critical value for the initiation of motion, the particles roll, slide and saltate along the underlying bed. At considerably higher values of the Shields parameter ($\theta > 1$), a regime of sheet-flow is entered in which the weight of the particles is balanced by gradients of the dispersive pressures arising from particle interactions.

Figure 5.1 shows the variation of the non-dimensional rate of bedload transport against the Shields parameter. The data covers over two orders of magnitude of the Shields parameter and exhibits a consistent trend. Meyer-Peter & Müller (1948) suggested the following empirical relationship between the rate of transport and the Shields parameter

$$q_{bl} / \sqrt{(s-1)gd^3} = 8\left(\theta - \theta_c\right)^{3/2} \tag{5.3}$$

There have been a number of subsequent empirical and semi-empirical models of bedload transport which exhibit the same fundamental dependence of the rate of bedload transport with $\theta^{3/2}$. For instance, Nielsen (1992) suggests that

$$q_{bl} / \sqrt{(s-1)gd^3} = 12\sqrt{\theta}\left(\theta - \theta_c\right) \tag{5.4}$$

The latter formula also matches the data collected at high values of the Shields parameter. In Figure 5.1, both of the relationships given by Equations 5.3 and 5.4 are shown.

Bagnold (1954) studied the normal and tangential stresses developed within a sheared suspension of particles. By repeating the experiment with a pure fluid and comparing results, he was able to deduce empirical relations for the inter-granular forces. He noted that the total shear stress of a suspension is a combination of stresses arising from the fluid and particulate phases. This led Bagnold (1954) to postulate that for an equilibrium state, the fluid shear stress at a vertically uniform bed must be reduced to the threshold value for motion; he argued that if the fluid shear stress exceeds the threshold, then particles would continue to be entrained from the bed. The application of this condition, together with an expression of a dynamic friction law which links the shear and normal stresses, implies that the concentration of particles within the bedload may be deduced without explicitly specifying the volume of entrained and deposited particles. This approach has permitted the prediction that the steady-state bedload transport varies with $\theta^{3/2}$ (Engelund & Fredsøe, 1976), which concurs with experimental studies. However, it has not been possible to extend this procedure to study non-equilibrium transport. In the course of this paper, a new framework for the analysis of bedload motion is presented which is driven by the transport arising from individual coherent turbulent motions.

TOWARDS A NEW MODEL OF BEDLOAD TRANSPORT

Within a turbulent boundary layer flow there are velocity fluctuations which lead to fluctuations of the local pressure field and influence the motion of particles in the bedload. Noting this effect led Fernandez-Luque (1974) to introduce a concept of bulk erosion, a local phenomenon where large, instantaneous and localized pressure gradients within the turbulent boundary layer can lead to entrainment of particles from the boundary. From consideration of the forces exerted on the granular bed by

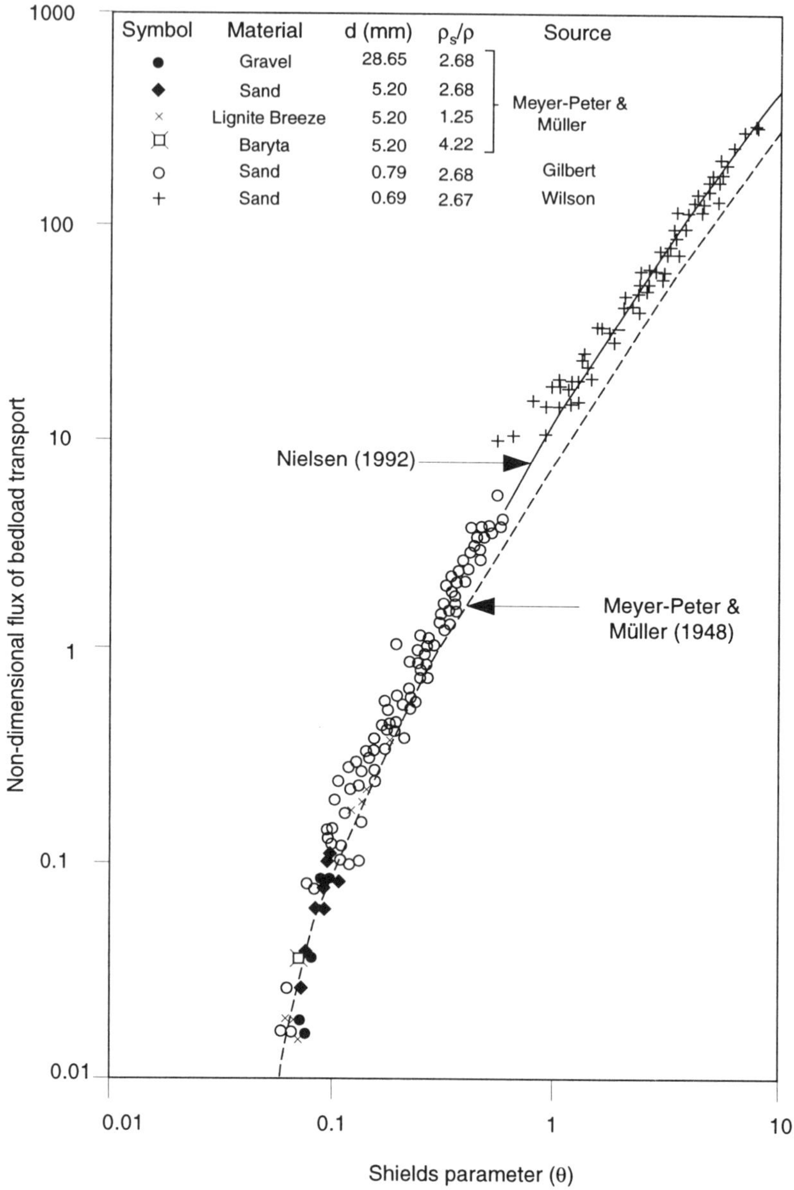

Figure 5.1. The steady-state rate of bedload transport. The dimensionless rate of transport ($q_{bl}/\sqrt{(s-1)gd^3}$) is plotted against the Shields parameter (θ), based on skin friction. The curves shown correspond to the equations proposed by Meyer-Peter & Müller (1948) ($q_{bl}/\sqrt{(s-1)gd^3} = 8(\theta-\theta_c)^{3/2}$) and Nielsen (1992) ($q_{bl}/\sqrt{(s-1)gd^3} = 12\sqrt{\theta}(\theta-\theta_c)$). Reproduced from Nielsen, 1992

the individual events, Fernandez-Luque (1974) defined a local and instantaneous bulk erosion boundary to represent the volume of grains which could be mobilized by a single event.

As noted above, recent research has indicated the existence and importance of coherent motions within turbulent boundary layers. Of these motions, the violent ejection of fluid away from the bed and the compensating sweep towards the boundary are thought to be of the most dynamical significance for mobilizing sediment. While ejection events advect particles far away from the boundary, sweep events impinge upon the bed. They lead to pressure fluctuations within the boundary layer and exert a shear stress on the surface of the bed. They are thus suitable candidates to initiate bulk erosion as envisaged by Fernandez-Luque (1974). Motivated by the observations of bedload transport driven by sweep events and armed with the concept of bulk erosion, the current paper suggests a new framework for modelling the mobilization of sediment particles by turbulent events which seeks to calculate the mass of particles mobilized by events of a given magnitude. Such a calculation yields an effective pick-up function for the bedload transport (*cf.* Nielsen, 1992) which in this study is linked to the impinging events.

This paper proposes four stages of a recipe for modelling turbulent event-driven bedload transport, based upon the study of the impingement of a jet on a granular bed (see below, Figure 5.2). The basic framework of this approach is now discussed and in subsequent sections details are presented of the mathematical modelling. The model acts as a basis from which further studies may elucidate more clearly the nature of the sweep events and their interaction with erodible granular beds.

An incoming, fast-moving sweep event may be treated as a jet flow impinging on a granular bed and locally this flow will be similar to a stagnation point flow. This impingement creates a pressure and shear stress distribution along the surface of the bed, the magnitudes of which are determined by the hydrodynamic characteristics of the jet. Within the bed the effects of the surface shear stress are rapidly attenuated, and so pressure gradients are the main forces responsible for mobilizing the sediment (as demonstrated below). A stability criterion can be defined which gives a minimum value of the ratio of the driving to resisting forces on a particle and allows calculation of a zone of erosion. This permits determination of the volume of particles mobilized by the impinging event. The rate of bedload transport can then be calculated by invoking a simple balance between the volume of particles entrained by the event and those deposited between events.

It should be emphasized that this analysis is relevant only to the parameter range $\theta_c < \theta < 1$; the flow is sufficiently energetic to mobilize particles, but not to launch them into suspension or initiate a sheet-flow layer. Furthermore, the time-scale of the inertial response of the particle must be assumed to be much less than the duration of the event. Smaller particles (sands and fine gravel) are rapidly accelerated by the flow, whereas larger particles (gravels) have a much longer inertial time-scale and are insensitive to the turbulent fluctuations. The transport of

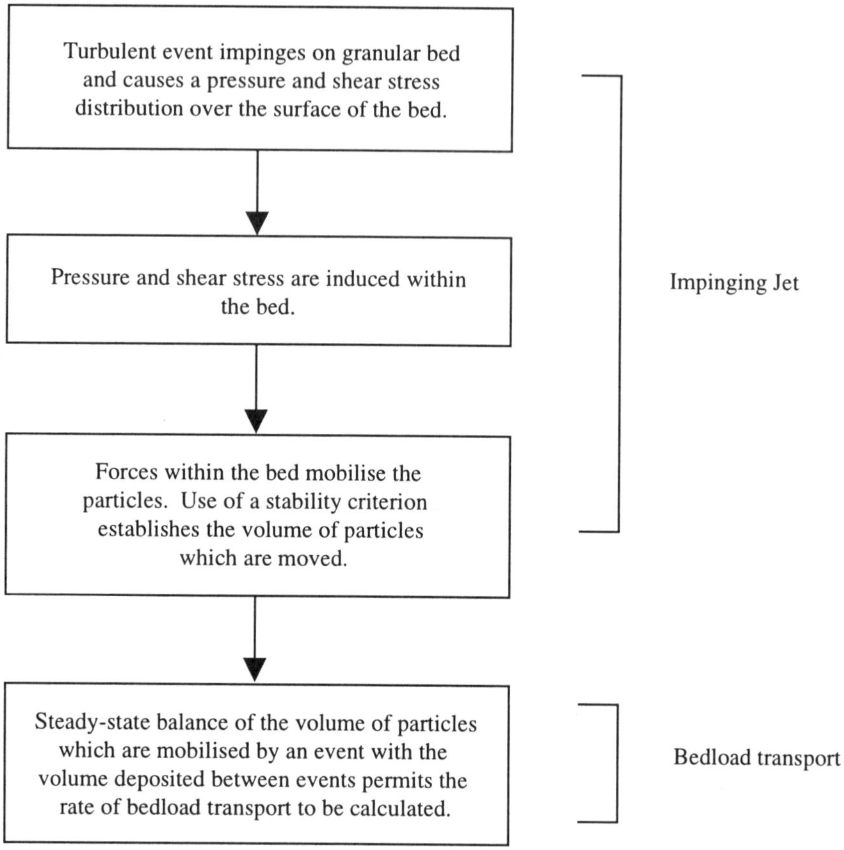

Figure 5.2. The four stages in the mathematical model of a jet impinging on a granular bed and its application to bedload transport

these larger particles is therefore poorly described by this model because they are incapable of responding to the fluctuating velocity field on a sufficiently rapid time-scale.

MATHEMATICAL MODEL OF THE IMPINGING JET FLOW AND BULK EROSION BOUNDARY

This paper considers the impact of an axisymmetric jet on an initially horizontal bed of cohesionless particles and calculates the erosion boundary, the concept of which was described above. The incoming jet is considered to be axisymmetric, because this is the simplest model of a three-dimensional jet flow. Additionally the axis of the jet is assumed to be normal to the granular bed, which is in accord with the viscous stagnation-point flow considered by Cleaver & Yates (1975) as a simple model of a sweep event. Furthermore, Raudkivi (1990) suggests that the angle of impingement has negligible influence upon the depth of erosion, although the erosion profiles become increasing asymmetric as the angle departs from the vertical.

IMPINGING JET FLOW

The impinging jet flow leads to distributions of shear stress and pressure on the surface of the granular bed. Beltaos & Rajaratnam (1974) and Rajaratnam (1976) performed experimental studies of jets impinging on a rigid plane, the configuration of which are shown in Figure 5.3. The source of the jet had velocity U_0, diameter a and was of a height H above the plane. The steady impingement of a jet onto an erodible bed exhibits substantially different behaviour from the impingement on a rigid boundary because the modification of the bed topography will influence the flow. However, the current model applies these experimental results for the rigid boundary, because, the jets considered here are active for only a short duration. Beltaos & Rajaratnam (1974) find experimentally that the pressure at the stagnation point is given by

$$p_0 = 25\rho U_0^2 / (H/a)^2 \tag{5.5}$$

and the surface pressure distribution is given by

$$p_s(r)/p_0 = \exp\left(-114(r/H)^2\right) \tag{5.6}$$

where r is the radial distance from the stagnation point. Furthermore, they find experimentally that the maximum shear stress exerted on the boundary is given by

$$\sigma_0 = 0.16\rho {U_0}^2 / (H/a)^2 \tag{5.7}$$

and note that far from the stagnation point, the shear stress decreases with the inverse-square of the radial distance. Hence the following equation can be proposed to describe the shear stress distribution:

$$\sigma_s(r)/\sigma_0 = 0.033 \frac{\left[1-\exp\left(-114(r/H)^2\right)\left(1+114(r/H)^2+11.09(r/H)^3\right)\right]}{(r/H)^2} \tag{5.8}$$

This relationship is a slight amendment of that presented by Beltaos & Rajaratnam (1974) to fit the experimental data and to account for the far-field relationship.

The use of the experimental data of Beltaos & Rajaratnam (1974) fixes the value of the ratio of the maximum stagnation pressure (p_0) to the maximum shear stress (σ_0), given by $p_0/\sigma_0 = 150$. This value is applicable to an impinging jet but may not be appropriate for an impinging turbulent event. Johansson *et al.* (1987) studied the impingement of turbulent events on a smooth boundary and found that this

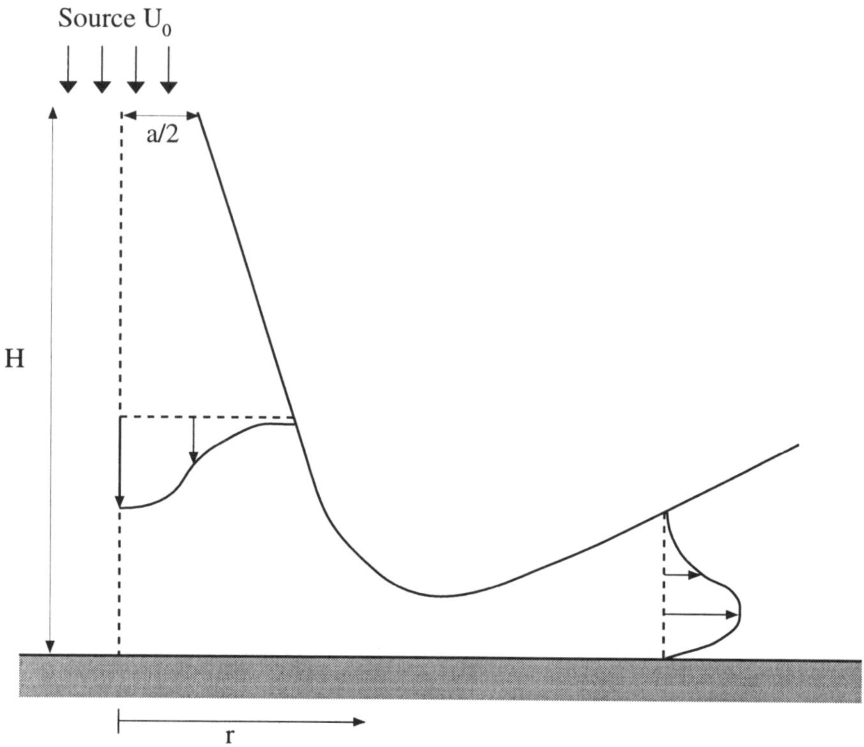

Figure 5.3. The radial wall jet arising from the impingement of a cylindrical jet

ratio is O(10). Nevertheless, the remaining stages of this analysis may be followed irrespective of the actual magnitude of p_0/σ_0. Furthermore, to a first approximation, it is shown below that the functional form of the relationship between the volume of mobilized particles and the hydrodynamic characteristics of the impinging event is independent of this ratio.

FLOW WITHIN THE BED

Flow within the bed is modelled in the present study by the Brinkman equation for filter flow through a porous media (Wiegel, 1980). This is a modification of Darcy's Law, which permits the modelling of flow when the porous region is bordered by a pure-fluid region (Koplik *et al.*, 1983). The bed is assumed to be semi-infinite and statistically homogeneous. The difference between the fluid pressure and the hydrostatic pressure is denoted by *p*, the permeability of the porous bed by Λ and the dynamic viscosity of the fluid by μ. Hence the equations which govern the flow within the bed may be given by:

$$\mu\nabla^2\boldsymbol{u} - \frac{\mu}{\Lambda}\boldsymbol{u} = \nabla p \tag{5.9}$$

$$\nabla\cdot\boldsymbol{u} = 0 \tag{5.10}$$

$$p \to 0,\ \boldsymbol{u} \to 0, \quad y \to -\infty \tag{5.11}$$

$$p = p_s(r), \quad \sigma = \sigma_s(r), \quad v = 0, \quad y = 0 \tag{5.12}$$

where $\boldsymbol{u} = (u, v)$ are the radial and vertical velocity components. The permeability of the bed is related to the grain size of which it is composed. Bear (1972) suggests an empirical relation for the permeability, measured in m^2

$$\Lambda = 0.617 \quad 10^{-3}\, d^2 \tag{5.13}$$

The following non-dimensional variables are introduced for the vertical and radial co-ordinates and the pressure and velocity fields

$$Q = p / p_0, \quad \boldsymbol{U} = \boldsymbol{u}\mu H / \Lambda p_0, \quad \xi = r / H, \quad \eta = y / H. \tag{5.14}$$

Noting that the pressure field satisfies Laplace's equation and providing that the surface pressure distribution is localized to the point of bed-impingement, the axisymmetric pressure distribution within the bed may be derived, via Fourier-Bessel transforms (see Appendix), as

$$Q(\xi, \eta) = \int_0^\infty \frac{k}{228} \exp\left(k\eta - k^2/456\right) J_0(k\xi)\, \mathrm{d}k \tag{5.15}$$

Isobars of this distribution are shown in Figure 5.4. Since the non-dimensional permeability $\lambda^{-2} \equiv \Lambda / H^2 << 1$, there is an 'effective' boundary layer within the flow through the porous media, adjacent to $y = 0$, within which viscous dissipation is non-negligible. Outside of this layer, Darcy stresses dominate and the velocity field is linearly related to the gradient of the pressure field (i.e. the first term of Equation 5.9 is negligible compared to the second). Within this boundary layer, variations in the vertical direction are more significant than those in the horizontal direction and so an approximate velocity field which satisfies the boundary conditions is given by

$$U = -\frac{\partial Q}{\partial \xi} + \left(\sigma_s(\xi)\lambda / p_0 + \frac{1}{\lambda} \left.\frac{\partial^2 Q}{\partial \xi \partial \eta}\right|_{\eta=0} \right) \exp(\eta\lambda) \tag{5.16}$$

$$V = -\frac{\partial Q}{\partial \eta} + \left.\frac{\partial Q}{\partial \eta}\right|_{\eta=0} \exp(\eta\lambda) \tag{5.17}$$

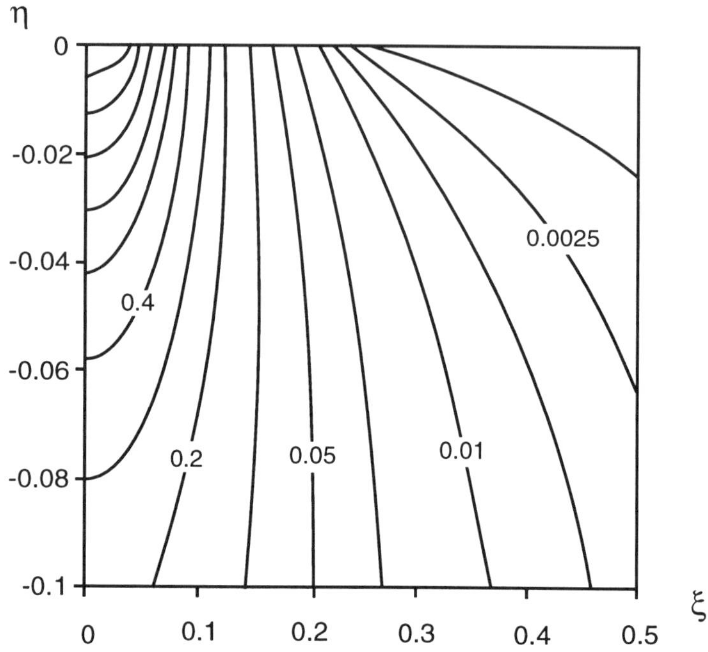

Figure 5.4. Isobars of $Q(\xi,\eta)$. The contours are shown for Q = 0.9, 0.8, 0.7, 0.6, 0.5, 0.4, 0.3, 0.2, 0.1, 0.05, 0.025, 0.01, 0.005, 0.0025, 0.001

Hence, the dimensionless force per unit volume acting on the particles, vertically averaged over a particle diameter, consists of radial and vertical components, given by

$$F_r = -\frac{\partial Q}{\partial \xi} + \frac{H}{d}\exp(\lambda\eta)\left(\sigma_s(\xi)/p_0 + \frac{1}{\lambda^2}\left.\frac{\partial^2 Q}{\partial \xi \partial \eta}\right|_{\eta=0} \right) \tag{5.18}$$

$$F_y = -\frac{\partial Q}{\partial \eta} + \frac{H}{d}\exp(\lambda\eta)\frac{1}{\lambda}\left.\frac{\partial Q}{\partial \eta}\right|_{\eta=0} \tag{5.19}$$

The extent of the boundary layer is $O(1/\lambda)$ which is also the extent of the region through which the influence of the surface stress is transmitted. It may be concluded, therefore, that the effect of the shear stress is absorbed by the granular bed over a relatively short, vertical distance (of the order of a grain diameter), whereas the pressure distribution is transmitted over a much greater length-scale. Similar conclusions are drawn by Fernandez-Luque (1974) and Wiegel (1980).

STABILITY CRITERION AND ZONE OF EROSION

This section considers the forces exerted on the particles which comprise the granular bed and invokes a stability condition to determine those which may be mobilized by the impingement of the incoming jet. This permits the definition of a zone of erosion and quantifies the mobilized particles as a function of the characteristics of the jet.

The forces acting on an individual particle in repose on a planar bed inclined at an angle β to the horizontal are considered; the forces which drive the motion in the radial and vertical direction are denoted by F_r and F_y respectively, while the weight of the particle is given by W. Motion is resisted by a frictional force which arises from the contact of the surface of the grains. The static angle of repose is denoted by α and is the angle at which the bed must be inclined before the grains start to move in the absence of any overlying fluid flow (for spherical, well-sorted sand grains, the angle of static repose is approximately given by $\alpha = 32°$). The configuration of the problem is shown in Figure 5.5.

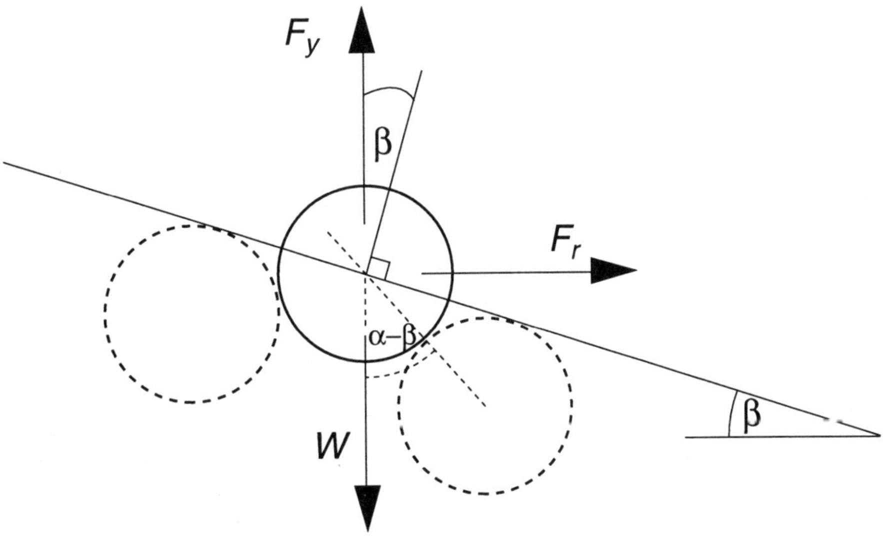

Figure 5.5. The forces acting on a particle in repose on a sloping bed. The horizontal and vertical components of the driving force are denoted F_r and F_y respectively, whilst the weight of the particle and the angle of static repose are given by W and $\tan\alpha$

An individual grain will start to move if the turning moment about its point of contact with a neighbouring grain exceeds the turning moment due to the weight of the particle. This condition is expressed by

$$F_r \cos(\alpha - \beta) + F_y \sin(\alpha - \beta) - W \sin(\alpha - \beta) > 0 \qquad (5.20)$$

Hence, motion occurs if

$$\frac{F_r}{-F_y + W} > \tan(\alpha - \beta) \tag{5.21}$$

Motion inevitably occurs if the inclination of the plane exceeds the angle of static repose (i.e. $\beta > \alpha$). The forces F_r and F_y are associated with the forces induced within the porous bed by the impingement of the jet, while the submerged weight is given by $W = C_b \rho(s-1)g$, where C_b is the concentration of particles in the bed. This allows the calculation of an erosion boundary by application of this stability criterion. This erosion boundary is denoted by $\eta = F(\xi)$ and hence in terms of dimensionless variables, the following relationship along the boundary is produced,

$$\left. \frac{F_r}{-F_y + \rho(s-1)C_b gH / p_0} \right|_{\eta = F(\xi)} \leq \tan(\alpha - \beta) \tag{5.22}$$

$$\text{where} \quad \frac{dF}{d\xi} = -\tan\beta \tag{5.23}$$

These equations introduce a new dimensionless parameter, given by ($\Theta = p_0 / \rho(s-1)C_b gH$), which is a measure of the scale of the pressure gradients to the gravitational force and is similar, but not identical, to the Shields parameter introduced above. The parameter Θ expresses the ratio of the driving to the stabilizing forces, which act on the bed of particles and is henceforth referred to as the 'effective' erosion parameter, noting that if $p_0 = \rho u_*^2$ then $\Theta = \theta d / HC_b$.

This analysis illustrates how a downward moving event can mobilize particles. The turning moment about the point of contact with a neighbouring particle, arising from the radial force (F_r), can be sufficient to overcome the resisting moment arising from the vertical force on the particle, even though the incoming event may augment the weight of the particle.

The erosion boundary is calculated using Equations 5.18, 5.19, 5.22 and 5.23. Far from the stagnation point there is no excess pressure and so this boundary must return to the bed surface. The point at which this occurs ($\xi = \xi_*$) is found by enforcing the equality of Equation 5.22 and solving for $\tan\beta = 0$. The range between the stagnation point and this position is the extent of the influence of the jet. In order for such a position to exist, it is a requirement that $\Theta > \Theta_c$. Hence this imposes a critical value of the parameter Θ, below which it is assumed there is no bulk erosion. For $0 < \xi < \xi_*$, the equality of Equation 5.22 is enforced, whereas beyond this point the bed is assumed flat ($\beta = 0$) which permits the ratio of the driving and stabilizing forces to be less than $\tan\alpha$. The condition that the erosion boundary must lie below the initial boundary of the bed is also imposed.

The critical erosion parameter varies with the permeability of the bed, which is related to the grain size (Figure 5.6). It is found empirically that

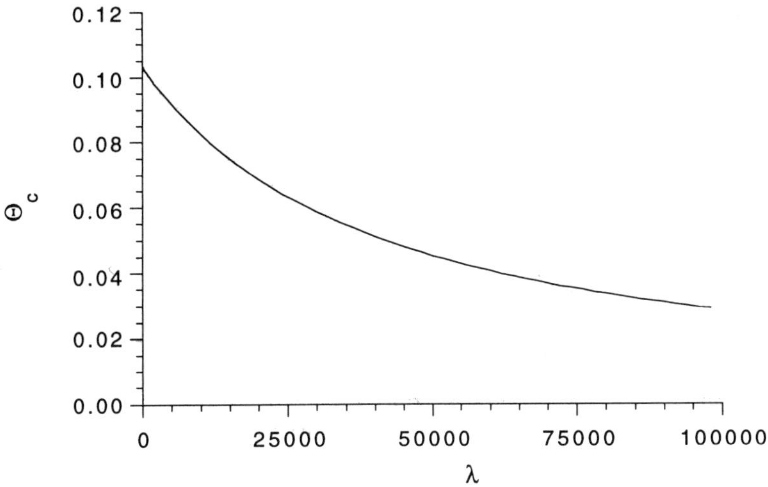

Figure 5.6. The variation of the critical erosion parameter (Θ_c) with the dimensionless permeability (λ)

$$\Theta_c \sim 0.1/((\lambda/40\,000)+1) \tag{5.24}$$

The following section discusses the connection between this critical value of the erosion parameter and the critical Shields parameter.

Profiles of the erosion boundary are calculated for various values of the erosion parameter and shown in Figure 5.7. For these profiles, λ is set equal to 1000. Two regions are elucidated by these erosion profiles. Close to the surface of the bed in a thin layer, the grains are mobilized by the action of shear. However, these effects are poorly transmitted through the bed and further from the surface the grains are mobilized by gradients of the induced pressure field.

If the eroded volume between the flat bed and the erosion boundary is calculated as a function of the effective erosion parameter (Figure 5.8), it is found that there is an approximately linear relation between the two, empirically given in dimensional form by

$$\text{Eroded Volume} = 0.001\pi(\Theta - \Theta_c)H^3 \tag{5.25}$$

This linear relationship may be rationalized by a simple scaling argument in each of the two regions of the flow within the porous bed. For the region in which the pressure gradients are the dominant force, $\partial Q/\partial\eta \sim \eta^{-3}$ and so by continuity $\partial Q/\partial\xi \sim \xi^{-3}$. Therefore, by enforcing the stability criterion (Equation 5.22), the erosion boundary $\xi, \eta \sim \Theta^{1/3}$ and hence the eroded volume is found to scale linearly with Θ. Conversely, within the boundary layer (of depth $1/\lambda$) adjacent to the bed

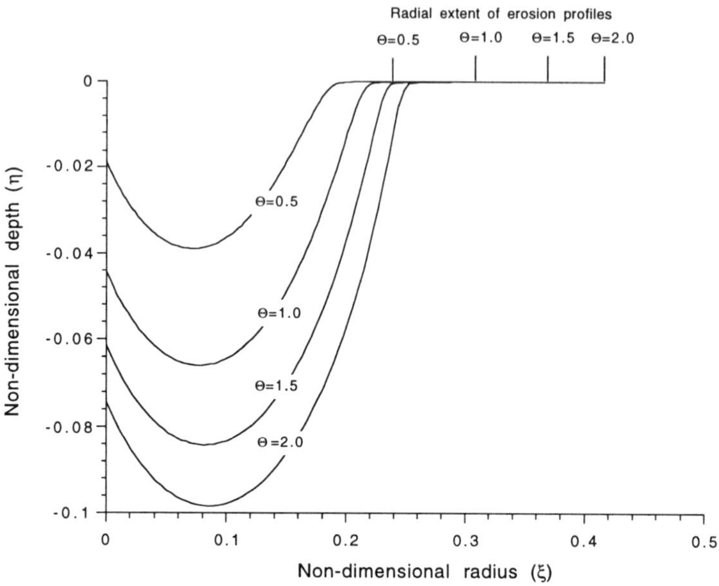

Figure 5.7. Theoretically calculated erosion boundary profiles for various values of the effective Shields parameter (Θ)

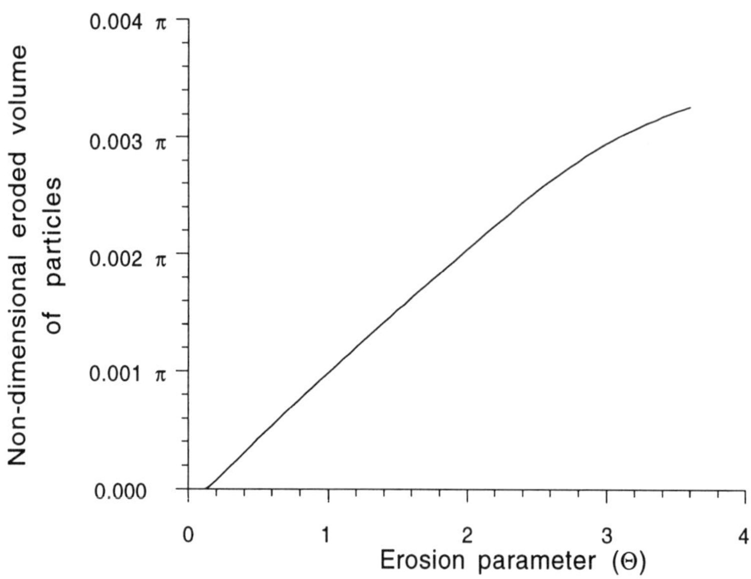

Figure 5.8. The variation of the dimensionless volume of particles with the erosion parameter (Θ)

surface, the shear stress decays as $O(1/\xi^2)$. Hence, the radius of the eroded, cylindrical volume is given by $\xi \sim \sqrt{\lambda\Theta}$ and so the volume within the boundary layer also scales linearly with Θ. These scaling arguments justify the claim that the functional dependence of the eroded volume is independent of the magnitude of the ratio of maximum pressure to shear stress (p_0/σ_0). Within both the pressure and shear stress dominated regions, the eroded volume of particles varies linearly with the erosion parameter (Θ). In the case shown here, $p_0/\sigma_0 >> 1$ and so the pressure dominated region is of more significance.

According to this simple reasoning, within neither the shear-dominated boundary layer, nor the pressure-dominated main body of the flow, does the eroded volume depend on the non-dimensional permeability. This fact is borne out by the numerical evaluation of the eroded volume for different permeabilities. A wide dispersion of grain sizes, however, may affect the friction angle ($\tan\alpha$), introducing a grain size effect in the mobility criterion and the shape of the eroded volume.

RATE OF BEDLOAD TRANSPORT

The above analysis is now used to develop an expression for the rate of bedload transport in equilibrium conditions and suggest a scaling relationship between the flux of bedload transport and the characteristics of the flow. More specifically, a relationship is proposed of the form:

$$q_{bl} = U_s C_s \Delta \tag{5.26}$$

where U_s, C_s are respectively the typical horizontal velocity and concentration of particles within the bedload layer of thickness Δ.

Most models for bedload transport (e.g. Engelund & Fredsøe, 1976) neglect an explicit description of the pick-up and deposition of particles and assume that the fluid shear stress on the boundary takes the critical value for incipient motion. The concentration of particles is then elucidated by use of an empirical dynamic friction condition which links the weight of the particles to the drag exerted on them. The present study instead proposes to assess the magnitude of the bedload transport in equilibrium conditions by assuming a balance between the volume of particles entrained and deposited.

The sweep event has been modelled by considering an axisymmetric jet of scale H and it is assumed that the jet is quasi-periodic with temporal period of recurrence T and areal frequency A, the area on the bed within which only one event may be found (Figure 5.9). The volume of particles picked up by a single event is simply related to the eroded volume from Equation 5.25 and is thus given by

$$V_e = 0.001\pi C_b H^3 (\Theta - \Theta_c) \tag{5.27}$$

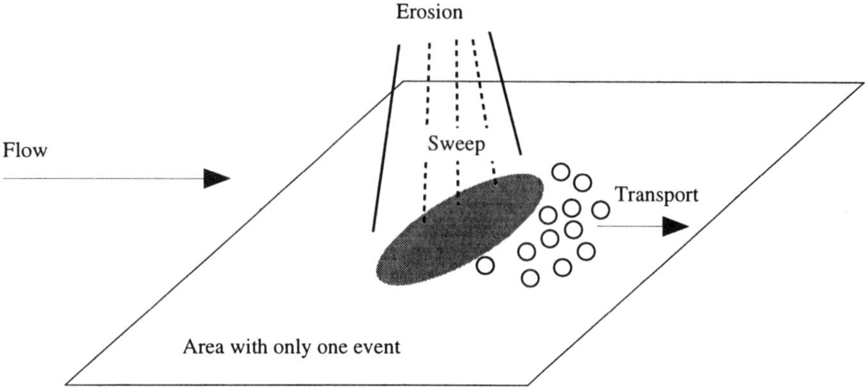

Figure 5.9. The turbulent, event-driven, picture of bedload transport. The sediment is mobilized by incoming sweep events which impinge upon the surface of the erodible bed, in-between which the particles are transported and deposited. The events are intermittent, yet repetitive flow structures within the turbulent boundary layer. Within an area of the bed (A), there is only one transport event

The volume of particles deposited during the period between events, throughout the area between events, is given by

$$V_d = \begin{cases} Aw_s TC_s, & \Delta \geq w_s T \\ A\Delta C_s, & \Delta < w_s T \end{cases} \tag{5.28}$$

where w_s is the settling velocity of the particles. These two scaling options correspond to whether or not all of the entrained particles are deposited before a subsequent event occurs. In equilibrium, the volume of entrained and deposited particles may be equated to find that

$$C_s = 0.001\pi C_b \frac{H^3}{Ah}(\Theta - \Theta_c) \tag{5.29}$$

where $h = \min\{w_s T, \Delta\}$. There has been considerable study to establish whether the scaling characteristics of sweep events depend upon viscous or inviscid processes (see the review of Cantwell, 1981). Blackwelder & Haritonidis (1983) found that the inter-event period is given by $T = 100\nu / u_*^2$, where ν is the kinematic viscosity of the fluid and u_* is the friction velocity. Conversely other investigators conclude that $T = 6\delta / u_\infty$, where δ is the extent of the boundary layer and u_∞ is the flow velocity at the edge of the boundary layer. The current study investigates a lower bound of the distance settled by a particle during the inter-event period using the scaling for this period proposed by Blackwelder & Haritonidis (1983) and the Stokes settling velocity of the particle $w_s = (s-1)gd^2/18\nu$. It is then found that

$$w_s T / d = 50/9\theta \tag{5.30}$$

The particle will therefore settle through many particle diameters during the inter-event period because the Shields parameter in the regime under consideration here is less than unity. However, observations suggest that the height of the bedload layer is only of the order of a few particle diameters. Hence it is concluded that $h = \Delta$, which is equivalent to assuming that all the mobilized particles settle back to the boundary before subsequent events occur. The rate of bedload transport is then given by:

$$q_{bl} = 0.001\pi \frac{U_s C_b H^3}{A}(\Theta - \Theta_c) \tag{5.31}$$

or in dimensionless form:

$$q_{bl} / \sqrt{(s-1)gd^3} = 0.001\pi \frac{U_s}{u_*} \frac{p_0}{\rho u_*^2} \frac{H^2}{A} \sqrt{\theta}(\theta - \theta_c) \tag{5.32}$$

where $\theta_c = \Theta_c H C_b \rho u_*^2 / p_0 d$. This expression is of the same functional form as that proposed by Nielsen (1992). The magnitude of the three ratios U_s / u_*, $p_0 / \rho u_*^2$ and H^2 / A, which represent the ratio of the horizontal velocity of the sediment to the friction velocity, the ratio of the stagnation pressure of the sweep event to the bed shear stress and the ratio of the square of the length-scale of the sweep event to the areal frequency, respectively, may now be estimated. Nielsen (1992) suggests that the typical velocity of the sediment within the bedload is approximately $5u_*$ and so it is assumed $U_s / u_* = 5$. Drake *et al.* (1988) calculated the area on the bed within their window of study which was affected by an individual sweep event, and assuming their study window corresponds to the area within which only one sweep event may be found, it can be estimated from their data that $H^2 / A = 0.3$. Finally from a field study of Start Bay, south-west England (Soulsby, 1983), it was found that sweep events contribute approximately 62% of the Reynolds shear stress in only 10% of the time and so the average shear stress contribution per event is approximately 6 times the overall time-averaged shear stress. The extent of a sweep event is fixed as when the shear stress exceeds the mean shear stress by a factor of 3 (Drake *et al.*, 1988) and hence an event is identified $0 < \xi < \xi_{sw}$, where $\sigma_s(\xi_{sw}) = 3\rho u_*^2$. The shear stress distribution (Equation 5.8) is then averaged over this interval and the average is required to be $6\rho u_*^2$ This sets $\sigma_0 / \rho u_*^2 = 12$ and thus $p_0 / \rho u_*^2 = 2 \times 10^3$. It is emphasized that both p_0 and σ_0 are the peak pressure and shear stress associated with the event. The pressure decays exponentially rapidly away from the point of impingement, while the shear stress decays algebraically. Using these estimates, Equation 5.30 is used to obtain the flux of bedload transport:

$$q_{bl} / \sqrt{(s-1)gd^3} = 9\sqrt{\theta}(\theta - \theta_c) \tag{5.33}$$

This formula is similar to the expressions for bedload transport which were semi-empirically derived by a number of other investigators (Engelund & Fredsøe, 1976; Nielsen, 1992). Furthermore, not only is the functional form of this relationship similar to previously suggested formulae, but also the crudely estimated constant, with a value of 9, compares favourably with previous studies.

These estimates may also be used to derive a value of the critical Shields parameter. Noting Equation 5.24 and that $\theta_c = \Theta_c H C_b \rho u_*^2 / p_0 d$, it is found that $\theta_c = 0.03$. This value is of the correct order of magnitude, but is slightly less than the usual critical value for the initiation of motion.

CONCLUSIONS

This paper has presented a framework for the analysis of turbulent, event-driven bedload transport by considering the impingement of individual sweep-like events upon a granular bed and calculating the volume of particles which are mobilized by an event. This volume depends both on the hydrodynamic characteristics of the incoming event and the size and density of the particle. Balancing the volume of particles which are mobilized with those deposited between events, allows formulation of an expression for the rate of bedload transport in equilibrium conditions. The expression which emerges from this new framework is in accord with those suggested by experimental studies and with those derived by existing theories. This result strongly suggests that models cannot neglect a description of event-driven transport. Indeed, the model presented here for event-driven bedload transport can account for a significant portion of material in transport for the range $\theta_c < \theta < 1$.

This framework has developed an expression for the pick-up function in bedload transport by appealing to the dynamics of the turbulent events and their interaction with the granular bed. The present study has not explicitly invoked the Bagnold assumption which requires the mean shear stress to be reduced to critical at the bed, nor has it utilized a dynamic coefficient of friction to relate the weight of the moving particles to the horizontal drag force exerted by the fluid. Rather these issues have been supplanted by an event-dominated view of the turbulent motion. The current model implicitly assumes that the flow is dominated by turbulent events close to the boundary and that the mean flow, onto which the event motions are superimposed, does not contribute to the mobilization of the sediment. Instead the mean flow does play a vital role in the advection of the mobilized sediment. This is analogous to sediment transport under combined waves and currents, in which the wave-motion mobilizes the particles and the current transports them. This study has focused on the most frequent turbulent events which render positive contributions to the Reynolds shear stress, namely ejection and sweep events. Critical conditions have been enforced at the bed by means of a stability criterion in which the contact friction is expressed through an angle of static repose and therefore elucidates the volume of particles mobilized as a function of the particle characteristics and the hydrodynamic characteristics of the incoming event. Consideration of the impact of sweep events on a granular bed can therefore potentially serve as a boundary condition for the turbulent event-driven suspension of particles, as described by Hogg *et al.* (1994).

It should be noted that the analytical framework presented here is based upon the notion of particle conservation; it has not considered the momentum transported by these downward moving, relatively fast events, nor the momentum associated with the mobilized particles. A more complete model would evaluate these contributions to the total shear stress within the flow and would enable the assessment of the particle velocity within the flow, rather than postulating that $U_s \sim u_*$. Furthermore, this model may be improved if it is driven statistically, taking into account the statistical distribution of the characteristic length and time scales of the impinging events. In addition, it may be possible to extend the analysis to non-equilibrium conditions or bedload transport driven by oscillatory flows. Nevertheless, the new model presented here has revealed how downward moving sweep events are capable of mobilizing particles in repose on the bed and has permitted derivation of an equilibrium transport formula which is in accord with experimental observations. Such turbulent event driven models can thus provide considerable insight to the process of bedload transport.

ACKNOWLEDGEMENTS

This research was financially supported by EPSRC, NERC and HR Wallingford. RLS was funded by the Commission of the European Communities Directorate General for Science, Research and Development as part of the MAST G8M Coastal Morphodynamics community research programme, under Contract No. MAS2-CT92-0027. We also acknowledge Albert Gyr and John Wells for their useful reviews of earlier drafts of this paper.

APPENDIX

This appendix presents a derivation of the pressure distribution within a porous bed of particles which is driven by a distribution of pressure along the surface of the bed. The equations to be solved are Equations 5.9–5.12. An axisymmetric variation of the pressure is considered ($p \equiv p(r, y)$) and hence the equations reduce to:

$$\frac{1}{r}\frac{\partial}{\partial r}\left(r\frac{\partial p}{\partial r}\right)+\frac{\partial^2 p}{\partial y^2}=0 \qquad (5.34)$$

$$p \to 0, \quad y \to -\infty \qquad (5.35)$$

$$p = p_s(r), \quad z = 0 \qquad (5.36)$$

The Fourier-Bessel transform of the pressure field is taken with respect to r, assuming that the surface pressure decays sufficiently rapidly away from the stagnation point, so that the Fourier-Bessel transform of $p_s(r)$ exists. The transformed variables are denoted by $\tilde{p} = (k, y)$ and hence

$$\tilde{p}(k, y) = \tilde{p}_s(k)\exp(ky) \qquad (5.37)$$

Inverting this transform, the standard result is found (Morse & Feshbach, 1953)

$$p(r, y) = \int_0^\infty k\exp(ky)J_0(kr)\int_0^\infty r'\, p_s(r')J_0(kr')\mathrm{d}r'\,\mathrm{d}k \qquad (5.38)$$

where $J_0(r)$ is the zeroth-order Bessel function of the first kind. Substitution of the surface pressure distribution into this expression $p_s = p_0\exp(-\gamma r^2)$ yields

$$p(r, y) = \int_0^\infty k\exp(ky)\, J_0(kr)\exp(-k^2 / 4\gamma)/2\gamma\,\mathrm{d}k \qquad (5.39)$$

This integral is then evaluated numerically to give the pressure distribution within the bed.

REFERENCES

Bagnold, R.A. 1954. 'Experiments on a gravity-free dispersion of large solid spheres in Newtonian fluid under shear', *Proc. Roy. Soc. London*, **A225**, 49–63.

Bear, J. 1972. *Dynamics of Fluids in Porous Media*, Elsevier, 764 pp.

Beltaos, S. & Rajaratnam, N. 1974. 'Impinging circular turbulent jets', *Proc. ASCE, J. Hydraul. Div.*, **100**, 1313–1328.

Blackwelder, R.F & Haritonidis, J.H. 1983. 'Scaling of the bursting frequency in turbulent boundary layers', *J. Fluid Mech.*, **132**, 87–103.

Cantwell, B.J. 1981. 'Organized motion in turbulent flow', *Ann. Rev. Fluid Mech.*, **13**, 457–515.

Cleaver, J.W. & Yates, B. 1975. 'A sublayer model for the deposition of particles from a turbulent flow', *Chem. Eng. Sci.*, **30**, 983–992.

Drake, T.G., Shreve, R.L., Dietrich, W.E., Whiting, P.J. & Leopold, L.B. 1988. 'Bedload transport of fine gravel observed by motion-picture photography', *J. Fluid Mech.* **192**, 193–217.

Engelund, F. & Fredsøe, J. 1976. 'A sediment transport model for straight alluvial channels', *Nordic Hydrol.*, **7**, 293–306.

Fernandez-Luque, R. 1974. *Erosion and Transport of Bedload Sediment*, Unpub. PhD Thesis, Delft Hydraulics Institute, 61 pp.

Gordon, C.M. 1974. 'Intermittent momentum transport in a geophysical boundary layer', *Nature*, **248**, 292–294.

Grass, A.J. 1971. 'Structural features of turbulent flow over smooth and rough boundaries', *J. Fluid Mech.*, **50**, 233–255.

Grass, A.J., Stuart, R.J. & Mansour-Tehrani, M. 1991. 'Vortical structures and coherent motions in turbulent flow over smooth and rough boundaries', *Phil. Trans. Roy. Soc. London*, **A336**, 36–65.

Heathershaw, A.D. 1974. 'Bursting phenomena in the sea', *Nature*, **248**, 294–295.

Heathershaw, A.D. & Thorne, P.D. 1985. 'Sea-bed noises reveal role of turbulent bursting phenomenon in sediment transport by tidal currents', *Nature*, **316**, 339–342.

Hogg, A.J., Huppert, H.E. & Soulsby, R.L. 1994. 'The dynamics of particle-laden fluid elements', in Bélorgey, M., Rajaona, R.D. & Sleath, J.F.A. (Eds) *Sediment Transport Mechanisms in Coastal Environments and Rivers, Proc. Euromech* **310**, 64–78.

Jackson, R.G. 1976. 'Sedimentological and fluid dynamical implications of the turbulent bursting phenomenon in geophysical flows', *J. Fluid Mech.*, **77**, 531–560.

Johansson, A.E., Her, J.Y. & Haritonidis, J.H. 1987. 'On the generation of high amplitude wall-pressure peaks in turbulent boundary layers and spots', *J. Fluid Mech.*, **175**, 119–142.

Kline, S.J., Reynolds, W.C., Schraub, F.A. & Runstadler, P.W. 1967. 'The structure of turbulent boundary layers', *J. Fluid Mech.*, **30**, 741–773.

Koplik, J., Levine, H. & Zee, A. 1983. 'Viscosity normalization in the Brinkman equation', *Phys. Fluids*, **26**, 2864–2870.

Meyer-Peter, E. & Müller, R. 1948. 'Formula for bedload transport', *Proc. Int. Assoc. Hydraul. Struct. Res.*, Stockholm, 39–64.

Moin, P. & Kim, J. 1985. 'The structure of the vorticity field in turbulent channel flow. Part 1: Analysis of instantaneous fields and statistical correlations', *J. Fluid Mech.*, **155**, 441–464.

Morse, P.M. & Feshbach, H. 1953. *Methods of Theoretical Physics*, McGraw-Hill, 1978 pp.

Nielsen, P. 1992. *Coastal Bottom Boundary Layers and Sediment Transport,* World Scientific, 324 pp.

Rajaratnam, N. 1976. *Developments in Water Sciences: Turbulent Jets,* Elsevier, 304 pp.

Raudkivi, A.J. 1990 *Loose Boundary Hydrodynamics,* Pergamon Press, 538 pp.

Robinson, S.K. 1991. 'Coherent motions in the turbulent boundary layer', *Ann. Rev. Fluid Mech.*, **23**, 601–639.

Shields, A. 1936. 'Anwendung der ähnlichkeitsmechanik und der turbulenzforschung auf die geschiebebewegung', *Mitt. Preuß. Versuchsanstalt für Wasserbau und Schiffbau, Berlin*, **26**, 1–26.

Smith, C.R., Walker, J.D.A., Haidari, A.H. & Sobrun, U. 1991. 'On the dynamics of near-wall turbulence', *Phil. Trans. Roy. Soc. London*, **A336**, 131–175.

Soulsby R.L. 1983. 'The bottom boundary layer of shelf seas', in Johns, B. (Ed) *Physical Oceanography of Coastal and Shelf Seas*, Elsevier, Amsterdam, 189–266.

Sumer, B.M. & Deigaard, R. 1981. 'Particle motions near the bottom in turbulent flow in an open channel – part 2', *J. Fluid Mech.*, **109**, 311–337.

Wiegel, F.W. 1980. *Fluid Flow through Porous Macromolecular Systems, Lecture Notes in Physics 121*, Springer-Verlag, 102 pp.

NOTATION

a	diameter of jet nozzle
A	areal frequency of events
C	particle concentration within flow
C_b	particle concentration within granular bed
C_s	representative particle concentration in bedload layer
d	particle diameter

D	depth of flow
F_r	non-dimensional radial force within bed
F_y	non-dimensional vertical force within bed
g	gravitational acceleration
H	distance from jet nozzle to boundary
J_0	zeroth-order Bessel function of the first kind
p	pressure within granular bed
p_s	pressure on surface of the bed
p_0	pressure at stagnation point
q_{bl}	bedload flux of particles
Q	non-dimensional pressure within granular bed
r	radial distance from stagnation point
s	ratio of solid to fluid densities
T	temporal period of event recurrence
u	longitudinal velocity
u_*	friction velocity
u_∞	velocity at edge of boundary layer
U_0	jet velocity
V_d	volume of particles deposited between events
V_e	volume of particles entrained by event
w_s	particle settling velocity
W	submerged weight of particles
y	vertical distance from stagnation point
α	angle of repose
β	angle of sloping bed
δ_∞	boundary layer extent
Δ	thickness of bedload layer
η	non-dimensional vertical distance from stagnation point
θ	Shields' parameter
θ_c	critical Shields' parameter
Θ	erosion parameter
Θ_c	critical erosion parameter
λ	non-dimensional permeability of granular bed
Λ	permeability of granular bed
μ	dynamic viscosity
ν	kinematic viscosity
ξ	non-dimensional radial distance from stagnation point
ξ_{sw}	non-dimensional radial extent of sweep event
σ_b	shear stress on the lower boundary
σ_s	distribution of shear stress on the lower boundary
σ_0	maximum shear stress on the lower boundary

6 Ratio of Lift and Shear Forces Over Rough Surfaces

ANDREAS DITTRICH, FRANZ NESTMANN
Institute of Hydraulic Structures and Agricultural Engineering, University of Karlsruhe, Kaiserstraße 12, D-76128 Karlsruhe, Germany

PETER ERGENZINGER
Institut für Geographische, Wissenschaften Freie Universität Berlin, Grunewaldstraße 35, Berlin 12165, Germany

ABSTRACT

Coherent turbulent structures like sweeps and ejections have been shown to generate the impulses that induce incipient motion and thus bed instability over smooth surfaces. The higher moments of the instantaneous streamwise velocities are related directly to the intensity, range of activity and frequency of occurrence of coherent turbulent structures. In order to investigate such structures over rough surfaces, an experimental study was conducted to measure flow close to artificial roughness elements and determine the first four statistical moments of the fluctuating velocities. These measurements indicate that flow close to the roughness elements is dominated by sweep-like structures.

Turbulence produced by these coherent turbulent structures gives rise to a dynamic lift. The ratio of dynamic lift to bed shear stress decreases with increasing roughness, i.e. the rougher the surface, the greater the shear stress compared to dynamic lift. Since the magnitude of the lift force and the temporal occurrence of sweeps is roughness dependent, the ratio of lift and shear force depends upon surface roughness. As a result, existing stability diagrams (e.g. the Shields diagram) must be extended by at least one additional parameter to account for the roughness density and parameterize correctly threshold conditions over rough surfaces.

INTRODUCTION

One of the central problems in hydraulic engineering concerns the ability of water flows to destabilize, entrain and transport solid material. A prerequisite to description of these processes is an understanding of the mechanisms causing incipient motion. For a long time this problem has been solved considering only the mean parameters of the solid and the liquid phases, for example, Shields (1936) linked incipient motion with the mean shear stress and mean grain diameter. However, this approach is unable to explain the great scatter of experimental data since it does not provide an insight into the mechanisms that determine bed instability. For this purpose it is necessary to consider the balance of forces acting

Coherent Flow Structures in Open Channels.
Edited by P.J. Ashworth, S.J. Bennett, J.L. Best and S.J. McLelland.

on a particle in a stream bed. Figure 6.1 shows that in addition to the usually considered shear force, F_τ, and the resisting force R which is proportional to the weight under buoyancy G', and the angle of repose φ, the vertically directed lift force, L, must be taken into account. The lift force arises owing to the pressure difference between the bottom and top of a particle (e.g. Bretting, 1958; Einstein & El-Samni, 1949). The dynamic lift is attributed to the wake zone at the rear of a particle and the higher velocities at its upper surface compared to its base. During burst events fluid passes over and around the particle (Grass *et al.*, 1991) and these coherent structures create favourable conditions for the generation of dynamic lift. These coherent structures form linked parts of a quasi-cyclical process of turbulence generation and give rise to the major features of turbulent shear stresses and pressure fluctuations. Importantly, the *sweep* events are believed to be the primary pulses inducing particle movement (Grass, 1971) and entrainment (Drake *et al.*, 1988).

As a consequence of the sweeps and ejections, the probability distributions of the instantaneous velocities are neither symmetric nor Gaussian. For instance, in the viscous sublayer and wake region of the boundary layer, strongly skewed probability distributions also show high kurtosis values. These deviations from Gaussian distribution profiles are usually attributed to the pronounced intermittent nature of the flow in these regions (Durst *et al.*, 1987). One way to characterize these distributions is by means of the first four statistical moments: the mean velocity U, the standard deviation of the velocity fluctuations u', the skewness S and the kurtosis F which are defined as follows:

$$U = \frac{1}{N}\sum_{i=1}^{N} u_i \qquad (6.1)$$

$$u' = \sqrt{\frac{1}{N-1}\sum_{i=1}^{N}(u_i - U)^2} \qquad (6.2)$$

$$S = \frac{1}{N-1}\,\frac{\sum_{i=1}^{N}(u_i - U)^3}{u'^3} \qquad (6.3)$$

$$F = \frac{1}{N-1}\,\frac{\sum_{i=1}^{N}(u_i - U)^4}{u'^4} \qquad (6.4)$$

where u_i is one value of the instantaneous streamwise velocity and N the number of samples.

The distributions of these four parameters have already been determined for smooth surfaces (e.g. Durst *et al.*, 1987). Figure 6.2 shows the distributions of the second (made dimensionless with the shear velocity u_*), third and fourth moments as a function of the dimensionless depth y^+ $(= yu_*/\nu)$ for flows over smooth

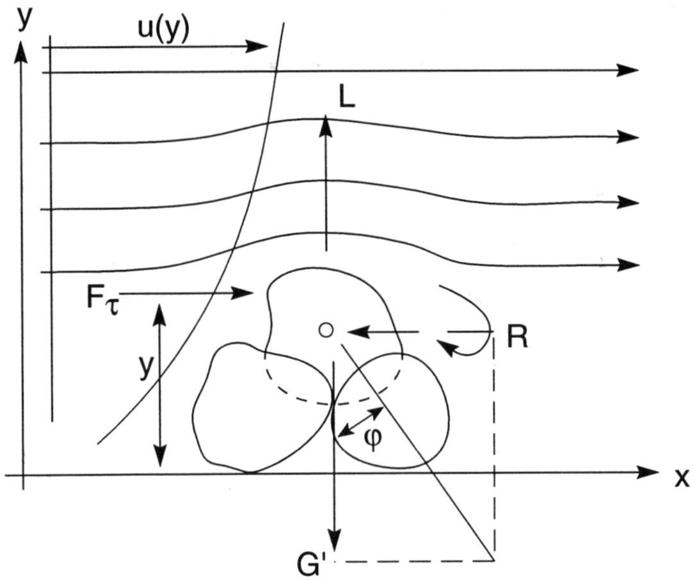

Figure 6.1. Forces acting on a particle

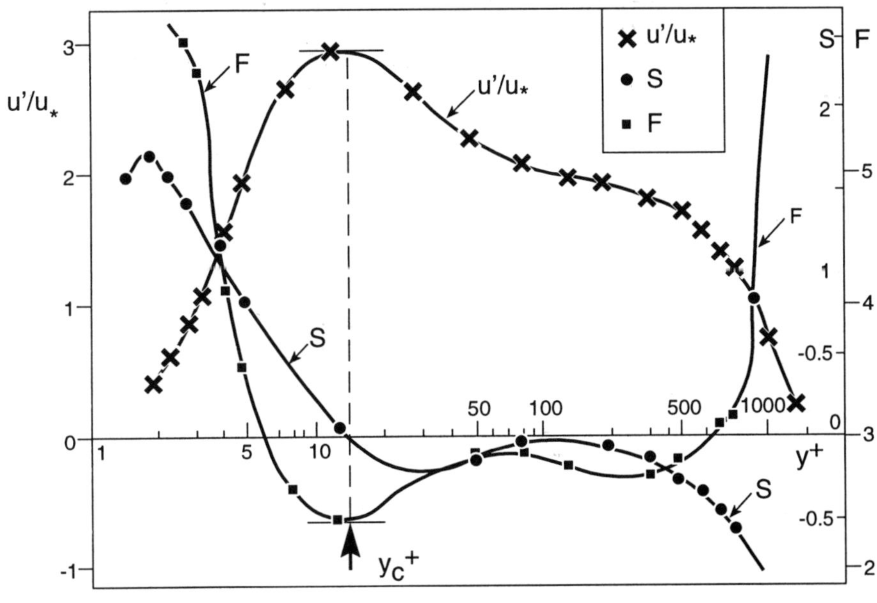

Figure 6.2. Second, third and fourth moments of the probability density distributions of the streamwise velocity over depth y^+ for smooth beds (from Durst *et al.*, 1987)

surfaces. This Figure illustrates that the values of skewness and kurtosis are much higher in the vicinity of the bed surface than the values obtained for the Gaussian distribution ($S = 0$ and $F = 3$) which means that the velocity distributions close to the boundary are peaked and positively skewed. As interest must focus on the mechanisms causing bed instability, the probability distributions or four statistical moments of the instantaneous velocities are required at the edge of a single particle (Figure 6.1). Furthermore, the maximum value of the turbulence intensities u'/u_*, the change of sign of S and the minimum value of F occur at a distance of $y_c^+ = 13$ (Figure 6.2). Since the lift force is likely to be caused by the maximum value of turbulence intensity in the wall region, a constant value of about three is obtained for $(u'/u_*)_{max}$ in the smooth bed case.

More information concerning determination of the lift force is available from measurements carried out over smooth surfaces by Emmerling (1973) and from assumptions made by Hinze (1975). Emmerling measured wall-pressure fluctuations with wall-microphones of different diameters and observed that pressure and velocity fluctuations were caused by the action of sweeps. The measurements of wall-pressure fluctuations yielded the following relationship between the intensity p'_w, $(=\overline{(p_w^2)}^{1/2})$, and the mean wall shear stress τ, and thus the dynamic lift L_1 (force per unit area):

$$L_1 = p_w' = C\tau \tag{6.5}$$

with C being a proportionality factor. In general a relationship of the form

$$p' = C_1 Re\rho\bar{u}^2 \tag{6.6}$$

is appropriate (Hinze, 1975), where $\bar{u}^2 = 1/3u'_i u'_i$ ($\bar{u}^2$ is the normal component of the Reynolds stress tensor, Re is the flow Reynolds number and u'_i represents the fluctuating velocities in $x(u'_1)$, $y(u'_2)$ and $z(u'_3)$ directions). The factor C_1 varies only slightly with Reynolds number, and values between 0.6 and 0.8 are found in the literature for this factor in flows with grid-generated turbulence. If a similar relationship for p'_w can be used and p'_w is determined mainly by the maximum values of turbulence intensities in the wall region, that is

$$(u'_1/u_*)_{max} \cong 3 \qquad (u'_2/u_*)_{max} \cong 1.1 \qquad (u'_3/u_*)_{max} \cong 1.7 \tag{6.7}$$

the following expression is obtained:

$$1/3\rho\overline{u'_i u'_i} = 4.3\rho u_*^2 = 4.3\tau \tag{6.8}$$

Consequently, with the above values for C_1:

$$L_1 = p'_w = C\tau = 4.3C_1\tau \tag{6.9}$$

and the value of C is in the range 2.6 to 3.4. Hinze (1975) suggested the following relationship as an appropriate average

$$L_1 = p'_w \cong 3\tau \tag{6.10}$$

On the basis of the buoyancy model, Einstein & El-Samni (1949) determined the average lift by measuring the pressure difference between the top and bottom of natural and artificial particles placed in an open channel. In one run they used the upper half of spheres by gluing them in a hexagonal pattern to the flume bottom, whilst in another experiment they repeated the measurements with natural sediment as the bottom texture. Using their data, Gessler (1965) calculated a mean value of $L_1 \cong 3\tau$ and Bretting (1958) obtained a maximum value of $L_1 \cong 5\tau$ for the lift force per unit area.

Recently, Jüpner (1993) measured the critical lift and shear forces directly under flume and natural conditions with a special device called COSSY (COBble Satellite SYstem) consisting of an artificial cobble (diameter > 120 mm) containing differential pressure sensors. Jüpner (1993) placed the cobble on the bottom of a flume among particles with $D_{50} = 55$ mm and in the very rough mountain torrent 'Lainbach' in upper Bavaria among particles with $D_{50} = 92$ mm, and measured the hydrodynamic forces (pressure differences) under supercritical flow conditions. Expressed in terms of the relative protrusion of the cobble, Jüpner obtained the following values for the ratio of lift force, L, and shear force, F_τ: $L/F_\tau = 0.5 - 1.6$.

Less reliable data and information concerning the physics of the skewness and kurtosis are available, especially in the case of rough boundaries. Almost all studies (e.g. Nowell & Church, 1979; Karlsson, 1980; Raemy, 1991) were confined to regions some distance from the roughness elements, owing to difficulties in obtaining data close to the grains.

However, measurements obtained by Raupach (1981) close to the roughness elements indicated that the third moments follow universal curves which are independent of surface roughness, except in a layer close to the surface. Raupach also obtained the following fractional difference ΔA between the contribution to the turbulent stress by sweeps and ejections:

$$\Delta A = A_{sweep} - A_{ejection} = 0.37S \tag{6.11}$$

where A characterizes the fraction of turbulent stress and S the third moment. According to Equation 6.11, quite a good description of the ejection-sweep character of boundary-layer turbulence can be obtained from consideration of the third moment of u. From Equation 6.11 and Figure 6.2, it is apparent that ejections dominate in the 'outer part' ($y_c^+ > 13$) and sweeps in the 'inner part' ($y_c^+ \leq 13$) of the turbulent flow field over smooth surfaces. Raupach (1981) conducted wind tunnel experiments over an initial smooth surface and over five cylindrical roughness elements in either square or diamond arrays and low roughness densities, c_k^*, (representing the ratio of plan area of elements to total bed area). According to the flow types described by Morris (1955), an 'isolated roughness' flow was investigated by Raupach.

The results reported by Raupach (1981) and Andreopoulos *et al.* (1984) reveal that the skewness factor can be related to the particular structure of coherent flow patterns whereas variations of the kurtosis factor across the boundary layer are inherently caused by variations of the frequency and duration of these structures. Due to the physics of flow and the schemes for detecting coherent structures, the kurtosis factor should be inversely related to the percentage of time occupied by the higher energy fluctuations of the local velocity field. This can be expressed by a so-called 'intermittency factor' which expresses the product of frequency of occurrence of flow structures and their time of duration (Durst *et al.*, 1987). The strong interrelation of the intermittency factor, γ, with the properties in the following expression:

$$\gamma = f_B T_B \quad \propto 1/F \tag{6.12}$$

can be related directly to the inverse of the kurtosis factor, F, close to the boundary. In Equation 6.12, f_B designates the mean frequency of occurrence and T_B the mean time duration of coherent structures.

The conclusion can be drawn from Figure 6.2 that the kurtosis factor is high close to smooth boundaries and low away from the boundary. Therefore, according to Equation 6.12, the percentage of time occupied by the higher energy fluctuations such as sweeps and ejections is low, whereas a short distance from the boundary this percentage of time increases rapidly and reaches a maximum at $y_c^+ \cong 13$. However, no reliable data are available in this region for flows over rough boundaries. Nevertheless, according to the results of the investigations of Raupach (1981) and Durst *et al.* (1987), the detection of coherent structures is possible by statistical analysis of the streamwise component of the velocity field using the third and fourth moments of the one-dimensional velocity field.

The visualization study of Grass *et al.* (1991) revealed that powerful vortical structures with a general horseshoe-type configuration occur over smooth surfaces as well as in the turbulent boundary layer near rough walls and that they are similarly linked to bursting events in both the smooth and rough bed cases. The nature of coherent structures and their processes of development over rough surfaces therefore appear to be similar to those originating over smooth walls. Furthermore, Grass *et al.* (1991) and Grass & Mansour-Tehrani (this volume) observed that a remarkable feature of the rough wall flow is its apparent ability to order itself very rapidly in a small vertical distance above the tops of the roughness elements. Their measurements indicated that for fully rough wall conditions, the near-wall turbulence structures are directly proportional to the bed roughness size for geometrically similar roughness elements and the packing used in their tests. The results were achieved using a range of different diameter glass spheres (D = 1.15, 6 and 12 mm), randomly packed in a single layer resting on a glass bed plate. According to Morris' (1955) flow types, skimming to wake interference flow is assumed to occur in the experiments of Grass *et al.* (1991).

Roughness density was varied by Nowell & Church (1979) to cover the whole range of 'skimming', 'wake interaction' and 'isolated roughness' flows. Their measurements were made in water in a tilting recirculating flume with the bed of the flume being covered with Lego™ baseboard onto which bricks were fixed in various arrays. Nowell & Church (1979) identified three flow regions of streamwise velocities over the flow: (i) for $y/d > 0.35$ (where d is the water depth and y is the vertical co-ordinate), turbulence intensities steadily decrease for all densities and approximately linearly toward the surface; (ii) for $0.35 > y/d > 0.20$, turbulence intensities are approximately constant at $u'/u_* \cong 2$; and (iii) for $y/d < 0.20$, values diverge according to roughness density. When densities (c_k^*) are equal to 1/8 and 1/12 (skimming flow), there is a maximum of turbulence intensity near $y/d = 0.25$ and a decrease toward the bed. Densities (c_k^*) of 1/16 and 1/20 (wake interaction) have $u'/u_* \cong 2$ down to the bed, whilst the lower densities (isolated roughness) show continued increases in turbulence intensity towards the bed. For intermediate roughness densities, wakes shed from the roughness elements interact to determine flow properties in this region with the wakes being related to the size of the elements.

In order to clarify the discrepancy in the data found in the literature and to improve knowledge of the physical processes that determine bed instability, an experimental study was conducted to describe the turbulent flow field close to artificial roughness elements. For this purpose the velocity field was measured by LDA in u direction, since the aim was to determine the impact of turbulence and describe the effect of coherent structures on the flow field by using the first four statistical moments of the fluctuating velocities in the streamwise direction. To make the results for artificial roughness arrangements comparable to natural surfaces, the roughness density of the artificial roughness elements, c_k, was approximated to that of natural river beds, i.e. $0.2 < c_k < 0.8$. The roughness density c_k was defined in accordance with O'Loughlin & MacDonald (1964) as the ratio of the upstream projected area of a roughness element to the floor area assigned to that element (Dittrich & Hammann de Salazar, 1993)

In the course of armouring, the roughness of the bed surface as well as bed resistance increase till the stage of maximum bed stability is approached. (Dittrich *et al.*, 1994). A value of $c_k \cong 0.8$ was assumed as the highest roughness density for bed surfaces of lowest roughness (see also O'Loughlin & MacDonald, 1964). Bertschler (1985) conducted experiments with spheres of the same height but with different roughness densities to simulate roughness configurations that cause maximum turbulent shear stress as observed in the armour layer experiments of Gessler (1965) and Günter (1971). A comparison of the configuration of spheres with $c_k \cong 0.26$ and the distribution of the particles used in the experiments of Gessler and Günter shows a good agreement between the artificial configuration causing a maximum shear stress and particles of the armour layers with diameters greater than D_{60}. Therefore, a value of $c_k \cong 0.2$ was assumed as the lowest roughness density for bed surfaces at the stage of maximum bed stability.

EXPERIMENTS

The experiments were conducted in a tilting flume, 13.2 m long and 0.3 m wide (Figure 6.3), in the Theodor-Rehbock laboratory at the University of Karlsruhe. More information on the flume and the instrumentation are given in Dittrich & Hammann de Salazar (1993). The flume base was covered with glass spheres which were fixed on perforated steel plates. The spheres had a diameter of 9 mm and an absolute height from the plate of 8.3 mm. The average flow velocity was held constant throughout the experiments at a value of $u_m = 0.58$ m s^{-1}.

In the course of the experiments the bed and the 2 m long measuring section were covered with four different artificial roughnesses, two comprising glass spheres and two being of PVC cubes. The diameter and absolute height of the spheres in the measuring section were the same as those fastened to the flume bed whilst the height of the PVC cubes was 8 mm. The roughnesses used were characterized by different shapes, arrangements and roughness densities. The main characteristics of the four roughnesses are summarized in Table 6.1.

One-dimensional laser Doppler anemometry, details of which are given in Hammann de Salazar & Dittrich (1994), was used to measure the horizontal components of the instantaneous velocities. The co-ordinate system used in the experiments is defined in Figure 6.4a with the x-axis being parallel to the flow, the y-axis being normal to the bed and the z-axis being transverse to flow.

To characterize the whole flow field close to the artificial roughnesses, three representative locations with respect to the roughness elements were chosen (Figures 6.4b–6.4d): (i) above the roughness elements; (ii) between the elements; and (iii) behind the elements. Within the important region close to the roughness elements, the measuring points had a vertical spacing of 1 mm from each other. The distributions of the statistical moments over the flow depth were determined at several points up to a depth of $y/d = 0.3$. Flow rate, Q, and mean velocity, u_m, were also measured. The measured and calculated flow parameters are summarized in Table 6.2.

Additional parameters for the presentation and interpretation of the higher moments (e.g. the shear velocity u_*, the equivalent sand roughness k_s and the zero

Table 6.1. Characteristics of the four roughnesses

Texture	shape	arrangement	roughness density c_k
1	spheres	in-line	0.70
2	spheres	staggered	0.70
3	cubes	staggered	0.28
4	cubes	staggered	0.58

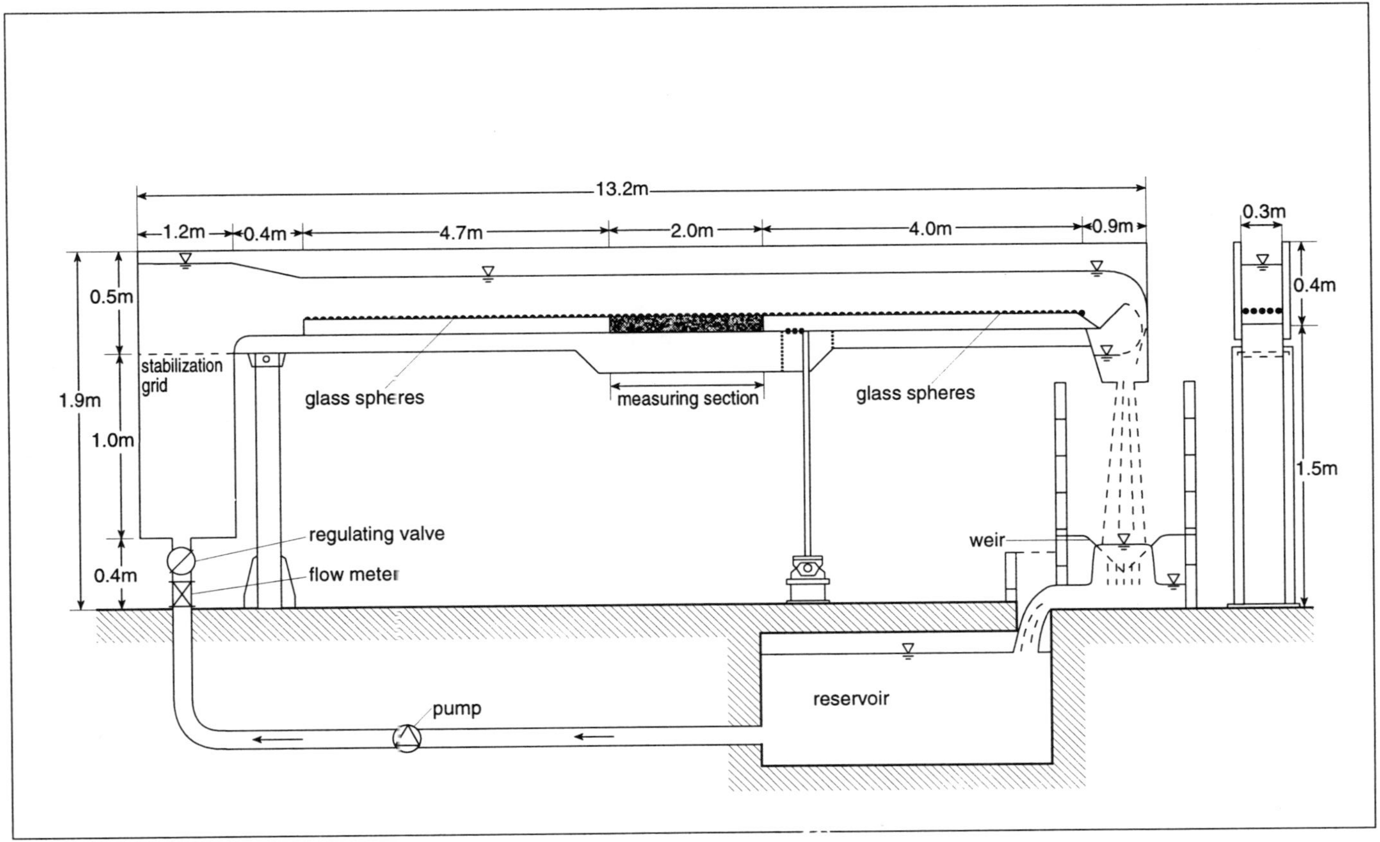

Figure 6.3. Tilting flume (exaggerated vertical scale)

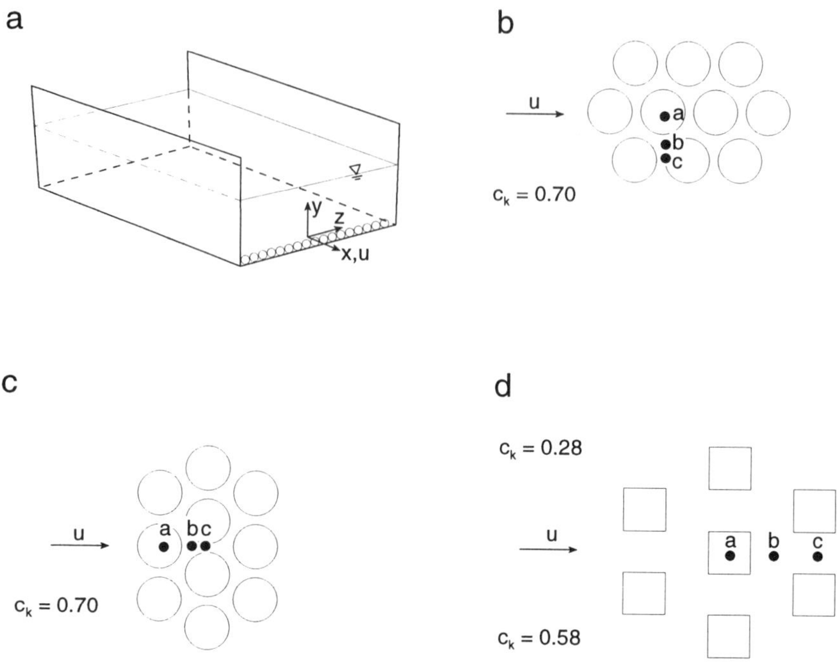

Figure 6.4. (a) Location of the *y*-axis; (b) glass spheres orientated parallel to the flow (in-line); (c) glass spheres staggered to the flow direction; and (d) cubes staggered to the flow direction

Table 6.2. Summary of flow parameters. I is the slope of the flume bottom, d is the water depth, $Fr = u_m/(gd)^{1/2}$ is the Froude number, $Re = u_m\delta/\nu$ is the flow Reynolds number and $Re_* = u_*k/\nu$ is the roughness Reynolds number (k is the geometric roughness height and ν is the kinematic viscosity of water)

Roughness elements	Q (l s^{-1})	I (‰)	d (m)	u_m (m s^{-1})	Fr	Re	Re_*
Spheres, in-line	18.0	3	0.1048	0.573	0.565	6.00×10^4	502.1
Spheres, staggered	18.3	3	0.1055	0.578	0.568	6.10×10^4	445.3
Cubes, widely spaced	18.0	3	0.1063	0.564	0.552	6.00×10^4	461.1
Cubes, closely spaced	27.4	2	0.1572	0.581	0.468	9.13×10^4	398.4

level y' of the velocity profiles) were also determined from the mean longitudinal velocity data. The procedure for the calculation of these parameters and a summary of the data obtained are given in Dittrich & Hammann de Salazar (1993).

RESULTS

The dimensionless mean velocities U/u_* are plotted as a function of the dimensionless depth y^+ in Figure 6.5 and show that the mean velocities follow the logarithmic law of the wall almost down to the top of the roughness elements (in Figure 6.5, the top of the roughness elements occur for values of y^+ between 50 and 230). Detailed analysis of the curves at the four roughnesses (Dittrich & Hammann de Salazar, 1993) shows that the separation point of the mean velocities from the logarithmic wall law is determined by one value of y^+ for each texture; this point, defined as y_c^+, is the upper limit of the roughness sublayer (as also found in Nakagawa *et al.*, 1988). This finding is illustrated more clearly for one texture in Figure 6.6.

For the in-line spheres, the distributions of the velocities U/u_* separate at $y_c^+ \cong 230$ (about 2.4 mm above the roughness elements), for the staggered spheres at $y_{c2}^+ \cong 100$ (almost identical with the top of the elements), for the widely spaced cubes at $y_{c3}^+ \cong 170$ (about 1 mm above the roughness elements), and for the closely packed cubes at $y_{c1}^+ \cong 50$ (roughly 1 mm below the top of the elements).

In Figure 6.7 the distributions of the turbulence intensities, skewness and kurtosis are plotted versus y^+ for the flow over the widely spaced cubes. This Figure shows that the level of maximum turbulence intensities coincides with the region of change in sign of the skewness and minimum in the kurtosis values at the characteristic point y_c^+, defined (in an extension to the definition given above) as the upper limit of the roughness sublayer. In addition to the roughness sublayer

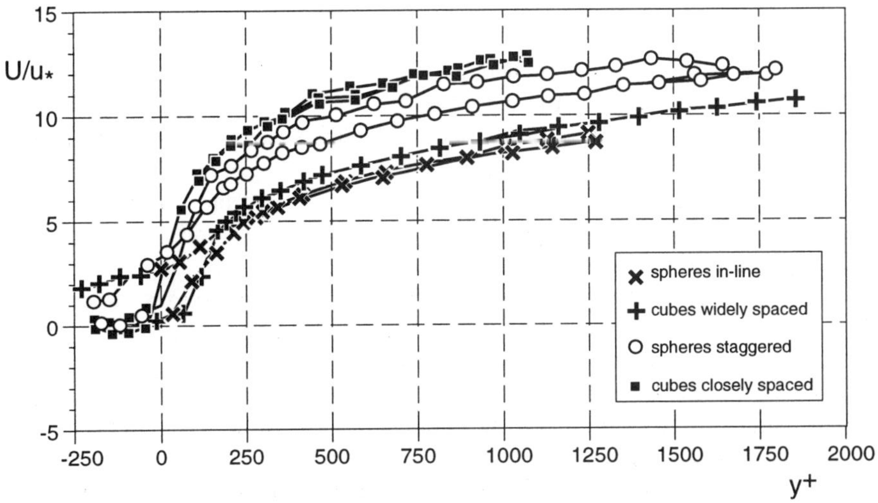

Figure 6.5. Distribution of U/u_* as function of y^+ for the four surface textures

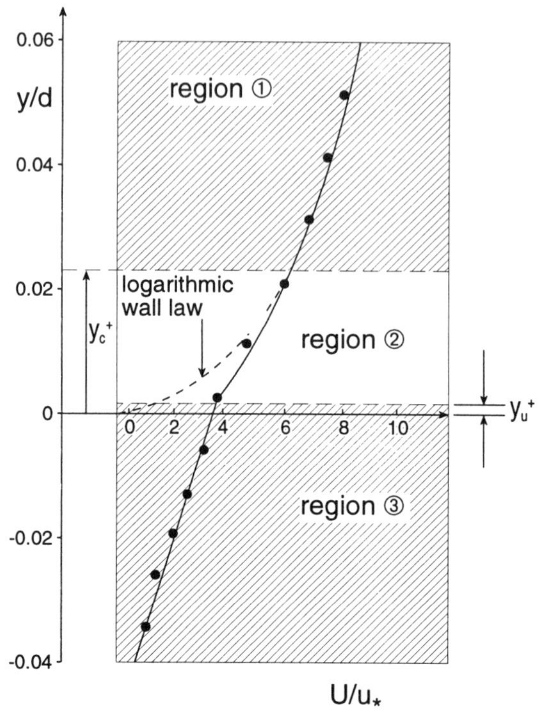

Figure 6.6. Three-layered, vertical velocity profile for one roughness texture close to the boundary

(marked by cross hatching), the height of the top of the cubes (marked by the double line) is also plotted on this diagram.

In the case of closely packed cubes, the maximum value of turbulence intensity u'/u_* is 2.75 whereas in the case of staggered spheres it is 2.6. In the case of widely spaced cubes the maximum value of 2.1 is not so well defined as for the other roughnesses. No value is given for the in-line spheres, due to operational difficulties with the equipment (Hammann de Salazar & Dittrich, 1994).

For all textures, the maxima of $u'/u*$ occurs at, or slightly above, the elevation of the top of the roughness elements, for the staggered spheres being at a the dimensionless depth of $y_{c2}^+ \cong 100$, for the widely spaced cubes at $y_{c2}^+ \cong 175$ and for the closely packed cubes at $y_{c1}^+ \cong 55$.

The change in sign of skewness, S, occurs in the case of the staggered spheres at $y_{c2}^+ \cong 90$, the widely spaced cubes at $y_{c3}^+ \cong 170$ and the closely packed cubes at $y_{c1}^+ \cong 70$. The minimum value of F is reached in the case of the staggered spheres at $y_{c2}^+ \cong 115$, for the widely spaced cubes at $y_{c3}^+ \cong 185$ and for the closely packed cubes at $y_{c1}^+ \cong 55$.

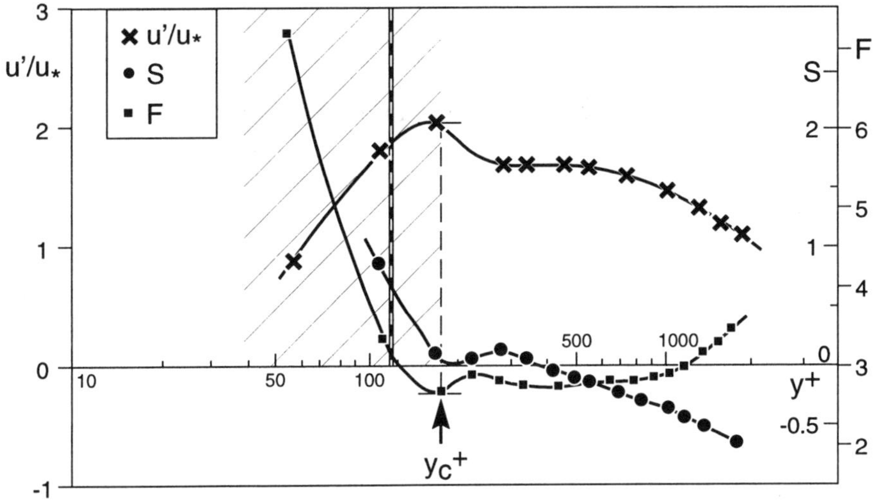

Figure 6.7. Second, third and fourth moments of the distributions of the streamwise velocity as a function of y^+ for the widely spaced cubes. The top of the roughness sublayer is marked by cross hatching and the height of the top of the cubes is shown by the double line

Figure 6.8 plots the turbulence intensities u'/u_*, skewness and kurtosis over the dimensionless depth y/y_c^+ for the staggered spheres, the widely spaced cubes, the more closely spaced cubes and the smooth bed case. The dimensionless depth y/y_c^+ was obtained by using $y_c^+ \cong 13$ in the smooth bed case and the mean values of y_{c1}^+ ($\cong 55$), y_{c2}^+ ($\cong 100$) and y_{c3}^+ ($\cong 175$) for the three different roughnesses. Figure 6.8 shows that the turbulence intensities, skewness and kurtosis are similarly distributed over $y/y_c^+ \geq 1$ for all roughnesses and are also similarly distributed over rough and smooth surfaces. However, a slightly stronger decrease seems to exist in the u'/u_* and S values with increasing water depth for the widely spaced cubes. If the distributions of u'/u_* over $y/y_c^+ > 1$ are described using the function of Nezu (1977; see Equation 6.13 below), the constants D_1 and the exponents λ_1 are almost identical.

Figure 6.9 shows the y_c^+ locations for the four surfaces as a function of k_s/k, based on the first four statistical moments for the closely packed and widely spaced cubes as well as the staggered spheres, and on the mean velocity data for the in-line spheres. The y_c^+ point for smooth surfaces is also plotted and values of y_c^+ increase almost linearly with increasing roughness.

Analysis of all this data (Dittrich & Hammann de Salazar, 1993) reveals that within the logarithmic wall region (region 1 in Figure 6.6) the turbulence

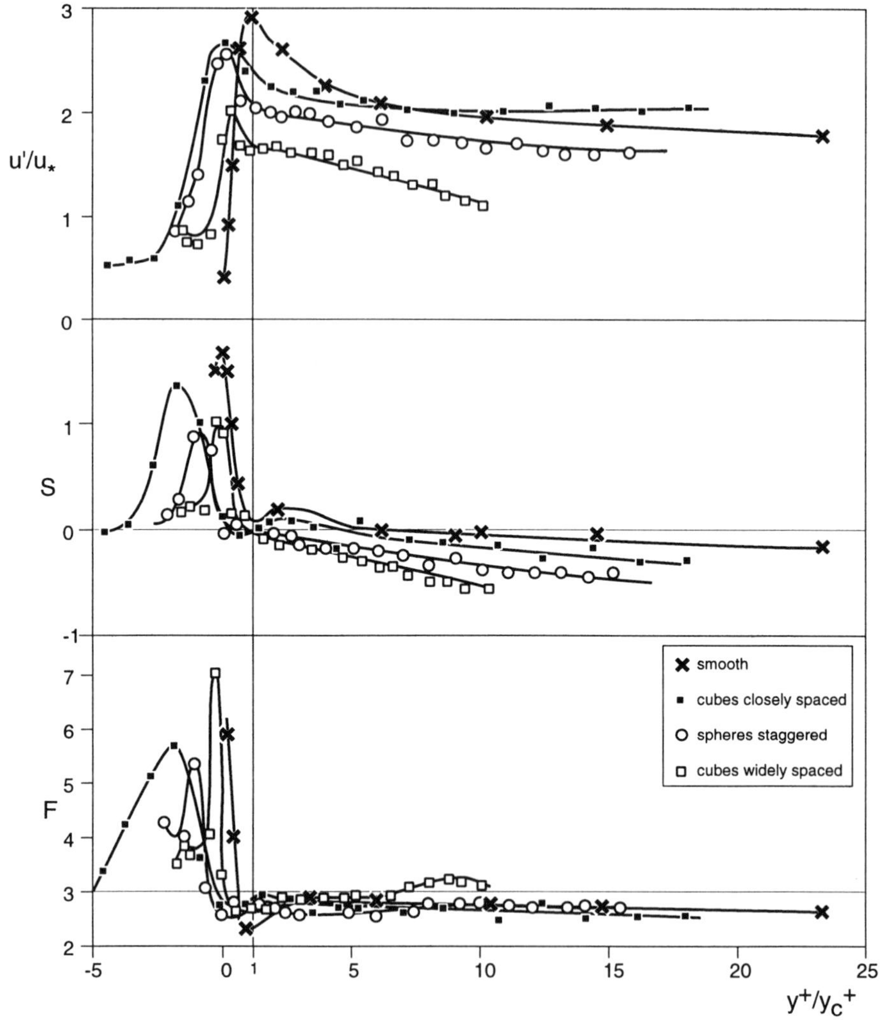

Figure 6.8. Second, third and fourth moments of the distributions of streamwise velocity for the staggered spheres (position b), the more widely spaced cubes (position b), the closely packed cubes (position c) and the smooth bed case as a function of y^+/y_c^+. Deviations of the locations of the maximum values of u'/u_* between 0 and 1 are in the order of 1 mm or less (compare also with Figure 6.9)

intensities follow the relationship recommended by Nezu (1977)

$$\frac{u'}{u_*} = D_1^{\left(-\lambda_1 y/d\right)} \tag{6.13}$$

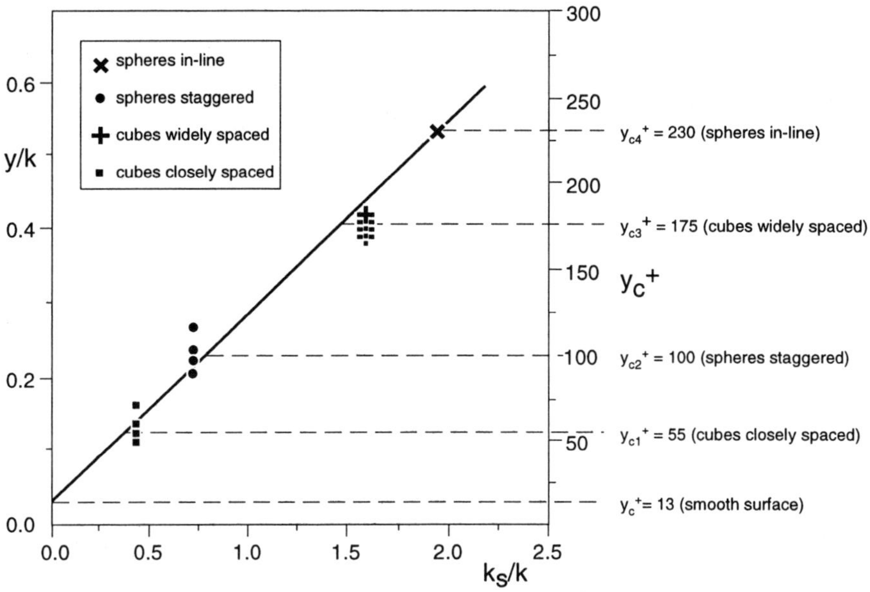

Figure 6.9. Upper limit of the roughness sublayer y_c^+ as a function of k_s/k for the four surfaces. The four points of y_c^+ for one roughness result from differences in the separation points, the maximum value of u'/u_* the change in sign of S and the minimum of F

with the following values for the constants D_1 and λ_1:

closely spaced cubes	$D_1 = 2.23$	$\lambda_1 = 1.22$	$(k_s/k = 0.44)$
staggered spheres	$D_1 = 2.09$	$\lambda_1 = 1.21$	$(k_s/k = 0.72)$
widely spaced cubes	$D_1 = 1.86$	$\lambda_1 = 1.50$	$(k_s/k = 1.58)$

The values for the constants D_1 and λ_1 are in good agreement with those obtained by Nezu (1977; $D_1 = 2.17$ and $\lambda_1 = 1.23$) for smooth surfaces and are independent of roughness for the staggered spheres and closely packed cubes, in contrast to values for the widely spaced cubes where the D_1 value is slightly lower and the λ_1 value is slightly greater. In the latter case, isolated roughness flow existed whereas in the other cases quasi-smooth to wake interference flow occurred (Dittrich & Hammann de Salazar, 1993).

A further similarity of the moments could be identified for depths less than y_c^+ (see Figure 6.8). The turbulence intensity decreases rapidly to reach minimum values whereas the skewness and kurtosis increase rapidly to reach maximum values. It is assumed that the lower limit of the roughness sublayer (region 2 in

Figure 6.6), y_u^+, lies approximately at the location of the maxima of S and F, this being in agreement with the higher moments determined over smooth surfaces which show a maximum of S in the viscous sublayer at $y_u^+ = 2$ (see Figure 6.2). The lower limit of the roughness sublayer is given as $y_{u1}^+(S) \cong -50$ and $y_{u1}^+(F) \cong -45$ for the closely packed cubes, $y_{u2}^+(S) \cong -10$ and $y_{u2}^+(F) \cong -5$ for the staggered spheres and $y_{u3}^+(S) \cong 50$ and $y_{u3}^+(F) \cong 55$ for the widely spaced cubes.

It was found (Dittrich & Hammann de Salazar, 1993) that in region 2 (Figure 6.6) the mean velocities follow the function recommended by Nakagawa *et al.* (1988):

$$\frac{U}{u_*} = \frac{u_0}{u_*} + \frac{1}{\kappa \varepsilon_R^+}\left(\frac{y}{d} - \frac{1}{2}\left(\frac{y}{d}\right)^2\right) \tag{6.14}$$

where u_0 is the mean velocity at the zero point of the logarithmic velocity profile, κ is the von Kármán constant and ε_R^+ is the dimensionless vortex viscosity within the roughness sublayer. This sublayer is further characterized by a similar correlation between skewness and kurtosis for both smooth and rough surfaces.

Below the roughness sublayer a further sublayer (region 3 in Figure 6.6) could be identified in which the mean velocities decrease almost linearly with decreasing depth and the four moments are dependent on shape, arrangement and density of the roughness elements. This sublayer is also characterized by no correlation between the skewness and kurtosis.

DISCUSSION

The distributions of the statistical moments over the flow depth are similar over different roughnesses and over rough and smooth surfaces down to the roughness sublayer (region 2 in Figure 6.6), a thin layer which extends in the present experiments from about 1 mm below the top of the roughness elements to about 3 mm above the elements. Although no direct dependence on shape, arrangement or roughness density could be identified, the effect of roughness is reflected in the value of y_c^+, the upper limit of the roughness sublayer. The flow field of rough surfaces can thus be characterized by the dimensionless distance y_c^+ that increases linearly with increasing roughness (Figure 6.9).

In contrast to the results of Nowell & Church (1979), the distributions of the statistical moments are independent of roughness density for quasi smooth to wake interference flow down to the roughness sublayer and the turbulence intensities do not decrease linearly toward the surface, but follow the relationship recommended by Nezu (1977). Grass *et al.* (1991) and Grass & Mansour-Tehrani (this volume) observed that rough wall flow is able to order itself very rapidly over a small vertical distance above the tops of the roughness elements, which is in agreement with the characteristic properties of the flow field in region 2 (see Figures 6.6–6.8).

As illustrated in Figures 6.8 and 6.10, the maxima of the turbulence intensities are roughness dependent. In Figure 6.10, the distributions of the turbulence intensities for the three different roughnesses are plotted as a function of y^+ with the top of the roughness elements being indicated by vertical bars. The maxima of the turbulence intensities are seen to be very closely located near the top of the roughness elements and these maxima decrease with increasing roughness k_s/k.

With the assumptions made in Equations 6.7 to 6.9 and the data of Figure 6.10, the following relationships are obtained of the lift force per area, L_1:

$$L_1 = (2.3 - 3.1)\tau \qquad \text{for} \qquad k_s/k = 0.40$$

$$L_1 = (2.1 - 2.9)\tau \qquad \text{for} \qquad k_s/k = 0.72$$

$$L_1 = (1.7 - 2.3)\tau \qquad \text{for} \qquad k_s/k = 1.58$$

These data are plotted in Figure 6.11 and show that the ratio of dynamic lift and shear stress decreases almost linearly with increasing dimensionless wall distance, y_c^+, and increasing k_s/k. Further, the comparison of the data with those found in the literature shows that the present data correspond well with those of Gessler (1965), Jüpner (1993) and Hinze (1975). However, Bretting's (1958) calculation results in much higher L_1 values than those from the present data.

By applying Equation 6.11 to Figures 6.7 and 6.8, it appears, as is the case with smooth surfaces, that ejection-like structures ($S < 0$, the velocity data are almost Gaussian distributed) dominate for $y^+ > y_c^+$ and sweep-like structures ($S > 0$, the velocity distributions are peaked and positively skewed) in region 2 ($y_u^+ < y^+ < y_c^+$) of the turbulent flow field over rough surfaces. Close to the top of the roughness elements, the turbulent flow field is characterized by the action of sweeps (Figure 6.7).

Furthermore, Figure 6.7 shows that close to the top of the roughness elements, the kurtosis factor is low and thus, according to Equation 6.12, the percentage of time occupied by higher energy fluctuations like sweeps is high. As with increasing roughness, the upper limit of the roughness sublayer, y_c^+, increases as does the distance between the top of the roughness elements and y_c^+, producing high kurtosis values close to the elements. High kurtosis values imply a low percentage of time occupied by sweeps. Thus with increasing roughness, not only do the maximum values of turbulence intensity decrease but also the temporal occurrence of sweeps decreases. The latter factor is another indication that with increasing roughness, the ratio L_1/τ, decreases; i.e. the rougher the surface, the greater the shear stress τ compared to the dynamic lift L_1. Since the roughest surface occurred at the lowest roughness density, $c_k = 0.28$, the result is in agreement with the measurements of Bertschler (1985) who found maximum turbulent shear stress to occur at a roughness configuration of $c_k = 0.26$.

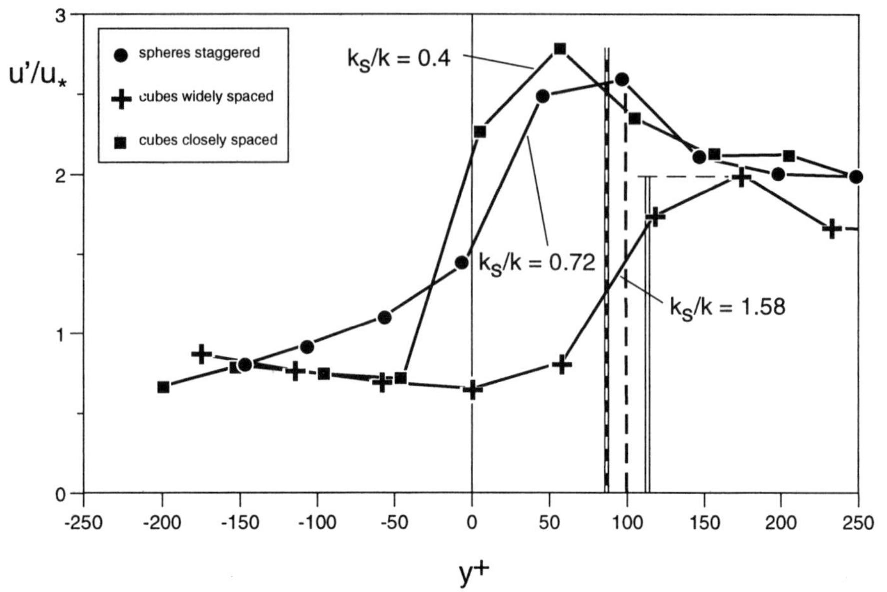

Figure 6.10. Distributions of the turbulence intensities over y^+ for three different roughnesses measured behind the elements

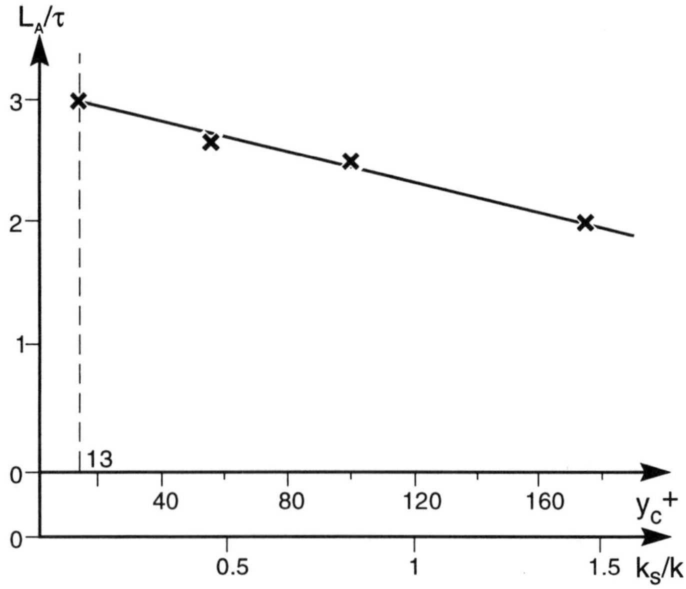

Figure 6.11. The ratio of dynamic lift and shear stress as a function of distance from the wall, y_c^+, and the dimensionless roughness, k_s/k

SUMMARY

The flow field over rough surfaces with quasi-smooth to wake interference flow can be characterized by a value of the dimensionless distance y_c^+ that increases linearly with increasing roughness and a value of the ratio of lift and shear force that decreases with increasing roughness. The point y_c^+ is defined as the location at which the mean velocities cease to follow the logarithmic law of the wall, turbulence intensity u'/u_* reaches a maximum, there is a change in the sign of skewness and the kurtosis reaches a minimum.

Sweeps dominate close to the top of the roughness elements, i.e. in the roughness sublayer (region 2 in Figures 6.6 and 6.7), and are therefore assumed to be the primary pulses inducing particle movement. The roughness sublayer is characterized by a rapid ordering of the flow. The thickness of the roughness sublayer ($y_c^+ - y_u^+$ in Figure 6.6) increases with increasing roughness and it shelters the roughness elements similar to the viscous sublayer associated with smooth beds. The effect of roughness on the flow can be expressed by a value of the distance y_c^+. Roughness is a function of particle shape, arrangement and roughness density as illustrated in the experiments of Grass *et al.* (1991). It has been further shown that the ratio of dynamic lift -to- shear stress decreases with increasing roughness or thickness of the roughness sublayer, y_c^+.

When considering the results of the present study with respect to the problems of incipient motion and the development of armoured beds, several conclusions may be drawn. In the case of non-uniform material such as gravels, the stability of the bed and its roughness are strongly time dependent. More or less well-arranged particles form the surface structure of a bed consisting of loosely packed non-uniform material shortly after a flow begins to impart forces on them and the roughness is low. With further development of the bed, a rearrangement of embedded particles and a change of the surface structure occurs with the new surface being characterized by higher roughnesses (Dittrich *et al.*, 1994). Thus, with an increase in time not only does the roughness change but also the ratio L_1/τ varies. Because the Shields diagram only includes the shear force as the active force, it can thus only be used to determine the end stage of maximum bed stability and no other stages (Graf & Pazis, 1977). Therefore, the implication is that in gravel bed rivers with similar shape and arrangement of gravels, the Shields diagram has to be extended by one additional parameter (Dittrich *et al.*, 1994), involving the roughness density, c_k, or an equivalent roughness parameter like the K_3 roughness coefficient (Ergenzinger, 1992; Ergenzinger & de Jong, this volume). Another possibility is use of the parameters c_k and y_c^+ and development of a stability diagram such as that proposed by Dittrich *et al.* (1994).

REFERENCES

Andreopoulos, J., Durst, F., Zaric, Z. & Jovanovic, J. 1984. 'Influence of Reynolds number on characteristics of turbulent wall boundary layers', *Exper. Fluids*, **2**, 7–16.

Bertschler, M. 1985. *Untersuchung der Turbulenzstruktur im offenen Gerinne über glatter Sohle und mit Rauhigkeitselementen*, Diss. **7648**, Institut für Hydromechanik und Wasserwirtschaft, ETH Zürich, (1-1)–(5-4).

Bretting, A.E. 1958. 'Stable channels', *Acta Polytechnica Scandinavica, Civil Engrg. and Building Construction Series*, **1**, 245/1958.

Dittrich, A. & Hammann de Salazar, K. 1993. 'Bed instability caused by clear water flow', *Bericht des Instituts für Wasserbau und Kulturtechnik*, Karlsruhe, FRG, 1–102.

Dittrich, A., Larsen, P. & Ergenzinger, P. 1994. 'Grain sorting and armouring leading to maximum bed stability', *J. Sediment. Res.*, **9**, 28–35.

Drake, T., Shreve, R., Dietrich, W., Whiting, P. & Leopold, L. 1988. 'Bedload transport of fine gravel observed by motion-picture photography', *J. Fluid Mech.*, **192**, 193–217.

Durst, F., Jovanovic, J. & Kanevce, L. 1987. 'Probability density distribution in turbulent wall boundary-layer flows', *Turbulent Shear Flows 5*, Springer Verlag, Berlin, 197–220

Einstein, H.A. & El-Samni, E.A. 1949. 'Hydrodynamic forces on a rough wall', *Rev. Modern Physics*. **21**, 520–524.

Emmerling, R. 1973. 'Die momentane struktur des wanddruckes einer turbulenten grenzschicht', *Mitt. Max-Planck-Institutes für Strömungsforschung*, Göttingen, **56**, 1–94.

Ergenzinger, P. 1992. 'River bed adjustment in a step-pool system: Lainbach, Upper Bavaria', in Billi, P., Hey, R.D., Thorne, C.R. & Tacconi, P. (Eds) *Dynamics of Gravel-bed Rivers*, John Wiley & Sons Ltd., Chichester, 415–430.

Gessler, J. 1965. 'Der geschiebetrieb bei mischungen untersucht an natürlichen abpflästerungserscheinungen in Kanälen', *Mitt. Versuchsanstalt für Wasser und Erdbau, ETH Zürich*, **69**, 1–67.

Graf, W.H. & Pazis, G.C. 1977. 'Les phenomenes de deposition et d'erosion dans un canal alluvionnaire', *J. Hydraul. Res.*, **15**, 151–166.

Grass, A.J. 1971. 'Structural features of turbulent flow over smooth and rough boundaries', *J. Fluid Mech.*, **50**, 233–255.

Grass, A.J., Stuart, R.J. & Mansour-Tehrani, M. 1991. 'Vortical structures and coherent motion in turbulent flow over smooth and rough boundaries', *Phil. Trans. Roy. Soc. London*, **A336**, 35–65.

Günter, A. 1971. 'Die kritische mittlere sohlenschubspannung bei geschiebemischungen unter berŸcksichtigung der deckschichtbildung und der turbulenzbedingten sohlen schubspannungsschwankungen', *Prom. Nr.* ***4649***, ETH Zürich, 1–70.

Hammann de Salazar, K. & Dittrich, A. 1994. 'Measuring systems to determine the velocity field in and close to the roughness sublayer', in Dittrich, A. (Ed) *Dynamics and Geomorphology of Mountain Streams*, Lecture Notes, Springer Verlag, 265–288.

Hinze, J.O. 1975. *Turbulence*, 2nd Edition, McGraw-Hill, 656–683.

Jüpner, R. 1993. *Die Erosion Grober Sohlpartikel: Untersuchung der Auftriebs- und Schubkräfte*, Diss. im Fachbereich Geowissenschaften, Freien Univ. Berlin, 1–69.

Karlsson, R.I. 1980. 'Studies of skin friction in turbulent boundary layers on smooth and rough walls', *Publikations Nr.* **80/4**, Dept. Applied Thermo and Fluid Dynamics, Chalmers Tekniska Högskola Göteborg, 1–65.

Morris, H.M. 1955. 'Flow in rough conduits', *ASCE Transactions*, **120**, 373–410.

Nakagawa, H., Tsujimoto, T. & Shimizu, Y. 1988. 'Velocity profile of flow over rough permeable bed', *Proc. 6th Congr. of APD-IAHR*, Kyoto, Japan, **2**, 449–456.
Nezu, I. 1977. 'Turbulence intensities in open channel flows', *Proc. Japan. Soc. Civil Eng.* **261**, 67–76.
Nowell, A. & Church, M. 1979. 'Turbulent flow in a depth-limited boundary layer', *J. Geophys. Res.*, **84**, 4816–4824.
O'Loughlin, E.M. & MacDonald, E.C. 1964. 'Some roughness-concentration effects on boundary resistance', *La Houille Blanche*, **7**, 773–782.
Raemy, F. 1991. 'Comparaison de quelques charactéristiques de la turbulence mesurée dans divers écoulements d'eau en canal et en conduite', *La Houille Blanche*, **1**, 13–21.
Raupach, M.R. 1981. 'Conditional statistics of Reynolds stress in rough-wall and smooth-wall turbulent boundary layers', *J. Fluid Mech.*, Vol. 108, 363–382.
Shields, A. 1936. 'Anwendung der ähnlichkeitsmechanik und der turbulenzforschung auf die geschiebebewegung', *Mitt. Preuß. Versuchsanstalt für Wasserbau und Schiffbau, Berlin*, **26**, 1–26.

NOTATION

A	fraction of turbulent stress
c_k	ratio of the upstream projected area of a roughness element to the floor area assigned to the element
c_k^*	ratio of plan area of elements to total bed area
C	proportionality factor
C_1	proportionality factor
d	water depth
D	diameter of glass spheres
D_{50} D_{60}	50th and 60th percentiles of the grain size distribution
D_1	constant
f_B	mean frequency of occurrence of coherent structures
F	kurtosis
Fr	Froude number ($u_m/(gd)^{1/2}$)
F_τ	shear force
g	acceleration due to gravity
G'	weight under buoyancy
I	slope of flume base
k	geometric roughness height
k_s	equivalent sand roughness
L	lift force
L_1	dynamic lift (force per unit area)
N	number of samples
p_w	intensity of wall-pressure fluctuations
Q	discharge
R	resisting force
Re	flow Reynolds number ($u_m d/\nu$)
Re^*	roughness Reynolds number ($u_* k/\nu$)

S	skewness
T_B	mean time duration of coherent structures
u_i	value of the instantaneous streamwise velocity
u_m	cross-section averaged mean velocity
u_0	mean velocity at the zero point of the logarithmic velocity profile
u_*	shear velocity
u'	velocity fluctuation in longitudinal direction
u_1' u_2' u_3'	fluctuating velocities in x, y and z-directions
$\overline{u^2}$	normal component of the Reynolds stress tensor
u'/u_*	turbulence intensity
U	time averaged mean
y	vertical co-ordinate
y^+	dimensionless depth yu_*/ν
y'	zero level of the vertical velocity profile
y_c^+	upper limit of the roughness sublayer
y_{c1}^+ y_{c2}^+ y_{c3}^+	upper limit of the roughness sublayer for closely packed cubes, staggered spheres and widely spaced cubes respectively
y_u^+	lower limit of the roughness sublayer
y_{u1}^+ y_{u2}^+ y_{u3}^+	lower limit of the roughness sublayer for the closely packed cubes, staggered spheres and widely spaced cubes respectively
γ	intermittency factor
ε_R^+	dimensionless vortex viscosity
κ	von Kármán constant
λ_1	constant
ν	kinematic viscosity
ρ	density of fluid
τ	shear stress
φ	angle of repose of sediment

7 Scales of Turbulent Coherent Flow Structures in a Gravel-bed River

ANDRÉ G. ROY, THOMAS BUFFIN-BÉLANGER & SYLVAIN DELAND
Département de géographie, Université de Montréal, C.P. 6128, Succursale Centre-Ville, Montréal, Québec, H3C 3J7, Canada

ABSTRACT

Although it is generally admitted that turbulent coherent flow structures over rough surfaces have signatures similar to those observed over smooth beds, few studies have characterized these structures in river flows over gravel beds. The objective of this paper is to quantify the turbulent structures in a depth-limited flow over an armoured gravel bed (D_{50} = 38 mm). Fluctuations in the longitudinal and vertical velocity components were sampled at 10 Hz for 2 minutes using Marsh-McBirney electromagnetic current meters. Two sensors of different sizes (13 and 38 mm) were simultaneously deployed first near the bed and then in the outer region of the flow at relative heights (y/Y) above the bed of 0.12 and 0.67. Seventeen replicates sampled at each location and for each sensor were used to derive average turbulence statistics. The detection and quantification of the coherent structures relied on temporal and spatial correlations and on burst detection techniques (Quadrant, U-Level, VITA and VITA with slope). Our results indicate that similar structural scales are present at both heights above the bed. Although burst frequency differs from one detection technique to another and decreases with sensor size, average burst frequency is not significantly different between the two heights above the bed. In the near-bed region, shear stress production appears to be dominated by strong sweep-like motions while the outer region is dominated by ejection events. The similarity between the scales of the structures in the near-bed and outer regions may indicate the presence of large wedges of low- and high-speed fluid.

INTRODUCTION

In the turbulent boundary layer over smooth beds, most of the turbulent energy is generated in the near-bed region where bursting processes generate intermittent high shear stress events. Despite the importance of these events for initiating particle motion and sediment transport (Drake *et al.*, 1988), the characteristics and structure of turbulence above natural gravel beds remain poorly quantified and understood. It is admitted, however, that the structural features observed in the turbulent boundary layer over smooth surfaces, most notably ejections and sweeps associated with the bursting process, are also observed over rough beds (Grass, 1971; Grass *et al.*, 1991, Defina, this volume; Grass & Mansour-Tehrani, this volume). Intermittent

Coherent Flow Structures in Open Channels.
Edited by P.J. Ashworth, S.J. Bennett, J.L. Best and S.J. McLelland.

events similar to ejections and sweeps are present over beds of homogeneous roughness and several researchers have confirmed the dominance of ejection- and sweep-like motions in turbulence production in flows over rough surfaces. Nakagawa & Nezu (1977) have documented in detail the changes in the structural features of turbulence when the bed surface roughness is gradually increased. They reported a switch from the well-documented domination of ejection events in shear stress production on smooth beds to a sweep-dominated stress production over rough surfaces. More recently, Krogstad *et al.* (1992) have shown clear differences between smooth and rough beds. In both cases, ejections and sweeps dominate shear stress production but events are stronger and more frequent over rough beds where sweeps tend to dominate shear stress production.

Over smooth beds, features present in the near-bed region interact with larger scale structures in the outer layer (Falco, 1977; Nezu & Nakagawa, 1981). Large-scale structures may take the form of a sequence of wedges of low- and high-speed fluid with shear zones at the margin of the wedges (Levi, 1983; Nezu & Nakagawa, 1981, 1993; Yalin, 1992). These larger scale structures may exhibit statistical similarities with ejections (low-speed fluid moving upward) and sweeps (high-speed fluid moving downward). The origin of large-scale structures is not clear (Robinson, 1990), but interactions between the structures composing the bursting process in the inner layer and those in the outer layer appear central to the generation and maintenance of the latter. The intensity of the structures in the inner layer appear to affect the outer-layer structures (Offen & Kline, 1974, 1975).

Very few turbulence studies have been conducted in flows over rough natural surfaces and insight into the turbulent coherent structures in these environments is limited. This is partly due to a lack of adequate instrumentation which prevents the examination of a whole range of structures, especially in the near-bed region. Often, flows over gravel beds are of very limited depth with respect to the size of the roughness elements composing the bed, thus exaggerating the effects of crude sensors on turbulence statistics. However, recent studies indicate that statistical signatures analogous to those of ejections and sweeps exist in depth-limited flows over gravel beds (Robert *et al.*, 1993; in press; Kirkbride, 1994). These events appear responsible for most of shear stress production and they tend to dominate the whole thickness of the flow. In rough environments, vortex shedding and advection of low-speed fluid from the lee of protruding particles play a crucial role in the dominating turbulent processes in the near-bed region and on the generation of larger scale structures (Robert *et al.*, 1993; Kirkbride, 1993, 1994). The presence of high-speed and low-speed wedges has also been reported over gravel beds (Kirkbride, 1994; Ferguson *et al.*, this volume). Despite these recent attempts, detailed studies of the scales and characteristics of the structural features of turbulent flows over natural rough beds are still lacking. There is a need for reliable turbulence statistics based on several replicated series of measurements.

The primary objective of this paper is to quantify the scales of the turbulent structures in the near-bed and outer regions of the flow over a natural gravel bed river. A combination of techniques are used to achieve this objective, ranging from correlation analysis to burst detection algorithms. It is also necessary to examine the effect of the size of the sensor on the turbulence statistics. Here, the performance of Electromagnetic Current Meters (ECMs) of two sizes will be compared.

METHODS

The selected site for this study is a portion of an armoured gravel bed in a low sinuosity reach of the Eaton North River, Québec. The bed surface clasts have a D_{50} of 38 mm and are tightly imbricated with some finer-grained material filling the interstices between the rougher elements. The measurement section is in a straight riffle where large protruding clasts are rare. Flow depth in the section remained constant at 0.3 m during the duration of the experiment. The measuring devices were deployed over a 1 m long section oriented in the streamwise direction. Three Marsh-McBirney ECMs were used on a horizontal rod clamped to vertical metal bars planted into the bed. Two sensors were attached at the upstream and downstream extremities of the 1 m long horizontal rod while the third one could be moved in between the two fixed sensors (Figure 7.1). The sensor heads were set at 0.2 m from the rod. The sensors were first positioned at a relative height (y/Y where y is height above the bed and Y is flow depth) above the bed of 0.12. Because the centre of the sensor head was used to position the ECMs, this is the closest to the bed that we could deploy the sensors. Seventeen velocity time-series were collected at that height which is referred to as the near-bed position. The horizontal rod was then moved to a position in the outer region of the flow at a relative depth of 0.67 above the bed. Again 17 time-series were collected at that position. All three sensors were operated in synchronicity to obtain concomitant

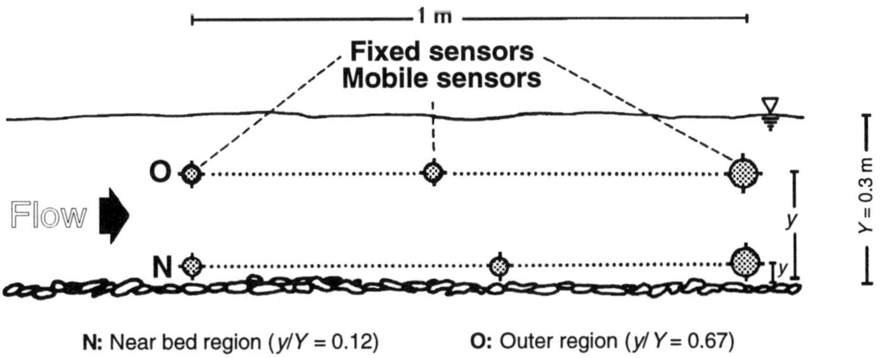

Figure 7.1. The experimental set-up for the velocity measurements using three ECMs of two different sizes

signals. Synchronicity was ensured by a master/slave configuration of the electronics linking the sensors.

Longitudinal (*U*) and vertical (*V*) velocity components were measured using ECMs of two sizes. The upstream fixed sensor and the mobile sensor had the smallest spherical head available for this type of instrument (13 mm). The fixed sensor at the downstream end had a 38 mm head which is the most common size for these instruments. The diameter of the sensing volume of these sensors is between 2.5 to 3 times that of the spherical head. Thus, the sensing volume of the largest current meter incorporates the bed while deployed in the near-bed region, a situation which may affect the signal quality. The response time of all ECMs is 0.05 s. The signals captured by the sensors are processed through a low-pass R/C filter with a half-power frequency of 3.18 Hz to remove noise. Each velocity time-series was sampled at 10 Hz during more than 120 s. The sampling frequency is low with respect to the response frequency of the sensor thus introducing some aliasing effects. However, these effects are small because of the severity of the filter in the high frequency range. The choice of the parameters of the sampling scheme is a compromise between data logger capacity and the lower frequencies that we intended to examine. A summary of the experimental conditions is given in Table 7.1.

Table 7.1. Experimental conditions (August 1994)

Site characteristics:	
Average flow depth (*Y*)	0.30 m
Average channel width	14.0 m
Slope of the reach	≈0.005
Bed material size (*b* axis)	
D_{25}	0.018 m
D_{50}	0.038 m
D_{75}	0.060 m
Reynolds Number	≈110 000
Sampling characteristics :	
Record length	120 s
Sampling frequency	10 Hz
Sampling height above the bed (*y*)	
Near-bed region (17 replicates)	0.035 m
Outer region (17 replicates)	0.200 m
Marsh-McBirney ECMs characteristics:	
Time constant	0.05 s
Half power	3.18 Hz
Diameter	
Small sensors	0.013 m
Large sensor	0.038 m

Table 7.2. Variance losses in the U and V components and in the Reynolds shear stress due to the sampling frequency (10 Hz) and to the record length (120 s) for two positions above the bed

		Variance losses in %		
y/Y	Frequency End	U	V	Reynolds Shear Stress
0.12	high	12.0	45.0	9.0
	low	0.2	0.9	0.5
0.67	high	5.0	17.0	2.0
	low	4.5	0.7	1.8

The experimental design bears important consequences on the losses in the velocity signals used in the analysis. Given the sensor sizes, the sampling parameters and the flow conditions, losses from the ideal velocity signal for both the high and low frequencies were estimated using the diagram presented by Soulsby (1980). Variance losses estimated for the U and V velocity components and for the Reynolds shear stress are given in Table 7.2. Due to the relatively long sampling records, the losses are small at the low frequency end of the spectrum. However, the relatively low sampling rate introduces important variance losses at the high frequency end of the V component, especially at the near-bed position. These losses do not affect the value of the Reynolds shear stress much because it is more dependent upon the fluctuations in the U component which is less sensitive to the low sampling rate. The variance losses in the high frequencies must be added to the effects of the filter and of the large sensing volume relative to flow depth, thus diminishing even further the quality of the response in these frequencies. It will therefore be difficult to resolve the finer-scale structures of the flow, especially with the large sensor close to the bed.

Before proceeding to the quantification of the structures contained within the velocity signals, all velocity series were examined for conspicuous defects or contamination problems. Visual scrutinization and spectral analysis were used to validate the data quality (Lapointe *et al.*, in press). In all, three signals taken in the near-bed region were found to contain odd spikes and these time-series were removed from subsequent analyses.

The means and standard errors of turbulence intensities of the Reynolds shear stresses and of the correlations between u' and v' were estimated from 15 replicates for the small sensor and 16 replicates for the large sensor in the near-bed region and from 17 replicates in the outer region for both sensors. To quantify the scales of the coherent flow structures, several techniques were also applied to these replicates. Correlation techniques both for the temporal and spatial domains were used to assess the integral time-scales and the degree of persistence in the velocity signals. The application of the temporal autocorrelation technique is based on Taylor's frozen flow hypothesis which should be restricted to low turbulence intensity environments (<10%) in which vortices are not rapidly deformed and sheared (see Clifford, this volume). However, the interpretation of structure scale based on the integral scales obtained from autocorrelation methods will be validated by spatial correlations and by other techniques. These techniques are derived from the literature on burst detection and the estimation of burst frequency. They focus on the ejection component of the bursting process and they are generally applied in studies of the near-wall region of a turbulent boundary layer. However, the well-known quadrant analysis which is part of such techniques is now often used in several environments far removed from the near-wall region of the turbulent boundary layer (Biron *et al.*, 1993). Several other techniques proposed to detect and scale the coherent structures have not yet been tested in the fluvial environment. The simpler techniques studied by Bogard & Tiederman (1986) will be examined in this paper. The techniques, Quadrant, *U*-Level, VITA and VITA with slope, are briefly presented in Table 7.3. These techniques identify the presence of an ejection if the velocity signals respond to specific conditions. The algorithms used in these techniques do not detect ejections composed of multiple events unless the signals fall below the detection threshold.

Because visualization could not be implemented simultaneously to the velocity measurements, optimal detection thresholds for each technique could not be determined and we decided to use those proposed by Bogard & Tiederman (1986). An example of burst detection and delineation using the four algorithms illustrates the differences between the techniques (Figure 7.2). The strongest events are

Table 7.3. Burst detection techniques used to characterize the scales of turbulent structures

Detection technique	Condition for positive detection
Quadrant	$u' < 0$ and $v' > 0$ $-u'v' > 1.2\ \sigma_u \sigma_v$
U-Level	$u' < -1.3\ \sigma_u$
VITA	$\sigma_u{}^2{}_k > 0.64\ \sigma_u{}^2$ window size: 7 time-steps
VITA with slope	$\sigma_u{}^2{}_k > 0.35\ \sigma_u{}^2$ $\partial u/\partial t < 0$ window size: 5 time-steps

u' is the fluctuation of the downstream velocity, v' is the fluctuation of the vertical velocity, σ_u is the RMS of the downstream velocity, σ_v is the RMS of the vertical velocity, σ^2 is the variance of a variable, $\sigma^2{}_k$ is the variance of a variable in a window of k time-steps and t is time

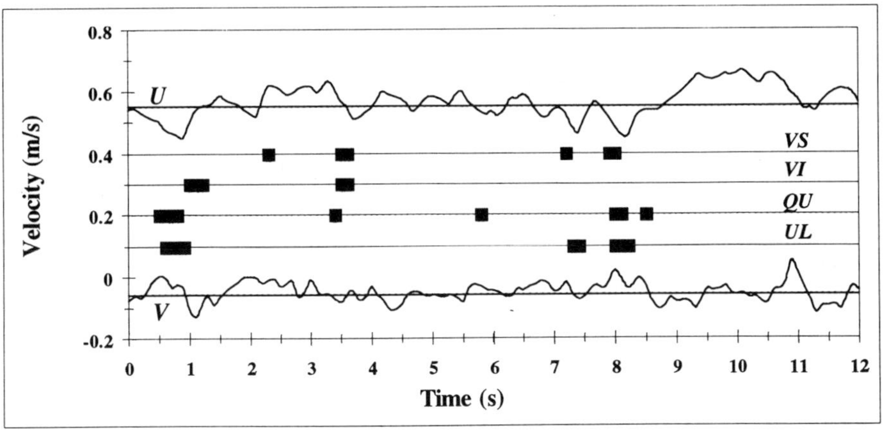

Figure 7.2. Example of positive detection by four burst detection techniques on a typical set of *U* and *V* velocity measurements

generally picked up by all techniques although the duration of the event may vary markedly. Smaller events can pass undetected, especially with the VITA technique. In the case of Quadrant analysis, a more complete examination of the shear stress contributions of each quadrant at various thresholds (hole sizes which are defined as a measure of deviation from the average stress) will be undertaken.

RESULTS

GENERAL TURBULENCE STATISTICS

Turbulence statistics presented in Table 7.4 are the overall means and standard errors of the means obtained from the replicate velocity series measured in the near-bed and outer regions. As expected, turbulence intensities of both velocity components, average Reynolds shear stress and the absolute value of the correlations between u' and v' all decrease from the near-bed position to the higher position. This observation holds true for both sensors although it is more pronounced for the most responsive small sensor. It is clear, when we examine the differences between the sensors, that all statistics associated with the amplitude of the velocity fluctuations are smaller for the large sensor. This effect is much more severe in the near-bed region where fluctuations of higher frequencies should contribute more to the velocity signal. This results in a reduction of the average Reynolds shear stress by as much as 50% and of the turbulence intensities by 20% for the *U* velocity component. The statistics for the vertical velocity component are less affected by the size of the sensor because the signals have been severely truncated for both sensors in the near-bed region. However, correlation estimates are very robust to the

Table 7.4. Averages and standard deviations or standard errors of turbulence statistics obtained from the replicated series of velocity measurements from two ECMs of different size for the near-bed and outer regions of a flow above a gravel bed

	Small sensor				Large sensor			
y/Y	0.12		0.67		0.12		0.67	
Component	U	V	U	V	U	V	U	V
Number of replicates	15	15	17	17	16	16	17	17
Average velocity ($m\ s^{-1}$)	0.41 [0.01]	-0.03 [0.01]	0.61 [0.01]	-0.03 [0.01]	0.39 [0.02]	0.02 [0.00]	0.56 [0.01]	-0.05 [0.00]
Turbulence intensity ($\langle \sigma_u / \overline{U} \rangle$ and $\langle \sigma_v / \overline{U} \rangle$, %)	19.23 (0.88)	8.48 (0.35)	10.60 (0.47)	6.09 (0.24)	15.56 (1.07)	6.90 (0.41)	9.10 (0.49)	5.89 (0.27)
Average Reynolds shear stress ($N\ m^{-2}$)	1.12 [0.12]		0.71 [0.11]		0.61 [0.15]		0.49 [0.11]	
Average correlation between u' and v'	-0.41 (0.04)		-0.30 (0.04)		-0.38 (0.06)		-0.29 (0.06)	

Standard deviations shown in normal brackets; standard errors shown in square brackets

sensor size effect. The near-bed correlations are close to -0.4, a value often reported as typical of turbulent boundary layers (Townsend, 1976; Lapointe, 1992; Deland, 1994). Finally, it is noticeable that in all cases the standard errors of estimates for the means are relatively small, thus indicating a very limited variability among the replicates.

TEMPORAL AUTOCORRELATIONS AND INTEGRAL TIME AND LENGTH SCALES

The average autocorrelation functions (ACF) for the U and V velocity components and the corresponding length scales are presented in Figure 7.3 and in Table 7.5. The similarity between the ACFs obtained in the near-bed and outer regions is striking. This similarity results in subtle differences in the integral time-scales between the two regions. In fact, the time scales estimated for the U component from the small sensor measurements are identical. Also, the difference between the ACFs for the two sensors is subtle. For the U velocity signals measured with the small sensor, the correlation persists until 20 time lags whilst for those measured with the large sensor it remains positive over 30 time lags. This increased persistence in the signal due to the larger sensing volume yields a longer time-scale. It is nonetheless counter-intuitive to have a longer time-scale at the bed than in the outer region. This is probably caused by the proximity of the sensor to the bed. The ratios between the time scales measured for the V to those measured for the U component vary from 28 to 37%. For both sensors, this ratio is larger in the near-bed region. When we examine the integral length-scales, the structures in the

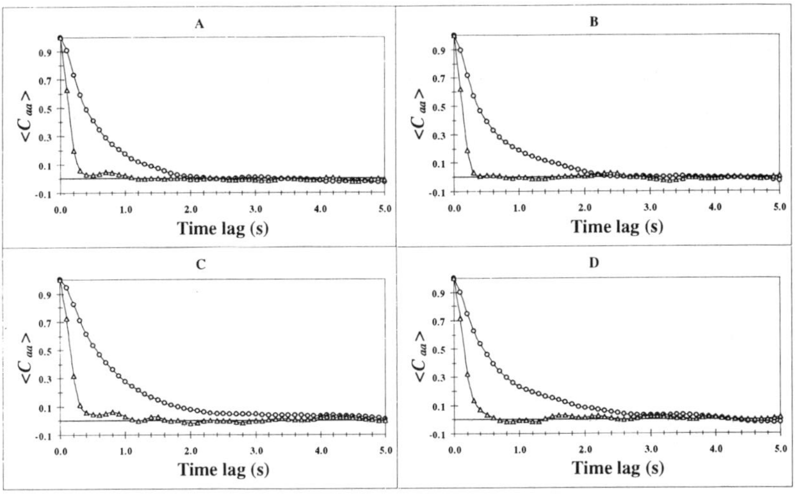

Figure 7.3. Average autocorrelation functions for the U velocity component (circles) and the V velocity component (triangles): (A) near-bed region, small sensor; (B) outer region, small sensor; (C) near-bed region, large sensor; and (D) outer region, large sensor

Table 7.5. Average integral time-scales (ITS) and length-scales (ILS) from U and V velocity replicates for the near-bed and outer regions above a gravel bed

	y/Y	Small sensor		Large sensor	
		U	V	U	V
⟨ITS⟩	0.12	0.64	0.20	0.95	0.26
(s)	0.67	0.66	0.19	0.87	0.25
⟨ILS⟩	0.12	0.26	0.08	0.37	0.10
(m)	0.67	0.40	0.12	0.49	0.14

$\text{ITS} = \int_0^k C_{xx}(t)\mathrm{d}t$ where k is the time step where C_{aa} is no longer significantly different from zero

$\text{ILS} = \text{ITS} \times \overline{U}$

U component are 1.3 to 1.5 times larger in the outer region than in the near-bed region. This increase in scale is obviously related to the increase in velocity from the bed to the water surface. On the other hand, the length of the structures of the V component appears quite stable throughout the flow. The ratio between the length scales of the V to those of the U component is slightly higher at the bed than in the outer region.

SPATIAL CORRELATION IN THE STREAMWISE DIRECTION

Spatial correlations were estimated in the streamwise direction for the *U* signals measured using the two fixed sensors at the upstream and downstream ends of the 1 m horizontal rod and the mobile sensor which could be set at varying distances between the upstream and downstream fixed sensors. In all, the mobile sensor was set at 17 locations but two records had to be eliminated from the analysis of the near-bed region. The sampling locations were set to be denser in the upstream portion of the flow. The first measurements from the mobile sensor were taken very close to the fixed upstream sensor. This could have an effect on the correlations between the signals as the downstream sensor is in the wake of the upstream one.

The results of the spatial correlation analysis are displayed in Figure 7.4. Again, there is a remarkable similarity between the signatures obtained for the near-bed and outer regions. In both cases, positive correlation is maintained over more than 0.7 m downstream from the initial fixed sensor. However, high positive values (>0.2) are slightly more persistent in the outer region. The correlations between the measurements of the mobile sensor and of the downstream sensor become close to zero at a separation distance of 0.80 m in the near-bed region and of 0.90 m in the outer region. This indicates that the structures are detectable in the streamwise direction on a distance between 2.5 and 3 flow depths and that they are slightly longer in the outer region, which is consistent with the integral length-scales.

BURST DETECTION AND BURST FREQUENCY ESTIMATED FROM SEVERAL TECHNIQUES

Burst detection algorithms were applied to all velocity time-series and the results were averaged for each region and each sensor. Despite the difference in the variability of the estimated burst frequencies within each of the four techniques, the results are highly consistent from one technique to another (Figure 7.5). Considering both the near-bed and outer regions, the average burst frequency ranges from 0.30 to 0.35 events per second for the small sensor (Table 7.6). In this case, there is no clear pattern from the near-bed to the outer region, some techniques (Quadrant and *U*-Level) showing a minute decrease in burst frequency while the VITA-based techniques indicate a small increase in burst frequency. In view of the variability within the replicates, these differences are negligible. The average burst frequencies show the effects of sensor size. Values estimated from the measurements taken with the large sensor are clearly always less than those obtained from the small sensor. This is especially true of the techniques based on variance estimates (VITA). The very low burst frequencies derived from the VITA algorithms for the near-bed region may be the result of the extreme proximity of the sensor bottom electrode to the bed. Frequencies computed for the outer region are more consistent from one technique to another and the overall average is 0.26 events per second, that is 80% of the average obtained for the small sensor in the

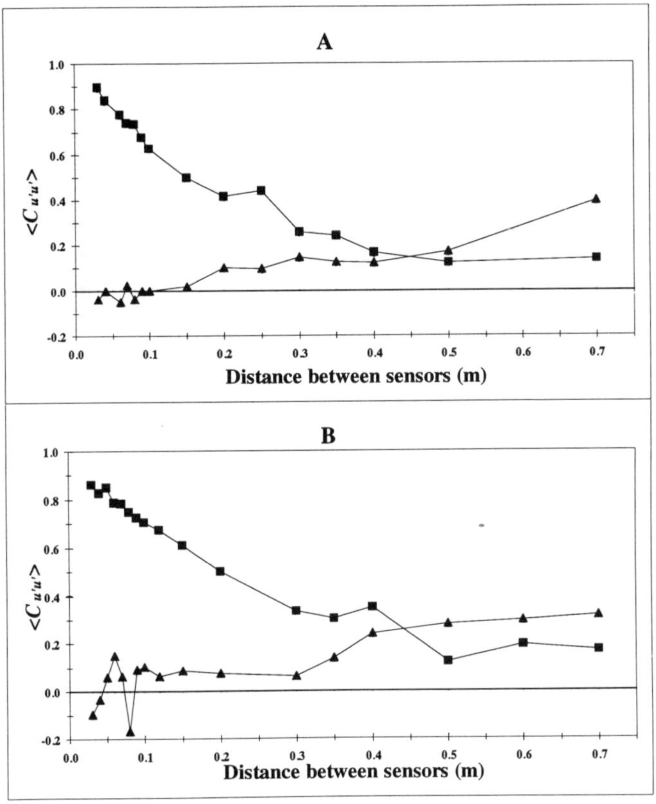

Figure 7.4. Spatial correlation for the *U* velocity component in the (A) near-bed region and (B) outer region. Squares: correlation between the upstream fixed sensor and the mobile sensor; Triangles: correlations between the mobile sensor and the downstream fixed sensor

near-bed region. Quadrant analysis appears to be the most consistent technique when applied to measurements obtained from sensors of different sizes. Using this technique, average burst frequency using the large sensor is 80% that from the small sensor for both regions.

When only the small sensor results are examined, the agreement among the techniques is remarkable. The thresholds used in this analysis were the same as those reported in Bogard & Tiederman (1986). This decision is based on the lack of concomitant visualization to guide the selection of threshold values appropriate for gravel bed environments. Despite the marked difference between the conditions studied by Bogard & Tiederman and those examined in this paper, the coincidence among the techniques suggests that the optimal thresholds proposed by the authors are robust. This coincidence is even more striking when the overall average burst

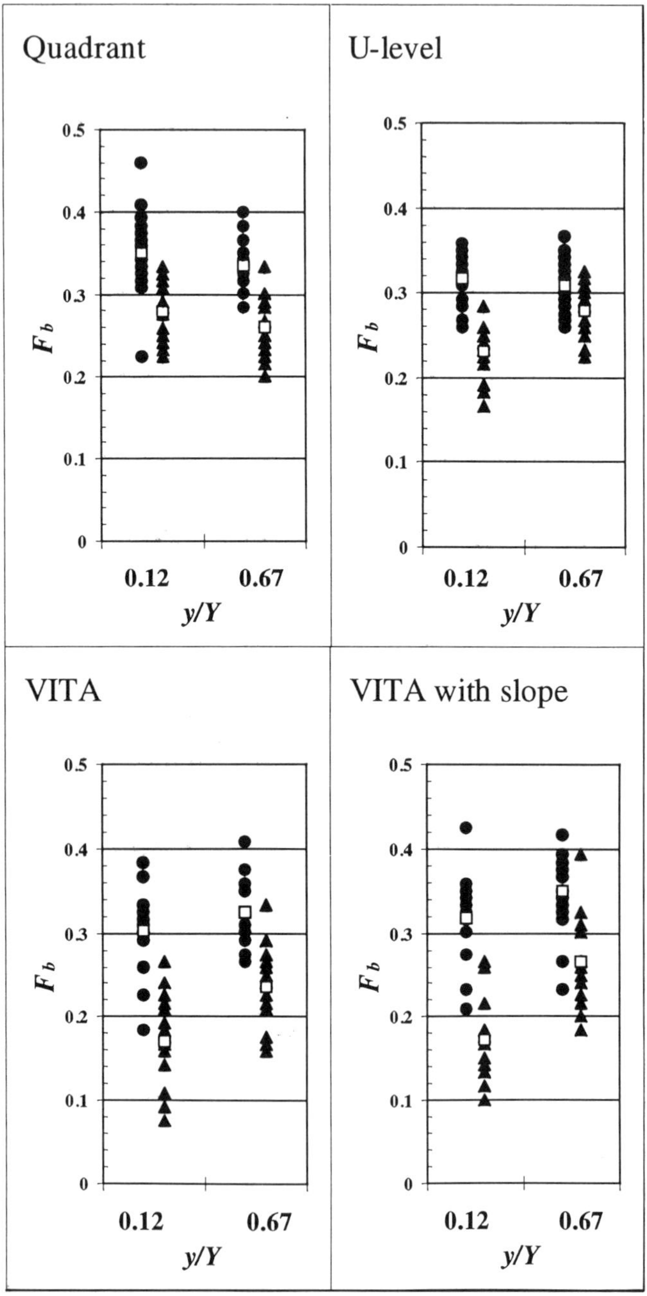

Figure 7.5. Burst frequencies obtained from four detection techniques. Circles : small sensor; Triangles: large sensor; Squares: mean

Table 7.6. Average burst frequency detected by different techniques for the near-bed and outer regions of flow above a gravel bed

	$\langle F_b \rangle$ = ⟨number of detected events / record length⟩							
	Small sensor				Large sensor			
y/Y	QD	UL	VI	VS	QD	UL	VI	VS
0.12	0.35	0.32	0.30	0.32	0.28	0.23	0.17	0.17
0.67	0.33	0.31	0.33	0.35	0.26	0.28	0.24	0.27

QD: Quadrant analysis; UL:U-Level technique; VI:VITA technique; and VS: VITA with slope technique

frequency (0.33) is compared to the expected burst frequency ($F_b = 0.36$) obtained from the equation $F_b = 5Y/\langle U \rangle$ first introduced by Laufer & Narayanan (1971).

QUADRANT ANALYSIS

The interpretation of structures will be completed through an analysis of quadrant dominance in terms of residence time and of shear stress production in each quadrant. For the examination of residence time, no threshold was applied and all events were taken into account (Table 7.7). As expected, quadrant-2 (the ejection quadrant) and quadrant-4 (the sweep quadrant) largely dominate the residence times. This is true irrespective of the sensor size and of the position above the bed. There is however a slight decrease in the dominance of the ejection and sweep quadrants from the near-bed to the outer region, confirming the tendency towards a less anisotropic turbulence away from the bed. The dominance of the ejection quadrant in the near-bed region is replaced by a subtle dominance of the sweep quadrant in the outer region. These changes are associated with a decline of the importance of the ejection events while the residence times in the sweep quadrant remain nearly constant.

Table 7.7. Average and standard deviations percentage of time in each quadrant from velocity replicates for near-bed and outer regions of flow above a gravel bed

y/Y	Small sensor				Large sensor			
	$Q1$	$Q2$	$Q3$	$Q4$	$Q1$	$Q2$	$Q3$	$Q4$
0.12	17.0	33.3	18.3	31.4	18.3	32.0	18.8	30.9
	(1.2)	(1.3)	(1.3)	(1.2)	(1.4)	(1.8)	(1.7)	(1.5)
0.67	20.4	29.3	19.5	30.7	20.6	29.9	19.2	30.3
	(1.2)	(1.5)	(1.1)	(1.2)	(1.5)	(1.8)	(1.6)	(1.4)

Standard deviations shown in brackets

Residence times only reveal part of the relevant information, and the contributions of each quadrant to shear stress production is probably more indicative of the intensity of turbulence activity associated with the dominant structures. To perform this analysis, we have used the concept of hole size (H) which is defined as a threshold of shear stress: $H = -u'v'/\sigma_u\sigma_v$. The proportions of the contributions to the Reynolds shear stress from each quadrant are plotted as a function of hole size. As hole size increases, only the stronger events remain as contributors. These strong events are the most crucial in particle entrainment and in sediment transport.

In all cases, hole size analysis reveals that events from quadrants-2 and 4 are the dominant contributors to Reynolds shear stresses (Figure 7.6). There are, however, some important nuances brought to light in this analysis. For the small sensor, residence time dominance of the ejection quadrant in the near-bed region is not matched with a shear stress production dominance (Figure 7.6a). In fact, sweep events appear to dominate shear stress production at all hole sizes. Conversely, sweep dominance in residence times in the outer region is associated with a quadrant-2 dominance in shear stress production (Figure 7.6b). For the large sensor, dominance of a single quadrant in shear stress production is tenuous especially in the outer region where relative stress contributions are equal for both quadrant-2 and 4 until $H = 2$ (Figure 7.6d). At the bed, there is a slight tendency for the ejection quadrant to dominate for all hole sizes considered (Figure 7.6c).

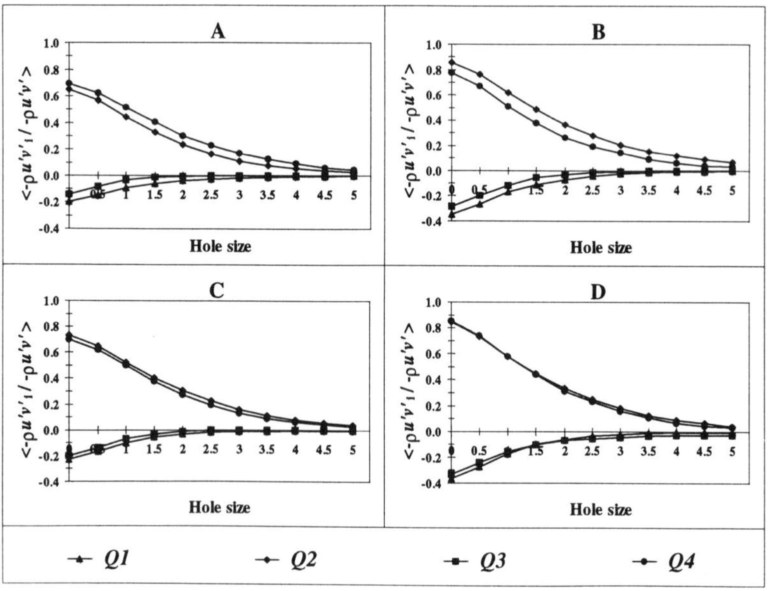

Figure 7.6. Hole size analysis of the relative contributions of each quadrant to the Reynolds shear stress: (A) near-bed region, small sensor; (B) outer region, small sensor; (C) Near-bed region, large sensor; and (D) outer region, large sensor

DISCUSSION AND CONCLUSION

The relevance of this paper is two-fold. Firstly, this study presents results from a number of replicated velocity measurements taken in a natural river with a rough bed. This approach allows the derivation of reliable statistics for several key variables in the analysis of turbulence structures in a boundary layer above a gravel bed. Such statistics provide background data to be used in further comparisons of turbulent boundary layers. Secondly, and more importantly, the data analysis of the replicated measurements yields the most detailed characterization of the turbulent boundary layer over natural gravel bed rivers published so far. The data collection was designed to quantify through a statistical approach the structural features of the turbulence in the near-bed and outer regions respectively and to compare the results obtained from sensors of different sizes. The main conclusions of this study are the following:

1. There is a tendency for the turbulence to become less anisotropic from the near-bed to the outer region. This tendency is revealed by a decrease in the absolute value of the correlation between the u' and v' velocities and by the more uniform distribution of the residence times within the quadrants. Turbulence intensities, especially of the U component, also decrease from the bed to the surface. Such results were expected and they confirm previous findings by Robert *et al.* (1992; in press) and Kirkbride (1994) for similar gravel bed environments.
2. For a given sensor size, the scales of coherent flow structures are remarkably similar in the near-bed and outer regions. This is suggested by correlation analyses in the temporal and spatial domains and by the burst frequency analysis. The structures detected by these techniques appear to be large with respect to flow depth and asymmetrical: they extend 1.3 to 1.7 times the depth in the streamwise direction, if not more, and consequently, they are 3 to 4 times longer for the longitudinal velocity component than for the vertical component. These figures agree well with those published by Gordon & Witting (1977) for the longitudinal extent of the structures and Heathershaw (1979) for the ratio of scales for both velocity components. The intense shearing in this relatively high turbulence intensity (> 10%) depth-limited flow may entrain and cause deformation of the structures. One interpretation of the similarity in size of the structures measured in both regions of the flow is that they correspond to large-scale wedges of high- and low-speed fluid. These wedges were observed in turbulent boundary layers over smooth beds but they may also be characteristic of gravel-bed environments (Kirkbride, 1994; Ferguson *et al.*, this volume). The smaller-scale structures in the shear zones between the wedges or advecting from the rough bed are either too weak to contribute significantly to the velocity signal or too small to be detected by the ECM sensors. Even the smallest of the two sensors with a time constant of 0.05 s does not allow full

characterization of the small structures. In this study, the sampling rate fixed at 10 Hz decreases even further the likelihood of detecting the small features. The relationships between sensor and sampling characteristics and the scales of the flow structures are crucial because they dictate the limitations of the analysis. This point may look obvious but it is often forgotten when comparisons are made between studies. In this case, the similarity between the scales of the structures in the near-bed and outer regions indicating the dominance of large wedges may be due to the temporal and spatial resolutions of the measurements. These scales could not reveal the smaller structures which may be associated with vortex shedding in the near-bed zone. The impossibility of setting the ECMs very near the bed without introducing a severe bias (which is perceptible in some statistics of the large sensor in the near-bed region) is also a critical limitation.

3. The various burst detection techniques yield consistent results despite the uncertainty of the thresholds used. In the estimation of burst frequency, quadrant analysis appears very consistent and reliable but did not show any significant difference between the near-bed and outer regions. The statistics obtained for the large sensor are equal to 80% of those for the small sensor.
4. Quadrant analysis showed a clear dominance of the ejection (quadrant-2) and sweep (quadrant-4) events in both the near-bed and outer regions. Although ejection motion appears to dominate in the percentage of time spent in each quadrant, shear stress production in the near-bed region is clearly sweep dominated. This is consistent with observations by Nezu & Nakagawa (1993) and it confirms results obtained by Robert *et al.* (in press) at a smooth-to-rough transition in a natural gravel bed river. Also, Krogstad *et al.* (1992) have shown from measurements over a rough bed in a flume experiment that sweep events dominate shear stress production. The sweep dominance at the bed is consistent with the motion of high-speed wedges (Kirkbride, 1994).

ACKNOWLEDGEMENTS

The authors are grateful for the financial support given to this research by NSERC of Canada and for the equipment grant given by FCAR to AGR and Jim Best. The comments of Sean Bennett and of two anonymous reviewers were helpful in improving the paper.

REFERENCES

Biron, P., De Serres, B., Roy, A.G. & Best, J.L. 1993. 'Shear layer turbulence at an unequal depth channel confluence', in Clifford, N.J., French, J.R. & Hardisty, J. (Eds) *Turbulence: Perspectives on Flow and Sediment Transport*, John Wiley & Sons Ltd., Chichester, 197–213.

Bogard, D.G. & Tiederman, W.G. 1986. 'Burst detection with single-point velocity measurements', *J. Fluid Mech.*, **162,** 339–413.

Drake, T.G., Shreve, R.L., Dietrich, W.E., Whiting, P.J. & Leopold, L. 1988. 'Bedload transport of fine gravel observed by motion picture', *J. Fluid Mech.*, **192**, 2193–2217.

Deland, S. 1994. *La structure de la Couche-limite Turbulente en Rivière à Lit Graveleux*, Unpub. MSc Thesis, Université de Montréal, 115 pp.

Falco, R.E. 1977. 'Coherent motions in outer region of turbulent boundary layers', *Phys. Fluids*, **20**, S124–S132.

Gordon, C.M. & Witting, J. 1977. 'Turbulent structure in a benthic boundary layer', in Nihoul, J.C. (Ed) *Bottom Turbulence*, Elsevier, Amsterdam, 57–81.

Grass, A.J. 1971. 'Structural features of turbulent flow over smooth and rough boundaries', *J. Fluid Mech.*, **50**, 233–255.

Grass, A.J., Stuart, R.J. & Mansour-Tehrani, M. 1991. 'Vortical structures and coherent motion in turbulent flow over smooth and rough boundaries', *Phil. Trans. Roy. Soc. London.*, **A336**, 35–65.

Heathershaw, A.D. 1979. 'The turbulent structure of the bottom boundary layer in a tidal current', *Geophys. J. Roy. Astr. Soc.*, **58**, 395–430.

Kirkbride, A. 1993. 'Observation of the influence of bed roughness on turbulence structure in depth limited flows over gravel beds', in Clifford, N.J., French, J.R. & Hardisty, J. (Eds) *Turbulence: Perspectives on Flow and Sediment Transport*, John Wiley & Sons Ltd., Chichester, 185–196.

Kirkbride, A. 1994. *Turbulence Structures in Straight Gravel-bed Channels*, Unpub. PhD Thesis, University of Sheffield, 176 pp.

Krogstad, P.-Å, Antonia, R.A. & Browne W.B. 1992. 'Comparison between rough- and smooth-wall turbulent boundary layers', *J. Fluid Mech.*, **245**, 599–617.

Laufer, J. & Narayanan, M.A.B. 1971. 'Mean period of the turbulent production mechanism in a boundary layer', *Phys. Fluids,* **14**, 182–183.

Lapointe, M. 1992. 'Burst-like sediment suspension events in a sand-bed river', *Earth Surf. Processes Landf.*, **17**, 253–270.

Lapointe, M., De Serres, B., Biron, P. & Roy, A.G. in press. 'Using spectral analysis to detect sensor noise and correct turbulence intensity and shear stress estimates from EMCM flow records', *Earth Surf. Processes Landf.*

Levi, E. 1983. 'Oscillatory motion for wall-bounded turbulence', *J. Eng. Mech.*, **109**, 728–740.

Nakagawa, H. & Nezu, I. 1977. 'Prediction of the contributions to the Reynolds stress from the bursting events in open-channel flows', *J. Fluid Mech.*, **80**, 99–128.

Nakagawa, H. & Nezu, I. 1981. 'Structure of space-time correlations of bursting phenomena in open channel flow', *J. Fluid Mech.*, **104**, 1–43.

Nezu, I. & Nakagawa, H. 1993. *Turbulence in Open-channel Flows*, IAHR/AIRH Monograph series, A.A. Balkema, Rotterdam, 281 pp.

Offen, G.R. & Kline, S.J. 1974. 'Combined dye-streak and hydrogen bubble visual observations of a turbulent boundary layer', *J. Fluid Mech.*, **62**, 223–239.

Offen, G.R. & Kline, S.J. 1975. 'A proposed model of the bursting process in turbulent boundary layers', *J. Fluid Mech.*, **70**, 209–228.

Robert, A., Roy, A.G. & De Serres, B. 1992. 'Changes in velocity profiles at roughness transitions in coarse grained channels', *Sedimentology*, **39,** 725–735.

Robert, A., Roy, A.G. & De Serres, B. 1993. 'Space-time correlations of velocity measurements at a roughness transition in a gravel bed river', in Clifford, N.J., French, J.R. & Hardisty, J. (Eds) *Turbulence: Perspectives on Flow and Sediment Transport*, John Wiley & Sons Ltd., Chichester, 165–183.

Robert, A., Roy, A.G. & De Serres, B. in press. 'Turbulence at a roughness transition in a depth limited flow over a gravel bed', *Geomorphology*.
Robinson, S.K. 1990. 'Coherent motions in the turbulent boundary layer', *Ann. Rev. Fluid Mech.*, **104**, 387–405.
Soulsby, R.L. 1980. 'Selecting record length and digitization rate for near-bed turbulence measurements', *J. Phys. Oceanogr.*, **10**, 208–219.
Townsend, A.A. 1976. *The Structure of Turbulent Shear Flow*, Cambridge University Press, 429 pp.
Yalin, M.S. 1992. *Rivers Mechanics*, Pergamon Press, Oxford, 220 pp.

NOTATION

C_{aa}	correlation coefficient between two velocity signals for component a
D_j	percentile *j* of the bed material size distribution of bed material for the *b*-axis
F_b	burst frequency as defined by the number of detected events divided by the record length
ITS	integral time-scale
ILS	integral length-scale
$Q1$	represents the quadrant where $u' > 0$ and $v' > 0$
$Q2$	represents the quadrant where $u' < 0$ and $v' > 0$
$Q3$	represents the quadrant where $u' < 0$ and $v' < 0$
$Q4$	represents the quadrant where $u' > 0$ and $v' < 0$
r	correlation coefficient
t	time
u	instantaneous velocity of U (m s^{-1})
u'	velocity fluctuation of U (m s^{-1})
U	velocity component in downstream direction
$\overline{U}$	mean velocity of U for an individual signal (m s^{-1})
v	instantaneous velocity of V (m s^{-1})
v'	velocity fluctuation of V (m s^{-1})
V	velocity component in vertical direction
Y	mean flow depth at sampling position (m)
y	height above bed (m)
ρ	fluid density (kg m^{-3})
$-\rho u'v'$	mean Reynolds shear stress for a velocity signal (N m^{-2})
$-\rho u'v'_i$	mean Reynolds shear stress for the velocity signal in quadrant *i* (N m^{-2})
σ_u	root-mean-square of U for an individual signal (m s^{-1})
σ_v	root-mean-square of V for an individual signal (m s^{-1})
σ^2	variance of a variable
σ^2_k	variance of a variable in a window of *k* time-steps
$\langle\ \rangle$	mean of a variable for all replicates

8 Markov Analysis of Velocity Fluctuations in Gravel-bed Rivers

R.I. FERGUSON
Department of Geography, University of Sheffield, Sheffield, S10 2TN, UK

A.D. KIRKBRIDE
Department of Geography, University of Lancaster, Lancaster, LA1 4YB, UK

A.G. ROY
Département de Géographie, Université de Montréal, Montréal, H3C 3J7, Canada

ABSTRACT

We present novel state-transition analyses of turbulence structure above a gravel bed, based on measurements in Canadian and Scottish rivers using electromagnetic current meters. The analysis is of states defined by the positive or negative signs of the fluctuating parts of the streamwise (u') and vertical (v') velocity components. At a single height these define the familiar quadrants, which we analyse for four sites; when data are available for several heights simultaneously, as at two sites, the signs of either u' or v' at each height define instantaneous velocity profiles. The results show that v' reverses more frequently than u', and time-scales are similar to those found in previous work. In two rivers for which instantaneous velocity profiles are available, state frequencies and transition probabilities show small but statistically significant deviations from randomness, with a tendency for sweep-like top-down change from the state in which u' < 0 at all heights. Quadrant transitions in three of the four rivers show a statistically significant preference near the bed for transitions in a clockwise direction in the quadrant diagram, suggesting the passage of coherent eddies. These findings cast doubt on the realism of random-walk models of turbulent dispersion and transport in shallow rivers, and are interpreted as supporting the existence of separate coherent structures in the outer flow (quasi-periodic sweep-like events) and near the bed (intermittent eddy shedding from particle wakes).

INTRODUCTION

Coherent flow structures in gravel-bed rivers can be studied in simplified flume models or in natural rivers. Working in the field is undeniably realistic but is more difficult. Possibilities of flow visualization are extremely limited, measurements of velocity series are generally restricted to one or at most a few points at a time, and unsteady flow limits the ability to build up a comprehensive spatial picture by

Coherent Flow Structures in Open Channels.
Edited by P.J. Ashworth, S.J. Bennett, J.L. Best and S.J. McLelland.

sequential measurements at different points. Inferences about coherent flow structures are therefore made from much more limited data than in most flume experiments. In this paper we outline a novel approach to the analysis of turbulent velocity series based on Markov-chain analysis of transitions between states defined using the signs of the fluctuating parts of the streamwise and vertical velocity components. The potential of this approach is illustrated using single- and multiple-height measurements from four gravel-bed rivers. The results reveal some features that were not obvious from conventional analyses of velocity series using correlation and quadrant methods, and other features that are consistent with such analyses as reported in Roy *et al.* (this volume).

VELOCITY SERIES AS MARKOV CHAINS

Velocity measurements at-a-point in a river are discrete in time, with interval Δt, but continuous in value at least to the digital resolution of the logging system. In contrast, a Markov chain is a series that is discrete in value as well as time. The discrete values are termed states. The normal Markov assumption is that the state at time t depends stochastically on the state at time t-Δt, and thus indirectly on the state at times t-$2\Delta t$, t-$3\Delta t$, etc. The nature of the dependence is quantified through a transition probability matrix (TPM). This is a square matrix in which element ij is the probability of change from state i to state j over one time step; elements on the diagonal are thus probabilities of staying in the same state over one time step. If the TPM can be specified *a priori*, or estimated from observations, it is possible to derive many other properties of series behaviour including probabilities of transitions over longer time intervals, relative frequencies of total time spent in each state, and mean durations of stays in a particular state (e.g. Cox & Miller, 1965).

In our work we define states according to the positive or negative signs of the fluctuating parts of the streamwise and vertical velocity components in the familiar Reynolds decomposition:

$$u' = u - U \qquad (8.1)$$

in the streamwise direction, where u is instantaneous velocity and U its time-average, and

$$v' = v - V \qquad (8.2)$$

in the vertical (positive upwards). The positive or negative sign of u' indicates whether streamwise velocity is temporarily faster or slower than the time-average; the sign of v' indicates what can loosely be termed upwards or downwards flow ('up' and 'down' are relative to the mean streamline, which may be non-horizontal in some circumstances).

The familiar turbulence quadrant diagram (e.g. Lu & Willmarth, 1973) can be viewed as consisting of four Markov states defined by combinations of positive and negative signs of u' and v'. A time-series of quadrant values is thus a Markov chain, and later in this paper we analyse such series in terms of transition probabilities. But if u and/or v are measured simultaneously at more than one location, other Markov series can be defined according to the simultaneous signs of fluctuating velocity components at different places. We have discussed elsewhere (Kirkbride & Ferguson, 1995) Markov chains based on the instantaneous velocity profile at three measurement heights. The instantaneous profile of the streamwise velocity component has eight states defined by different combinations of $u' > 0$ or $u' < 0$ ('fast' or 'slow' flow relative to the time-average) at each of the three heights, and a time-series of these states is a Markov chain. Similarly, it is possible to define and analyse eight states of the instantaneous vertical profile of the vertical component of velocity ($v' > 0$ or $v' < 0$ at each height). The digital resolution of our logging system was such that u' and v' were never exactly zero. Another possibility, not pursued here, is to define states by the sign of the change in velocity over each time step (S. Bennett, pers. comm., 1995). In principle one could extend these approaches to more than three heights, or to quadrants at two or more heights simultaneously. However, as more states are defined the frequency of occurrence of each decreases if the number of observations in the series remains the same. This reduces the power of statistical tests involving state frequencies or counts of transitions between states. Markov analysis using many states therefore requires a long high-frequency record, as with more conventional correlation and spectral analyses. In this paper we restrict attention to 3- or 4-point instantaneous velocity profiles (8 or 16 states) and to quadrant series at single heights (4 states), generally using series of 10^3–10^4 values.

Expressing a velocity series as a Markov chain involves loss of quantitative information as velocity is reduced from a continuous-valued measurement to a discrete state. This is clearly only worthwhile if the subsequent analysis reveals features which are obscured, inaccessible, or overlooked in a conventional analysis. A Markov-chain approach affords simple but powerful statistical tests of deviation from randomness for a variety of definitions of what might constitute a 'random' turbulent flow. The results from this approach are therefore a potentially useful complement to conventional methods such as correlation analysis (Clifford, this volume; Roy *et al.*, this volume) in the phenomenological investigation of coherent flow structures, and also provide a test of assumptions about flow randomness made in random-walk models of Fickian turbulent transport and dispersion (e.g. Allen, 1982; Wallis, 1994).

FIELD MEASUREMENTS AND DATA SETS

All the data analysed here are field measurements from natural gravel-bed rivers using electromagnetic current meters (ECMs). A few measurements were made with a single Marsh-McBirney model 511 ECM with a 38 mm diameter probe attached to a wading rod and aligned to measure streamwise (u) and vertical (v) components of velocity at a height which could be adjusted, as sketched in Figure 8.1a. This ECM had a switchable time constant of 0.2 s and the output signal was logged at 4 Hz through an analogue-to-digital converter to a laptop computer. Most measurements were made with an array of up to four Marsh-McBirney probes mounted at different heights on a single wading rod (Figure 8.1b). One probe was of 38 mm diameter and the others 13 mm; the smaller probes have a smaller sensing volume and thus have greater ability to detect small flow structures, but all probes had the same signal conditioning (time constant 0.05 s, half-power frequency 3.2 Hz) and were logged synchronously at 10 or 20 Hz into a single digital data logger. One set of measurements used only three probes since one small probe was temporarily unserviceable. The use of small probes, and several probes at once, means these data are amongst the best available using ECMs in rivers.

Measurements were made in four gravel-bed rivers, three in Canada and one in Scotland (Table 8.1). Beauty Creek is a non-glacial tributary of the proglacial Sunwapta River in the Rocky Mountains, Alberta, Canada. The study site was in the centre of an almost straight reach about 20 m wide (see also Kirkbride, 1994, and Kirkbride & Ferguson, 1995). An array of three probes was used, with replicate measurements using two different sets of probe heights (BC1 and BC2 in Table 8.1). The Eaton North River in the Eastern Townships, Québec, Canada, is a slightly sinuous gravel-bed channel about 30 m wide; the study reach is the same as in Robert *et al.* (1993) and the measurements analysed here were made with the aim of comparing the response characteristics of large (EN1) and small (EN2) probes. Replicate measurements were made with each probe at each of two heights. The Mastigouche River drains part of the southern Laurentian hills, Québec. The study reach was at a 30 m wide riffle in a slightly sinuous reach. An array of four ECM probes was used to obtain a single, long series of simultaneous velocity measurements at four heights. The Allt Dubhaig is in the Scottish Highlands and has been described by Ferguson & Ashworth (1991); the turbulence measurements reported here were made in a straight reach about 10 m wide in the distal part of the stream (Kirkbride, 1994). A single ECM was used to make measurements at several different heights at each of two locations (AD1 and AD2) which had similar flow depth but different bed roughness.

In each river, measurements were made near the channel centreline at a relatively low discharge, when flow was shallow enough for easy wading (depths 0.24 to 0.55 m) but deep enough relative to the median bed-surface grain size (25 to 42 mm) that the flow structure was not dominated by individual obstacle wakes. The time-averaged, vertical velocity-profile was approximately logarithmic in each

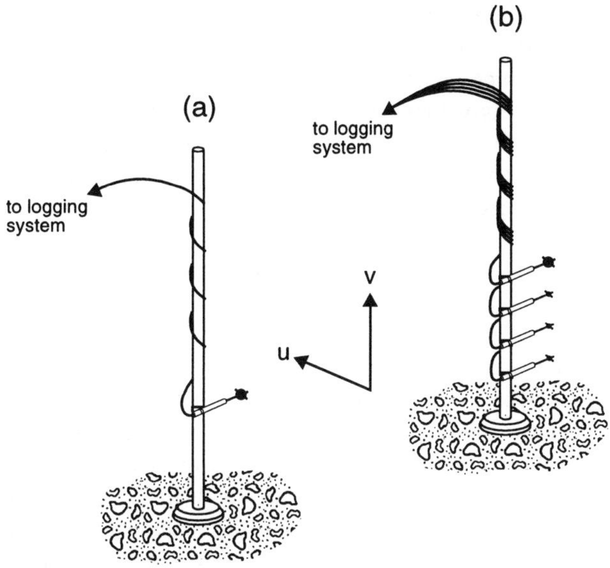

Figure 8.1. Schematic of instrument rig for turbulence measurements using (a) single electromagnetic current meter; and (b) array of synchronized electromagnetic current meters

Table 8.1. Field site characteristics and details of electromagnetic current meter and logger configuration used at each site

Data code	Name of river	Flow depth (m)	Bed D_{50} (mm)	ECM probe size	Probe heights above bed (m)	Type of analysis	Time interval Δt (s)	Series length (s)
BC1	Beauty Creek	0.55	26	SSL	0.03, 0.10, 0.30	V, Q	0.1	4 × 130
BC2	Beauty Creek	0.55	26	SSL	0.14, 0.20, 0.30	V, Q	0.1	4 × 130
EN1	Eaton North River	0.30	33	L	0.035, 0.20	Q	0.1	15 × 120
EN2	Eaton North River	0.30	33	S	0.035, 0.20	Q	0.1	15 × 120
MR	Mastigouche River	0.41	26	SSSL	0.04, 0.15, 0.25, 0.35	V, Q	0.05	1920
AD1	Allt Dubhaig	0.24	25	L	0.04, 0.06, 0.10, 0.14	Q	0.25	3 × 100
AD2	Allt Dubhaig	0.24	42	L	0.04, 0.06, 0.10, 0.14	Q	0.25	3 × 100

Probe sizes are S: small (13 mm), L: large (38 mm). Types of analysis are V: instantaneous velocity profile using simultaneous measurements at multiple heights, Q: quadrant transitions at single heights case (velocity was not measured at sufficient heights to detect small deviations), there were no protruding obstacle clasts or obvious clusters, and there was no obvious secondary circulation. The Mastigouche data consist of a single, very long record. In the other rivers velocity was logged for about 2 minutes at a time but with several replicates at each position (Table 8.1). The replicate series differed little and in some of our analyses the replicates are combined to create more reliable transition probability matrices.

RESULTS

Since the four study sites are all fairly similar in having quasi-uniform rough turbulent boundary layers (near-straight channels with width-to-depth ratio greater than 20, depth -to- grain-size ratio of about 10), we treat them collectively with the emphasis on common features. We discuss first the reversal probabilities of u' and v' in the various data sets, then the extent to which fluctuations in instantaneous velocity profiles show definite patterns and the nature of the coherent flow structures implied, and finally the non-random tendencies revealed by analysis of quadrant transitions.

REVERSAL PROBABILITIES

Following Kirkbride & Ferguson (1995) we define as p and q the probabilities of reversal in sign of u' and v' respectively over one time step of length Δt. Thus p is the probability that the streamwise velocity component changes from faster than its time-average value to slower, or from slower to faster, while q is the probability that the vertical component changes from up to down or down to up. Evidently the values of p and q may depend on the sampling interval Δt. We expect that both p and q tend to zero as Δt tends to zero, and that both p and q tend to 0.5 as Δt becomes very long so that there is no memory in the process. Between these extremes p and q should increase monotonically with Δt, but at a less than linear rate since the number of recorded reversals per second, $p/\Delta t$ or $q/\Delta t$, must decrease through loss of high frequency resolution. The possibility exists that p and q differ for a given sampling interval, although this has not been allowed for in random-walk models for turbulent transport and dispersion which have generally assumed temporal independence and spatial homogeneity (Wallis, 1994). The size of an ECM probe influences the high-frequency response of the measurement system and thus the inferred reversal probabilities. Roy *et al.* (this volume) found a 20% difference in turbulence intensity recorded by 13 and 38 mm probes near the bed, but the difference was negligible higher in the flow and there was no significant difference between probes in quadrant counts.

Table 8.2 summarizes the reversal probabilities estimated from each data set by counting the number of reversals in u' or v' and dividing by the total number of

Table 8.2. Sign-reversal probabilities in different data sets (see notes for Table 8.1); p and q denote probability of reversal of sign of u' and v' respectively over time Δt. Last two columns show mean intervals between reversals

Data set	Probe diameter (mm)	Δt (s)	p	q	$\Delta t/p$ (s)	$\Delta t/q$ (s)
BC1	38	0.1	0.15	0.26	0.7	0.4
	13	0.1	0.15	0.27	0.7	0.4
BC2	38	0.1	0.14	0.27	0.7	0.4
	13	0.1	0.13	0.27	0.8	0.4
EN1	38	0.1	0.11	0.23	0.9	0.4
EN2	13	0.1	0.14	0.28	0.7	0.4
MR	38	0.05	0.17	0.18	0.3	0.3
	13	0.05	0.16	0.18	0.3	0.3
AD1	38	0.25	0.16	0.29	1.6	0.9
AD2	38	0.25	0.26	0.35	1.0	0.7

time steps. Since the sampling intervals vary, the results are also expressed as the mean time interval in seconds between reversals. Three features are apparent. First, q is usually much larger than p: the vertical velocity component reverses about twice as often as the streamwise component. With data collection at 10 Hz ($\Delta t = 0.1$ s; BC and EN data sets), reversals of u' and v' occur at roughly 1 and 2 Hz respectively. The greater persistence in the u series is consistent with the findings of Robert *et al.* (1993) using correlation analysis in one of the same rivers (EN). However, the difference diminishes at slow sampling rates (AD data sets) and is small at the fastest sampling rate (MR). The second feature is that the mean time between reversals of both u' and v' decreases with increasing sampling rate, as expected when progressively higher-frequency turbulence is resolved. Thirdly, at a given sampling interval there is very little difference in reversal probabilities for small and large probes; this indicates that little bias will be introduced by including data from both sizes of probe in analyses of instantaneous velocity profiles or comparisons of quadrant sequences.

The typical reversal frequency of 1–2 Hz in our measurements is compatible with turbulence time-scales found by other workers using more conventional techniques to analyse gravel-bed river velocity series. Clifford *et al.* (1992) found a characteristic time-scale of 1–2 s by visual inspection and Box-Jenkins analysis, and Robert *et al.* (1993) found modal values of 1.2 s for u and 0.6 s for v, which implies a higher reversal rate for v' than u' as found in the present study.

INSTANTANEOUS VELOCITY PROFILES

The concept of the instantaneous velocity profile is described by Kirkbride & Ferguson (1995) with illustrations and inferences based on the Beauty Creek data set. Here we summarize the main results from Beauty Creek and discuss the extent to which similar features are apparent in our other multi-height data set from the Mastigouche River.

In order to derive testable null hypotheses for further analysis it is necessary to define a 'random' structure for Markov states of the instantaneous velocity profiles of u and v. We base it on three postulates:

1. the probability p that u' reverses sign over time Δt is constant in time and space, and in particular is the same at all heights;
2. the reversal probability q for v' is likewise constant; and
3. state transitions at different heights are independent.

The third postulate accords with random-walk models for turbulent transport, in which the random deflections at each point and time are spatially uncorrelated; however, these models are generally more restrictive in assuming temporally uncorrelated (e.g. white noise) disturbances, which amounts in our terms to $p = q = 0.5$. The other postulates are also very general and permit autocorrelation in the u and v series, possibly with different time and length scales for each.

This random model is introduced as a statistical null hypothesis which yields testable predictions. Three such predictions for an n-height profile are that:

1. the probabilities of reversal at $r = 0, 1, 2, ..., n$ heights during one time step are given by the binomial distribution: $\text{prob}(r) = \binom{n}{r} p^r (1-p)^{n-r}$ for the u' profile, and similarly for v' but with q instead of p;
2. each of the 2^n profile states has equal probability of occurrence; and
3. patterns of transitions should be spatially symmetric, e.g. if at one moment u' is positive at all n heights and a single reversal occurs over the next time step, it is equally likely to occur at each of the n heights.

The Beauty Creek results, described more fully in Kirkbride & Ferguson (1995), show systematic departures from each aspect of this random model. The binomial distribution gives a good fit to the number of heights at which u' or v' reverses in a time step (Figure 8.2), with differences of not more than 2% between observed and expected frequencies. However, given the large sample size these small discrepancies are statistically significant for the u' series, though not for v' (here, and subsequently, testing is by means of one-sample chi-square tests of observed versus expected counts, using a 5% significance level).

Secondly, the eight profile states differ significantly in frequency. The most frequent u' states are all-positive and all-negative, i.e. flow either faster than average at each height, or slower than average at each height; in the BC2 data these are occupied no less than 55% of the time, the other six states only 45% of the time. The most frequent v' states in the BC2 data are again all-positive or all-negative, which occurred 39% of the time with the other six states totalling 61%, but this tendency is less clear in the BC1 v' data. The least frequent states are those where the sign reverses twice along the profile.

Thirdly, transitions from the all-negative state of the u' profile are not spatially symmetric but show a statistically significant tendency towards top-down change

(Figure 8.3). Transitions from the all-positive (up) state of the *v'* profile also show a top-down tendency, which may indicate a sweep-like combination of acceleration and downwards declination of the velocity vector, but in this case the tendency is not statistically significant. Transitions in the opposite direction, from the all-positive (faster than time-average) state back towards all-negative, do not show such a clear pattern (Figure 8.3): the counts are again significantly unequal, but with

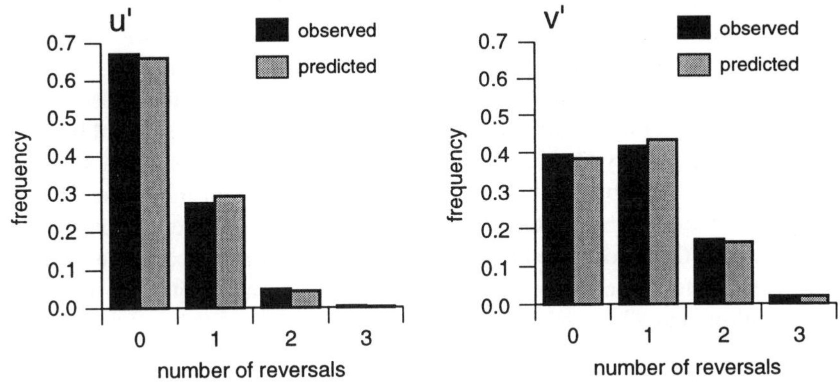

Figure 8.2. Binomial distributions of frequency with which *u'* and *v'* reverse sign at 0, 1, 2 or all 3 heights in the BC2 data set. Observed frequencies are based on measurements at 0.14, 0.20, 0.30 m above bed and time interval of 0.1 s. Expected frequencies are based on binomial model described in text with constant reversal probability $p = 0.129$ for *u'*, $q = 0.273$ for *v'*

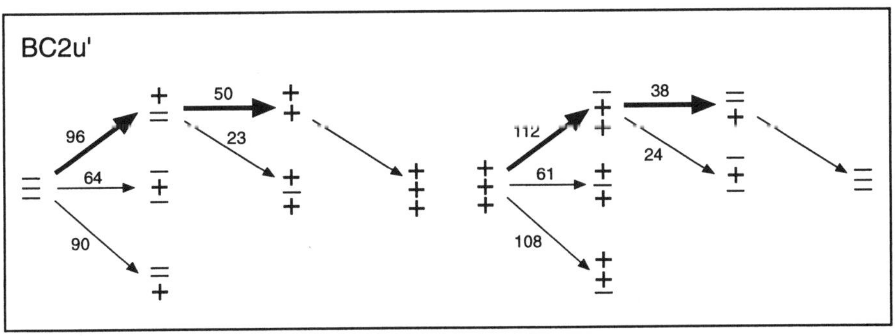

Figure 8.3. Tree diagrams showing numbers of observed transitions in BC2 three-height *u'* data. Left: changes from all-negative towards all-positive values of *u'*; right: from all-positive towards all-negative. Total number of transitions marked out of each state varies because alternatives shown exclude null transitions (state unchanged) and reverse transitions (return to previous state). Bold arrows are used for the most frequent transition where the frequencies of the alternatives differ significantly; they all show preference for top-down change

almost equal tendencies for top-down and bottom-up change rather than the bottom-up preference that might be expected if the classic burst-sweep cycle occurred in these rivers.

The Mastigouche River data are for four heights not three, so there are 16 profile states instead of 8, but analysis on the same lines is possible because long series were measured. The results resemble those from Beauty Creek in several respects but with some differences. The binomial model is again a very good fit for the number of heights at which reversal occurs during a time-step, in this case with no significant difference between observed and predicted frequencies. The state frequencies of both *u*' and *v*' are significantly different from a uniform distribution, but the pattern of deviation is less clear than in Beauty Creek; the all-negative and all-positive states of *u*' are amongst the commonest but the *v*' states differ less in frequency. As in Beauty Creek, the instantaneous profiles of *u*' show a statistically significant preference for top-down sequences of change from the all-negative state (Figure 8.4); the probability of the top probe turning from slower to faster than the time-average is 0.31, compared to only 0.20 for the bottom probe. Once again there are no clear preferred sequences of transitions from the all-positive *u*' profile, nor from the all-up and all-down states of the *v*' profile.

QUADRANT TRANSITIONS

Any coherent flow structure is likely to be represented in both the *u* and *v* components of velocity. Our analysis of instantaneous velocity profiles from Beauty Creek showed a clear sweep-like tendency in the *u*' data, and less clearly also for *v*', but the two velocity components were analysed separately so there is no

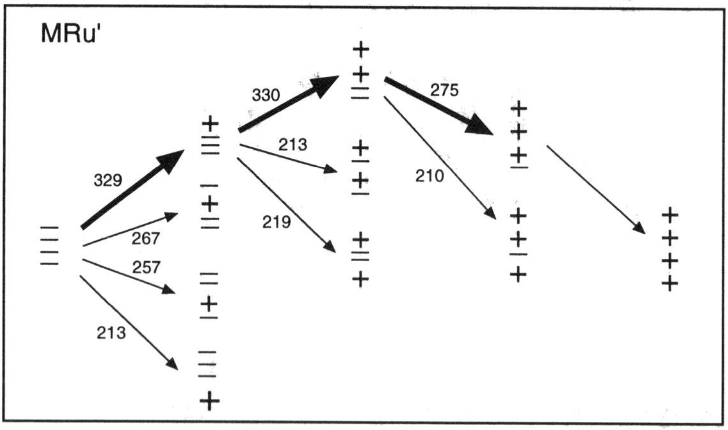

Figure 8.4. Tree diagram (as in Figure 8.3) of transition counts for 4-point instantaneous velocity profile of *u*' in Mastigouche River. As in Figure 8.3 there is a significant tendency for top-down change from the all-slow state

guarantee that the top-down preferred sequences in u' and in v' relate to synchronous events and a single coherent structure. Also, the higher reversal frequency of v' than u' implies that any structures identified in Markov chains of u' must contain additional structures involving shorter-term fluctuations in v'. In an attempt to learn more about coherent covariation in u and v we have investigated transition sequences of the quadrant states defined by simultaneous signs of u' and v'. Analysis of quadrant transitions at a single point allows detection of structured behaviour, and any difference in results from different heights in the same flow may indicate further systematic tendencies.

The simplest possible random model for quadrant transitions would be to assume not only that the reversal probabilities p and q for the signs of u' and v' are constant over time (postulate 1 of the previous section) but also that p and q act independently, with no correlation between the behaviour of u' and v'. This model is consistent with some random-walk models but it ignores the negative correlation that generally exists between u' and v', associated with the existence of a positive Reynolds shear stress. Testable predictions by this simplistic model include:

1. the four states (quadrants) are equally probable so should be approximately equally frequent in data series;
2. there should be only four different values of the transition probabilities, according to whether the transition involves reversal of the sign of u', v', both, or neither, as shown in Figure 8.5; and
3. symmetry in the transition probabilities in the sense that the probability of moving from quadrant i to j should be the same as that from j to i. This implies that probabilities of moving clockwise round the quadrant diagram (e.g. from quadrant 1 to 4, 4 to 3) are the same as probabilities in the anticlockwise direction (e.g. quadrant 3 to 4, 4 to 1); in view of prediction (1) this also means the total numbers of clockwise and anticlockwise transitions should on average be the same. In practice the negative correlation between u and v is associated with a tendency for the flow to occupy quadrants 2 and 4 more often than quadrants 1 and 3.

Inspection of quadrant transition counts for each data set confirms that quadrants 2 and 4 tend to be occupied more often than quadrants 1 and 3 (Table 8.3), though the difference is very small for the Mastigouche. The averages listed in Table 8.3 conceal a small but systematic vertical change in quadrant frequencies; this is discussed in Roy *et al.* (this volume). Analysis of quadrant transitions reveals that, again with the exception of the Mastigouche, there is substantial asymmetry in numbers of clockwise and anticlockwise transitions (Figure 8.6). Measurements made low in the flow show an excess of clockwise over anticlockwise transitions, whereas those higher in the flow show a more symmetric pattern, or in the case of the EN data an anticlockwise preference. Chi-square tests of total numbers of clockwise and anticlockwise transitions show significant departures from symmetry at both heights in the EN1 and EN2 data, the four lowest heights in the AD1 data, the two lowest for AD2, and the two lowest for BC1. Where there is a clear

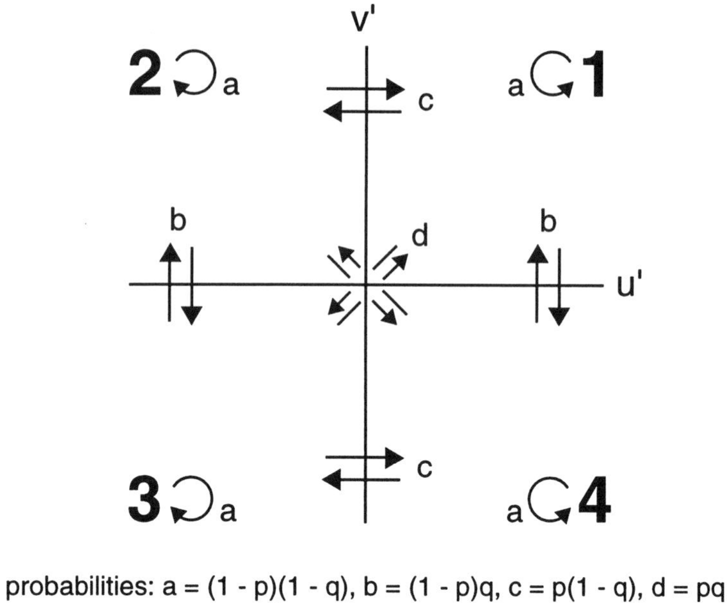

Figure 8.5. Quadrant diagram showing the four $u'v'$ states and all possible transitions between them, including null transitions where flow remains in same state. If probability p of reversal of sign of u', and probability q of reversal of sign of v', are constant and independent each transition has one of four probabilities as indicated by a, b, c or d

Table 8.3. Percentage of time spent in each quadrant in each river. Figures are aggregates over all measurement depths; in all cases differences between quadrants are more pronounced near the bed, less pronounced higher in the flow

River	Quadrant 1	Quadrant 2	Quadrant 3	Quadrant 4
BC1 & 2	22	27	21	30
EN1 & 2	18	33	18	31
MR	23	27	24	26
AD1 & 2	22	29	21	29

clockwise tendency it applies to all four pairs of quadrants but is more pronounced for transitions involving reversal of the sign of u' (between quadrants 1 and 2, or 3 and 4) than of v' (quadrants 1 and 4 or 2 and 3).

This pattern of asymmetry could be interpreted either as supporting the existence of coherent flow structures with different characteristics at different heights within the flow, or as evidence of the existence of two different sets of structures, one

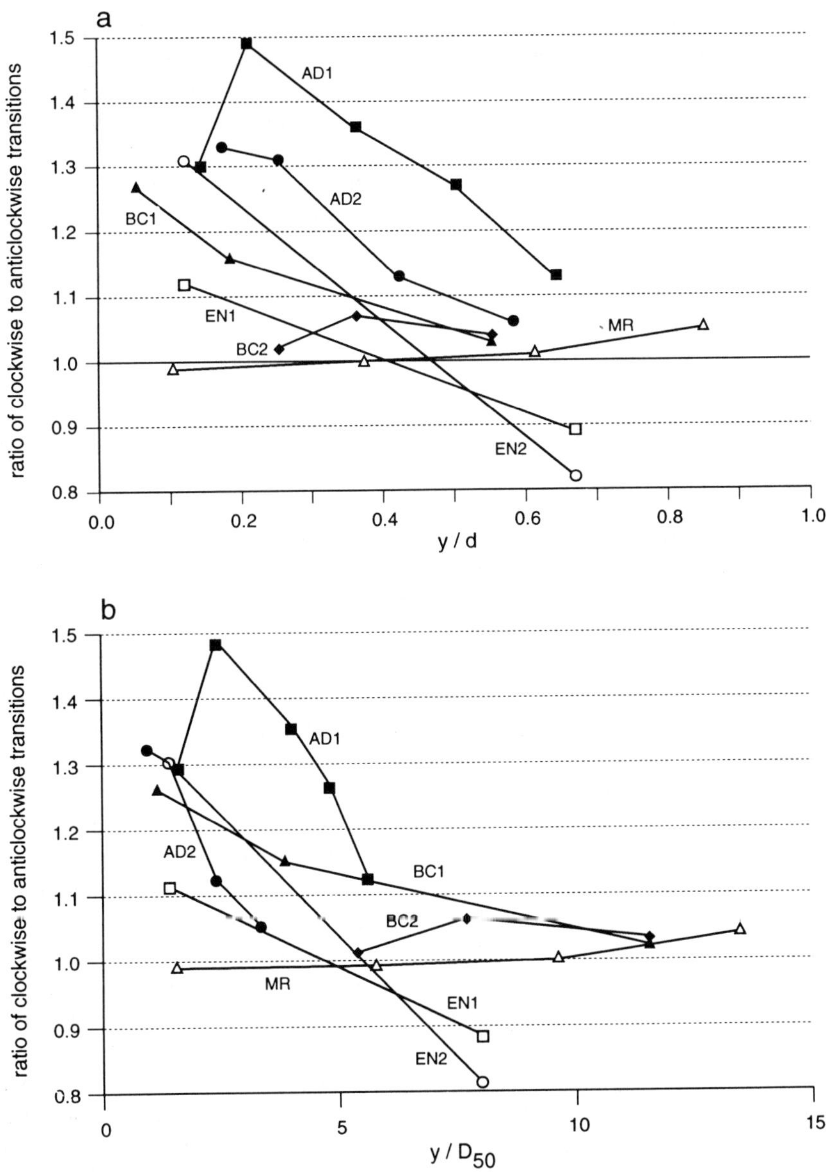

Figure 8.6. Asymmetry in quadrant transitions in measurements from four gravel-bed rivers. Ratio of clockwise to anticlockwise transitions is plotted against height above bed, scaled by (a) flow depth or (b) median bed-material diameter. Tendency for more clockwise transitions near bed is discussed in text

with little or no quadrant transition asymmetry in the outer flow and another near the bed which is biased in favour of clockwise transitions. In the first case symmetry characteristics would presumably alter with absolute height (y) above the bed or height relative to flow depth (y/d); in the second case the scaling should be with height relative to bed obstacle size, e.g. y/D_{84} or y/D_{50}. Figures 8.6a and 8.6b are scaled by d and D_{50} respectively. Because our study sites do not differ greatly in relative roughness it is not clear which scaling is better. Statistically significant clockwise tendencies are restricted to the lower one third or so of the flow depth, and to heights up to about 3 to 5 times the D_{50} diameter, or about twice D_{84}. We discuss possible explanations below.

DISCUSSION

Analysis of the fluctuating velocity profile reveals structures which decay in coherence towards the bed, whereas analysis of the quadrants suggests (different) structure near the bed which decays higher in the flow. These results are not contradictory. Both are consistent with the conceptual model of coherent flow structures over spatially complex rough beds that was suggested by Kirkbride (1993, 1994) and Kirkbride & Ferguson (1995). In this conceptual model there is one coherent structure in the outer flow and another, wake-dominated, near the bed, with only limited interaction, e.g. when sweep-like outer-flow structures temporarily squash near-bed wakes, or eddies are intermittently shed from large clasts giving an opportunity for outer-flow eddies to approach the bed. The existence of a near-bed wake zone in gravel-bed streams had previously been suggested by Nowell & Church (1979), Clifford *et al.* (1992), and Nakagawa *et al.* (1993). We suggest that the two different techniques of Markovian analysis reported in this paper are complementary and both bring out aspects of the difference in structure between the outer and near-bed flow. This discussion will focus on what the results of these two types of analyses can tell us about the spatial and temporal characteristics of the structures, and the mechanisms by which they form.

The main result from the analysis of transitions in instantaneous velocity profiles in Beauty Creek and the Mastigouche River is the significant preference for top-down change as the flow speeds up from the state with $u' < 0$ at all heights. In Beauty Creek there is also a significant tendency for top-down change from the state with $v' > 0$ at all heights. These findings suggest sweep-like behaviour in which relatively slow and upwards-directed flow is progressively replaced downwards through the profile by faster and downwards-directed flow. Kirkbride & Ferguson (1995) interpreted this as supporting a streamwise alternation of faster-than-average and slower-than-average flow zones occupying the full flow depth with the possible exception of a wake-dominated near-bed layer, the front of each fast zone being inclined with its upper part leading its lower part. Such a 'wedge' structure has been suggested by many previous authors (e.g. Nakagawa & Nezu, 1981; Levi, 1983;

Yalin, 1992), although these works do not relate to flows over complex rough gravel beds. The present results, in conjunction with those of Kirkbride & Ferguson (1995) and Roy *et al.* (this volume), suggest that wedge structures are common in flows over hydraulically rough, spatially complex, mixed-size gravel beds. The main characteristics of wedge fronts were described by Kirkbride & Ferguson (1995) as: (i) progressive acceleration of streamwise velocity from the top downwards in a sweep-like manner; (ii) fairly incoherent decay over time to the lower speed quasi-stable state; (iii) a progressive bedwards declination of the $u'v'$ vector which again is sweep-like in character; and (iv) a decay in coherence towards the bed. The sweep-like part of the process is probably important for bed entrainment. The pattern of velocity speed-up is clearer in the BC2 data than for BC1 and MR, where the bottom ECM probe was much nearer to the bed. This suggests the high-speed fronts are only coherent in the outer flow zone, and tend to be suppressed near the bed by eddy shedding from obstacle clasts. There is inevitably some transfer of mass, momentum and vortical structure between the zones, even though it may be intermittent. The nature of the interaction between the zones is discussed further later.

The other main result of our Markov-chain analyses is the tendency in the near-bed zone for more clockwise than anticlockwise transitions between quadrants. This asymmetry has an interesting parallel with the increased near-bed asymmetry in quadrant contributions to Reynolds stress that was found by García & López (1995) in rough-bed flume experiments, and by Roy *et al.* (this volume) in gravel-bed rivers analysed by methods complementary to those used here. Hydrodynamically, a clockwise transition sequence starting from quadrant 2 and passing through quadrants 1, 4, 3 back to 2 implies a streamwise acceleration followed in turn by declination, deceleration, then inclination. The hypothesis that streamwise acceleration is associated with the input of higher momentum fluid from higher in the velocity profile would, by definition, result in a diagonal transition in the quadrant diagram (as is the tendency in the high-speed front zone). Rotation in the quadrant diagram implies a more complex two-dimensional transfer of momentum, and is more likely to be related to the passage of eddies.

A tendency for clockwise, or for that matter anticlockwise, transitions between quadrants does not inevitably imply non-zero spanwise vorticity (since this vorticity depends on the magnitude as well as the sign of u' and v'), but it strongly suggests it. Our result therefore invites comparison with conceptual models of spanwise eddies in turbulent velocity profiles, for example the classic Prandtl mixing-length model. If momentum is transferred down the profile by eddies which are advected with the flow and have downwards flow at their leading edges, the passage of these eddies past a fixed current meter will give a pseudo-periodic sequence of quadrant transitions. If such an eddy is advected horizontally past a current meter, the sequence will be clockwise if the current meter is below the centre of the eddy but anticlockwise if above the centre of the eddy. This could be seen as an explanation of a clockwise tendency close to the bed. However, for several reasons this does not appear to explain our findings in a complete and consistent way. If the velocity profile is maintained by a series of horizontally-advected eddies one above another, there would not be a simple distinction between clockwise transitions near the bed and anticlockwise higher up, but a vertical alternation between a clockwise tendency in the lower half of each eddy and an anticlockwise tendency in the upper half of each eddy. If the profile is maintained by eddies which are advected upwards, some should pass just above and others just below a current meter so that no net clockwise or anticlockwise tendency should be observed. And it is not obvious how a simple Prandtl-type model can explain why v' reverses sign almost twice as often as u' in several of our data sets. However, both this and the clockwise tendency can be explained if there are spanwise eddies within the slow zones beneath the high speed fronts discussed earlier; Figure 8.7 shows schematically how this might work.

The temporal pattern of change of $u'v'$ signatures (i.e. pathways through the quadrant diagram) can be predicted for advecting eddies of any type, not just simple Prandtl eddies. This requires a knowledge of the direction of eddy advection/convection, the position of the probe relative to the eddy, and the speed of the eddy relative to the ambient fluid velocity. It is possible to derive $u'v'$ signatures for various permutations of a frozen eddy passing a monitoring probe: probe above, within and below the eddy; eddy advection velocities above, equal to and below streamwise velocity; and the eddy being advected in the mean streamwise direction and at 30° in the vertical. The combinations we have tried show that clockwise quadrant pathways would only be recorded by probes *below* eddies being advected in the streamwise direction or being ejected. All other permutations result

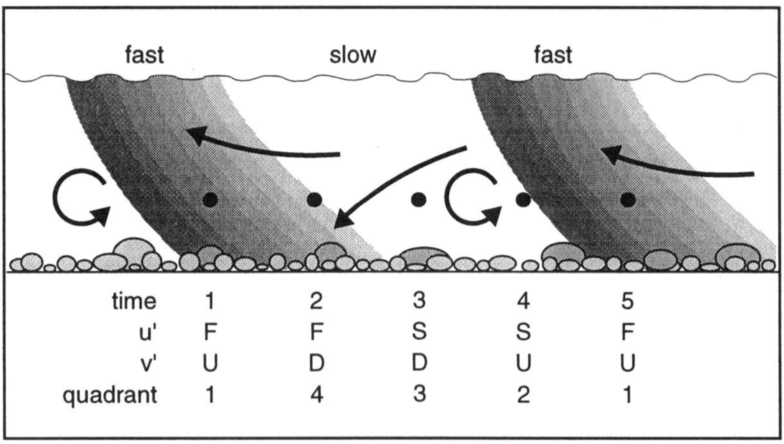

Figure 8.7. Schematic diagram of possible generation of clockwise quadrant transitions near the bed in turbulent flow in gravel-bed rivers. Outer flow is dominated by inclined high-speed fronts, but near-bed flow contains spanwise eddies beneath these fronts. See text for discussion of wake-shedding or other source of these eddies

in anti-clockwise or diagonal quadrant transitions, and the passage of a high-speed front produces diagonal transitions.

This supports the proposal that the near-bed region is dominated by the passage of structures with spanwise vorticity. Nearer the bed, the increased tendency for clockwise transitions suggests that those eddies resolved by the 13 mm sensor are mostly passing over the probes. If the intense mixing near a rough bed destroyed structures, quadrant transitions ought to be symmetric. If the dominant eddies were those shed by obstacles, we would expect evidence of symmetric or anticlockwise tendencies as the eddies either envelope the probe or pass beneath. The evidence does not support either of these hypotheses, so we conclude there is an extra factor: spanwise vortical structures which have origins not directly related to eddy shedding. Proponents of the 'wedge' or high-speed front model outlined above (e.g. Yalin, 1992) have suggested that shear associated with the front induces the roll-up of spanwise vorticity. The increased tendency for clockwise quadrant transfer towards the bed could therefore be interpreted either as an increased tendency for roll-up of the front nearer to the bed, or, more probably, the preferential "seeding" of roll-up of the front by the sub-coherent vortical remnants of eddies shed from the bed obstacles. Either way, it seems likely that the zone within which there is a tendency for clockwise quadrant transitions extends to a height which scales with bed roughness rather than flow depth.

As pointed out at the start of this paper, inferences about coherent flow structure in natural rivers have to be made from very limited evidence. The use of an array of ECMs, as in the work of Robert *et al.* (1993) and the Beauty Creek and Mastigouche studies reported here, yields additional information which cannot be obtained by deploying a single instrument at different heights one after another. Reducing the u and v time-series at each height to Markov states for analysis of transitions in instantaneous velocity profiles, and in quadrant states, has revealed some fairly consistent large-scale coherent motion in flow over gravel beds. The results are concordant with other recent work on such flows. It would nonetheless be valuable to do similar analyses of multi-height measurements in rivers of different relative roughness, and in a laboratory situation in which statistical properties of velocity could be compared with evidence from flow visualization. Another avenue of research would be to subdivide each quadrant at a certain hole value and analyse 8-state quadrant transitions to single out coherent structures making important contributions to Reynolds stress.

ACKNOWLEDGMENTS

Fieldwork was supported by grants from the Royal Society (to Ferguson for Alberta fieldwork), NSERC (to Roy for Canadian fieldwork), and NERC (studentship to Kirkbride including Scottish fieldwork). We also thank NSERC and FCAR (Roy), and Sheffield University (Ferguson and Kirkbride), for equipment grants and computing facilities. Pascale Biron (Montréal) assisted with some of the fieldwork. Sean Bennett made helpful review comments on the presentation of our methods and interpretations. We thank Paul Coles (Department of Geography, University of Sheffield) for the diagrams.

REFERENCES

Allen, C.M. 1982. 'Numerical simulation of contaminant dispersion in oscillatory flows', *Proc. Roy. Soc. London,* **A381**, 179–194.

Clifford, N.J., Robert, A. & Richards, K.S. 1992. 'Estimation of flow resistance in gravel-bedded rivers: a physical explanation of the multiplier of roughness length', *Earth Surf. Proc. Landf.*, **17**, 111–126.

Cox, D.R. & Miller, H.D. 1965. *The Theory of Stochastic Processes*, Methuen, London, 398 pp.

Ferguson, R.I. & Ashworth, P.J. 1991. 'Slope-induced changes in channel character along a gravel-bed stream: the Allt Dubhaig, Scotland', *Earth Surf. Proc. Landf.*, **16**, 65–82.

García, M. & López, F. 1995. 'Turbulent coherent structures in cobble-bed open channel flow', in Ashworth, P., Best, J.L., Bennett, S.J. & McLelland, S.J. (Eds) *Coherent Flow Structures in Open Channels: Abstract Volume*, Leeds, 16.

Kirkbride, A.D. 1993. 'Observations of the influence of bed roughness on turbulence structure in depth limited flows over gravel beds', in Clifford, N.J., French, J.R. & Hardisty, J. (Eds) *Turbulence: Perspectives on Flow and Sediment Transport*, John Wiley & Sons Ltd., Chichester, 185–196.

Kirkbride, A.D. 1994. *Turbulence Structure in Straight Gravel-bed Channels*, Unpub. PhD thesis, Univ. of Sheffield, 176 pp.

Kirkbride, A.D. & Ferguson, R.I. 1995. 'Turbulent flow structure in a gravel-bed river: Markov chain analysis of the fluctuating velocity profile', *Earth Surf. Proc. Landf.*, **20**, 721–733.

Levi, E. 1983. 'Oscillatory motion for wall-bounded turbulence', *J. Eng. Mech.*, **109**, 728–740.

Lu, S.S. & Willmarth, W.W. 1973. 'Measurements of the structure of the Reynolds stress in a turbulent boundary layer', *J. Fluid Mech.*, **60**, 481–511.

Nakagawa, H. & Nezu, I. 1981. 'Structure of space-time correlations of bursting phenomena in open channel flow', *J. Fluid Mech.*, **101**, 1–43.

Nakagawa, H., Tsujimoto, T. & Shimizu, Y. 1993. 'Turbulent flow with small relative submergence', in Armanini, A. & di Silvio, G. (Eds) *Fluvial Hydraulics of Mountain Regions*, Springer, Berlin, 33–44.

Nowell, A.R.M. & Church, M. 1979. 'Turbulent flow in a depth-limited boundary layer', *J. Geophys. Res.,* **C84**, 4816–4824.

Robert, A., Roy, A.G. & De Serres, B. 1993. 'Space-time correlations of velocity measurements at a roughness transition in a gravel-bed river', in Clifford, N.J., French, J.R. & Hardisty, J. (Eds) *Turbulence: Perspectives on Flow and Sediment Transport*, John Wiley & Sons Ltd., Chichester, 165–184.

Wallis, S.G. 1994. 'Simulation of solute transport in open channel flow', in Beven, K. J., Chatwin, P.C. & Millbank, J.H. (Eds) *Mixing and Transport in the Environment*, John Wiley & Sons Ltd., Chichester, 89–111.

Yalin, M.S. 1992. *River Mechanics*, 2nd edn., Pergamon, Oxford, 220 pp.

NOTATION

d	depth of water
D D_{50} D_{84}	diameter of bed sediment, where subscript indicates percentile of grain-size distribution
n	number of heights in flow at which velocity is simultaneously measured
p	probability of reversal in sign of u' over one time step
q	probability of reversal in sign of v' over time Δt
r	number of heights at which u' or v' reverses over time Δt
u	instantaneous flow velocity in streamwise direction
u'	fluctuating part of flow velocity in streamwise direction
U	mean flow velocity in streamwise direction
v	instantaneous flow velocity in vertical direction, positive upwards
v'	fluctuating part of flow velocity in vertical direction
V	mean flow velocity in vertical direction
y	height above bed
Δt	time interval between velocity measurements

9 The Application of Particle Image Velocimetry to the Study of Coherent Flow Structures over a Stabilizing Sediment Bed

S.J. TAIT
Department of Civil and Structural Engineering, University of Sheffield, Sir Frederick Mappin Building, Mappin Street, Sheffield, S1 4DU, UK

B.B. WILLETTS & M.W. GALLAGHER
Department of Engineering, University of Aberdeen, Fraser Noble Building, Aberdeen, AB9 2UE, UK

ABSTRACT

Research into the fundamental structure of turbulent boundary layers has revealed the existence of organized vortical-type structures. The aim of the work reported here was to conduct an experiment which would attempt to identify, measure and then examine the influence of these structures on the entrainment of grains from a non-cohesive sediment bed.

It was believed that one of the factors which may control the grain entrainment rate is the temporal and spatial distribution of these flow structures close to the bed. In order to examine the influence of the spacing and persistence of these structures on sediment transport rate it was necessary to obtain simultaneous, spatially distributed measurements of the turbulent flow field close to the bed. Particle image velocimetry was used to produce quantitative measurements of changes in the near-bed velocity field over a specified flow area whilst a sediment bed stabilized.

Observations are reported from a single experiment in which the spatial organization of the flow region adjacent to a sediment bed was measured. Measurements were made at two different times but at the same position above the bed. This was done in order to investigate the differences in the near-bed flow organization when there was: (i) significant bedload transport; and (ii) essentially no bedload transport. Vector maps of the near-bed flow field indicated the presence of flow features which were believed to be important in the grain entrainment process. Although a limited amount of data have been examined, it did suggest differences in the distribution and frequency of flow features of a particular size as the sediment transport rate dropped and the sediment surface stabilized.

Coherent Flow Structures in Open Channels.
Edited by P.J. Ashworth, S.J. Bennett, J.L. Best and S.J. McLelland.

INTRODUCTION

Change in the form and character of any river channel is a consequence of bedload transport. The problem of defining the critical bed and flow conditions associated with the threshold of motion and subsequent sediment transport is therefore of some significance. In natural rivers, gravel beds with non-uniform grain sizes have been observed to exhibit wide variations in the threshold conditions for bedload transport (Reid *et al.*, 1985). Hence, there does not appear to be a direct and unique relationship between the prevailing flow conditions and the initial movement of sediment

Traditionally, bedload transport rates and incipient motion have been determined by the comparison of the imposed time-averaged fluid shear stress with empirically measured values of a reference critical shear stress for all grains within a particular grain-size fraction (Shields, 1936; Meyer-Peter & Müller, 1948). These empirically calibrated relationships of mean time-averaged values of the imposed flow and sediment bed variables have been insufficient to explain the wide scatter in the experimental data observed. The aim of the work presented here has been to examine the degree of variability of the fluid's physical processes within the near-bed flow environment and its possible bearing on the prediction of the entrainment of sediment grains from a non-cohesive sediment bed.

Sediment entrainment occurs when the mobilising fluid forces exceed the grain forces resisting motion. Direct measurements have indicated that mixed grain-size sediment beds exposed to the action of water organize their surface topography to enhance their stability. In channels with mixed grain-size beds, the resulting bedforms have been described as coarse grain threads arranged in an irregular diamond pattern and have been observed in the field and in the laboratory (Wolcott, 1990; Tait & Willetts, 1992). This organization of surface topography is believed to give rise to spatial variations in the local critical shear stress required to affect grain entrainment. Kirchner *et al.* (1990) measured the surfaces of mixed grain-size sediment beds with different surface grain layouts and reported that different bed topographies with comparable grain-size distributions had a marked effect on the value of the critical shear stress calculated using a relatively simple analytical model. They realized that a major assumption in their analysis was the use of mean values of flow velocity and fluid shear stress to represent the effect of the near-bed flow region. Although the force-balance relationship aimed to study the variability in the resistance forces of grains against entrainment, it did not account for any variability within the flow. Kirchner *et al.* stated that there was an urgent need for a more complete understanding of the flow field close to rough sediment beds since the complex topography of water-worked mixed grain-size sediment beds bear little resemblance to the more uniform and ordered spacing of roughness elements and sediment normally found in fluid dynamics studies (Grass, 1971; Nowell & Church, 1979). Comparison of field and laboratory data collected by Robert *et al.* (1992) indicated similarities in the response of the mean velocity gradient and roughness length to a sudden increase in the surface roughness in both the field and laboratory.

It is clear that the size and range of the turbulent fluctuations in flow are important in determining the entrainment of sediment. Grass (1970) conceptualized the process of grain entrainment as the intersection between two probability distributions. These distributions represent: (i) the resisting forces of the grains on a bed surface; and (ii) the time-varying fluid forces applied to the bed by the near-bed turbulent flow field. Kirchner *et al.* (1990) believed that the first distribution was related not only to the grain-size distribution but also to grain arrangement within different surface geometries. Changes in the nature of either or both of these distributions can cause large differences in their areal overlap representing the grain entrainment rate.

In previous work using fine sediment, beds of low relief were formed with a characteristic pattern and scale suggesting that the entrainment process is influenced by both temporal and spatial variability in the turbulent flow fluctuations (Best, 1992). Analysis of a large number of moving grains in successive plan view images of a gravel-bed river by Drake *et al.* (1988) produced detailed quantitative information on the typical rates and modes of grain entrainment. They found that grain entrainment was predominantly caused by fluid forces rather than by grain impact with the bed. Analysis of the collective motions of particle groups indicated that most grain entrainment events occurred almost simultaneously in small localized areas of the bed. Although no detailed measurement of the near-bed flow field was conducted, Drake *et al.* postulated that these transport events, which accounted for 70% of the total bedload transported, were caused by turbulent fluid sweeps and termed this type of entrainment 'sweep-transport'. Gyr *et al.* (1989) described various surface textures of a self-stabilizing sand surface in terms of differences in the turbulent flow fluctuations measured at a single point and attempted to relate these changes to the scales of the evolving sediment patterns that were formed as the bed stabilized. These observations have indicated that the spatial organization of the sediment bed surfaces may be related to a spatial pattern found in the near-bed flow environment.

THE NEAR-BED FLOW FIELD AND GRAIN ENTRAINMENT PROCESSES

The dynamics of sweeps and bursts within turbulent boundary layers over rough walls reported by Grass (1970) were made by observing their interaction with a highly mobile sand-bed, using sediment entrainment as an indicator of the presence of coherent flow structures. The entrainment of fine sediment was therefore believed to be linked to the sweep impact with the bed (Grass, 1970; Gyr *et al.*, 1989). The characteristic pattern created in the sediment indicated a streak spacing similar to the fluid streak spacing. From flow visualization experiments, Best (1992) suggested that the characteristic dimensions of the bed defects may be larger than the fluid streak spacing and that groups of related sweeps and ejections may be responsible for the pattern of sediment entrainment. It was also noted that the long temporal persistence of low-speed streaks was greater than the more frequent bursts and sweeps.

While descriptions of turbulence over smooth plates may be useful when investigating the significant flow features in that idealized situation, questions arise as to whether a similar hierarchy of flow patterns can be found in natural river channels. It may be assumed that the structure of the inner-wall region is different for the smooth and rough fixed-wall cases. Over smooth walls, the momentum exchange is produced by the variability within the viscous-dominated layer close to the boundary whilst over a rough surface the momentum exchange is promoted by the pressure field created by roughness elements. Therefore, the energetic near-bed flow structures associated with the roughness elements would be expected to scale with the roughness elements rather than the inner-layer parameters (ν kinematic fluid viscosity and u_* shear velocity at wall). Thus, the structures within the boundary layer may appear similar but their formation processes are quite different. The influence of sediment moving close to a solid boundary on the formation, movement and dissipation of flow structures has not been rigorously quantified. Best (1993) stated that two of the key issues requiring research when considering the interaction between flow and sediment transport were the influences of surface roughness and mobile sediment on the turbulence structure of the flow.

The generally accepted model of the turbulent boundary layer has as its foundation the presence of streamwise streaks of relatively slow flow next to the bed which are then uplifted away from the boundary into an outer-flow region in which there are larger coherent flow structures. Observations of change in averaged flow variables such as shear velocity, hydraulic roughness height and Reynolds stress have indicated that the near-bed flow can dramatically change in response to variations in surface roughness (Mulhearn & Finnigan, 1978; Nowell & Church, 1979; Robert *et al.*, 1992). The influence of distinct surface roughness patterns in changing near-bed flow structure and, therefore, the ability of flow to entrain sediment at localized sites needs to be investigated. Best (1992) reported on experimental measurements of streak spacing and persistence over a smooth bed with longitudinal ridges at a lateral spacing of approximately 80 wall units (wall unit = yu_*/ν, where y is the distance from solid boundary). These observations indicated that the fixed ridges influenced the near-bed flow sufficiently to concentrate the low speed streaks either over or close to the ridges. The organization of turbulence in rivers is likely to differ from that in laboratory studies involving surfaces of uniform roughness. It is necessary to examine if the roughness pattern as well as the roughness height are important in determining the extent of turbulent flow organization.

A detailed description of both the temporal and spatial variation of near-bed turbulence promises to explain the spatial variability previously observed in the entrainment of sediment as described earlier. It is valuable to be able to classify flow features by size and frequency in order to estimate which type has most influence on sediment entrainment. The study of entrainment processes at the grain scale is important if numerical models are to be developed to predict sediment movement using descriptions of physical processes without strong dependence on empirical data. Numerical models have been developed to simulate the transport of sediment by examining the forces found during the trajectories of numerous grains using the fundamental equations of motion, but these models still rely on simplified descriptions of grain interactions and entrainment (Sekine & Kikkawa, 1992; Haff *et al.* 1993). The work reported here attempts to obtain an accurate description of the flow environment at entrainment in order to explore the physical processes which effect entrainment and explain the errors found in current grain transport models.

EXPERIMENTAL WORK

Turbulent flow patterns are not fully random, but consist in some part of flow structures with a spatial and temporal coherence. Earlier studies have observed these flow patterns using flow visualization techniques and attempts have been made to measure their size and influence using single-point velocity measurement techniques such as hot-wire and laser Doppler anemometry. However, these attempts have been hindered by interpretation difficulties encountered when trying to relate single-point measurements to spatial flow patterns. Therefore, it was decided to use a flow measurement technique which would measure simultaneously the flow velocity at many points and be able to accumulate these measurements over a period of time through successive coherent structures. Particle image velocimetry (PIV) is such a technique. This technique yields quantitative information on the instantaneous flow pattern in a plane which produces direct information about the spatial coherence of any flow pattern or structure. The spatial distribution of various quantities can be calculated, such as the flow velocity and the out of plane component of vorticity which is relevant to the identification and the dynamics of coherent flow structures (Hussain, 1986). Gyr (1982) believed that the frequency of large flow structures which pass a fixed point in the boundary layer is related to the frequency of suitably defined high-stress events at that point. It is proposed to use this measurement technique to examine the size and persistence of the coherent flow structures which form during a period of substantial transport and then those during a period of minimal bedload transport. This comparison will determine whether change in the spatial structure of the near-bed flow contributes significantly to the decrease in sediment mobility.

EXPERIMENTAL SET-UP

A flow area above a sediment bed was examined and the instantaneous flow velocities over that area quantified at two points in time: (i) during a phase of intense sediment transport; and (ii) later when the bedload transport rate had dropped dramatically.

The experiment was carried out in a recirculating, tilting glass-sided flume 12.5 m long and 0.3 m wide. The slope of the flume was set at 0.0013 and the flow rate was held constant at 7.0 l s^{-1}. The water depth was controlled by an adjustable, sharp-edged weir at the downstream end. The bedload transport rate was measured using a slot-type trap at 10.5 m from the flume inlet. The base of the flume was covered with a level surface layer of thoroughly mixed, poorly-sorted sediment which had a median diameter of 1.64 mm and a sorting coefficient ($\sigma_g = (D_{84}/D_{16})^{1/2}$, where D_{84} and D_{16} are the 84th and 16th percentiles of the grain-size distribution) equal to 2.67. A steady flow was introduced and the tail weir was adjusted to extend the uniform depth of flow to as large a reach as possible. During the experiment measurements were made of the water depth, bedload transport rate and of the flow velocity field close to the bed on the centreline at 8.5 m from the inlet. Initially the average water depth was 63.6 mm and the wall shear velocity was 0.011 m s^{-1}. The flow field measurements were taken at 25 and 275 minutes, corresponding to periods of high and low transport rate respectively (Figure 9.1). The bed exhibited no prominent bed morphology during the experiment.

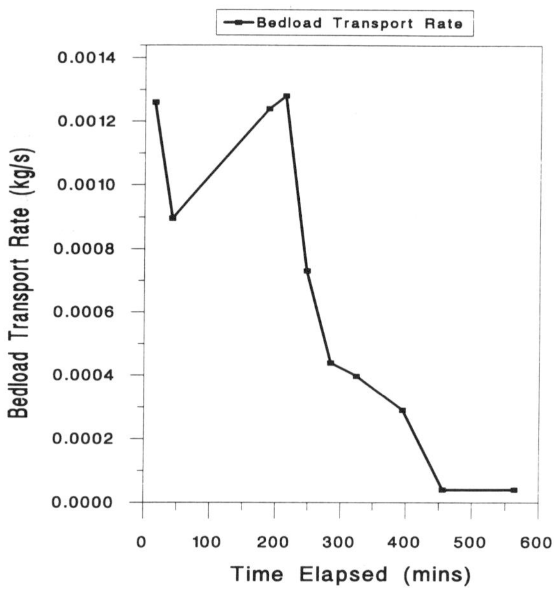

Figure 9.1. Changes in the bedload transport rate with time

PARTICLE IMAGE VELOCIMETRY

PIV is a measurement technique which relies on the successive illumination of an area of flow which is seeded with reflective particles. These particles need to be scatter light efficiently, be small but visible and be approximately neutrally buoyant so that they can accurately follow the flow. Each single image contains a series of successive illuminations of the observed flow area. Images are processed in order to determine the local displacement of the seeding particles. The velocity of the flow at various points can be determined if the time interval between the illuminations of the seeding particles is known.

A schematic layout of the measurement equipment is shown in Figure 9.2. In this experiment an area of flow approximately 45 mm in the streamwise direction and 30 mm in the vertical direction was illuminated using a scanning beam system and an Argon-ion laser. The illuminated area was on the centreline of the flume and approximately 8.5 m from the flume inlet, and it corresponded to approximately the lower 50% of the overall flow depth. The scanning beam box system produced an intense vertical beam of light which traversed the measurement flow area at a constant streamwise velocity, controlled by the speed of the rotating polygon mirror which repeats scans at a time interval of 2.0 ms. This was taken to be the time interval between successive illuminations of the seeding particles. The seeding particles used were grains of conifer pollen approximately 50 μm in size with approximately neutral buoyancy.

In order to study the dynamics of coherent flow structures as they undergo cycles of formation, growth and decay, the measurement of the temporal evolution of the flow field was important. A time-series of flow images needed to be recorded. A CCD array camera was used to record the flow images digitally. The CCD array had a resolution of 752 by 582 pixels and was able to record images at 25 Hz with a shutter speed of 1/125 s. The images were time coded and recorded on video tape for subsequent retrieval. Thus, a continuous time-series of images was available for subsequent processing.

Image processing and analysis

Two series of images were used from the recorded video tape to represent two periods of flow, each 25 successive frames long, which represented the sediment bed in a highly mobile and a stabilized state, respectively. The digitized images were analysed by splitting up each frame into a grid of small interrogation areas 32×32 pixels in size. Each interrogation area corresponded to a physical flow area of approximately 1.6×1.6 mm. There was a 50% overlap of the neighbouring interrogation areas so a grid of 47×36 velocity vectors was produced for the flow field for each image. The mean displacement of the seeded particles within an interrogation area was found using correlation techniques (Gray, 1994). A two dimensional cross-correlation was conducted on each interrogation area with a flow area displaced a small distance upstream. This calculated the displacement which

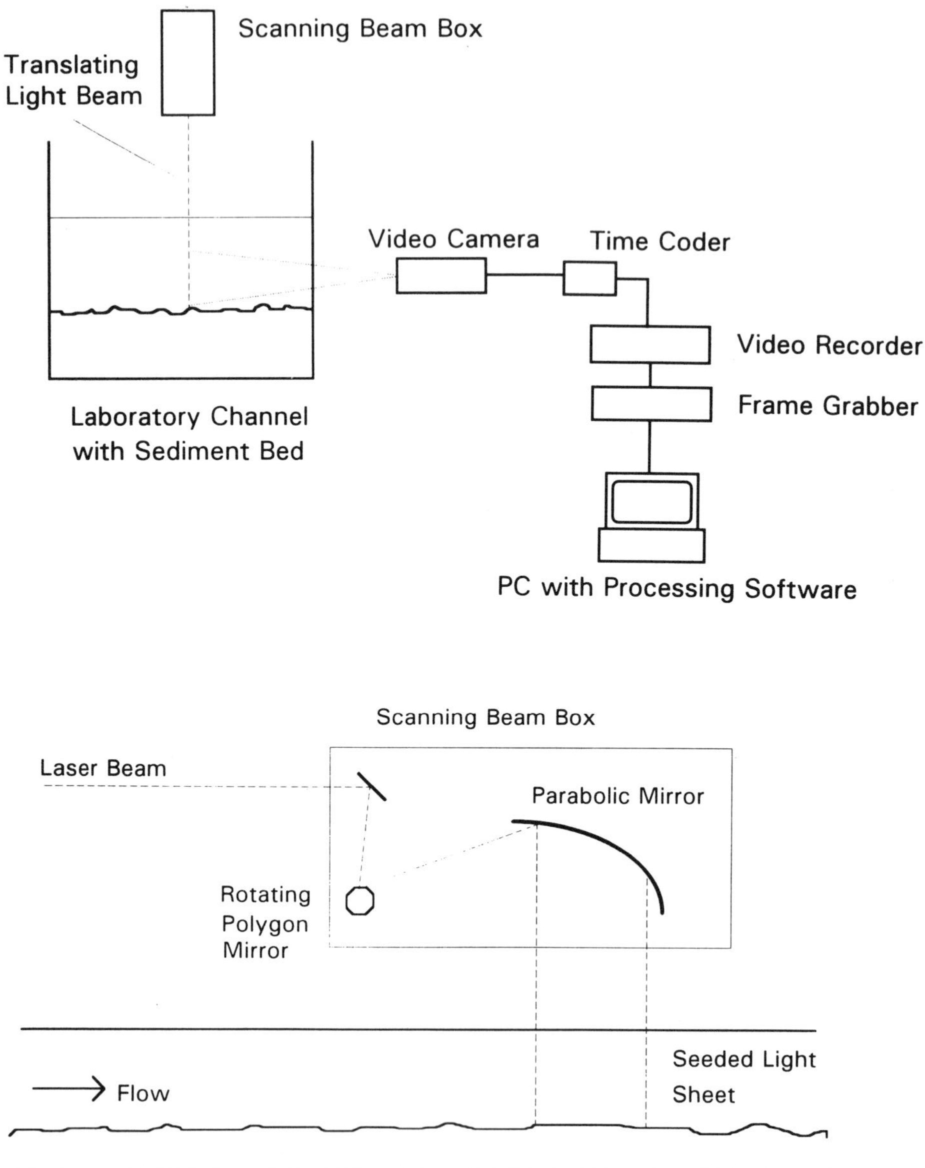

Figure 9.2. Schematic layout of the particle image velocimetry system

best described the movement of the seeding particle pattern within an interrogation area during successive illuminations. The trajectories of individual particles were not tracked using this technique. Once the mean displacement is calculated for each interrogation area, division by the scanning beam time interval allows the mean local flow velocity vector to be calculated which is assumed to be located at the centre of each interrogation area.

Post-processing: erroneous vectors

The raw PIV vector map contained some erroneous vectors created by the inability of the correlation routines to determine the displacement of the seeding particles either due to poor image quality or low seeding density in a part of the image. For a large number of frames it would be a laborious task to identify and delete these spurious vectors manually. Moreover such inspection of the raw PIV results is subjective. Therefore a method has been used based on the work of Westerweel (1994) which rejects vectors as spurious based on a deviation from a local median. In this case, nine neighbouring vectors are examined, and vectors which deviate greatly from the median value are defined as spurious and deleted. The proportion of rejected vectors per image was approximately 5–6% per frame. Particular care was taken with the treatment of the vectors adjacent to the bed as a proportion of these interrogation areas were occupied by the bed and not seeded fluid, so reducing the robustness of the correlation technique used. These vectors were also examined visually.

LABORATORY RESULTS

The mean velocity profiles were obtained by examining 25 consecutive frames from both sediment transport states. The values of mean flow velocity were obtained by averaging the streamwise velocity data along lines of constant height in each frame and then calculating the line averages over each set. The values of the mean streamwise velocity were non-dimensionalized with the wall shear velocity u_*, which was calculated from the slope of the semi-logarithmic velocity profile of the flow within the bottom 10% of the overall water depth.

The normalized mean velocity profile plotted in a semi-logarithmic form indicated a distribution with two segments. The distribution became more uniform closer to the bed (Figure 9.3). This 'kinked' form of velocity distribution is characteristic of turbulent flow over rough surfaces and the variability in the gradient of the lower segment has been linked to local changes in surface roughness (Nowell & Church, 1979).

Turbulent flow structures

Visual inspection of the series of instantaneous velocity flow maps reveals organized flow patterns. This is illustrated in Figures 9.4 and 9.5 which are vector maps describing the flow field at an instant in time during a period of high sediment transport. Figure 9.4 shows the measured instantaneous flow velocities

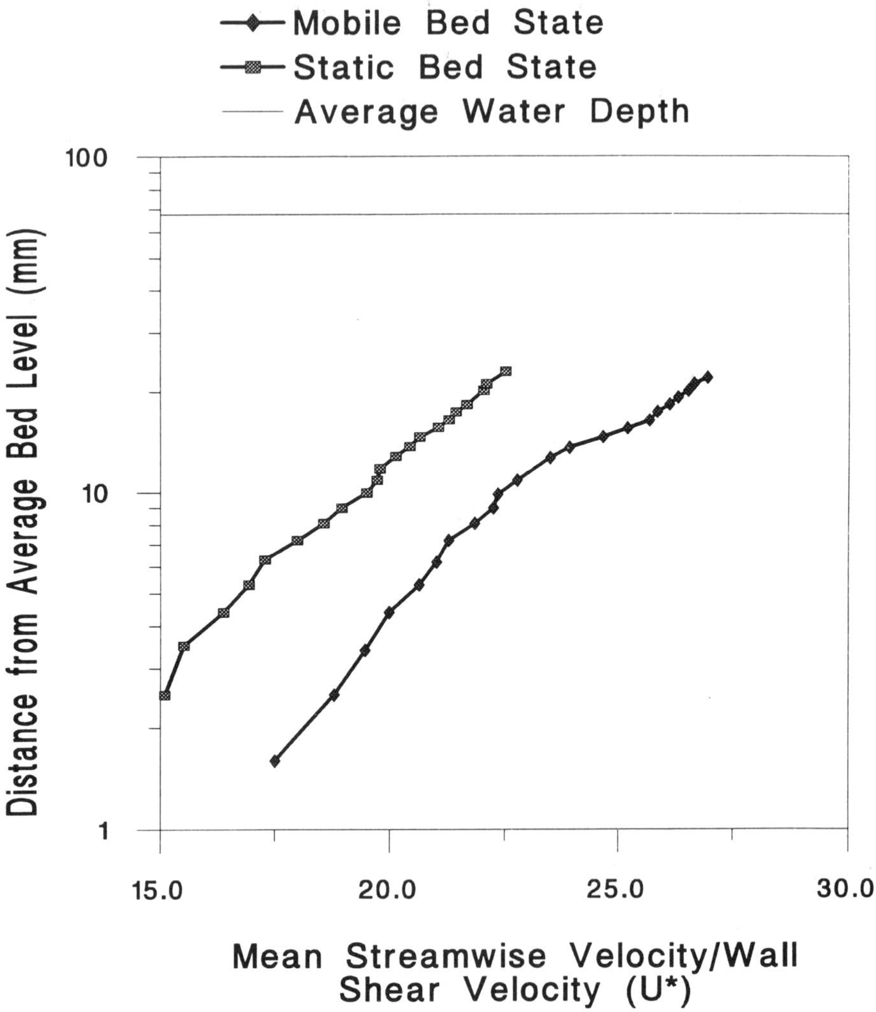

Figure 9.3. Mean streamwise velocity with depth, during a periods of high and low sediment transport

and illustrates two downward moving flow areas of differing sizes, annotated A and B. Figure 9.5 is a plot of the instantaneous velocity fluctuations of the same flow area at the same point in time. The fluctuating velocity field was produced by subtracting the local mean flow velocity calculated with respect to depth within the single frame from the measured instantaneous flow velocities. Observations indicated that two scales of flow feature were present: a larger-scale structure (A)

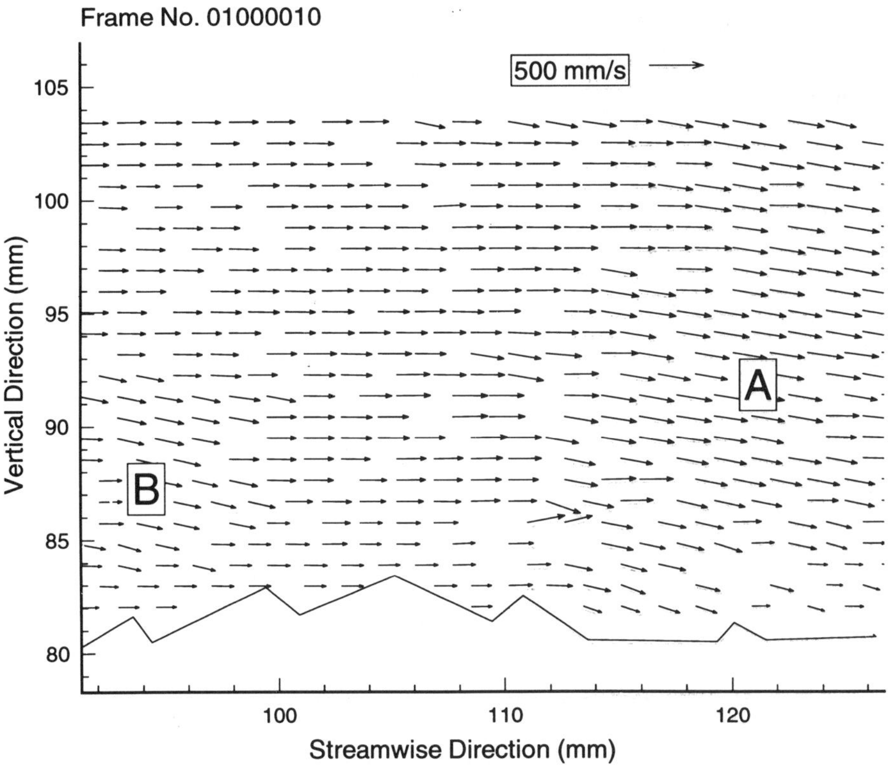

Figure 9.4. Vector map of the instantaneous velocity field, frame no. 01000010

with a size of the order of several centimetres and a smaller-scale structure (B) with a size of the order of several millimetres.

A typical sequence of formation and development starts with the rise of a small upward moving flow area, such as that labelled 'C' in Figure 9.6. This Figure illustrates the flow pattern that is present approximately 0.25 s before the appearance of the downward moving flow features in Figures 9.4 and 9.5. This upward moving feature expanded as it rose into the flow. Typically upward flow features formed separately close to the bed, rose, and then amalgamated as they rose into the flow. Small downward moving packets of fluid are then observed in the measurement area, typically followed by a large downward moving area of fluid which may occupy the majority of the measured flow area (Figure 9.4). Once this structure has passed, several smaller flow features trail the larger one. The flow features tend to move at approximately the local mean flow velocity although it was seen that the upward moving flow generally was slightly slower than the local mean flow velocity whilst the downward moving flow features moved faster. Several instances were observed of these flow features interacting with each other.

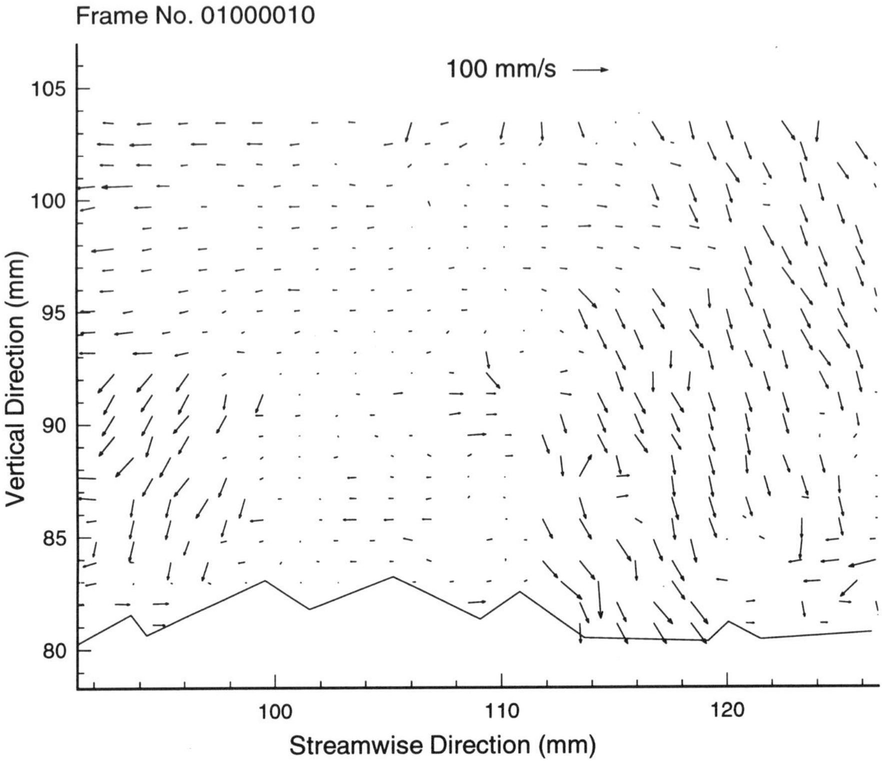

Figure 9.5. Vector map of the instantaneous fluctuations in the velocity field, frame no. 01000010

The smaller flow features (both the upward and downward moving ones) appeared to be approximately 4 to 5 times more prevalent than the larger downward moving flow features. Examining the two time periods corresponding to high and low sediment transport, it was seen that the average deviation, for the upward and downward moving flow, from the mean flow velocity was larger on average during the period of high sediment transport than during the period in which the bed had stabilized. The average angular deviation from the streamwise direction was approximately 15° for the upward moving fluid and around 10° for the downward moving fluid when the bedload movement was intense. Both these angles increased to approximately 18 and 12° respectively when the sediment bed had stabilized.

Examination was concentrated on the downward moving areas of flow as they were thought to be the more important in the grain entrainment process. These areas of flow tended to be associated with times in which the streamwise velocity was significantly above that expected at a particular flow depth. The distribution of the flow velocities was determined for several images at a flow depth close to the

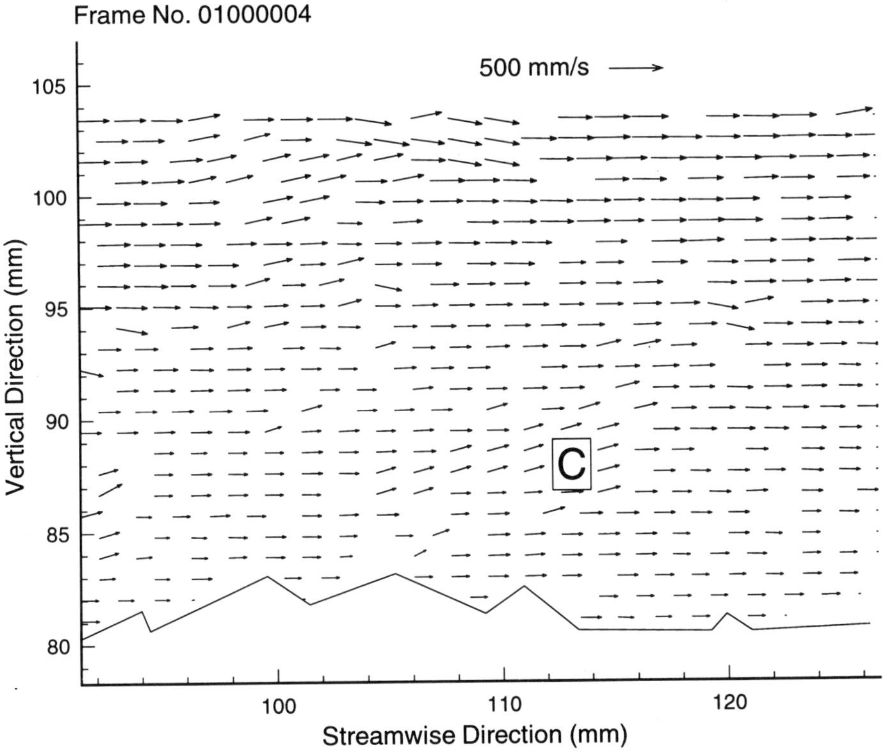

Figure 9.6. Vector map indicating rise of fluid close to the bed, frame no. 01000004

bed (3.5 mm above average bed surface). The aim was to correlate increases in the streamwise velocities with the passage and collision with the bed of both large and smaller downward moving flow features. Figure 9.7 indicates the occurrence of peaks in the streamwise velocity. These peaks corresponded to image frames in which the flow was generally down. The highest peaks in this Figure corresponded to frames in which large downward moving areas of flow were dominant, and the troughs to frames in which the flow was generally upwards. The distribution of the individual velocity fluctuations close to the bed within each frame was then examined and it was seen that during the periods in which the bed was exposed to the highest mean streamwise velocities, the distributions of the streamwise and vertical velocity encompassed the largest fluctuations (Figures 9.8 and 9.9). The widest range of velocity fluctuations occurred exclusively when the large moving flow features impacted with the bed. It seems probable that these features are the most important when considering the processes of grain entrainment. They may correspond with the 'sweep-like' transport events which were observed by Drake *et*

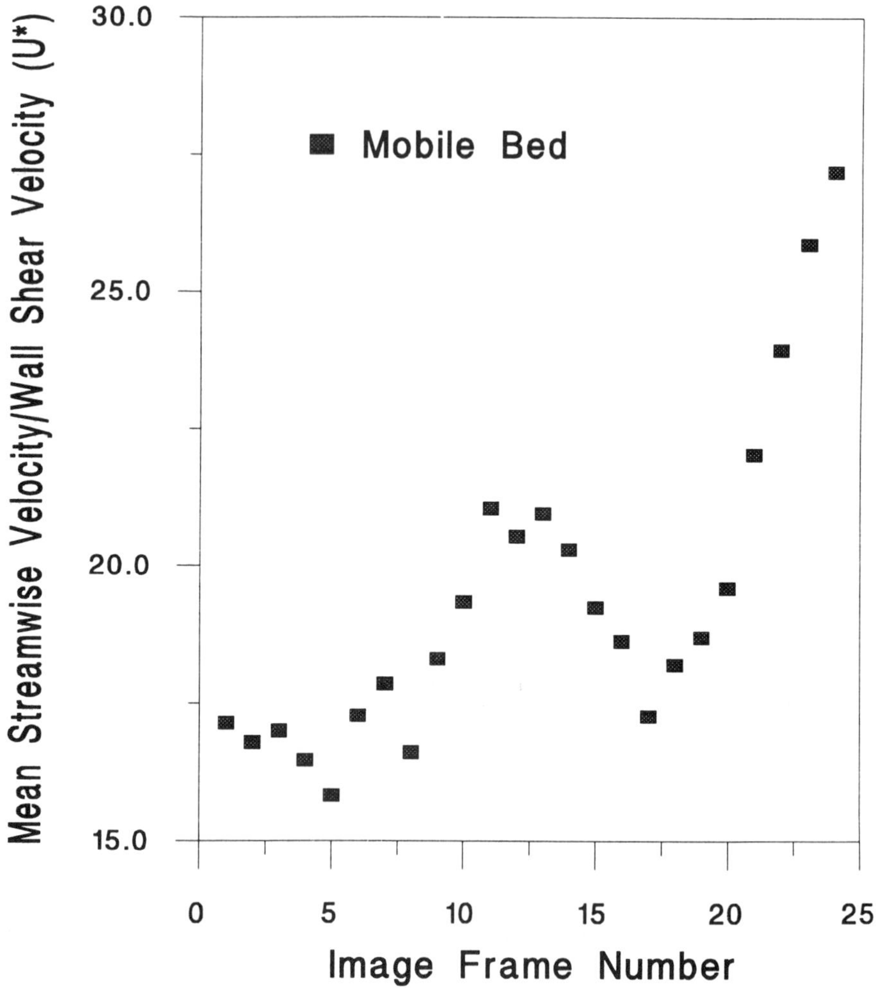

Figure 9.7. Variation of mean streamwise velocity, calculated for each image, at 3.5 mm from average bed surface during a series of 25 consecutive frames. Frame numbers 01000000 to 01000024. Sediment bed is highly mobile

al. (1988) whose observations indicated that the flow features responsible for entrainment had a size equivalent to several grain diameters.

There are insufficient data at present to identify conclusively any significant changes in the relative frequency of occurrence of large-scale flow features over a mobilized and stabilized sediment bed. This small data set did suggest that the larger coherent flow structures are less frequent over a stabilized bed, the smaller flow features being slightly more common and interacting more frequently with the

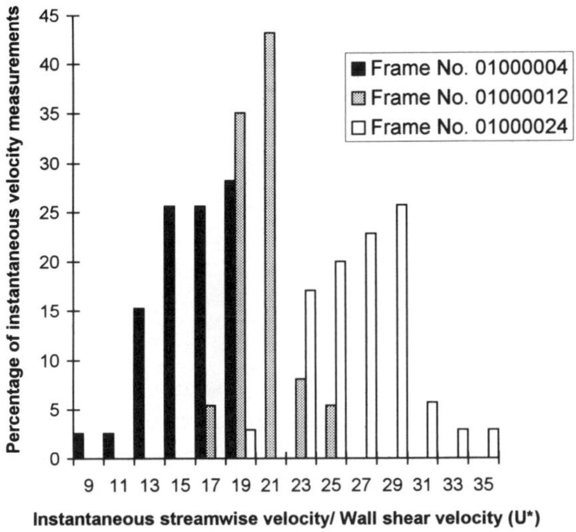

Figure 9.8. Distribution of the instantaneous streamwise velocity measurements at 3.5 mm from average bed surface for frame numbers 01000004, 01000012 and 01000024

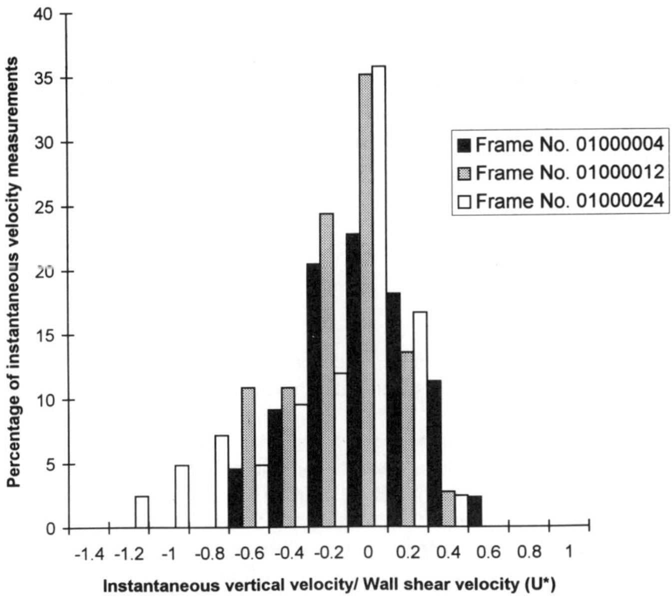

Figure 9.9. Distribution of the instantaneous vertical velocity measurements at 3.5 mm from average bed surface for frame numbers 01000004, 01000012 and 01000024. Negative values of velocity indicate flow towards the bed

upward moving flow structures. Examining the percentage area covered by the upward and downward moving fluid in each frame did reveal differences between the two transport stages. At high transport rates the range of the variations in the area covered between the frames in a time series was larger: overall approximately 48% of the velocity vectors indicated flow away from the bed and 52% indicated flow towards the bed. Once the bed had stabilized the percentage variation between the images during this measurement period was reduced and now only 45% of the vectors were towards the bed.

CONCLUSIONS

The measurements presented here begin to quantify the turbulent coherent flow structures which appear important in the entrainment of grains from a sediment bed. Flow features with different scales and dynamic characteristics (rate of rise/descent, and speed through the flow) have been identified from the visual inspection of consecutive vector maps of the measured instantaneous flow velocities. These different types of flow features appear to have dissimilar characteristics in that rising fluid tends to move more slowly in the streamwise direction but to move with greater vertical velocity than descending fluid.

The distribution of the instantaneous near-bed velocities changes as different coherent flow features either impact or pass over a point on the bed suggesting that the presence of different types of flow features is important in determining the probability of grain entrainment. It appears that the large downward moving flow features may be the most significant in terms of grain entrainment due to the wide range of velocities found within them. More images at situations of both high and low transport rate need to be examined in order to obtain a more complete description of the frequency of formation of these flow features and their interaction with each other and the bed.

REFERENCES

Best, J. 1992. 'On the entrainment of sediment and the initiation of bed defects: insight from recent developments within turbulent boundary layer research', *Sedimentology*, **39**, 797–811.

Best, J. 1993. 'On the interactions between turbulent flow structure, sediment transport and bedform development: Some recent considerations from recent experimental research', in Clifford, N.J., French, J.R. & Hardisty, J. (Eds) *Turbulence: Perspectives On Flow and Sediment Transport*, John Wiley & Sons Ltd., Chichester, 61–92.

Drake, T.G., Shreve, R.L., Dietrich, W.E., Whiting, P.J. & Leopold, L.B. 1988. 'Bedload Transport observed by motion-picture photography' *J. Fluid Mech.*, **192**, 193–217.

Grass, A.J. 1970. 'Initial instability of fine bed sand', *J. Hydraul. Div., ASCE*, **96**, 619–632.

Grass, A.J. 1971. 'Structural features of turbulent flow over smooth and rough boundaries', *J. Fluid Mech.*, **50**, 233–255.
Gray, C. 1994. *VidPIV for Microsoft Windows: Analysis tools for Particle Image Velocimetry*, Univ. of Edinburgh, 63 pp.
Gyr, A. 1982. 'Towards a better definition of the three types of sediment transport', *J. Hydraul. Res.*, **21**, 1–15.
Gyr, A., Müller, A. & Schmid, A. 1989. 'Observations of self stabilisation processes in sediment transport linked to scales of coherent structures', *Proc. XXIII Congr. IAHR*, Ottawa, A31–A38.
Haff, P.K., Jiang, Z. & Forrest, S.B. 1993. 'Transport of granules by wind and water: Micromechanics to macromechanics in geology and engineering', *Mechanics of Materials*, **16**, 173–178.
Hussain, A.K.M.F. 1986. 'Coherent structures and turbulence', *J. Fluid Mech.*, **173**, 303–356.
Kirchner, J.W., Dietrich, W.E., Iseya, F. & Ikeda, H. 1990. 'The variability of critical shear stress, friction angle, and grain protrusion in water worked sediments', *Sedimentology*, **37**, 647–672.
Meyer-Peter, E. & Müller, R. 1948. 'Formulas for bedload transport', *Proc. Int. Congr. IAHR Stockholm*, Appendix 2, 39–64.
Mulhearn, P.J. & Finnigan, J.J. 1978. 'Turbulent Flow over a very rough random surface', *J. Boundary Layer Metrology*, **15**, 109–132.
Nowell, A.R.M. & Church M. 1979. 'Turbulent flow in a depth-limited boundary layer', *J. Geophys. Res.*, **84**, 4816–4824.
Reid, I., Frostick, L.E. & Layman, J.T. 1985. 'The incidence and nature of bedload transport during flood flows in coarse-grained alluvial channels', *Earth Surf. Processes Landf.*, **10**, 33–44.
Robert, A., Roy, A.G. & De Serres, B. 1992. 'Changes in velocity profiles at roughness transitions in coarse grained channels', *Sedimentology*, **39**, 725–735.
Sekine, M. & Kikkawa, H. 1992. 'Mechanics of saltating grains II', *J. Hydraul. Eng. ASCE*, **118**, 536–558.
Shields, A. 1936. 'Application of similarity principles and turbulence research to bedload movement' *Mitt. Preuß Versuchsanstalt für Wasserbau und Schiffbau, Berlin*, **26**, 1–26. (translation by Ott, W.P. & van Uchelen, J.C., Californian Inst. of Technology Pasadena).
Tait, S.J. & Willetts, B.B. 1992. 'Characteristics of armoured bed surfaces', in *Proc. Int. Grain Sorting Seminar*, Mitteilungen **117**, ETH Zürich, 207–225.
Westerweel, J. 1994. 'Efficient detection of spurious vectors in particle image velocimetry data', *Exper. Fluids*, **16**, 236–247.
Wolcott, J.F. 1990. 'Flume studies of gravel bed surface response to flowing water', Unpub. PhD Thesis, Univ. of British Columbia, 111 pp.

NOTATION

D	grain size, where subscript value denotes 'percentage finer than'
U_*	shear velocity
y	distance above boundary
ν	kinematic viscosity of fluid
σ_g	grain-size sorting coefficient

10 Flow-sediment Interactions in Separating Flows over Bedforms

STEPHEN R. McLEAN
Mechanical and Environmental Engineering Department, University of California, Santa Barbara, CA 93106, USA

JONATHON M. NELSON
National Research Program, Water Resources Division, U.S. Geological Survey, Box 25046, MS 413, Lakewood, CO 80225, USA

RONALD L. SHREVE
Department of Earth and Space Sciences and Institute of Geophysics and Planetary Physics, University of California, Los Angeles, CA 90024, USA

ABSTRACT

Flow separation occurs over many naturally occurring bedforms, such as ripples and dunes. The resulting wake that forms in the lee of the separation zone has a profound effect on the development of the underlying boundary layer. Detailed measurements of turbulence in separated flow over bottom features with different bottom slopes downstream of the reattachment region provide insight into the nature of such flows. The turbulence characteristics in the near-bed region near reattachment are significantly different than those in a fully-developed, uniform boundary layer. For example, in the developing boundary layer downstream of a reattachment point, high turbulence levels are not necessarily accompanied by high Reynolds shear stress estimates, as demonstrated by the anomalously low correlation coefficients in this region. This arises because of more even partitioning of the contributions to the turbulent stress among all the quadrants of the joint probability distribution. Simultaneous flow and sediment transport observations indicate that, because outward interactions have characteristics that are conducive to transport of erodible bed material (high downstream velocity and upward vertical velocity), sediment fluxes are higher than expected near reattachment. A favourable pressure gradient, arising as a result of the upward sloping topography on the stoss side of a ripple or dune, alters the nature of the turbulence field and also increases the rate of acceleration in the mean flow. The interaction of the turbulence adjustments and the accelerating flow field provide a feedback mechanism that controls the shape and evolution of bedforms.

INTRODUCTION

Analysis of sediment-transporting flows can be carried out at different levels of detail, ranging from the prediction of the mean transport rate over a channel section

Coherent Flow Structures in Open Channels.
Edited by P.J. Ashworth, S.J. Bennett, J.L. Best and S.J. McLelland.

(sediment load) to the prediction of local transport at many points in a complex, highly non-uniform flow over features such as bars or bedforms. In the latter case, the accuracy of local fluxes is frequently poor because of our lack of understanding of the details of the interaction of fluid and sediment in complex flows. Even in the simpler case of predicting spatially averaged transport, flow non-uniformity or unsteadiness can produce significant errors even in the best applied techniques. These errors point to the need for a more comprehensive view of the coupled fluid-sediment system. Predicting the behaviour of an erodible bed acted upon by an overlying flow requires an accurate understanding of three different aspects of that system: (i) the adjustment of the flow to the underlying topography; (ii) the response of sediment particles to the surrounding flow; and (iii) the evolution of the bottom topography in response to spatial and temporal gradients of the local sediment flux. These three processes are interdependent and thus must be considered together. With the advent of instrumentation that has allowed us to observe turbulent flows in detail, it has become evident that near-bed velocity fluctuations play a key role in determining the response of sediment on the bed (e.g. Nelson, *et al.*, 1995). In this paper, the coupling between turbulent fluid and sediment grains moving as bedload is examined in the context of separating flows over bedforms. This examination yields significant insight into the mechanics of transport on the stoss sides of typical bedforms, and also yields general information on the mechanics of sediment entrainment and motion in non-uniform and/or unsteady flows.

In the past, prediction of transport in non-uniform flows, including those over bedforms, have been predicated on the extension of formulae and equations developed for steady, uniform flows to these more complex cases (Engelund & Skovgaard, 1973; Fredsøe, 1982; McLean & Smith, 1986; Nelson & Smith, 1989). Thus, engineers and scientists wishing to predict local sediment transport rates within a ripple or dune field begin by estimating the local boundary shear stress, τ_b. This quantity is inserted, along with sediment parameters, into a variety of empirical formulae developed for uniform flows relating bottom stress and transport. For erodible beds that are devoid of topographic features (i.e. uniform flows), observations suggest that there is indeed a consistent relationship between bedload transport and boundary stress (e.g. Meyer-Peter & Müller, 1948; Yalin, 1963; Bagnold, 1973; Wiberg & Smith, 1989). For such flows, both the mean velocity and the standard deviation of the turbulent fluctuations scale with the shear velocity $u_* = \sqrt{\tau_b / \rho}$ where ρ is the density of the fluid. Therefore, it is not surprising that a strong correlation between the transport rate and the boundary shear stress is found for this special case. However, separated flows, such as those that occur over ripples and dunes, exhibit characteristics that are distinctly different from those of uniform turbulent flow. Beyond the difficult task of even determining the local skin friction, the existence of a simple relationship between local shear stress and local transport rates is questionable. The interaction between the highly turbulent wake, created by separation, and the underlying boundary layer creates turbulence characteristics that not only are different than uniform boundary layers, but vary significantly with position downstream. Therefore, other factors such as spatial variation in the statistics and frequency distribution of turbulence fluctuations must also be considered when trying to predict sediment response.

For many years, investigators have argued that coherent structures are critical to our understanding of suspended and bedload transport, respectively (e.g. Heathershaw, 1974; Jackson, 1976). The relative frequency and strength of these coherent structures are significantly affected by the process of separation and subsequent flow acceleration near the bed. Consequently, it is not surprising that the response of sediment in the vicinity of separated flow is significantly different from that which is observed over flat beds. It is imperative for our understanding of bedform shape and stability that the sediment response be characterized.

SEDIMENT-FLUID INTERACTION

McLean *et al.* (1994) provided substantial evidence that the local sediment flux may not be directly related to the local boundary shear stress by comparing transport rates calculated with the Meyer-Peter & Müller (1948) equation

$$\Phi = \frac{q_s}{\left[\dfrac{\rho_s - \rho}{\rho g D^3}\right]^{1/2}} = \alpha\left(\tau^* - \tau^*_{cr}\right)^{3/2}$$

using actual measurements (see Figure 10.1a). Here q_s is the transport rate, ρ_s is the density of the sediment, g is the acceleration of gravity, D is the diameter of the sediment, $\tau^* = \tau_b / (\rho_s - \rho) g D$ is the non-dimensional boundary shear stress, τ^*_{cr} is its value at incipient motion, and α is a constant taken to be 8 by Meyer-Peter & Müller. The calculations were done using boundary shear stress estimates derived in two ways: (i) from the near-bed Reynolds stress measurements; and (ii) from calculating the shear velocity u_* from the slope of a logarithmic profile intercepting the measured near-bed velocity and passing through zero velocity at y_0, the roughness parameter suggested by the measurements of Nikuradse (1933) for smooth, transitional or rough flows. The error in estimating the stress is of the order of 20% for both methods. The calculations were performed specifying $\tau^*_{cr} = 0.045$ and also $\tau^*_{cr} = 0$ (the latter requires reducing α by a factor of four to produce reasonable values of transport). In no case do the calculations capture the total structure of the measured transport. There are two possible reasons for this lack of agreement: (i) the equation relating the flux to the stress is inappropriate for this situation; and (ii) the estimates of stress are inaccurate. It is likely that both are true. For example, the observations of sediment transport near the point of reattachment clearly indicate that turbulence levels are sufficiently large in this

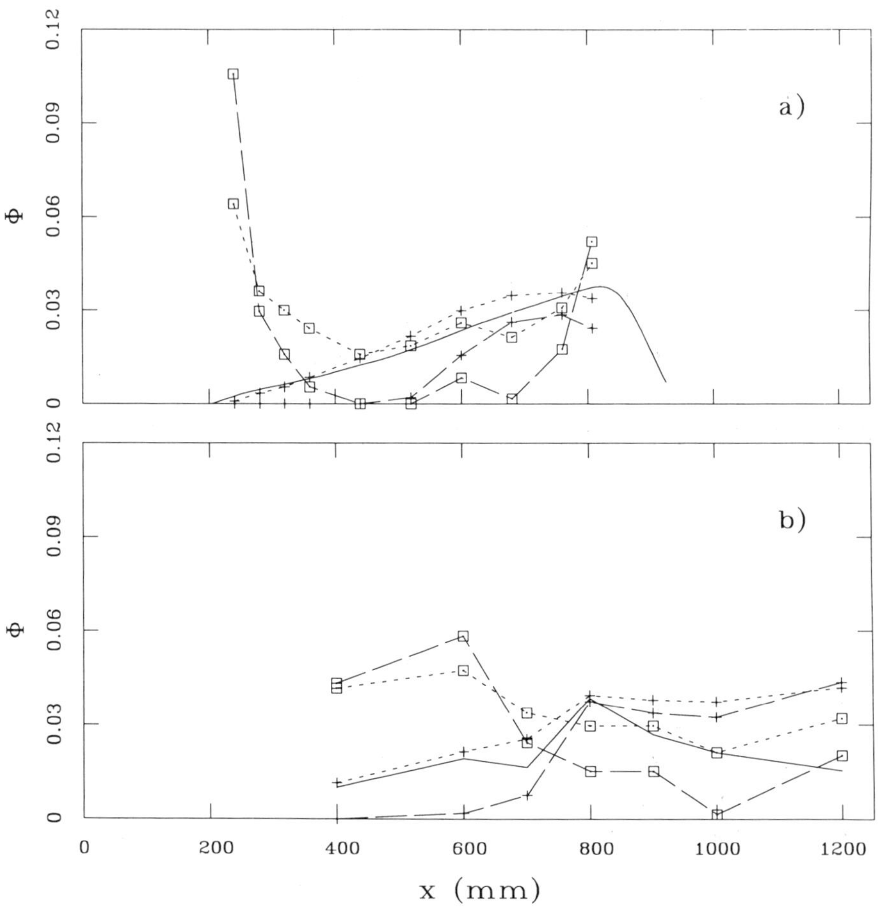

Figure 10.1. Plots of non-dimensional sediment transport rate versus distance downstream of the point of separation for: (a) a bedform 40 mm high and 800 mm long comprising 0.5 mm sand; and (b) for a flat bed comprising 1 mm sand downstream of a 40 mm high backward facing step. The mean reattachment point was estimated to be at 200 mm for (a) and at 240 mm for (b). The solid lines in these plots are estimates made by: (a) measuring the local elevation of the bed as a function of time and using the erosion equation to calculate the transport rate; and (b) counting the number of grains passing a 10 mm long line perpendicular to the flow in 0.1 s using a high-speed film camera. The lines with symbols are estimates of the transport from the Meyer-Peter & Müller equation, the square represents calculations using the Reynolds stress measured about 3 mm from the bed, and the plus represents calculations using estimates of boundary shear stress inferred from logarithmic velocity profiles near the bed. The long dashed lines are the results using $\tau^*_{cr} = 0.045$, the short dashed lines are for $\tau^*_{cr} = 0$. Note that in no case is there good agreement between the calculations and the more direct measurements of transport

region that even where the mean velocity (and presumably the stress) are zero, there are times when the instantaneous near-bed velocity is sufficient to induce sediment motion. Therefore, as long as the frequency distribution of these events is such as to favour one direction over another, net transport can occur even for zero mean velocity (or stress).

Local mean sediment transport rates are found by time-averaging the instantaneous local fluxes. Because of the highly non-linear nature of sediment-fluid interaction, simultaneous measurements of near-bed flow and sediment flux at relatively high-frequency are needed in developing relationships between flow and sediment transport rates for a variety of conditions. With this in mind, an experiment was carried out in the flume at the Friday Harbor Laboratories of the University of Washington wherein measurements were made in the simplest possible separated flow and in a uniform boundary layer. Flow separation was created using a downstream-facing step, thus producing a wake flow with minimal, if any, pressure gradient downstream of the reattachment point. High-speed films of the sand bed in the lee of the step were made simultaneously with velocity measurements at 5 mm above the bed using a laser Doppler velocimeter. Details of this experiment and analysis can be found in Nelson *et al.* (1995). From the resulting data, 10 Hz time-series of the flux and velocity were generated. The measured mean transport rates appear in Figure 10.1b along with estimates using the measured velocities and Reynolds stresses as was done for Figure 10.1a. Again, there is generally poor agreement between the estimates and the measurements; the measured transport exhibits a maximum at 800 mm whereas the estimates using the Reynolds stress tend to decrease with distance downstream and those from the velocity tend to increase in the downstream direction.

Time-series of the various quantities that were measured are shown in Figures 10.2a and 10.2b. The former depicts results under uniform conditions, the latter presents results from a position $10H$ downstream of the step, where H is the step height. The downstream and vertical velocity components and the product of their instantaneous, fluctuating values are included. Comparison of the downstream velocity records between the two Figures clearly shows that there is more turbulent energy in the case near reattachment (Figure 10.2b), even though the velocity is significantly lower owing to the wake structure downstream of the step. The second panel from the bottom relates to the force a particle on the bed would experience and is discussed in detail below. In the bottom panel, the instantaneous transport rate is shown. As Nelson *et al.* (1995) note, the character of transport is markedly different between the two cases; that from the separated flow exhibits more intermittency, with fewer, larger transporting events. These events generally coincide with periods of high downstream velocity, but the correlation between the flux and the velocity is usually less than 0.5.

In an effort to understand the response of a sediment particle to the ambient flow, a logical starting point is to investigate the forces that act on a sediment particle lying on a bed of similar particles. Let $\vec{F}$ be the force exerted by the flow on the

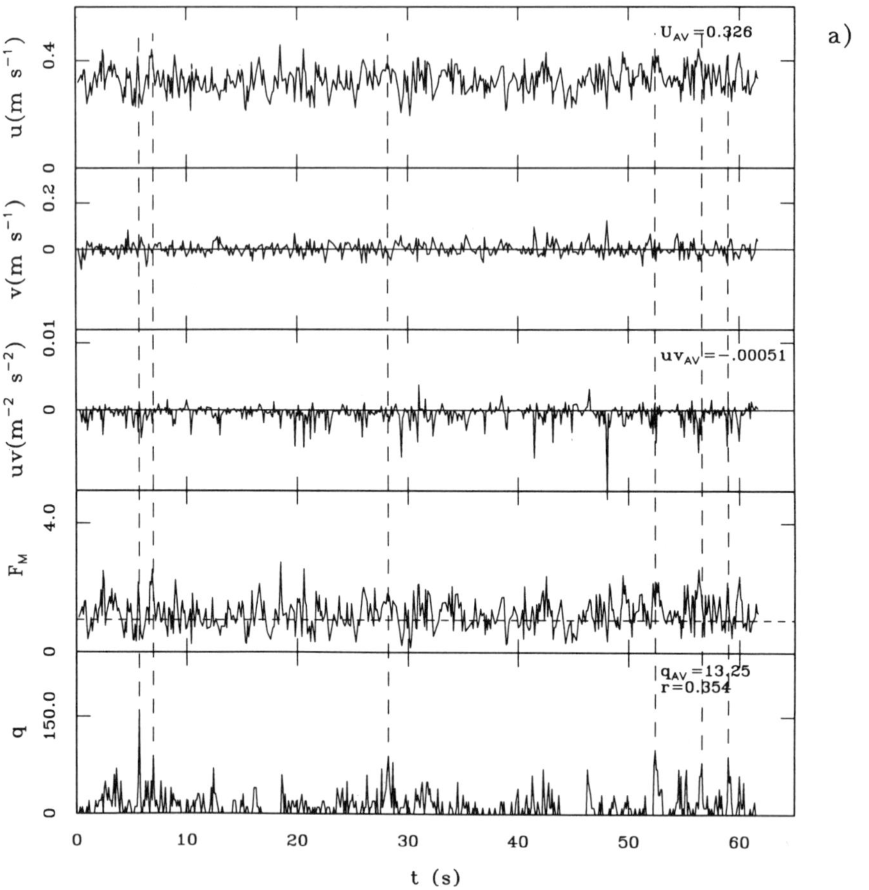

Figure 10.2. Time-series of the downstream and vertical velocity components, the instantaneous product of u' and v' (the fluctuating velocity components), F_M, which is related to the force acting on a particle resting on the bed, and the sediment transport rate, in number of grains per centimetre per second passing below the LDV measuring volume. The time-series are from: (a) a uniform boundary layer; and (b) a point $10H$ downstream of a backward-facing step of height $H = 40$ mm. Note the highly intermittent nature of the sediment transport rate in (b) and that the occasional high transport rates are generally associated with peaks in the downstream velocity. *Continued on facing page*

particle and let the line connecting the centre of mass of the particle and its point of contact with the underlying grain on the downstream side be of length $D/2$ with an angle ϕ_0 from vertical. Incipient motion occurs when the fluid forces just begin to roll the particle forward so that the only forces acting on the particle are the drag, lift and wetted weight. These forces are assumed to act at the centre of mass, and

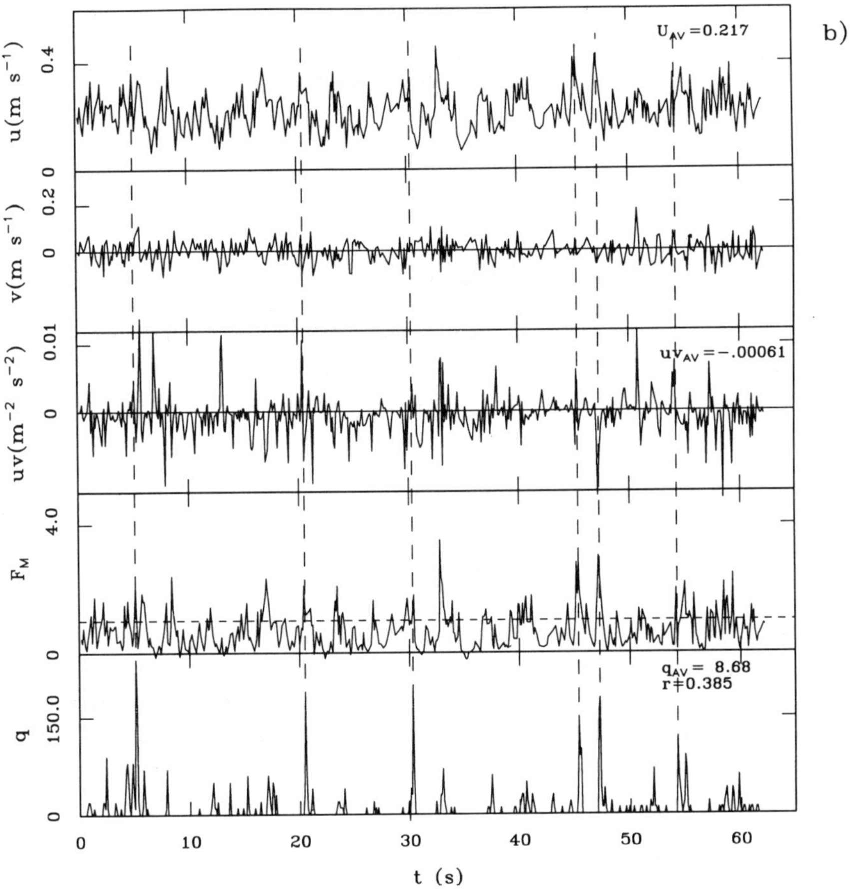

Figure 10.2. *continued*

normal and friction forces exerted by the downstream underlying grain act at the point of contact. If the particle were just suspended in this state, the sum of the moments about the point of contact is zero, therefore

$$F_x \frac{D}{2} \cos \phi_0 + \left(F_y - F_g'\right) \frac{D}{2} \sin \phi_0 = 0 \qquad (10.1)$$

where F_g' is the wetted weight and x and y represent the downstream and vertical directions, respectively. Using the equation for conservation of momentum in the fluid, the force exerted on a particle moving with velocity $\vec{u}_s$ can be written as

$$F_x = \rho V_s \frac{Du}{Dt} + \rho C_m V_s \frac{Du_\Delta}{Dt} + \frac{1}{2}\rho C_D A|\vec{u}_\Delta| u_\Delta + \frac{1}{2}\rho C_L A\left(|\vec{u}_\Delta|_t^2 - |\vec{u}_\Delta|_b^2\right)(\vec{n}_L)_x \tag{10.2a}$$

and

$$F_y = \rho V_s \frac{Dv}{Dt} + \rho C_m V_s \frac{Dv_\Delta}{Dt} + \frac{1}{2}\rho C_D A|\vec{u}_\Delta| v_\Delta + \frac{1}{2}\rho C_L A\left(|\vec{u}_\Delta|_t^2 - |\vec{u}_\Delta|_b^2\right)(\vec{n}_L)_y \tag{10.2b}$$

where V_s is the volume of the particle, u and v are the downstream and vertical velocity components of the fluid, C_m, C_D and C_L are coefficients of added mass, drag and lift, respectively, A is the cross-sectional area of the particle, the relative velocity is

$$\vec{u}_\Delta = \vec{u} - \vec{u}_s = (u_\Delta, v_\Delta)$$

and the unit vector in the direction of the lift force is

$$(\vec{n}_L)_x = -\frac{v_\Delta}{|\vec{u}_\Delta|};\ (\vec{n}_L)_y = -\frac{u_\Delta}{|\vec{u}_\Delta|}$$

The total derivatives in Equations 10.2a and 10.2b denoted with an upper case D are rates of change seen moving with the fluid velocity, whereas those denoted with a lower case d are changes seen moving with the sediment particle. It should be noted that the fluid velocity, as defined above, would be the average velocity over the particle volume of the fluid velocity that would exist if the particle were absent. Also the subscripts t and b in the lift terms above denote the 'top' and 'bottom' of the grain as defined by the unit vector $\vec{n}_L$. For the case of incipient motion, $\vec{u}_s = 0$. The fluid velocity for this experiment was measured 5 mm from the bed. In order to calculate the lift and drag on a particle resting on the bed, $\vec{u}_\Delta = \vec{u}$ in Equations 10.2a and 10.2b are assumed to be $(u_D + u'$ (5 mm), v' (5 mm)) where u_D is the mean velocity extrapolated, using a logarithmic profile, to the level $D/2$ from the bed (D = 1 mm) and u' and v' are the fluctuating velocities, having zero mean. This is done because both the downstream and vertical fluctuating velocities exhibit little variation with distance from the bed beyond y^+ ($= u_* y/\nu$, where ν is the kinematic viscosity) of a few tens. For the 10 Hz velocity data, the drag and lift terms are by far the dominant terms in Equations 10.2a and 10.2b. It is entirely possible that velocity data sampled at higher frequency might reveal time derivatives that would indicate that added mass effects may sometimes be significant.

Defining

$$F_M = \frac{\left(\dfrac{F_x}{\tan\phi_0} + F_y\right)}{F_g'} \qquad (10.3)$$

we see from Equation 10.1 that the condition for incipient motion is $F_M = 1$. In the second panel from the bottom in Figures 10.2a and 10.2b F_M is plotted where ϕ_0 has been assumed to be 60° (Wiberg & Smith, 1987). This non-dimensional force is strikingly similar to the downstream velocity because the drag force is dominant. This is clearly seen in the top two panels of Figures 10.3a and 10.3b which plot the normalized drag and lift forces for the uniform boundary layer and the separated flow respectively. The drag force is significantly larger than the lift force and the variance is also much greater.

McLean *et al.* (1994) and Nelson *et al.* (1995) point out that outward interactions ($u' > 0$ and $v' > 0$), although contributing negatively to the Reynolds stress, are conducive to the transport of sediment. Upon inspection of Figure 10.2a, it is apparent that many of the largest transporting events in the bottom panel coincide with periods when the vertical velocity is positive. In Figures 10.3a and 10.3b, the middle panel,

$$\Delta F_M = \begin{cases} F_M - 1 & \text{if } F_M > 1 \\ 0 & \text{if } F_M \leq 1 \end{cases}$$

is plotted. This can be thought of as excess force that is available to move a sediment particle. Especially for Figure 10.3b, this quantity bears a striking resemblance to the time-series of the flux q. Directly below this panel, ΔF_M is plotted if $u' > 0$ and $v' > 0$ (quadrant-1 of the $u' - v'$ joint distribution) and $-\Delta F_M$ is plotted if $u' > 0$ and $v' < 0$ (quadrant-4). Almost all of the non-zero values of ΔF_M are accounted for in these two quadrants. This is related to the transport rates in Figures 10.4a and 10.4b where each panel is associated with a different quadrant in the joint distribution of u' and v'. Quadrants 2 and 4 represent bursts and sweeps, respectively, and quadrants 1 and 3 represent outward and inward interactions, respectively. In each panel, the instantaneous product $u'v'$ is plotted if u' and v' fall in the appropriate quadrant and it is set to zero if they fall in another quadrant. In addition, either q or $-q$ is plotted when u' and v' fall in the appropriate quadrants and it is set to zero if they do not. The sign is chosen so that the plot of q always appears on the opposite side of the time axis from $u'v'$. The numbers appearing on the left side of the figures are the fraction of q or $\overline{u'v'}$ which is accounted for by each quadrant. Two points are demonstrated in these plots. First, although the Reynolds stress for these two cases differ by only about 20%, the fraction of the total that is contributed by each of the quadrants is much greater for the case near reattachment. The higher overall turbulence intensity does not result in

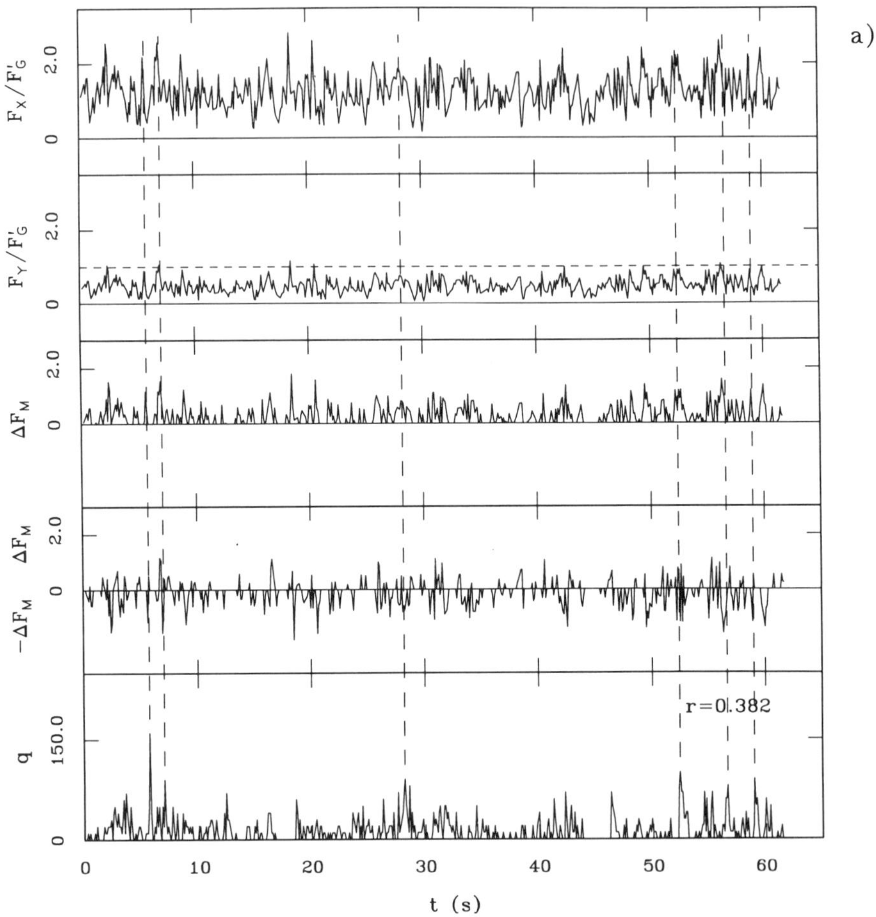

Figure 10.3. Time-series of the normalized drag and lift forces acting on a particle resting on the bed, and the excess moment producing fluid force ΔF_M that would tend to propel a particle at rest into motion (the transport rate is also plotted again for reference). In the third panel from the bottom, ΔF_M is plotted normally; in the second panel from the bottom, if $u' > 0$ and $v' > 0$ (quadrant-1), ΔF_M is plotted, if $u' > 0$ and $v' < 0$ (quadrant-4), $-\Delta F_M$ is plotted, otherwise a zero value is plotted. The time-series are from: (a) a uniform boundary layer; and (b) a point $10H$ downstream of a backward-facing step of height H = 40 mm. Note that almost all non-zero occurrences of ΔF_M are represented by one of these two quadrants and that the large transporting events are generally associated with quadrant-1. *Continued on facing page*

concomitantly higher Reynolds stresses because of relatively poor correlation between the horizontal and vertical velocity fluctuations or, in other words, because of the increased importance of quadrants 1 and 3 relative to quadrants 2 and 4. This

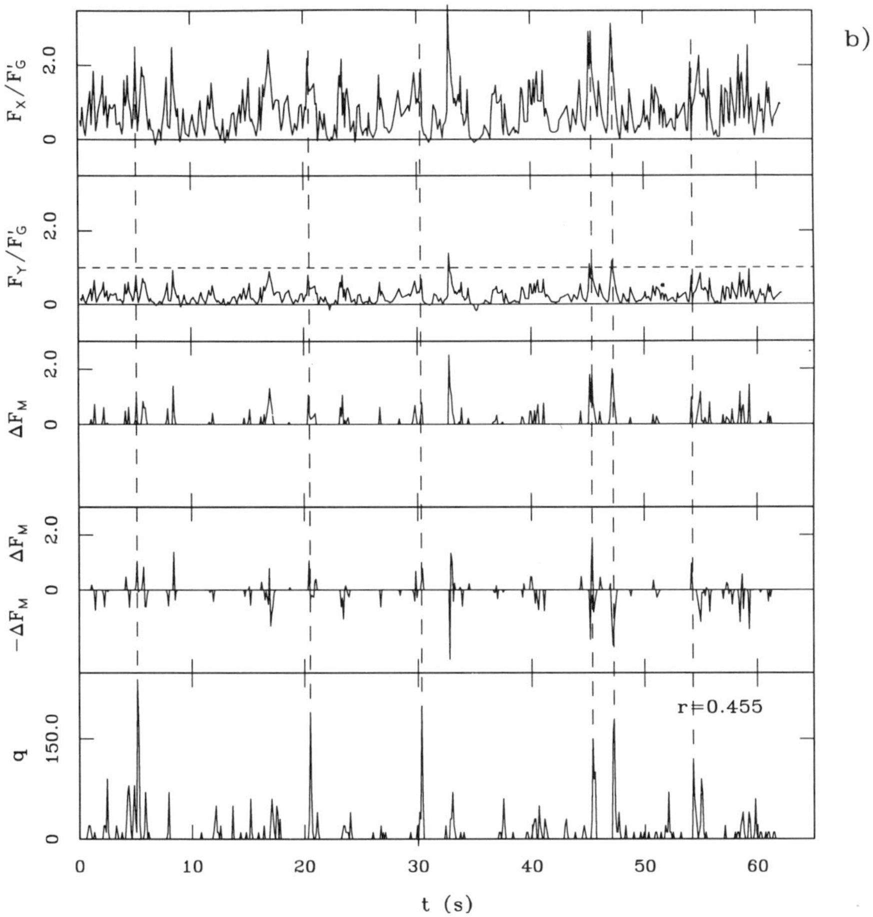

Figure 10.3. *continued*

trend is clearly seen in comparing the first and third panels in Figure 10.4a with those in Figure 10.4b. Second, the ratio of the flow velocity downstream of the step (see Figure 10.2b) to that of the uniform flow (see Figure 10.2a) and the ratio of the transport rates are about 0.66. Commonly used transport relations would predict a much smaller transport ratio given the ratio of the velocities. The discrepancy arises because of differences in the turbulence statistics between the two cases. Over 25% of the sediment transport occurs while the fluctuating velocities are in quadrant-1 for both of the cases shown here, even though quadrant-1 represents a negative contribution to the flux of momentum toward the bed. However, for the separated flow, quadrant-4 accounts for about 54% of the transport rate, but only 37% in uniform flow. The difference lies in the fact that quadrant-2

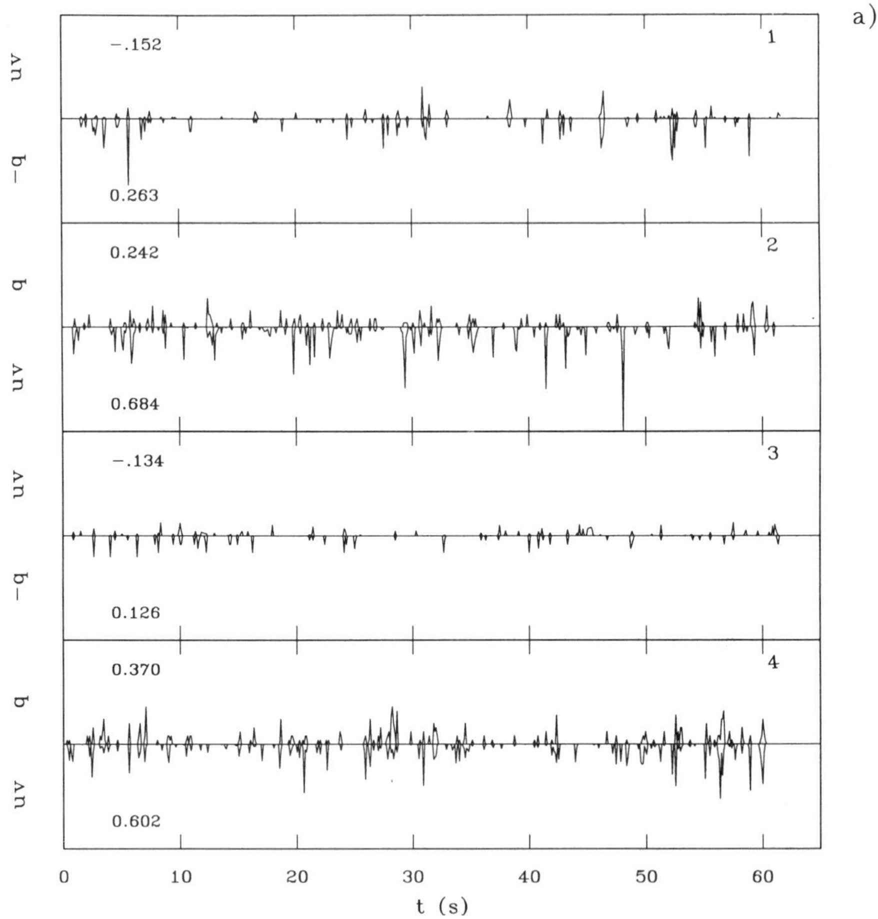

Figure 10.4. Time-series of the instantaneous product $u'v'$ and q. In the top panel $u'v'$ is only plotted if u' and v' fall in quadrant-1, in the second panel it is only plotted if they are in quadrant-2 and so on. Similarly, either q or $-q$ are plotted so that they are always on the opposite side of the time axis. On the left, the fraction of the Reynolds stress, or the total transport, associated with each quadrant is shown. For example, in (a) -15.2% of the Reynolds stress is accounted for by quadrant-1 and 26.3% of the sediment is transported while the instantaneous velocity is within quadrant-1. The time-series are from: (a) a uniform boundary layer; and (b) a point $10H$ downstream of a backward-facing step of height H = 40 mm. In both cases quadrant-4 is associated with the highest transport, but quadrant-1 is the next highest. For (a) quadrant-2 is associated with almost 25% of the transport but in (b) it is less than 10%. *Continued on facing page*

contributes almost 24% of the transport rate in uniform flow, where the velocity is high enough that the mean velocity is probably above critical conditions.

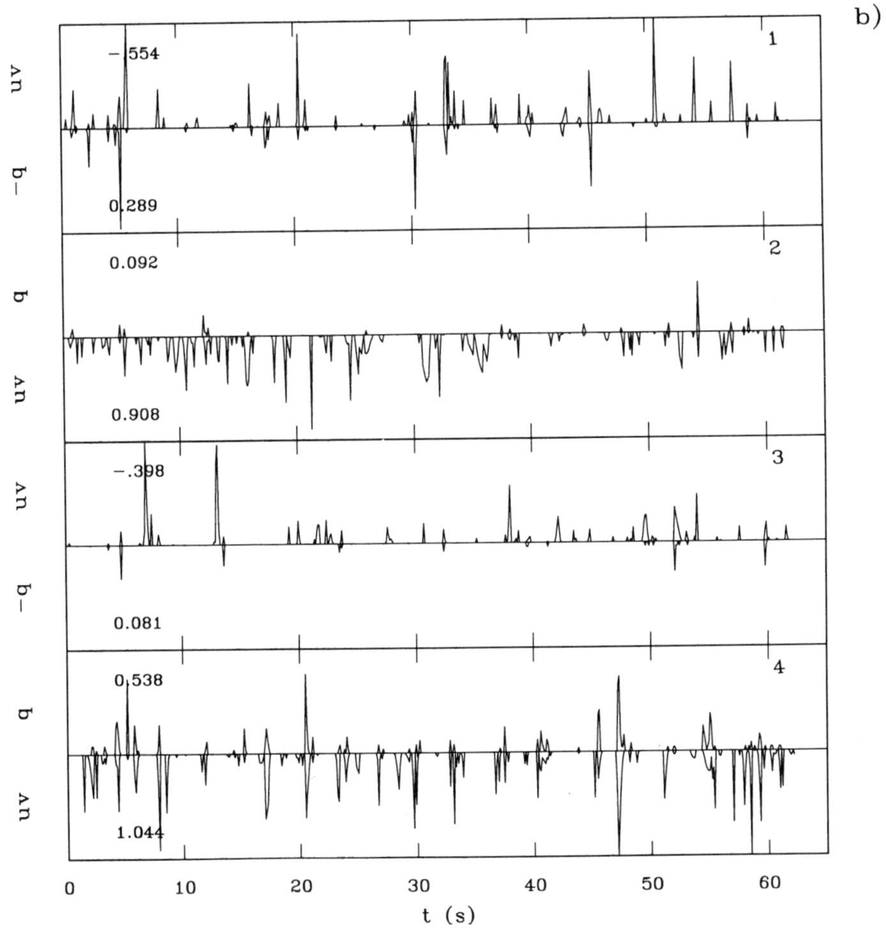

Figure 10.4. *continued*

SHEAR STRESS STRUCTURE OVER BEDFORMS

McLean *et al.* (1994) presented detailed measurements of mean flow and turbulence over fixed forms of height 40 mm and length 800 mm. Some of these data are presented in Figure 10.5. Here the water depth was approximately 160 mm and the mean velocity was 0.42 m s^{-1}. In the bottom panel, the bed topography is shown and in the next panel up, the velocity at 3 mm from the bed is shown. The Reynolds stress at that level appears in the next panel and is repeated in the top panel. These data actually represent averages over 512-point LDV burst measurements made at 2, 3 and 4 mm from the bottom. Also, plotted in the top

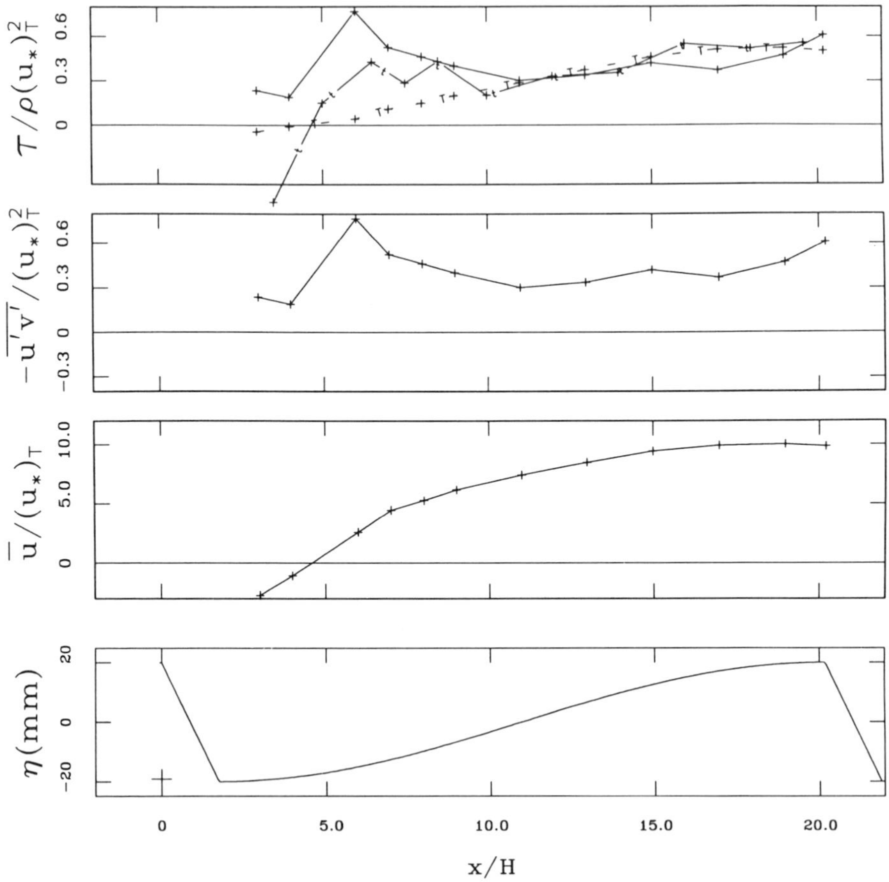

Figure 10.5. Plot of normalized stress and mean velocity as a function of position over a fixed bedform, 40 mm high and 800 mm long (shown in bottom panel). The Reynolds stress is shown in the second panel from the top and repeated in the top panel along with estimates of the boundary shear stress found using a control volume approach, incorporating measured pressure, mean velocity and turbulent stresses (marked with a lower case *t*), and from the slope of a logarithmic velocity profile near the bed (marked by an upper case *T*). Note that toward the crest of the bedform, all the stress estimates yield similar values

panel are estimates of the boundary shear stress made by: (i) assuming a logarithmic profile below 3 mm and assuming a roughness coefficient, y_0, from the results of Nikuradse (1933); and (ii) using a control volume analysis of the volume within 3 mm of the bed and bounded on the bottom by the bed itself and at the ends by the locations of two consecutive measurement profiles. For the latter, both mass and momentum are conserved assuming a logarithmic velocity profile within 3 mm of the bed and setting y_0 using Nikuradse (1933). The pressure is specified from measurements of the bottom pressure from a series of small holes drilled into the bedforms and connected through a manifold of valves to a pressure transducer. The roughness Reynolds number near the crest was in excess of 10 and the y^+ of these measurements is nearly 100, therefore, one would expect the Reynolds stress to be approximately equal to the boundary shear stress. And indeed, toward the crest all three estimates of the bottom stress are nearly the same. This is not the case near the point of reattachment because the developing internal boundary layer there is quite thin and even measurements at 3 mm from the bed actually are from within the wake of the separation zone. Therefore, below 3 mm there can be large gradients in the flow that are not included in the control volume calculations.

Subsequent to the investigation of the flow over the fixed, dune-shaped forms, additional measurements were made over steeper, more ripple-like forms, having height 40 mm and length 400 mm. In this case, 2048-point data sets were gathered at 2 mm from the bed in addition to vertical profiles of 512-point measurements at 1 mm spacing near the bed. In Figure 10.6, both the 2048-point, 2 mm data and the 512-point, 2, 3 and 4 mm data are shown for this steeper bedform case. The control volume calculation for both data sets are also included in the top panel, but for clarity, the Reynolds stress which appears in the second panel from the bottom has been omitted from the top panel. The 2 mm data, which appear here as the larger symbols, clearly indicate that very near the bed the Reynolds stress is virtually zero or even negative. The Reynolds stress taken from 2, 3 and 4 mm is somewhat larger, but still smaller than that expected from the velocity field. The control volume analysis yields disparate results depending on the data set that is input. The results using the 2 mm data yield bottom stresses that are, for the most part, at or below zero, while the data from further off the bed yields a bottom stress that is virtually the same as the stress predicted from the velocity field. This is a puzzling result suggesting that great care is needed in interpreting Reynolds stress measurements in separated flow.

At this juncture, an important point must be emphasized. The process of fitting a logarithmic profile so that it passes through a near-bed velocity measurement and intercepts zero velocity at a roughness length in keeping with Nikuradse (1933), and relating the slope of this profile to the boundary shear stress, is only valid for uniform boundary layers. In flows that are accelerating, there is as yet no concrete evidence that such a methodology yields an accurate estimate of the boundary shear stress. Just because the near-bed mean velocity varies logarithmically does not necessarily mean that the slope of that profile is related to the local boundary shear stress.

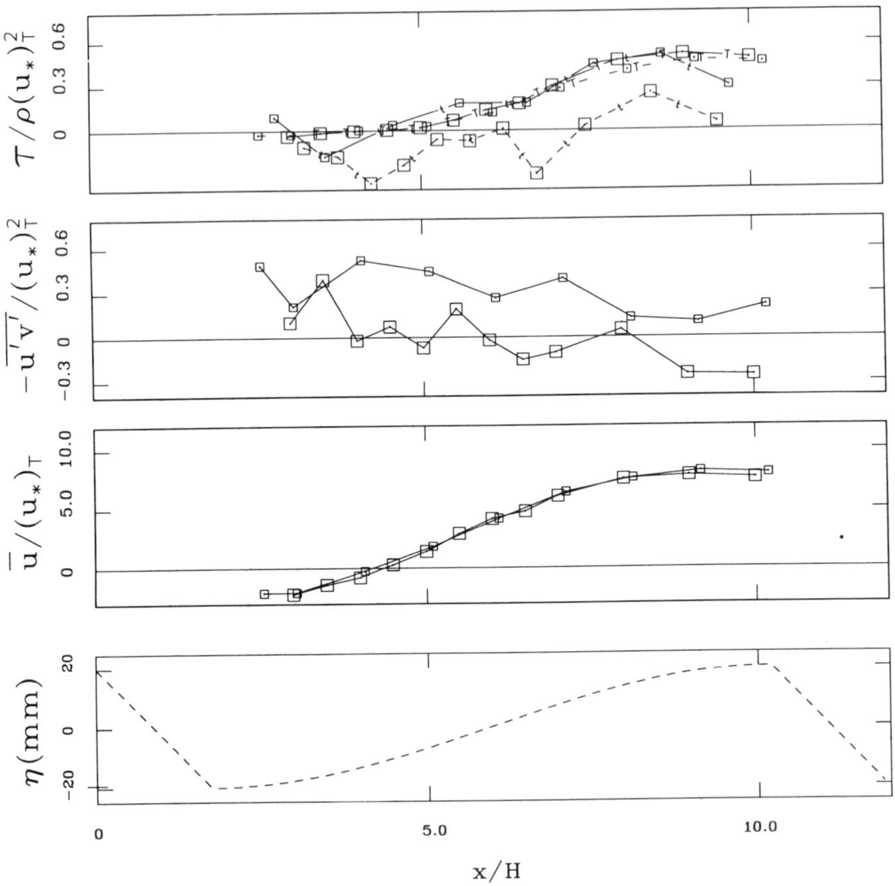

Figure 10.6. Plot of normalized stress and mean velocity as a function of position over a fixed bedform, 40 mm high and 400 mm long (shown in bottom panel; see caption from Figure 10.5). In this case data from a single 2048-point LDV sample at 2 mm (large symbols) as well as combined data from three 512-point samples from 2, 3 and 4 mm (small symbols) are plotted. The Reynolds stress is omitted from the top panel for clarity

PRESSURE GRADIENT EFFECTS DOWNSTREAM OF REATTACHMENT

Models that accurately predict the shape and stability of ripples and dunes must take into account spatial variation of flow characteristics. The size of the separation zone as well as the turbulence field downstream of the reattachment point are strongly affected by the pressure gradient in the flow. The data from the backward step and the bedforms with two different aspect ratios form a set of measurements from flows of similar velocity and water depth that have progressively larger favourable pressure gradients, and hence larger acceleration. This provides an opportunity for comparison. In Figure 10.7a, the statistics of the downstream component of velocity for the three cases are plotted versus position downstream from the point of reattachment. These measurements are all from an average distance of 5 mm from the bed in order to make the comparisons as accurate as possible. Consequently, some of the results shown here differ somewhat from data presented in previous Figures that were centred at a distance of 3 mm from the bed. Also

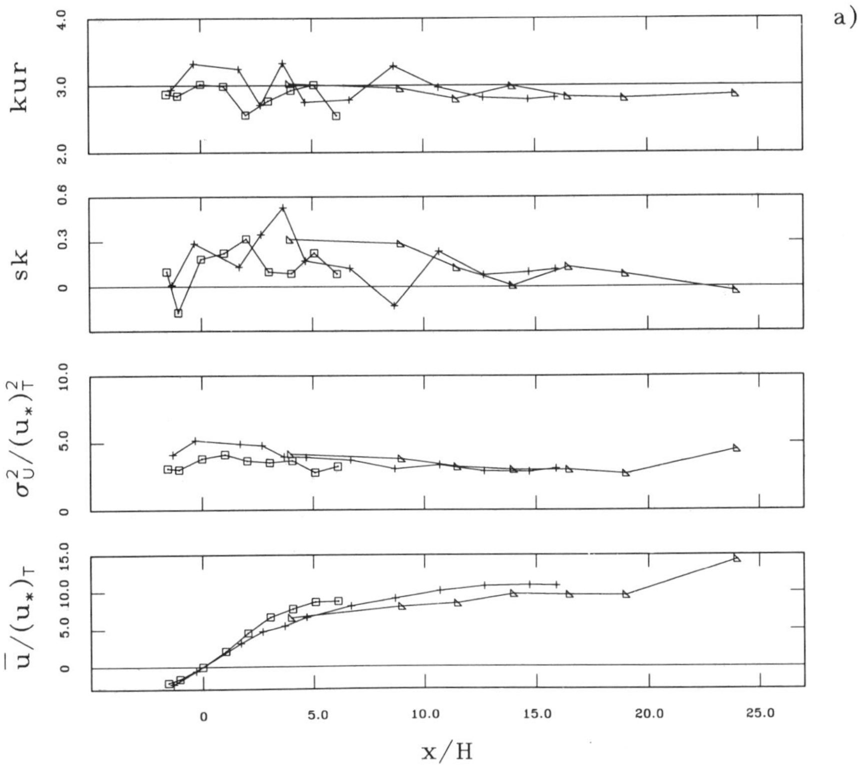

Figure 10.7. Plot of: (a) downstream velocity; and (b) 'vertical' velocity statistics as a function of position downstream of the point of reattachment for three different topographic conditions: (1) flat bed downstream of a 40 mm high backward step (triangles); (2) dune-shaped feature 40 mm high, 800 mm long (pluses); and (3) steeper bedform 40 mm high and 400 mm long (squares). The farthest triangle on the right represents data from a uniform boundary layer. *Continued on next page*

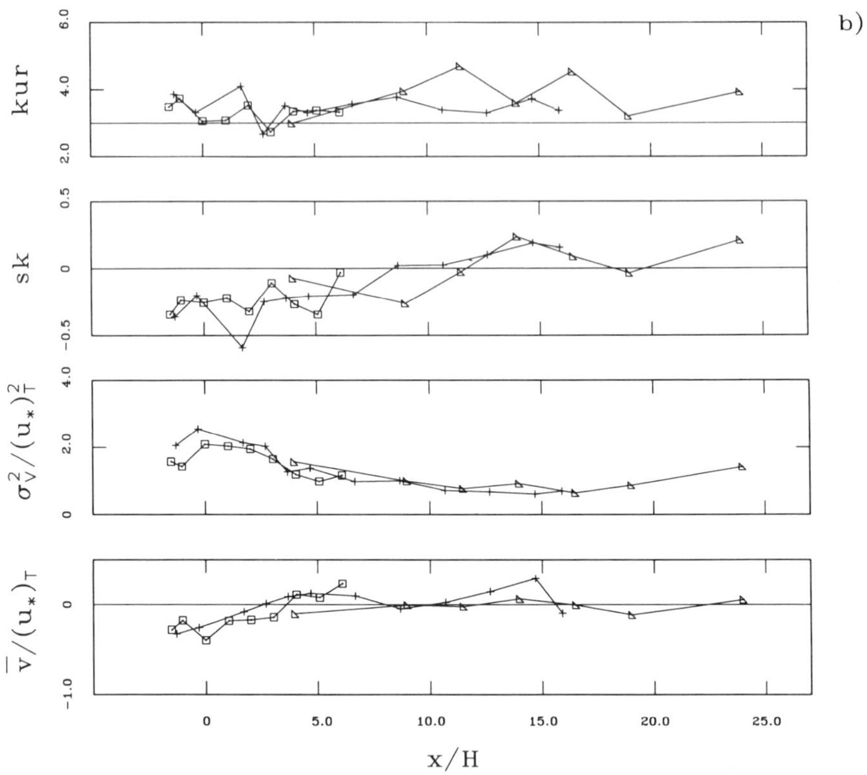

Figure 10.7. *continued*

included are data from the uniform boundary layer; these appear as the farthest downstream point of the backward step data (triangles). The mean and variance are normalized using the shear velocity associated with the total boundary shear stress. For the two bedform cases, the total boundary shear stress was estimated using pressure measurements and from the slope of the free surface. For the backward step, a drag coefficient was used to calculate the form drag which was combined with the measured Reynolds stress to arrive at a reasonable total shear stress estimate. For the uniform case, the data were normalized by the shear velocity derived from the measured Reynolds stress.

The mean velocity, shown in the bottom panel of Figure 10.7a, clearly displays the differences in acceleration. The steepest bedforms, which provide the most obstacles to the flow, exhibit the smallest normalized maximum velocity, in keeping with the fact that the maximum velocity is inversely related to the friction velocity. The variance of the downstream velocity, shown in the second panel from the bottom in Figure 10.7a, displays surprisingly little structure for all three cases.

Interestingly, because of the higher total stress due to the added form drag and because the actual turbulence levels are much higher in the separated cases, the normalized variance for the separated cases near reattachment is about the same as for the uniform boundary layer case. Except for a tendency for the skewness to be positive near reattachment, decreasing toward zero for the uniform case, it is difficult to draw strong conclusions from the higher moments presented in Figure 10.7a. The positive skewness indicates that there are infrequent large positive fluctuations; this is in keeping with the observed large transporting events near reattachment in the lee of the negative step.

In Figure 10.7b, the statistics of the vertical velocity component are shown. It should be noted that this is actually the velocity perpendicular to the local bed topography, rather than that defined by gravity. In this case, there is more structure in the variance, with a marked peak in the normalized variance near reattachment, decreasing to values less than that of the uniform case. The negative mean velocity near reattachment is in keeping with the downward velocity associated with the flow expansion caused by the bedforms. Unlike the downstream velocity, the vertical skewness is negative near reattachment, increasing toward slightly positive values with distance downstream. Negative skewness indicates infrequent, but large, downward velocities, in keeping with the dominance of quadrant-4 relative to the other quadrants, as seen in Figure 10.2b.

The average of the product of u' and v' from the first and fourth quadrants, normalized by the square of the shear velocity associated with the total stress, is plotted in Figure 10.8. Here the effects of acceleration are clearly seen. In all cases, the absolute value of contributions from both quadrants is larger than average near reattachment, but then they decrease with distance downstream. However, the rate of decrease is much greater for the steeper bedform as compared to the backward step case. These effects are further seen in Figure 10.9 which includes both the correlation coefficient and the ratio of the contributions from quadrants 1 and 4. The trends are different for each case; the correlation coefficient for the backward step and the dune-shaped bedform is small near reattachment, increasing with distance downstream toward the expected uniform value of about 0.4 (Lu & Willmarth, 1973). The increase is more gradual for the backward step case. On the other hand, for the steep bedform correlation decreases downstream, and for measurements closer to the bed, r is even negative near the crest. The ratio of the two quadrants also exhibits different characteristics for the different cases; (i) increasing downstream for the steeper bedform; (ii) varying randomly without a noticeable trend for the dune shape; and (iii) decreasing for the backward step.

Nelson *et al.* (1993) suggested that adjustments in turbulence characteristics due to topographic forcing are crucial to the evolution of bedforms on an erodible bed. This can be illustrated by considering the development of an isolated bedform downstream of an obstacle. Initially we assume that the obstacle is sitting on an otherwise flat bed, reminiscent of the backward step case. The bottom panel in Figure 10.1 indicates that there will be a maximum in the transport rate

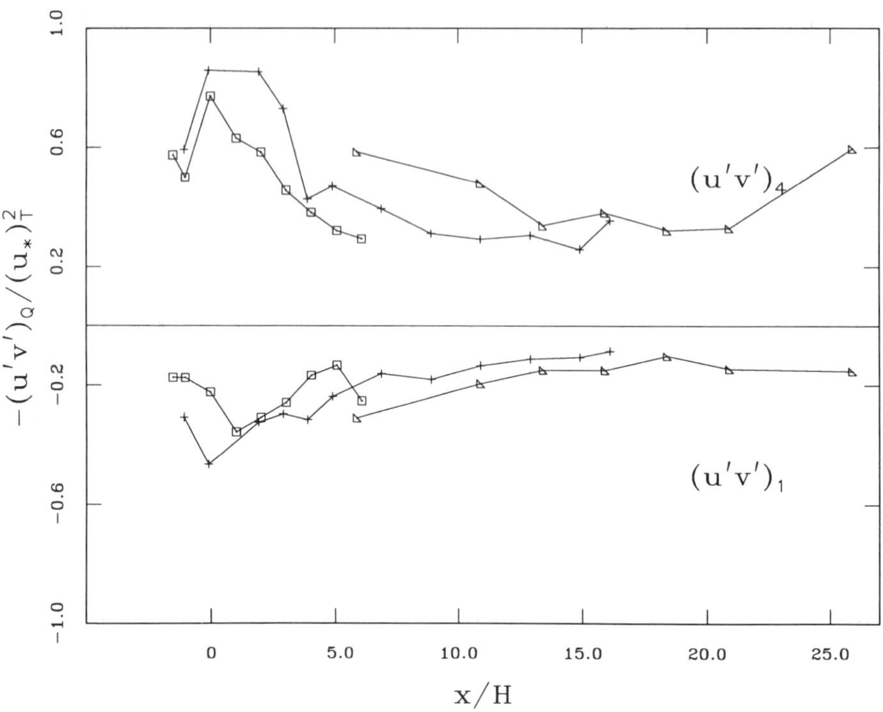

Figure 10.8. Plot of the average product $u'v'$ for quadrants 1 and 4, normalized by the square of the shear velocity associated with the total boundary shear stress, for three topographic conditions (see Figure 10.7)

downstream of the point of reattachment that will lead to erosion upstream of this point and deposition downstream. The maximum exists not simply because of a maximum in the velocity or the boundary shear stress, but also because of gradients in turbulent characteristics. Just downstream of reattachment, the correlation coefficient is low, there are high turbulence levels and there is enhanced quadrant-1 activity. Even though the mean velocity and stress are small (perhaps below critical conditions), this situation is especially efficient for transporting sediment. Although the velocity increases with distance downstream, thus increasing the transport potential, anomalous turbulence characteristics decay, reducing the associated transport enhancement, resulting in the observed transport maximum. For the initial stages, the variation of the turbulence characteristics is slow as is the increase in the mean velocity. Therefore, the feature that forms initially will be long and low. As it grows, however, a favourable pressure gradient will develop that accelerates the flow more rapidly and causes the adjustments in the turbulence to occur closer to the reattachment point. The net effect is to move the point of maximum growth closer to the point of reattachment. If the obstacle that began this process were in fact the crest of a bedform that was itself moving, the feature

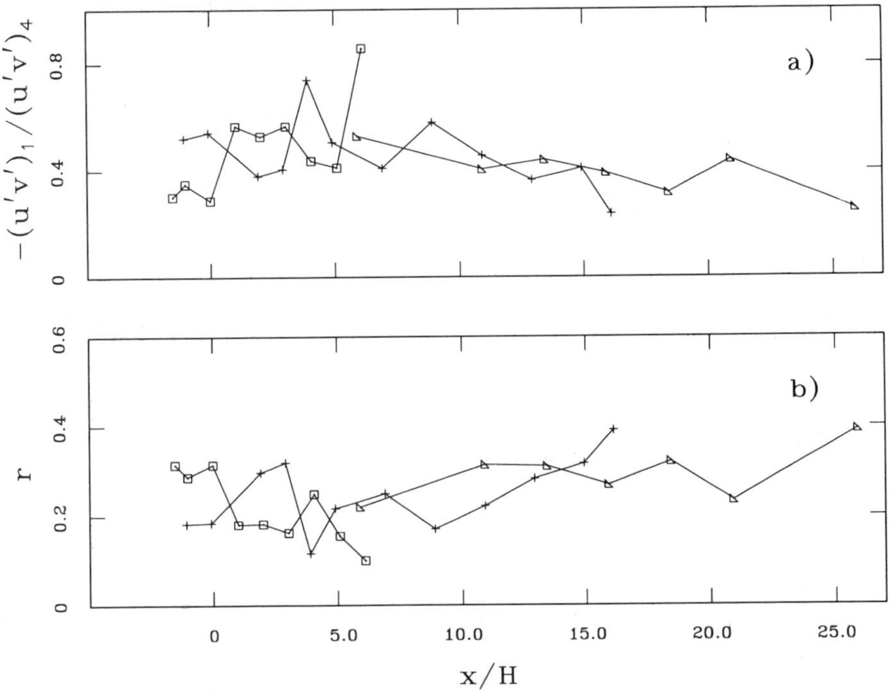

Figure 10.9. Plot of the ratio of the contributions from quadrants 1 and 4 to the Reynolds stress as well as the correlation coefficient as a function of position downstream of the point of reattachment for three topographic conditions (see Figure 10.7)

that grows downstream would begin as a long, low, fairly symmetrical form that would migrate and gradually become asymmetrical, leading to separation in its lee. Eventually, the acceleration caused by the increased height of this downstream bedform will cause the transport maximum to occur at its crest and the wavelength will then be set. Thus, the interaction between the accelerating flow and the decaying turbulence in the separation wake determines the shape and evolution of the bedforms.

SUMMARY AND CONCLUSIONS

Separated flow, common to many naturally occurring bedforms, produces turbulence that is significantly different from that occurring in normal, uniform boundary layers. Both the magnitude and the variability in these differences have many ramifications, especially for the evolution of an erodible bed in the vicinity of such separation. Observations include:

1. abnormally low correlation coefficients that are indicative of larger than normal contributions from quadrants 1 and 3 relative to quadrants 2 and 4 of the $u' - v'$ distribution;
2. although quadrant-1 represents a negative contribution to the Reynolds stress, fluctuations occurring in this quadrant are quite efficient at transporting sediment;
3. although the frequency of occurrence of these quadrant-1 events does not appear to be heightened significantly, the average magnitude is greater. As a result, the transport is higher than would be expected for uniform flow;
4. near the point of reattachment, where the velocity fluctuations are very large, there are occasional events that are sufficient to move sediment, even though the mean velocity is well below critical conditions for incipient motion; and
5. the interaction of the enhanced transport due to the anomalous turbulence characteristics and the acceleration downstream of reattachment, due to both wake decay and topographic effects, create a transport maximum that depends strongly on the topography and thus provides a feedback mechanism that sets the size and shape of the bedforms.

REFERENCES

Bagnold, R.A. 1973. 'The nature of saltation and of bed-load transport in water', *Proc. Roy. Soc. London*, **A332**, 473–504.

Engelund, F. & Skovgaard, O. 1973. 'On the origin of meandering and braiding in alluvial streams', *J. Fluid Mech.*, **57**, 289–302.

Fredsøe, J. 1982. 'Shape and dimensions of stationary dunes in rivers,' *J. Hydraul. Div., ASCE*, **108**, 932–947.

Heathershaw, A. D. 1974. '"Bursting" phenomena in the sea', *Nature*, **248**, 394–395.

Jackson, R.G. 1976. 'Sedimentological and fluid-dynamic implications of the turbulent bursting phenomenon in geophysical flows', *J. Fluid Mech.*, **77**, 531–560.

Lu, S.S. & Willmarth, W.W. 1973. 'Measurements of the structure of the Reynolds stress in a turbulent boundary layer', *J. Fluid Mech.*, **60**, 481–511.

McLean, S.R. & Smith, J.D. 1986. 'A model for flow over two-dimensional bed forms', *J. Hydraul. Engrg.*, **112**, 300–317.

McLean, S.R., Nelson, J.M. & Wolfe, S.R. 1994. 'Turbulence structure over two-dimensional bedforms: Implications for sediment transport', *J. Geophys. Res.*, **99**, 12729–12747.

Meyer-Peter, E. & Müller, E. 1948. 'Formulas for bed-load transport', *IAHR, 2nd Mtg.*, Stockholm, 39–64.

Nelson, J.M. & Smith, J.D. 1989. 'Evolution and stability of erodible channel beds', in Ikeda, S. & Parker, G. (Eds) *River Meandering*, AGU Water Resources Monograph **12**, Washington, 321–377.

Nelson, J.M, McLean, S.R. & Wolfe, S.R. 1993. 'Mean flow and turbulence fields over 2-dimensional bedforms', *Water Resour. Res.*, **29**, 3935–3953.

Nelson, J.M., Shreve, R.L., McLean, S.R. & Drake, T.G. 1995. 'Role of near-bed turbulence structure in bedload transport and bedform mechanics', *Water Resour. Res.*, **31**, 2071–2086.

Nikuradse, J. 1933. *Strömungsgesetze in Rauhen Rohren*, Kaiser Wilhelm-Institut für Stromungsforschung, Göttingen, Forschungsheft **361**, 22 pp.

Wiberg, P.L. & Smith, J.D. 1987. 'Calculations of the critical shear stress for motion of uniform and heterogeneous sediments', *Water Resour. Res.*, **23**, 1471–1480.

Wiberg, P.L. & Smith, J.D. 1989. 'Model for calculating bed load transport of sediment', *J. Hydraul. Eng.*, **115**, 101–123.

Yalin, M.S. 1963. 'An expression for bed-load transportation', *J. Hydraul. Div., ASCE*, **89**, 221–250.

NOTATION

A	cross-sectional area of sediment particle
C_D	drag coefficient
C_L	lift coefficient
C_m	added mass coefficient
D	mean sediment diameter
$\vec{F}$	force exerted by flow on sediment particle
F_g'	wetted weight of a sediment particle
F_M	ratio of the moments acting on a particle on the bed
F_x F_y	downstream and vertical components of $\vec{F}$
g	acceleration of gravity
H	bedform/step height
$\vec{n}_L$	unit vector in the direction of the lift force
q	instantaneous transport rate (number of grains cm^{-1} s^{-1})
q_s	bedload transport rate (m^2 s^{-1})
$\vec{u}$	instantaneous fluid velocity
$\vec{u}_s$	instantaneous sediment particle velocity
u v	downstream and vertical components of instantaneous fluid velocity
u' v'	downstream and vertical components of fluctuating fluid velocity
u_D	downstream velocity extrapolated to height $D/2$ above bed
$\vec{u}_\Delta$	velocity of fluid relative to moving sediment particle
u_Δ v_Δ	downstream and vertical relative velocity components
V_s	sediment particle volume
x y	downstream and vertical co-ordinates
y_0	roughness parameter
y^+	non-dimensional distance from bed $(= u_* y / \nu)$
u_*	shear velocity $(= \sqrt{\tau_b / \rho})$
α	constant
ν	kinematic viscosity
ρ	fluid density
ρ_s	sediment density
τ_b	boundary shear stress

τ^* non-dimensional boundary shear stress ($= \tau_b/(\rho_s - \rho)gD$)
τ^*_{cr} non-dimensional critical boundary shear stress ($= \tau_{cr}/(\rho_s - \rho)gD$)
Φ non-dimensional bedload transport rate
ϕ_0 angle of repose of particle lying on bed of similar particles

11 The Role of Coherent Structures in Developing Bedforms During Sediment Transport

ALBERT GYR & ANDREAS MÜLLER
Institute of Hydromechanics and Water Resources Management, Federal Institute of Technology, CH-8093 Zürich, Switzerland

ABSTRACT

In modelling sediment-laden flows using numerical or physical methods, the scaling laws of both the flow and the transport have to be known. It is shown herein how bedforms reflect the interaction of 'coherent' flow structures with the size of the grains, D, or the scales of the previous bedforms specified by their height h. The criteria for the development of different bedforms depend on the influence of viscosity on these interactions. The following ranges are found: $D^+ < 5$ *for the maintenance of a smooth bed;* $5 < D^+ < 12$ *or* $h^+ < 30$ *for ripples and their preform;* $D^+ < 12$ *or* $h^+ < 30$ *for three-dimensional bedforms; and* D^+, $h^+ < 70$ *for rough wall conditions. Dunes can exist when* $h^+ > 100$. *In the intermediate range, sediment transport requires a time-resolved description.*

INTRODUCTION

Scales are important when a flow has to be understood or modelled by numerical or physical methods. Here, the importance of such scaling parameters for a sediment-laden turbulent flow continuously interacting with a bed of movable sand grains will be discussed. The aim of this study is to investigate the dependence of sediment transport on turbulent flow, especially in the case of sand grains which are small compared to the size of the so-called coherent structures present.

It is well known that the conditions for dynamic similarity of the flow cannot be achieved with those under which the sediment transport is similar. However, this fact is often irrelevant, especially for flows where the friction factor becomes independent of the Reynolds number, Re; this condition is achieved when $Re > 70$ where $Re = u_* D / \nu$, $u_* = \sqrt{\tau_w / \rho}$, D is the grain diameter, τ_w is the wall shear stress, ρ is the fluid density and ν is the kinematic viscosity. All parameters, if non-dimensionalized by the wall parameters (ν, u_*), are marked by the superscript,

Coherent Flow Structures in Open Channels.
Edited by P.J. Ashworth, S.J. Bennett, J.L. Best and S.J. McLelland.

+. For rough walls, it is possible for similarity given by the Shields parameter $\theta = \rho u_*^2/(\rho_s - \rho)gD$, where ρ_S is sediment density. However, modelling is problematic for the transport of fine sand; in this case, the flow becomes dependent on the Reynolds number. The rough wall assumption often holds, but not when the development of bedforms can be observed. The reason for this failure of the rough wall assumption is that, for fine sand, the roughness criterion is often not satisfied. This problem will be elucidated since it emphasizes the significance of the so-called 'coherent structures' for sediment transport.

At a first view it is surprising that a Reynolds-number-independent similarity can be achieved for the frictional behaviour of the flow but not for sediment transport. As we show below, the reason is due to the time-scales of the two processes. The time-scales of the turbulent structures are irrelevant for the friction factor because the friction factor is the time-average force per unit area scaled with u_*^2 and the contributions of all time-scales are of quasi-equivalent importance. This is not the case for sediment transport, for which Grass (1970) showed that sediment transport occurs whenever the intensity of the sweeps is higher than a certain threshold value in the local and instantaneous wall shear stress. In other words, to describe sediment transport, the intermittent character of the turbulent flow cannot be neglected.

Intermittency is synonymous with structured flow. We avoided the term 'coherent' since it is associated with two different meanings: (i) as an inner coherence, i.e. the structure has a collective motion; and (ii) as an outer coherence, i.e. a synchronization of the events, which implies the existence of a feedback system. We will see that such systems exist, but it seems rather inappropriate to use the term 'coherent' for every kind of structure present in a turbulent flow. We prefer to use the description: 'the flow has structures belonging to multifractal classes'.

The best known structures are ejections and sweeps such as those observed over smooth, flat walls. These are the result of a viscous-dominated, near-wall instability. However, the time-scales for ejections and sweeps are independent of viscosity and occur due to triggering disturbances located in the outer flow. Grass *et al.* (1991) have shown that ejections and sweeps can be interpreted as two features belonging to a more universal structure, namely a generalized 'horseshoe vortex', the natural shear instability. This explains why ejections and sweeps are also encountered as the main structures on rough walls. In that case, however, the eddies shed from the roughness elements would be the main disturbance of the viscous instabilities. This shows up in the lateral scaling, λ, of the structures. Grass *et al.* found an interesting result that $\lambda/k \cong$ constant, where k is the surface roughness scale.

As Grass (1971) showed, sweeps are the relevant structures for sediment transport and are responsible for grain motion. Therefore, it seems important to study the correspondence between the structures of the flow and sediment transport mechanisms.

SWEEP DOMINATED PROCESSES

In time-resolved descriptions of sediment transport, an attempt is made to describe this transport by means of micromechanical processes. They depend on the ratios of the scales of various turbulent structures and the sizes of the sand grains. Table 11.1 shows how the different bedforms correlate with the different parameters.

Let us start with incipient motion. A variety of publications attempt to explain incipient motion by considering the interaction of the flow field with a single grain. However, observations of this process, including those by Grass, clearly show that there is no such thing as the motion of single grains. Incipient motion consists of the movement of whole batches of grains by sweeps. This is evident from the scales of the grains and the sweeps (Table 11.1a). Sweeps have a lateral dimension of $\sim 50(\nu/u_*)$. For grains with a typical $D^+ < 10$, a lateral area of about five grains is simultaneously exposed to the incoming sweep. In the direction of flow, this area is much longer. The study of incipient motion of single grains therefore only makes sense for fairly large grains. As Grass *et al.* (1991) have shown, the lateral scaling of sweeps over a rough wall increases and therefore, even for larger grains, motion in patches is observed (Drake *et al.*, 1988).

Experimental results show that sweeps transport batches of grains intermittently at random locations. Because of the viscous loss of energy, the fluid moving the grains loses its transport capacity and the sand settles, producing a pattern which looks like a horseshoe. The mechanism of the interaction of a sweep with the bed is similar to that described by Falco (1979) in the context of the evolution of so-called pockets.

These horseshoe-like patterns are continuously deformed by new, incoming sweeps. This kind of incipient motion can stop after the relic forms an orange peel-like bedform. For the first time a certain feedback seems to be present and the sweeps are triggered by the grain-scale topographic elevations, which become comparable in size with the lateral scale of the sweeps (Table 11.1b).

At higher sediment transport, which means at higher velocities, the grains start to move again while the sweeps remain fairly stable laterally. The pertinent bedform is streak-like, with sand stripes of an elevation of about one grain diameter and a lateral spacing of 100 viscous units. This scale corresponds to that of two counter-rotating pairs of vortices, well-known in boundary layers over smooth flat walls. The transport between these stripes shows that it is maintained by a series of locally stabilized sweeps. It also shows that sweeps are the origin of the counter-rotating vortices, a fact already recognized in the process of incipient motion. It is important that the time-scale by which the stripes form differs from that of a single sweep. It is an iterative mechanism by a feedback between the forming bedforms and the channelling of the sweeps (Table 11.1c).

At higher wall shear stress, the sweeps become laterally destabilized. A single, non-channelled sweep, can destroy a stripe. It disturbs the pattern, transporting grains sideways, so that a new arrow-like pattern is immediately created. The sand

Table 11.1. Sweep dominated processes

<table>
<tr><th></th><th>Processes</th><th>Interaction</th><th>Dimension</th></tr>
<tr><td>a</td><td>Incipient motion</td><td>Sweep/grains</td><td>$D^+_{xs} \approx 50$ and $D^+_g \approx 10$</td></tr>
<tr><td colspan="4">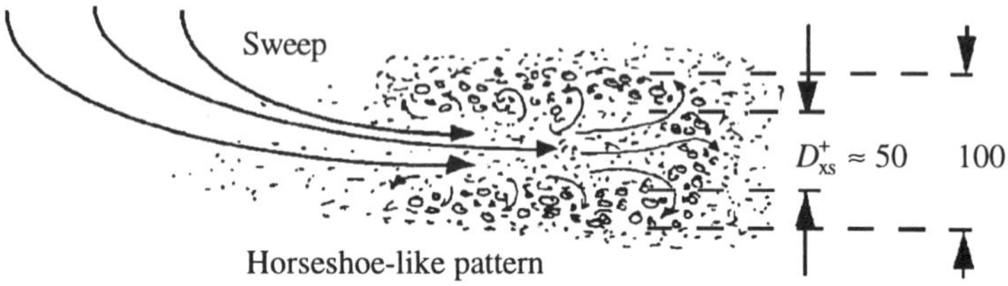
</td></tr>
<tr><td>b</td><td>1st Stabilization (orange peel)</td><td>Sweep/diameter of the elevation of the relics</td><td>$D^+_{xs} \approx 50$ and $D^+_r \approx 10$</td></tr>
<tr><td colspan="4">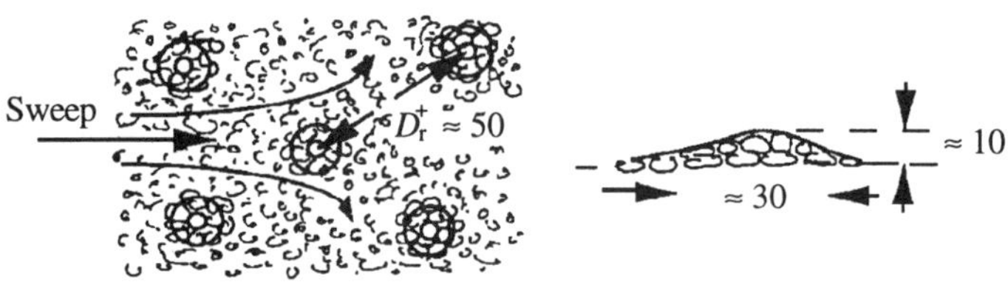
</td></tr>
<tr><td>c</td><td>Sand riblets</td><td>Sweep/λ_z riblets</td><td>$D^+_{xs} \approx 50$ and $D^+_{zr} \approx 100$</td></tr>
<tr><td colspan="4">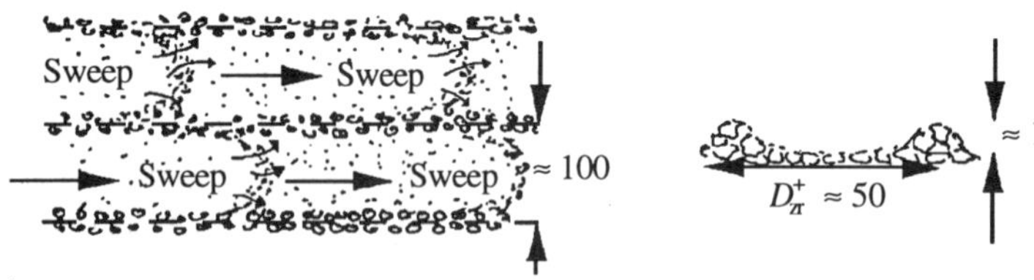
</td></tr>
<tr><td>d</td><td>Rhomboidal pattern</td><td>Sweep/λ_x, λ_z</td><td>$D^+_{xs} \approx 50$, $D^+_x \approx 500$ and $D^+_z \approx 100$</td></tr>
<tr><td colspan="4">
</td></tr>
</table>

elevations form the corners of these patterns. The averaged sweep frequency can be transformed into a length-scale by using the convective velocity in the near-wall region. This scale is identical to the distance between the subsequent arrow pattern (Table 11.1d). The ratio between the scales in flow and lateral directions is about five, which corresponds to the dimensional ratio, λ_x / λ_z, found for single sweeps. This enables one to formulate the arrow pattern as the footprint of two oblique waves, the so-called shear waves, with a five-to-one velocity ratio, producing a 12° arrow pattern. In this situation, sweeps are again organized by a triggering mechanism and can be called coherent.

SWEEP AND VISCOSITY DOMINATED SEPARATIONS FORMING RIPPLES

Ripple formation is described in several papers as a two-dimensional instability which occurs as an interaction of the flow and continuous sediment transport. It seems futile to encourage the stability theoreticians to look at a flow in which ripples start to form. Ripples are formed by a very complex mechanism, starting from the arrow-like pattern, and by no means from a two-dimensional disturbance. This is of utmost interest since it is one of the rare mechanisms by which a three-dimensional pattern turns into a two-dimensional one.

The main mechanism is a viscosity-dominated separation at topographic elevations, i.e. the separation of flow over the arrow-like bedforms and not over single grains. Experimentally it was found that for $D^+ > 12$, the single grain separation overcomes the separation at the pattern and ripple formation is no longer possible.

The length-scale of ripples, at least in the first stage, is identical to the sweep scale obtained by using the Taylor hypothesis. This scale can be interpreted as being associated with large structures. Sweeps are then the instabilities of the near-wall flow due to outer inviscid structures which scale with the whole boundary layer thickness. In this terminology sweeps have a time-scale determined by the diameter of the large structures of the flow. The relation is roughly given by $T_E U_\infty / \delta \approx 5 — 6$, where T_E is the mean time between the events, U_∞ is the outer velocity and δ the depth of the flow, considered to represent the boundary layer thickness. The reproduction process preserves the scale of the arrow-like pattern. The complex micromechanical reproduction mechanism of forming the two-dimensional pattern can be found in Schmid (1985) and a short description of it was given in Müller & Gyr (this volume). The two-dimensional lateral bedforms trigger these large inviscid structures and, therefore, the appearance of the sweeps becomes laterally synchronized. In fact, here we can call them coherent in the sense of an outer coherence. However, although the time-scale becomes inviscid, the flow separation remains viscous dominated. This is why the arrow-like pattern transforms into a two-dimensional one called ripples. We used the so-called outer scaling, although we are aware that it would probably be better to use mixed variables because sweeps and ejections are the features responsible for the momentum exchange between the wall and the outer flow. On the other hand, measurements show that the time-scales based on outer variables are quite good and the results found here lend support for this kind of scaling.

DUNES AND THEIR SCALES

Large flow structures interact with the wall as well as with the free surface. The large scale of the flow can therefore be triggered by an additional scale: the wavelength of gravity waves. As found in nature, water surface waves become dominant when ripples transform into dunes. This metamorphosis was described by stability calculations, which in this case make more sense since the starting instability is really quasi-two-dimensional with a given scale. A mechanistic description, however, has not yet been given and observations show that dunes can develop from a two-dimensional bedform, though only very rarely, while they usually develop from a three-dimensional ripple pattern. This three-dimensional pattern develops when the elevation of the ripples increases and, hence, the viscous separation assumption is no longer valid. When the separation becomes inertia-dominated, a free shear layer forms with its inherent instabilities which introduces new scales. The three-dimensional pattern can again organize into a two-dimensional bedform called dunes. The mechanism by which this organization occurs is still unknown.

The free shear layer developed at both two- and three-dimensional sand patterns form vortical loops which can interfere with the free surface or the trough. From observations, it became evident that sediment transport is due to the interaction of the downward oriented loops with the bed. The flow pattern together with sediment transport play an essential role in dune formation. The two-dimensional case is described by Müller & Gyr (this volume). The two interacting flow structures are the vorticity of separation and the surface waves which trigger the instabilities of the shear layer and with it a spanwise intermittency of sediment transport. The interesting point is that this mechanism does not require very fine sand. Rough sediment can also form dunes if a collective inertial separation occurs. This mechanism has not been studied in detail and it is possible that suspended material plays a significant role under such circumstances. If it is essential, scales like grain settling velocity have to be introduced into the model. On the other hand, the surface waves scale with the Froude number and, therefore, give the scales of the relevant bedforms (Kennedy, 1969). Hence, the coherent structures dominating dune formation and transport are those associated with the free shear layer.

An additional coherent motion is often encountered in sediment transporting channels: so-called secondary motions. They scale with water depth and their occurrence is strongly related to the aspect ratio of the channel. They give rise to so-called 'sand ridges', a bedform not discussed here since the terminology 'coherent structure' is used in its restricted form as small-scale features of the turbulent flow.

THE INFLUENCE OF THE ROUGHNESS

It is known that ejections and sweeps can also be observed above rough walls. The investigations of Grass (1971) showed an ejection-sweep cycle for $D^+ \leq 85$. Grass *et al.* (1991) gave an explanation of this phenomena by introducing a generalized instability process by which roughness changes only the scale in the cascade but not the main resulting structural features of the turbulent flow. The structure over rough walls has some similarity with the original sweeps and ejections because these too are triggered by the so-called large structures of the flow, hence, sediment transport remains intermittent (Drake *et al.*, 1988). However, this is only one aspect of the influence of roughness, that introduced by the sand roughness. As we showed earlier, form roughness starts to play a role and these two roughness are not independent of each other.

One of the main results of this investigation is that we have to distinguish between form roughness with a viscous-dominated separation and forms with a rather inviscid separation, i.e. shear layers. Viscous separation occurs only as long as $D^+ < 12$ and only those bedforms occurring under this condition are described.

The so-called rough wall condition where the friction factor becomes independent of Reynolds number is achieved when $D^+ > 70$. For $12 < D^+ < 70$, three-dimensional bedforms can be observed. They are either relics of ripples or they are similar to the arrow-like patterns probably due to the generalized ejection-sweep cycle. In this context, knowledge of the structures of a turbulent flow over a rough wall and the pertinent bedforms is very limited.

The structures of the flow at the wall are a combination of turbulent outer-flow structures and the separation at single grains and their interaction. As long as the separation at single grains is small, bedforms with viscous-dominated separations can occur. Shed vortices become increasingly more important for vorticity transport from the bed into the outer flow region. With increasing roughness, k^+, the longitudinal turbulent intensity decreases relative to the vertical turbulent intensity. The sweeps become more homogeneous in intensity and therefore, sediment transport also becomes more homogeneous. The ejections also include the separation structure, some of which become so strong that they are capable of sediment transport. The sweeps are, however, still the important features for sediment transport and the change in structures is important. Turbulence remains intermittent, yet the intensity of the intermittent features becomes increasingly more homogeneous and practically every event is more intense than the threshold value needed for the sediment transport. We encounter a saturation in sediment transport limited by the transport capacity of the flow events. Before this state is achieved not every sweep is strong enough, that means the intermittency of the transport is lower than that of the structure. With increasing Reynolds number, the intermittency loses its significance for sediment transport since the wall shear becomes increasingly more homogeneous. At $D^+ \approx 70$ this saturation is achieved.

CONCLUSIONS

The purpose of this study was to investigate how useful the so-called 'coherent' structures are for understanding bedforms in a turbulent flow with sediment transport. We found that it is important to know the intermittent character of the turbulence due to these structures, which also leads to an intermittency in the sediment transport. The results are listed in Table 11.2. For small roughness, if the grains are $12 > D^+ > 5$, we have a regime with turbulent, coherent structures dominating sediment transport and the separation on the forming bedforms is viscous dominated. Only in this small region is it adequate to assume that the structures are similar to those found on flat, smooth walls. Interestingly, it is sediment transport and the proper bedforms which make the turbulent structures even more coherent. It is evident that this regime is of interest for rivers with very fine-grained sediment only.

Table 11.2. The roughness ranges given by their interaction criteria with the flow structures

Roughness type	Inviscid representation	Intermittency considered
$D^+ < 5$	no (smooth)	yes (sweeps)
$5 < D^+$ or $h^+ < 12$ ripples and their preform	no	yes (sweeps)
$12 < D^+ < 70$ or $h^+ < 70$ three-dimensional bedforms	mixed inviscid-viscid	yes (sweeps, bedforms and roughness elements)
D^+, $h^+ > 70$ rough wall	yes	intermittency existing, but not considered to describe the sediment transport
$D^+ \approx 10 — 100$ and $h^+ > 100$ dunes can exist	yes	yes (free shear instabilities)

Coherent structures produced by separation at single roughness elements and their interactions are poorly understood. The intermittency of turbulence due to these structures is no longer a significant parameter for sediment transport and sediment transport can be modelled by a Froude similarity as well. This holds as long as no single roughness element is extremely dominant.

For large bedforms, such as dunes, the class of coherent structures as they appear in a free shear layer are of importance. In this case the intermittency becomes important again; the mechanism, however, is different.

Therefore, we can conclude by citing (Grass *et al.*, 1991): "the breakdown of wall similarity means that differences must be expected between laboratory based observations and corresponding behaviour at geophysical scales".

REFERENCES

Drake, T.S., Shreve, R.L., Dietrich, W.E., Whiting, P.J. & Leopold, L.B. 1988. 'Bedload transport of fine gravel observed by motion-picture photography', *J. Fluid Mech.*, **192**, 193–217.

Falco, R.E. 1979. 'Structural aspects of turbulence in boundary layer flows', in Patterson, G.K. & Zakin, J.L. (Eds) *Turbulence in Liquids*, University of Missouri-Rolla, 1–15.

Grass, A.J. 1970. 'Initial instability of fine sand', *J. Hydraul. Div., ASCE*, **96**, 619–632.

Grass, A.J. 1971. 'Structural feature of turbulent flow over smooth and rough boundaries', *J. Fluid Mech.*, **50**, 233–255.

Grass, A.J., Stuart, R.J. & Mansour-Tehrani, M. 1991. 'Vortical structures and coherent motion in turbulent flow over smooth and rough boundaries', *Phil. Trans. Roy. Soc. London*, **A336**, 35–65.

Kennedy, J.F. 1969. 'The formation of sediment ripples, dunes, and antidunes', *Ann. Rev. Fluid Mech.*, **1**, 147–168.

Schmid, A. 1985. *Wandnahe Turbulente Bewegungsabläufe und ihre Bedeutung für die Riffelbildung*, Unpub. PhD Thesis, ETHZ Nr. 7697, Zürich, 295 pp.

NOTATION

d	boundary layer thickness
D	diameter of the grain
h	height of the bedform
k	roughness scale
Re	Reynolds number
T_E	the mean time between events
u_*	wall friction velocity
U_∞	the mean outer velocity
θ	Shields' parameter
λ	length-scale of the structures
ν	kinematic viscosity of water
ρ	density of fluid
ρ_s	density of sediment
τ_w	wall shear stress

Subscript

g	grain
r	relics

Superscript

+	in viscous units (u_*, ν)

12 Geometrical Analysis of the Feedback Between Flow, Bedforms and Sediment Transport

ANDREAS MÜLLER & ALBERT GYR
Institute of Hydromechanics and Water Resources Management, Federal Institute of Technology, CH-8093 Zürich, Switzerland

ABSTRACT

The interaction between flow, bedforms and sediment transport can be described by a feedback loop. In this loop, the flow changes the bedforms by means of sediment transport, and bedforms represent a boundary condition for the flow with its coherent structures. In this paper, an attempt is made to use this concept for the description of flow structures when ripples and dunes are present. The geometry of the flow field is checked for consistency using topological methods.

INTRODUCTION

Traditional sediment transport formulae relate the wall shear stress, averaged in space and time, to sediment transport. Since the wall shear stress and the shear velocity are efficient scaling parameters, this concept is partially successful, but it cannot describe the local processes in space and time which provide the feedback between flow, bedforms and sediment transport (Figure 12.1). For this description

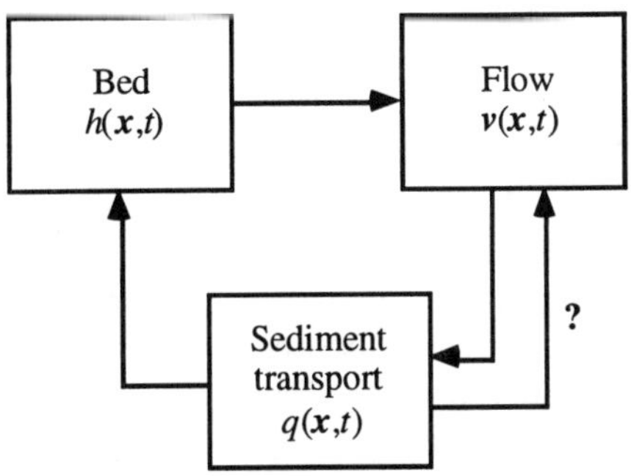

Figure 12.1. Feedback loop between flow, bedforms and sediment transport

Coherent Flow Structures in Open Channels.
Edited by P.J. Ashworth, S.J. Bennett, J.L. Best and S.J. M^c^Lelland.

it is necessary to understand the relevant instantaneous and local flow processes which can be characterized by coherent flow structures. The bedform topography ($h(\boldsymbol{x},t)$, where $\boldsymbol{x}$ is space and t is time) determines the three-dimensional flow field ($\boldsymbol{v}(\boldsymbol{x},t)$) with its coherent structures as a boundary condition, and the flow changes the bedforms by means of sediment transport ($q(\boldsymbol{x},t)$) only. A direct influence of the moving sediment on the flow structures is likely but needs to be explored further.

The concept of coherent flow structures implies a collective motion of fluid which is typical for the flow field under investigation. These structures have a certain spatial extent and life-span and often are characterized by their vortex skeleton, i.e. a conceptual concentration of vorticity in vortex tubes. They have a repetitive, quasi-periodic appearance in time. In this concept the flow field is decomposed into: (i) a flow within a structure given by the vortex skeleton; (ii) a transient evolution of the vortex skeleton from its formation to its decay; and (iii) a downstream convection of the whole structure. It is obvious that there is no unique description of the three-dimensional time-dependent flow and pressure field, $\boldsymbol{v}(\boldsymbol{x},t)$, and $p(\boldsymbol{x},t)$, by coherent structures. They have to be conceived and selected depending on their relevance for the investigation. Therefore, there are 'many animals in this zoo' for two reasons: (i) a variety of different flow structures are present in a given flow; and (ii) different aspects of the same flow feature are emphasized by different researchers, partly depending on their experimental methods.

There are several basic laws of fluid mechanics which are obeyed by flow fields and, therefore, also by coherent structures:

1. The vorticity field ω is solenoidal (i.e. $\nabla \bullet \omega = 0$). Vortex lines never end within the flow and, due to the no-slip condition, they also do not end at the wall.
2. In typical flows with sediment transport, vorticity is only produced by pressure gradients at the wall or by horizontal density gradients.
3. The topology of a flow field is characterized by the nodes and saddles within the flow and half- nodes and half-saddles at the wall (Figure 12.2) in every plane. The number of nodes and saddles fulfils the following condition (Hunt *et al.*, 1978)

$$\left(\sum N+\tfrac{1}{2}\sum N'\right)-\left(\sum S+\tfrac{1}{2}\sum S'\right)=1-n$$

where ΣN is the number of nodes, $\Sigma N'$ is the number of half-nodes, ΣS is the number of saddles, $\Sigma S'$ is the number of half-saddles and n is the connectivity of the flow section. In a two-dimensional flow field a streamline divides the flow in two parts and there is no exchange between them. In three dimensions the division requires a stream-surface. As a consequence, a three-dimensional separation is characterized by a stream-surface bifurcation. Three-dimensional topology is discussed by Hornung & Perry (1984) and Perry & Hornung (1984).

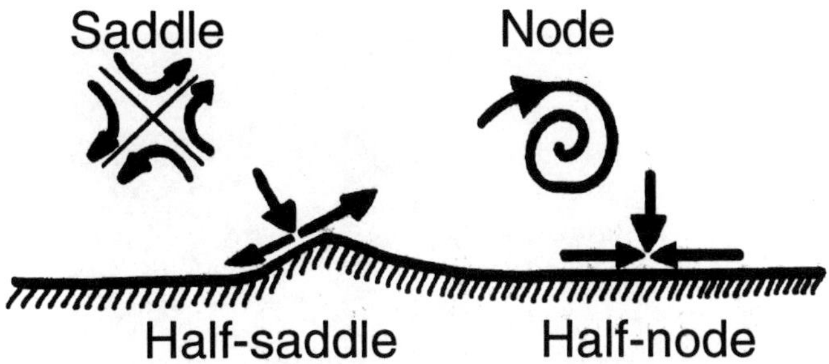

Figure 12.2. Definition sketch of nodes and saddles within the flow, and half-nodes and half-saddles at the wall

COHERENT STRUCTURES OF THE FLOW FIELD ABOVE AN ISOLATED RIPPLE

When the wall shear stress is increased above the entrainment threshold on a uniform, flat sand bed, sediment transport is initiated. Where the bed is deformed by more than 4 viscous units ($d \cong 4\nu/u_*$, where d is the sand grain diameter, ν is the kinematic viscosity of the fluid and u_* is the shear velocity), sediment transport is interrupted just downstream of the ripple which is formed. The deformation of the bed causes a flow separation which is dominated by viscosity.

The flow above an isolated ripple (Figure 12.3a) and its topology was studied by visualization in an experiment (Schmid, 1985). A plastic replica of the ripple showing the typical indentation of the crest (Figure 12.3b) was fixed on a smooth flat wall in a developed channel flow. The local pressure gradients above the ripple lead to the generation of vorticity in the wall layers as indicated in Figure 12.3b. The fluid originating from these layers was rolled in on the two sides leading to the vortex tubes (1) and (3), shown in Figure 12.4, and was convected downstream. The separation vortex (2) behind the ripples was fed from the wall layers on the two sides and from the shear layer on the crest. The fluid entrained into this vortex is flowing out laterally. It was observed that this fluid then flowed first downward at the two edges of the crest inside the vortex tube (2). At the wall it reversed its flow direction and was later convected downstream. The vortex tube (2) was pairing with the vortex tube (1) and (3) of the corresponding side. This bending of the vortex tube (2) toward the wall and the flow reversal causes the vorticity vector to change its direction.

The topology can be completed by drawing the wall stream lines (Figure 12.5). They have two nodes: one where the vortex tube (2) touches the wall and another at the downstream end of the separation zone. In-between there are two saddle points, and three half-saddle points are located at the edge of the ripple. Together with the half-saddle at the upstream stagnation point of the ripple there is an excess of two half-saddles corresponding to the condition of Hunt *et al.* (1978) for a flow domain

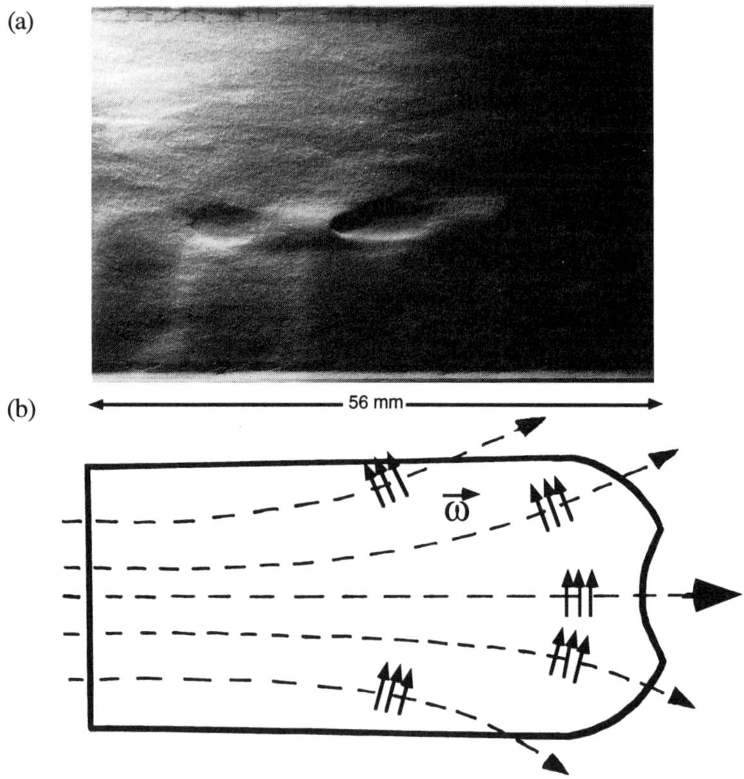

Figure 12.3. (a) Photograph of an isolated ripple; and (b) plastic replica of the ripple with a sketch of streamlines and vorticity vectors

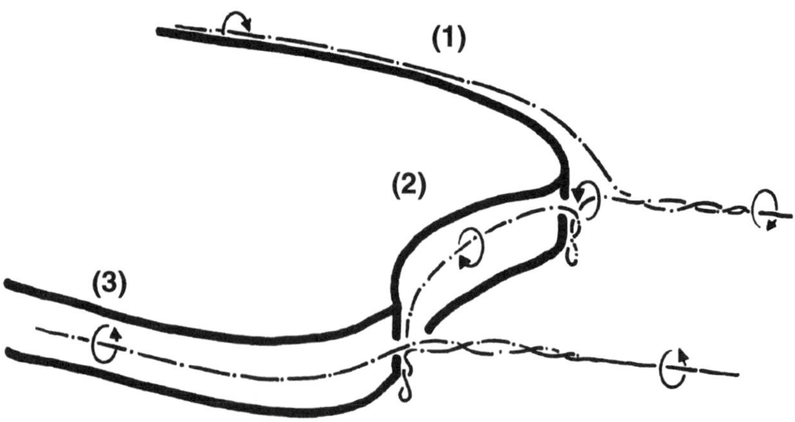

Figure 12.4. Vortex skeleton of the flow above an isolated ripple

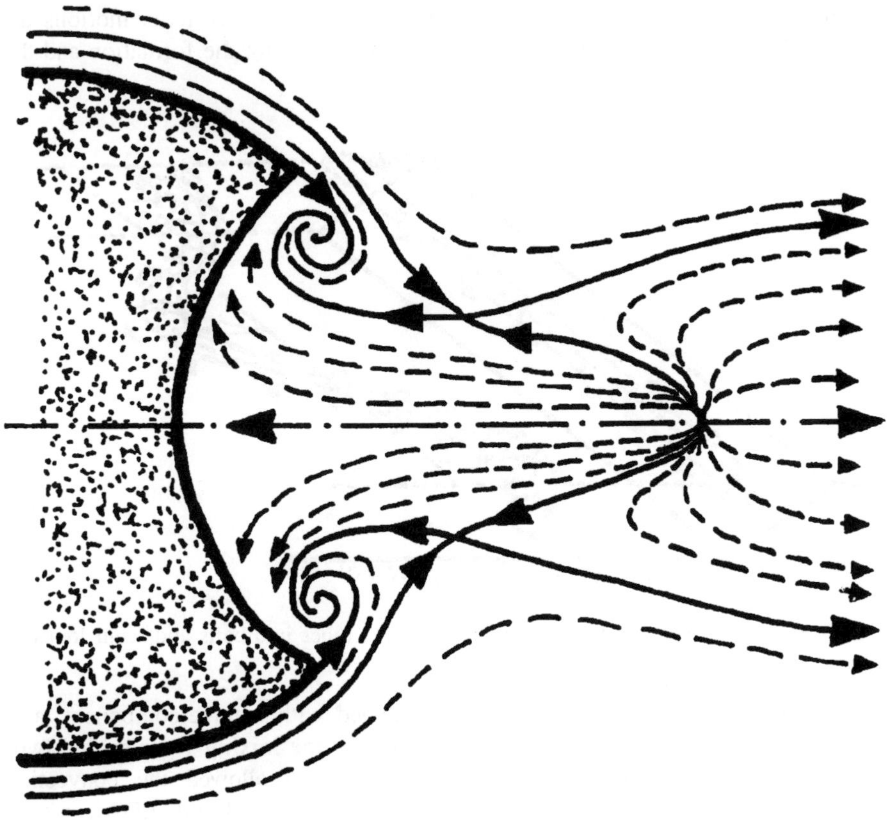

Figure 12.5. Topology of the wall streamlines in the separation zone of an isolated ripple

with $n = 2$. When the equilibrium between the interacting vortex tubes is disturbed by a neighbouring ripple, the fluid of vortex (2) is no longer diverted in the mean flow direction. It flows out laterally which leads to a synchronization of the two ripples and to the formation of an apparently two-dimensional bedform (Schmid, 1985).

DUNES

Dunes appear when the deformation of the bed is large enough such that a free shear layer develops in its lee, as the vorticity of the upstream boundary layer is concentrated in vortex lines. These lines are laterally unstable and form boils which transport low-speed fluid upwards, and scour loops which transport high-speed fluid into the trough (Figure 12.6; Müller & Gyr, 1983, 1986). These latter motions are responsible for the scour in the trough which is necessary for the formation and the migration of the bedform.

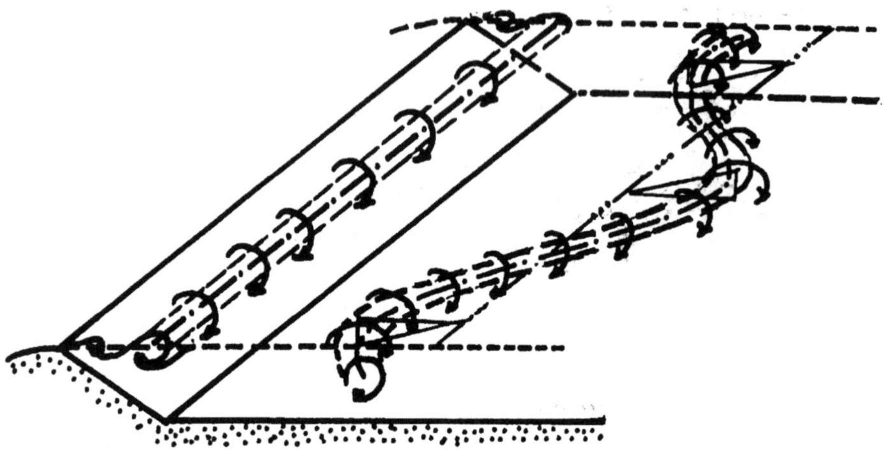

Figure 12.6. Sketch of the evolution of vortex tubes into boils and scour loops

Flow visualization with a movie was combined with a three-component laser Doppler anemometer (LDA) for the investigation of the mixing layer behind two-dimensional dunes (Müller, 1985). The visualization allowed the convection velocity of the boils to be estimated and the velocity measured with LDA can then be transformed into the frame of reference moving with the boil. The streamline pattern in a horizontal plane (Figure 12.7a) shows the two nodes of the vortex tube

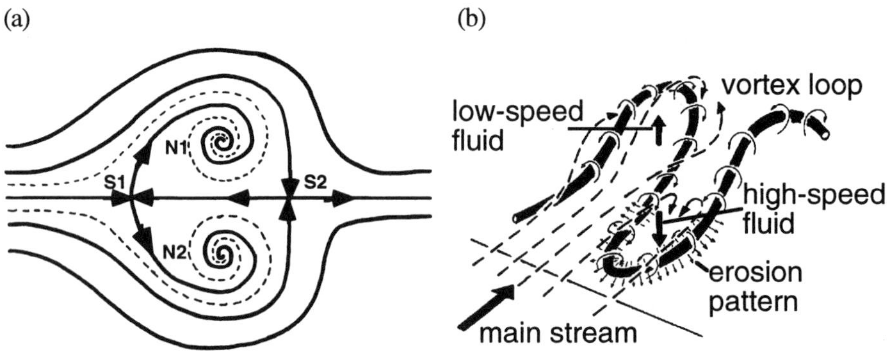

Figure 12.7. (a) Streamline pattern of a horizontal cut of the vortex loop of a boil; and (b) sketch of the three-dimensional flow around a boil

and the upstream stagnation point of the structure. The streamline pattern can be completed by an additional downstream saddle point.

The following observation of the evolution in time of an initially two-dimensional vortex is an indication that the vorticity of opposite sign produced by the return flow in the separation zone also has to be considered to understand the dynamics of the process. Here, the flow was visualized with the aid of a photochromic substance which turns into a dye when illuminated by a strong ultraviolet laser beam (Müller & Wiggert, 1988).

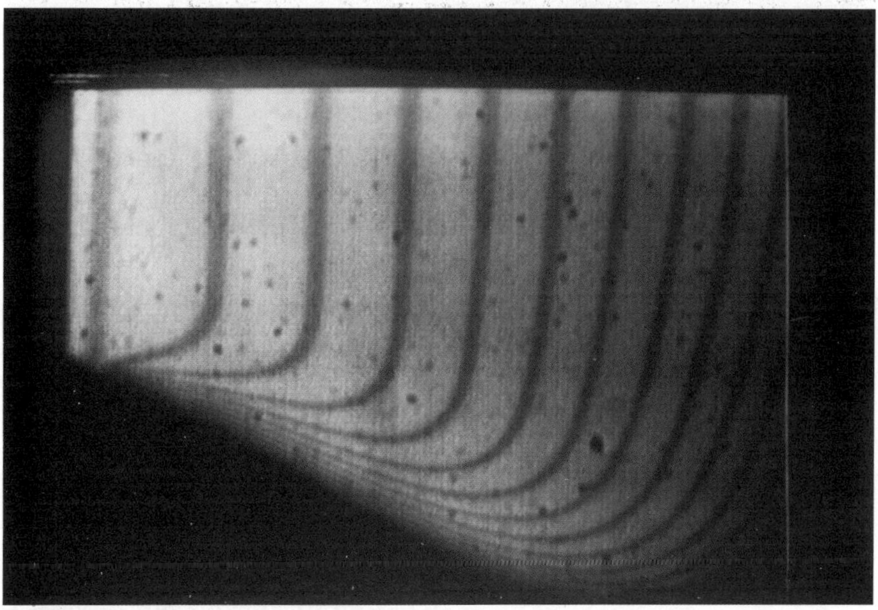

Figure 12.8. Phase 1 of the time evolution of an evolving vortex: growth of the boundary layer

In phase 1 (Figure 12.8), there are no saddles and nodes, as long as the flow does not separate. At separation (phase 2) a half-saddle, S1, forms at the crest (Figures 12.9a and 12.9b) and the reattachment point forms the second half-saddle, S2, which moves downstream. The node, N1, of the nascent vortex compensates the two half-saddles. In phase 3 (Figures 12.10a and 12.10b) the return flow starts to separate, which leads to the half-saddles, S4 and S5, a node, N2, and a saddle, S3. The topological rule requires an additional node, which appears as, N3, and connects the wall layer with the region of concentrated vorticity.

(a)

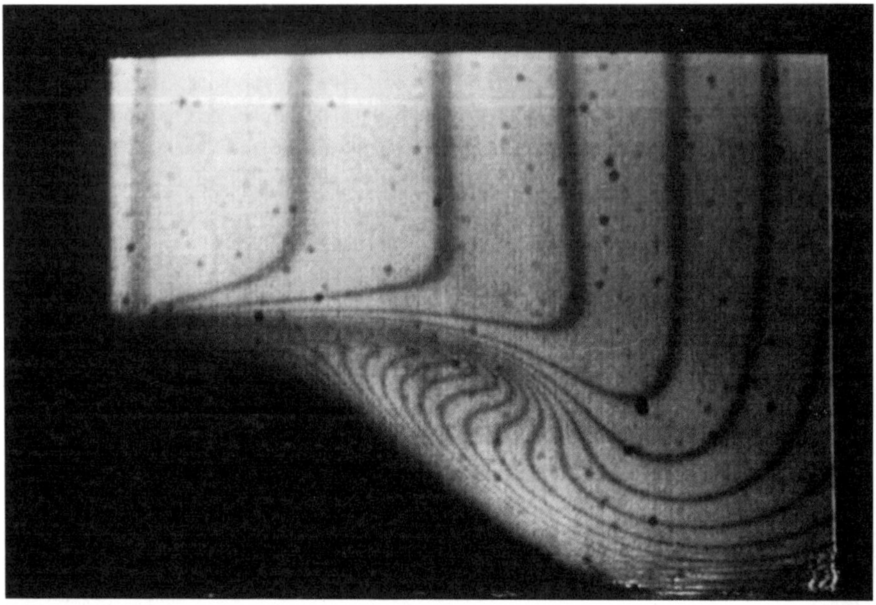

(b)

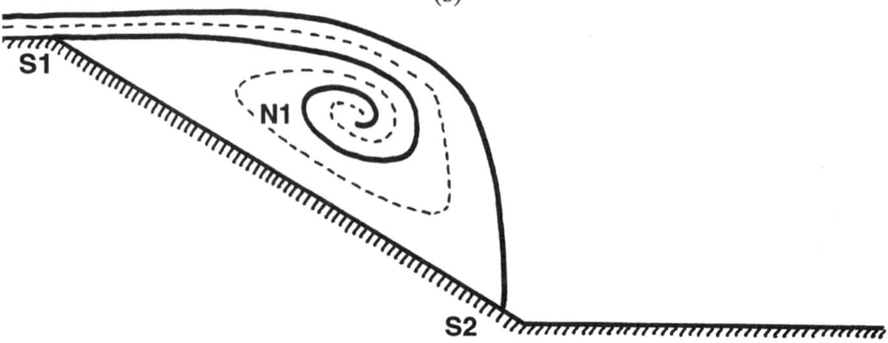

Figure 12.9. Phase 2 of the time evolution of an evolving vortex: (a) separation and formation of the first vortex; and (b) topological streamline pattern of phase 2

CONCLUSIONS

1. Bed topography is related to the fluid flow field via a feedback loop in which the flow changes the bedforms by means of sediment transport and the bedforms represent a boundary condition for the flow with its coherent structures.

(a)

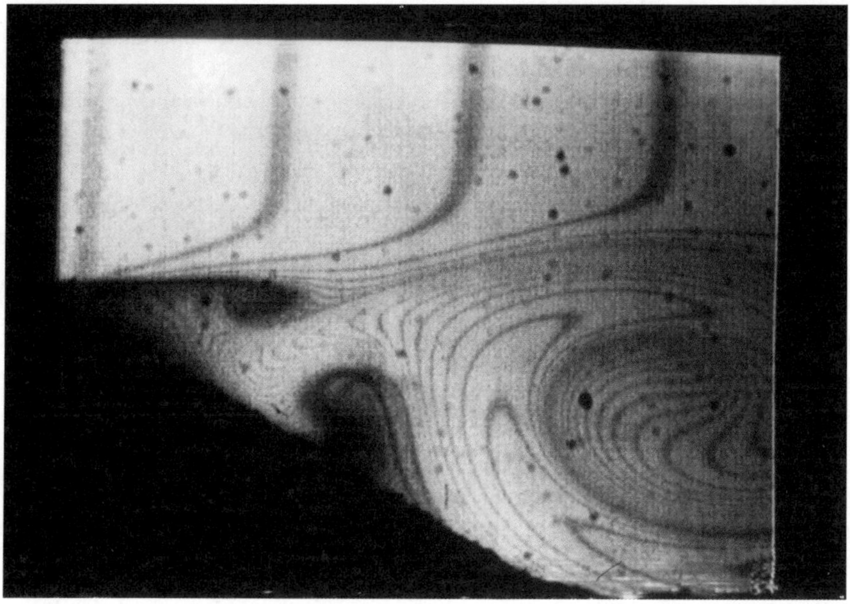

(b)

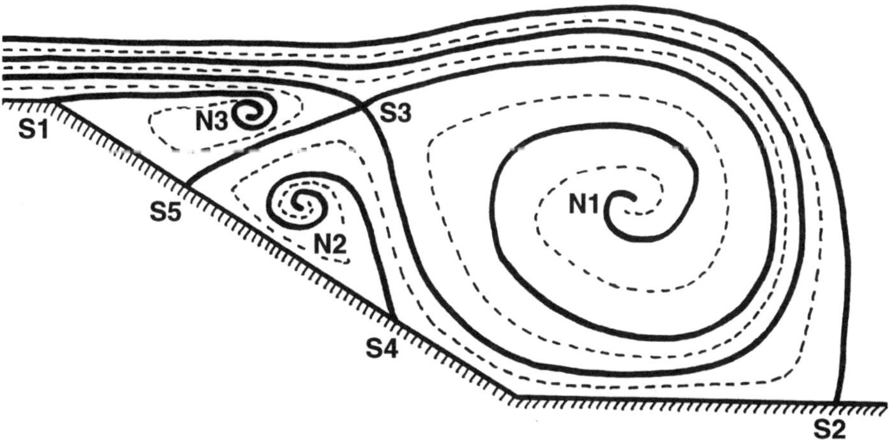

Figure 12.10. Phase 3 of the time evolution of an evolving vortex: (a) separation of the reverse flow, formation of a stagnation point in the flow and two new saddles and two new nodes; and (b) topological streamline pattern of phase 3

2. Ripple formation is linked to a complex three-dimensional flow separation at low Reynolds numbers.
3. The formation and migration of dunes can be linked to coherent structures in the mixing layer behind dunes.
4. Topology is an efficient tool for verifying the consistency of flow field models which incorporate coherent flow structures.

REFERENCES

Hornung, H.G. & Perry, A.E. 1984. 'Some aspects of three-dimensional separation, Part I: Streamsurface bifurcations', *Z. Flugwiss. Weltraumforsch.*, **8**, 77–87.

Hunt, J.C.R., Abell, C.J., Peterka, J.A. & Woo, H. 1978. 'Kinematical study of the flow around free or surface-mounted obstacles; applying topology to flow visualization', *J. Fluid Mech.*, **86**, 179–200.

Müller, A. 1987. 'An experimental investigation of the mixing layer behind dunes with combined flow visualization and velocity measurement', in Chen, C.J., Chen, L.-D. & Holly, F.M., Jr. (Eds) *Turbulence Measurements and Flow Modelling*, Hemisphere Publishing Company, Washington, 229–238.

Müller, A. & Gyr, A. 1983. 'Visualisation of the mixing layer behind dunes', in Sumer, B.M. & Müller, A. (Eds) *Mechanics of Sediment Transport*, Proc. Euromech **156**, A.A. Balkema, Rotterdam, 41–45.

Müller, A. & Gyr, A. 1986. 'On the vortex formation in the mixing layer behind dunes', *J. Hydraul. Res.*, **24**, 359–375.

Müller, A. & Wiggert, D.C. 1989. 'Analysis of turbulent shear flows using photochromic visualization', in Adrian, R.J., Asanuma, T., Durao, D.F.G., Durst, F. & Whitelaw, J.H. (Eds) *Laser Anemometry in Fluid Mechanics*, Vol. IV, Springer-Verlag, Heidelberg, 518–534.

Perry, A.E. & Hornung, H.G. 1984. 'Some aspects of three-dimensional separation, Part II: Vortex skeletons', *Z. Flugwiss. Weltraumforsch.*, **8**, 155–160.

Schmid, A. 1985. *Wandnahe Turbulente Bewegungsabläufe und ihre Bedeutung für die Riffelbildung*, *Bericht R* **22-85**, Inst. für Hydromechanik und Wasserwirtschaft, ETH Zürich.

NOTATION

d	diameter of sand grains
$h(\boldsymbol{x},t)$	bed topography
n	connectivity of the flow section
$p(\boldsymbol{x},t)$	pressure field
$q(\boldsymbol{x},t)$	sediment transport
t	time
u_*	shear velocity
$\boldsymbol{v}(\boldsymbol{x},t)$	velocity vector
$\boldsymbol{x}$	space
ν	kinematic viscosity of water
ΣN	number of nodes

$\Sigma N'$	number of half-nodes
ΣS	number of saddles
ΣS	number of half-saddles
ω	vorticity vector

13 Genesis and Morphology of Erosional Shapes Associated with Turbulent Flow over a Forward-facing Step

A. POLLARD, N. WAKARANI
Department of Mechanical Engineering, Queen's University, Kingston, Ontario, K7L 3N6, Canada

J. SHAW
Department of Geography, University of Alberta, Edmonton, T6G 2H4, Canada

ABSTRACT

Landforms in deserts, in areas scoured by catastrophic outburst floods from Pleistocene glaciers and in areas once covered by continental ice sheets include erosional grooves extending down-flow from the edges of forward-facing escarpments (steps). These grooves define intervening ridges which are known by a variety of names depending on environment: yardangs, fluting, rat tails and rock drumlins. They range in scale from millimetres to tens of kilometres. The coincidence of grooves and steps facing into the flow suggests that the steps may be responsible for flow structures which form the grooves.

In this paper, we consider laminar, transitional and fully turbulent flow over a step (height, h) inside a plain channel (height, H) of aspect ratio h/H = 0.47 for Reynolds number, based on H, between 1514 and 6724. The flow is investigated using: (i) flow visualization by placing a patch (6h×6h) of erodable material (plaster of Paris) in and at the leading edge of the step; and (ii) through the time-dependent numerical modelling of the Navier-Stokes equations which reveals complex three-dimensional motion near and distal to the step. Surface erosion was obtained under a variety of leading edge modifications that mimic possible landforms, which give rise to different erosion morphologies that can be traced to the turbulence structure. Computations confirm that the structure induces a time-dependent wall shear stress.

INTRODUCTION

Erosional grooves or furrows are noted in nature in a number of environments. They define yardangs, residual ridges between grooves, found in deserts where they are eroded by wind (McCauley *et al.*, 1977). Longitudinal grooves in mud beds are

Coherent Flow Structures in Open Channels.
Edited by P.J. Ashworth, S.J. Bennett, J.L. Best and S.J. McLelland.

mainly meandering although Allen (1982) produced remarkably straight grooves in flume experiments in transitional flow. Grooves are also noted in areas scoured by water released as a result of catastrophic lake drainages (Baker & Nummedal, 1978). In this case, the grooves are scoured into basaltic bedrock and are of relatively large scale with widths and depths measured in metres. Spectacular furrows are noted on the ground where scour resulted from high velocity mixtures of vapour and particulate matter ejected by explosive volcanic eruptions (Kieffer & Sturtevant, 1988).

Fluted bedrock and surficial deposits in formerly glaciated areas are well illustrated by Smith (1948; Figure 13.1). Fluting occurs over seven orders of magnitude in scale (Shaw, 1994). Whereas the fluid responsible for other grooves in nature is not in doubt, there is controversy as to whether the sub-glacial grooves were formed by meltwater or ice (Shaw, 1994). Some argue that the grooves are products of glacial abrasion (e.g. Smith, 1948; Boulton, 1974; Laverdiére *et al.*, 1985). Others use field evidence to suggest that erosion was by turbulent meltwater (e.g. Murray, 1988; Shaw, 1988). Lucchitta (1982) takes glacial grooves on Earth as analogues to argue that similar grooves on Mars were also produced by glacial erosion.

Detailed study of the morphology of grooves, the internal architecture of associated ridges and surface markings on bedrock led Shaw (1988, 1994) to argue that the grooves result from meltwater erosion rather than glacial abrasion. The distribution of striae is telling in this respect: whereas striae are commonly weakly developed or absent in grooves, adjacent bedrock surfaces are strongly striated (Figure 13.2). This argues against formation of the grooves by glacial abrasion which produces striae. Truncation of the internal structure of remnant ridges and the presence of a gravel lag deposit on the erosional surface also point to fluvial rather than glacial erosion: there is no documented glacial process that would create a gravel lag.

Clearly, a better understanding of the genesis of such grooves would enable confident interpretation of their environment and mode of formation. Their topographic positions may give an important clue as to how they form. In this respect, the common appearance of grooves downstream from upstream or forward-facing escarpments or steps is probably significant to their formation (Figure 13.2).

The flow over a forward-facing step is a simplified analogue to flow over escarpments and is often considered to be two-dimensional, especially when the aspect ratio of the flow bounded by an upper wall (or underside of a glacier) is greater than, say, 10:1. However, as shown by Neto *et al.* (1993), the shear layer that develops from the cousin of this flow, that is the flow over a backward-facing step, is highly three-dimensional.

Dimaczek *et al.* (1989) investigated flow over two-dimensional, surface-mounted obstacles at high Reynolds numbers and found that, for all obstacles considered, certain vortical structures could be identified upstream of the obstruction and that these propagated over the obstruction. These structures had a regular spanwise

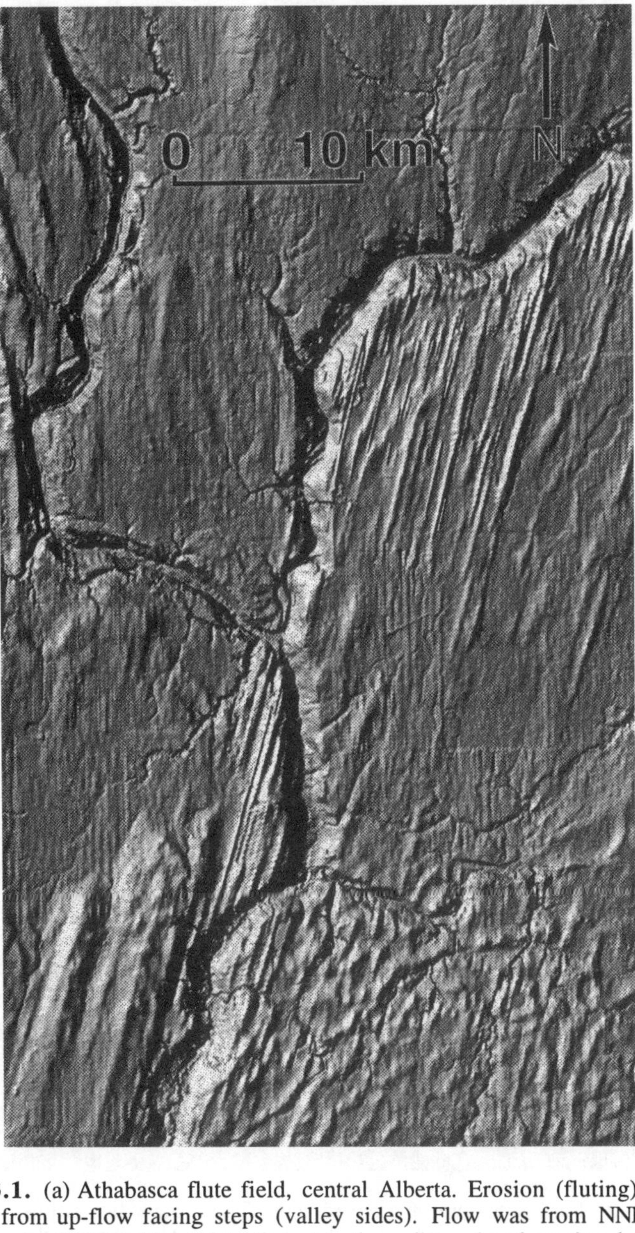

Figure 13.1. (a) Athabasca flute field, central Alberta. Erosion (fluting) that extend downstream from up-flow facing steps (valley sides). Flow was from NNE. Individual ridges may be up to 30 m high. Note the more three-dimensional erosional marks to the south. *Continued on next page*

Figure 13.1. *continued* (b) Grooves cut in hard bedrock (Palaeozoic carbonate) at Wilton Creek, near Kingston, Ontario. The grooves define ridges (light colour) and extend downstream from a low (about 10 m high) escarpment. Note the noses on the escarpment which may have served as obstacles for horseshoe vortices. Flow from NE

Figure 13.2. Surface texture of grooved rock. The groove lies below the prominent rim running across the photograph. Note that the rock surface outside the groove (top) is intensely striated. The surface of the groove (bottom), particularly immediately downflow of the rim, is unstriated in places and only weakly striated elsewhere. Scale 0.13 m. Flow was towards the right

spacing that ranging from 0.4 to 0.8 channel depths. No correlation could be established between the spacing and the obstacle height. The flow in the region up to 1.5 step heights upstream of the obstacle displayed significant, organized three-dimensionality and Dimaczek *et al.* concluded that side-wall effects are not responsible for creating these structures. This flow case was also calculated using large eddy simulation, but with little resolution of factors responsible for the formation of the structures (Werner & Wengle, 1989; Wengle & Werner, 1993).

The separated flow found at the juncture of the downstream wall and the cavity floor (downstream secondary eddy) in a shear-driven cavity flow exhibits similar characteristics to the pre-step separated flow over a forward-facing step. The flow visualization studies of Koseff & Street (1984) in a three-dimensional cavity reveal: (i) at the point of separation, dye forms into discrete spirals (vortices) at fairly regular intervals; (ii) the vortices occur in pairs; (iii) they bear a strong resemblance to Taylor-Görtler vortices; and (iv) the end walls appear to affect the distribution of the vortices.

Freitas & Street (1988) extended the earlier work of Koseff & Street (1984) by carrying out a numerical investigation of the shear-flow in a three-dimensional cavity. This study confirmed that the vortices observed by Koseff & Street were indeed Taylor-Görtler-like.

The present paper addresses issues relevant to transitional phenomena in two-dimensional channel flow. In the study of laminar to turbulent transition in channel flow, Patel & Head (1969) found that the laminar mean velocity profiles display periodic explosions in the hot-wire signal located at the centre plane of the channel at $Re = 1500$ (see their Figure 12), where $Re = \rho UH/\mu$ and U is mean flow velocity in the channel, H is the channel depth and ρ and μ are the density and molecular viscosity of the fluid, respectively. They refer to these explosions as bursts. Further work clarifies the periodic turbulence-like activity (Singer *et al.*, 1989). In transitional channel flow, K- or H-type transition modes can appear. These modes are precursors to breakdown of the flow into broad-spectrum turbulence and occur at Reynolds numbers significantly less than that predicted by linear stability theory (Nishioka *et al.*, 1975). Important to the present work is that the K- or H-modes are symptomatic of the presence of longitudinal streamwise vortices. Singer *et al.* (1989) used Dean vortices to produce spanwise distributions of vorticity in an otherwise laminar channel flow. These vortices fill the whole channel cross-section. They postulated that these vortices, generated either in the boundary layer or as remnants of the inlet conditions, "shift the dominant mode of breakdown from the H-type to the K-type". Interestingly, both types are symptomatic of vortices with axes aligned in the longitudinal flow direction. Also, Aihara *et al.* (1984) argue that streamwise vorticity can be generated on planar surfaces, in a manner similar to Görtler vortices on curved surfaces, if the boundary layer is forced to thicken, as it would, for example, approaching a step change in channel depth (contraction or escarpment).

In this paper, flow over a forward-facing step is studied experimentally and computationally. Experiments include flow visualization and the erosion of a patch of plaster of Paris embedded in the step. A link between the wall structures upstream of the pre-step separation point and the streamwise vortices downstream of the contraction plane is presented. The erosion patterns in the erodable step are found to be very similar to those observed in the field.

EXPERIMENTS

Water in a closed, return flow channel containing a sharp-edged step was used for the flow visualization experiments. It was supplied from a constant head tank and the water discharge was controlled by a calibrated valve. The flow was conditioned as it entered the channel using screens and honeycomb. The channel is 12.7 mm high (H), 0.18 m wide (W) and 2 m long. The step is 6 mm high (h) and 0.9 m long; thus, the forward face of the step is located about $60H$ from the flow conditioning devices. The aspect ratio is $h/H = 0.47$. Flow Reynolds numbers used ranged between $1514 < Re < 6724$, the larger Re being the limit of the rig.

The fluid was marked with fluorescein dye, which was carefully injected into the flow using a 0.635 mm diameter syringe at and parallel to the juncture of the step and the wall containing the step. Standard equipment was used for photography and video recording (see WaKarani, 1992, for details).

A 60×60 mm square by 3 mm deep cavity was inserted 1 mm downstream of and central to the step. It was filled with plaster of Paris and brought to a smooth surface before immersion in the flow. The time-dependent formation and evolution of erosion patterns was recorded by video and still photography.

THE COMPUTATIONS

The unsteady flow of an isothermal, incompressible fluid is fully described mathematically through the basic laws of conservation of mass and, for the case considered here, the Navier-Stokes equations. They need not be presented here.

DISCRETIZATION AND DIFFERENCING SCHEME

The equations are integrated over non-uniformly distributed staggered control volumes (e.g. Patankar, 1980). The convection terms are discretized using a variant of the third-order *QUICKER* scheme of Pollard & Siu (1982), but in a more robust form (see Xiao, 1992). The diffusion terms are discretized using central differencing. Time advancement is accomplished using a fully implicit method. The effect of time step size will be discussed below. The equations were solved using standard pressure-coupling solution algorithms (e.g. Rubini *et al.*, 1992).

SOME COMPUTATIONAL DETAILS AND BOUNDARY CONDITIONS

The computational domain is a section of the channel extending approximately $1.5H$ upstream and downstream of the contraction plane, which was deemed acceptable to capture the relevant flow phenomena. The physical dimensions were 12.7 mm high and with a width (W) of either 30 or 90 mm. The co-ordinate axes are oriented such that the downstream flow direction is along the x-axis, the y direction is perpendicular to the channel floor and the z-axis is across the flow in the spanwise direction.

The step/channel was non-uniformly covered with 28×21×32 control volumes, clustered near the step. Grid dependency tests showed that this grid density was adequate. Further reference to the resolution of the calculations will be made below. The flow was calculated over a time period of 3.4 s. Using a mean time-scale of $H/U = 0.11$ s, ($Re = 1500$), this corresponds to about 32 non-dimensional, turn-over time periods. Time advancement is achieved using 232 time steps to cover the first 0.72 s of flow development. 268 time steps are required to advance the remaining 2.68 s using a fixed time step of 0.01 s. Tests reveal that this number of time steps is more than sufficient to resolve the flow accurately (WaKarani, 1992).

End-view hydrogen-bubble visualization studies of Smith & Shwartz (1983) and Kasagi *et al.* (1986) provide extensive information on streamwise or quasi-streamwise vortices in the near-wall region of a turbulent boundary layer. The term quasi-streamwise is used here to include those streamwise vortices that may possess a significant spanwise and/or wall-normal component. The above mentioned studies confirm the common occurrence of these vortices in the near-wall region, including frequent observation of counter-rotating pairs. "In the simultaneous top and end views by Smith and Shwartz, counter-rotating vortex pairs in the near-wall region were always associated, in space and time, with low-speed streak formation" (Robinson, 1991). The mean spanwise streak spacing (l^+) is about 100 (in non-dimensional wall units; Kasagi *et al.*, 1986).

Kasagi *et al.* (1986) found that solitary quasi streamwise vortices are more common than vortex pairs in the near-wall region. This observation was also made by Guezennec *et al.* (1989). Diameters of the quasi-streamwise vortices (d^+) range between 15 and 50 wall-units with centres occurring predominantly between 20 and 30 wall units away from the wall. From statistics computed by Robinson (1991), the average location of quasi-streamwise vortices from the wall is 30 wall units. The vortex Reynolds number (Re_v) varies between 0 and 80 with an average of 22. However, the histogram of vortex Reynolds number density is significantly skewed, hence, two values, 17.7 and 45, were chosen for the present study.

At the inlet boundary, spanwise (w) and vertical (v) components of velocity are specified such that $w = v$. At each location, spaced 100 wall units apart in the spanwise direction and at 30 wall units away from the wall containing the step, the velocity varies linearly from zero to v at the circumference of a circle. Everywhere else these velocities are set to zero.

A parabolic inlet profile, as opposed to some other profile, is used for convenience. Making the profile uniform, for example, did not alter the flow in such a way that would affect our conclusions. Shear velocity was calculated assuming the flow is turbulent at that Reynolds number.

It must be stressed that the vortices used here produce results that are remarkably consistent with the observations. These vortices were chosen because they are well described. Other types or arrangements of vortices could create the three-dimensional secondary flow found here. Moreover, with increasing Reynolds number the spectrum of sizes of vortices will grow. Consequently, this study is limited by use of only one vortex size. The inlet boundary conditions are summarized in Table 13.1.

Table 13.1. Summary of boundary conditions for simulated turbulence structure

Re	$W/2H$	l/h	Re_v	d^+	l^+
1500	4.72	1.633	17.7	18	62
1500	4.72	1.633	45.0	18	62
1500	14.17	2.833	17.7	20	100
3200	4.72	1.633	45.0	18	62

Using these values, the grid density used, especially in the plane normal to the main flow direction, is approximately three, which compare favourably to those used by Robinson (1991). Note, also, that the non-dimensional time-step ranges from 2×10^{-5} to 2×10^{-4}.

EXPERIMENTAL OBSERVATIONS

Figures 13.3 and 13.4 show plan views of the channel, the step and the fluorescein dye at $Re = 1514$ and 2100 respectively. The step is at the 231 cm scale mark (arbitrary location). The blotches in the Figures are air bubbles trapped beneath the transparent step plate (removable) and the lower wall of the channel. The pictures shown were taken well after the dye had been slowly injected into the corner of the step.

In both figures and upstream of the step, the dye shape or outline is scalloped, with ends that curve towards the step face; these shapes are reminiscent of 'sausages' and will be so referred to here. Dye is ejected at the ends of the sausages and it is only this dye that is seen to pass over the step. Of course, unmarked fluid may also be flowing over the step. These lines of dye will be referred to as 'threads'. An interesting feature in the figures is the alteration in the apparent length and surface or edge definition of the sausages and the wavelength between the threads, which all decrease with increasing Re. The sausages and threads move

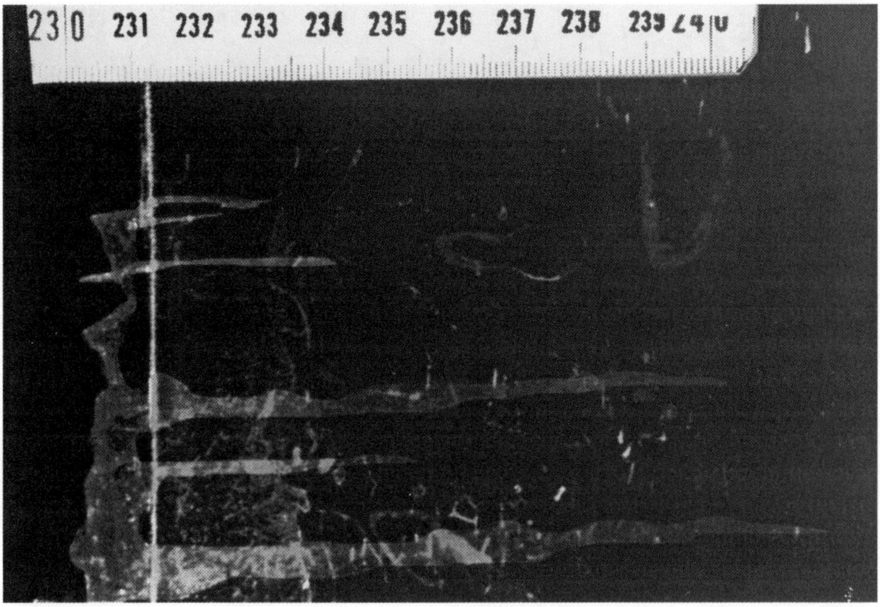

Figure 13.3. Flow visualization of flow over forward-facing step. Dye appears light in colour. Step appears as a line at about the 231 cm mark. Flow from left to right. $Re = 1514$

transverse to the main stream direction. That is, the sausages rotate, ejecting fluid from their ends and, all the while, both sausages and threads oscillate transverse to the main flow. Video records indicate no apparent periodicity in this movement. Sometimes there is a 'gulping' phenomenon where two sausages amalgamate, following which the flow re-establishes the conditions shown in the accompanying Figures. Careful measurement of the transverse wavelength of the threads from photographs reveal that, on average, they decrease with increasing *Re*, as given in Table 13.2. When plotted, the thread spacing asymptotically approaches about 0.9.

COMPUTATIONAL RESULTS

Results obtained from implementing a parabolic inlet profile with no embedded structure are in excellent agreement with previously reported results (WaKarani, 1992).

PARABOLIC INLET PROFILE

Dimaczek *et al.* (1989) interpreted their observations using the notion that for turbulent boundary layer flow Taylor-Görtler-type vortices exist on the surface of

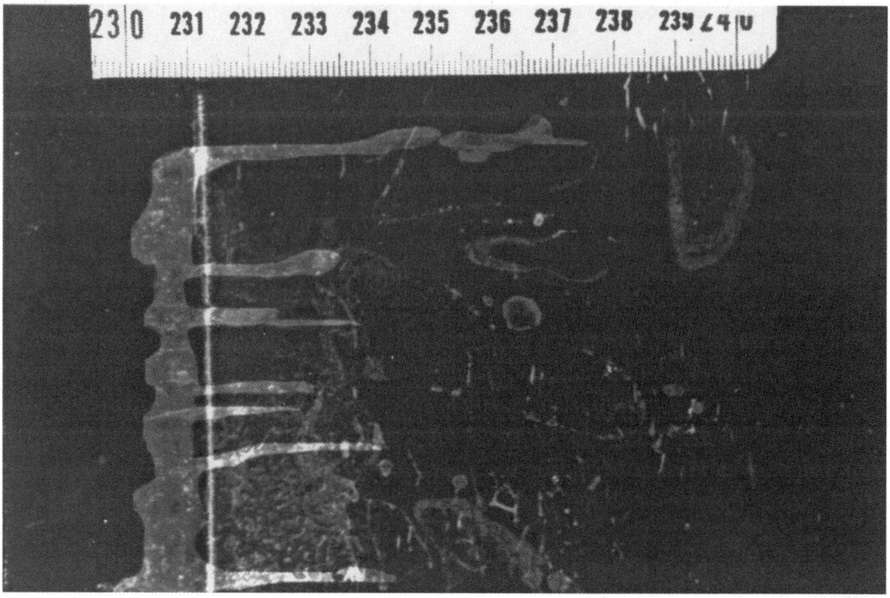

Figure 13.4. Flow visualization of flow over forward-facing step. Dye appears light in colour. Step appears as a line at about the 231 cm mark. Flow from left to right. $Re = 2100$

Table 13.2. Experimental mean and standard deviation of 'thread' wavelength (l/h) downstream of step

l/h	Re
2.81 ± 0.94	1514
2.03 ± 0.65	2100
1.09 ± 0.45	3500
0.95 ± 0.38	6724

the recirculation bubble upstream of the step. We apply similar interpretation of the flow visualization (WaKarani, 1992). Our observations lead us to believe that for this flow there must be a 'trigger' mechanism for Taylor-Görtler-type vortices. Computationally we tried to simulate this trigger by introducing perturbations, that were random in both space and time, onto the parabolic mean velocity profile and by altering the aspect ratio of the channel. While it was observed that these disturbances affected the flow, they were washed out of the calculation domain and the flow re-established its pre-disturbed state. It is important to note that excessively fine discretization in both time and space did not alter these findings.

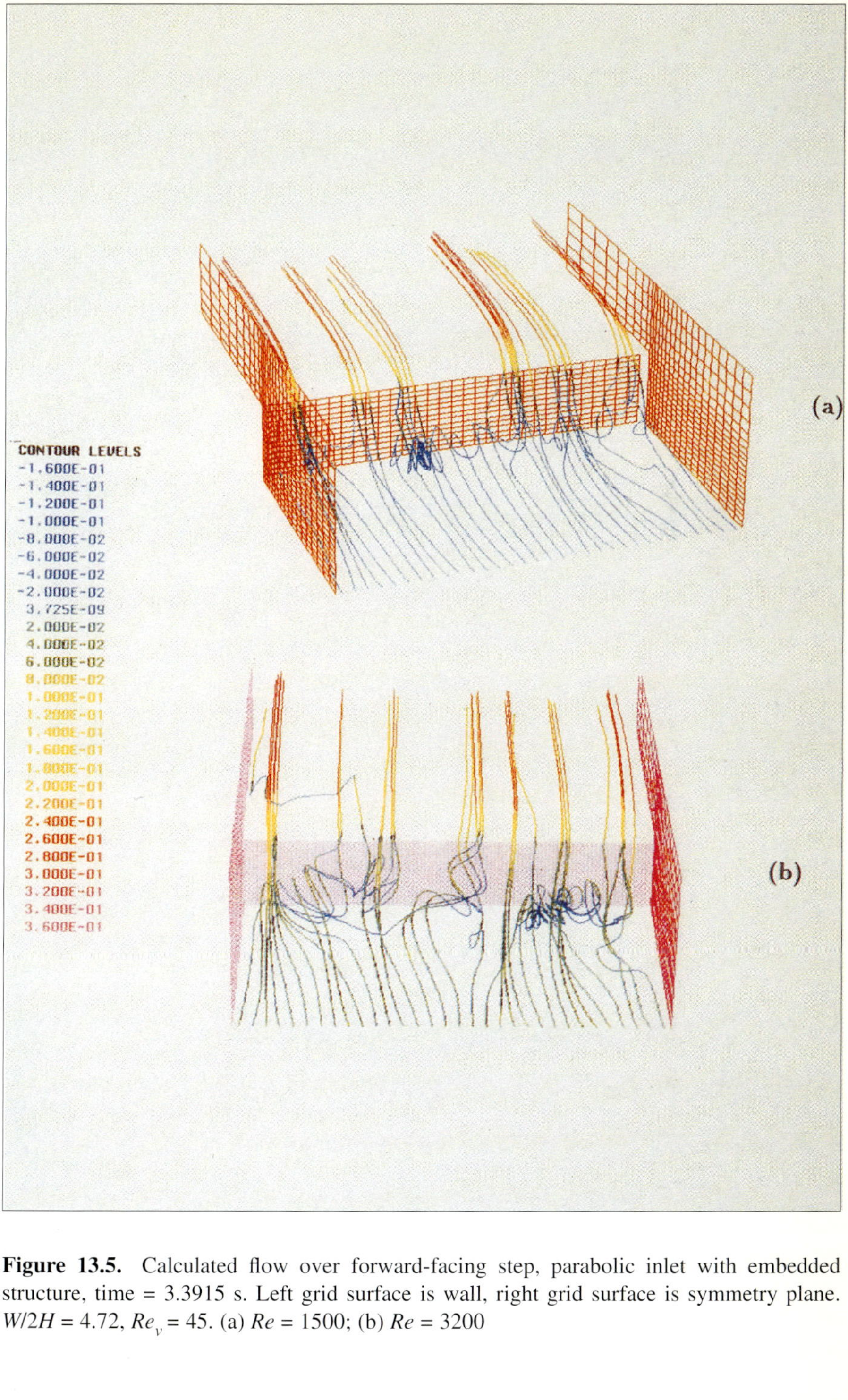

Figure 13.5. Calculated flow over forward-facing step, parabolic inlet with embedded structure, time = 3.3915 s. Left grid surface is wall, right grid surface is symmetry plane. $W/2H = 4.72$, $Re_v = 45$. (a) $Re = 1500$; (b) $Re = 3200$

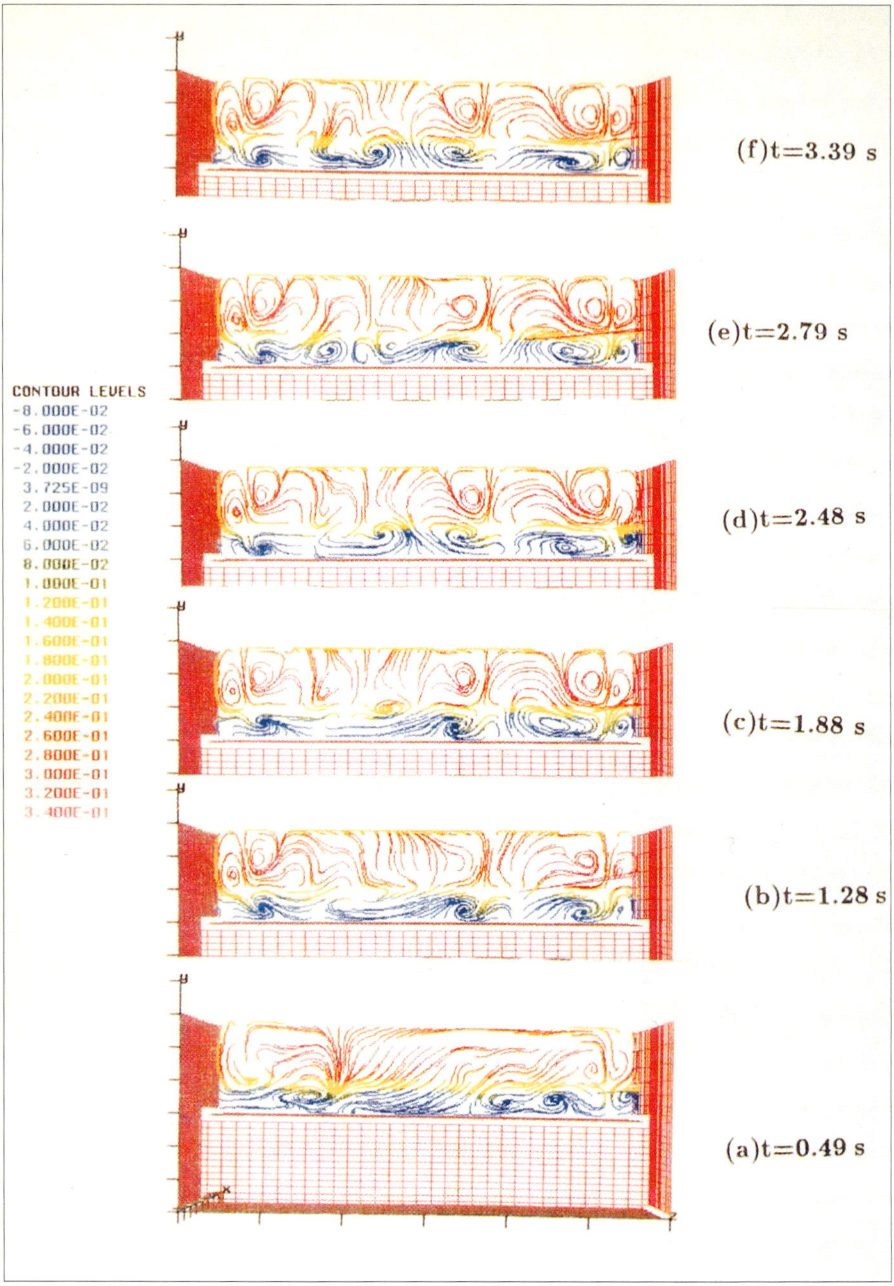

Figure 13.6. Time evolution of vortices, transverse plane, x/h = 1.66 downstream of step, 0.49 s < t < 3.39 s, Re = 1500. Left surface is wall, right surface is symmetry plane

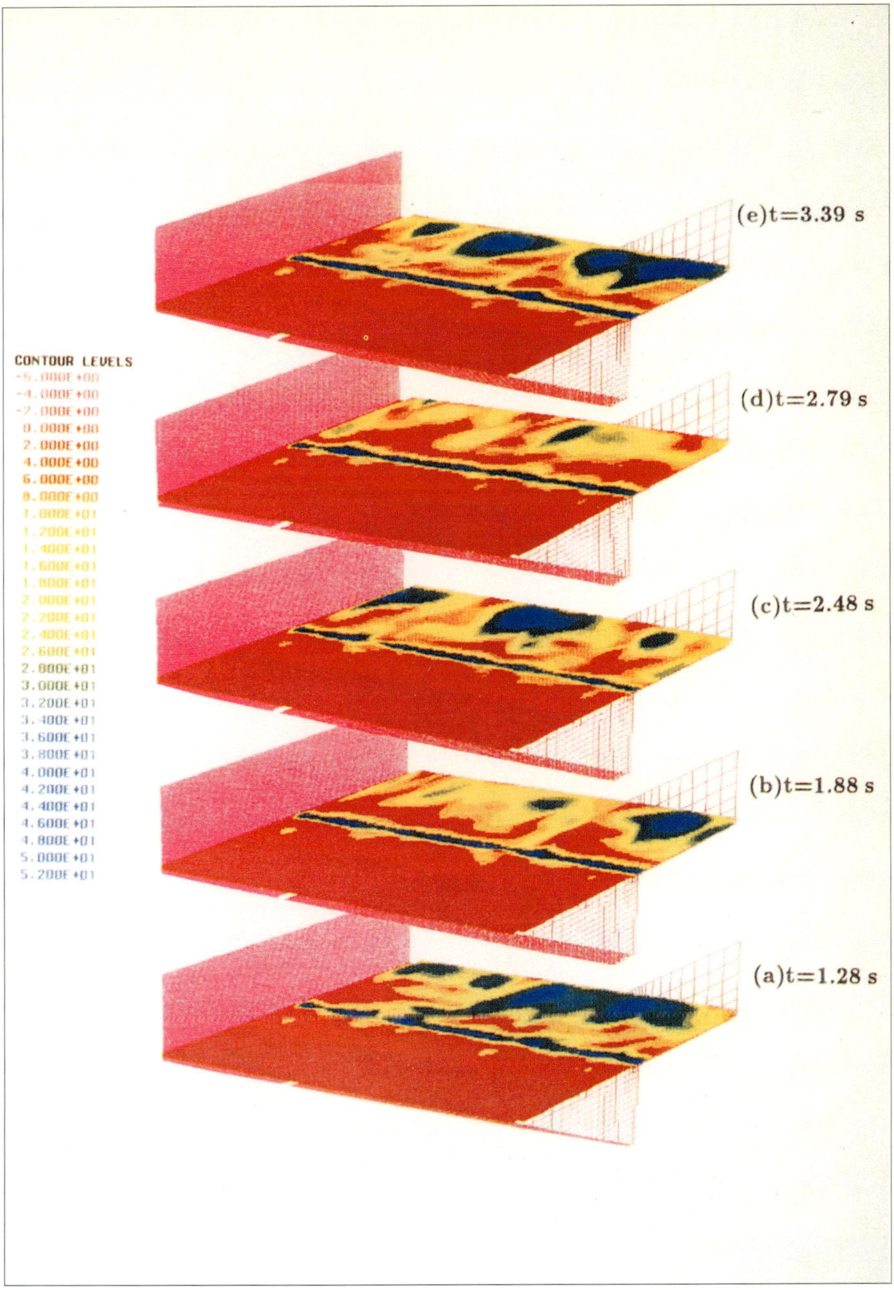

Figure 13.7. Time evolution of surface shear stress Re = 3200. Left surface is wall, right surface is symmetry plane. y/h = 1.0, on wall containing the step. Re_v = 45, $W/2H$ = 4.72 and t = 3.39 s

PARABOLIC INLET PROFILE WITH EMBEDDED VORTEX STRUCTURE

Artificially produced distributed disturbances were introduced into the boundary layer upstream of the step. These distributions are longitudinal low-speed streaks, as given in Table 13.1. Note that all other boundary conditions remain the same as used previously.

Figures 13.5a and 13.5b show particle traces for particles released uniformly along a constant transverse line after about 3.4 s for $Re = 1500$ and 3200, respectively. In addition to pre-step and on-step separation regions, there is a variety of streamwise vortices on the face of the step. Figures 13.5a and 13.5b both show the resulting complicated paths taken before the particles emerge concentrated together distal to the step. The 'sausage' and 'threads' identified in the flow visualization are evident in both Figures 13.5a and 13.5b. Some of the fluid particles are seen to flow into the recirculation regions, which exist only under the emerging threads. The vortices are located between threads. It is interesting to note that as Re increases, for the same number of particles introduced at the inlet plane, there is an increase in the number of threads distal to the step; still more encouraging, the wavelength between the threads decreases with increasing Re. The time-average mean and standard deviation of the thread wavelength deduced from the calculations are given in Table 13.3.

Table 13.3. Calculated mean and standard deviation of 'thread' wavelength (l/h) downstream of step

Re	1500	1500	3200	1500
Re_v	17.7	45.0	45.0	17.7
$W/2H$	4.72	4.72	4.72	14.17
l/h	0.7±0.12	0.75±0.11	0.61±0.10	0.55±0.14

The experimental mean wavelengths of the 'threads' are 2.81 and 1.09 for Re of 1514 and 3500 respectively (Table 13.2). These values are about twice those deduced computationally. The poor agreement probably results from specifying a single inlet length-scale (see Table 13.1). However, it is encouraging that, while there is little sensitivity to changes in vortex Reynolds number, there is some sensitivity to increasing Re. Similar sensitivity is also apparent as W/h changes.

Figure 13.6 shows a sequence in time of particle traces in the plane of the step, $x/h = 1.66$ distal to the step. This time sequence clearly demonstrates the transient nature of the structures both on the wall containing the step and the genesis and temporal evolution of others along the flat upper wall.

The dynamics of the flow can also be illustrated using carpet maps of the wall shear stress, Figure 13.7 on the wall distal to the step, $y/h = 1$. As time increases the high shear near the leading edge of the step produces a variegated separation region and intense, variegated shear stress distribution in the reattachment zone that clearly indicates not only oscillations transverse to but also in the main flow direction. This is similar to the findings of Latornell & Pollard (1986) for flow over a backward-facing axisymmetric pipe expansion.

While it can be said that the general features of the flow visualization have been re-created, it is quite clear that the current simulations do not contain the proper distribution of disturbances at the inlet plane. This can only be done using large eddy or direct numerical simulation of the fully turbulent equivalent of this flow.

EXPERIMENTAL OBSERVATIONS USING AN ERODABLE STEP

The calculations do not include provision for the time-dependent shear to alter the surface geometry. The experimental observations are shown in Figures 13.8–13.11 depicting erosion surfaces after about 4–5 hours of exposure to flow. The Plexiglas lip (1 mm thick) separating the leading edge of the step from the cavity was removed for the experiment illustrated in Figure 13.10 and the plaster at the step was exposed to the flow. Small pins were inserted to model the effects of obstructions along an escarpment rim (Figure 13.11).

Longitudinal erosional marks in the plaster distal to the step are very clearly seen in all figures. Their spatial distribution has not been subjected to statistical analysis. The random movement of the turbulence structure over the step clearly plays a role in the genesis of these bedforms, however, the link between the

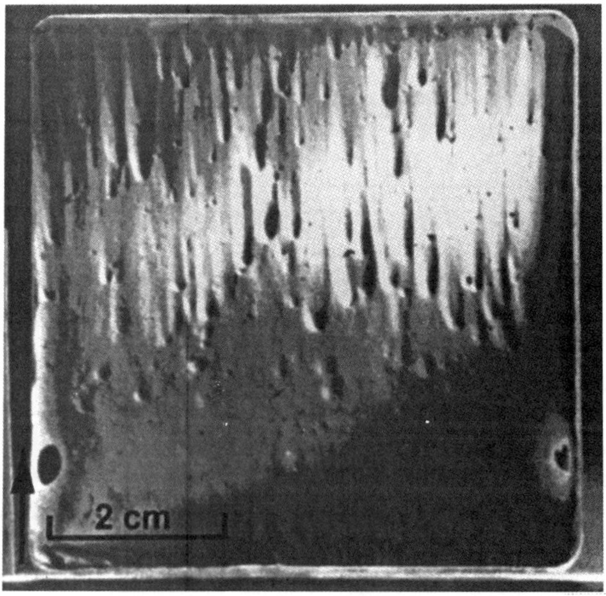

Figure 13.8. Erosion of plaster insert after 10 hours, $Re = 2800$. Step at bottom of photograph

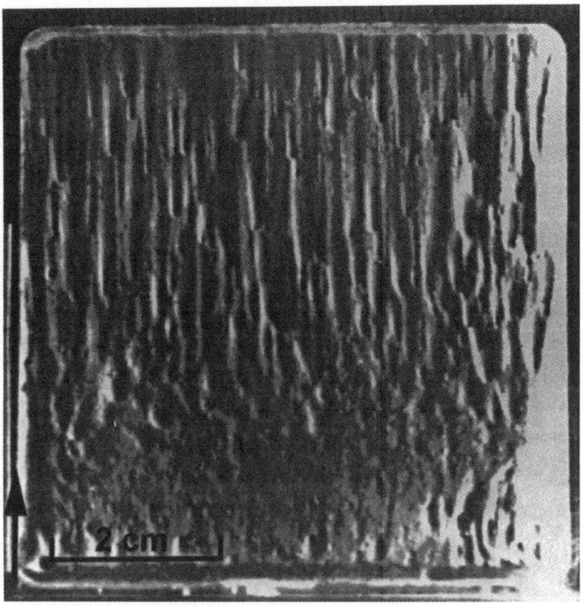

Figure 13.9. Erosion of plaster insert after 5 hours, *Re* = 3500. Step at bottom of photograph.

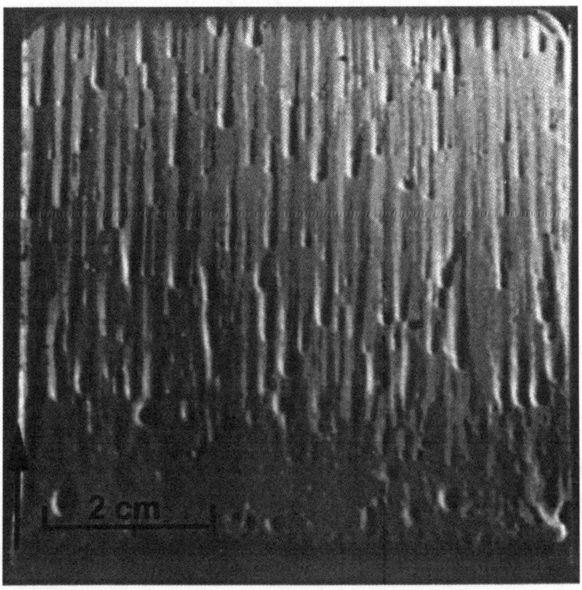

Figure 13.10. Erosion of plaster insert without protected lip after 4 hours, *Re* = 3500. Step at bottom of photograph

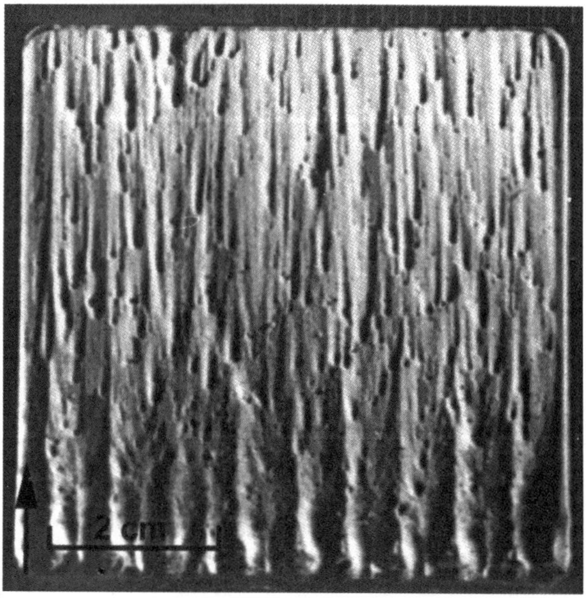

Figure 13.11. Erosion of plaster insert with prominent remnant ridges down flow from obstacles (pins), *Re* = 3500. Step at bottom of photograph

nucleation site and the high-shear regions has yet to be established. The pins appear to cause obstacles for the generation of horseshoe-type vortices and have little effect on the apparent random distribution of depressions downstream. The removal of the protective Plexiglas lip thereby exposing the leading edge of the plaster to the very high shear as identified in the computations also does little to alter the distribution pattern of depressions. However, the leading edge appears to have become preferentially eroded across its span.

Erosional features in nature are very similar to those produced experimentally. Particularly strong features are found immediately downstream from the step (river valley rim, Figure 13.1a, or bedrock escarpment, Figure 13.1b). Grooves initiated beyond the steps are less well-defined and their erosional forms become more three-dimensional (Figure 13.1a). It appears that horseshoe-type vortices as well as vortices generated at the steps and vortices carried across the step from upstream produced the striking, natural bedforms.

SUMMARY

The flow over a forward-facing step has been used to investigate the relation between the turbulent boundary layer structures and the erosion downstream from the step. The streaky structures of the turbulent boundary layer are included as part of the inlet boundary condition in the computational portion of the work. This study suggests a relation between the wall region streak spacing and the spanwise spacing of the vortices both in the vicinity of and distal from the step.

The importance of this paper to the formation of erosional marks downstream from steps is that it provides a mechanism whereby vortex action may be concentrated on parts of the bed. However, remembering that the streaks shift laterally with high frequency, this explanation does not account for the fixed position of erosional marks. A feedback mechanism is required whereby, once formed, the erosional marks themselves fix the positions of intense erosion at the bed. This is straightforward and well-understood for horseshoe vortices generated at obstacles. By contrast, it is unclear how grooves or furrows attract the action of vortices generated upstream of the step. Further work on this aspect of the erosional mark problem and experiments at Reynolds numbers well into the turbulent regime would enhance the results of this study.

ACKNOWLEDGEMENTS

This work is supported by the Natural Sciences and Engineering Research Council of Canada. The resources and friendly working atmosphere of the Centre D'Etudes Aerodynamiques et Thermiques, Université de Poitiers, France and in particular, Dr. J.P. Bonnet, assisted greatly in the preparation of this paper while AP was there on leave.

REFERENCES

Aihara, Y., Tomita, Y. & Ito, A. 1984. 'Generation, development and distortion of longitudinal vortices in boundary layers along concave and flat walls', in Kozlov, V.V. (Ed) *IUTAM Symp. Laminar and Turbulent Transition*, Springer Verlag, 447–454.

Allen, J.R.L. 1982. *Sedimentary Structures: Their Character and Physical Basis*, Vol. I, Elsevier, Amsterdam, 593 pp.

Baker, V.R. & Nummedal, D. (Eds) 1978. *The Channelled Scabland: a Guide to the Geomorphology of the Columbia Basin*, Planetary Geology Program, NASA, Washington, 186 pp.

Boulton, G.S. 1974. 'Processes and patterns of glacial erosion', in Coates, D.R. (Ed) *Glacial Geology*, State Univ. of New York, Binghampton, 123–148.

Dimaczek, G., Kessler, R., Martinuzzi, R. & Tropea, C. 1989. 'The flow over two-dimensional surface-mounted obstacles at high Reynolds numbers', in F. Durst, F., Launder, B.E., Schmidt, F.W. & Whitelaw, J.H. (Eds) *Proc. of the Seventh Symp. on Turbulent Shear Flows*, Stanford Univ., 10.1.1–10.1.6.

Freitas, C.J. & Street, R.L. 1988. 'Non-linear transient phenomena in a complex recirculation flow: a numerical investigation', *Int. J. Numer. Methods Fluids*, **8**, 769–802.

Guezennec, Y., Piomelli, U. & Kim, J. 1989. 'On the shape and dynamics of wall structures in turbulent channel flow', *Phys. Fluids*, **A4**, 764–766.

Kasagi, N., Hirata, H. & Nishino, K. 1986. 'Streamwise pseudo-vortical structures and associated vorticity in the near-wall region of a wall bound turbulent shear flow', *Exper. Fluids*, **4**, 309–318.

Kieffer, S.W. & Sturtevant, B. 1988. 'Erosional furrows formed during the lateral blast of Mount St. Helens, May 18, 1980', *J. Geophys. Res.*, **93**, 14793–14816.

Koseff, J.R. & Street, R.L. 1984. 'Visualization studies of a shear driven three-dimensional recirculating flow', *J. Fluids Engrg.*, **106**, 21–29.

Latornell, D.J. & Pollard, A. 1986. 'Some observations on the evolution of shear layer instabilities in laminar flow through axisymmetric sudden expansions', *Phys. Fluids*, **29**, 2828–2835.

Laverdière, C., Guiment, P. & Dionne, J-C. 1985. 'Les formes et les marques de l'érosion glaciaire du plancher rocheux; signification, terminologie, illustration', *Paleogeogr., Paleoclim. Paleoecol.*, **51**, 365–387.

Lucchitta, B.K. 1982. 'Ice sculpture in the Martian outflow channels', *J. Geophys. Res.*, **B87**, 9951–9973.

McCauley, J.F., Grolier, M.J. & Breed, C.S. 1977. *Yardangs of Peru and Other Desert Regions*, USGS, Interagency Report, Astrogeology, **81**, 177 pp.

Murray, E.A. 1988. *Subglacial Erosional Marks in the Kingston, Ontario, Canada Region: Their Distribution, Form and Genesis*, Unpub. MSc Thesis, Queen's Univ., Kingston, 171 pp.

Neto, A.S., Grand, D., Metias, O. & Lesieur, M. 1993. 'A numerical investigation of the coherent structures of turbulence behind a backward-facing step', *J. Fluid Mech.*, **256**, 1–25.

Nishioka, M., Iida, S. & Ichikawa, Y. 1975. 'An experimental investigation of the stability of plane Poiseuille flow', *J. Fluid Mech.*, **72**, 731–751.

Patankar, S.V. 1980. *Numerical Heat Transfer and Fluid Flow*, Hemisphere Publishing Corp., New York, 197 pp.

Patel, V.C. & Head, M.R. 1981. 'Some observations on skin friction and velocity profiles in fully developed pipe and channel flows', *J. Fluid Mech.*, **83**, 181–201.

Pollard, A. & Siu. A.L-W. 1982. 'The calculation of some laminar flows using various discretization schemes', *Comp. Methods Appl. Mech. and Engrg.*, **35**, 293–313.

Robinson S.K. 1991. *The Kinematics of Turbulent Boundary Layer Structures*, NASA Technical Memorandum (**103859**), 490 pp.

Rubini, P., Becker, H.A., Grandmaison, E.W., Pollard, A., Sobiesiak, A. & Thurgood, C.P. 1992. 'Multi-grid acceleration of three dimensional turbulent variable density flows', *Numer. Heat Transfer*, **B22**, 163–177.

Shaw, J. 1988. 'Subglacial erosional marks, Wilton Creek, Ontario', *Can. J. Earth Sci.*, **25**, 1256–1267.

Shaw, J. 1994. 'Hairpin erosional marks, horseshoe vortices and subglacial erosion', *Sediment. Geol.*, **91**, 269–283.

Singer, B.A., Reed, H.L. & Ferziger, J.H. 1989. 'The effects of streamwise vortices on transition in the plane channel', *Phys. Fluids*, **A1**, 1960–1971.

Smith, C.R. & Shwartz, S.P. 1983. 'Observation of streamwise rotation in the near-wall region of a turbulent boundary layer', *Phys. Fluids*, **26**, 641–652.

Smith, H.T.U. 1948. 'Giant glacial grooves in northwest Canada', *Am. J. Sci.*, **246**, 503–514.

WaKarani, N. 1992. *Flow Over an Upstream-Facing Step Including the Effects of Erosion*, Unpub. MSc Thesis, Queen's Univ., Kingston, 175 pp.

Wengle, H. & Werner, H. 1993. 'Large eddy simulation of turbulent flow over sharp edged obstacles in a plate channel', in Gersten, K. (Ed) *Numerical Fluid Dynamics*, Vol. **40**, 192–199.

Werner, H. & Wengle, H. 1989. 'Large eddy simulation of turbulent flow over a square rib in a channel', in Fernholz, H.H. & Fiedler, H.E. (Eds) *Advances in Turbulence*, Vol. 2, Springer Verlag, 418–423.

Xiao, D.D. 1992. *Experimental and Computational Investigation of Multiple Jets in a Duct Cross Flow*, Unpub. PhD Thesis, Queen's Univ., Kingston, 258 pp.

NOTATION

d	vortex diameter
H	channel depth
h	step height
l	spanwise spacing of threads
L	mean spanwise spacing of threads
Re	Reynolds number, based on mean velocity and channel depth
Re_v	vortex Reynolds number
t	time
U	mean flow velocity
$v\ w$	vertical and spanwise components of velocity
W	width of channel
$x\ y\ z$	downstream, vertical and spanwise flow directions
μ	fluid molecular viscosity
ρ	fluid density

Superscripts

+	normalized based on shear velocity

14 Coherent Flow Structures in Open-channel Slot Flow

D.P. KEOGH & P.S. ADDISON
Department of Civil and Transportation Engineering, Napier University, 10 Colinton Road, Merchiston Campus, Edinburgh, EH10 5DT, UK

ABSTRACT

The use of benchmark testing is an important and fundamental tool of many engineering problems, non more so than in the investigation of fluid dynamics. It is known that the addition of a backward-facing step and in some cases a forward-facing step, in a prismatic channel produces flow separation and subsequent growth of a free-shear layer. This study presents the main flow characteristics of turbulent flow over backward-facing steps, focusing on the zones of flow separation and reattachment, and the turbulent free-shear layer. Many studies have quantified the length of the recirculation zone, bounded by the separation and reattachment points, downstream of a backward-facing step. It is widely accepted that for turbulent flows the reattachment point occurs at a length of approximately 6–8 step heights, downstream of the step. This study concentrates on slot flow of various aspect ratios, defined as the slot length divided by slot height. At large aspect ratios the reattachment length is less than the slot length and the flow is strongly influenced by the backward-facing step, reattaching at the slot base and subsequently impinging on the forward-facing step at some distance further downstream. This is in contrast to the work of Manson (1994) which suggested the addition of a forward-facing step would cause extension of the length at which reattachment occurs.

An experimental study was undertaken in a laboratory flume using laser Doppler anemometry to acquire the mean and fluctuating components of velocity in the streamwise and vertical direction in open-channel slot flow. The results illustrate that distinct coherent flow structures of high turbulence intensity may develop in slot flows despite the flow reversal and stagnant water. This may have important ramifications for erosion and sedimentation in natural or engineered cross-channel trenches and slots.

INTRODUCTION

Many different processes operating within alluvial channels can be addressed by simple hydraulic modelling (e.g. sediment transport, flow around bends, flow around obstructions, floodplain flow and pollutant dispersion) whereby each variable is systematically isolated and controlled. This approach is termed *benchmark testing* and has played an important role not only in river hydraulics but also in many other scientific and engineering subjects. Benchmark tests are generally simplistic in design, but have the ability to produce the fundamental characteristics essential for the understanding of the system under consideration. To

Coherent Flow Structures in Open Channels.
Edited by P.J. Ashworth, S.J. Bennett, J.L. Best and S.J. McLelland.

this end, their use has been widely accepted in the development of our understanding of fluid dynamics.

Much literature exists on the study of compound channels, whereby the dominant direction of flow follows the down-valley floodplain gradient, whilst a smaller proportion flows within the meandering river channel (see Naish & Sellin, this volume; Willetts & Rameshwaran, this volume). The shearing which occurs as these flows interact produces highly three-dimensional patterns of flow circulation, which can lead to significant erosion of the river banks. Investigation of this flow phenomena has concentrated on the flow structures which develop when the river channel is skewed relative to the floodplain, the angles of which have varied between 0–90°, with the smaller angles receiving more attention (Jasem, 1990). When the angle is increased to 90° (where the channels intercept at right angles), the conditions produce a maximum cross-over shear flow. The configuration of this cross-over flow is a vertical expansion in the channel followed by a vertical contraction, and is generally called *slot, trench* or *cavity* flow (Alfrink & van Rijn, 1983; Fujita & Komura, 1992). This situation is similar to flow over backward- and forward-facing steps which has been the subject of much work (e.g. Bradshaw & Galea, 1966; Etheridge & Kemp, 1978; Quraaishi & Al-Brahim, 1993; McLean *et al.*, this volume; Pollard *et al.*, this volume).

The turbulent flow mechanisms which occur in these channel geometries are of direct application to the study of sediment transport and pollutant diffusion. Turbulence measurements in open-channels have been greatly improved with the advent of laser Doppler anemometry (LDA). This method superseded hot-film anemometry as the favoured method of data acquisition in open-channel flow, mainly due to its greater accuracy and ability to be non-intrusive which is especially suited to velocity measurements in water (see Steffler *et al.*, 1985; Nezu & Rodi, 1986). This paper reports on a series of flume experiments with varying slot configurations which use LDA to measure the two-dimensional flow structure and turbulence intensity in the downstream and vertical directions.

CHARACTERISTICS OF BACKWARD-FACING STEP AND SLOT FLOW

Investigations into the flow characteristics of separated flow has been extensive over the last two decades, primarily due to the increased computational power available to researchers. This has provided the opportunity for direct comparison between numerical and experimental data, and thus provided greater confidence in numerical models. It is evident from the literature that the favoured condition for investigating separated flow is that over a backward-facing step (BFS), owing to its geometrical simplicity and direct engineering implications.

The numerical approach to modelling the effects of free-surface flows has proved to be a major hurdle in the development of open-channel flow models. This has led to many investigations on separated flow being conducted in closed-channels, a process which has been replicated experimentally using both air and water as the fluid medium.

From Figure 14.1, it can be seen that the flow separates as it reaches a step, producing a separation streamline which reattaches at some point along the step floor. The length of this reattachment point has received much attention, as it is one of the most important parameters characterizing the global feature of a separated flow. The dimensionless distance at which reattachment occurs is generally defined by X_s/H_s, where X_s is the distance along the step floor and H_s is the step height. For laminar flow, the value of X_s/H_s is found to increase directly with Reynolds number (*Re*). In an experimental study undertaken in a closed channel, Armaly *et al.* (1983) found that the flow changed from laminar to transitional flow at *Re* greater than 1250. At this stage, it was observed that the reattachment length suddenly decreased, and continued to decrease further with increasing Reynolds number. This continued until the turbulent state was achieved. A secondary recirculation zone can be seen to develop on the upper wall which has been shown to develop when *Re* is greater than 500 (Armaly *et al.*, 1983; Sohn, 1988). As indicated in Figure 14.1, a second recirculation bubble along the step floor was observed by Armaly *et al.* (1983). This bubble had not been reported before in the literature and no other evidence has been found to substantiate it. However, it appeared in the early transitional stage, where the reattachment length experiences a sudden drop in magnitude and is believed to be caused by the development of vortex shedding at the step edge. An increase in Reynolds number, still within the transitional range, leads to the disappearance of this separation bubble. Similarly, the secondary recirculation bubble at the upper surface was found to move upstream as the Reynolds number increases, before disappearing when the flow becomes turbulent.

With increasing Reynolds number the flow becomes unstable and breaks down into a turbulent state. When this occurs, it has been found that the reattachment

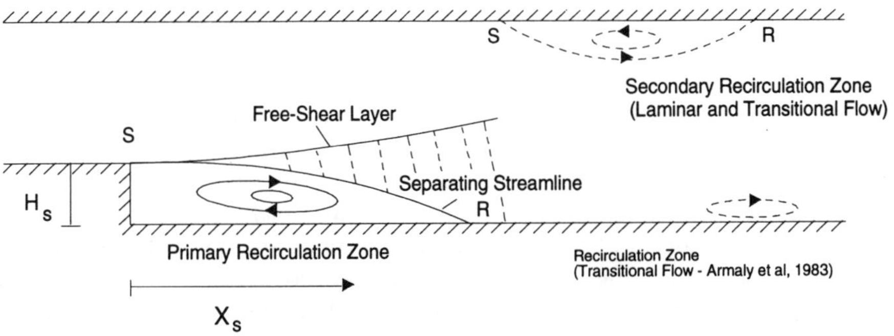

Figure 14.1. Recirculation pattern as a function of Reynolds number (S shows the separation point and R shows the point of reattachment)

length stabilizes at a length varying between 6–9 step heights. Shih & Ho (1994) have shown that the flow in the immediate vicinity of the step is highly three-dimensional. As a result, the reattachment length is less in the centreline and greater towards the sidewalls.

An important feature observed in the transitional and turbulent phase is the development of a turbulent free-shear layer. This occurs as the faster main channel flow shears across the top of the slower fluid contained within the primary recirculation area. It is characterized by high turbulence intensities generated by shearing of the two flows along the separation streamline, although, it should be remembered that there will be small fluctuations of this separation streamline along its defined time-averaged position. This area of high turbulence intensity convects downstream producing a turbulent plume which gradually dissipates as the flow recovers from the sudden expansion. It has been documented that the effects of this shear layer can still be measured at a distance of 50 step heights downstream of the reattachment point (Bradshaw & Wong, 1972).

The Reynolds stresses have been shown to reach a maximum within the shear layer, especially at points along the separation streamline. Etheridge & Kemp (1978) have shown that the Reynolds shear stress in the streamwise-vertical (u-v) plane increases rapidly with distance from the step, X_s, up to a maximum value which occurs near the point of reattachment.

Due to the hydrodynamic instability of this free-shear layer, the flow develops finite amplitude oscillations. Perturbations over a wide range of frequencies develop as the flow separates. This takes the form of wavy patterns which develop into vortices that are small at first and amplify as they convect downstream at rates appropriate to their wavelengths (see Betts, 1972; Ethembabaoglu, 1978).

There is not much information in the literature regarding flow over a backward-facing step for open-channels, this is due to the particular difficulty of modelling this type of flow numerically. However, the flow characteristics are similar to those which develop in closed-channel BFS flow, where the main flow feature is still the primary recirculation zone. Due to the free-surface, the secondary recirculation zone appearing in the closed-channel does not occur, and there is no evidence of a second recirculation zone at the base of the slot.

Similar investigations have also been undertaken for lateral expansions as opposed to vertical expansions as already discussed. These have been chosen primarily to investigate the cross-shear flows which develop in harbours and behind islands (see Babarutsi *et al.*, 1989). In this situation, the reattachment length is approximately eight step heights which is consistent with the findings for vertical expansions. A slot (trench or cavity) flow is essentially made up of a forward-facing step inserted downstream of the backward-facing step. Slot flow is applicable to a wide range of engineering applications, such as floodplain flow and dredged channels for sediment traps, traffic tunnels and pipelines. Slot flow exhibits several phenomena which are highly dependant on the flow conditions and the size of the slot, defined by its aspect ratio (A_s):

$$A_s = L_s / H_s$$

where L_s is the length of slot.

Figure 14.2 illustrates the general flow characteristics that may develop in slot flow. When the aspect ratio is small, generally less than between 5–7, flow is characterized by a recirculating eddy driven by the main flow which dominates the slot region (see Figure 14.2a). When the aspect ratio is increased to a value greater than approximately 5–7, the main flow produces a separation streamline which reattaches somewhere along the bottom of the slot, as illustrated in Figure 14.2b. This produces a zone of flow recirculation driven by the main flow. The flow beyond the reattachment point continues along the slot bed until it is rapidly accelerated over the forward-facing step (FFS) due to a sudden contraction of the channel, producing a secondary recirculating region on top of the FFS. The distance at which the separation streamline reattaches itself to the main flow has been found to occur in the slot at a length of approximately 5–7 slot heights by various authors. However, Manson (1994) suggests that in some cases, reattachment does not occur until 10 slot heights due to the acceleration of the flow over the FFS dragging the reattachment point downstream.

The interaction of the main flow and the slower moving recirculating region in the slot causes the generation of a free-shear layer. This shear layer is similar to that described above for BFS flow. The main influence of the FFS is that it interferes with the development of the shear layer and also generates a secondary

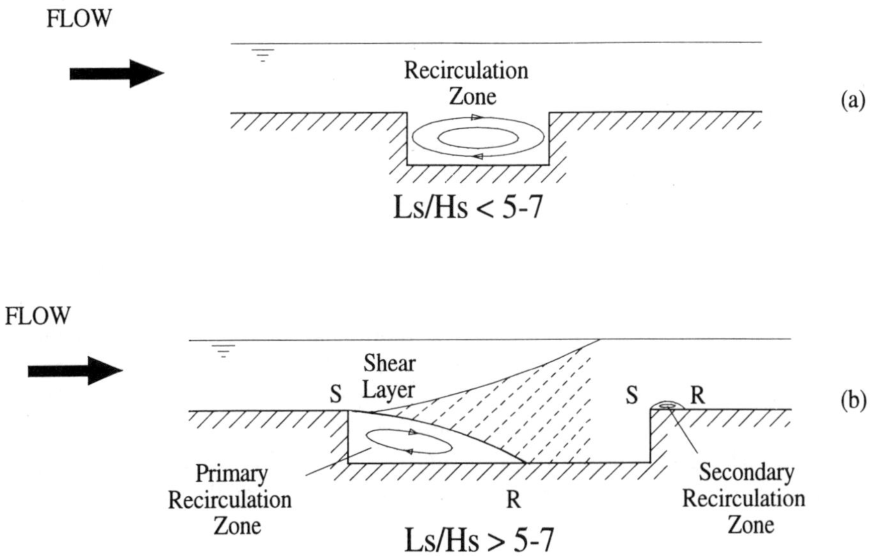

Figure 14.2. Slot flow characteristics in open channels

shear layer beyond the FFS. Associated with this will be a feedback mechanism when the aspect ratio is small, caused by the interaction of the shedding vortices with the FFS. This gives rise to alternating pressure increases within the slot, causing smaller scale vorticity. When these disturbances are transmitted back to the origin of the shear layer, which is very sensitive to external disturbances, new perturbations are produced and eventually the system reaches a state of dynamic equilibrium (see Betts, 1972; Ethembabaoglu, 1978).

EXPERIMENTAL APPARATUS

This study was undertaken in a glass panelled, 4 m long, tilting laboratory flume, adjusted to acquire two-dimensional open-channel slot flow data. The flume was set to a slope of 1×10^{-5}. Flume width is 0.1 m and a controlled flow depth (d) of 0.1 m was maintained in the main channel by use of depth gauges mounted on top of the flume. A slot is located 2.6 m downstream of the inlet. Different aspect ratios of 2, 5, 10, 15 and a backward-facing step allow investigation of the flow characteristics associated with varying configurations of slot and step flow. The mean and fluctuating components of velocity were acquired along the centreline of the flume for the streamwise (u) and vertical (v) directions of flow using a one-component Dantec Flowlite LDA system. Supporting software processed the flow data and calculated the instantaneous velocity and its variance.

The Reynolds number of the flow was 10 000, based on the hydraulic radius of the upper channel. This corresponds to an average flow velocity of 0.29 m s^{-1}. A logarithmic profile was obtained at the inlet and outlet and regulated conditions were maintained throughout the experiments. The position of the laser was controlled using a purpose built traverse/support accurate to ±0.1 mm in all three planes of movement.

The LDA sampling rate was approximately 40 Hz, depending on the location of the laser within the fluid. It was found that the upstream corner of the slot yielded a sampling frequency as low as 5 Hz owing to the reduced velocities experienced in this region. On average, however, these frequencies were maintained for a period of 60 s, thereby producing approximately 2400 samples in the faster flow regions, and reducing to approximately 300 samples in the regions of slower flow. There was no deliberate particle seeding of the fluid as it was observed that naturally occurring particles within the fluid greatly improved the rate of data acquisition.

Data were acquired in each configuration using a grid defined by $\Delta x = L_s/20$ and $\Delta y = d/20$, where x and y are the streamwise and vertical directions respectively. This produced a grid of 400 defined points within the slot domain. Measurements were also taken at selected profiles at 500 mm upstream of the slot and 700 mm downstream of the slot.

RESULTS

The flow structure is presented as a series of vector plots which illustrate the time-averaged velocities for both the streamwise and vertical directions (Figure 14.3). From these plots, reference is made to the separating streamline which is defined as the separation line between the main flow and primary recirculation zone, identified by a change in velocity direction. Positive flow direction is from left to right and bed to water surface for the streamwise and vertical components respectively. Contour plots of turbulent kinetic energy (k), are also presented where:

$$k = \frac{1}{2}\rho\left(\overline{u'^2} + \overline{v'^2}\right)$$

ρ is the fluid density and $\overline{u'^2}$ and $\overline{v'^2}$ are the mean square values of the velocity fluctuations in the streamwise and vertical directions respectively (Figure 14.4).

In this study, it must be remembered that the results are two-dimensional, neglecting the mean and fluctuating component of the cross-stream velocity. Although flow of this nature has been shown to be highly three-dimensional in the step region (Shih & Ho, 1994), the flow conditions used in this study limit a true analysis to two-dimensional flow. The analysis of the shear layer makes reference to k_{slot} which is defined as the maximum k that occurs in the main free-shear layer, and is not to be confused with the higher k values which are shown to occur in the secondary recirculation zone.

SLOT ASPECT RATIO 2

As expected, with the aspect ratio of 2, a recirculating eddy fills the slot. This eddy rotates due to the action of the main flow shearing across the fluid contained within the slot. The time-averaged velocity vectors are presented in Figure 14.3a. The centre of the recirculation zone and a small area of essentially stagnant flow in the bottom upstream corner of the slot can be observed. As the flow approaches the FFS, it is deflected towards the bed of the slot where it then moves upstream with relatively high velocities.

The turbulent kinetic energy of the flow can be seen in the contour plot of Figure 14.4a. Before the flow enters the slot it exhibits the characteristic behaviour for a turbulent boundary layer in open-channel flow. Turbulence intensity is zero at the bed and increases rapidly away from the laminar sub-layer. A zone of higher turbulence intensity then appears which develops due to the friction of the flow with the channel bed. The maximum turbulent kinetic energy occurring in this region is approximately 9% of k_{slot}. Moving vertically through the flow, the turbulent activity decreases until the free-surface is reached.

As the flow enters the slot, the creation of the free-shear layer can be observed which develops as it progresses downstream. k_{slot} occurs near the end of the slot in this shear layer. Close to the forward-facing step there is a rapid downward

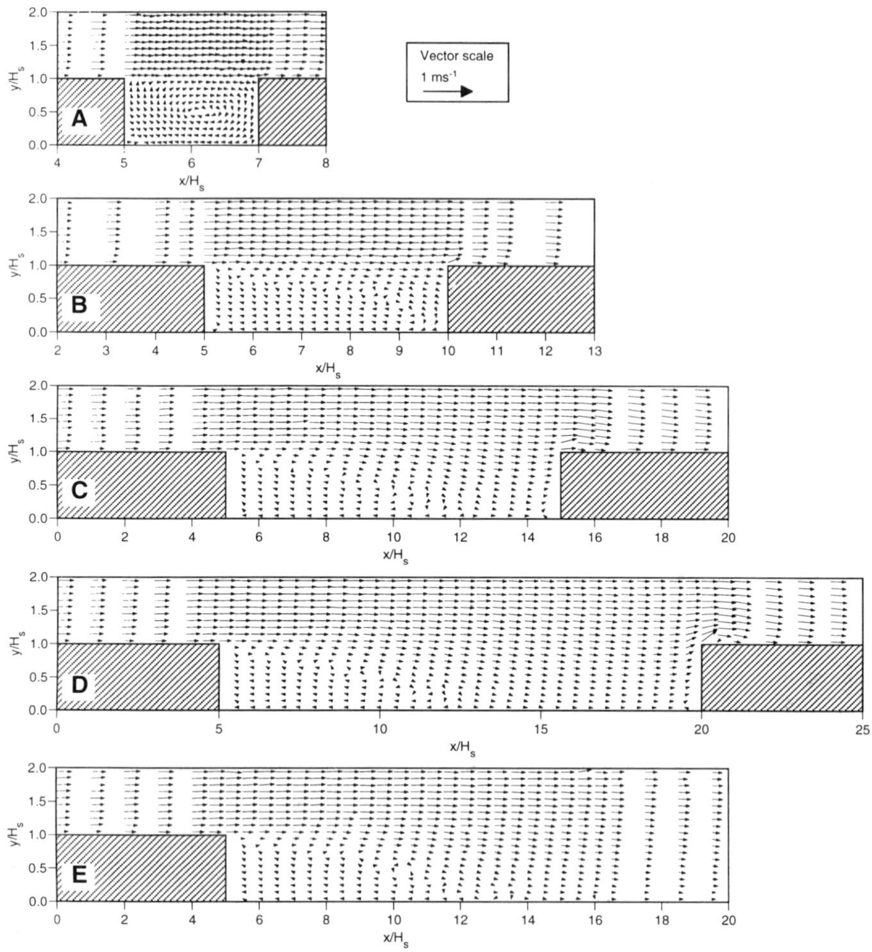

Figure 14.3. Vector plots showing the distribution of flow velocities for the following experimental configurations: (A) $A_s = 2$; (B) $A_s = 5$; (C) $A_s = 10$; (D) $A_s = 15$; and (E) backward facing step

deflection of the shear layer into the slot. This is caused by the nature of the recirculation which is associated with such low aspect ratios. However, it is believed that other factors include the feedback mechanism caused by vortex development, and the suppression of the shear layer by the flow in the main channel. The shear layer can also be seen to 'spill over' the forward-facing step and begin to dissipate quickly beyond this point. In this region there is a significant reduction in the turbulent kinetic energy from the shear layer towards the bed of the channel. This may be caused by the deflection upwards of the shear layer by the

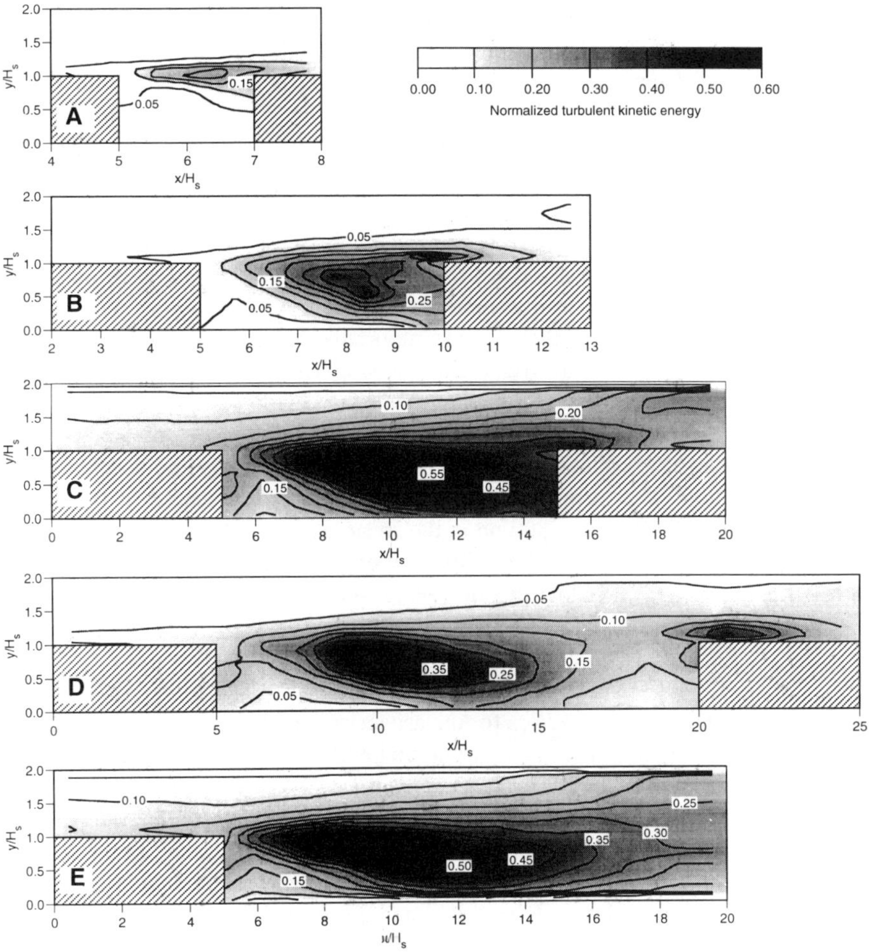

Figure 14.4. Contour plots showing the distribution of normalized turbulent kinetic energy for the following experimental configurations: (A) $A_s = 2$; (B) $A_s = 5$; (C) $A_s = 10$; (D) $A_s = 15$; and (E) backward facing step

secondary recirculation region, where turbulent kinetic energy levels reach a value equal to 44% of k_{slot}.

SLOT ASPECT RATIO 5

The time-averaged velocity vectors plotted in Figure 14.3b indicate that the main feature at this aspect ratio is still a recirculating eddy within the slot. This time however, the eddy does not completely fill the slot. The centre of the recirculating region can again be identified, as can the small region of essentially stagnant water at the bed near the upstream corner of the slot. As flow leaves the slot, there is an increase in flow velocity. This is due to an upward deflection of the flow as it meets the forward-facing step and merges with the faster moving main flow. This process is responsible for the secondary recirculating region which though not apparent at this scale, is evident on the turbulent kinetic energy plot where the development of the free shear layer occurs in the slot region (Figure 14.4b). The increased length of the slot promotes the growth of the free-shear layer. It appears to grow within the slot with its top boundary suppressed into the slot by the faster moving main flow. The shear layer then impacts onto the FFS, at which point most of the flow deflects downwards into the recirculation region. With aspect ratios in the order of 4–7 this is an important feature, whereby a feedback mechanism confines the shear layer within the slot. This produces a bulging of the shear layer which can be seen to attach to the bottom of the slot whilst there is no reattachment of the main flow. The high turbulent kinetic energy of the secondary recirculation zone can be clearly seen and is primarily caused by the acceleration of flow over the step. The turbulent kinetic energy values experienced in this region are approximately 212% of k_{slot} found in the shear layer. Another feature which can be observed is a small region of high turbulent energy occurring near the surface of the flow just downstream of the forward-facing step.

SLOT ASPECT RATIO 10

The time-averaged velocity vectors for the aspect ratio of 10 are presented in Figure 14.3c. Again the main feature is a recirculating eddy in the slot, although, the flow reattaches along the slot bed. The reattachment length is approximately 7–8 step heights. Downstream from the point of reattachment, flow encounters the FFS and deflects upwards and over the step, as can be seen by the high velocities at the top of the FFS. The primary recirculation zone is clearly illustrated by the velocity vectors, characterized by a well-defined core and strong reverse flows along the slot bed. An area of essentially stagnant water can again be observed at the bed in the upstream section of the slot. The turbulent kinetic energy plot, Figure 14.4c, illustrates the growth of the free-shear layer, again suppressed in the slot by the faster moving main flow. The shear layer can be seen to attach to the bottom of the slot at a point close to the reattachment point of the main flow. As the shear layer moves downstream, early signs of its dissipation are evident before impacting on the FFS. The secondary recirculation zone can also be seen clearly where values of turbulent kinetic energy are approximately 94% of k_{slot}. Unfortunately, the measured flow field does not extend far enough downstream to yield the high turbulent kinetic energy experienced at the surface of the flow. However, there are early signs of this feature developing with k values at 20% of k_{slot}.

SLOT ASPECT RATIO 15

The largest aspect ratio of 15 yields similar results to that of the aspect ratio of 10, whereby a separation streamline reattaches at the bottom of the slot at a distance of 7–8 step heights (Figure 14.3d). The flow continues along the slot until it deflects upwards as it impacts against the FFS. Associated with this are very high velocities experienced as the flow from the slot rejoins the main flow. Figure 14.4d shows the distribution of turbulent kinetic energy, where the full development of the free-shear layer and its dissipation within the slot are clearly noticeable. The shear layer attaches to the bottom of the slot at a point close to the reattachment point of the main flow. Turbulence intensity dissipates more rapidly from the bottom of the slot, possibly caused by the faster main flow dragging the shear layer at the top of the slot further forward. Due to the very high velocities developed at the lip of the forward-facing step, the associated turbulent kinetic energy is very high. This increase in k is of the order of 3–10 times the k_{slot}. A marginal increase in the turbulent kinetic energy, 14% of k_{slot}, can be seen where the original boundary layer interacts with the free shear layer and surface of the flow.

BACKWARD-FACING STEP

This configuration allows the flow to develop at an expanded depth ratio of 2. Figure 14.3e shows the vector plot, which highlights the expected recirculation region. The reattachment length can be seen to occur at approximately 7–8 step heights, suggesting that the effect of a forward-facing step has no difference on the reattachment length. As the flow progresses downstream it recovers from the sudden expansion and settles down to a log-profile at its new depth. The k plot can be seen in Figure 14.4e. The main feature is the shear layer, which develops without any adverse effects produced by the forward-facing step. This shear layer can be seen to attach at a distance of 6 step heights, with maximum values occurring along the separation streamline. Early signs of its dissipation can be seen to occur close to the channel bed at a distance of 12 slot heights.

CONCLUSIONS

The coherent flow structures which develop due to sudden expansions and contractions of turbulent open-channel flow have been documented through a series of flume experiments. The results have focused on defining the separation streamline, its associated reattachment point, and the development of the turbulent free-shear layer in the wake of backward-facing steps. The following conclusions can be made from the study:

1. Time-averaged velocities have illustrated that the reattachment length of the separated streamline is consistently 7–8 step heights for aspect ratios exceeding 10. This is in good agreement with the findings from various

authors using different flow conditions, suggesting there is some consistency in the reattachment length. In the case of the backward-facing step configuration, the reattachment length is also 7–8 step heights, suggesting that the influence of the forward-facing step on the reattachment length is negligible. Although the slot length was the only parameter changed in this study, it is suggested that future work may focus on the flow depth. An investigation with the FFS at the time-averaged point of reattachment has not been performed, however, it is anticipated that some instability might exist in the bottom downstream corner of the slot and the characteristic flow features that develop will not be dissimilar to those for aspect ratios greater than 10.

2. The primary recirculation zones have been clearly illustrated for each case of varying slot length. For small aspect ratios (less than 5–7) high velocities have been shown to develop at the face of the forward-facing step, sweeping downwards and along the step floor maintaining strong reverse flows. In all cases the bottom upstream corner of the slot consists of essentially stagnant water. A combination of these effects has direct implications for the morphological development of rivers. An example is a trench cut perpendicular to the main channel (e.g. a sediment trap) or irrigation ditch on the adjacent floodplain. In both cases it is expected that sediment will accumulate in the bottom upstream corner and hence, alter the shape and possibly the efficiency of the trench. The same effect will also occur in sections of a river which are perpendicular to the direction of the floodplain when in flood.
3. The development of a turbulent free-shear layer over various slots has been investigated. Maximum turbulence intensity within this shear layer occurs along the separating streamline. For aspect ratios less than 5–7, a feedback mechanism is responsible for the bulging of the shear layer within the slot. For the aspect ratios greater than 5–7, the shear layer is seen to develop and attach to the slot floor at a length similar to that of the separating streamline.
4. The vector plots illustrate the acceleration of the flow as it passes over the forward-facing step, causing the development of a secondary recirculation zone on top of the forward-facing step. This zone is not clearly defined by the vector plots, owing to the large spacing of the measurement grid, however, the high turbulence intensity associated with this is clearly seen in the contour plots. This high turbulence intensity is seen to increase with increasing aspect ratio, with the exception of when the aspect ratio is 10. The aspect ratio of 15 produces the largest increase of turbulence intensity in this secondary recirculation zone, with k values reaching 10 times k_{slot}.
5. A small zone of increased turbulent energy has also been seen to occur at the free-surface with the slot flows. This zone generally occurs at the interface between the original boundary layer, free-shear layer and the free-surface.

This zone does not experience any change in mean velocity, thus cannot be confused with the recirculation zone found at the top surface experienced in closed-channel flow. Its existence appears to be the result of these different layers interacting, however, it is suggested that this feature requires more thorough investigation.

Preliminary comparisons with a free-surface flow model developed at the University of Glasgow (Pender *et al.*, 1995) are in good agreement with the development and location of flow features found in this study. Future work will seek to investigate the vortex development which occurs due to the hydrodynamic instability of the free-shear layer, an important characteristic of this type of flow. Analysis of the frequency spectra of points within the free-shear layer may provide some information as to the nature of this flow phenomenon. An initial flow visualization study has been undertaken (Keogh, 1995) which clearly illustrates qualitatively the development of these vortices.

ACKNOWLEDGEMENTS

Stuart M[c]Lelland is thanked for helping reproduce Figures 14.3 and 14.4.

REFERENCES

Alfrink, B.J. & Van Rijn, L.C. 1983. 'Two-equation turbulence model for flow in trenches', *J. Prof. Issues in Engrg., ASCE*, **109**, 941–958.

Armaly, B.F., Durst, F., Pereira, J.C.F. & Schonung, B. 1983. 'Experimental and theoretical investigation of a backward-facing step flow', *J. Fluid Mech.*, **127**, 473–496.

Babarutsi, S., Ganoulis, J. & Chu, V.H. 1989. 'Experimental investigation of shallow recirculating flows', *J. Hydraul. Engrg.*, **115**, 906–924.

Betts, P. L. 1972. 'Self induced oscillations in an open water-channel with slotted walls', *J. Fluid Mech.*, **55**, 401–417.

Bradshaw, P. & Galea, P.V. 1967. 'Step-induced separation of a turbulent boundary layer in incompressible flow', *J. Fluid Mech.*, **27**, 111–130.

Bradshaw, P. & Wong, F.Y.F. 1972. 'The reattachment and relaxation of a turbulent shear layer', *J. Fluid Mech.*, **52** 113–135.

Demuren, A.O. & Wilson, R.V. 1994. 'Estimating uncertainty in computations of two-dimensional separated flows', *J. Fluid Mech.*, **116**, 216–220.

Ethembabaoglu, S. 1978. 'Some characteristics of unstable flow past slots', *J. Hydraul. Div., ASCE*, **104**, 649–666.

Etheridge, D.W. & Kemp, P.H. 1978. 'Measurements of turbulent flow downstream of a rearward-facing step', *J. Fluid Mech.*, **86**, 3, 545–566.

Fujita, I., Komura, S. & Kanda, T. 1993. 'Measurements of turbulent flow in a trench using an image processing technique', *Proc. 5th Intl. Symp., Refined Flow Modelling and Turbulence Measurements,* Paris, France, 309–316.

Jasem, H.K. 1990. *Flow in Two Stage Channels with the Main Channel Skewed to the Flood Plain Direction*, Unpub. PhD Thesis, Univ. of Glasgow, 455 pp.

Keogh, D.P. 1995. 'Flow Visualisation: Open-Channel Slot Flow', Video Presentation, Napier University, Edinburgh.

Manson, J.R. 1994. *The Development Predictive Procedure for Localised Three-dimensional River Flows*, Unpub. PhD Thesis, Univ. of Glasgow, 135–218.

Nezu, I. & Rodi, W. 1986. 'Open channel flow measurement with a laser Doppler anemometer', *J. Hydraul. Engrg, ASCE,* **112**, 335–355.

Pender, G., Keogh, D.P., Manson, J.R. & Addison, P.S. 1995. 'Verification of a three-dimensional flow model with experimental data', *Hydra 2000, Proc. 24th IAHR Congr.*, Thomas Telford, London, **1**, 290–295.

Quraaishi, A.A. & Al–Brahim, A.M. 1992. 'Hydraulic jump in sloping channel with positive or negative step', *J. Hydraul. Res.*, **30**, 769–782.

Shih, C. & Ho, C.-M. 1994. 'Three-dimensional recirculation flow in a backward facing step', *J. Fluids Engrg*, **116**, 228–232.

Sohn, J.L. 1988. 'Evaluation of FIDAP on some classical laminar and turbulent benchmarks', *Intl. J. Numer. Methods Fluids*, **8**, 1469–1490.

Steffler, P.M., Rajaratnam, N. & Peterson, A.W. 1985. 'LDA measurements in open channels', *J. Hydraul. Engrg, ASCE,* **111**, 119–130.

NOTATION

A_s	slot aspect ratio
d	slot flow depth
H_s	step height.
k	turbulent kinetic energy
L_s	slot length
Re	Reynolds number
u v	streamwise and vertical flow directions
$\overline{u'^2}$ $\overline{v'^2}$	mean square values of the velocity fluctuations in the streamwise and vertical directions
x y	streamwise and vertical directions
X_s	distance along the step floor
ρ	fluid density

15 Mean Flow and Turbulence Structure over Fixed Ripples and the Ripple-dune Transition

SEAN J. BENNETT† & JAMES L. BEST
Department of Earth Sciences, University of Leeds, Leeds, LS2 9JT, UK
†Current address: USDA-ARS, National Sedimentation Lab., PO Box 1157, Oxford, MS, 38655, USA

ABSTRACT

Detailed measurements of downstream and vertical flow velocities and their turbulent fluctuations in a unidirectional current were obtained using laser Doppler anemometry over two fixed ripples in a laboratory channel. High-density sampling allows construction of contour maps of flow and turbulence parameters which illustrate that: (i) time-averaged downstream and vertical flow velocities are affected by bed morphology through topographically-induced acceleration and deceleration in near-bed regions only ($y/d \leq 0.3$); (ii) the maximum root-mean-square values of the downstream and vertical velocity components and Reynolds stress occur within the flow separation zone, along the separation cell free shear-layer and at reattachment; (iii) positive vertical skewness values and high-magnitude quadrant-2 events (ejections) occur both along the free shear layer and in mid-depth regions; and (iv) low-magnitude quadrant-4 events (sweeps) occur near flow reattachment. Whilst ripples and dunes may be considered morphologically and dynamically similar, when compared to dunes, ripples have: (i) a greater degree of flow uniformity; (ii) a separation zone shear layer which is restricted to near-bed regions and has much lower associated turbulence intensities; and (iii) much lower Reynolds stresses at flow reattachment. A schematic model is presented detailing the physical and fluid dynamic mechanisms responsible for the transition from ripples to dunes and the changing dynamics of the flow separation zone.

INTRODUCTION

A unidirectional turbulent shear flow acting on a nominally flat bed of cohesionless sediment grains less than ~0.6 mm in diameter and just sufficient to move the sediment will cause the bed to deform, creating a quasi-random distribution of bed undulations (see discussion in Gyr & Schmid, 1989; Best, 1992; Gyr & Müller, this volume). These bed defects promote flow separation and perturbations further downstream that increase in size to become ripples (e.g. Liu, 1957; Raudkivi, 1963; Grass, 1970, 1983; Williams & Kemp, 1971, 1972; Gyr & Schmid, 1989; Best, 1992). Ripples only form in turbulent boundary layers that are initially

Coherent Flow Structures in Open Channels.
Edited by P.J. Ashworth, S.J. Bennett, J.L. Best and S.J. McLelland.

hydraulically smooth where the sediment grains are immersed in the viscous sub-layer (e.g. Simons & Richardson, 1961; Yalin, 1977; Leeder, 1980, 1983). Once formed, ripples are generally no higher than 0.03 m and no longer than 0.6 m and their length, L, scales approximately with grain diameter, D, such that $L \sim 1000D$ (e.g. Yalin, 1977, 1985; Richards, 1980; Allen, 1982), but there is much scatter in this relationship (see Baas, 1993). Although several studies have sought to characterize flow and turbulence over ripples (e.g. Raudkivi, 1963, 1966; Wiberg & Nelson, 1992; Lyn, 1993; Li, 1994; see references in Bennett & Best, 1995), the interaction between fully-developed ripples, sediment transport mechanics and the outer flow (greater than 20% of the flow depth) is still poorly understood.

Ripples formed in sediment larger than ~0.15 mm diameter increase dramatically in length and height with increasing transport stage and transform into dunes (see Allen, 1982). Dune length has been found to scale not with grain size but with boundary layer thickness such that $L \sim 5\text{–}6d$ where d is flow depth (see Jackson, 1976; Allen, 1982). The transition from ripples to dunes has been linked to the formation of 'rogue' ripples (Leeder, 1983) which, due to their relatively larger heights, cause deeper lee-side scour, thus producing greater downstream sediment fluxes and bedform heights. Little quantitative information, however, has been presented to support this proposition and detail the differences in flow and turbulence associated with ripples and dunes. The aims of the present paper are: (i) to describe the structure of mean flow and turbulence over fixed ripples in a laboratory channel; (ii) to compare these data with the mean flow and turbulence fields observed over dunes; and (iii) to discuss some of the fluid dynamic mechanisms responsible for the transition from ripples to dunes.

EXPERIMENTAL EQUIPMENT AND PROCEDURE

The experiments were conducted in the Sedimentological Fluid Dynamics Laboratory, Department of Earth Sciences, University of Leeds, using a tilting recirculating flume 10 m long, 0.3 m wide and 0.3 m deep. An equilibrium rippled bed was formed using glass spheres 0.22 mm in diameter, a mean flow depth of 0.1 m and mean flow velocity of ~0.4 m s^{-1}. The ripples ranged from 5 to 25 mm in height (h; mean ~13 mm) and 70 to 120 mm in length (L; mean ~100 mm) and had sinuous crest-lines. Once equilibrium conditions had been achieved, the flow was carefully stopped to preserve the ripples on the bed surface, the flume was slowly drained of water and the bed was allowed to dry. The sediment surface was then sprayed with polyurethane varnish in order to harden it and then a 1 m long rubber mould of the bed surface was prepared using Vinamould™ moulding rubber. Five concrete bed inserts were cast from this mould, each encompassing about 10 ripple lengths. The surfaces of two inserts were painted and covered with glass beads while the paint was still wet in order to obtain the appropriate surface grain roughness.

The concrete bed inserts were placed in the flume and flow conditions similar to the mobile-bed experiment were established (Figures 15.1a and 15.1b). Achievement of quasi-equilibrium flow over fixed-ripples required adjustment of the pump discharge and flume slope to obtain the mean flow velocity and depth (±2 mm) over successive bed-wave crests as observed in the mobile bed experiment. Flow over artificial bedforms cannot replicate mobile bed conditions exactly because of the absence of a sediment load and a decrease in bed porosity (see Bennett & Best, 1995).

(a)

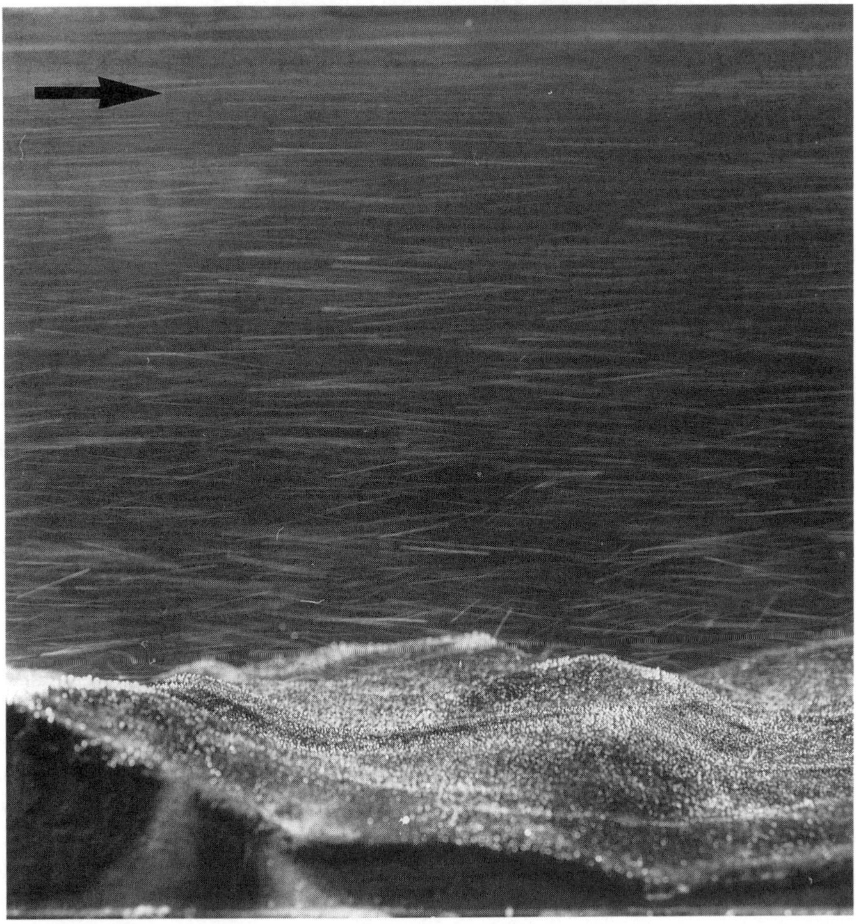

Figure 15.1. Experimental fixed ripple configuration. (a) Long-exposure (1/30 s) photograph of flow over fixed ripples. Flow is from left to right (as indicated by arrow) and the white flow tracers are neutrally-buoyant Pliolite particles. Note that most particle paths are parallel to the bed except in near-bed regions. Field of view is 0.15 m and the bed surface is covered with air bubbles. *Continued on next page*

(b)

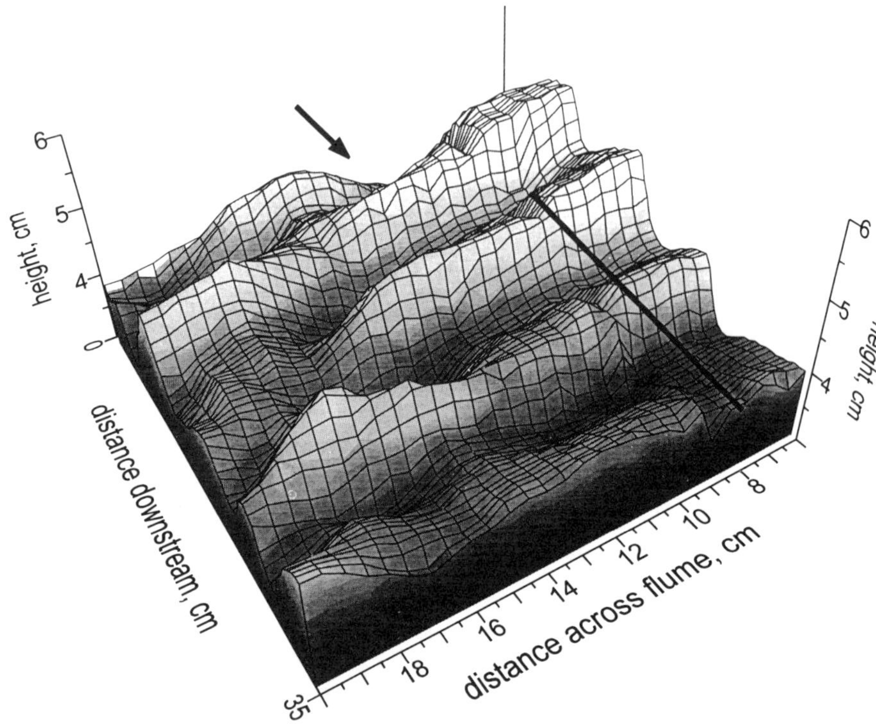

Figure 15.1. *continued* (b) Three-dimensional view of the fixed ripple bed insert. The line denotes the orientation of the section used for the LDA traverse (see Figure 15.1c) and arrow indicates flow direction. *Continued on facing page*

The flume test section was 5.0 to 6.25 m downstream of the flume entrance and the dynamic (i.e. moving) and static (i.e. still) water surface elevations were measured to within ±1 mm along the flume centre-line using a vernier point gauge mounted on a movable carriage. Water surface elevations relative to a horizontal datum were then calculated from the difference between the dynamic and static water surface elevations. The mean water surface slope, $\overline{S}$, was obtained from a least-squares linear regression of these elevations within the test section and this slope was used to calculate the spatially-averaged mean bed shear stress, $\overline{\tau}_0$, from

$$\overline{\tau}_0 = \rho g \overline{dS} \qquad (15.1)$$

where ρ is fluid density, g is gravitational acceleration and $\overline{d}$ is the mean flow depth. This shear stress was adjusted for smooth sidewalls using Williams' (1970) correction method, requiring a reduction of ~6%.

(c)

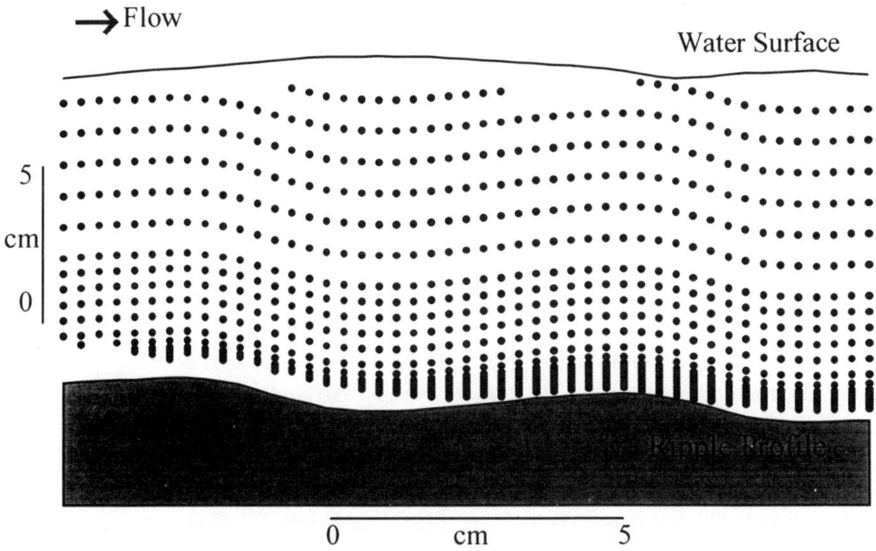

Figure 15.1. *continued* (c) Vertically exaggerated (×0.5) bed and water surface profiles of the fixed ripple insert showing all velocity sample locations. Profiles are numbered 1 to 47 from left (upstream) to right (downstream).

At-a-point velocity measurements were obtained using a DANTEC two-component fibre-optic laser Doppler anemometer (LDA) used in back-scatter mode with a 161 mm focal length lens and 100 mW argon-ion laser. The LDA was operated with: (i) a 40 MHz frequency shift to enable detection of positive and negative flow directions; and (ii) a DANTEC Particle Dynamics Analyser processor that operates on a correlation-type process (DANTEC, 1991). The flow was seeded with a small amount of a water-based white paint that provided an excellent source of scattering centres, yielding data rates between 50 and 200 Hz, in excess of those desirable to ensure full characterization of the turbulence spectra (see Nezu & Nakagawa, 1993, p. 30). The probe was mounted on a high-precision, electronically-controlled traverse system that allowed positioning of the laser measurement volume to within ±0.1 mm.

Forty-seven vertical profiles of velocity, spaced 3 mm apart in the downstream direction, were measured 0.085 m from the flume sidewall over two successive ripples (Figures 15.1b and 15.1c). Downstream and vertical velocities were collected simultaneously at 13 to 23 points in each profile (in some cases, only the downstream component was measured nearest the bed; Figure 15.1c). This distance from the wall was chosen to maximize the number of points that could be measured near the bed in the presence of three-dimensional ripples (see Figure 15.1a and 15.1b). The flow was sampled for 60 s at each point and a grid of over 950 points was established, with the lowest point in each profile being 1 to 5 mm above the bed depending on the local ripple morphology.

For each point in the profile, the following moments were determined for both the downstream and vertical velocities (where these calculations represent at-a-point time averages). The mean downstream ($\overline{U}$) and vertical ($\overline{V}$) velocities (arithmetic particle means) were calculated using

$$\overline{U} = \frac{1}{n}\sum_{i=1}^{n} u_i \tag{15.2a}$$

$$\overline{V} = \frac{1}{n}\sum_{i=1}^{n} v_i \tag{15.2b}$$

where n is the total number of velocity measurements and u_i and v_i are the instantaneous downstream and vertical velocities. No attempt was made to correct for a velocity biasing effect (when the particle measuring rate is correlated to the magnitude of the instantaneous velocity vector within the flow; Adams & Eaton, 1988, p. 276), since such bias is significant only in highly turbulent flows (in excess of several metres per second; see Gould & Loseke, 1993). The root-mean-square (RMS) fluctuation about the mean of each component (u'_{rms}; v'_{rms}) was calculated using

$$u'_{rms} = \left[\frac{1}{n}\sum_{i=1}^{n}(u_i - \overline{U})^2\right]^{0.5} \tag{15.3a}$$

$$v'_{rms} = \left[\frac{1}{n}\sum_{i=1}^{n}(v_i - \overline{V})^2\right]^{0.5} \tag{15.3b}$$

where $u' = u_i - \overline{U}$ and $v' = v_i - \overline{V}$. The skewness of each velocity distribution (U_{sk} and V_{sk}) was calculated from

$$U_{sk} = \frac{1}{n}\sum_{i=1}^{n}\left(\frac{u_i - \overline{U}}{u'_{rms}}\right)^3 \tag{15.4a}$$

$$V_{sk} = \frac{1}{n}\sum_{i=1}^{n}\left(\frac{v_i - \overline{V}}{v'_{rms}}\right)^3 \tag{15.4b}$$

The Reynolds stress, τ_R, was determined using

$$\overline{u'v'} = \frac{1}{n}\sum_{i=1}^{n}(u_i - \overline{U})(v_i - \overline{V}) \tag{15.5a}$$

$$\tau_R = -\rho\overline{u'v'} \tag{15.5b}$$

Velocity profiles constructed from the at-a-point, time-averaged downstream velocities (Equation 15.2a) were used to calculate the depth-averaged velocity. These values were then spatially averaged over one ripple length to determine a mean flow velocity ($\overline{U}_d$). The spatially-averaged bed shear stress, $\overline{\tau}_W$, was calculated from the Kármán-Prandtl law of the wall (using those measurements above ~0.07d; see discussion below):

$$\frac{U}{u_*} = \frac{1}{\kappa}\ln\left(\frac{y}{y_0}\right) \tag{15.6a}$$

$$\overline{\tau}_W = \rho u_*^2 \tag{15.6b}$$

where u_* is shear velocity, κ is the von Kármán coefficient, y is the distance from the bed where U is measured and y_0 is the zero-velocity roughness height. The equivalent sand roughness height, k_s, is calculated from

$$k_s = 30.2 y_0 \tag{15.7}$$

for hydraulically rough flows where $u_* k_s / \nu \geq 70$ and ν is the kinematic viscosity of water (see Bridge & Bennett, 1992).

Quadrant analysis was employed to discriminate important instantaneous turbulent events. Plotting the horizontal and vertical velocity fluctuations about the mean defines four quadrants with quadrants 2 ($+v'$, $-u'$) and 4 ($-v'$, $+u'$) describing 'ejection' and 'sweep' events and quadrants 1 ($+v'$, $+u'$) and 3 ($-v'$, $-u'$) defining 'outward' and 'inward' interactions respectively (e.g. Lu & Willmarth, 1973; Bogard & Tiederman, 1986). Each velocity pair may be investigated by examining the entire signal or only those events lying above a certain threshold value (H; or hole size) defined as:

$$H = |u'\,v'| / (u'_{rms}\,v'_{rms}) \tag{15.8}$$

The contribution of important high-magnitude events within each quadrant to the total turbulent signal is assessed by using different threshold values. Typical threshold values associated with burst detection (quadrant-2 events) range from 1.07 (Bogard & Tiederman, 1986) to 4.0 to 4.5 (Lu & Willmarth, 1973; Willmarth & Lu, 1974) whereas values of 2–3 have been used for quadrant-4 events. Because of the lack of data and reliable guidelines concerning event thresholds in boundary layers which are not flat nor hydraulically smooth (the 'non-canonical' boundary layers of Robinson, 1991), no attempt was made to impose limits on the time between successive measurements in order to differentiate between single-event and multiple-event ejections (see Luchik & Tiederman, 1987).

RESULTS

MEAN FLOW AND TURBULENCE STRUCTURE OVER FIXED RIPPLES

All spatially-averaged flow values are summarized in Table 15.1 and contour plots for all mean flow and turbulence parameters are presented in Figure 15.2. Figure 15.3 presents profiles of mean downstream and vertical flow velocities, their RMS

Table 15.1. Summary of experimental flow parameters

Flow Parameter	Spatially-Averaged Values
width, $\overline{w}$ (m)	0.30
mean depth, $\overline{d}$ (m)	0.104
mean velocity, $\overline{U}_d$ (m s^{-1})	0.398
maximum velocity, U_{max} (m s^{-1})	0.483
height of maximum velocity, δ (m)	0.085
discharge, $\overline{Q}$ (m^3 s^{-1})	0.0124
water surface slope, $\overline{S}$	0.0013±0.0002
bed shear stress, $\overline{\tau}_0 = \rho g \overline{dS}$ † (Pa)	1.25
shear velocity, u_* (m s^{-1}) using $\overline{\tau}_0$	0.0354
Darcy-Weisbach friction factor, $f = 8\overline{\tau}_0 / \rho \overline{U}_d^2$	0.063
Froude number, $Fr = \overline{U}_d / \sqrt{g\overline{d}}$	0.39
From Reynolds stress profile	
$\overline{\tau}_R$ (Pa)	1.36
u_* (m s^{-1})	0.0369
f	0.069
From Law of the Wall for $0.07\overline{d} \leq y \leq \delta$ (averaged along ripple stoss)	
For $\kappa = 0.410$:	
$\overline{\tau}_W$ (Pa)	1.60
u_* (m s^{-1})	0.040
f	0.081
zero-velocity roughness height, y_0 (mm)	0.57
equivalent sand roughness height, $k_s = 30.2y_0$ (mm)	17.21
For $u_* = 0.0369$ (m s^{-1}):	
von Kármán coefficient, κ	0.379

† corrected for smooth sidewalls using the method of Williams (1970)

values and Reynolds stress at four locations over a ripple (back, crest, separation zone and trough).

Maximum time-averaged downstream velocity (~0.4 m s^{-1}) occurs slightly below the water surface (at 0.82*d*; Figure 15.2a; Table 15.1) due to minor secondary flow. Small negative downstream velocities (up to −0.1 m s^{-1}) occur within the flow separation zone on the lee-side of the bedforms (Figures 15.2a and 15.3). The wake region associated with flow separation is restricted in vertical height to less than 0.2*d* (see Figure 15.3); above this zone the downstream velocity contours are essentially parallel to the average bed surface (Figure 15.2a), indicating that flow is approximately uniform. This flow uniformity is also evidenced by the time-averaged vertical velocities (Figures 15.2b and 15.3) which are close to zero for most of the flow field. Minor flow expansion occurs over the ripple troughs (up to -0.03 m s^{-1}, Figure 15.2b) and slight topographic forcing (positive vertical velocities) occurs over the ripple back and within the separation cell (up to 0.03 m s^{-1}). The length of the flow separation zone determined for the downstream ripple shown in Figure 15.2a is 0.021 m (~1.8*h*), smaller than that observed previously for both fixed ripples and dunes (~3.5–4.5*h*; Engel, 1981; Nelson & Smith, 1989; Wiberg & Nelson, 1992; Bennett & Best, 1995). This smaller separation length may be due to the three-dimensional shape of the ripples (see below).

The maximum RMS values for the downstream and vertical velocity components (up to 0.1 and 0.06 m s^{-1}, respectively) occur near the ripple crest, along the flow separation zone free shear layer and at flow reattachment (Figures 15.2c, 15.2d and 15.3). These maxima, however, are restricted to near the free shear layer and within the lowest 20% of the flow depth. Above this region, both u'_{rms} and v'_{rms} decrease linearly with distance from the boundary (Figures 15.2c, 15.2d and 15.3), similar to the distribution of turbulence in flat-bed boundaries (e.g. Nezu & Rodi, 1986; Nezu & Nakagawa, 1993). The maximum Reynolds stresses (up to 3–3.5 Pa, Figure 15.2e; or about $2-3\overline{\tau}_0$, Table 15.1) occur within the flow separation zone, along the free shear layer and at flow reattachment (Figures 15.2e and 15.3). Above this near-bed wake-influenced region, τ_R also decreases linearly with distance from the boundary.

The skewness values for the downstream velocity component are, in general, close to zero for most of the flow field, although positive skewness values occur within the flow separation cell and just above the free shear layer (Figure 15.2f). Maximum positive vertical skewness values occur in two regions over these fixed ripples (Figure 15.2g): (i) along the free shear layer originating at the bedform crest-line and extending the entire length of the bedform; and (ii) in the mid-depth flow regions, an area distinct and separate from that encompassing the free shear layer. The positive vertical skewness values present in both of these regions match well with the location of frequent, high-magnitude quadrant-2 events (ejections; Figure 15.2h–j) which originate from the shedding of Kelvin-Helmholtz instabilities (Müller & Gyr, 1983, 1986; Levi, 1991; Bennett & Best, 1995). The free shear layer extends for the length of the bedform, its thickness being approximately equal to the ripple (step) height and its attitude being essentially parallel to the bed. Like the mid-depth regions observed here, a similar signature of high-magnitude quadrant-2 events dominating the region away from the wall has also been observed in flat-bed flows (e.g. Willmarth & Lu, 1974). An explanation for this zone of quadrant-2 dominance may lie in the advection of the higher-magnitude coherent structures, either near the boundary or along the shear layer, into the outer part of the flow which, as they pair, can interact and move upwards through the region of highest shear and increase in relative magnitude when compared to the other three quadrants. This process of ejection growth, advection and dissipation may be linked to the presence of outer flow 'bulges' and their associated shear layers (e.g. Falco, 1977, 1991; Thomas & Bull, 1983; Allen, 1984). It should be noted that ejections associated with ripples are confined in their vertical extent to approximately the lower 30% of the flow depth.

The greatest number of high-magnitude quadrant-4 events over a threshold value of 2 (Figure 15.2k) occur near the ripple crest, within the flow separation cell and at reattachment, and mark the inrush of fluid both in response to flow expansion in the bedform lee-side and to replace fluid advected in quadrant-2 events as required by continuity. Near the wall, quadrant-4 events dominate Reynolds stress production (see also Wallace *et al.*, 1972; Willmarth & Lu, 1974), and relatively few quadrant-4 events are present at higher threshold values ($H = 4$; Figure 15.2l).

BOUNDARY SHEAR STRESS AND ROUGHNESS HEIGHT OVER FIXED RIPPLES

Independent determinations of mean boundary shear stress acting on the rippled bed surface incorporating both form drag and skin friction components can be made using a number of different techniques. First, a mean boundary shear stress of 1.25 Pa is calculated using mean flow depth and a sidewall corrected water surface slope (Equation 15.1, Table 15.1). Second, the spatial average of the direct stress measurements well above the bedform (those unaffected by flow separation) can also be used (see Lyn, 1993, Nelson *et al.*, 1993; Bennett & Best, 1995). This is accomplished by averaging the values of τ_R (Equation 15.5) from flow

---

Figure 15.2. Contour plots for all mean flow and turbulence parameters for the fixed ripple bed showing: (a) mean downstream velocity; (b) mean vertical velocity; (c) RMS of the downstream velocity; (d) RMS of the vertical velocity; (e) Reynolds stress; (f) skewness of the downstream velocity distribution; and (g) skewness of the vertical velocity distribution. Contour plots of the normalized number (given as % of the total) of quadrant-2 events (Q_2) for a threshold value (H) of: (h) 2; (i) 4; and (j) 6. Contour plots of the normalized number of quadrant-4 events (Q_4) for a threshold value of: (k) 2; and (l) 4. In all maps, flow is from left to right, colour bar indicates contour intervals and vertical exaggeration is ×0.4. *See facing page*

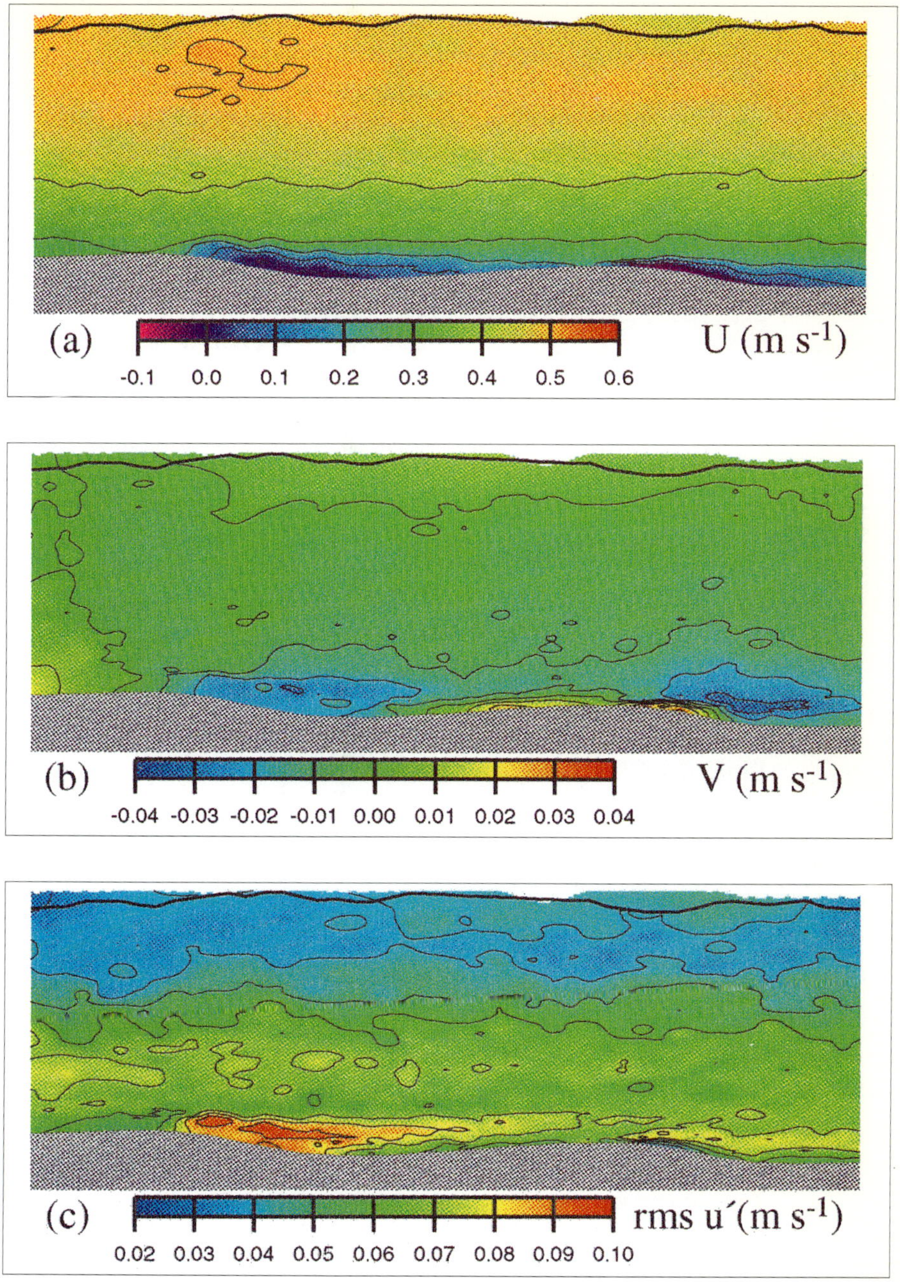

Figure 15.2. *Continued on next page*

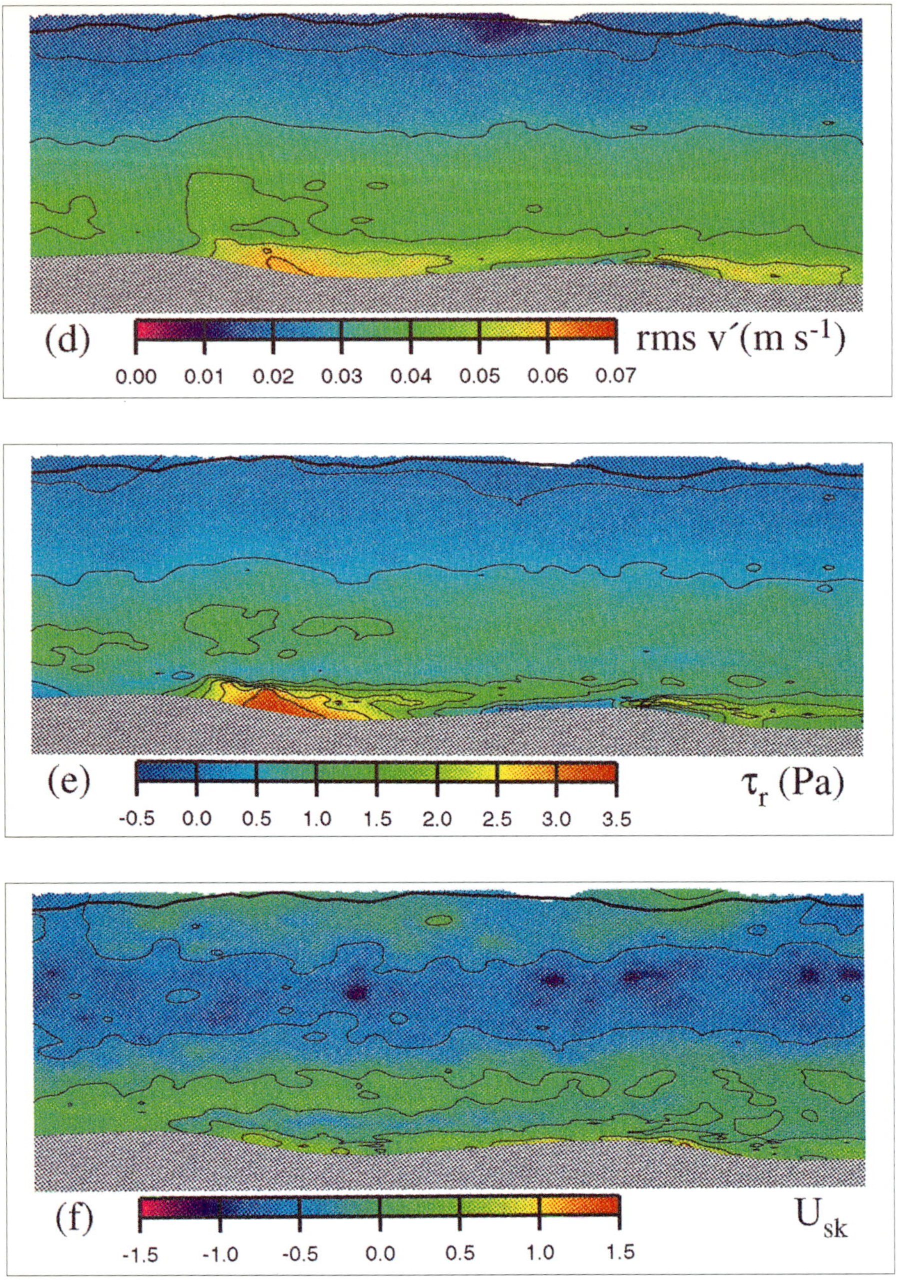

Figure 15.2. *Continued on facing page*

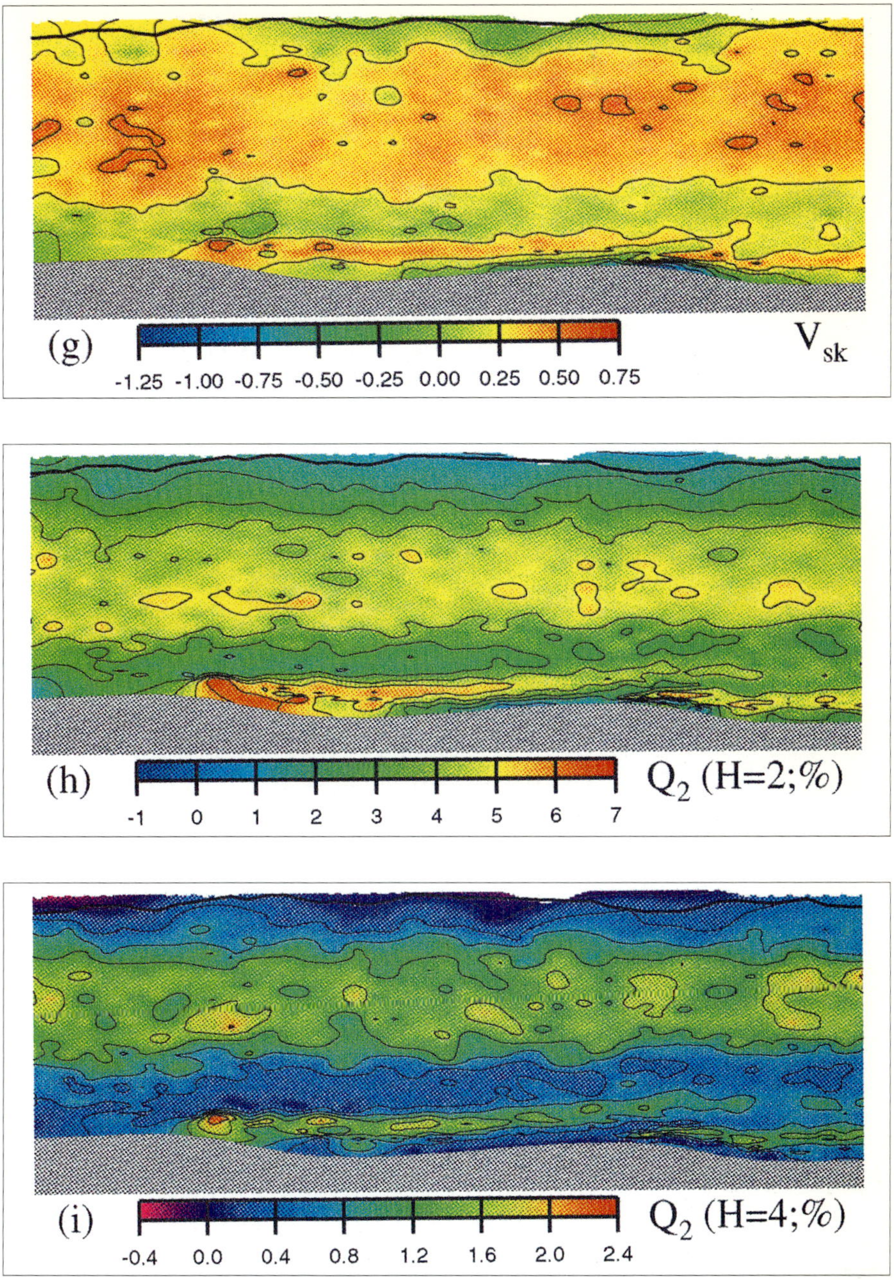

Figure 15.2. *Continued on next page*

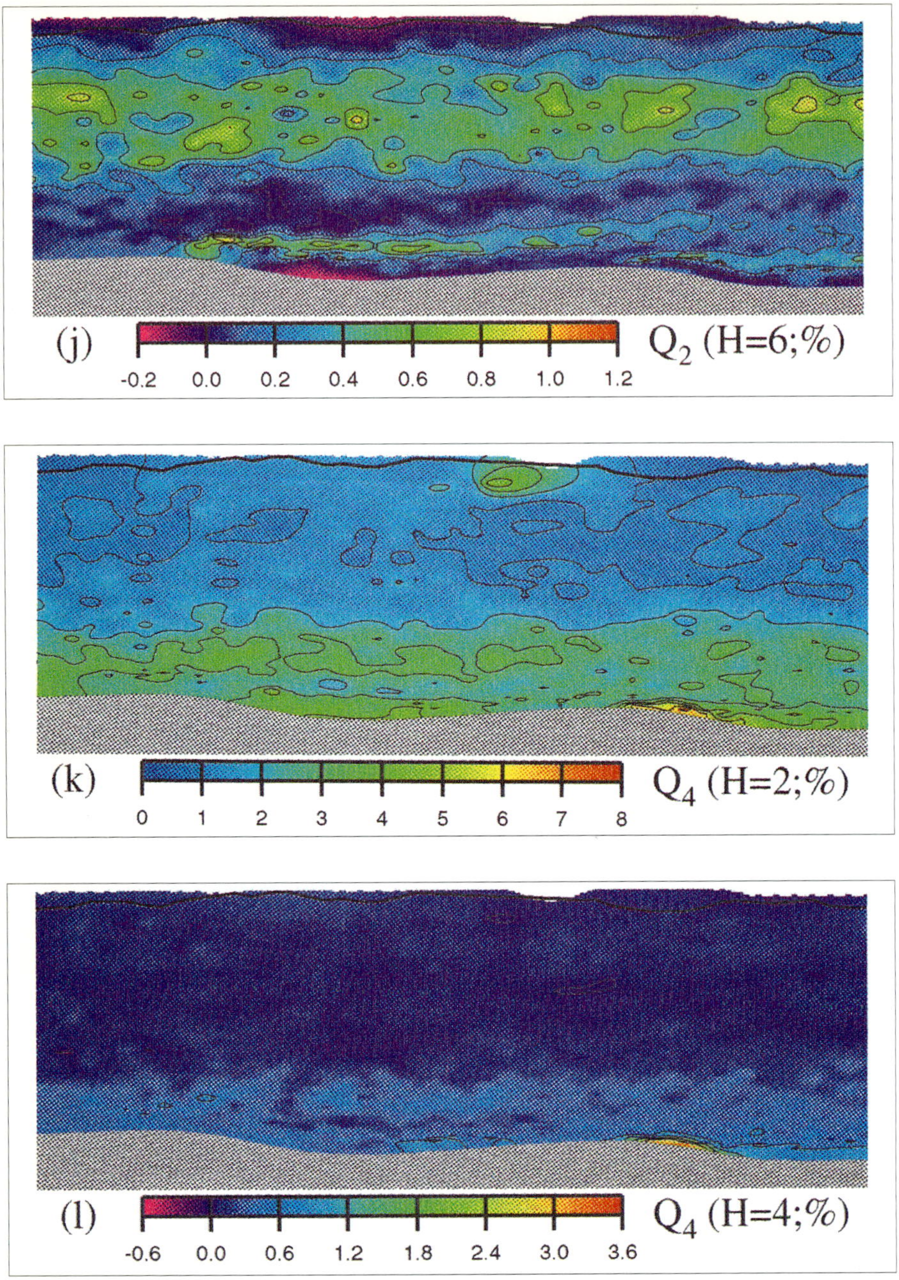

Figure 15.2. *continued*

reattachment to the ripple crest along a constant height about the mean bed elevation. Because the measurements were staggered in space, each stress was sorted into 5 mm bins, averaged and the mid-point of each bin plotted (Figure 15.4). A region of linear stress exists between 0.03 and 0.09 m above the mean bed elevation (see also Figure 15.3). A least-squares linear regression through these points projected to the zero-mean bed height, corresponding to a hypothetical equivalent flat-bed flow, gives $\overline{\tau}_R = 1.36$ Pa. Third, the spatially-averaged flow velocities along the stoss of the ripple, from 0.07 to 0.82d (i.e. the outer velocity profile; see Wiberg & Nelson, 1992; Li, 1994), can be used in the law of the wall (Equation 15.6 with $\kappa = 0.410$) to calculate a boundary shear stress, $\overline{\tau}_W$, of 1.60 Pa. If u_* derived from $\overline{\tau}_R$ is used in Equation 15.6, then $\kappa = 0.379$. The zero-velocity roughness height, y_0, is 0.57 mm and the equivalent sand roughness height, k_s, is 17.2 mm (Equation 15.7). This latter roughness height is approximately equal to the largest ripples present on the bed surface or, in the case of the ripples where the measurements were obtained, ~2h, in close agreement to those determined previously (e.g. Grant & Madsen, 1982; van Rijn, 1982; Wiberg & Nelson, 1992; Li, 1994).

Although downstream velocity measurements were obtained very close to the boundary, thereby offering the potential to calculate directly skin friction shear stresses (as discussed by Nelson & Smith, 1989; using Equation 15.6), the roughness heights determined from these velocities were much larger than the expected values based on surface grain size (>>0.008 mm; Nikuradse, 1933). This was due to the three-dimensional shape of the ripple stoss-side (Figure 15.1b) which produced transverse near-bed velocities, thus causing the observed downstream velocities orthogonal to the laser to be unexpectedly low (see Figure 15.3). Hence, no reliable measure of skin friction shear stress could be made. Furthermore, near-bed Reynolds stresses (those below the shear layer) decrease dramatically towards the boundary (Figure 15.3c) due to increased viscous damping of turbulence and could not be used for Reynolds stress projections as outlined above.

COMPARISON OF FLOW OVER FIXED RIPPLES AND DUNES

Bennett & Best (1995) conducted experiments similar to those presented herein on flow and turbulence structure over fixed two-dimensional dunes. Profiles of mean flow velocity, their RMS values and Reynolds stress at four localities over these two-dimensional dunes are plotted in Figure 15.3 for comparison with the present study. In general, similarities in flow over both ripples and dunes include: (i) separation of flow at the bedform crest and development of a lee-side separation zone; (ii) a free shear layer above the separation zone whose length and thickness are approximately equal to the bedform length and height, and characterized by positive vertical skewness values and frequent, high-magnitude ejection events (quadrant-2) arising from vortex shedding; and (iii) a turbulent wake region

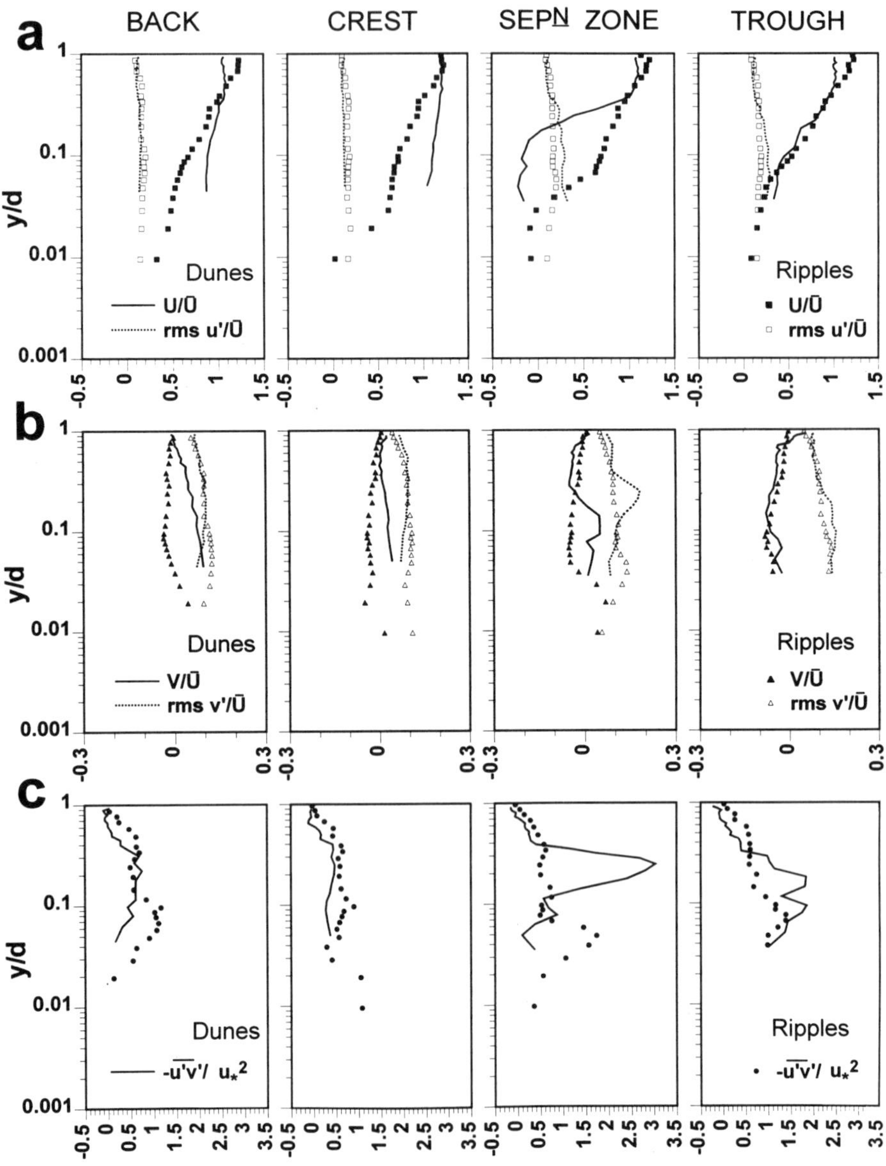

Figure 15.3. Profiles of flow velocity, turbulence and Reynolds stress at four locations (back, crest, separation zone and trough; profile numbers 27, 34, 36 and 43 on Figure 15.1c, respectively) over the fixed ripple bedform (symbols). Profiles shown are for mean and RMS values of the (a) downstream and (b) vertical components of velocity and (c) Reynolds stress. Also plotted are the profiles obtained over back, crest, separation zone and trough (profile numbers 52, 6, 21 and 31, respectively) of the two-dimensional fixed dunes (lines) reported by Bennett & Best (1995)

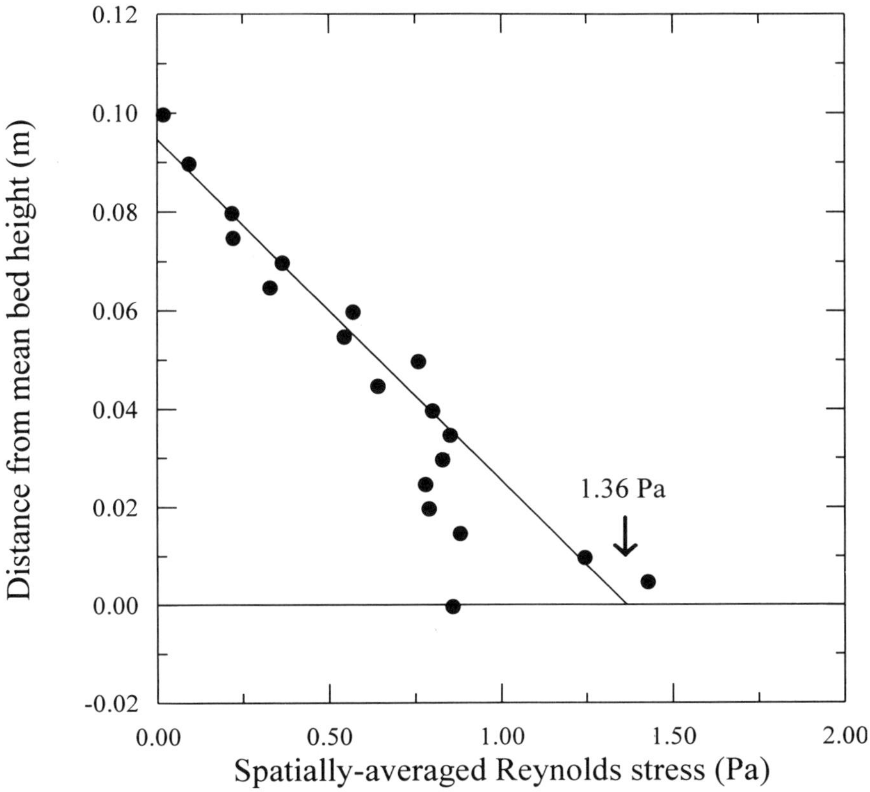

Figure 15.4. Vertical variation of the spatially-averaged Reynolds stress, τ_R, from flow reattachment to the ripple crest plotted about the mean bed elevation. A least-squares regression through the linear segment of the stress profile projected to the zero-mean bed height is used to determine $\overline{\tau}_R$

encompassing the free shear layer where u'_{rms}, v'_{rms} and τ_R are relatively high (Figure 15.3).

Several important differences in flow over ripples and dunes are apparent and these are summarized in Table 15.2. Flow is essentially uniform over ripples because the wake-induced effects of the crests are restricted to near-bed regions and complete mixing is achieved within the lower 20–40% of the flow. While both ripples and dunes possess separation zone shear layers, the shear layer present over ripples is essentially parallel to the bed whereas the shear layer over dunes is topographically forced up into the flow by the stoss side of the next dune downstream. The ejection events associated with the ripple shear layer dissipate rapidly and are therefore limited to the near-bed regions. In contrast, the ejection events associated with the shear layer over dunes can extend to 0.6*d* and to the water surface as 'boils' in large events. Bennett & Best (1995) suggested that, because the

Table 15.2. Summary of primary differences between flow over ripples and dunes

Flow parameter	Flow over ripples	Flow over dunes
Uniformity	weakly uniform	strongly non-uniform
Separation cell shear layer:		
Attitude	bed parallel	stoss parallel
Vertical extent	limited to 0.2d	extends to 0.6d (and to water surface)
Quadrant-2 events	restricted vertically and of low-magnitude	unrestricted vertically and of high-magnitude
Velocity gradient	small	large
High-speed fluid inrushes from outer regions	no	yes
At flow reattachment:		
Quadrant-4 events	infrequent, low-magnitude	frequent, high-magnitude
Reynolds stresses	2–3 $\bar{\tau}_0$	6–9 $\bar{\tau}_0$

velocity gradient across the separation zone shear layer is relatively large over dunes as compared to ripples (see also Figure 15.3), the frequency and magnitude of ejection events are also greater. This difference is further illustrated in Figure 15.5 which shows the maximum turbulence intensity of the downstream velocity component (as normalized by the free-stream approach velocity), either along the shear layer or at reattachment, is significantly larger in flows over dunes as compared to ripples and negative steps (in both air and water flows). The maximum value of u'_{rms} / U_0 is approximately constant at ~0.2 for negative steps over a range of flow depth -to- step height values (Figure 15.5), whereas bedforms show some variation in their RMS values with the absolute size of the bedform. This may be due to: (i) significant flow convergence downstream from reattachment associated with bedforms, a feature often excluded in studies of negative steps but important in determining reattachment lengths (see Eaton & Johnston, 1980); and (ii) the presence of a free water surface and its strong effect on flow above larger bedforms (a factor excluded in wind tunnel studies). The step angle may also influence turbulence generation along the free shear layer (see inset, Figure 15.5). Ruck & Makiola (1987) suggested that, for given mean flow conditions, step angles of 25° to 30° (i.e. those often associated with bedform slip-faces) produced maximum turbulence intensities. Transverse flow generated by the three-dimensionality of bedforms may also complicate the shear layer turbulence signatures when compared to two-dimensional steps. Regardless, it is clear that turbulent velocities associated with flow separation over dunes will be much larger than observed for ripples. At flow reattachment, time-averaged Reynolds stresses are 2–3 $\bar{\tau}_0$ for ripples and 6–9 $\bar{\tau}_0$ for dunes. These significantly higher Reynolds stresses are the result of

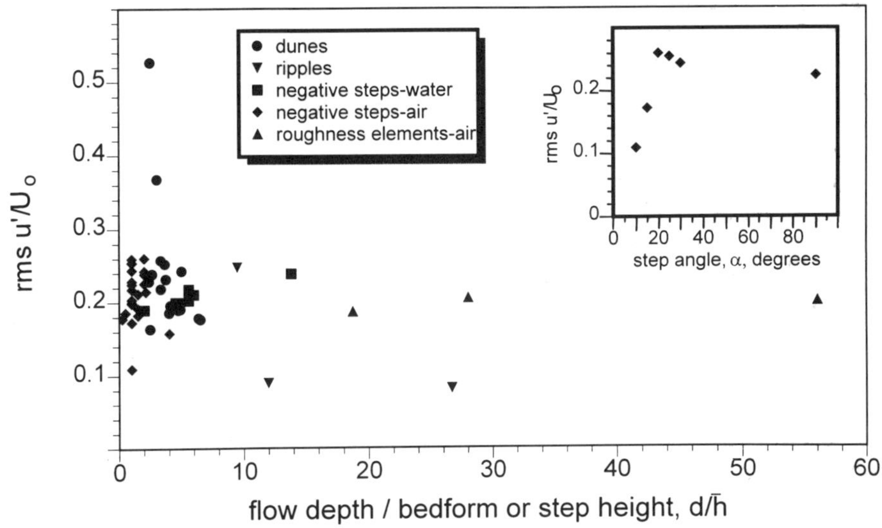

Figure 15.5. Variation in the maximum turbulence intensity of the downstream velocity component observed along the shear layer (either along the shear layer or at reattachment) as a function of the step height -to- flow depth ratio for water flows over dunes and ripples and for water and air flows over negative steps. Data sources: *Ripples* – McCorquodale & Giratalla, 1973; this study; *Dunes* – Raudkivi, 1966; McQuivey, 1973 (bedform height estimated from turbulence profiles); van Mierlo & de Ruiter, 1988; Lyn, 1993; Bennett & Best, 1995; Kornmann, 1995; *Negative steps in water* – Rifai & Smith, 1971; Etheridge & Kemp, 1978; Nakagawa & Nezu, 1987; Best & Bennett (unpub.); *Negative steps in air* – Bradshaw & Wong, 1972; Eaton & Johnston, 1980; Stevenson *et al.*, 1984; Vogel & Eaton, 1985; Ruck & Makiola, 1987, 1988; Berbee & Ellzey, 1989; Ötügen, 1991; Ötügen *et al.*, 1992; *Roughness elements* – Mueller & Robertson, 1963. Inset shows variation in maximum turbulence intensity over a negative step in a wind tunnel under identical mean flow conditions but with variable step angle (data from Ruck & Makiola, 1987)

frequent, high-magnitude quadrant-4 events ('sweeps'; positive contributors to the Reynolds stress) which are notably absent at flow reattachment in ripples (Figure 15.2k and 15.2l).

Since quadrant-2 (ejection) events penetrate further into the outer flow over dunes than ripples, the return flow required for continuity (including quadrant-4 events) must also originate from higher in the boundary layer, and therefore these events possess higher downstream velocities. Although the nature of the interaction between quadrant-2 events and the free surface has not been investigated (i.e. 'boils'; Matthes, 1947; Coleman, 1969; Jackson, 1976; Müller & Gyr, 1986; Kostaschuk & Church, 1993; Babakaiff & Hickin, this volume), such interactions may produce accelerated flow near the surface which are then directed back towards the bed (similar to the observations of Rashidi & Banerjee (1988) over flat beds in shallow flows). The possible enhanced mixing of quadrant-2 events caused by acceleration beneath the free surface, or by 'blocking' of the mean flow by the ejection as observed by Kirkbride (1993) over gravel clasts in depth-limited flows, may serve to increase the magnitude of return flows towards the bed, thereby increasing local bed shear stresses and transport rates.

THE PHYSICAL AND FLUID DYNAMIC MECHANISMS FOR THE RIPPLE-DUNE TRANSITION

It is well known that as transport stage increases and ripples transform into dunes, several important morphological changes occur (e.g. Guy *et al.*, 1966): (i) both bedform height and length become significantly larger but at an increasingly slower rate; (ii) a maximum value exists for ripple length and height relative to the flow depth before they are transformed to dunes; (iii) the bedform aspect ratio (L/h) does not change appreciably; (iv) bedform migration and bedload transport rates increase but their rates of change decrease significantly across the transition; and (v) the increase in bedload transport rate can be related directly to the increase in bedform size and migration rate (e.g. Simons *et al.*, 1965).

The data presented in the current study demonstrate the clear distinction between flow over ripples and dunes. Flow over dunes is markedly non-uniform: the bed topography causes large-scale spatial flow accelerations and decelerations which are absent in flow over ripples. The shear layer present over dunes is much more energetic (greater frequency of high-magnitude quadrant-2 and quadrant-4 events and higher Reynolds stresses) and it extends further into the outer flow as compared to ripples. These differences may be related to the pattern of flow non-uniformity generated by bed topography – water surface interactions and flow convergence/divergence and the complex turbulence patterns caused by the 'stacking' of individual wake regions generated upstream (Nelson & Smith, 1989).

As flow velocity, bed shear stress and sediment transport rate increase over ripples, the following sequence of events is suggested as a descriptive model of how ripples are transformed to dunes (Figure 15.6):

1. Ripples increase in mean height, length and migration rate, although a range of ripple dimensions will exist in an equilibrium flow (see Baas, 1993, 1994). Turbulence associated with the separation zone shear layer is restricted to the lower 20–30% of the flow, quadrant-2 events do not extend through the entire flow depth and quadrant-4 events originate in near-bed regions (Figure 15.6a);
2. A bedform of greater height than the other ripples present (a 'rogue' ripple; Leeder, 1983) may be created by:

   (i) the amalgamation of ripples of different height and celerity (see Raudkivi & Witte, 1990; Ditchfield & Best, 1992; Coleman &

Melville, 1994);

(ii) smaller-scale fluctuations in sediment transport rate, for example as a result of a local increase in τ_0 due to ripple three-dimensionality; and/or

(iii) 'flapping' of the separation zone shear layer (see Simpson, 1989) as mean flow velocity over the ripples is increased. Such 'flapping' will cause the point of reattachment to vary, exposing a larger bed area to the relatively high instantaneous Reynolds stresses, thereby increasing erosion and downstream sediment transport rates. The rapid increase in lee-side erosion with increasing ripple height has been demonstrated in the theoretical model of Asaeda *et al.* (1989); and

3. Once formed, this larger bedform causes a series of critical changes to the flow structure (Figure 15.6b):

(i) the absolute magnitude of the turbulent fluctuations in velocity increase (Figures 15.2–15.4), thus generating higher Reynolds stresses near reattachment (Figures 15.6b and 15.6c) and increasing bed erosion; and

(ii) the magnitude and frequency of ejections associated with vortex shedding along the free shear layer not only increase (Bennett & Best, 1995) but penetrate further into the outer flow and interact with the water surface. Such interactions may induce higher velocity inrushes of fluid back towards the bed (Figure 15.6b).

All of these factors serve to increase the erosion rate at reattachment and downstream sediment transport rate, thereby increasing deposition at the next crest downstream. Additionally, since higher bedforms possess greater flow separation lengths, the bedform length also increases. This process, once initiated, causes the growth of larger bedforms (dunes) that replace ripples. The critical stage in this process lies in alteration of the existing flow field by a larger-than-average bedform, and especially the influence of a larger, more energetic free shear layer associated with flow separation (Figures 15.6b and 15.6c). The importance of the causative association between ejection events and dunes, noted long ago by Jackson (1976) and others, may lie not in the enhanced potential for suspension (see below) but in promoting higher-magnitude return flows towards the bed combined with water surface interactions. These turbulent events will significantly affect bedload transport and provide a physical explanation for the presence of dunes in both gravel- and sand-sized sediment (see Dinehart, 1989, 1992; Best, 1996).

The rate of suspended load transport increases significantly as ripples are transformed to dunes (e.g. Guy *et al.*, 1966). It has been suggested that high-magnitude quadrant-2 events occurring along the shear layer and the quadrant-4 events occurring at the dune crest and at flow reattachment provide a mechanism for the initiation and maintenance of a suspended load (see discussions in Bennett & Best, 1994, 1995). High concentrations of suspended sediment in the outer flow and the occurrence of macroturbulent events such as 'kolks' or 'boils' (see references above), notably absent in flow over ripples, are also a direct consequence of the

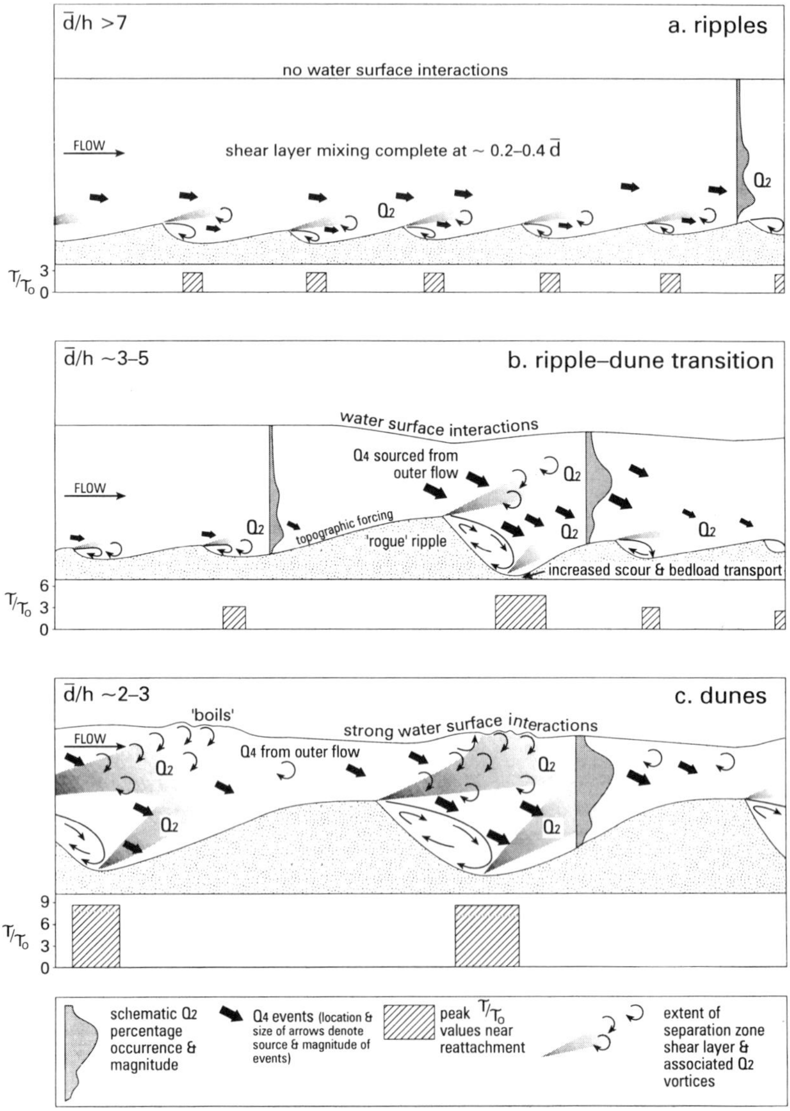

Figure 15.6. Schematic model of changing flow and turbulence structure across the ripple-dune transition (see text for further explanation): (a) equilibrium flow over a rippled bed; (b) non-equilibrium flow (increasing transport stage) over a rippled bed. Note the formation of a larger-than-average bedform ('rogue' ripple); and (c) equilibrium flow over a dune-covered bed. Note that figures are schematic and not drawn to scale in order to highlight the turbulent flow processes present. τ/τ_0 denotes the approximate ratio of the time-averaged Reynolds stress near reattachment to the spatially-averaged boundary Reynolds stress over the entire bedform

enhanced growth and development of the separation cell shear layer or a shear layer created by flow expansion downstream of a dune (Kostaschuk & Villard, this volume).

The size of dunes is ultimately limited by: (i) the applied shear stress where the increased shear has been balanced by increased sediment transport and bedform migration rates together with increased form drag associated with the bedform dimensions; and (ii) the depth of flow such that no further deposition can occur on the bedform crests (where conditions on the crest are similar to an upper-stage plane bed). However, the influence of greater sediment transport rates and sediment concentrations on both turbulence intensities and the dynamics of the flow separation zone are critical in the subsequent transition to upper-stage plane beds (e.g. Bridge & Best, 1988).

ACKNOWLEDGEMENTS

This work was supported by grant GR3/8235 from the UK Natural Environment Research Council. JLB gratefully acknowledges receipt of a Nuffield Foundation Science Research Fellowship which allowed this work to be completed. We thank Jaco Baas, John Bridge, Mike Leeder and an anonymous referee for their many constructive comments on improving the manuscript and Neil Woodhouse, Phil Fields and Tony Windross for technical support. The processed data for the fixed ripple experiments are available from the authors on provision of a 3.5" disk.

REFERENCES

Adams, E.W. & Eaton, J.K 1988. 'An LDA study of backward-facing step flow, including the effects of velocity bias', *J. Fluids Engrg.*, **110**, 275–282.

Allen, J.R.L. 1982. *Sedimentary Structures: their Character and Physical Basis*, Elsevier, Amsterdam, 539 pp.

Allen, J.R.L. 1984 'Parallel lamination developed from upper stage plane beds: a model based on the larger coherent structures of the turbulent boundary layer', *Sediment. Geol.*, **39**, 227–242.

Asaeda, T., Nakai, M., Manandhar, S.K. & Tamai, N. 1989.'Sediment entrainment in channel with rippled bed', *J. Hydraul. Engrg.*, **115**, 327–339.

Baas, J.H. 1993. *Dimensional analysis of current ripples in recent and ancient depositional environments*, Unpub. PhD Thesis, Utrecht Univ., The Netherlands, 199 pp.

Baas, J.H. 1994. 'A flume study on the development and equilibrium morphology of small-scale bedforms in very fine sand', *Sedimentology*, **41**, 185–209.

Bennett, S.J. & Best, J.L. 1994. 'Structure of turbulence over two-dimensional dunes', in Bélorgey, M., Rajaona, R.D. & Sleath, J.F.A. (Eds) *Sediment Transport Mechanisms in Coastal Environments and Rivers EUROMECH 310*, World Scientific Publishing Corp., Singapore, 3–14.

Bennett, S.J. & Best, J.L. 1995. 'Mean flow and turbulence structure over fixed, two-dimensional dunes: implications for sediment transport and dune stability', *Sedimentology*, **42**, 491–514.

Berbee, J.G. & Ellzey, J.L. 1989. 'The effect of aspect ratio on the flow over a rearward-facing step', *Exper. Fluids*, **7**, 447–452.

Best, J.L. 1992. 'On the entrainment of sediment and initiation of bed defects: insights from recent developments within turbulent boundary layer research', *Sedimentology*, **39**, 797–811.

Best, J.L. 1996. 'The fluid dynamics of small-scale alluvial bedforms', in Carling, P.A. & Dawson, M. (Eds) *Advances in Fluvial Dynamics and Stratigraphy,* John Wiley & Sons Ltd., Chichester, 67–125.

Bogard, D.G. & Tiederman, W.G. 1986. 'Burst detection with single-point velocity measurements', *J. Fluid Mech.*, **162**, 389–414.

Bradshaw, P. & Wong, F.Y.F. 1972. 'The reattachment and relaxation of a turbulent shear layer', *J. Fluid Mech.*, **52**, 113–135.

Bridge, J.S. & Bennett, S.J. 1992. 'A model for the entrainment and transport of sediment grains of mixed sizes, shapes, and densities', *Water Resour. Res.,* **28**, 337–363.

Bridge, J.S. & Best, J.L. 1988. 'Flow, sediment transport and bedform dynamics over the transition from dunes to upper-stage plane beds: implications for the formation of planar lamination', *Sedimentology*, **35**, 753–763.

Coleman, J.M. 1969. 'Brahmaputra River: channel processes and sedimentation', *Sediment. Geol.*, **3**, 129–239.

Coleman, S.E. & Melville, B.W. 1994. 'Bed form development', *J. Hydraul. Engrg.,* **120**, 544–560.

DANTEC 1991. *Particle Dynamics Analyzer: Users Manual*, DANTEC ELECTRONIK, 107 pp.

Dinehart, R.L. 1989. 'Dune migration in a steep, coarse-bedded stream', *Water Resour. Res.,* **25**, 911–923.

Dinehart, R.L. 1992. 'Evolution of coarse gravel bed forms: field measurements at flood stage', *Water Resour. Res.,* **28**, 2667–2689.

Ditchfield, R. & Best, J. 1992. 'Development of bed features: discussion', *J. Hydraul. Engrg.,* **118**, 647–650.

Eaton, J.K. & Johnston, J.P. 1980. *Turbulent Flow Reattachment: an Experimental Study of the Flow and Structure Behind a Backward-facing Step*, Report **MD-39**, Thermosciences Div., Dept. Mech. Engrg., Stanford Univ., 164 pp.

Engel, P. 1981. 'Length of flow separation over dunes', *J. Hydraul. Div., ASCE,* **107**, 1133–1143.

Etheridge, D.W. & Kemp, P.H. 1978. 'Measurements of turbulent flow downstream of a rearward-facing step', *J. Fluid Mech.*, **86**, 545–566.

Falco, R.E. 1977. 'Coherent motions in the outer region of turbulent boundary layers', *Phys. Fluids*, **20**, S24–S32.

Falco, R.E. 1991. 'A coherent structure model of the turbulent boundary layer and its ability to predict Reynolds number dependence', *Phil. Trans. Roy. Soc. London,* **A336**, 103–129.

Gould, R.D. & Loseke, K.W. 1993. 'A comparison of four velocity bias correction techniques in laser Doppler velocimetry', *J. Fluids Engrg.*, **115**, 508–514.

Grass, A.J. 1970. 'Structural features of turbulent flow over smooth and rough boundaries', *J. Fluid Mech.*, **50**, 233–255.

Grass, A.J. 1983. 'The influence of boundary layer turbulence on the mechanics of sediment transport', in Sumer, B.M. & Müller A. (Eds) *Mechanics of Sediment Transport*, A.A. Balkema, Rotterdam, 3–18.

Grant, W.D. & Madsen, O.S. 1982. 'Movable bed roughness in unsteady oscillatory flow', *J. Geophys. Res.,* **87**, 469–481.

Guy, H.P., Simons, D.B. & Richardson, E.V. 1966. 'Summary of alluvial channel data from flume experiments, 1956-1961', *US Geol. Surv. Prof. Paper* **462-I**, 96 pp.

Gyr, A. & Schmid, A. 1989. 'The different ripple formation mechanism', *J. Hydraul. Res.*, **27**, 61–74.

Jackson, R.G. 1976. 'Sedimentological and fluid-dynamic implications of the turbulence bursting phenomenon in geophysical flows', *J. Fluid Mech.*, **77**, 531–560.

Kirkbride, A. 1993. 'Observations of the influence of bed roughness on turbulence structure in depth limited flows over gravel-beds', in Clifford, N.J., French, J.R. & Hardisty, J. (Eds) *Turbulence: Perspectives on Flow and Sediment Transport*, John Wiley & Sons Ltd., Chichester, 185–196.

Kostaschuk, R.A. & Church, M.A. 1993. 'Macroturbulence generated by dunes: Fraser River, Canada', *Sediment. Geol.*, **85**, 25–37.

Kornmann, B.A. 1995. *The Effect of Changes in the Lee Side Shape of Dunes on the Flow Field, Turbulence and Hydraulic Roughness,* Report **R95-1**, Inst. Marine and Atmospher. Res., Utrecht Univ., The Netherlands, 20 pp.

Leeder, M.R. 1980. 'On the stability of lower stage plane beds and the absence of current ripples in coarse sands', *J. Geol. Soc. London*, **137**, 423–429.

Leeder, M.R. 1983. 'On the interactions between turbulent flow, sediment transport and bedform mechanics in channelised flows', in Collinson, J.D. & Lewin, J (Eds) *Modern and Ancient Fluvial Systems*, Intl. Assoc. Sediment. Spec. Pub. **6**, 5–18.

Levi, E. 1991. 'Vortices in hydraulics', *J. Hydraul. Engrg.*, **117**, 399–414.

Li, M.Z. 1994. 'Direct skin friction measurements and stress partitioning over movable sand ripples', *J. Geophys. Res.*, **99**, 791–799.

Liu, H.K. 1957. 'Mechanics of sediment ripple formation', *J. Hydraul. Div., ASCE*, **83**, 1–23.

Lu, S.S. & Willmarth, W.W. 1973. 'Measurements of the structure of the Reynolds stress in a turbulent boundary layer', *J. Fluid Mech.*, **60**, 481–511.

Luchik, T.S. & Tiederman, W.G. 1987. 'Timescale and structure of ejections and bursts in turbulent channel flows', *J. Fluid Mech.*, **174**, 529–552.

Lyn, D.A. 1993. 'Turbulence measurements in open-channel flows over artificial bedforms', *J. Hydraul. Engrg.*, **119**, 306–326.

Matthes, G.H. 1947. 'Macroturbulence in natural stream flow', *Trans. AGU*, **28**, 255–262.

McCorquodale, J.A. & Giratalla, M.K. 1973. 'Flow over natural and artificial ripples', in *Proc. 15th Congress, Inter. Assoc. Hydraul. Res.*, Istanbul, **1**, A22-1–A22-6.

McQuivey, R.S. 1973. 'Summary of turbulence data from rivers, conveyance channels and laboratory flumes', *US Geol. Surv. Prof. Paper* **802-B**, 66 pp.

Mueller, T.J. & Robertson, J.M. 1963. 'A study of the mean motion and turbulence downstream of a roughness element', *Dev. Theor. App. Mech.*, **1**, 326–240.

Müller, A. & Gyr, A. 1983. 'Visualisation of the mixing layer behind dunes', in Sumer, B.M. & Müller A. (Eds) *Mechanics of Sediment Transport*, A.A. Balkema, Rotterdam, 41–45.

Müller, A. & Gyr, A. 1986. 'On the vortex formation in the mixing layer behind dunes', *J. Hydraul. Res.*, **24**, 359–375.

Nakagawa, H. & Nezu, I. 1987. 'Experimental investigation on turbulent structure of backward-facing step flow in an open channel', *J. Hydraul. Res.*, **25**, 67–88.

Nelson, J.M., McLean, S.R. & Wolfe, S.R. 1993. 'Mean flow and turbulence fields over two-dimensional bed forms', *Water Resour. Res.*, **29**, 3935–3953.

Nelson, J.M. & Smith, J.D. 1989. 'Mechanics of flow over ripples and dunes', *J. Geophys. Res.*, **94**, 8146–8162.

Nezu, I. & Nakagawa, H. 1993. *Turbulence in Open-channel Flows*, IAHR Monograph Series, A.A. Balkema, Rotterdam, 281 pp.

Nezu, I. & Rodi, W. 1986. 'Open-channel flow measurements with a laser Doppler anemometer', *J. Hydraul. Engrg.*, **112**, 335–355.

Nikuradse, J. 1933. *Strömungsgesetze in Rauhen Rohren*, Kaiser Wilhelm-Institut für Stromungsforschung, Göttingen, Forschungsheft 361, 22pp. (Translation: *Nat. Advisory Comm. Aero.*, Tech. Mem. **1292**, 1950).

Ötügen, M.V. 1991. 'Expansion ratio effects on the separated shear layer and reattachment downstream of a backward-facing step', *Exper. Fluids*, **10**, 273–280.

Ötügen, M.V., Papadopoulos, G., Vradis, G. & Muckenthaler, G. 1992. 'Spanwise characteristics of the separated flow in a suddenly expanding duct', *Exper. Fluids*, **14**, 216–214.

Rashidi, M. & Banerjee, S. 1988. 'Turbulence structure in free-surface channel flows', *Phys. Fluids*, **31**, 2492–2503.

Raudkivi, A.J. 1963. 'Study of sediment ripple formation', *J. Hydraul. Div., ASCE*, **89**, 15–33.

Raudkivi, A.J. 1966. 'Bed forms in alluvial channels', *J. Fluid Mech.*, **26**, 507–514.

Raudkivi, A.J. & Witte, H.H. 1990. 'Development of bed features', *J. Hydraul. Engrg.*, **116**, 1063–1079.

Richards, K.J. 1980. 'The formation of ripples and dunes on an erodible bed', *J. Fluid Mech.*, **99**, 597–618.

Rifai, M.F. & Smith, K.V.H. 1971. 'Flow over triangular elements simulating dunes', *J. Hydraul. Div., ASCE*, **97**, 963–975.

Robinson, S.K. 1991. 'Coherent motions in the turbulent boundary layer', *Ann. Rev. Fluid Mech.*, **23**, 601–639.

Ruck, B. & Makiola, B. 1987. 'Flow over a single-sided backward facing step with step angle variations', in Turner, J.T. (Ed) *Proc. 3rd Inter. Conf. Laser Anemometry*, 369–378.

Ruck, B. & Makiola, B. 1988. 'Particle dispersion in a single-sided backward-facing step flow', *Int. J. Multiphase Flow*, **14**, 787–800.

Simons, D.B. & Richardson, E.V. 1961. 'Forms of bed roughness in alluvial channels', *J. Hydraul. Div., ASCE*, **87**, 87–105.

Simons, D.B., Richardson, E.V. & Nordin, C.F. 1965. 'Bedload equation for ripples and dunes', *US Geol. Surv. Prof. Paper*, **462-H**, 9 pp.

Simpson, R.L. 1989. 'Turbulent boundary-layer separation', *Ann. Rev. Fluid Mech.*, **21**, 205–234.

Stevenson, W.H., Thompson, H.D. & Craig, R.R. 1984. 'Laser velocimeter measurements in highly turbulent recirculating flows', *J. Fluids Engrg.*, **106**, 173–180.

Thomas, A.S.W. & Bull, M.K. 1983. 'On the role of wall-pressure fluctuations in deterministic motions in the turbulent boundary layer', *J. Fluid Mech.*, **128**, 283–322.

van Mierlo, M.C.L.M. & de Ruiter, J.C.C. 1988. *Turbulence Measurements above Artificial Dunes*, Report **TOW A55 Q789**, Delft Hydraulics, The Netherlands, 42 pp.

van Rijn, L.C. 1982. 'Equivalent roughness of alluvial bed', *J. Hydraul. Div., ASCE*, **108**, 1215–1218.

Vogel, J.C. & Eaton, J.K. 1985. 'Combined heat transfer and fluid dynamic measurements downstream of a backward-facing step', *J. Heat Transfer*, **107**, 922–929.

Wallace, J.M., Eckelmann, H. & Brodkey, R.S. 1972. 'The wall region in turbulent shear flow', *J. Fluid Mech.*, **54**, 39–48.

Wiberg, P.L. & Nelson, J.M. 1992. 'Unidirectional flow over asymmetric and symmetric ripples', *J. Geophys. Res.*, **97**, 12745–12761.
Williams, G.P. 1970. 'Flume width and water depth effects in sediment transport experiments', *US Geol. Surv. Prof. Paper* **562-H**, 37 pp.
Williams, P.B. & Kemp, P.H. 1971. 'Initiation of ripples on flat sediment beds', *J. Hydraul. Div., ASCE*, **97**, 505–522.
Williams, P.B. & Kemp, P.H. 1972. 'Initiation of ripples by artificial disturbances', *J. Hydraul. Div., ASCE*, **98**, 1057–1070.
Willmarth, W.W. & Lu, S.S. 1974. 'Structure of the Reynolds stress and the occurrence of bursts in the turbulent boundary layer', in Frenkiel, F.N. & Munn, R.E. (Eds) *Turbulent Diffusion in Environmental Pollution*, Academic Press, New York, 287–314.
Yalin, M.S. 1977. 'On the determination of ripple length', *J. Hydraul. Div., ASCE*, **103**, 439–442.
Yalin, M.S. 1985. 'On the determination of ripple geometry', *J. Hydraul. Engrg.*, **111**, 1148–1155.

NOTATION

d $\overline{d}$	flow depth and mean value
D	grain diameter
f	Darcy-Weisbach friction factor
Fr	Froude number
g	gravitational acceleration
h $\overline{h}$	bedform or step height and mean value
H	threshold (hole size) used in quadrant analysis
k_S	equivalent sand roughness height
L	bedform length
n	number of measurements
$\overline{Q}$	mean flow discharge
Q_2 Q_4	quadrant-2 and quadrant-4 events
u'_{rms} v'_{rms}	root-mean-square of the downstream and vertical velocity fluctuations
$\overline{S}$	mean water surface slope
u_i v_i u' v'	instantaneous downstream and vertical velocity and their fluctuations about the mean
u_*	shear velocity
U $\overline{U}$ $\overline{U}_d$	downstream velocity, its time-averaged value and its depth- and spatially-averaged value
U_{max}	maximum downstream velocity
U_0	mean downstream velocity approaching step or bedform
U_{sk} V_{sk}	skewness of downstream and vertical velocity distribution
$\overline{V}$	time-averaged vertical velocity
$\overline{w}$	flume width
y	height above bed
y_0	zero-velocity roughness height

α	angle of step lee-side
δ	height of maximum velocity above bed
κ	von Kármán coefficient
ν	fluid kinematic viscosity
ρ	fluid density
τ_R	Reynolds stress
$\bar{\tau}_R$	boundary shear stress calculated from linear interpolation of Reynolds stress profile
$\bar{\tau}_W$	bed shear stress using Kármán-Prandtl law of the wall
$\bar{\tau}_0$	boundary shear stress determined from momentum equation

16 Turbulent Sand Suspension Events: Fraser River, Canada

RAY KOSTASCHUK & PAUL VILLARD
Department of Geography, University of Guelph, Guelph, Ontario, N1G 2W1, Canada

ABSTRACT

Time-series of turbulent suspended-sediment events over dunes were gathered with acoustic profilers deployed in Fraser River, Canada, and tested for an underlying Poisson process. A Poisson process implies a random distribution of inter-event periods over time. The hypothesis of an underlying Poisson process could not be rejected for seven of the eight time-series examined and the model appears to apply to time-series generated from both single and multiple dune sources. The only time-series that did not fit the Poisson model was measured close to the bed and is significantly clustered. A physical explanation for this result is that clusters of low-period events close to the bed merge to form larger structures that are randomly distributed further above the bed. It is also possible that only the largest structures, which are random rather than clustered in pattern, are sufficiently coherent and energetic to be visible a significant distance above the bed. Smaller structures near the flow surface may represent events generated upstream or they could be fragments of larger events. Although suspended-sediment events are randomly distributed in time, they are tied to generation sites on dunes and are spatially non-random.

INTRODUCTION

Turbulence is a fundamental characteristic of most fluid flows and is linked to a variety of sedimentary processes including the entrainment and transport of sediment (Drake *et al.*, 1988; Rood & Hickin, 1989; Thorne *et al.*, 1989; Lapointe, 1992), the genesis and maintenance of bedforms (Jackson, 1976; Best, 1992) and the development of stream meanders (Levi, 1991; Rhoads & Welford, 1991). In geophysical flows, turbulence is evident at the surface as 'boils', circular patches of upwelling fluid (e.g. Matthes, 1947; Jackson, 1976; Rood & Hickin, 1989). Descriptions of boils in rivers indicate that they originate behind backward-facing steps such as dunes (Matthes, 1947; Coleman, 1969; Jackson, 1976). Matthes (1947) proposed that boils resulted from slowly-rotating vortices that he called 'kolks', an interpretation supported by Jackson (1976). Our understanding of the processes of boil formation by dunes, however, has been advanced primarily by laboratory investigations (for detailed reviews see Best, 1993; Nezu & Nakagawa,

Coherent Flow Structures in Open Channels.
Edited by P.J. Ashworth, S.J. Bennett, J.L. Best and S.J. McLelland.

1993). Müller & Gyr's (1986) flow visualizations of coherent structures over fixed dunes show that boils are related to eddy shedding along the shear layer above the lee-side flow separation zone. Detailed velocity measurements over fixed dunes (e.g. Nezu & Nakagawa, 1989; Nelson *et al.*, 1993; McLean *et al.*, 1994; Bennett & Best, 1995) reveal high levels of turbulence at the point of reattachment of flow separation, suggesting strong potential for sediment suspension. Iseya & Ikeda's (1986) mobile-bed flume experiments confirm that bed sediment is suspended in coherent structures generated at reattachment and appears at the surface in boils. Rood & Hickin's (1989) field study shows that boils in Squamish River contain much higher suspended bed-material concentrations than the ambient flow.

Several studies (e.g. Thorne & Hardcastle, this volume) have employed acoustic backscattering of suspended sediment to 'visualize' turbulent sediment suspension events. Recently, Kostaschuk & Church (1993) used acoustic flow visualizations to link sediment suspension and coherent flow structures generated by large dunes in Fraser River. They found that acoustic 'clouds', interpreted as representing high suspended-sediment concentrations, originate on the lower stoss-sides of dunes and appear at the surface as boils. Fraser River dunes are commonly low, symmetrical features with gentle lee-side slope angles and no lee-side flow separation bubble (Kostaschuk & Ilersich, 1995). A well defined zone of decelerated, downstream flow does occur close to the bed over the lee-side, however. Kostaschuk & Church (1993) suggest that the coherent structures evident on the acoustic records are the product of Kelvin-Helmholtz instabilities generated along the shear zone between the high speed outer flow and the decelerated flow on dune lee-sides. Villard & Kostaschuk (1995) show that high levels of turbulence and suspended-sediment concentration occur on lower stoss-sides of unseparated flow dunes in Fraser River, supporting Kostaschuk & Church's interpretation of acoustic clouds as turbulent sediment suspension events.

The quasi-periodic nature of turbulence has led investigators to focus on the statistical characteristics of the phenomenon, most notably the time interval between events (Luchik & Tiederman, 1987). An early flow-visualization flume study by Kim *et al.* (1971) indicated a log-normal distribution for burst inter-event periods. Jackson (1976) measured inter-event periods of surface boils in Wabash River, plotted his data on probability paper, and concluded that boils followed a log-normal distribution "fairly well" (pp. 542–543). Flow visualizations over fixed dunes by Nezu *et al.* (1980) also support a log-normal distribution for inter-event periods of surface boils. A flow visualization flume investigation by Bogard & Tiederman (1986), however, found that ejection and burst periods follow an exponential distribution rather than a log-normal distribution. This result has theoretical implications because an exponential distribution of inter-event periods indicates an underlying Poisson process (e.g. Giffin, 1978; Haan, 1979), a model generally accepted for atmospheric turbulence (e.g. Raupach *et al.*, 1986). A Poisson process represents discrete events on a continuous time scale, an appealing interpretation for time-series of coherent structures such as suspended-sediment events.

A Poisson process is a discrete parameter stochastic process that has the inherent Markov property of 'memoryless-ness', where the present is independent of the past (Gross & Harris, 1985). Events generated by a Poisson process are therefore discrete and random. The Poisson process stochastic model has been applied to problems as diverse as the arrival of jobs at a machine shop (Giffin, 1978) and rainstorm events (Haan, 1979). In addition, such processes can result from the superposition of any number of mutually independent, component Poisson processes (Cooper, 1972). Three probability distributions describe the Poisson process (Haan, 1979): (i) the number of events in a given time period follows a Poisson distribution; (ii) the time between events follows an exponential distribution; and (iii) the time to the nth event is a gamma distribution.

The purpose of this paper is to examine time-series of acoustic events in Fraser River for an underlying Poisson process. The acoustic events are interpreted as coherent turbulent flow structures containing high concentrations of suspended bed-material (e.g. Kostaschuk & Church, 1993).

SETTING

Fraser River drains 250 000 km^2 of mountainous terrain on the west coast of Canada and discharges into Strait of Georgia, a semi-enclosed marine basin with a mean tidal range of 3 m and spring tides approaching 5 m. Mean annual river discharge is 3400 $m^3 s^{-1}$ near the mouth. The lower reach of the river (Figure 16.1) is an estuary affected by tidal variations and salinity intrusions (Hodgins *et al.*, 1977; Kostaschuk & Atwood, 1990). Sandy bed-material transport is predominantly downstream-directed and is greatest under conditions of high river discharge, ebb tides and unstratified flow (Kostaschuk *et al.*, 1989a, 1989b). Median grain-size in the estuary is consistently 0.25–0.30 mm (Kostaschuk *et al.*, 1989a).

Dunes in Fraser River estuary range from 0.1 m to greater than 4 m in height, and 2 m to greater than 100 m in length (Kostaschuk *et al.*, 1989a). Multi-track surveys of large dunes (Kostaschuk & MacDonald, 1988) reveal a curved, concave-downstream planform with crests that are continuous for at least 300 m across the channel. This planform geometry indicates a two-dimensional shape (Ashley, 1990), at least for the larger features. Changes in dune height and length follow, but lag behind, seasonal variations in river discharge and weekly variations in tidal conditions (Kostaschuk *et al.*, 1989a; Kostaschuk & Ilersich, 1995). Dunes actively develop and migrate around low tide when flow is unstratified and bed-material is in transport (Kostaschuk *et al.*, 1989a, 1989b; Kostaschuk & Church, 1993).

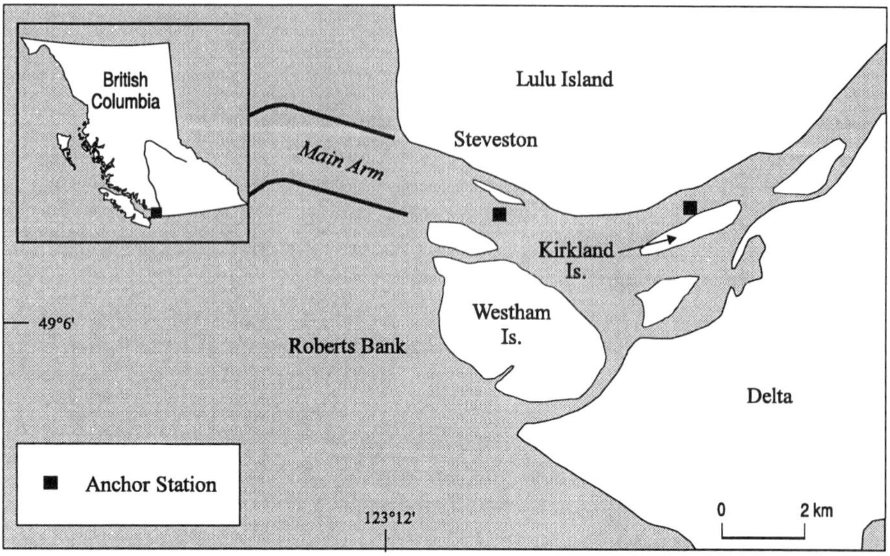

Figure 16.1. Lower Fraser River with anchor stations used in this investigation

METHODS

Time-series were gathered in Main Arm of Fraser River estuary (Figure 16.1) during high river discharge in June of 1990, 1992 and 1994 (the 1990 data are discussed in detail in Kostaschuk & Church, 1993). Measurements were made during the 2–3 hour period around low tide when flow was unstratified and bed sediment was in suspension. Acoustic data were obtained from an anchored research launch (Figure 16.1) using a 25 kHz Datasonics profiler in 1990 and a 200 kHz Apelco sounder in 1992 and 1994. Surface and velocity profiles were measured with mechanical and electromagnetic current meters and flow depth with a digital sounder. Dune geometry near anchor stations was monitored with the 200 kHz sounder. Surveyed shore markers provided position. The Steveston anchor station was employed in 1990 and 1994, and the Kirkland Island station was employed in 1992 (Figure 16.1).

Two distinct approaches were used to identify acoustic events on sounder records. Events were identified visually on 1990 and 1992 analogue records (e.g. Figure 16.2) and time of arrival at the anchor station measured 3–5 m above the bed. Some subjectivity is inherent in defining individual events using this procedure, but the events are sufficiently distinct to allow confident identification. The 1994 analogue records were digitized with a grey-scale scanner (range 0–255 units, temporal resolution 2 Hz) to examine detailed variations in acoustic reflectivity. Time-series were extracted at fixed heights above the bed and in this investigation

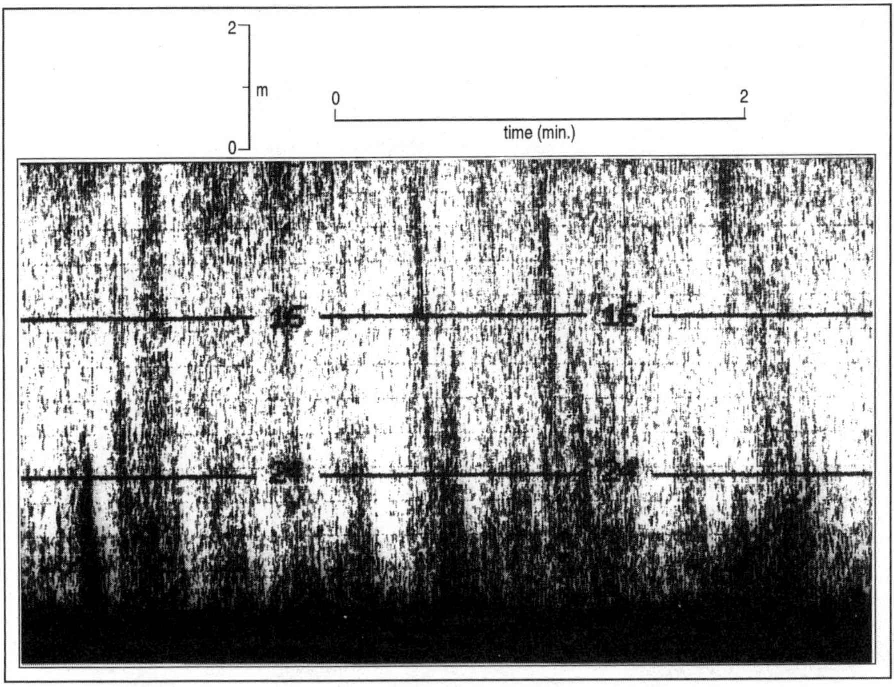

Figure 16.2. 200 kHz analogue chart record from 17 June 1995

we examine series at 1 m. A 2.5 s running-mean was used to filter background noise. Point data of grey-scale intensity 1 m above the bed was divided by the mean grey-scale intensity within known events to provide a grey-scale ratio (Figure 16.3). The mean grey-scale ratio is 1, so observations with a grey-scale ratio greater than 1 are defined as events. Events identified with this method show good correspondence with those identified visually, but the scanner has the advantage of a reproducible, less subjective methodology and allows comparison of event characteristics at different heights above the bed. Time of arrival of acoustic events at the launch is defined by the downstream edge of the structure in all data sets, and inter-event period is the time between arrivals of subsequent events.

RESULTS

Flow and dune characteristics vary considerably over the three field seasons (Table 16.1). Depth and surface velocity were generally largest in 1990 and 1992, although the lowest surface velocity was recorded in 1990. The largest dunes were in 1990, the smallest in 1992. The small dunes in 1992 may be controlled by a cohesive substrate that restricts dune development near Kirkland Island (Figure 16.1), though

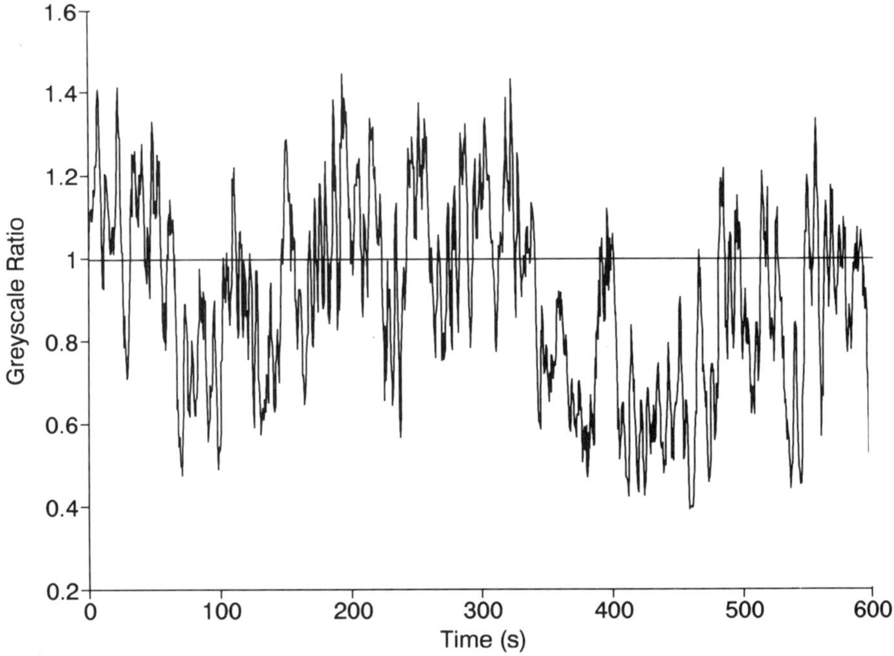

Figure 16.3. Digitized grey-scale scanner record 1 m above the bed for 19 June 1994. See text for explanation

Table 16.1. Flow characteristics and dune geometry

Date	$\bar{h}$ (m)	$\overline{U_s}$ (m s^{-1})	$\bar{H}$ (m)	$\bar{L}$ (m)
18/06/90	10.1	0.72	2.21	38.3
25/06/90	9.8	1.59	1.63	47.2
17/06/92	10.1	1.43	0.12	1.9
25/06/92	10.2	1.25	0.14	2.1
25/06/92	10.1	1.23	0.14	2.1
19/06/94	9.4	0.99	0.72	22.1
21/06/94	8.8	1.31	1.14	28.9
22/06/94	9.8	1.11	1.02	31.6

$\bar{h}$ is mean flow depth, $\overline{U_s}$ is mean surface velocity, $\bar{H}$ is mean dune height, $\bar{L}$ is mean dune length. $\bar{H}$ and $\bar{L}$ are means of 5–24 dunes around anchor stations we have no core or geophysical evidence to test this hypothesis.

The analogue record of Figure 16.2 illustrates several distinct acoustic events. Most features appear tied to the bed and a few of these extend towards the water surface at the top of the chart. A limited number of structures appear at the top of the record and are not linked to the bed. In some cases the larger events are composed of smaller structures close to the bed.

Figures 16.4 and 16.5 illustrate the two main time-series patterns observed. The 25 June 1992 record is very irregular and 'random' in appearance, although there is a cluster of small period events centred at an arrival time of 1200 s. In contrast, the 19 June 1994 record displays a distinctly 'clustered' structure, with groups of small-period events centred around arrival times of 250, 600, 1100 and 1500 s.

Table 16.2 summarizes descriptive statistics for inter-event periods for the eight data sets examined and Figure 16.6 provides frequency distributions for the seven long series. Inter-event periods generally are smaller for the 1994 data, a result consistent with the near-bed location of the measurements (e.g. Figure 16.2). All of the data are positively skewed, although skewness values are consistently larger in 1994. The 1994 observations are concentrated in the smallest class on Figure 16.4, with very few observations in larger classes. The Shapiro-Wilk test (Wilkinson, 1990) confirms that none of the data sets are normally-distributed at the 95% level of significance.

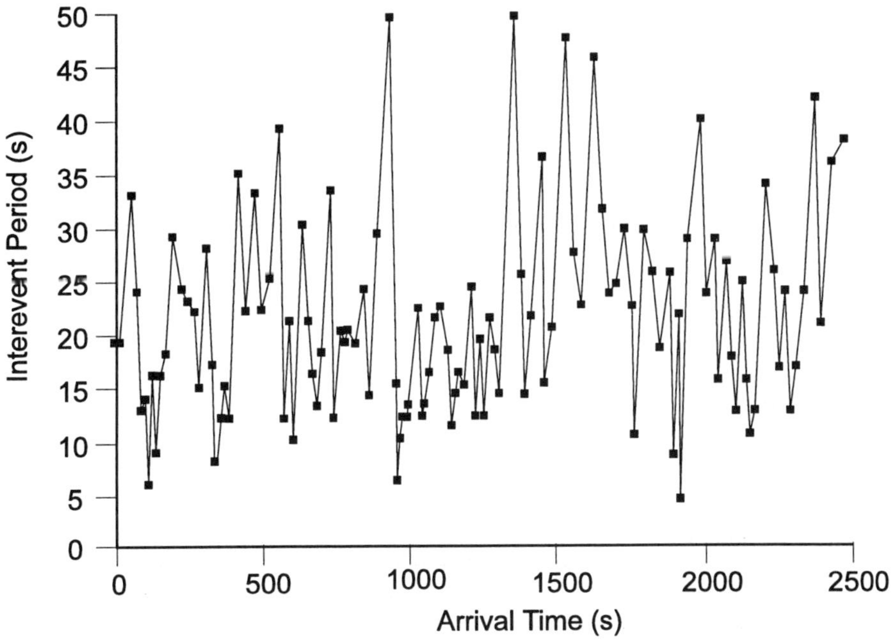

Figure 16.4. Time-series of inter-event period versus arrival time for 25 June 1992

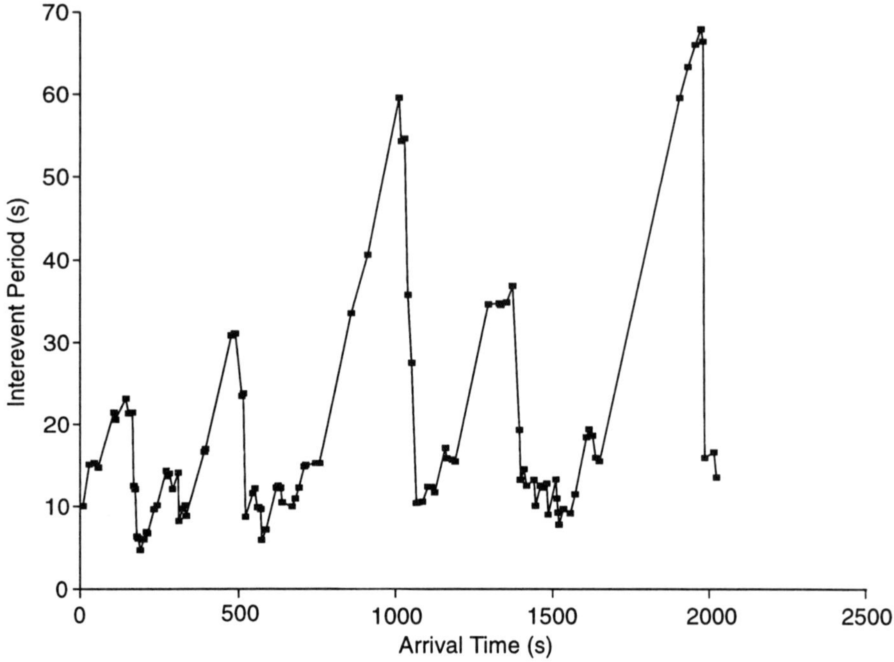

Figure 16.5. Time-series of inter-event period versus arrival time for 19 June 1994

Table 16.2. Descriptive statistics of time between acoustic events

Date	n	$\overline{t_i}$ (s)	σ (s)	Sk
18/06/90	76	46	22	0.9
25/06/90	10	22	8	0.63
17/06/92	98	41	26	1.59
25/06/92	160	22	10	0.7
25/06/92	120	30	16	1.83
19/06/94	107	19	30	5.37
21/06/94	108	19	22	3.54
22/06/94	73	26	33	4.31

n is number of events, $\overline{t_i}$ is the sample mean inter-event period, σ is sample standard deviation, Sk is sample skewness

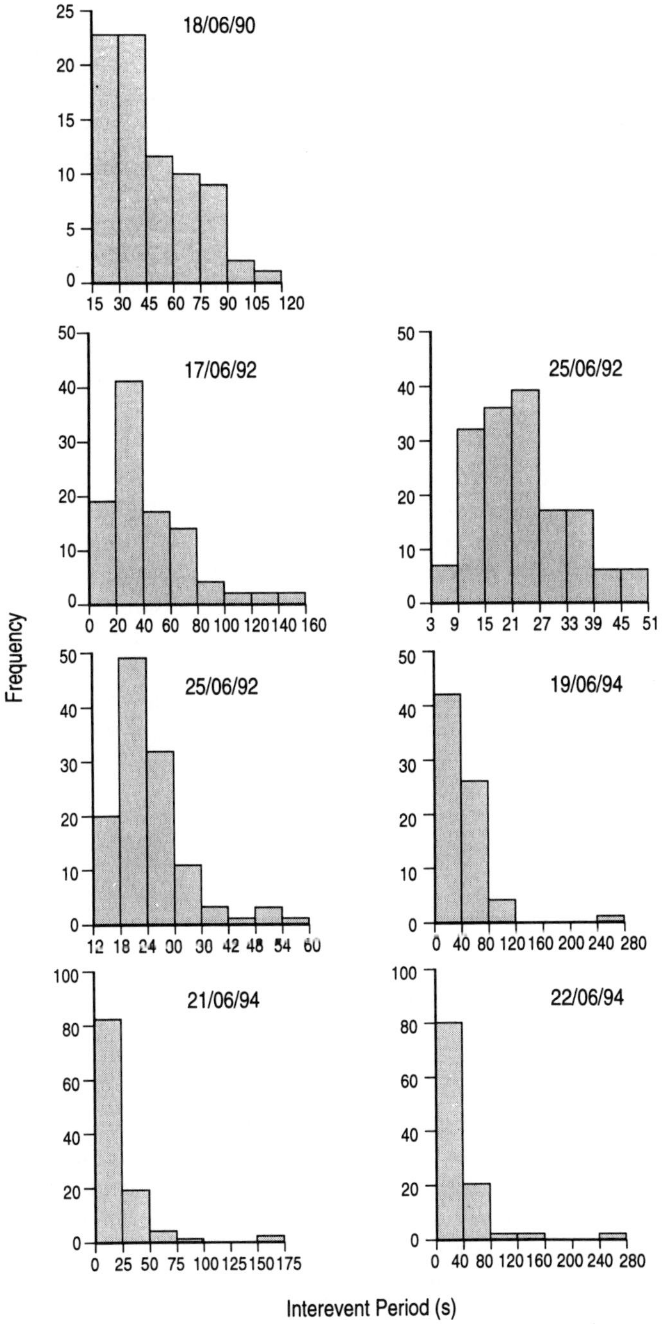

Figure 16.6. Frequency distributions for inter-event periods

DISCUSSION

Inter-event periods of turbulent events have been scaled to several flow and dune variables in both laboratory (e.g. Luchik & Tiederman, 1987) and field (e.g. Kostaschuk & Church, 1993) investigations. Such relations typically scale inter-event period directly with a length parameter (e.g. flow depth or dune height) and inversely with a measure of velocity (e.g. surface velocity, mean velocity or shear velocity). Although a detailed test of these relations is beyond the scope of this paper and we have very few observations (n = 8), a brief examination of our data may provide useful insight into the factors that influence inter-event period in our study. We have selected correlation analysis for this purpose. Since our data clearly are not normally distributed (Table 16.2), the non-parametric Spearman coefficient (90% level of significance) was chosen (Table 16.3). The only significant correlation is a positive result between inter-event period and flow depth, a correlation consistent with predictions from scale relations. Surface velocity may scale inversely with inter-event period, but the result is not statistically significant. Correlations with dune height and length are very low and highly insignificant.

One of the fundamental assumptions of the Poisson-process model is that events occur independently. We tested event independence for our time-series using the autocorrelation procedure described by Wilkinson (1990). None of these series were autocorrelated, suggesting event-independence for our data and allowing us to proceed with a formal test for a Poisson process. We use Giffin's (1978) 'normal test' to examine our acoustic data for an underlying Poisson process. The test relies on the argument that for r events in a predetermined time interval T, the sum of arrival times t is normally distributed. The mean μ and variance γ^2 of the distribution are:

$$\mu = rT/2\mu \qquad (16.1)$$

$$\gamma^2 = rT^2/12 \qquad (16.2)$$

The sample statistic for the sum of arrival times S is:

$$S = \sum t \qquad (16.3)$$

The 95% acceptance interval for the test statistic S_A is:

$$S_A = \mu \pm 1.96\gamma \qquad (16.4)$$

If S falls within this interval we cannot reject the hypothesis that the underlying process is Poisson. Table 16.4 summarizes the normal test results for our data. With the exception of 19 June 1994, all values of S fall well within the 95% acceptance levels of S_A so we cannot reject the hypothesis of an underlying Poisson process. This implies that turbulent sediment-suspension events are randomly distributed over time for seven of eight of our data sets.

Kostaschuk & Church (1993) report that the 25 June 1990 data appear to have been generated from a single source dune near their research vessel and our analysis shows a random distribution for this process. The remaining data sets probably result from multiple dune sources, as revealed on soundings along survey lines in the centre of the channel (Kostaschuk & Church, 1993). If our interpretations are correct, then these results provide physical evidence that a Poisson process from single dune sources can be superimposed to create a Poisson process from multiple dune sources.

The non-random nature of the 19 June 1994 record could be caused by the clustered pattern suggested on Figure 16.5. We tested this hypothesis with the 'linear nearest-neighbour' procedure of Hammond & McCullough (1980). The test is a variant of the two-dimensional (spatial) nearest neighbour analysis and identifies patterns as regular, random or clustered. The linear nearest-neighbour index is L_{calc}:

$$L_{calc} = \frac{T\overline{t_{obs}}}{0.5[T/(n-1)]} \qquad (16.5)$$

in which t_{obs} is the mean time between each event and its nearest neighbour and n is the number of events. This calculated value is then compared to a critical (95% level of significance) value L_{crit}. For the 19 June 1994 record $L_{calc} = 0.79$ and

Table 16.3. Spearman sign-rank correlation coefficients

	$\bar{h}$ (m)	$\overline{U_s}$ (m s^{-1})	$\bar{H}$ (m)	$\bar{L}$ (m)
$\bar{t_i}$ (s)	0.652 (0.080)	-0.205 (0.627)	-0.067 (0.875)	-0.091 (0.831)

see Tables 16.1 & 16.2 for explanatory text. Values in brackets are probabilities of chance occurrence

Table 16.4. Normal test results for a Poisson process

		S_A (s)	
Date	S (s)	lower	upper
18/06/90	133 988	116 595	151 381
25/06/90	1 180	786	1 602
17/06/92	214 900	179 252	230 507
25/06/92	289 800	263 955	315 644
25/06/92	207 228	196 209	241 207
19/06/94	95 468	96 537	107 519
21/06/94	18 388	12 418	20 562
22/06/94	66 067	59 641	77 851

S is the sample statistic for sum of arrival times, S_A is the test statistic for sum of arrival times (95% acceptance level).

$L_{crit} = 0.78$, placing the data almost exactly on the boundary between clustered and random. This result implies that we cannot reject the hypothesis that the data are clustered. The clustered pattern for 19 June 1994 may be related to methodology used in identifying events, although the other 1994 data are randomly distributed. A possible physical explanation is that 'clusters' of low-period events close to the bed (e.g. Figure 16.2) merge to form larger structures that are randomly distributed further above the bed. Alternatively, it could be argued that only the largest structures, which are random rather than clustered in pattern, are sufficiently coherent and energetic to propagate upwards from the bed. The structures at the top of Figure 16.2 that are not linked to the bed probably represent events generated upstream that have risen toward the surface and propagated past the sensor. They could also be fragments that have separated from larger structures.

Many investigations of turbulence in open channels have shown that large coherent eddies contain most of the turbulent energy in the flow and should have a major impact on the transport of bed-material in suspension (Nezu & Nakagawa, 1993). Our acoustic images (e.g. Figure 16.2) and those of Kostaschuk & Church (1993) suggest that such structures occupy a significant portion of the flow. Rood & Hickin (1989) found that suspended sediment concentration within coherent structures is much higher than the surrounding flow. Most sediment transport formulae in common use, however, do not account for the influence of coherent structures (Nezu & Nakagawa, 1993), but rather rely on an empirical approach using time-averaged 'bulk' flow properties such as shear velocity. It is generally accepted that suspension of coarse sediment requires upward vertical velocities greater than the settling velocities of the particles. Coherent structures are characterized by strong upward velocities (e.g. Kostaschuk & Church, 1993) and a time-averaged measure of flow cannot capture their effect on sediment suspension. Adequate prediction of sediment transport will only be achieved by models that effectively account for the role of coherent flow structures.

SUMMARY

We tested for an underlying Poisson process in time-series of turbulent suspended sediment events gathered with acoustic profilers deployed in Fraser River. A Poisson process interpretation implies a random distribution of inter-event periods over time. The hypothesis of an underlying Poisson process could not be rejected for seven of the eight time-series examined. The model appears to apply to time-series generated from both single and multiple dune sources. The only time-series that did not fit the Poisson model was measured close to the bed and is significantly clustered. Clusters of low-period events close to the bed may merge to form larger structures that are randomly distributed higher in the flow. It is also possible that only the largest structures, which are random rather than clustered in pattern, are visible above the bed. Smaller structures near the flow surface may represent events generated upstream, or they could also be fragments that have separated from larger structures.

A Poisson process interpretation for turbulent sediment suspension indicates temporally-independent, random events. The physical mechanisms generating turbulence, however, are not random. Scaling of inter-event periods to parameters such as flow depth and surface velocity (e.g. Levi, 1983, 1991), or to flow parameters controlled by dunes (Kostaschuk & Church, 1993), implies underlying physical controls to turbulence generation. On flat beds, interaction between high- and low-speed regions in the flow may generate coherent vortices (Levi, 1983, 1991; Yalin, 1992). Dunes control the physical location of vortex generation and reduce inter-event periods compared to flat beds (Yalin, 1992; Kostaschuk & Church, 1993). So although suspended sediment events are randomly distributed in time, they are spatially non-random.

ACKNOWLEDGEMENTS

Thanks are due John Luternauer and Mike Church for their ongoing interest and support, Ray Sanderson for piloting the research launch in Fraser Estuary, and Shannon Sterling and Michel Lapointe for field assistance in 1992. Funding was provided by a Natural Sciences and Engineering Research Council Research Grant to Kostaschuk, a Natural Sciences and Engineering Research Council Postgraduate Scholarship to Villard, and Luternauer's Geological Survey of Canada Project 790006.

REFERENCES

Ashley, G.M. 1990. 'Classification of large-scale subaqueous bedforms: a new look at an old problem', *J. Sediment. Petrol.*, **60**, 160–172.

Bennett, S.J. & Best, J.L. 1995. 'Mean flow and turbulence structure over fixed, two-dimensional dunes: implications for sediment transport and bedform stability', *Sedimentology*, **42**, 491 513.

Best, J. 1992. 'On the entrainment of sediment and initiation of bed defects: insights from recent developments within turbulent boundary layer research', *Sedimentology*, **39**, 797–811.

Best, J.L. 1993. 'On the interactions between turbulent flow structure, sediment transport and bedform development: some considerations from recent experimental research', in Clifford, N.J., French, J.R. & Hardisty, J. (Eds) *Turbulence: Perspectives on Flow and Sediment Transport*, John Wiley & Sons Ltd., Chichester, 61–93.

Bogard, D.G. & Tiederman, W.G. 1986. 'Burst detection with single point velocity measurements', *J. Fluid Mech.*, **162**, 389–413.

Coleman, J.M. 1969. 'Brahmaputra River channel processes and sedimentation', *Sediment. Geol.*, **3**, 129–239.

Cooper, R.B. 1972. *Introduction to Queuing Theory*, MacMillan Co., New York, 359 pp.

Davis, J.C. 1986. *Statistics and Data Analysis in Geology*, John Wiley & Sons, Toronto, 646 pp.

Drake, T.G., Shreve, L.R., Dietrich, W.E., Whiting, P.J. & Leopold, L.B. 1988. 'Bedload transport of fine gravel observed by motion picture photography', *J. Fluid Mech.*, **192**, 193–218.

Giffin, W.C. 1978, *Queuing: Basic Theory and Applications*, Grid Inc., Columbus, Ohio, 356 pp.

Gross, D. & Harris, C.M. 1985. *Fundamentals of Queuing Theory*, John Wiley & Sons, Toronto, 556 pp.

Haan, C.T. 1979. *Statistical Methods in Hydrology*, Iowa State University Press, 378 pp.

Hammond, R. & McCullough, P. 1980. *Quantitative Techniques in Geography*, Clarendon Press, Oxford, 364 pp.

Hodgins, D.O., Osborn, T.R. & Quick, M.C. 1977. 'Numerical model of stratified flow', *J. Waterways, Ports and Coastal Div., ASCE,* **1**, 25–42.

Iseya, F. & Ikeda, H. 1986. 'Effect of dune development on sediment suspension under unsteady flow conditions', in *Proc. 30th Japanese Conf. Hydraulics (JSCE)*, Tokyo, Japan, 505–510. (in Japanese).

Jackson, R.G. 1976. 'Sedimentological and fluid-dynamic implications of the turbulent bursting phenomena in geophysical flows', *J. Fluid Mech.*, **77**, 531–560.

Kim, H.T., Kline, S.J. & Reynolds, W.C. 1971. 'The production of turbulence near a smooth wall in a turbulent boundary layer', *J. Fluid Mech.*, **50**, 133–169.

Kostaschuk, R.A. & Atwood, L.A. 1990. 'River discharge and tidal controls on salt-wedge position and implications for channel shoaling: Fraser River, British Columbia', *Can. J. Civ. Engrg.*, **17**, 452–459.

Kostaschuk, R.A. & Church, M.A. 1993. 'Macroturbulence generated by dunes: Fraser River, Canada', *Sediment. Geol.*, **85**, 25–37.

Kostaschuk, R.A. & Ilersich, S.A. 1995. 'Dune geometry and sediment transport: Fraser River, British Columbia', in Hickin, E.J. (Ed) *River Geomorphology*, John Wiley & Sons Ltd., Chichester, 19–36.

Kostaschuk, R.A. & MacDonald, G.M. 1988. 'Multitrack surveying of large bedforms', *Geo-Mar. Lett.*, **8**, 57–62.

Kostaschuk, R.A., Church, M.A. & Luternauer, J.L. 1989a. 'Bedforms, bed-material and bed load transport in a salt-wedge estuary – Fraser River, British Columbia', *Can. J. Earth Sci.*, **26**, 1440–1452.

Kostaschuk, R.A., Luternauer, J.L. & Church, M.A. 1989b. 'Suspended sediment hysteresis in a salt-wedge estuary, Fraser River, Canada', *Marine Geol.*, **87**, 273–285.

Lapointe, M.F. 1992. 'Burst-like sediment suspension events in a sand bed river', *Earth Surf. Processes Landf.*, **17**, 253–270.

Levi, E. 1983. 'A universal Strouhal law', *J. Eng. Mech., ASCE,* **109**, 718–727.

Levi, E. 1991. 'Vortices in hydraulics', *J. Hydraul. Engrg.*, **117**, 399–413.

Luchik, T.S. & Tiederman, W.G. 1987. 'Timescale and structure of ejections and bursts in turbulent channel flows', *J. Fluid Mech.*, **174**, 529–552.

Matthes, G.H. 1947. 'Macroturbulence in natural stream flow', *Trans. AGU*, **28**, 255–262.

McLean, S.R., Nelson, J.M. & Wolfe, S.R. 1994. 'Turbulence structure over two-dimensional bedforms: implications for sediment transport', *J. Geophys. Res.*, **99**, 12729–12747.

Müller, A. & Gyr, A. 1986. 'On the vortex formation in the mixing layer behind dunes', *J. Hydraul. Res.*, **24**, 359–375.

Nelson, J.M., McLean, S.R. & Wolfe, S.R. 1993. 'Mean flow and turbulence fields over two-dimensional bedforms', *Water Resour. Res.*, **29**, 3935–3953.

Nezu, I., Nakagawa, H., Tominaga, A. & Yoshikawa, M. 1980. 'Visual study of large-scale vortical motions in open channel flows', *Annual Conf. JSCE*, Kansai-Branch, **II-10**. (in Japanese).

Nezu, I. & Nakagawa, H. 1989. 'Turbulent structure of backward-facing step flow and coherent vortex shedding from reattachment in open channel flows', in *Turbulent Shear Flows 6*, Springer Verlag, Berlin, 313–337.

Nezu, I. & Nakagawa, H. 1993. *Turbulence in Open-channel Flows*, IAHR Mono. Series, A.A. Balkema, Rotterdam, 281 pp.

Raupach, M.R., Coppin, P.A. & Legg, B.J. 1986. 'Experiments on scalar dispersion within a model plant canopy, Part 1: the turbulence structure', *Boundary Layer Meteorology*, **35**, 21–52.

Rhoads, B.L. & Welford, M.R. 1991. 'Initiation of river meanders', *Prog. Phys. Geog.*, **15**, 127–157.

Rood, K.M. & Hickin, E.J. 1989. 'Suspended sediment concentration and calibre in relation to surface flow structure in Squamish River estuary, southwestern British Columbia', *Can. J. Earth Sci.*, **26**, 2172–2176.

Thorne, P.D., Williams, J.L. & Heathershaw, A.D. 1989. 'In situ acoustic measurement of marine gravel threshold and transport', *Sedimentology*, **36**, 61–74.

Villard, P. & Kostaschuk, R. 1995. 'Turbulence and sediment suspension over unseparated flow dunes', in *Can. Assoc. Geographers Annual Meeting, Program and Abstracts*, University of Quebec at Montreal, Montreal, 76.

Wilkinson, L. 1990. *SYSTAT: The System for Statistics*, Evanston IL: SYSTAT Inc., 676 pp.

Yalin, M.S. 1992. *River Mechanics*, Pergamon Press, New York Inc., 220 pp.

NOTATION

$\bar{h}$	mean flow depth
$\bar{H}$	mean dune height
$\bar{L}$	mean dune length
L_{calc}	calculated linear nearest neighbour index
L_{crit}	critical linear nearest neighbour index
n	sample size
r	number of events passing a reference site
S	normal test sample statistic for the sum of arrival times of events
S_A	normal test statistic for the sum of arrival times of events (95% acceptance level)
Sk	sample skewness of inter-event periods
t	arrival time of events at reference site
$\bar{t_i}$	sample mean inter-event period
$\overline{t_{obs}}$	mean time between each event and its nearest neighbour
T	predetermined time interval for events passing reference site
$\overline{U_s}$	mean surface velocity
μ	normal distribution mean of t
γ^2	normal distribution variance of t
σ	sample standard deviation of inter-event periods

17 Coherent Flow Structures in Squamish River Estuary, British Columbia, Canada

C.S. BABAKAIFF & E.J. HICKIN
Department of Geography, Simon Fraser University, Burnaby, British Columbia, V5A 1S6, Canada

ABSTRACT

Macroturbulent features that appear as coherent flow structures (e.g. boils) in Squamish River estuary, Canada, are related to hydraulic and bedform conditions. Two types of boils occur in the estuary. Type-1 boils are produced in regions where the talweg crosses the channel and relate to bedform parameters and water depth. As bedform amplitude exceeds 0.5 to 0.6 m, the relation between amplitude and bedform wavelength becomes strongly linear and Type-1 boil intensity and sediment entrainment increase sharply. This suggests that bedforms become three-dimensional as scour holes form on the lee side. Sediment entrainment within Type-1 boils depends on water depth and boil period decreases rapidly as relative roughness increases. Type-2 boils are produced where the talweg moves along the outside banks, and relate to hydraulic conditions in the estuary. Time-series of streamwise current-speed display patterns also noted in simultaneous video imaging of Type-2 boil eruptions at the water surface, but there are no consistent relations. Type-2 boil periodicity is not governed exclusively by speed; there may be a feedback relation in which intense boil activity within the water column increases flow resistance, diminishes current speed and affects boil production.

INTRODUCTION

River and estuarine flows are characterized by two kinds of high-frequency fluctuations. Turbulence involves continuous, small-scale, random motions throughout the flow which are isotropic and well described by Gaussian statistics. Macroturbulence, however, involves large-scale, intermittent and distinctly non-random anisotropic perturbations to the flow.

Matthes (1947) proposes a classification of macroturbulence which includes discontinuous and intermittent vortex action (e.g. boils). He concludes that the process creating boils at the surface is the most powerful form of energy at work on stream beds. Jackson (1976) argues that features described in previous studies (e.g. Matthes, 1947; Korchokha, 1968) are analogous to the boils described by Coleman (1969). Coleman (1969) and Jackson (1976) present diagrams of a macroturbulent

Coherent Flow Structures in Open Channels.
Edited by P.J. Ashworth, S.J. Bennett, J.L. Best and S.J. McLelland.

structure with a single upwelling centre. It assumes a circular shape as water flows away from the centre towards a sharp boundary marked by vortices and, if sediment is present, a distinct decrease in suspended sediment. The structure then grows radially, flattening as its energy diminishes, and dissipates quickly. This is known henceforth as the 'Coleman/Jackson model'.

Boil generation has been linked to dunes by several researchers (Lane, 1944; Matthes, 1947; Coleman, 1969), but others suggest that bedforms are not required for boil generation (Jackson, 1976; Allen, 1986). Recent flume studies (Dyer, 1986; Müller & Gyr, 1986) suggest that boils result from flow separation in the lee of obstacles such as dunes. However, Kostaschuk *et al.* (1991) found that sub-surface boil generation occurs in the absence of flow separation and is consistently associated with only a few bedforms.

Levi (1983) concludes that 'if a fluid body of thickness d, which is restrained in its motion, is excited by an external flow of velocity, U, a tendency exists for it to oscillate with a frequency, f, given approximately by the universal Strouhal Law':

$$St = \frac{fd}{U} \tag{17.1}$$

This general law of instability has been used to describe many quasi-periodic features of turbulent activity, including boil generation in rivers and estuaries. The thickness of the restrained fluid body may be given by a length scale (L) of some obstacle in the flow such as the height of a bedform. Kostaschuk & Church (1993) suggest that a Strouhal relation may adequately describe the periodicity (T) of boil events:

$$T = \frac{2\pi d}{U} \tag{17.2}$$

Kostaschuk & Church (1993) argue that there may be several boil-generation mechanisms and that the length and velocity scales selected should depend on the particular mechanism. The disparities that they note between some predicted and measured boil periods may be related to the inability to attribute boils to a specific production site. In fact, Rood & Hickin (1989) question the applicability of the scaling equation to a single production site as opposed to a broad water-surface area. Other processes considered to influence scaling equations include secondary-flow patterns, secondary upwellings within a boil structure (Rood & Hickin, 1989) and multiple boil morphologies (Korchokha, 1968). There is little supporting quantitative analysis for these proposals.

An additional complexity to the application of a scaling equation is the choice of an appropriate averaging duration. Kostaschuk & Church (1993) present a time-series plot of periods of sonar-visualized boils which display repeated cyclicity on scales of several minutes. The temporal lags and response functions that may exist between observed macroturbulent phenomena and bedform or hydraulic parameters may render a short-term mean value a gross oversimplification. Lapointe (1989) notes that data regarding the frequency and intensity of turbulent events as a function of bedform geometry and flow parameters are needed. These concerns are directly addressed in our research.

The paucity of boil studies may be related to the intermittent appearance of boils in rivers. Fortunately, if a significant tidal flux is superimposed on the river flow, boil production is concentrated at certain stages in the diurnal tidal cycle (Rood & Hickin, 1989; Kostaschuk *et al.*, 1991) and observational difficulties are considerably reduced. It is postulated that consideration of wide ranges in hydraulic, tidal and bedform parameters will be most appropriate for elucidating scaling parameters for boil properties. Squamish River estuary in south-west British Columbia is an example of a high-energy, sand-bed river with significant tidal effects superimposed on large fluctuations in discharge.

The main objective of this study is to relate properties of boil activity to bedform and hydraulic parameters in Squamish River estuary. Rather than simply inspecting boil frequency, boil morphology and evolution were also observed to determine if there is evidence of more than one type of boil in Squamish River estuary.

FIELD SITE

Squamish River drains into Howe Sound approximately 40 km north of Vancouver, British Columbia (Figure 17.1). The drainage basin covers an area of 3600 km^2, much of which is mountainous and glaciated terrain. The region has a modified-marine climate, with cool, wet winters and warm, dry summers. Autumnal and winter floods associated with intense precipitation and rain-on-snow events are superimposed on a seasonal pattern of runoff characterized by a late spring-early summer snowmelt freshet. During the study period (May–August, 1992), mean hourly fluvial discharge in the estuary varied between 230 and 850 m^3 s^{-1}.

Tidal regime in Squamish River estuary is of the mixed semi-diurnal type. The tidal heights are not equal: there is a 'higher high tide' (HHT) and a 'lower high tide' (LHT), and corresponding low tides (HLT and LLT). The current is non-reversing, and near HHT (particularly during spring tides), current speeds greatly slacken and the river surface becomes mirror-like. Tidal range exceeds 4 m during spring tides and water surface slope ranges from 0 to 9.5×10^{-4}. The lower 5 km of Squamish River estuary, where the tidal effects on hydraulic parameters are most pronounced, constitutes the study reach (Figure 17.1).

METHODS

An NBA Controls DNC 3 current meter equipped with a 37 kg weight was deployed from a boat to obtain the current speed. Current-speed data collection

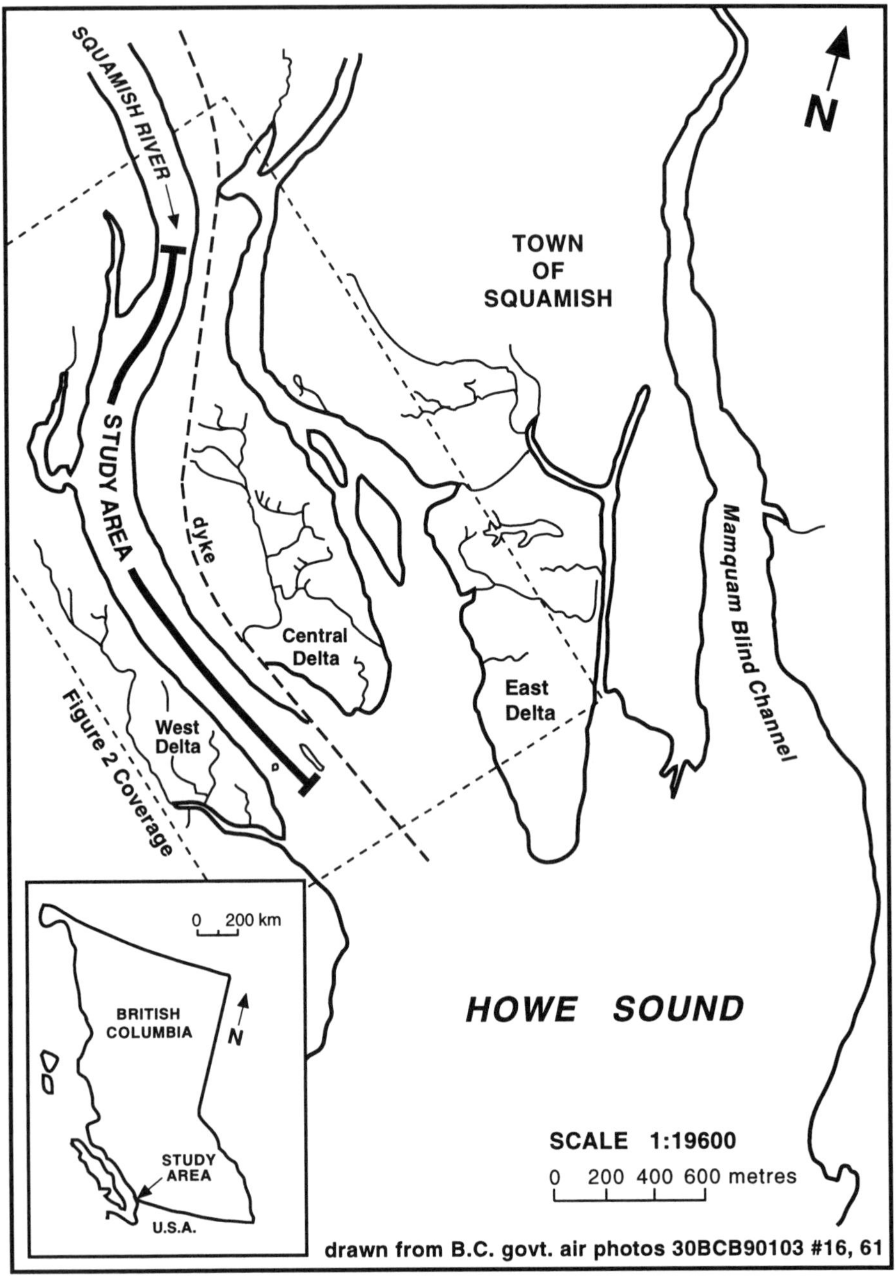

Figure 17.1. Study area

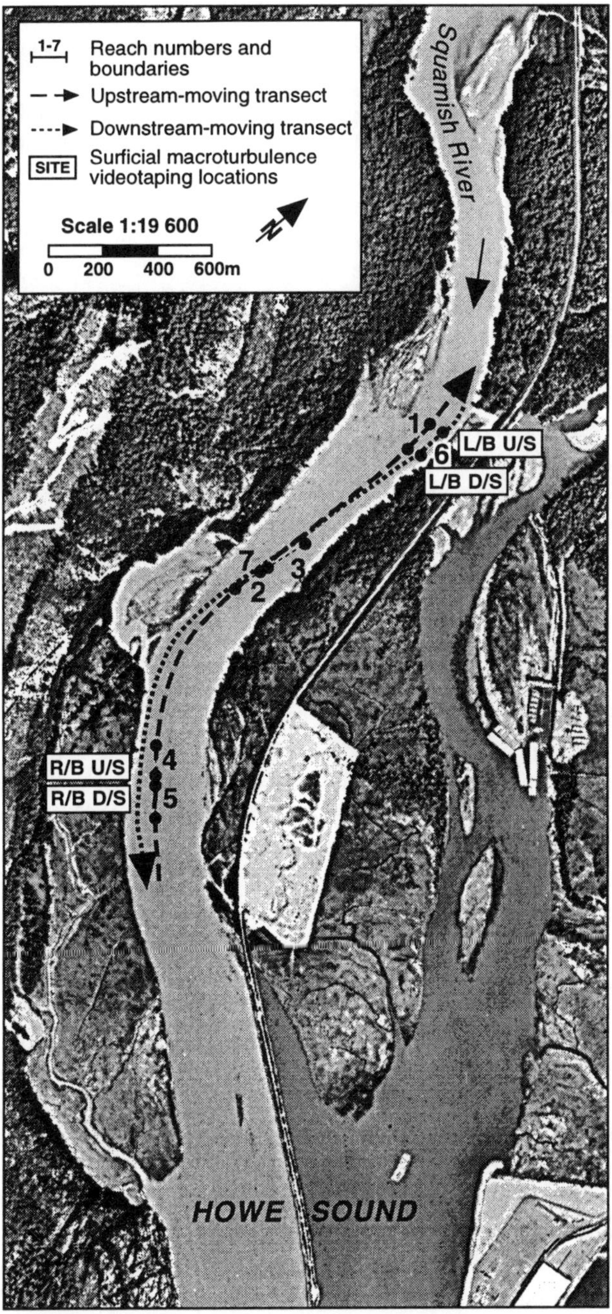

Figure 17.2. Sonar transects, surficial macroturbulence videotaping sites and study reaches

consisted of mean values collected over a duration of 30 s at the surface (U_{sfc}), and 10 minutes at 0.7 of total depth ($U_{0.7d}$). Water depth was measured with a Lowrance X-16 depth sounder.

Collection of simultaneous current-speed and boil-period data required filming the boil activity for later digitization. Only a relatively narrow range of boil intensity could be digitized. Bounding conditions had either such intense activity that structures could not be attributed to a single generation site, or such weak activity that the upwellings could not be recognized reliably. About 25% of the more than 100 hours of video record was usable.

The bedform parameters were measured along the talweg because the high-speed filament was associated with the most intense macroturbulence. Bedform profiles were obtained along a 2 km path length utilizing a graphic output Lowrance X-16 depth-recording sonar. Each profile of the path-length is known as a 'transect'; over 600 such transects were run during the field season. Most were run in pairs, first upstream, then downstream. The transect was divided into 12 'reaches' of approximately homogeneous average velocity, width and depth. Mean reach values of water depth, bedform height and wavelength were obtained from the graphic output. Observations of boil morphology and intensity in the talweg were also made during collection of bedform data. Five reaches were digitized from the upstream transects, and two from the downstream transects (Figure 17.2). 100 transects were selected for each of the seven reaches, chosen to cover the full range of tidal and fluvial conditions encountered. Details regarding methodology are described in Babakaiff (1993).

In order to determine individual bedform properties from a continuous profile, a bedform must be operationally defined. This definition is somewhat subjective depending on the scale of the features along the profile, but this is acceptable as long as consistency within the analysis is maintained. If the height of the secondary features exceeded 20% of the height of the 'parent' bedform, or if the secondary features were better defined than the parent feature, each secondary feature was digitized as a separate bedform.

Similarly, there is not a commonly accepted technique for measuring the morphometric parameters of bedforms. Inconsistencies among these techniques (described in Babakaiff, 1993) can be attributed to the bedforms often being symmetrical and two-dimensional, thus it being irrelevant whether height is measured on the upstream or downstream side. The importance of the wavelength measures depends upon whether individual bedforms are the study focus or if means of several bedforms are examined. In our analysis, reach-means of bedform wavelength (w) and height (h) and flow depth (d) are used, and the results are relatively insensitive to the measurement technique.

The bedforms in the estuary are asymmetrical (ebb-tidal asymmetry), and bedform height is measured on the upstream side. Wavelength is measured as the trough-to-trough distance, and depth is measured as the distance from the water surface to one-half the bedform height on the downstream side (Figure 17.3).

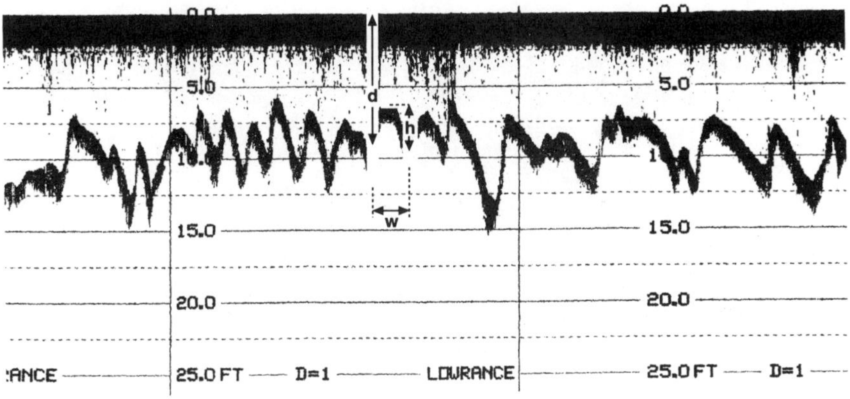

Figure 17.3. Bedform parameters and water depth from graphical output of sonar transect

BOIL MORPHODYNAMICS

TYPE-1 AND 2 BOIL DESCRIPTIONS

Field observations of boils often suggest a common morphology and evolution, but the structures in Squamish River estuary are complex (Rood & Hickin, 1989). Throughout the study reach, the boils are strongly influenced by tidal conditions.

Approaching LLT in the lower estuary, the water surface often exhibits streamwise parallel bands of alternating smooth and rough patterns. As the depth decreases, these bands become differentiated by the sediment and organics entrained by macroturbulent features. These lanes of boil activity become most strongly developed in the late ebb, particularly during spring tides. The lanes increase both in number and width 60 to 90 minutes before LLT. As LLT approaches, the bands begin to decrease in width, but they increase in number. This trend continues to the flood tide and all bands dissipate a few hours after LLT.

Within these bands, sites of repeated boil production were noted, but the duration rarely exceeded 5 minutes and never exceeded 10 minutes. Once boil production at a site stopped, another site was typically activated within a 10 m radius. The structures observed in the lower estuary frequently, but not always, conformed to the Coleman/Jackson model. Secondary features could also be present on the upstream side (Figure 17.4). As LLT approached, two distinct structures not fitting the Coleman/Jackson model were observed in the lower estuary. During spring tides, some boils erupting at the surface did not display one or two point sources, but had many small point sources of upwelling, each entraining much sediment and organics. These violently erupting boils did not have specific preferential downwelling regions and the edges were poorly defined. In plan view, the chaotic

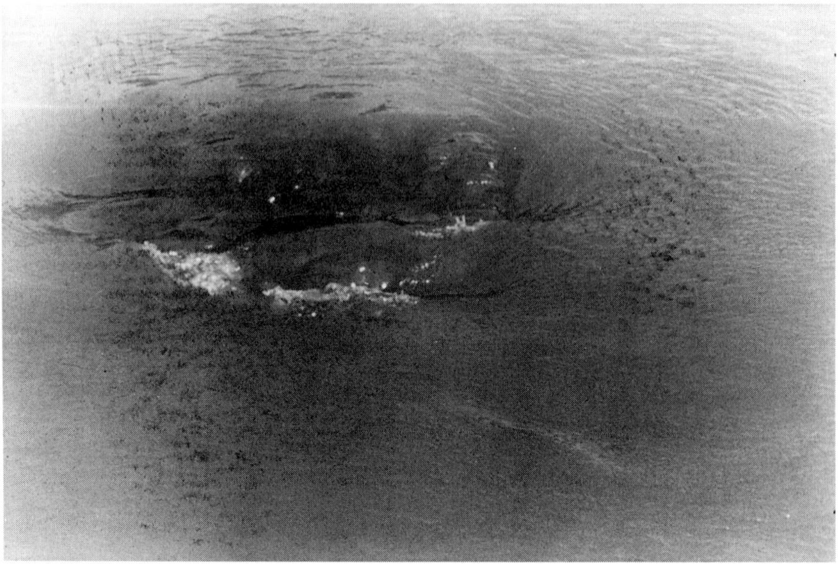

Figure 17.4. Secondary upwellings on upstream side of Coleman/Jackson structure (flow is towards bottom of photo)

pattern of upwellings created a structure similar in appearance to a head of cauliflower (Figure 17.5) and is thus termed a 'cauliflower' morphology.

Upwelling was also noted to occur along a linear source, with rotation along a horizontal axis, and downwelling along a linear front (Figure 17.6). This downwelling was frequently oriented normal to the flow on the downstream side of the upwelling, but could also be aligned parallel with the flow. These boils have been termed 'rollers'. The ends of the roller were either straight or curved, and the curvature seemed to determine how the structure would evolve. After the initial upwelling, there was a distinct cycle of events leading to dissipation. As illustrated in Figure 17.7, the ends began to curl towards the centre of the feature as it spread and rained out sediment. As the structure thinned, the ends looped strongly, and often formed strong vortices on one side or both. If the axes were initially straight, the structure simply dissipated. If they were curved, however, a secondary feature frequently upwelled either just between the vortices, or on the upstream side of the structure. These secondary features often spawned several smaller features, but all were poorly defined and short-lived. Although the secondary features represent a later phase of activity, there were no additional upwellings beyond 5 to 8 s after the original boil broke the surface and the structure dissipated as it was advected downstream.

Where the talweg crossed from the left bank to the right bank (i.e. reaches 2, 3 and 7 in Figure 17.2), all three structures described above were noted with

Figure 17.5. 'Cauliflower' boil structure (flow is from right to left)

intensities similar to those in the lower estuary. Although the production sites seemed to occur randomly rather than in lanes, their duration (less than 5 minutes) was also similar to those in the lower estuary site. These structures were collectively termed Type-1 boils.

Further upstream, the observed morphology, evolution and production sites of the boils were quite different. The most extensive boil activity in the upstream portions of the study area occurred where the talweg impinged on the banks. Wherever the talweg moved along the outer bank of the curved channel, there were numerous slump blocks and undercut trees in the channel which produced turbulent wakes. There were also several sites along both inner and outer banks that were the production sites for a more persistent form of macroturbulence, termed Type-2 boils. The intermittent upwelling of fluid produced a structure that would grow in surface area and stretch downstream as the upwelling continued. After detaching, the

Figure 17.6. Roller structure with curved long axis (flow is towards bottom of photo)

patch of smooth water moved downstream, acquiring a rotation in accordance with the lateral velocity gradient. Over several hundred metres downstream of the production site, the boil also underwent continual change, stretching, expanding and contracting, as numerous upwellings continually rejuvenated the structure. The contribution of many individual upwellings produced an agglomeration that typically had a surface area of 10–20 m^2, but occasionally exceeded 60 m^2. The upwellings within the boil agglomeration were frequently bounded by foam, vortices, or areas of downwelling, but precise delineation of the boundaries required ideal lighting conditions. As the boils advected past, the upwelling sites appeared random, but with the passage of several boil structures, it was apparent that they remained persistent over at least several tens of minutes. These boil agglomerations typically diffused 200 to 300 m downstream of the original production site. Within this region of boil dissipation, another production site was inevitably activated, and the cycle of production, advection and dissipation would repeat again. There were three to five distinct, persistent Type-2 boil production sites per bank, each separated by regions of transportation (with contributory upwellings) and dissipation.

Qualitative estimates of boil intensity were noted during the transect surveys, and are based on the following scale: *level 1*, strong 'eruption', sediment and organics entrained, audible; *level 2*, weak 'eruption', sediment and organics entrained; *level 3*, strong 'upwelling', organics but no sediment entrained; *level 4*, weak

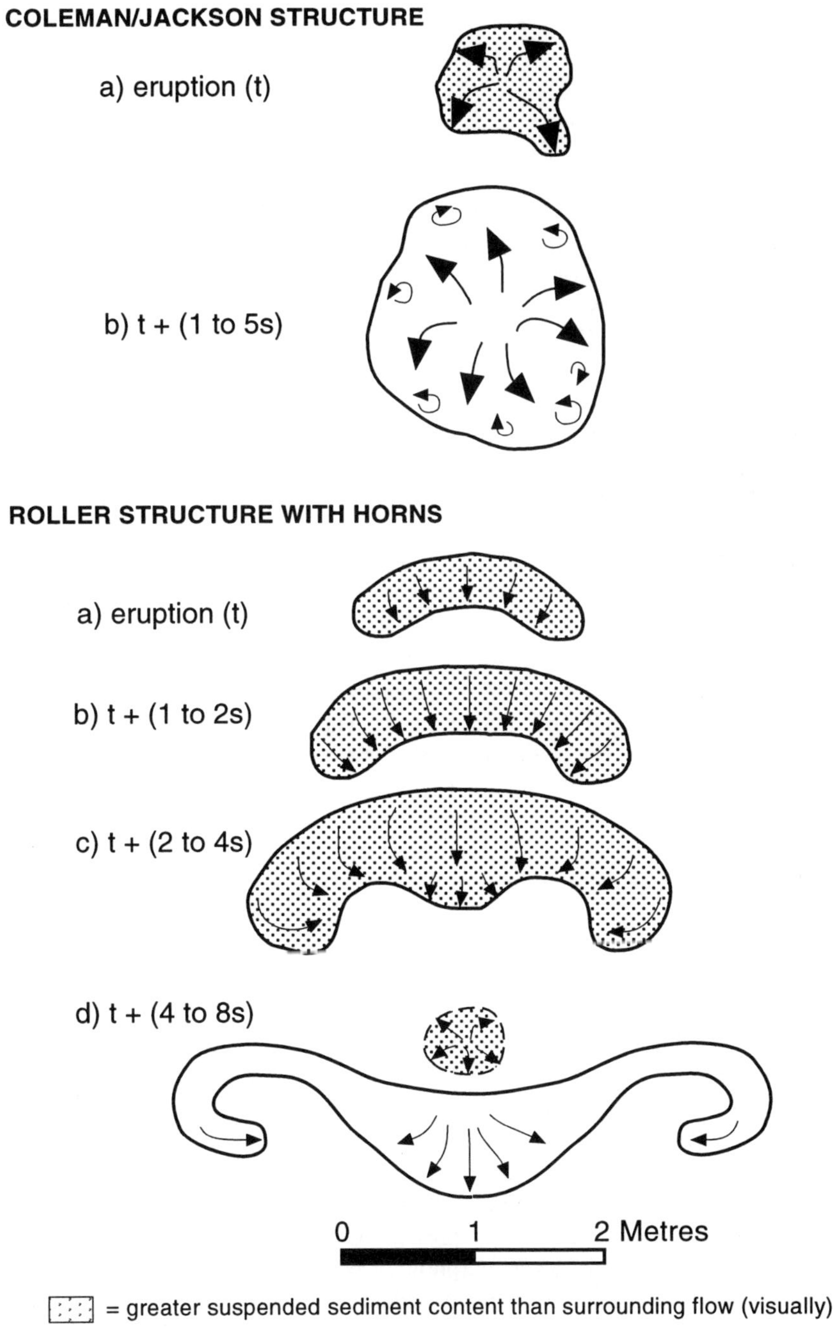

Figure 17.7. Evolution of boil structures: Coleman/Jackson and roller (with horns)

'upwelling', bit of organics; *level 5*, very weak upwellings, no organics; and *level 6*, calm, placid surface.

TYPE-1 BOIL ANALYSIS

To examine the dependence of Type-1 boils on bedform or hydraulic parameters, direct examination of the relations between bedform h and w and water depth was undertaken. The parameter used to differentiate such plots was the boil intensity observed at the water surface during the transect survey for each reach. Although the appearance of the features within a reach was temporally intermittent, the intensity was generally quite spatially homogenous within the talweg.

Plots of w versus h (for reaches 2, 3 and 7), display two distinct populations with a line of separation at $h = 0.5–0.6$ m (Figure 17.8). The data display a greater variance with h less than 0.5–0.6 m in all three reaches, and a linear pattern in reaches 2 and 7 once h exceeds 0.6 m. The sediment-laden boils (intensity levels 1 and 2) are associated with the latter population. In reach 2, the sediment-laden boils were associated with the higher values of dune steepness within this latter population (Figure 17.8a).

Examination of the w versus d plots for reaches 2, 3 and 7 (Figure 17.9) indicate that sediment entrainment within the macroturbulent features may depend on the relative roughness spacing ($RRS = w/d$, after Morris, 1955). A line of separation at $RRS = 0.8$ for reaches 2 and 7 is suggested, but most of the sediment-laden features occurred in depths less than 3.25 m. The RRS dependence seemed more strongly developed in reach 3, where the line of separation arguably may fall between $RRS = 0.25$ to 0.35 (Figure 17.9b).

From the plots of d versus h, reach 2 and 7 data indicate that sediment-laden features dominated when relative roughness, $RR = h/d$, is greater than 0.2 (Figure 17.10). As RR decreased, the features with intensity levels 1–4 concentrated with $d < 3$m, and the proportion of lower intensities increased. Reach 3 displayed a dominance of sediment-laden features with $RR > 0.25$, and as RR decreased, the gradation into progressively weaker boil intensities was again evident (Figure 17.10b).

Bedform conditions and water depth corresponded with Type-1 boil intensity well, but relations with boil morphology were not as apparent. Typically, there was a single dominant boil morphology and intensity noted over a surface area of tens of square metres at any particular time. Of the 300 digitized transect portions for reaches 2, 3 and 7, only a few displayed surface macroturbulence patterns that were not dominated by the simple quasi-circular upwellings described in the Coleman/Jackson model; they were present over the full range of hydraulic and bedform conditions. Cauliflower morphology occurred only when the mean depth was less than 2.5 m and the relative roughness exceeded 0.17. When rollers were noted, mean depth did not exceed 3.0 m and the relative roughness was always greater than 0.15.

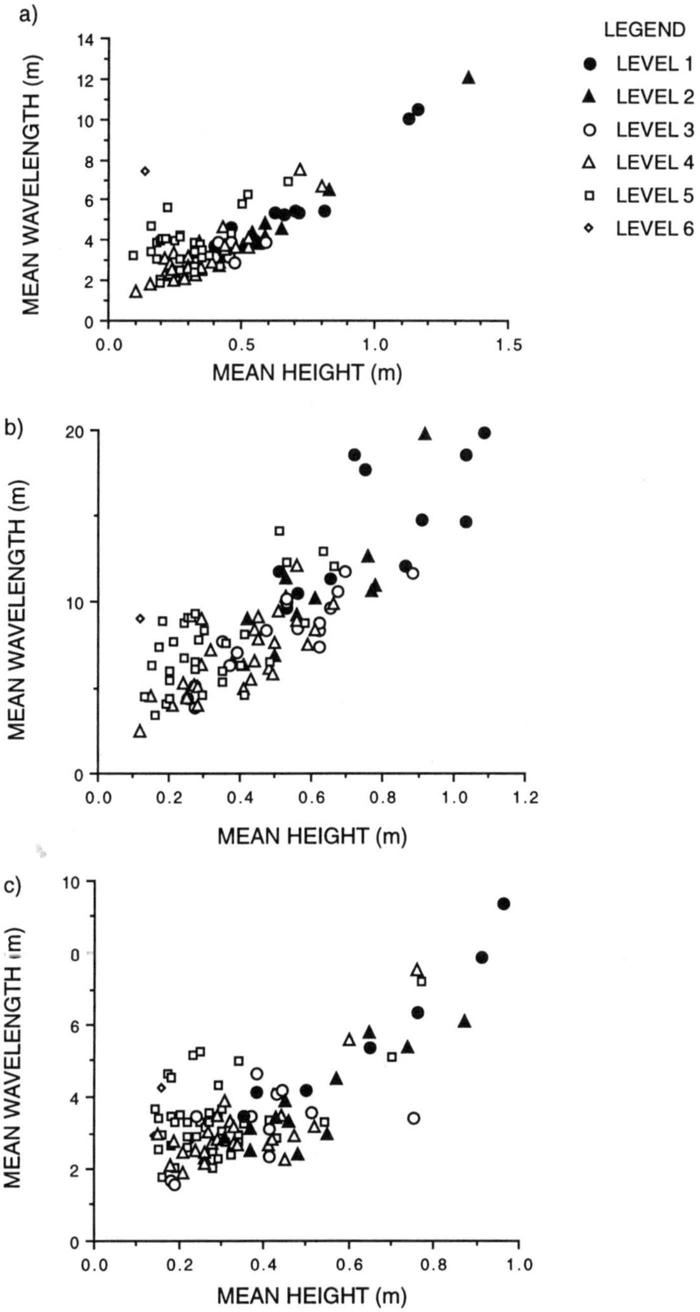

Figure 17.8. Mean bedform wavelength (*w*) versus height (*h*) for (a) Reach 2, (b) Reach 3 and (c) Reach 7

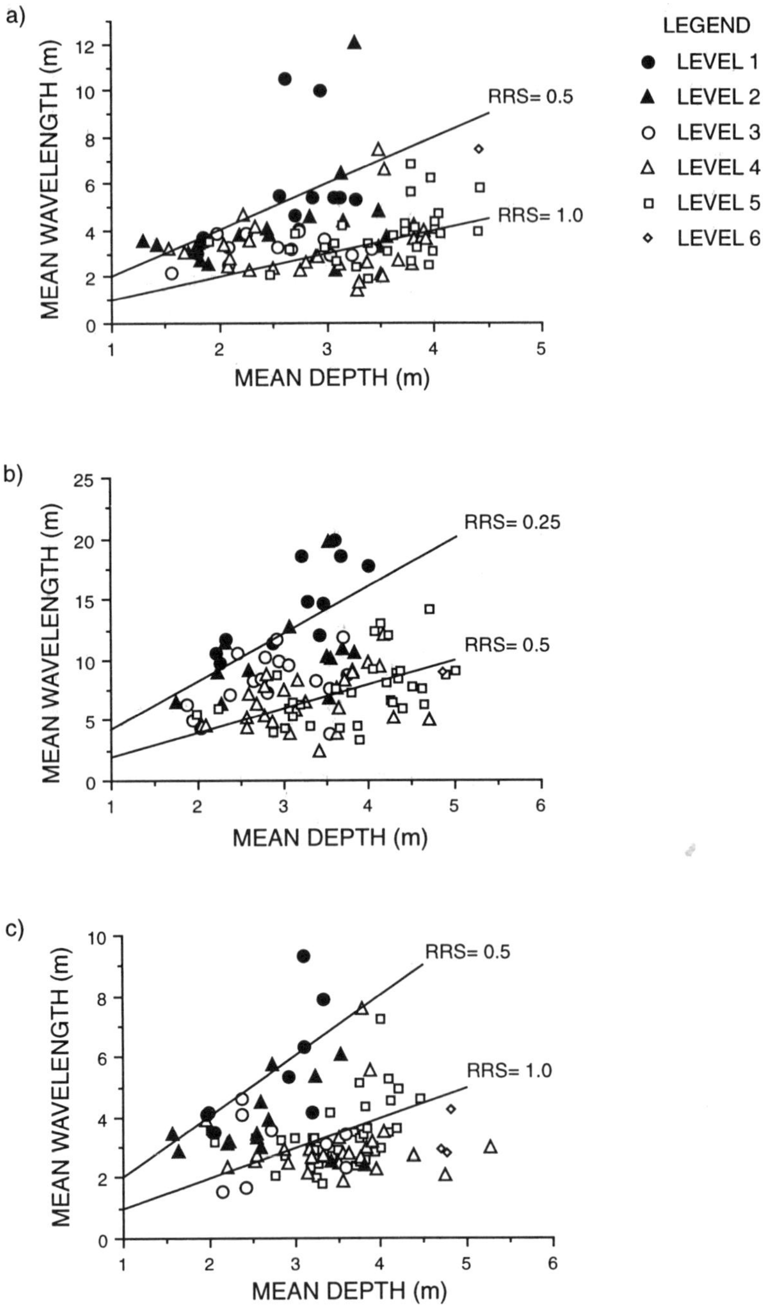

Figure 17.9. Mean bedform wavelength (w) versus water depth (d) for (a) Reach 2, (b) Reach 3 and (c) Reach 7

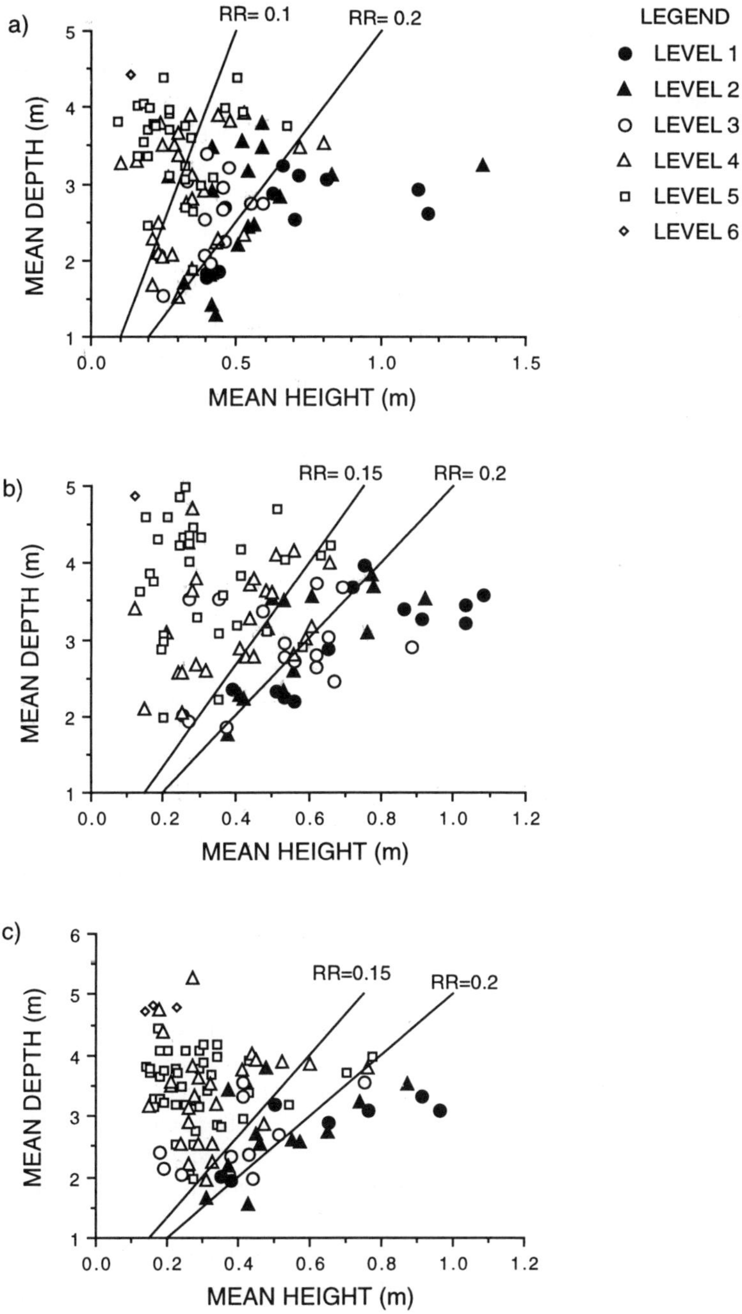

Figure 17.10. Mean water depth versus (d) bedform height (h) for (a) Reach 2, (b) Reach 3 and (c) Reach 7

Table 17.1. Relations between Type-1 boil period, current speed, water depth, bedform height, relative roughness and predicted length-scales

Date and time	Boil period (s)	U_{sfc} (m s^{-1})	$U_{0.7d}$ (m s^{-1})	Water depth (m)	Bedform height (m)	Relative roughness	Predicted length-scales (m) from U_{sfc}	from $U_{0.7d}$
Speed site #2								
29 July, 12:39	4	1.70	1.40	1.9	0.4	0.21	1.08	0.89
29 July, 12:45	4	1.70	1.30	1.9	0.4	0.21	1.08	0.82
29 July, 12:51	4	1.70	1.35	1.9	0.4	0.21	1.08	0.86
30 July, 14:23	7	1.45	1.10	2.1	0.3	0.14	1.62	1.23
Lower estuary								
3 May, 14:51	2	1.00	-	1.0	-0.2	0.20	0.32	-
3 May, 13:12	26	0.60	-	1.5	0.1	0.07	2.48	-
5 May, 16:00	14	0.85	-	1.3	0.1	0.08	1.89	-
5 May, 16:17	20	0.85	-	1.3	0.1	0.08	2.71	-
5 May, 16:30	3	1.50	-	1.6	-0.6	0.38	0.71	-
15 June, 12:48	2	1.40	-	1.5	-0.5	0.36	0.45	-
11 Aug., 10:12	17	0.65	0.35	1.4	0.1	0.06	1.75	0.95
11 Aug., 11:14	28	0.75	0.50	1.4	0.1	0.09	3.34	2.23
11 Aug., 12:21	23	0.75	0.45	1.5	0.1	0.04	2.75	1.65

Initial observations in reaches 2, 3 and 7 suggested that in the hours surrounding LLT when Type-1 boil activity was strong, the macroturbulence was intermittent and relatively periodic when present. The period between boils shed by a production site was similar to the period of nearby production sites, but this similarity often diminished as the separation increased. To test the applicability of Strouhal's Law to the mean Type-1 boil periodicities displayed, the predicted length-scales were calculated based on Equation 17.2 and compared to the mean water depth and the bedform height (Table 17.1).

For a speed measurement site within reach 2, the predicted length-scales fall between the dune heights and water depths. For the lower estuary sites, the predicted length-scales exceed the water depth for all time-series except those obtained when the bed was planar with occasional deep scour holes. The negative dune heights in Table 17.1 represent scour holes identified on the sonar record. Strong boils erupted at the water surface a few metres downstream of the scour hole as it progressed past the survey boat.

Table 17.1 also includes the calculated relative roughness; negative dune heights represent the elevation difference between the base of the scour hole and the surrounding river bed. Data in Kostaschuk & Church (1993), Jackson (1976) and Korchokha (1968) allow calculation of the relative roughness. Although there is

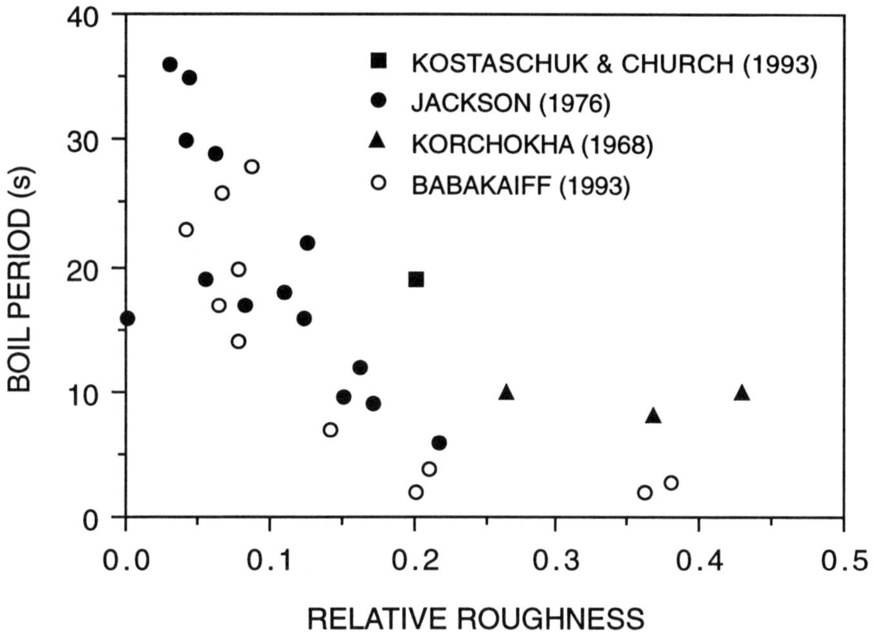

Figure 17.11. Boil period (*T*) versus relative roughness (*RR*)

some scatter in Figure 17.11, it is evident that, as the relative roughness increased, the boil period diminished rapidly and became constant as *RR* exceeded 0.2.

TYPE-2 BOIL ANALYSIS

Type-2 boils were very rarely observed to entrain organics or sediment, and the graphical comparison of boil intensity to bedform parameters and water depth did not reveal notable relations in reaches 1 and 4 to 6. Observations suggested that both the frequency and intensity of Type-2 boil shedding varied periodically over several minutes. Previous research describes strong periodic fluctuations in the current speed of Squamish River (e.g. Rood, 1980), therefore current-speed time-series were collected just downstream (within 10 m) of the boil videotape sites to examine the correlation.

Over 100 current-speed time-series were collected, and most displayed considerable deviations from the 10 minute mean value. In many instances there were apparent periodic fluctuations or repetitive patterns: several time-series displayed high-frequency fluctuations ranging from tens of seconds to several minutes. These fluctuations were often superimposed on longer-scale oscillations which exceeded 50% of the record length (Figure 17.12a). The tendency for current speed to accelerate unsteadily up to some maximum speed and then rapidly decrease

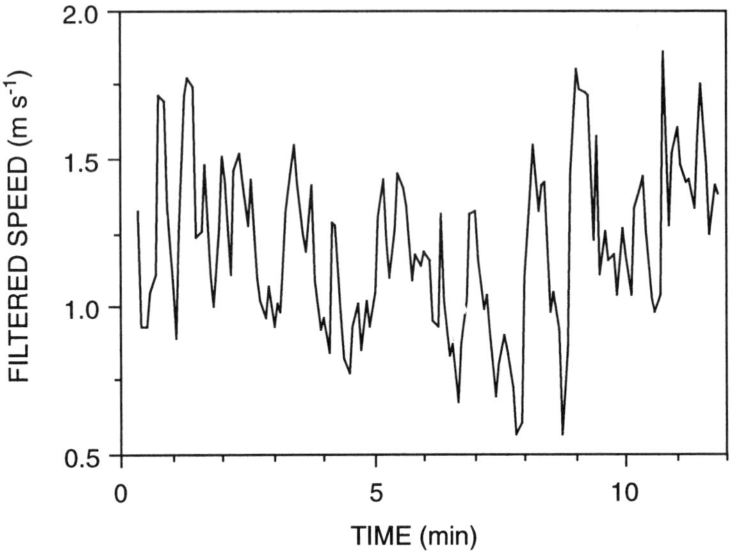

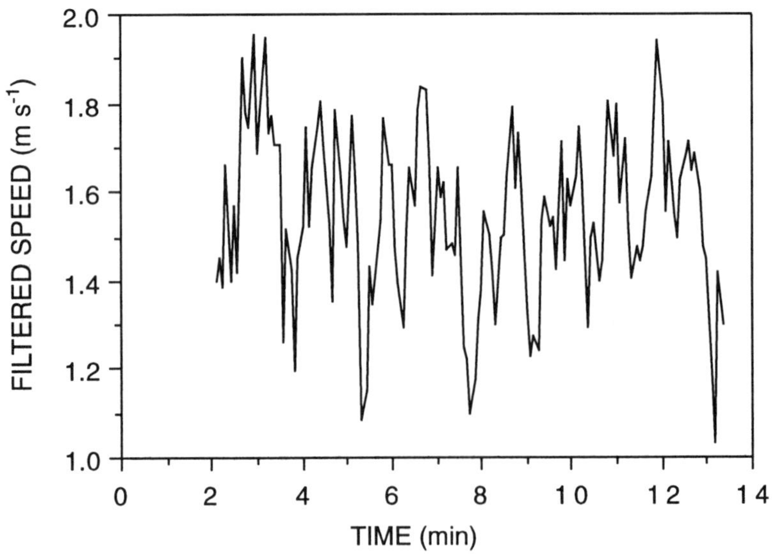

Figure 17.12. Current speed records (Binomial filter applied)

was noted in many time-series (e.g. Figure 17.12b, minutes 8–12). Rapid increases followed by slow unsteady decreases were also noted, but far less frequently.

All video records of boil activity with simultaneous current-speed time-series were digitized to facilitate comparisons. The reliability of the digitization technique from video was tested by repeating the digitization, and then comparing the derived frequency distributions using a chi-square test (99% level of significance). Violations of the test assumptions disqualified 20% of the comparisons. Of those that could be tested, just over 40% were concluded to have frequency distributions that were not similar in the repeat digitization. After removal of these time-series, and others which were fouled by wind and wave disturbance, twenty simultaneous current-speed and Type-2 boil-period time-series remained for analysis.

Simultaneous boil period and current-speed time-series displayed no consistent patterns. Vigorous boil activity often corresponded to high current speeds, but not exclusively. Often, the time-series varied on such short scales that it was difficult to discern whether the vigorous boil activity corresponded to the current speed maxima or minima (e.g. Figure 17.12a). A notable feature in many time-series was the tendency for current speed to accelerate unsteadily up to some maximum speed and then rapidly decrease (e.g. Figure 17.12b). A Strouhal relation was applied to Type-2 boil data, but the predicted length-scales were an order of magnitude less than the water depth and consistently less than the bedform height and wavelength.

DISCUSSION AND CONCLUSION

In the lower 2 km of the estuary, boil production was typically isolated to streamwise linear bands whose width and lateral spacing fluctuated with hydraulic and tidal conditions. Coleman (1969) related streamwise parallel bands of boil activity to longitudinal scours which may have been a product of secondary circulation cells. Due to the large amounts of sediment entrained in the boils of the lower estuary, it seems more likely that the limited duration of the production sites is related to local changes in the bedforms, rather than to the shifting of secondary cells. The stationarity of boil production sites observed by several researchers (e.g. Lane, 1944; Coleman, 1969) may in fact be related to stable bedform parameters.

Type-1 boil intensity was related to bedform parameters, which in turn were influenced by the hydraulic and tidal variables. Attempts to relate the dominant Type-1 structure observed within a reach to the bedform parameters and water depth were not conclusive. The cauliflower and roller structures were limited to specific relative roughness and depth domains which were consistent in all three reaches. The Coleman/Jackson structures were also noted within these domains, so there may be some degree of randomness involved, or perhaps dependence on variables not measured such as secondary circulation or bedform shape. Gabel (1993) related the size and vigour of boil eruptions to the curvature of dune crest-lines. Secondary upwellings, inevitably weaker and more poorly defined than the original upwelling, are also associated with the Coleman/Jackson model and the cauliflower structure. This suggests that there is some additional or more general causative mechanism. For example, Korchokha (1968) related paired structures to bedform shape.

For the w versus h plots (Figure 17.8), the line of separation of the data at $h = 0.5\text{–}0.6$ m was strongly apparent in reaches 2, 3 and 7. There was a strong linear structure and decreased variance in the data once the bedform height exceeded 0.5–0.6 m. It is postulated that once bedform height exceeds this critical value, troughs begin to be scoured and steepness increases markedly. Once the three-dimensional bedforms create separation-zone vortices and detached rollers, sediment is entrained in the macroturbulent features. Previous studies have related increased steepness to the formation of lee scour pits, producing increased three-dimensionality (Green, 1975; Dalrymple *at al.*, 1978). Scour pits have also been found to increase with high spring tides (Terwindt & Brouwer, 1986).

The plots of w versus d (Figure 17.9) suggested that sediment entrainment within the macroturbulent features may have been related to $RRS > 0.8$ for reaches 2 and 7 and $RRS > 0.25 - 0.35$ for reach 3. Furthermore, it seems that most sediment-laden features occurred when the water depth was less than some critical value. The apparent hysteresis between sediment content and current speed observed by Rood & Hickin (1989) may, in fact, be attributed to flow depth. Current speed peaked two hours before LLT, but water depth continued to diminish in the 2 hours following peak boil sediment content.

The plots of d versus h for reaches 2, 3 and 7 (Figure 17.10) displayed a gradation of decreasing boil intensities with decreasing relative roughness. Lapointe (1992) qualitatively relates boil intensity to relative roughness, and there may also be a relation between boil periodicity and relative roughness. Figure 17.11 suggests an inverse relation between the relative roughness and the boil period. Invariance in the relation beyond $RR = 0.2$ may explain why Korchokha (1968) concluded that boil period is invariant with depth. In fact, Rood (1980) postulates that Korchokha's observations may have been related to bedform properties.

This general relation has not been reported previously and deserves future study. Our findings indicate that the critical parameter for explaining Type-1 boil intensity and periodicity is relative roughness, although the dune steepness may also be important.

Simultaneous time-series of current speed and Type-2 boil periodicity displayed no consistent relations. The observed phenomena of a slow and unsteady increase in current speed followed by a rapid decrease may be associated with macroturbulent activity. Perhaps there is some critical level of boil activity which, once exceeded, creates such resistance to flow within the water column that the mean current-speed is diminished. Once the current speed decreases to the point that the boil activity declines and offers less resistance to flow, the current speed will begin to increase again, and the cycle repeats. Such a feedback relation would be highly complex over a variety of temporal and spatial scales.

The production sites of Type-2 boils are limited to the concave bank during strong currents, and it is under these conditions that a strong helical flow component may be expected. Several figures presented in Thompson (1986) based on flume studies and field observations indicate secondary flow and bed morphology patterns similar to those observed in Squamish River estuary. Tamai *et al.*(1986) conclude that a shear layer in the lateral velocity profile will shed eddies periodically, but the production sites of Type-2 boils here were always in the wake of an obstruction. This obstruction could presumably deflect flow towards the surface, particularly if the obstruction has trapped sediment and organic debris on its upstream face. The rotation within the Type-2 boils that had been attributed to lateral shear could also be imposed by helical flow at the production site.

Further study of the boils of Squamish River estuary may lead to a more complete differentiation of boil morphologies and a greater understanding of production mechanisms, but a more pressing issue is how these Type-1 and 2 boils influence sediment transport. Injection of sediment in discrete packages (such as boils) to levels above the bed creates suspended sediment transport that simply is not described by general diffusion theories of sediment transport. In rivers and estuaries displaying boils, study of sediment transport must include consideration of boil activity. Empirical relations between boil size or intensity and sediment content may exist, but such associations are likely linked to bedform or hydraulic conditions, and could be exceedingly complex.

ACKNOWLEDGEMENTS

This work is part of an ongoing project on the morphodynamics of rivers in coastal British Columbia, Canada, funded by the Natural Science and Engineering Research Council (NSERC) of Canada.

REFERENCES

Allen, J.R.L. 1986. *Physical Sedimentology*, George Allen & Unwin, London, 272 pp.

Babakaiff, C.S. 1993. *Flow Hydraulics, Bedforms, and Macroturbulence of Squamish River Estuary, British Columbia*, Unpub. MSc Thesis, Simon Fraser Univ., 298 pp.

Coleman, J.M. 1969. 'Brahmaputra River: Channel processes and sedimentation', *Sediment. Geol.*, **3**, 129–239.

Dalrymple, R.W., Knight, R.J. & Lambiase, J.J. 1978. 'Bedforms and their hydraulic stability relationships in a tidal environment, Bay of Fundy, Canada', *Nature*, **275**, 100–104.

Dyer, K.R. 1986. *Coastal and Estuarine Sediment Dynamics*, John Wiley & Sons, New York, 342 pp.

Gabel, S.L. 1993. 'Geometry and kinematics of dunes during steady and unsteady flows in the Calamus River, Nebraska, USA', *Sedimentology*, **40**, 237–269.

Green, C.D. 1975. 'A study of hydraulics and bedforms at the mouth of the Tay estuary, Scotland', in Cronin, L.E. (Ed) *Estuarine Research: Vol. 2 Geology and Engineering*, Academic Press, New York, 323–344.

Jackson, R.G. 1976. 'Sedimentological and fluid-dynamic implications of the turbulent bursting phenomenon in geophysical flows', *J. Fluid Mech.*, **77**, 531–561.

Korchokha, Y.M. 1968. 'Investigation of the dune movement of sediments on the Polomet River', *Soviet Hydrol.: Selected Papers*, **6**, 541–559.

Kostaschuk, R.A. & Church M.A. 1993. 'Macroturbulence generated by dunes: Fraser River, Canada', *Sediment. Geol.*, **85**, 25–37.

Kostaschuk, R.A., Church, M.A. & Luternauer J.L. 1991. 'Acoustic images of turbulent flow structures in Fraser River estuary, British Columbia', *Current Research Part E, Geol. Surv. Can.*, 83–90.

Lane, E.W. 1944. 'A new method of sediment transportation', *Trans. AGU*, **25**, 900.

Lapointe, M.F. 1989. *Sediment Suspension by Turbulent Bursting Over Advancing River Dunes*, Unpub. PhD Thesis, Univ. of British Columbia, 164 pp.

Lapointe, M.F. 1992. 'Burst-like sediment suspension events in a sand bed river', *Earth Surf. Processes Landf.*, **17**, 253–270.

Levi, E. 1983. 'Oscillatory model for wall-bounded turbulence', *J. Eng. Mech.*, **109**, 728–740.

Matthes, G.H. 1947. 'Macroturbulence in natural stream flow', *Trans. AGU*, **28**, 255–265.

Morris, H.M. 1955. 'Flow in rough conduits', *Trans. ASCE*, **120**, 373–398.

Müller, A. & Gyr, A. 1986. 'On the vortex formation in the mixing layer behind dunes', *J. Hydraul. Res.*, **24**, 359–375.

Rood, K.M. 1980. *Large Scale Flow Features in Some Gravel Bed Rivers*, Unpub. MSc Thesis, Simon Fraser Univ., 180 pp.

Rood, K.M. & Hickin E.J. 1989. 'Suspended-sediment concentration and calibre in relation to surface-flow structure in Squamish River estuary, southwestern British Columbia', *Can. J. Earth Sci.*, **26**, 2172–2176.

Tamai, N., Asaeda, T. & Ikeda, H. 1986. 'Study on generation of periodical large surface eddies in a composite channel flow', *Water Resour. Res.*, **22**, 1129–1138.

Terwindt, J.H.J. & Brouwer, M.J.N. 1986. 'The behaviour of intertidal sandwaves during neap-spring tide cycles and the relevance for palaeoflow reconstructions', *Sedimentology*, **33**, 1–31.

Thompson, A. 1986. 'Secondary flows and the pool-riffle unit: a case study of the processes of meander development', *Earth Surf. Processes Landf.*, **11**, 631–641.

NOTATION

d	mean water depth
f	frequency
h	mean bedform height
L	length scale
RR	relative roughness
RRS	relative roughness spacing
St	Strouhal number
t	time-scale
T	mean boil period
U	streamwise current speed
U_{sfc}	mean streamwise current speed at water surface
$U_{0.7d}$	mean streamwise current speed at 0.7 of total depth
w	mean bedform wavelength

18 Observations of Near-bed Suspended Sediment Turbulence Structures using Multifrequency Acoustic Backscattering

PETER D. THORNE, PETER J. HARDCASTLE
Proudman Oceanographic Laboratory, Bidston Observatory, Birkenhead, Merseyside, L43 7RA, UK

ANDREW HOGG
Institute of Theoretical Geophysics, Department of Applied Mathematics & Theoretical Physics, Silver Street, Cambridge, CB3 9EW, UK

ABSTRACT

The mechanisms of sand suspension and transport in turbulent tidal flows over sandwaves are complex, with the suspended concentration varying over periods of the order of seconds and having spatial dimensions which at present are not well-known. To investigate the process of sand suspension requires detailed measurements of the flow and suspended sediment which can be compared with mathematical developments in the modelling of suspended sediment transport. Acoustic techniques have the capability to non-intrusively provide these detailed observations. In the present work results are reported on a multifrequency acoustic backscatter system which has been used to measure suspended sediments in an estuary. Using recent developments in sediment scattering theory, the backscattered acoustic signal has been analysed to yield profiles of both suspended sediment concentration and particle size. The accuracy of the results have been evaluated by comparison with in-situ sampling. Using the high resolution capability of this acoustic technique, detailed multifrequency images of suspended sediments have been obtained which illustrate the dynamic nature of the suspension process. These images, coupled with measurements of the flow, are beginning to indicate the detailed links between turbulent flow and suspension dynamics.

INTRODUCTION

The utilization of sound for non-invasively revealing detailed structures, via conventional or tomographic techniques, has been increasingly employed in a number of fields (e.g. medical and non-destructive testing) in recent years. One of these applications has been in the study of suspended sediments (Hay & Bowen, 1994; Vincent & Downing, 1994; Lynch *et al.*, 1994; Thorne & Hardcastle, 1994).

Coherent Flow Structures in Open Channels.
Edited by P.J. Ashworth, S.J. Bennett, J.L. Best and S.J. McLelland.

The non-intrusive capability of acoustics, coupled with the facility to provide high spatial and temporal resolution, allows for unperturbed visualization of suspension processes and provides new opportunities to examine sediment dynamics. In the present investigation a multifrequency acoustic backscatter approach is used to probe suspended sediment structures generated by turbulent tidal currents over a sandy bed.

ACOUSTIC TECHNIQUE AND INSTRUMENTATION

The technique is based on propagating sound, typically of megahertz frequencies, through the water column and measurements of the suspended sediments are obtained by analysis of the backscattered signal. Since the method provides profile measurements which can be rapidly updated, the suspension processes can be imaged and these results, when coupled with the flow data, can be used to study detailed suspension mechanisms. In the present work an acoustic backscatter system (Thorne & Hardcastle, 1994) using three transceivers operating at 1.0, 2.5 and 5.0 MHz was employed. The set-up provided 128 backscattered profiles, at each of the three frequencies each second. The spatial resolution was 0.01 m and the operating range 1.28 m. Data at each range were averaged using on-line processing over 32 records to provide measurements at 0.25 s intervals which were used to compute profiles of both particle size and suspended sediment concentration. Such temporal averaging was necessary to mitigate against configurational noise arising from variations in the backscattered signal, due to the random phase distribution of the echoes, originating from the varying location of the particles within the insonified volume.

ACOUSTIC ANALYSIS

Over the past decade the application of acoustics for studying suspended sediments has been validated by a series of laboratory, marine, and theoretical studies (Sheng & Hay, 1988; Thorne & Campbell, 1992; Hay & Sheng, 1992; Thorne *et al.*, 1993). These have shown that it is possible to obtain measurements of suspended sediment concentration and particle size using acoustics. The development of acoustic theory is therefore not presented here, and only the final expressions are provided to illustrate how the acoustic data was converted to suspension parameters. The suspended sediment particle size, $a_s(r)$, and concentration, $c(r)$, profiles, can be written as

$$a_s(r) = F^{-1}\left(\frac{f_i(r)}{f_j(r)}\right) \qquad (18.1)$$

$$\frac{f_i(r)}{f_j(r)} = \frac{\left(P_i(r)/K_{t_i}\right)e^{2r(\alpha_i - \alpha_j)}}{\left(P_j(r)/K_{t_j}\right)} \qquad (18.2)$$

$$c(r) = \left\{ \frac{rP_i(r)}{\left(f_i(r)/\sqrt{a_s(r)r_s}\right)K_{ti}} \right\}^2 e^{4\alpha_i r} \tag{18.3}$$

where

$$\alpha_i = \alpha_{wi} + \alpha_{si} \tag{18.4}$$

$$\alpha_{si} = \frac{1}{r} \int_0^r \zeta_i(r) c(r) \mathrm{d}r \tag{18.5}$$

a_s can be associated with an inverse transform of the ratios of f, the form function and this parameter describes the scattering properties of the particles in suspension. P is the root-mean-square value of the backscattered pressure, r is range , α_w is the absorption due to water, α_s is the attenuation due to the suspension, ζ describes the attenuation properties of the suspended particles, and ρ_s is the density of the sediment particles. K_t is a calibration constant for the system and $i \neq j$ where $i, j = 1, 2, 3$ refers to the three frequencies. Using an iterative inversion algorithm with the above expressions, estimates of $a_s(r)$, and $c(r)$ can be obtained at each of the three frequencies.

MEASUREMENT SITE AND EXPERIMENTAL ARRANGEMENT

The field experiment was conducted during Spring tides in the River Taw estuary in North Devon, UK, in April 1991. No surface waves were present and the water was unstratified. The bed consisted of well-sorted fine to medium sand, having a Gaussian size distribution with a mean radius of $a_s \approx 80\ \mu\text{m}$ and standard deviation of $\sigma(a_s) \approx 20\ \mu\text{m}$. The area was dominated by sandwaves which typically had wavelengths of around 20 m and amplitudes of the order of 0.8 m, although there was substantial variability. The estuary was subject to strong rectilinear turbulent tidal currents which generated high concentrations of suspended sediments. At low water, the site was dry and readily accessible for arranging the experiment whilst at high water, the depth was approximately 6 m. Measurements were recorded over five flood periods at different locations along the sandwave. The data presented in the present paper comes from one of the locations where the surface profile was relatively uniform. The position of the instrumentation, and the surveyed surface profile is shown in Figure 18.1a. Although the sandwaves typically have an aspect ratio of approximately 0.04 and their presence strongly influences the flow, the rig position was chosen to be far downstream of a crest in a relatively flat location. Therefore, it was expected that the mean flow would be parallel with the boundary and the fluctuating velocity and concentration fields would not be dominated by the intermittent detachment of eddies. The experimental arrangement is shown in Figure 18.1b. The instrumentation was composed of a triple frequency backscatter system, pumped sampling to assess the acoustic

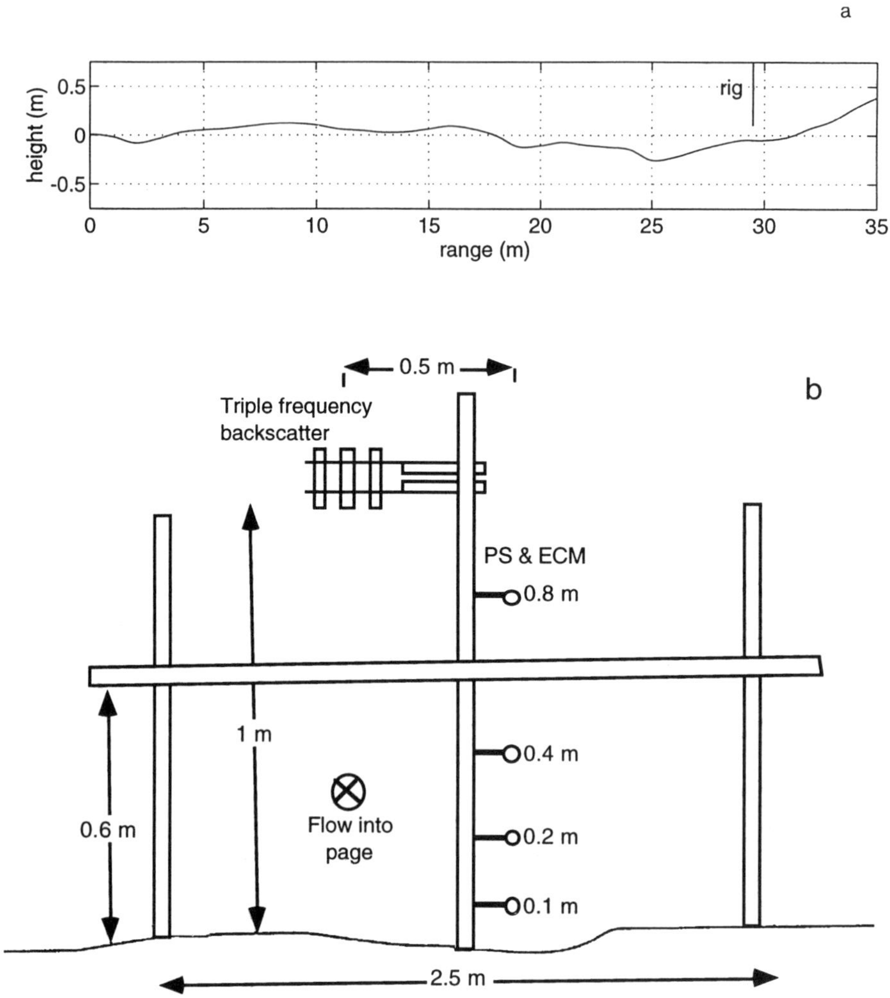

Figure 18.1. (a) location of the rig on the sandwave. (b) experimental arrangement: PS–Pumped samples; and ECM–Electromagnetic current meter

inversions and electromagnetic current meters (ECMs) to monitor the turbulent flow. The ECMs used were manufactured by Colnbrook and discoidal in shape with a 0.05 m diameter. Pumped samples (PS) were taken at the four heights shown in Figure 18.1b at 10 minute intervals and the horizontal and vertical velocities were logged at 4 Hz at the same heights.

ANALYSIS OF THE OBSERVATIONS

MEAN RESULTS

Before examining acoustic observations of detailed suspension processes, the accuracy of the acoustic inversions was examined. 10 minute averages of both the suspended sediment concentration and particle size were computed, and the results assessed by making comparisons with in-situ samples obtained using pumped sampling.

To illustrate the general features of the acoustic results obtained, Figure 18.2 shows three-dimensional plots of the particle radius and suspended sediment concentration over the flood. The particle size was derived from the ratio of the 2.5 and 1.0 MHz backscattered signals. There were some difficulties with using the 5.0 MHz signal for computing particle size because of the very high propagation attenuation of the signal due to the suspension. However, the suspended concentration was derived from the mean of the concentrations obtained at each of the three frequencies, calculated using the particle size data derived from the ratio of the 2.5 and 1.0 MHz backscattered signals. The inversion used to calculate the particle size and suspended concentration was conducted without reference to the pumped sample data or to laboratory calibrations using sediment from the site. This theoretical inversion uses an acoustic calibration of the system and recent descriptions of the scattering properties of non-cohesive sediments (e.g. Crawford & Hay, 1993; Thorne *et al.*, 1995; Schaafsma, 1994). The results for the particle radius, presented in Figure 18.2a, show a relatively uniform particle size, both with height above the bed, and over the duration of the experiment. The jump in size at zero height is simply due to the bed echo. The mean and standard deviation for the data shown in Figure 18.2a are $\bar{a}_s = 75\ \mu\text{m}$, and $\sigma(\bar{a}_s) = 10\ \mu\text{m}$ respectively, this low standard deviation demonstrating the temporal and spatial consistency of the suspended particle radius. The variability in particle size towards the end of the observational period is due to system noise and some acoustic echoes from the rig, both of which became noticeable at low concentrations. As for the absolute error in the data: (i) the system calibration error was approximately 10%; and (ii) the error in describing the scattering properties of the sediment would conservatively be of the same order. Hence, the estimated standard error for $\bar{a}_s$ is approximately 20%. Figure 18.2b shows the variation in suspended sediment concentration over the flood period, extending just over three orders of magnitude, 0.001–2 kg m^{-3}. A peak in concentration is seen at approximately 30 minutes into the record, reducing to low levels towards the end of the record at slack high water. The anomalously relatively high values seen towards the end of the record at particular ranges are due to echoes from the arrangement of the rig. Since three estimates of concentration were used to compute the results in Figure 18.2b, an estimate for the internal consistency of the data can be obtained. If the last two records contaminated by the rig echoes are ignored, then the standard error in the mean suspended concentration derived from the three data set is 10%. This provides a measure of the internal consistency, however, for the absolute accuracy, the concentration standard error is considered to be of the same order as the acoustic size measurements, i.e. approximately 20%.

To assess the accuracy of the acoustic measurement of particle size and suspended sediment concentration, pumped sample data were collected. Pumped samples on known water volumes were taken at 0.1, 0.2, 0.4 and 0.8 m above the bed, and each height was sampled simultaneously every 10 minutes. The filtered samples were dry weighed to give suspended concentration and, on a selection of the samples, where sufficient material was collected, a sieve analysis was conducted. Figure 18.3a provides an assessment of the particle size measurements. This shows a time-series of the ratio of the acoustic to the pumped sample measurements of the particle size at the four heights above the bed. Ideally this ratio should be unity and constant with time. Figure 18.3a shows values at all four heights close to unity, apart from the first and last ratios which over-estimate the particle size. Calculation of the mean ratio for the whole data set gives a value of $A_s/P_s = 1.23 \pm 0.27$, but this reduces to 1.1 ± 0.1 if the first and last values are excluded. Therefore the majority of the acoustic estimates of particle size are within 10% of the pumped sample data, and since the standard deviation of the ratio is 10% there is no significant difference between the majority of the two data sets. The result for the comparison between the acoustic and pumped sample measurements of suspended sediment concentration is shown in Figure 18.3b. The data cover just over three orders of magnitude, and as can be seen the data generally clusters about the line $A_c = P_s$ although there is some divergence at the lower concentration. This divergence at low concentration is due in part to the variability of the particle size estimates at lower concentration levels. Linear regression on the data gives a correlation coefficient of 0.97 which is significant at the 99.95 level (Bendat & Piersol, 1986). The mean ratio of $A_c/P_c = 0.93 \pm 0.3$, and so there is no significant difference from unity. If concentration levels below 0.01 kg m^{-3} are neglected, owing to the inversion being somewhat noisy, the result is $A_c/P_c = 1.02 \pm 0.18$, this ratio is closer to unity, and the standard deviation has reduced by 40%. Therefore at times of greatest interest, when significant quantities of suspended sediment are being transported, both the acoustic estimates of particle size and suspended concentration levels are in good agreement with the pumped sample data.

DETAILED RESULTS

Although the mean results provide an assessment of the acoustic technique, the reason for developing the system is to provide high temporal and spatial resolution measurements of suspended sediments. To illustrate the output which can be obtained, and the internal consistency achieved, high resolution images have been generated. Owing to the uniformity of the particle size with range and height above the bed, the high resolution analysis presented here was conducted with a fixed

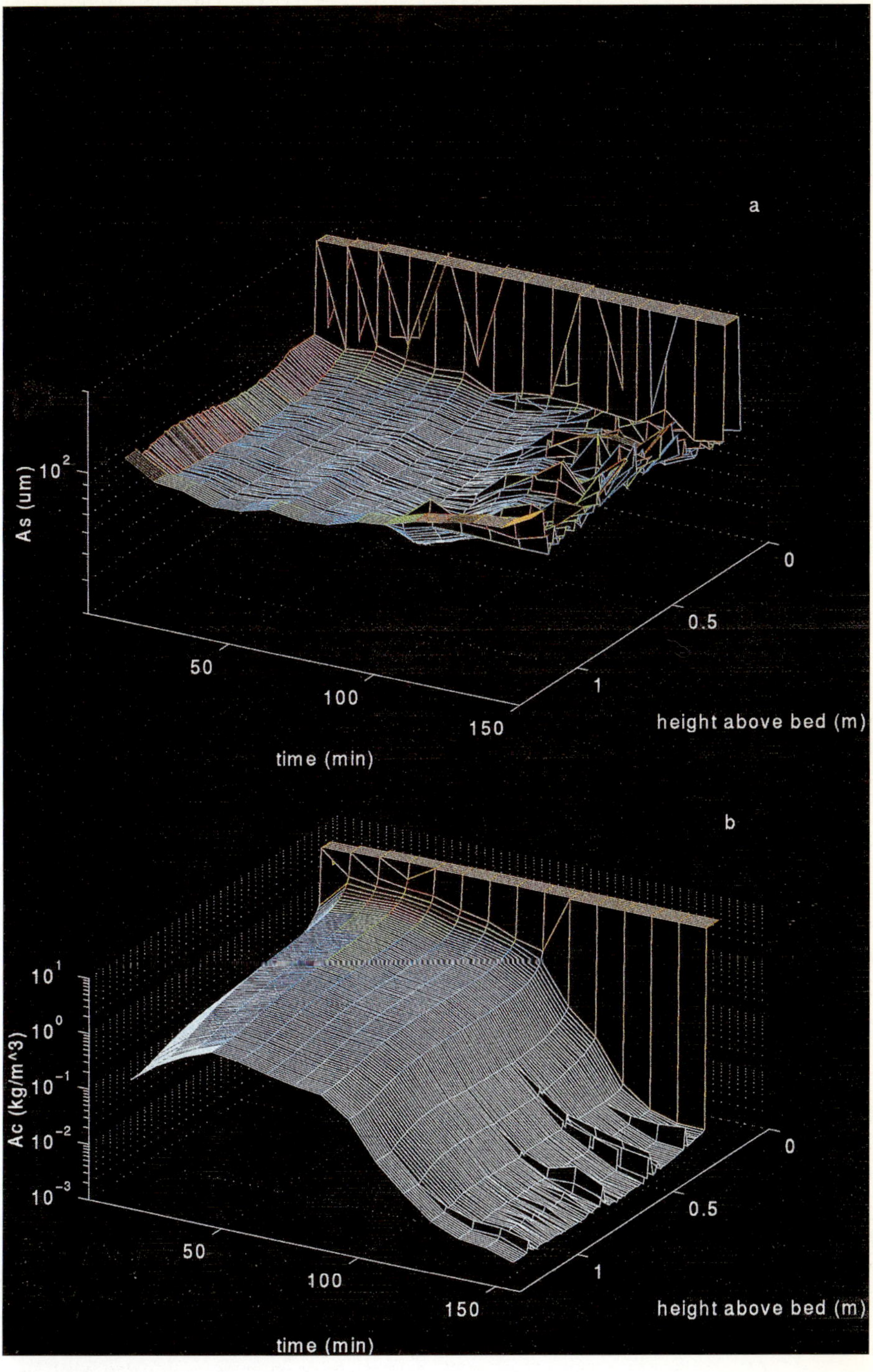

Figure 18.2. (a) Three-dimensional plot of acoustic measurement of particle size. (b) Three-dimensional plot of acoustic measurement of suspended sediment concentration

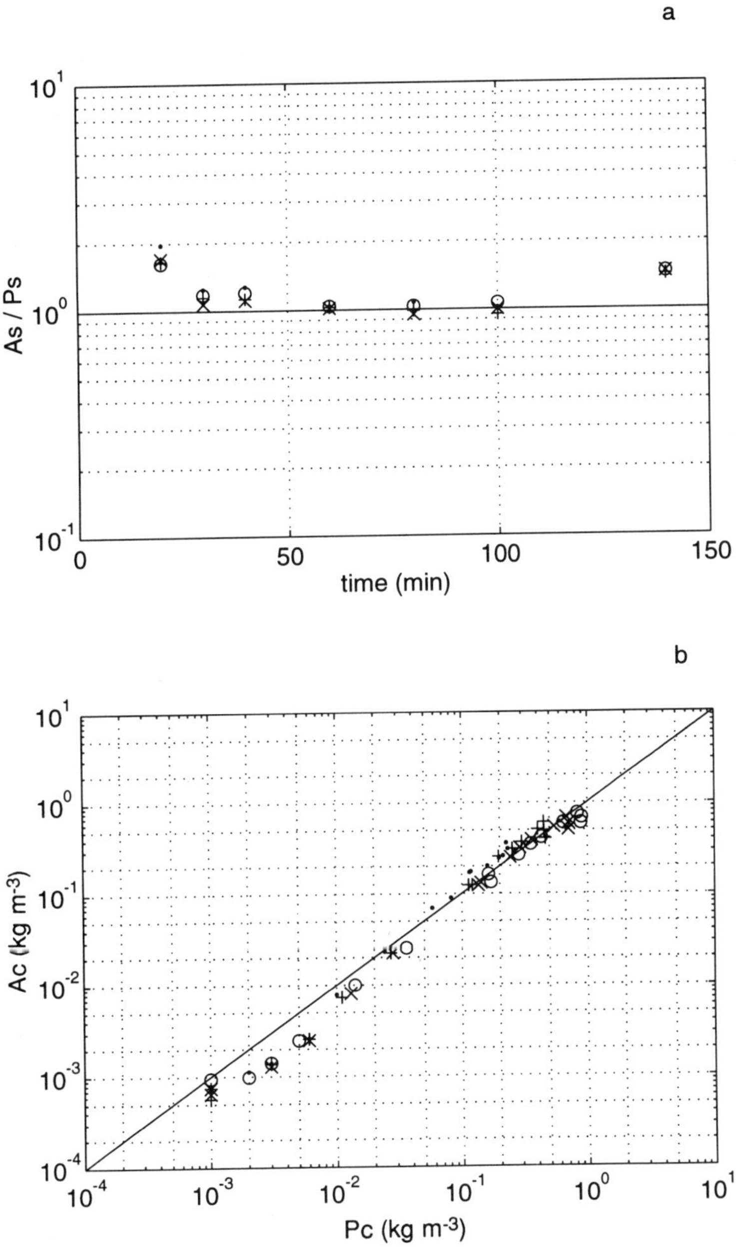

Figure 18.3. (a) Ratio of the acoustic measurement of particle size, A_s, to the pumped sample measurement, P_s, at ○0.1, ×0.2, +0.4, and •0.8 m above the bed. (b) Plot of acoustic measurement of concentration, A_c, against pumped sample measurements, P_c, at ○0.1, ×0.2, +0.4, and •0.8 m above the bed

particle radius of 65 μm. This approximation was valid as pumped samples at 0.1 and 0.8 m above the bed showed mean particle sizes within 2% of one another for the record examined here. Use of a constant input size simplifies and reduces the time to compute the inversion and, as long as the particle size measurement is accurate, the inversion can be obtained with higher precision. An illustration of the acoustic output is shown in Figures 18.4a and 18.4b, which show respectively the concentration field obtained at 1.0 and 2.5 MHz. The bed echo is again seen at zero height. These figures illustrate the suspended sediment images that can be developed using acoustics, and shows the technique provides a notable improvement over more conventional approaches, since it is possible to visualize and map out features in the suspension field as the material advects below the backscatter sensors. To form the images shown in Figure 18.4, a 2 s temporal filter and 0.04 m spatial filter have been applied in order to reduce the 'salt and pepper' type noise associated with both configurational noise and real high frequency fluctuations in the suspended concentration. The two images are comparable although some differences can be observed. The differences are due both to the fact that the two transceivers were not collocated, but were 0.11 m apart, as well as the effect of the configurational noise on the inversion which will introduce some fluctuations. However, the figures do show clearly a consistent image of a series of relatively high concentration events which penetrate well up into the water column. These events are interspersed with intervals which are relatively quiescent. This type of pattern is not unique, but common throughout the records, and is indicative of the intermittency of the suspended concentration.

SEDIMENT ANALYSIS

The analysis of the backscattered data, to date, has focused primarily on development of a framework to describe the interaction of sound with a marine suspension, the application of this description to the development of inversion techniques and an appraisal of the acoustic approach to deliver accurate results. The interpretation of the data in terms of suspended sediment dynamics is in its early stages, however, some interim results are available and these are presented to illustrate the opportunities provided by the acoustic technique. In this discussion of preliminary results, the focus is upon the mean suspended concentration profile, eddy diffusivity, and possible links between the kinematic Reynolds stress and suspension fluctuations. A single record block (2048×0.25 s, 128×0.01 m) has been chosen to illustrate the results being obtained.

Using the suspended sediment concentration time-series, it is trivial to extract the vertical variation of both the mean, C, and root-mean-square (rms) concentration, c'_{rms}, with height above the bed. These are shown in Figures 18.5a and 18.5b and are characterized by a smooth reduction in concentration with height above the bed. The ratio c'_{rms}/C also reduces with height above the bed, attaining a near constant value of ~0.5 (Figure 18.5c). The magnitude of this ratio expresses the inherent

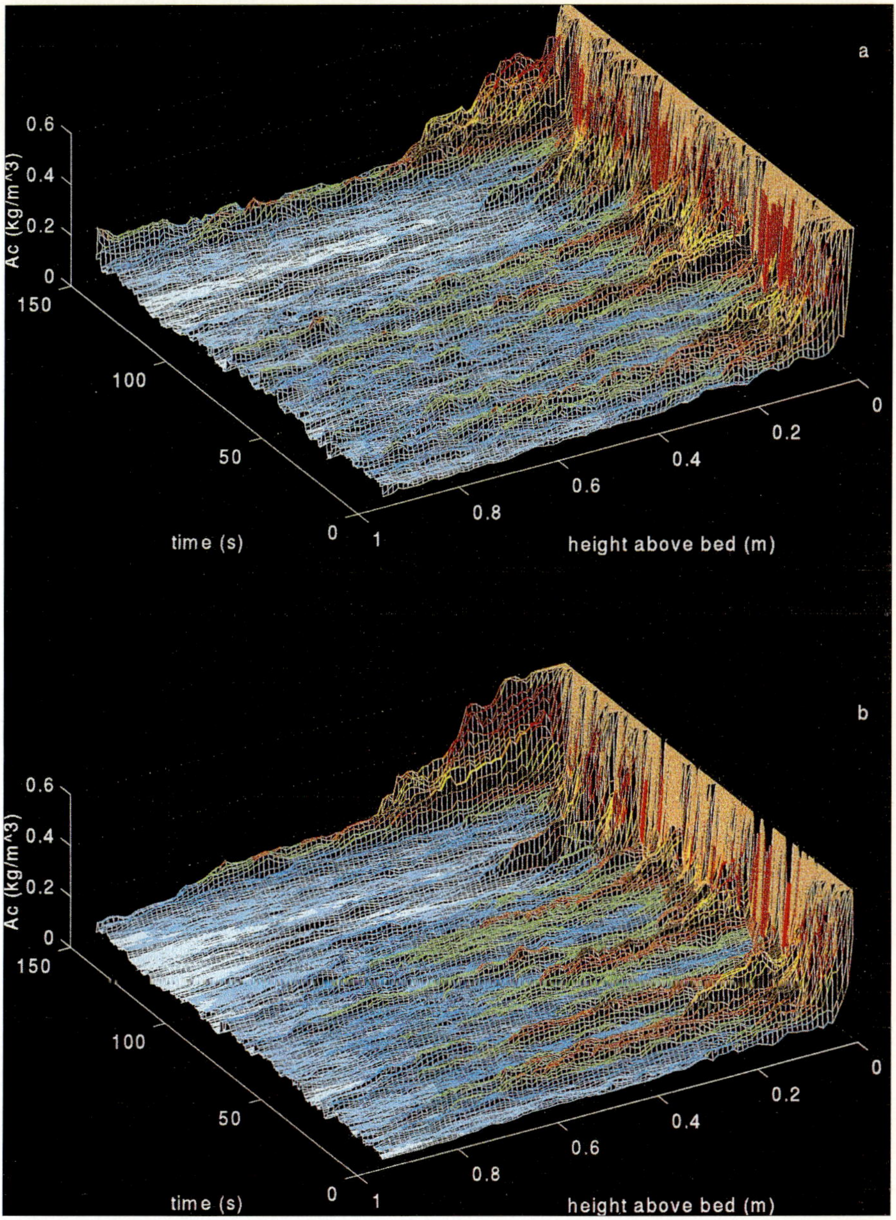

Figure 18.4. (a) High resolution measurement of suspended sediment concentration obtained at 1 MHz. (b) High resolution measurement of suspended sediment concentration obtained at 2.5 MHz

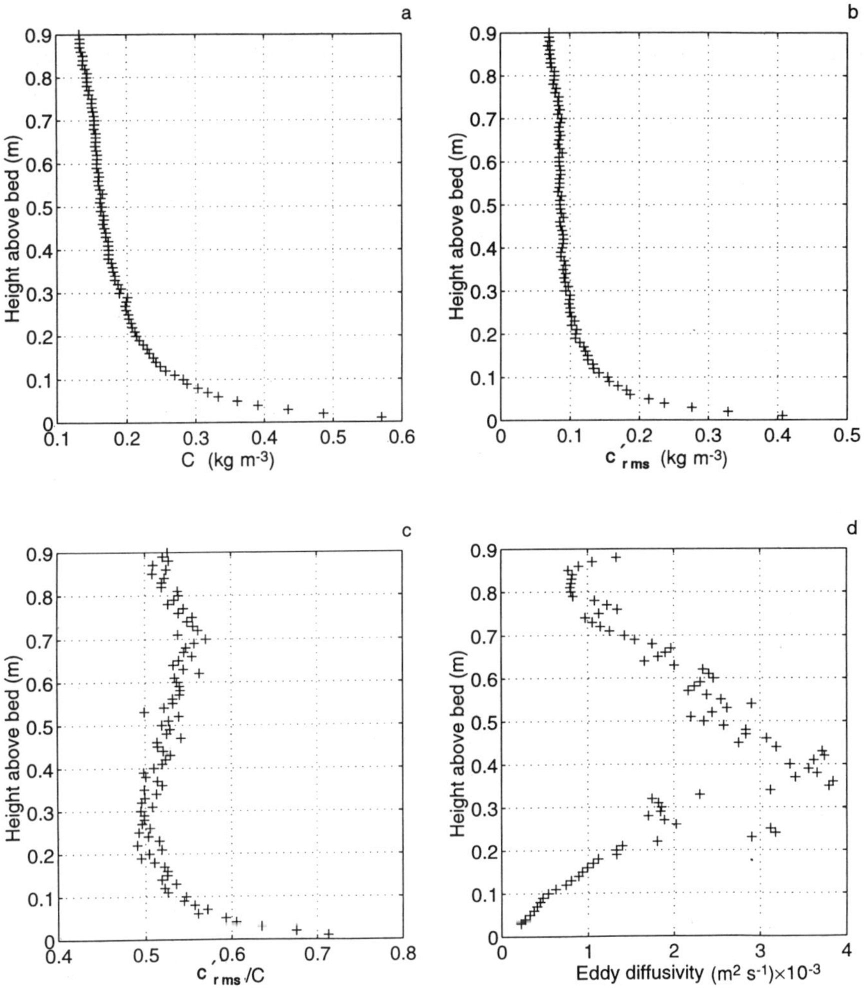

Figure 18.5. Acoustic measurements of: (a) mean concentration, C; (b) root-mean-square concentration c'_{rms}; (c) ratio of c'_{rms}/C; and (d) eddy diffusivity

variability of the concentration field with height above the bed and shows that closer to the bed the relative magnitude of the concentration fluctuations are larger. A possible interpretation of this observation is that close to the bed the suspension of particles is mainly driven by intermittent coherent events (*cf.* Figure 18.4). These discrete clouds carry fluid, of a substantially higher particle concentration than the surrounding fluid, upwards away from the boundary. As the cloud ascends, it expands, material settles out and the concentration within the cloud is reduced to average conditions. Hence, the relative magnitude of the concentration fluctuations also reduce with increasing distance from the boundary.

Many models of the steady vertical distribution of suspended sediment are based upon a balance between the upward motions of the particles, arising from the turbulent eddy motions, and the gravitational settling of the particles. It is conventional to invoke a gradient-diffusion model of the turbulent flux of particles by equating this flux of particles with the product of an eddy diffusivity ($K_s(y)$) and the vertical gradient of the concentration field. Hence, this balance is expressed by

$$K_s(y)(dC/dy) + v_s C = 0 \qquad (18.6)$$

where C is the mean concentration, v_s is the settling velocity of the particles and y is the height above the bed. The eddy diffusivity varies with the distance from the boundary, reflecting the typical size of the turbulent eddies responsible for the vertical transport of the particles. There is little consensus on the functional form of this eddy diffusivity over the entire flow depth (see review in Dyer & Soulsby, 1988) and it has proven difficult to measure experimentally. Using the present data we may calculate directly the functional form of the effective eddy diffusivity for this flow by evaluating

$$K_s(y) = -v_s C/(dC/dy) \qquad (18.7)$$

In this calculation, the eddy diffusivity is evaluated in dimensional form; it could be suitably non-dimensionalized by the product of the friction velocity, u_* and the flow depth (Y). Irrespective of the non-dimensionalization, the eddy diffusivity shown in Figure 18.5d initially increases linearly with distance from the boundary, and subsequently decreases with height. Such a variation is broadly in accord with theoretical considerations of the functional form of the eddy diffusivity; virtually all theories predict a linear variation with height adjacent to the boundary. It is noteworthy that the visualization of the coherent cloud of particles (*cf.* Figure 18.4) questions the assumption that the suspension of particles may be adequately described by a diffusive process. Rather, in accord with the studies of Nielsen (1992) and Hogg *et al.* (1994), an improved description of the process may be formulated by the inclusion of the advective transport associated with these discrete events. Nevertheless, irrespective of these theoretical questions and doubts, an effective eddy diffusivity for the flow can be evaluated using Equation 18.7.

Time-series of the horizontal and vertical flow fluctuations u' and v' respectively, the kinematic Reynolds stress, $-u'v'$, and the fluctuating concentration, c', are presented in Figure 18.6. The measurements were taken at 0.2 m above the bed (there were problems with the ECM data at 0.1 m above the bed). The velocity components were rotated into a frame in which the mean vertical velocity is zero, and the fluctuating velocity and concentration were calculated by removing means and linear trends throughout the data series. The Reynolds stress is a highly intermittent signal with peaks in excess of 20 times the mean. Positive contributions to the Reynolds stress arise from upward flows, moving slower than the mean ($u' < 0$, $v' > 0$ or 'ejections'; shown by the bolder line in Figure 18.6c),

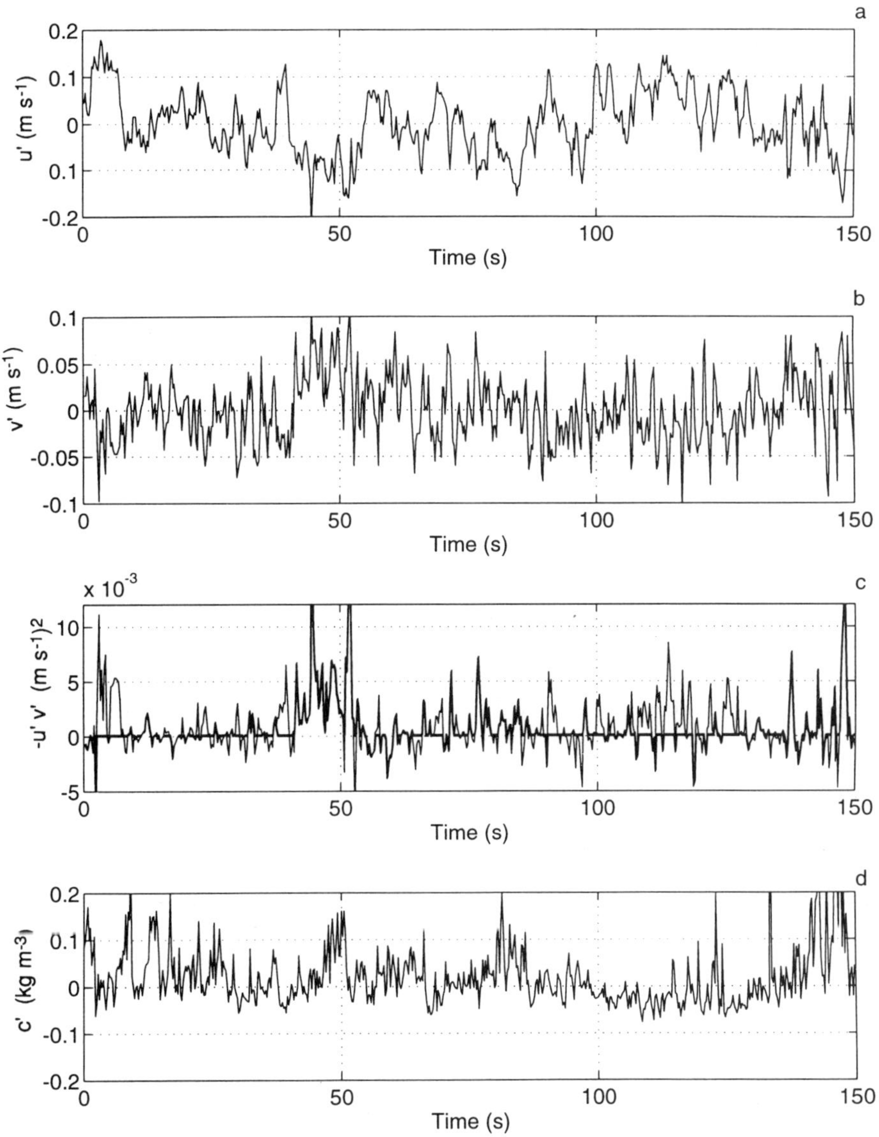

Figure 18.6. Time-series at 0.2 m above the bed of: (a) horizontal flow fluctuations *u'*; (b) vertical flow fluctuations *v'*; (c) kinematic Reynolds stress, *u'v'* (broad line represents $v' > 0$ events); and (d) concentration fluctuations *c'*

and from relatively fast downward flows ($u' > 0$, $v' < 0$ or 'sweeps'). Many studies, (Grass, 1983; Jackson, 1976; Sumer & Deigaard, 1981; Soulsby *et al.*, 1994) have sought to link ejection events with the suspension of sediment. Hence the expectation was to find some correlation between the peaks in Reynolds stress and the peaks in concentration. However, as can be seen in Figures 18.6c and 18.6d there is limited correspondence between ejection peaks and concentration peaks, possibly due to the 0.5 m lateral separation between the ECMs and acoustic transducers (Figure 18.1b). The lack of this correlation is illustrated by comparing the upward turbulent flux of particles, $\overline{v'c'}$, with the mean settling flux, v_sC which, as the flow is broadly in equilibrium, should be of the same order of magnitude. However, it is found that $\overline{v'c'} = 0.0014$, while $v_sC = 0.015$. Since it seems unreasonable to conclude that there is considerable deposition during this block of data, it seems most likely that there is an error in the calculation of the turbulent flux, and this error arises from the problems of cross-correlating the concentration and velocity measurements, owing to the spatial separation between the points at which they were measured.

It is can readily be observed in Figure 18.6d that the fluctuating concentration field is not symmetric about zero. There seem to be intermittent events in which the concentration is positive ($c > 0$) and attains relatively high values. In between these events the fluctuating concentration tends to be negative ($c < 0$) and smaller in magnitude. For instance, the peak positive c' attained in Figure 18.6d is 0.2 kg m^{-3}, while the peak negative c' is only -0.07 kg m^{-3}. This asymmetry is typical of all of the time-series at each height above the bed. A statistical measure of the asymmetry in the concentration is the skewness factor given by $c'^3/(c'^2)^{1.5}$. If the concentration fluctuations were purely Gaussian then the skewness factor would vanish. However, as seen in Figure 18.7a, the skewness is positive at each height above the bed, which is consistent with the observation of the time-series and may be explained in terms of the event-driven view of suspension. If the particles are advected by intermittent, coherent events, then these would leave just this signature in the statistics of the flow and lead to intermittent, yet high, peaks in the concentration fluctuations in-between which the concentration is lower. The positive skewness supports this event-driven view of suspended sediment. Given this model of suspension, it is possible to further develop a means of analysing these clouds of sediment. This technique is based upon that used to analyse Reynolds stress time-series, (*cf.* Gordon & Witting, 1977), but here is applied to the concentration field; the aim is to establish the average characteristics of the sediment-laden clouds. The technique focuses on the c'^2 (using c'^2 instead of $|c|$ as an analogy to $-u'v'$ for Reynolds stress) time-series, and finds the maximum value during the data block. This peak is then classified according to the sign of the c'. The time-series of c'^2 is then stepped forwards and backwards in time until either c'^2 has decreased to a given proportion of the peak value, in the present case 10%, or c' has changed sign. This permits calculation of the duration of an event and an average concentration within the event. Once identified, the event is entirely removed from the time-series and the process repeated until a certain proportion of the sum of the c'^2 measurements is accounted for, 50% being chosen in this study. This cloud analysis has been applied to the time-series at each height above the bed.

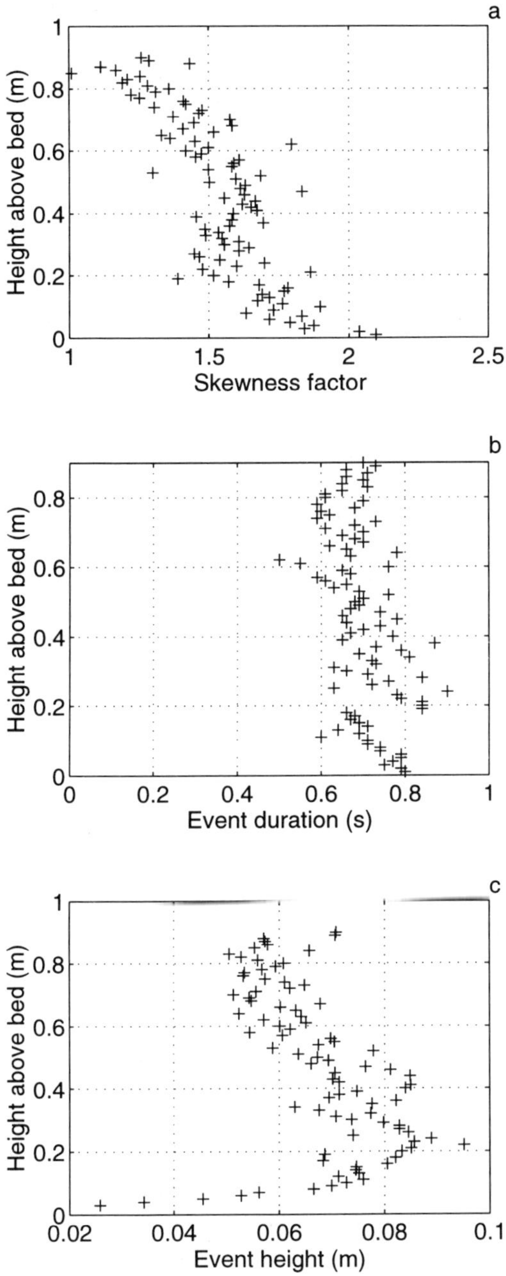

Figure 18.7. Measurements of: (a) skewness factor; (b) event duration; and (c) event height for the sediment concentration

Using the proportions of 10 and 50% for the critical parameters in the analysis, it was found that the ratio of positive ($c > 0$) to negative ($c < 0$) events is approximately 20, which is entirely in accord with the positive skewness. The average duration of positive events is approximately constant with distance from the boundary, and equal to 0.7 s, as shown in Figure 18.7b. Therefore the streamwise dimension, given by the product of the streamwise velocity and this duration, increases with height. The events are thus growing with distance from the boundary. The vertical extent of the events may also be assessed using a development of this cloud analysis. Having found a peak concentration at a certain time and an event duration at a certain height, the time-series for heights above and below this location are noted until c'^2 has decayed to 10% of the peak value or until c' has changed sign. This permits the vertical extent associated with the event to be estimated (Figure 18.7c) and shows that the events appear to grow during an initial phase before attaining an approximately constant value (~0.06 m in the present example).

Although the analysis of the detailed data is provisional, the picture emerging from the analysis of the acoustic measurements of suspension profiles in a turbulent tidal flow is one in which the suspended sediment is highly variable, with notable positive skewness values and fluctuations in concentration which decrease from the bed. These events are intense, transport highly concentrated clouds of sediment and remain coherent over considerable distances, although their transporting capacity reduces as they expand and mix with the main body of the flow.

CONCLUSION

This paper has outlined the application of multifrequency acoustics to the measurement of suspended sediment concentration and particle size, and validated its accuracy by comparison with pumped sample data. With the confidence that acoustics can yield accurate values for suspension parameters, advantage has been taken of the technique to generate high resolution images of the suspended concentration field. This visualization shows an intermittency in the suspended concentration, with high concentration events being projected well up from the bed and interspersed with relatively low concentration periods. The preliminary analysis indicates a number of ways in which data from acoustic backscatter devices may be used and how they reveal new details of the suspension processes. These results highlight the important role of coherent fluid motions in the process of sediment suspension and suggest that subsequent, more complete studies of this data will provide useful insights into the fundamental processes of suspending and transporting sediment.

ACKNOWLEDGEMENTS

Glenn Holdaway and Thomas Wylie are thanked for their input into the data processing. The HR Wallingford, UK, field team, in particular Roy Atkins, are thanked for their help in collecting the data. This work was funded by the Natural Environmental Research Council, and the Ministry of Agriculture Fisheries and Food, under the latter's Flood Protection commission with the Proudman Oceanographic Laboratory.

REFERENCES

Bendat, J.S. & Piersol, A.G. 1986. *Random Data*, John Wiley & Sons Ltd., Chichester, 566 pp.

Crawford, A.M. & Hay, A.E. 1993. 'Determining suspended sand size and concentration from multifrequency acoustic backscatter', *J. Acoust. Soc. Am.*, **94**, 3312–3324.

Dyer, K.R. & Soulsby, R.L. 1988. 'Sand Transport on the continental shelf', *Ann. Rev Fluid Mech.*, **20**, 295–324.

Grass, A.J. 1983. 'The influence of boundary layer turbulence on the mechanics of sediment transport', in Sumer, B.M. &. Müller, A. (Eds) *Mechanics of Sediment Transport*, Proc. Euromech **156**, A.A. Balkema, Rotterdam, 3–17.

Gordon, C.M. & Witting, J. 1977. 'Turbulent structure in a benthic boundary layer', in Nihoul, J.C.J. (Ed) *Bottom Turbulence*, Elsevier, Amsterdam, 59–81.

Hay, A.E. & Bowen, A.J. 1994. 'Coherence scales of wave-induced suspended sand concentration fluctuations', *J. Geophys. Res.*, **99**, 12749–12765.

Hay, A.E. & Sheng, J. 1992. 'Vertical profiles of suspended sand concentration and size from multifrequency acoustic backscatter', *J. Geophys. Res.*, **97**, 15661–15677.

Hogg, A.J., Huppert, H.E. & Soulsby, R.L. 1994. 'The dynamics of particle-laden fluid elements', in Belorgey, M., Rajaona, R.D. & Sleath, J.F.A. (Eds) *Sediment Transport Mechanisms in Coastal Environments and Rivers*, Proc. Euromech **310**, 64–68.

Jackson, R.G. 1976. 'Sedimentological and fluid dynamic implications of the turbulent bursting phenomenon in geophysical flows', *J. Fluid Mech.*, **77**, 531–560.

Lynch, J.F., Irish, J.D., Sherwood, C.R. & Agrawal Y.C. 1994. 'Determining suspended sediment particle size information from acoustical and optical backscatter measurements', *Continental Shelf Research*, **14**, 1139–1165.

Nielsen, P. 1992. *Coastal Bottom Boundary Layers and Sediment Transport*, World Scientific, 324 pp.

Schaafsma. A.S. 1994. 'Acoustic attenuation spectroscopy of suspended sediments', in Bjorno, L. (Ed) *Proc. 2nd European Conf. on Underwater Acoustics*, European Commission, Vol. **II**, 863–868.

Sheng, J. & Hay, A.E. 1988. 'An examination of the spherical scatterer approximation in aqueous suspensions of sand', *J. Acoustic Soc. Am.*, **83**, 598–610.

Soulsby, R.L., Atkins, R. & Salkield, A.P. 1994. 'Observations of the turbulent structure of a suspension of sand in a tidal current', *Continental Shelf Research*, **14**, 429–435.

Sumer B.M. & Deigaard, R. 1981. 'Particle motions near the bottom in turbulent flow in an open channel – part 2', *J. Fluid Mech.*, 109, 311–337.

Thorne, P.D. & Campbell, S.S. 1992. 'Backscattering by a suspension of spheres,' *J. Acoust. Soc. Am.*, **92**, 978–986

Thorne, P.D. & Hardcastle, P.J. 1994. 'Multifrequency acoustic imaging of suspended sediment processes', in *Proc. Oceans '94*, Brest, France, 13-16 Sept. 1994, Vol. **III**, 263–268.

Thorne, P.D., Hardcastle, P.J. & Soulsby, R.L. 1993. 'Analysis of acoustic measurements of suspended sediments', *J. Geophys. Res.*, **98**, 899–910.
Thorne, P.D., Waters, K.R. & Brunder, T.J. 1995. 'Acoustic measurements of scattering by objects of irregular shape', *J. Acoust. Soc. Am.*, **97**, 242–251.
Vincent, C.E. & Downing, A. (1994) 'Variability of suspended sand concentration, transport and eddy diffusivity under non-breaking waves on the shoreface', *Continental Shelf Research*, **14**, 223–250.

NOTATION

a_s	sediment radius (m)
A_c	acoustic measurement of suspended sediment concentration (kg m^{-3})
A_s	acoustic measurement of particle radius (m)
c	suspended sediment concentration (kg m^{-3})
c'	concentration fluctuations (c-C) (kg m^{-3})
c'_{rms}	root-mean-square concentration (kg m^{-3})
C	mean concentration (kg m^{-3})
f	form function
K_s	eddy diffusivity (m^2 s^{-1})
K_t	transceiver constant (Pa m$^{3/2}$)
P	backscattered root-mean-square pressure (Pa)
P_c	pumped sample measurement of suspended sediment concentration (kg m^{-3})
P_s	pumped sample measurement of particle radius (m)
r	range
u'	horizontal flow fluctuations (m s^{-1})
u_*	friction velocity (m s^{-1})
$-u'v'$	kinematic Reynolds stress (m^2 s^{-2})
v'	vertical flow fluctuations (m s^{-1})
v_s	settling velocity (m s^{-1})
y	height above the bed (m)
Y	flow depth (m)
α_s	suspended sediment attenuation (Nepers m^{-1})
α_w	water absorption (Nepers m^{-1})
ρ_s	sediment density
σ	standard deviation
ζ	sediment attenuation constant (Nepers kg^{-1} m^2)

19 Interaction Between Cellular Secondary Currents and Lateral Alternate Sorting

TETSURO TSUJIMOTO
Department of Civil Engineering, Kanazawa University, 2-40-20, Kodatsuno, Kanazawa, 920, Japan

TADANORI KITAMURA
School of Global Environmental Engineering, Kyoto University, Sakyo-ku, Kyoto, 606, Japan

ABSTRACT

In a stream with a bed composed of sand and gravel, longitudinal stripes of sorting, or the lateral alternation of relatively coarse- and fine-grain stripes of sediment, present a very interesting problem with regard to their formation. In the present study, the interaction between sediment transport with the prevailing flow regime is investigated theoretically to provide a model of the formation of cellular secondary currents and the associated longitudinal sorting stripes.

The flow is analysed numerically by employing an algebraic stress model for turbulence which can describe the secondary flow induced by spatial heterogeneity of the bed roughness. The flow analysis clarifies the flow direction near the bed. The bedload motion of graded bed material is described by a model employing pick-up rate and step length which are evaluated for each grain size. The behaviour of the bedload as affected by secondary flow and selective motion due to sediment size are then described, and the subsequent changes of bed surface composition are predicted.

The stability of the proposed model is investigated by applying a small disturbance to the system in a linearized scheme. As a result, the secondary currents are induced most intensively on short wavelength sorting stripes, but the predominant wavelength of sorting stripes which develops at the final stage is around twice the flow depth. This is consistent with both field observations and mobile bed experiments.

INTRODUCTION

In streams composed of graded bed materials, several kinds of sorting take place. Besides the formation of armour coats and pavements as monotone sorting processes, 'longitudinal stripes of sorting' and 'diffuse gravel sheets' (longitudinal alternate grain sorting patterns) are important from the viewpoint of both river morphology (Parker, 1991; Tsujimoto, 1991) and river engineering.

Coherent Flow Structures in Open Channels.
Edited by P.J. Ashworth, S.J. Bennett, J.L. Best and S.J. McLelland.

In a stream with a bed composed of sand and gravel, longitudinal stripes of sorting, or the lateral alternation of relatively coarser- and finer-grained stripes, are often observed (see Figure 19.1). Such a bed condition promotes a lateral flow from the rough to smooth part near the bed and the formation of paired cellular secondary currents (see Figure 19.2). In particular, when the lateral wavelength of alternation

Figure 19.1. Longitudinal stripes of sorting formed in an irrigation canal of 0.5 m wide ($d \cong 0.2\sim0.3$ mm; $\beta \cong 5\sim10$; $p_0 \cong 0.5$; $h \cong 0.15$ m; $U_m \cong 40$ m s^{-1}). Flow direction from bottom to top of picture.

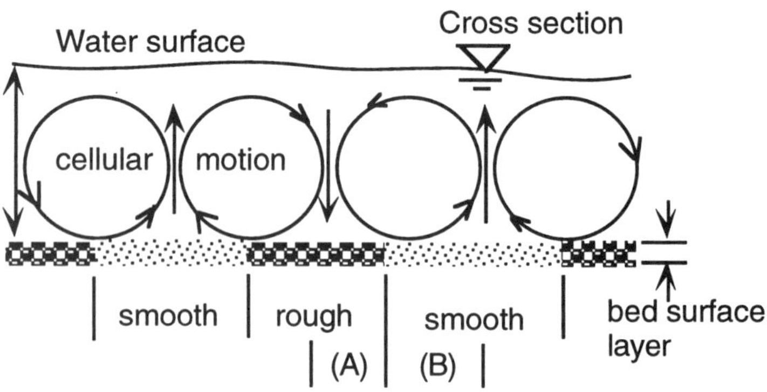

Figure 19.2. Schematic diagram of cellular secondary currents and sorting stripes

of sorting is roughly twice the flow depth, cellular secondary currents, often called longitudinal vortices, develop. However, cellular secondary currents and longitudinal sorting stripes may promote or suppress one another. Such an interaction was investigated in the earlier studies mostly from the viewpoint of secondary current initiation but rarely from the viewpoint of the mechanics of graded material transport.

Müller & Studerus (1979) and Nakagawa *et al.* (1981) made artificial longitudinal stripes of different roughness in order to produce a lateral heterogeneity of bed shear stress and successive cellular secondary currents, and conducted detailed measurements of such stable cellular currents. According to their experiments, a lateral secondary-flow component, the intensity of which near the bottom is 2–5% of the maximum primary velocity, was confirmed. Figure 19.3, which is from an exhibit at ETH-Hönggerberg, shows that secondary flow is induced by the difference of the primary velocity profile depending on the bed roughness.

Such secondary currents often form longitudinal stripes of bed undulation called 'sand ridges'. Kuroki & Kishi (1981) and Colombini (1993) performed a theoretical analysis on their formation mechanism, and Hirano & Ohmoto (1988) and Nezu *et al.* (1988) conducted turbulence measurements of flow over sand ridges.

Figure 19.3. Exhibit demonstrating the heterogeneity of primary velocity profile and cellular secondary currents (from the laboratory of ETH-Hönggerberg)

On the other hand, Tsujimoto (1989) explained the formation mechanism of longitudinal stripes of sorting as an instability of lateral distribution of bed-surface composition. He stated that a secondary current has a lateral component originating at a relatively rough part (*region A* in Figure 19.2), where coarser particles dominate, towards the relatively smooth part (*region B* in Figure 19.2), and this causes lateral bedload transport from the rough to the smooth. Finer particles are more likely to be picked-up from the bed, but are transported a shorter distance (step length) compared with coarser particles. Hence, the difference in bed-surface composition may be either amplified (unstable) or suppressed (stable). Tsujimoto clarified a criterion for the occurrence of longitudinal stripes of sorting by applying a non-equilibrium transport model of bedload for each grain size. In his analysis, the magnitude of secondary current was assumed to be independent of bed-surface composition. In this study, the magnitude of the secondary current is estimated by a numerical analysis of the flow based on an algebraic stress model of turbulence, coupled with a model of heterogeneous bedload transport.

INTERRELATING SYSTEM TO FORM LONGITUDINAL STRIPES OF SORTING

The formation mechanism for longitudinal stripes of grain sorting can be explained by analysing the following interrelating system: (i) the secondary flow induced by the heterogeneity of bed roughness as determined by the bed surface composition; (ii) bedload transport of graded material by secondary currents; and (iii) bed deformation and change of bed-surface composition due to bedload motion. In this study, a longitudinally uniform flow ($\partial/\partial x = 0$, where x is the downstream direction) is assumed and the change of each quantity in the lateral (z) direction is examined. The bed-surface composition is represented by the volumetric ratio of the i-th sand fraction to the total bed material in the surface layer (often called the 'exchange layer') represented as $p_i(z)$. For convenience, the bed roughness is represented by the mean diameter, $d_m(z)$, which is here defined as follows:

$$d_m = \sum_{i=1}^{N} p_i d_i \qquad (19.1)$$

where d_i is the diameter of the i-th sand fraction.

The existence of a secondary component in the flow influences the direction of bedload motion. Furthermore, if sand ridges are formed, the lateral bed slope also effects the direction of bedload motion. In other words, the direction of bedload motion, $\phi_i(z)$, is affected by the direction of flow, $\gamma(z)$ and the lateral bed slope $\theta_z(z) \equiv \partial y_b/\partial z$ (where y_b is the bed elevation) which have a decisive effect on the lateral flux of bedload, $q_{Bzi}(z)$. According to mass conservation, the lateral gradient of the lateral transport rate results in bed deformation ($\partial y_b/\partial t$) and the size class difference brings about the sorting ($\partial p_i/\partial t$).

FLOW OVER A BED WITH HETEROGENEOUS ROUGHNESS

When the rough and smooth zone are neighbouring each other in the longitudinal direction continuously, the upward secondary flow appears at the centre of the smooth zone while the downward secondary flow appears at the centre of the rough zone, causing the secondary flow from the rough to the smooth zone. This causes cellular motions or longitudinal vortices to form. Such a flow structure with secondary component induced by the heterogeneity of bed roughness is analysed using an algebraic stress model.

The heterogeneity of bed roughness is taken into account in the boundary condition at the bottom: the log-law velocity distribution, in which equivalent sand roughness is linked to the mean grain diameter $d_m(z)$, is assumed at the lowest grid points ($y = y_p$) of the calculation using

$$\frac{U_b(z)}{u_*(z)} = \frac{1}{\kappa}\ln\left[\frac{y_p}{d_m(z)}\right] + B_r \qquad (19.2)$$

where U_b is the primary velocity near the bed ($y = y_p$), u_* is the shear velocity, κ is von Kármán's coefficient and $B_r = 8.5$.

When a two-equation turbulence model is employed, the governing equations of flow are written as follows after a parabolic flow approximation where the longitudinal flow is dominant.

$$\frac{\partial U}{\partial x} + \frac{\partial V}{\partial y} + \frac{\partial W}{\partial z} = 0 \qquad (19.3)$$

$$\frac{\partial}{\partial x}(U^2) + \frac{\partial}{\partial y}\left(VU - \Gamma\frac{\partial U}{\partial y}\right) + \frac{\partial}{\partial z}\left(WU - \Gamma\frac{\partial U}{\partial z}\right) = g\sin\theta - \frac{d}{dx}\left(\frac{P}{\rho}\right) \qquad (19.4)$$

$$\frac{\partial}{\partial x}(UV) + \frac{\partial}{\partial y}\left(V^2 - \nu\frac{\partial V}{\partial y}\right) + \frac{\partial}{\partial z}\left(WV - \nu\frac{\partial V}{\partial z}\right) =$$

$$g\cos\theta - \frac{d}{dx}\left(\frac{P''}{\rho}\right) + \frac{\partial}{\partial y}\left(\frac{\tau_{yy}}{\rho}\right) + \frac{\partial}{\partial z}\left(\frac{\tau_{yz}}{\rho}\right) \qquad (19.5)$$

$$\frac{\partial}{\partial x}(UW) + \frac{\partial}{\partial y}\left(VW - \nu\frac{\partial W}{\partial y}\right) + \frac{\partial}{\partial z}\left(W^2 - \nu\frac{\partial W}{\partial z}\right) =$$

$$-\frac{\partial}{\partial z}\left(\frac{P''}{\rho}\right) + \frac{\partial}{\partial y}\left(\frac{\tau_{yz}}{\rho}\right) + \frac{\partial}{\partial z}\left(\frac{\tau_{zz}}{\rho}\right) \qquad (19.6)$$

$$\frac{\partial}{\partial x}(Uk) + \frac{\partial}{\partial y}\left[Vk - \left(\nu + \frac{\nu_t}{\sigma_k}\right)\frac{\partial k}{\partial y}\right] + \frac{\partial}{\partial z}\left[Wk - \left(\nu + \frac{\nu_t}{\sigma_k}\right)\frac{\partial k}{\partial z}\right] = P_k - \varepsilon \qquad (19.7)$$

$$\frac{\partial}{\partial x}(U\varepsilon)+\frac{\partial}{\partial y}\left[V\varepsilon-\left(\nu+\frac{\nu_t}{\sigma_\varepsilon}\right)\frac{\partial \varepsilon}{\partial y}\right]+\frac{\partial}{\partial z}\left[W\varepsilon-\left(\nu+\frac{\nu_t}{\sigma_\varepsilon}\right)\frac{\partial \varepsilon}{\partial z}\right]=$$

$$\frac{\varepsilon}{k}(C_{\varepsilon 1}P_k - C_{\varepsilon 2}\varepsilon) \qquad (19.8)$$

$$P_k=\nu_t\left[\left(\frac{\partial U}{\partial y}\right)^2+\left(\frac{\partial U}{\partial z}\right)^2\right] \qquad \nu_t = C_\mu\frac{k^2}{\varepsilon} \qquad (19.9)$$

in which x, y and z are the longitudinal, vertical and transverse axes (y is measured from the bed), U, V and W are the x, y and z components of time-averaged velocity, k is the turbulence energy, ε is the dissipation rate of turbulence energy, P and P'' are the static pressure and pressure deviation, ρ is the mass density of fluid, g is the gravitational acceleration, θ is the longitudinal bed slope, ν is the kinematic viscosity, ν_t is the kinematic eddy viscosity, P_k is the energy production, $\Gamma \equiv \nu + \nu_t$ is the effective viscosity, and σ_k, σ_ε, $C_{\varepsilon 1}$, $C_{\varepsilon 2}$ and C_μ are empirical parameters.

The secondary flow due to the heterogeneity of bed roughness must be explained by applying an anisotropic model of turbulence. Hence, an algebraic stress model by Launder & Ying (1973) is employed here and its applicability to open-channel flows as developed by Kawahara & Tamai (1988). Then, the Reynolds stresses are expressed as follows:

$$\frac{\tau_{xy}}{\rho}=\nu_t\left(\frac{\partial U}{\partial y}+\frac{\partial V}{\partial x}\right) \qquad \frac{\tau_{zx}}{\rho}=\nu_t\left(\frac{\partial U}{\partial z}+\frac{\partial W}{\partial x}\right) \qquad \frac{\tau_{yz}}{\rho}=C\,\nu_t\frac{k}{\varepsilon}\left(\frac{\partial U}{\partial y}\right)\left(\frac{\partial U}{\partial z}\right)$$

$$\frac{\tau_{yy}}{\rho}=C\,\nu_t\frac{k}{\varepsilon}\left(\frac{\partial U}{\partial z}\right)^2-C_k k \qquad \frac{\tau_{zz}}{\rho}=C\,\nu_t\frac{k}{\varepsilon}\left(\frac{\partial U}{\partial z}\right)^2-C_k k \qquad (19.10)$$

where C' and C'_k are empirical parameters.

The empirical parameters involved in the governing equations were determined by adopting the standard values as follows: $C_\mu = 0.09$; $C'_k = 0.522$; $C_{\varepsilon 1} = 1.44$; $C_{\varepsilon 2} = 1.92$; $\sigma_k = 0.9$; and $\sigma_\varepsilon = 1.3$. Meanwhile, since C' is subject to the magnitude of the secondary current of the Prantdl's second kind, Launder & Ying (1973) gave a standard value as $C' = 0.018$ from comparing the result with experiments in duct flows. However, for flow in a compound channel where secondary flow dominates, Kawahara & Tamai (1988) used twice the standard value.

The governing equations are discretized on a staggered grid and the *SIMPLE* algorithm developed by Patanker & Spalding (1972) is employed. A part of the cross-section of the stream is extracted as shown in Figure 19.4, with symmetric boundary conditions imposed. The boundary condition at the bed is given by the log-law as already mentioned. In the longitudinal direction, a periodic boundary condition is imposed at the upstream and downstream ends of the control volume to obtain a quasi-uniform flow solution.

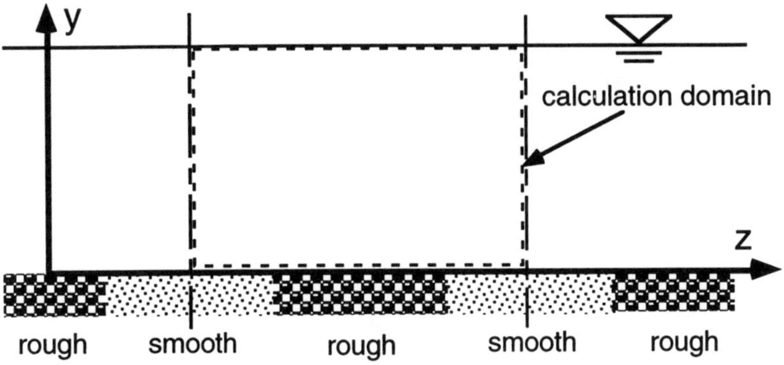

Figure 19.4. Control volume for calculation

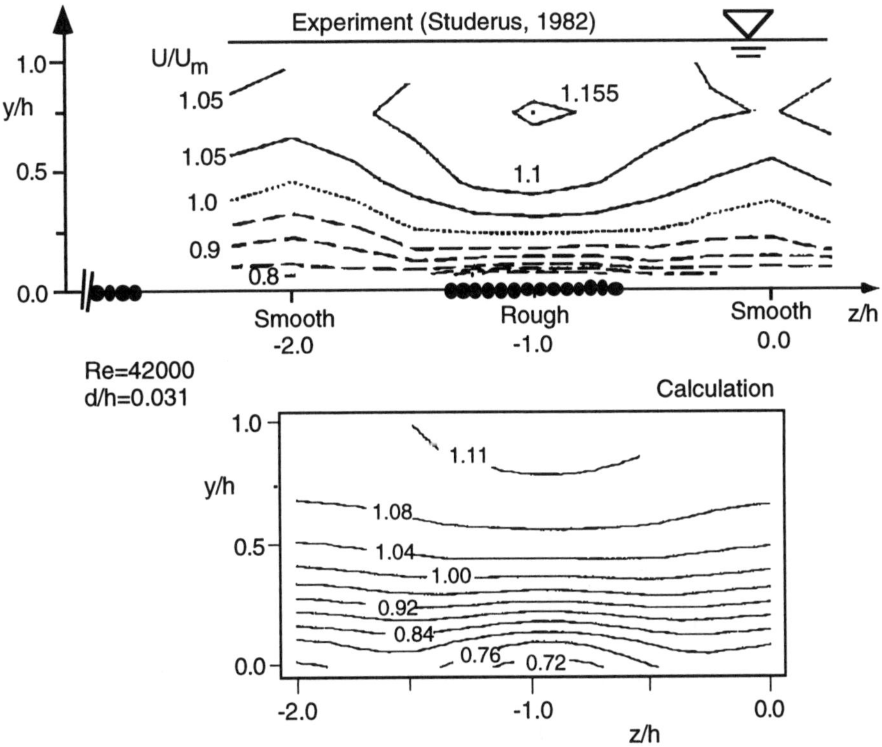

Figure 19.5. Comparison between experiment (Studerus, 1982) and calculation (this paper): contours of primary velocity (U). U_m is the cross-section average of the primary velocity

Figures 19.5 and 19.6 compare the experimental data by Studerus (1982) with the calculated result based on the above model. Figure 19.5 shows contours of the primary velocity, while Figure 19.6 is the vector expression of secondary components. From a qualitative comparison, it is verified that the numerical model can describe the essential characteristics of secondary currents induced by the heterogeneity of bed roughness.

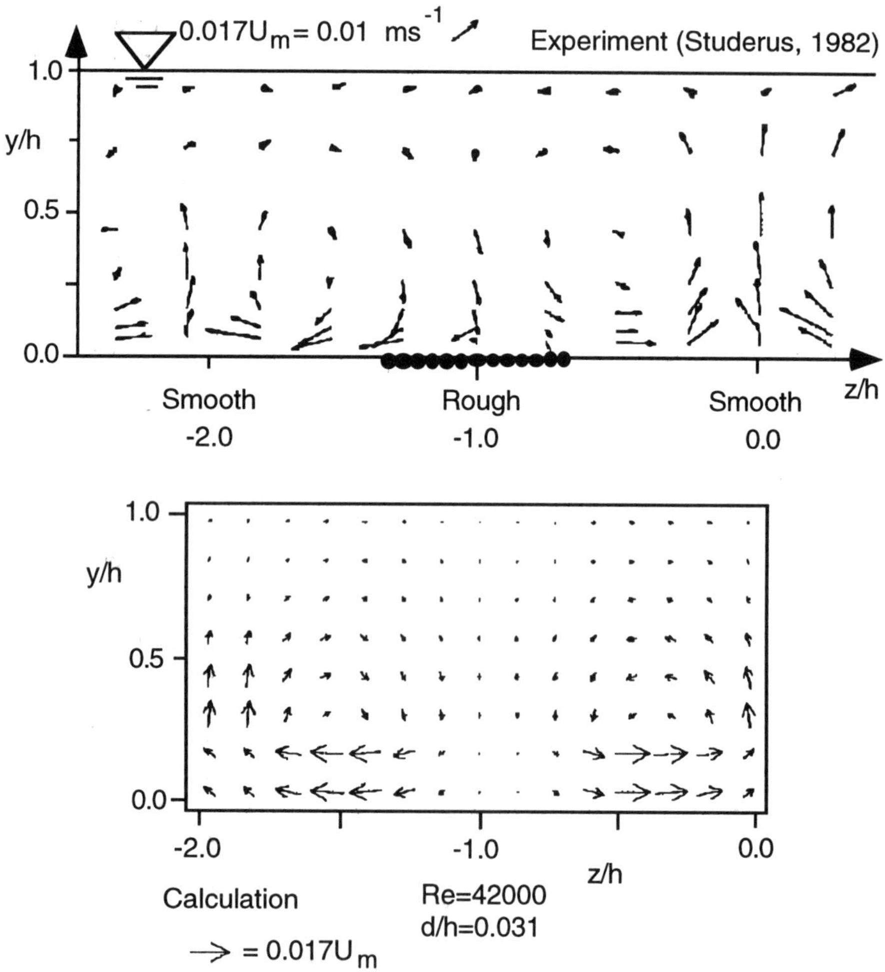

Figure 19.6. Comparison between experiment (Studerus, 1982) and calculation (this paper): secondary flow components. U_m is the cross-section average of the primary velocity

BEDLOAD TRANSPORT MODEL

The bedload transport model incorporates the pick-up rate and step length for each grain size of graded bed material. The fractional transport rate due to primary flow is given as follows:

$$q_{Bi} = (1-\rho_0)\theta_e p_i p_{si} \Lambda_i \tag{19.11}$$

where q_{Bi} is the bedload transport rate of the i-th fraction (substantial volume per unit width), ρ_0 is the porosity of sand, θ_e is the thickness of the exchange layer and p_{si} and Λ_i are the pick-up rate (probability density of a sand particle to be dislodged from a bed per unit time) and the mean step length of the i-th sand fraction. The lateral bedload flux of the i-th sand fraction is given as follows:

$$q_{Bzi} = q_{Bi} \tan \phi_i \tag{19.12}$$

where the direction of bedload motion (ϕ_i) is given by

$$\tan \phi_i \equiv -\alpha \sqrt{\frac{\tau_{*ci}}{\tau_{*i}}} \tan \theta_z + \tan \gamma \tag{19.13}$$

where $\tau_{*i} \equiv u_*^2 /[(\sigma/\rho - 1)gd_i]$; τ_{*ci} is the dimensionless critical tractive force for each grain size with σ representing the sediment density, $\alpha \equiv \sqrt{1/\mu_d}$, μ_d is the friction coefficient of sand, and $\gamma \equiv W/U|_{y\equiv 0}$ which is the deflection of the flow direction (Nakagawa *et al.*, 1986).

The dimensionless critical tractive force for each grain size is expressed as follows (Nakagawa *et al.*, 1982):

$$\frac{\tau_{*ci}}{\tau_{*cm}} = \left(\frac{d_i}{d_m}\right)^{-q} \tag{19.14}$$

where τ_{*cm} is the critical tractive force of sand with the diameter d_m in the mixture. Here, τ_{*cm} is assumed to be identical to the dimensionless critical tractive force for uniform size sand. The exponent of Equation 19.14, q, is 1.0 or slightly smaller than 1.0.

According to Nakagawa *et al.* (1982), the pick-up rate for each grain size of graded materials, p_{si}, is given by the following equation.

$$\frac{p_{si}}{\sqrt{(\sigma/\rho - 1)\, g/d_i}} = 0.03\tau_{*i}\left(1 - \frac{0.7\tau_{*ci}}{\tau_{*i}}\right)^3 \tag{19.15}$$

Furthermore, the mean step length for each grain size is written as

$$\Lambda_i = \lambda d_i \tag{19.16}$$

where $\lambda \cong 20$ (Nakagawa *et al.*, 1982).

When the thickness of the surface layer ('exchange layer') is assumed to be constant, the temporal change of the volumetric ratio of the i-th sand fraction is written as follows (Tsujimoto, 1991):

$$\frac{\partial p_i}{\partial t} = \frac{1}{\theta_e(1-\rho_0)}\left[p_i \sum_{i=1}^{N}\left(\frac{\partial q_{Bzi}}{\partial z}\right) - \frac{\partial q_{Bzi}}{\partial z}\right] \qquad (19.17)$$

The change of bed elevation is written as

$$\frac{\partial y_b}{\partial t} = \frac{-1}{\theta_e(1-\rho_0)}\sum_{i=1}^{N}\left(\frac{\partial q_{Bzi}}{\partial z}\right) \qquad (19.18)$$

LINEARIZED ANALYSIS FOR SMALL DISTURBANCE

In order to investigate the instability of the interrelating system among flow, bedload transport and bed-surface composition, a bed composed of heterogeneous material with two fractions, fine sand and coarse sand, is postulated. The diameters and the volumetric ratios of the respective fractions are represented by d and βd ($\beta > 1$), and 1-p and p. In order to make a linearized analysis possible, a small disturbance of the volumetric ratio of coarse sand is assumed as follows:

$$\frac{p(z)}{p_0} = 1 + a_p \cos Kz^* \qquad (19.19)$$

where $a_p p_0$ is the amplitude of the disturbance of p, $K \equiv 2\pi h_0/L_z$, L_z is the lateral wavelength of sorting stripe, $z^* \equiv z/h_0$, h_0 is the undisturbed flow depth, and the subscript 0 implies the undisturbed values. Then, the lateral change of the mean diameter d_m of the surface layer is written as

$$\frac{d_m(z)}{d_{m0}} = 1 + r_d a_p \cos Kz^* \qquad r_d = \frac{p_0(1-\beta)}{1+p_0(1-\beta)} \qquad (19.20)$$

where $r_d a_p d_{m0}$ is the amplitude of the disturbance of d_m.

The result of the secondary flow analysis is written as

$$\tan\gamma = r_w r_d a_p \sin Kz^* \qquad (19.21)$$

where $r_w r_d a_p d_{m0}$ is the amplitude of the disturbance of $\tan\gamma$. The lateral bed profile would be written as

$$\frac{y_b}{h_0} = -a_y \cos Kz^* \qquad (19.22)$$

where y_b is the bed-surface elevation from the mean bed level.

In a linear mode, the following pair of differential equations with respect to a_p and a_y are deduced:

$$\frac{da_p}{dt_*} = Aa_p - Ba_y \tag{19.23}$$

$$\frac{da_y}{dt_*} = Ca_p - Da_y \tag{19.24}$$

where A, B, C and D are defined below, together with the auxiliary parameters P_1, P_2 and Ω. A, B, C and D are all positive.

$$A \equiv r_w r_d (1 - \Omega\beta^{-1/2})K \tag{19.25}$$

$$B \equiv \alpha\sqrt{\frac{\tau_{*cm}}{\tau_{*m}}}\sqrt{[(1-p_0)+\beta p_0]^{q-1}}(1-\Omega\beta^{-q/2})K^2 \tag{19.26}$$

$$C \equiv r_w r_d\left(1 + \frac{p_0}{1-p_0}\Omega\beta^{-1/2}\right)\theta_e^* K \tag{19.27}$$

$$D \equiv \alpha\sqrt{\frac{\tau_{*cm}}{\tau_{*m}}}\sqrt{[(1-p_0)+\beta p_0]^{q-1}}\left(1 + \frac{p_0}{1-p_0}\Omega\beta^{-q/2}\right)\theta_e^* K^2 \tag{19.28}$$

$$P_1 \equiv 1 - \frac{0.7\tau_{*cm}}{\tau_{*m}}\left[\frac{\beta}{(1-p_0)+\beta p_0}\right]^{1-q}$$

$$P_2 \equiv 1 - \frac{0.7\tau_{*cm}}{\tau_{*m}}\left[\frac{1}{(1-p_0)+\beta p_0}\right]^{1-q} \tag{19.29}$$

$$\Omega = (P_1/P_2)^3 \tag{19.30}$$

in which $\theta_e^* \equiv \theta_e/h_0$, $\tau_{*m} \equiv u_*^2/[(\sigma/\rho - 1)gd_m]$, and $t_* \equiv (1-p_0)(h_0/d)p_{s0}\lambda t$. In Equations 19.23 and 19.24, A and C are factors to promote the disturbance, while B and D act as stabilizing factors.

If $q = 1$ in Equation 19.14, AD-$BC = 0$, and the following pair of differential equations are deduced from Equations 19.23 and 19.24.

$$\frac{d^2 a_p}{dt_*^2} - (A - D)\frac{da_p}{dt_*} = 0 \tag{19.31}$$

$$\frac{d^2 a_y}{dt_*^2} - (A - D)\frac{da_y}{dt_*} = 0 \tag{19.32}$$

As an initial condition, only a small disturbance of p is imposed. The bed is assumed to be initially flat. In other words, $a_p = a_p(0) \neq 0$ but $a_y = 0$ at $t_* = 0$. Then, the solutions are written as follows:

$$\frac{a_p}{a_p(0)} = \frac{-D}{A-D} + \frac{A}{A-D}\exp[(A-D)t_*] \tag{19.33}$$

$$\frac{a_y}{a_p(0)} = \frac{-C}{A-D}\{1-\exp[(A-D)t_*]\} \tag{19.34}$$

These solutions behave as follows (see Figure 19.7): (i) when $(A\text{-}D) < 0$, the disturbances of bed configuration and surface bed composition show an asymptotic behaviour; and (ii) when $(A\text{-}D) > 0$, the disturbances grow monotonously. In both cases, sand ridges and sorting stripes are formed.

Figure 19.8 shows the ratio of the amplitude of the fluctuation of $\tan\gamma$, r_w, to the amplitude of d_m/d_{m0} obtained as the solution of the secondary flow analysis. According to this figure, the maximum deflection of the flow direction appears at $K \cong 12$. The wavelength of this sorting stripe is unexpectedly small.

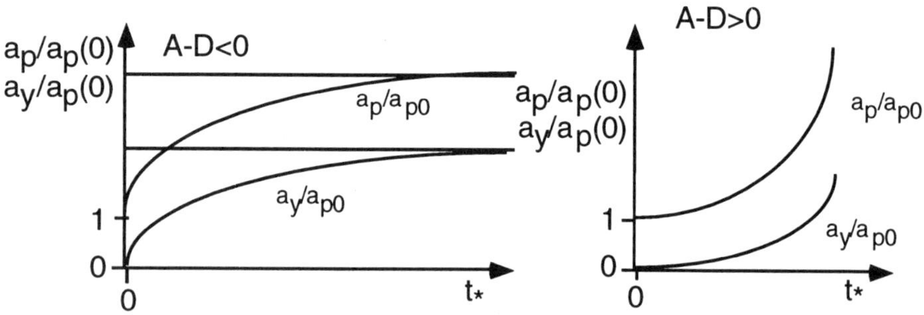

Figure 19.7. Illustrative relation of $a_p/a_p(0)$ and $a_y/a_y(0)$ with t_*

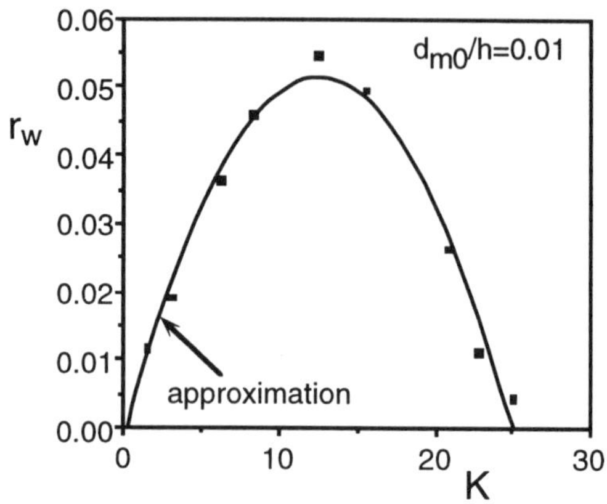

Figure 19.8. Amplitude of disturbance of flow direction by secondary currents relative to the value of K

However, the solution of the analysis coupled with bedload motion of heterogeneous bed material is shown in Figure 19.9, where the relations of the asymptotic values of $a_p/a_p(0)$ and $a_y/a_p(0)$ with K are depicted for the condition that $(A\text{-}D) < 0$. This result suggests that the predominant equilibrium state for sorting stripe development occurs when the stripes have a wavelength around twice the flow depth ($K \cong 2\text{–}3$) which is consistent with the phenomena we often observe in the field and movable bed experiments.

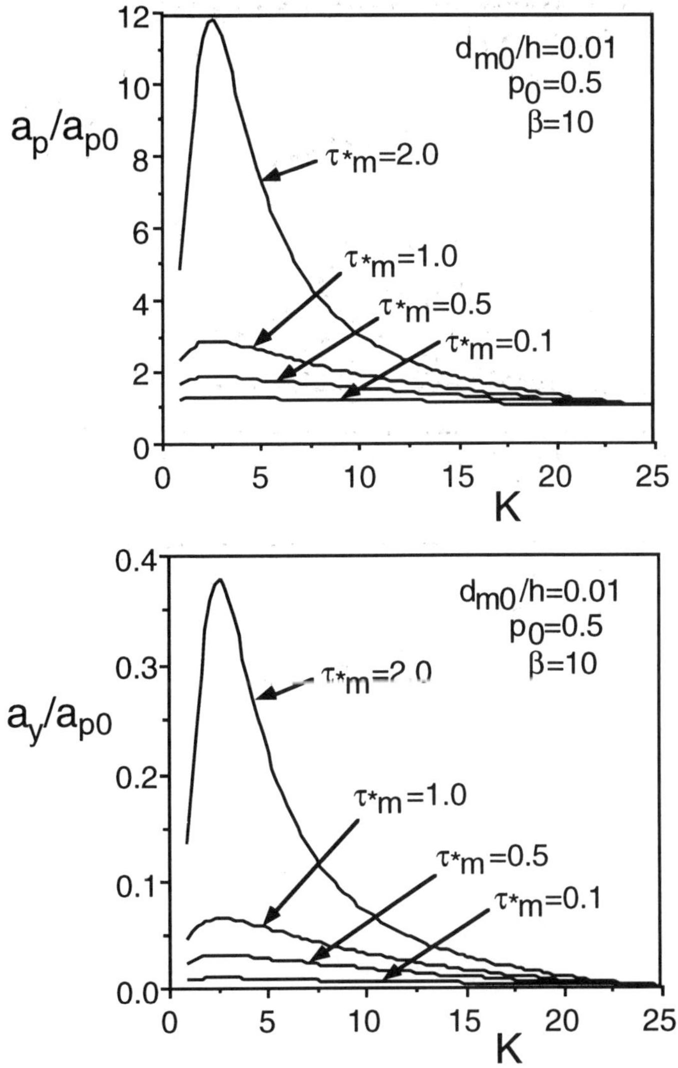

Figure 19.9. Magnitude of sorting and height of sand ridge

CONCLUSION

In this study, the complex system which incorporates flow, bedload transport and bed-surface composition has been investigated in order to understand the formation mechanism of cellular secondary currents and longitudinal sorting stripes. The secondary flow has been analysed numerically by employing an algebraic stress model of turbulence. The flow analysis has clarified the secondary flow behaviour near the bed which affects the bedload direction. The bedload motion of heterogeneous material has been described by a model employing pick-up rate and step length which have been evaluated for each grain size. Based on these models, the bedload behaviour, as determined by secondary flow and selective motion due to sediment size, has been described and the subsequent change of bed-surface composition has been predicted. As a result, the secondary currents are induced on the sorting stripes with rather shorter wavelength but the wavelength of sorting stripes which develop most predominantly at the final stage is around twice the flow depth and this is consistent with both field and experimental observations.

REFERENCES

Colombini, M. 1993. 'Turbulence-driven secondary flows and formation of sand ridges', *J. Fluid Mech.*, **254**, 701–719.

Hirano, M. & Ohmoto, T. 1988. 'Experimental study on the interaction between longitudinal vortices and sand ribbons', *Proc. 6th Congr. APD-IAHR*, Kyoto, Japan, **II**, 59–66.

Kawahara, Y. & Tamai, N. 1988. 'Numerical calculation of turbulent flows in compound channels with algebraic stress turbulence model', *Proc. 3rd Int. Symp. Refined Flow Modelling & Turbulent Measurements*, Tokyo, Japan, 527–536.

Kuroki, M. & Kishi, T. 1981. 'Sediment transport and dispersion of pollutant in flow with longitudinal vortices', *Proc. 25th Japanese Conf. on Hydraulics, JSCE*, 433–438. (in Japanese).

Launder, B.E. & Ying W.M. 1973. 'Prediction of flow and heat transfer in ducts of square cross-section', *Heat and Fluid Flow*, **3**, 115–121.

Müller, A. & Studerus, X. 1979. 'Secondary flow in an open channel', *Proc. 18th Congr. IAHR*, Cagliari, Italy, **B.a.3**, 142–144.

Nakagawa, H., Nezu, I. & Tominaga, A. 1981. 'Turbulent structure with and without cellular secondary currents over various bed configurations', *Annuals*, Disast. Prev. Res. Inst., Kyoto Univ., **24B/2**, 315–338. (in Japanese).

Nakagawa, H., Tsujimoto, T. & Murakami, S. 1986. 'Non-equilibrium bed load along side bank', *Proc. 3rd Int. Symp. River Sedimentation*, Jackson, Mississippi, 1029–1065.

Nakagawa, H., Tsujimoto, T. & Nakano, S. 1982. 'Characteristics of sediment motion for respective grain sizes of sand mixtures', *Bulletin*, Disast. Prev. Res. Inst., Kyoto Univ., **32**, 1–32.

Nezu, I., Nakagawa, H. & Kawashima, N. 1988. 'Cellular secondary currents and sand ribbons in fluvial channel flows', *Proc. 6th Congress APD-IAHR*, Kyoto, Japan, **II**, 51–58.

Parker, G. 1991, 'Some random notes on grain sorting', in *Proc. Int. Grain Sorting Seminar*, Mitteilungen **117**, ETH Zürich, 19–76.

Patankar, S.V. & Spalding, D.B. 1972. 'A Calculation Method for Heat, Mass and Momentum Transfer in three-dimensional Parabolic Flow', *Int. J. Heat Mass Transfer*, **15**, 1787–1806.

Studerus, X. 1982. *Sekundärströmungen im offenen Gerinne über Rauhen Längsstreifen*, Unpub. PhD Thesis, Institut für Hydromechanik und Wasserwirtschaft, ETH Zürich, 185 pp.

Tsujimoto, T. 1989. 'Longitudinal stripes of sorting due to cellular secondary currents', *J. Hydroscience Hydraul. Eng., JSCE*, **7**, 23–34.

Tsujimoto, T. 1992. 'Fractional transport rate and fluvial sorting', in *Proc. Int. Grain Sorting Seminar*, Mitteilungen **117**, ETH Zürich, 227–249.

NOTATION

a_p	amplitude of the disturbance of p/p_0
a_y	amplitude of the disturbance of y_b/h_0
$a_p p_0$	amplitude of the disturbance of p
A B C D	constants
B_r	experimental constant
C' C'_k	empirical parameters
C_{e1} C_{e2} C_μ	empirical parameters
d_i	diameter of i-th fraction sand
d_m	mean diameter of the bed material in the surface layer
d_{m0}	cross-sectional average of d_m
d	diameter of fine sand
g	gravitational acceleration
h_0	undisturbed flow depth
k	turbulence energy
K	dimensionless wave number($\equiv 2\pi h_0/L_z$)
L_z	lateral wavelength of sorting stripe
N	number of grain size fractions
P_k	energy production
p_i	the volumetric ratio of i-th fraction sand to the total bed materials in the surface layer
p_{i0}	cross-sectional average of p_i
p_0	undisturbed volumetric ratio of the coarse sand
p_{si}	pick-up rate (probability density of a sand particle to be dislodged from the bed per unit time) of the i-th fraction sand
P P''	static pressure, pressure deviation
P_1 P_2	auxiliary parameters
q	constant
q_{Bi}	bedload transport rate of i-th sand fraction
q_{Bzi}	the lateral flux of bed load for i-th sand fraction
$r_d a_p p_0$	amplitude of the disturbance of d_m

$r_w r_d a_p p_0$	amplitude of the disturbance of $\tan\gamma$
t	time
t_*	dimensionless time($\equiv (1-p_0)(h_0/d)p_{s0}\lambda t$)
$U\ V\ W$	x, y and z components of time-averaged velocity
U_b	near-bed primary velocity
U_m	cross-sectional average of longitudinal mean velocity
U_p	longitudinal mean velocity near the bottom
u_*	shear velocity
$x\ y\ z$	longitudinal, vertical and lateral axes
y_b	bed elevation
y_p	y at the lowest grids in numerical calculation
z^*	z/h_0
α	$\alpha \equiv \sqrt{1/\mu_d}$
βd	diameter of coarse sand
γ	direction of flow near the bottom;
Γ	effective viscosity ($= \nu + \nu_t$)
ε	dissipation rate of turbulence energy
θ	longitudinal bed slope
θ_z	lateral bed slope
θ_e	thickness of the exchange layer
κ	von Kármán's constant
Λ_i	the mean step length of i-th sand fraction
μ_d	friction coefficient of sand
ν	kinematic viscosity
ν_t	kinematic eddy viscosity
ρ	mass density of fluid
ρ_0	porosity of sand
σ	mass density of sand
$\sigma_k\ \sigma_\varepsilon$	empirical parameters
τ	Reynolds shear stress, where pairs of the subscripts x, y and z denote the plane in which the force acts
τ_{*i}	dimensionless tractive force of i-th sand fraction
τ_{*ci}	dimensionless critical tractive force of i-th sand fraction
τ_{*cm}	dimensionless critical tractive force of sand with the diameter d_m in the mixture
τ_{*m}	dimensionless tractive force of sand with the diameter d_m in the mixture
ϕ_i	direction of bedload motion for i-th sand fraction
Ω	auxiliary parameter

20 Coherent Fluctuations in a Vegetated Zone of Open-channel Flow: Causes of Bedload Lateral Transport and Sorting

TETSURO TSUJIMOTO
Department of Civil Engineering, Kanazawa University, 2-40-20, Kodatsuno, Kanazawa, 920, Japan

ABSTRACT

Vegetated zones in rivers are often preserved for their environmental contribution to floodplain dynamics and their use as a control of flow, sediment transport and general fluvial processes. It is therefore critical to understand the relationship between flow and sediment transport in streams with vegetation.

The coexistence of slower flow in a vegetated zone and faster flow in the main neighbouring channel promotes lateral diffusion which causes the lateral flux of suspended sediment resulting in river morphological change. Lateral mixing is not only contributed to by simple turbulent diffusion but also by more organized behaviour associated with water-surface fluctuations. Such an organized motion is characterized by low frequency alternations of the lateral flow velocity component whose amplitude remains strong near the bed. Hence, it influences bedload motion and leads to alternation of the lateral flux of bedload transport. Since the bedload transport rate in the main channel is visually higher than that in the vegetated zone, the net flux of lateral bedload transport is towards the vegetated zone and the sediment deposit forms a longitudinal sand ridge. Such a process is observed in a laboratory flume, and is described herein by coupling a model for non-equilibrium fractional bedload transport with a model for the organized fluid motion.

INTRODUCTION

The role of vegetation in Japanese rivers with floodplains is being assessed as a positive measure to help preserve a diverse ecological habitat and control river morphology and channel change. It is therefore important to understand how flow and sediment transport processes are affected in streams with vegetation.

Figures 20.1 and 20.2 show the contrast between a coarse gravel/boulder bed stream bordered by *Salix* (willow) bushes and a sand-bed stream bordered by reeds. In the more tranquil sand-bed case, the flow near the vegetated zone is retarded and the lateral gradient of velocity induces lateral mixing. Sand is transported

Coherent Flow Structures in Open Channels.
Edited by P.J. Ashworth, S.J. Bennett, J.L. Best and S.J. McLelland.

predominantly as suspended load and the lateral flux of sediment is probably driven by turbulent mixing. In general, faster flow entrains more sediment, increasing its sediment load and thus causing a sediment flux towards the vegetated zone where sand deposition occurs. On the other hand, in a more turbulent and roughness-dominated stream as shown in Figure 20.1, bed materials are transported as bedload, and one cannot expect a lateral flux of sediment due to turbulent mixing. However, even in such a dynamic stream, sediment deposition occurs in vegetated zones and contributes to the morphology of the floodplain.

In this paper, the following hypothesis is proposed based on an experiment of flow and bedload transport in a flume with a zone of idealized vegetation. In the flow with a vegetated zone, an organized fluctuation will build up which is associated with water-surface elevation changes. This flow structure affects the direction of bedload motion because the fluctuation is of low-frequency.

The low-frequency organized motions associated with water-surface fluctuations were observed by Fukuoka & Fujita (1988) and Tsujimoto & Kitamura (1992). Chu *et al.* (1991) and Ikeda *et al.* (1992) explained the appearance of such organized motions as an instability of horizontal shear flow with an inflection point. Tsujimoto & Kitamura (1992) described the organized motion as an instability of an interrelating system consisting of the water-surface and lateral velocity fluctuation.

FLUME EXPERIMENTS ON FLOW WITH A VEGETATED ZONE

The hydraulic characteristic of a vegetation zone is represented by the parameter K_s which is obtained from Equation 20.1 using measurements of the characteristic velocity U_{s0} and the energy gradient S when the bed is totally covered by vegetation and the depth is smaller than the vegetation height:

$$K_s = U_{s0}\sqrt{S} \tag{20.1}$$

The vegetation is herein idealized by a group of cylinders with constant diameter R and spacing s, for which K_s is given as follows:

$$K_s = \sqrt{2gs^2/(C_D R)} \tag{20.2}$$

where C_D is the cylinder drag coefficient and g represents gravitational acceleration.

Figure 20.3 shows the configuration of the vegetation in the laboratory flume model. The laboratory flume was 0.4 m wide and 12 m long. The zone $z < 0$ is vegetated, while the zone $z > 0$ has a smooth bed. Three kinds of vegetation models were employed in the present laboratory tests. The K_s values of them were as follows: Case A, K_s = 2.95 m s^{-1}; Case B, K_s = 2.34 m s^{-1}; and Case C, K_s = 1.20 m s^{-1}. The experiments were conducted under a condition where the water depth was smaller than the vegetation height. The flow under such a

Figure 20.1. A gravel-bedded reach of the Tedori river (Japan) bordered by *Salix* bush

Figure 20.2. A sand-bedded reach of the Tedori river (Japan) bordered by reeds

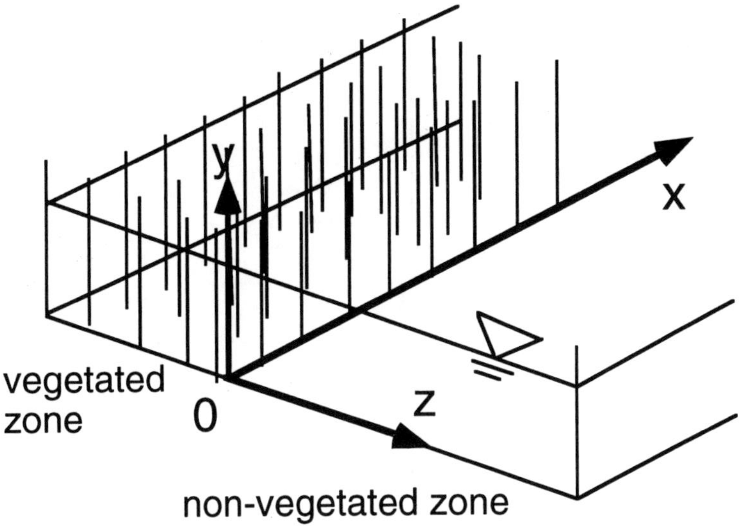

Figure 20.3. Schematic diagram of flume with model vegetation (bamboo rods) the value of K_s was changed by adjusting the rod spacing (s) and its diameter (R) (see Equation 20.2 in text)

condition appears to be two-dimensional.

This study deals with the situation where the flow in the reach undergoes minimal change in the downstream direction. Velocity measurements were obtained by a miniature propeller current meter (diameter 3 mm) and a small electromagnetic current meter (ECM). The probe of the ECM is cylindrical, 30 mm long and 8 mm in diameter. Its reliability for measurements of fluctuating velocities lower than 20 Hz was tested by comparing the data with those obtained with a hot-film anemometer. Fluctuations in the water-surface were measured by a capacity limnimeter.

Figure 20.4 shows the lateral distributions of the depth-averaged velocity ($\langle U \rangle$), the standard deviation of longitudinal and lateral velocities ($\langle u_{rms} \rangle$ and $\langle w_{rms} \rangle$) and the Reynolds stress in the lateral direction ($-\langle u'w' \rangle$). Near the boundary of the vegetation zone ($z \cong 0$), there are large gradients in lateral velocity, turbulence intensities, and the Reynolds stress. These characteristics can be partly explained by the kinematic eddy viscosity, although this is not fully evaluated here. The lateral flux of suspended sediment may be described in a similar manner.

Appreciable fluctuations appear in all components of velocity and water-surface elevation. They must be correlated with one another and the lateral mixing must be effected by such organized fluctuations.

Figure 20.5 shows the lateral variation in the standard deviation of water-surface fluctuation (d_{rms}). The maximum d_{rms} appears at the boundary of the vegetated zone ($z = 0$). With distance from the boundary, d_{rms} decreases rapidly in the vegetated

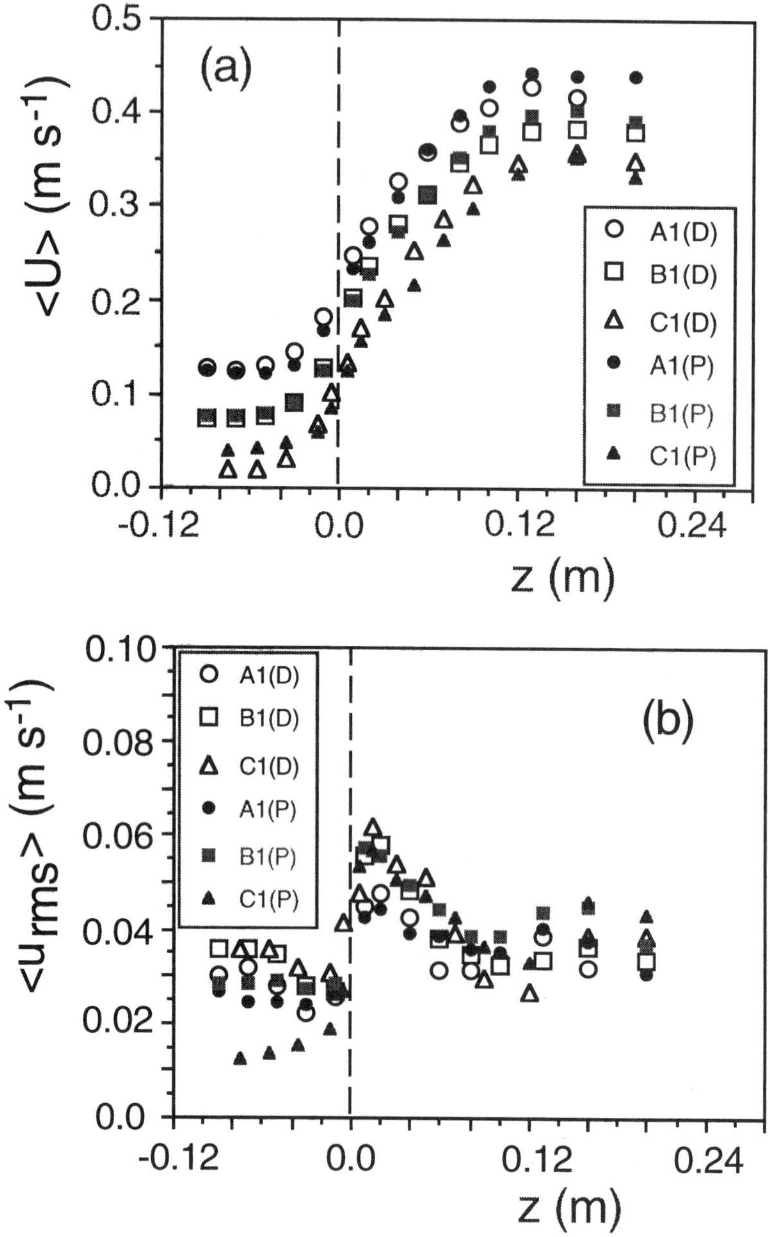

Figure 20.4. (a) Lateral distributions of depth-averaged primary velocity. (b) Standard deviation of downstream velocity fluctuations). *Continued on facing page*

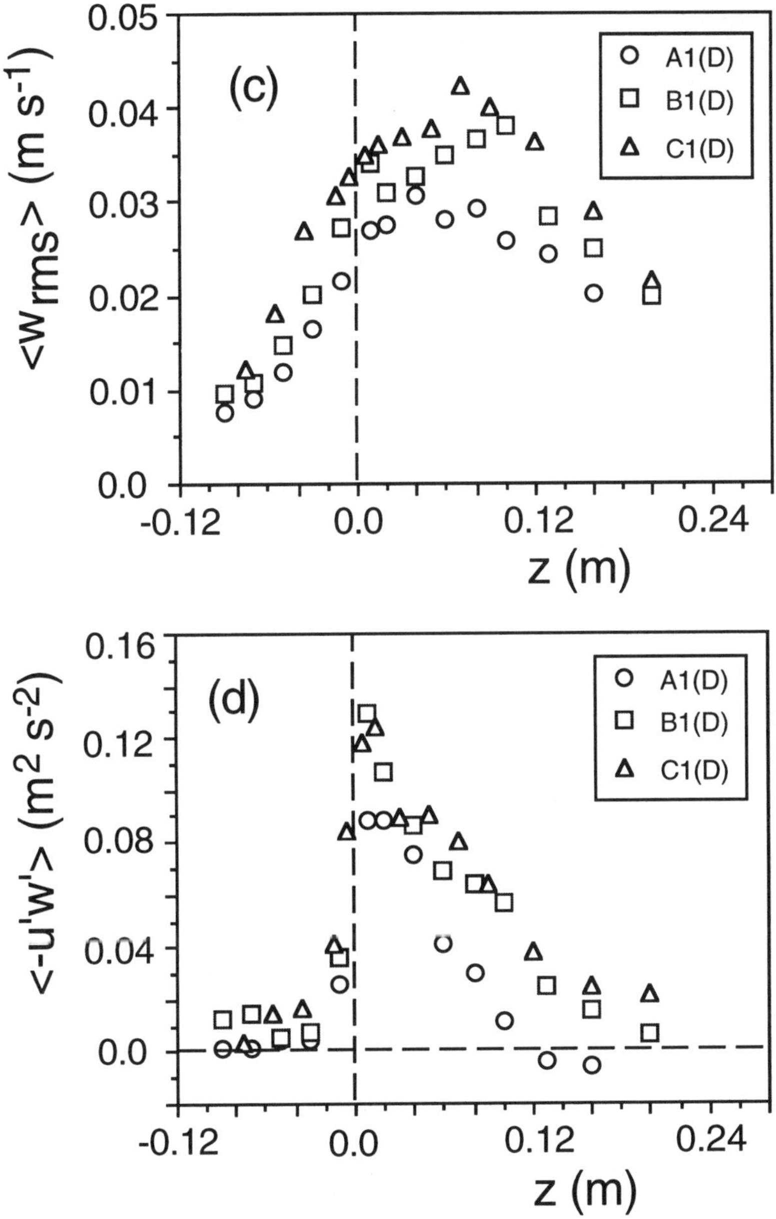

Figure 20.4. *continued* (c) standard deviations of cross-stream velocity fluctuations and (d) the Reynolds stress. The symbols D and P refer to depth-averaged velocity data obtained by ECM and propeller meter respectively. The vertical dashed lines represent the boundary between vegetated (negative z) and non-vegetated zones (positive z)

zone but only gradually toward the main channel. The simultaneous measurement of water-surface fluctuations at the different locations in the same cross-section imply that they are in phase as shown in Figure 20.6, where the water-surface fluctuations are normalized by their standard deviations. This fact implies an alternation of the water-surface lateral slope near the boundary, and it may be responsible for driving the alternating lateral motion of water.

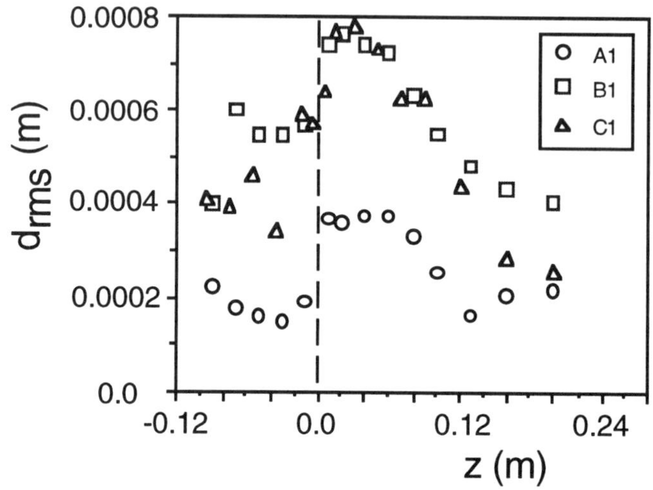

Figure 20.5. Lateral change in the standard deviation of water-surface fluctuations

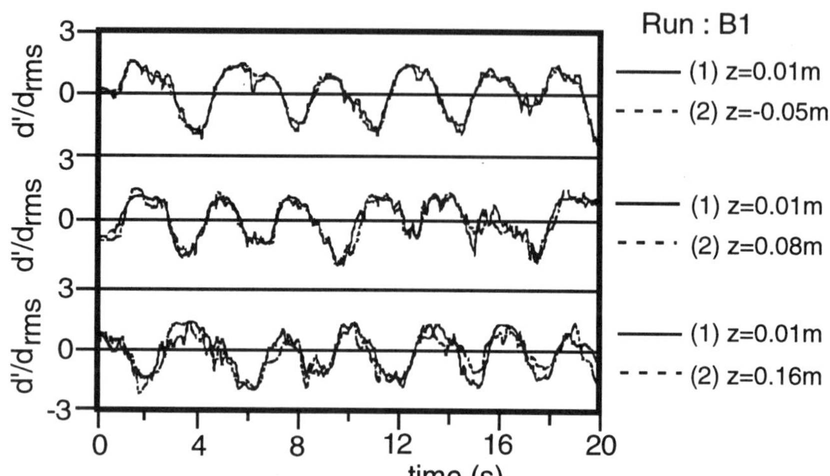

Figure 20.6. Time-series of water-surface fluctuations at several points in the same cross-section

Simultaneous measurement of the fluctuations of water-surface and longitudinal and lateral components of velocity were conducted using a limnimeter and an ECM, as shown in Figure 20.7. The time-series of the obtained fluctuations are shown in Figure 20.8 which clearly illustrate their close cross-correlation. An analysis of these correlations is shown in Figure 20.9 which reveals phase shifts among them. The results of the phase shifts are summarized in Figure 20.10. The phase shift between d and $\langle w \rangle$ is approximately $3\pi/2$, and that between $\langle w \rangle$ and $\langle u \rangle$ is approximately π near the boundary. The former suggests that the alternation of the lateral slope of water-surface causes the lateral motion of water, and the latter implies an efficient momentum exchange in the lateral direction.

Simultaneous measurements of the water-surface fluctuations were also conducted at two different points on a longitudinal axis. The cross-correlation analysis of the changes in water-surface level (Figure 20.11) illustrates that the fluctuation is

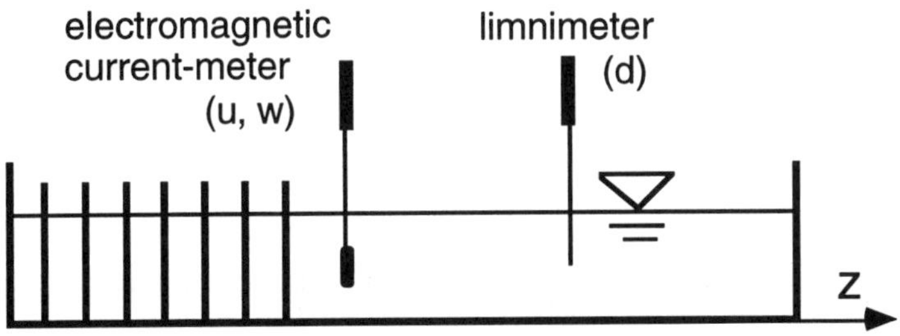

Figure 20.7. Experimental configuration of current meter and limnimeter

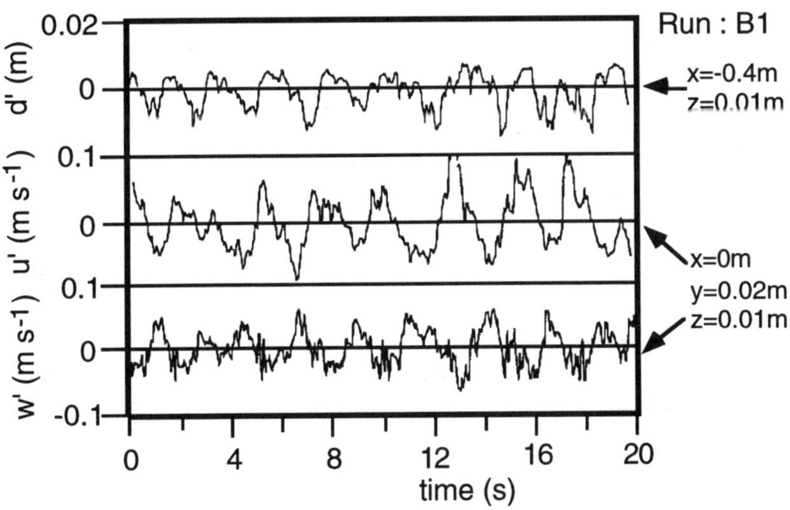

Figure 20.8. Time series of fluctuations of velocity and water-surface elevation

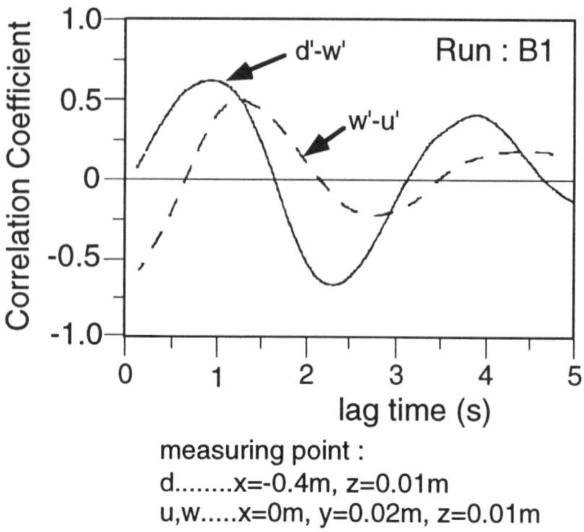

Figure 20.9. Cross-correlation coefficients of velocity fluctuations. Measurement points for d at x = -0.4 m and z = 0.01 m and for u and w at x = 0 m, y = 0.02 m and z = 0.01 m

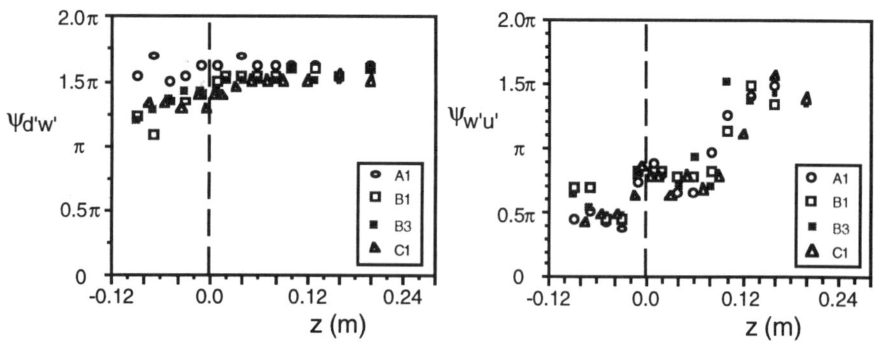

Figure 20.10. Phase shifts among the fluctuations of U and W velocity components and water-surface elevation

convective with the propagation celerity almost equal to the mean velocity near the boundary of the vegetated zone.

Furthermore, the simultaneous measurements of water-surface and local velocity clarify that the correlation characteristics among them, hardly vary in the vertical near the vegetated zone boundary (see Figure 20.12). This suggests that low-frequency organized fluctuations still exist near the bed.

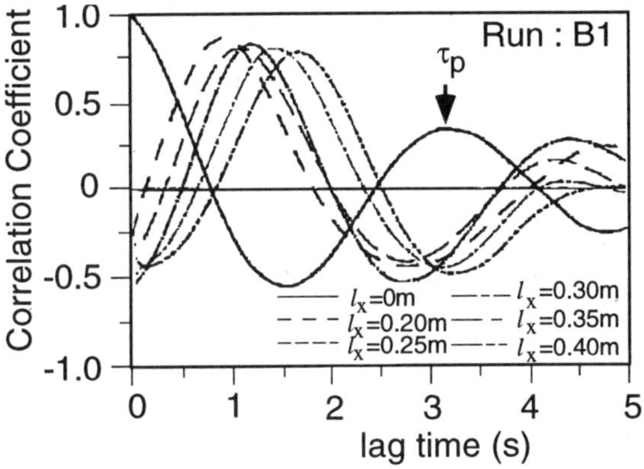

Figure 20.11. Cross-correlation of water-surface fluctuations at several points along the longitudinal axis near the vegetated zone. l_x represents the longitudinal distance between the two measuring points. τ_p is the time lag corresponding to the second peak of the correlation coefficient

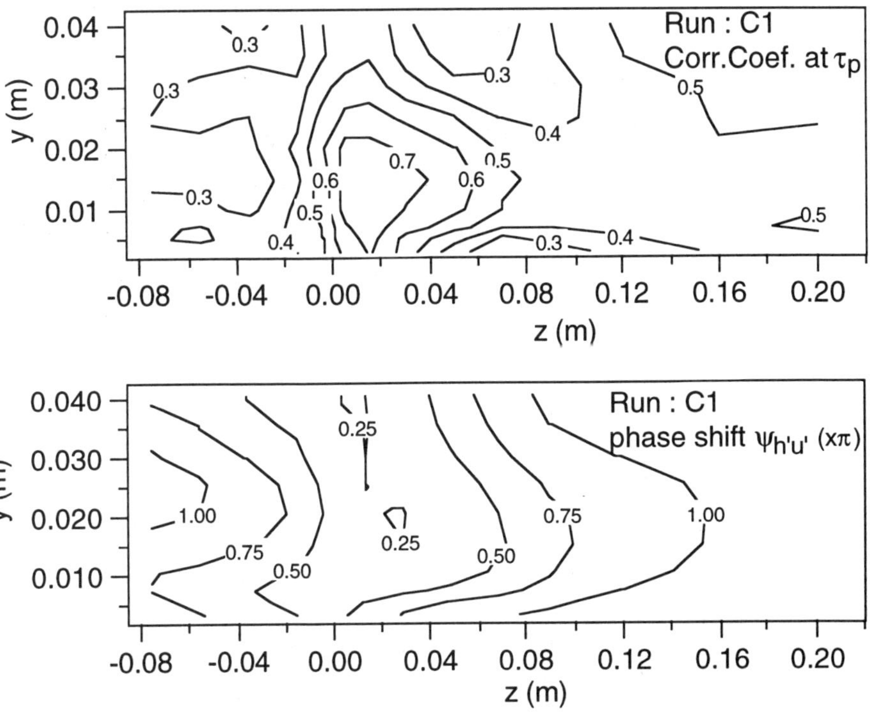

Figure 20.12. Coherence and in-phase cross correlation between local velocity and water-surface fluctuations

LATERAL BEDLOAD TRANSPORT AND SAND RIDGE FORMATION

IDEALIZED LOW-FREQUENCY FLUCTUATIONS AND LATERAL BEDLOAD TRANSPORT

In order to describe the bedload motion near the boundary of the vegetated zone, the organized fluctuations need to be simplified. The role of small-scale turbulence characterized by high-frequency fluctuations on the long-term time-series of velocity is schematically illustrated in Figure 20.13. Such small-scale turbulence is unlikely to have a major impact on bedload entrainment and transport direction. Hence, the time series of the lateral component of velocity, $\langle w \rangle$, is simulated as a rectangular wave by extracting low-frequency fluctuation as illustrated in Figure 20.13, or a periodic alternation of $\langle w_{rms} \rangle$ and $-\langle w_{rms} \rangle$. The period ΔT is large enough for bedload motion to respond to it. The lateral component of velocity affects the direction of bedload motion whose angle, ϕ, alternates as ϕ_1 and ϕ_2 according to the change in the sign of $\langle w \rangle$. The subscripts 1 and 2 imply the duration when $\langle w \rangle$ is positive and negative, respectively.

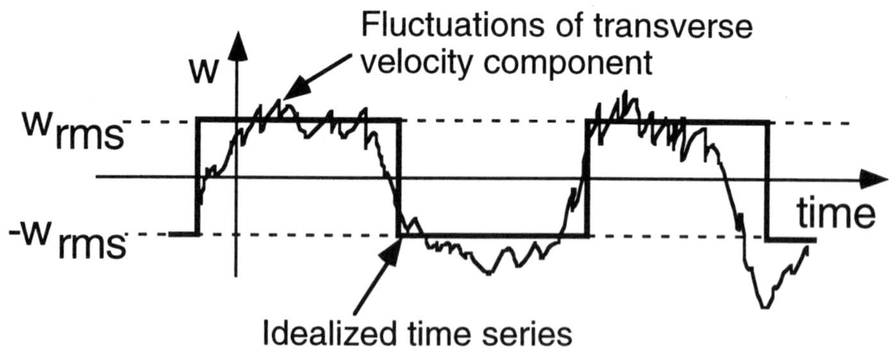

Figure 20.13. Lateral velocity fluctuation and its simplification into a series of idealized steps

The lateral change of the primary velocity suggests an increase in the sediment pick-up rate with z. The lateral flux of bedload at $z = 0$ in the duration $\langle w \rangle > 0$ is contributed by particles picked up in the region $z < 0$, while that in the duration $\langle w \rangle < 0$ by particles picked up in the region $z > 0$. It is obvious that the former is smaller than the latter, but such a fact cannot be described when one employs an equilibrium transport formula. Therefore, the non-equilibrium transport formula developed by Nakagawa & Tsujimoto (1980) is employed here along the trajectory of bedload motion, which is defined by the pick-up rate and the step length of Einstein's model (1942). If the lateral component of bedload transport is analysed, the bedload transported laterally during a half period of the fluctuation ΔT is written as follows:

$$q_{Bzj}(z)\frac{\Delta T}{2} = -(-1)^j \frac{A_3 D}{A_2}\int_0^\infty p_s(z+(-1)^j l_x)\exp\left[\frac{l_x}{\Lambda \sin\left|\phi_j(z)\right|}\right]dl_x \frac{\Delta T}{2} \qquad (20.3)$$

where A_2 and A_3 are geometrical coefficients of sand related to its projected area and the volume respectively, D is sand size, p_s is the sediment pick-up rate (the probability density per unit time for a bed material particle to be dislodged), j indicates the orientation of the lateral flow ($j = 1$: $w > 0$, $j = 2$: $w < 0$), and the step length is assumed to follow an exponential distribution, the mean of which is Λ. Because the pick-up rate at $z - l_x$ and that at $z + l_x$ are different from each other, $q_{Bz1}(z)$ and $q_{Bz2}(z)$ are not in balance. The net lateral transport rate of bedload, $q_{Bz}^{net}(z)$, is given as follows:

$$q_{Bz}^{net}(z) = \frac{A_3 D}{2A_2}\int_0^\infty \left\{ \begin{array}{l} p_s(z-l_x)\exp\left[\dfrac{l_x}{\Lambda \sin\left|\phi_1(z)\right|}\right] \\ -p_s(z+l_x)\exp\left[\dfrac{l_x}{\Lambda \sin\left|\phi_2(z)\right|}\right] \end{array} \right\} dl_x \qquad (20.4)$$

Since $p_s(z - l_x)$ is always smaller than $p_s(z + l_x)$ near the boundary of the vegetated zone, $q_{Bz}^{net}(z)$ is always negative in this zone, if the lateral bed-slope is neglected.

When sediment transport is in longitudinal equilibrium, bed deformation is governed by the net flux of the lateral bedload transport rate as follows:

$$\frac{\partial y_b}{\partial t} = -\frac{1}{1-\rho_0}\frac{\partial q_{Bz}^{net}}{\partial z} \qquad (20.5)$$

where y_b is the bed elevation and ρ_0 is the porosity of sand.

Since $q_{Bz}^{net}(z)$ is negative near the boundary of the vegetated zone, $\partial q_{Bz}^{net}/\partial z$ is negative and has a minimum near $z = 0$. This implies the formation of a sand ridge near the boundary of vegetation.

FORMULATION OF FLOW CHARACTERISTICS AND BEDLOAD TRANSPORT

Based on the flume experiments described earlier (see Figure 20.4), the following exponential-type distribution is adopted for $U(z)$, which is deduced from an analysis of horizontal shear flow with constant eddy viscosity.

$$\langle U_* \rangle = 1 + (\sqrt{\gamma} - 1)\exp\left(-\frac{z}{B_s}\right) \qquad (z \geq 0)$$

$$\langle U_* \rangle = \gamma + (\sqrt{\gamma} - \gamma)\exp\left(-\frac{z}{B_m}\right) \qquad (z \le 0) \tag{20.6}$$

$$\gamma \equiv \frac{U_{s0}}{\langle U_{m0} \rangle} = \sqrt{\frac{C_f}{C + \Omega}} \tag{20.7}$$

$$\frac{B_m}{d_0} = \sqrt{\frac{\varepsilon_*}{2}} C_f^{-1/4} \qquad \frac{B_s}{d_0} = \sqrt{\frac{\varepsilon_*}{2}} (C_f + \Omega)^{-1/4} \tag{20.8}$$

where $\langle U_* \rangle \equiv \langle U \rangle / \langle U_{m0} \rangle$, $\langle U_{m0} \rangle \equiv \left(\sqrt{1/C_f}\right) u_{*0}$ is the flow velocity of the zone unaffected by vegetation, C_f represents the resistance coefficient of a sand bed, $u_{*0} = \sqrt{g d_o S_e}$, d_0 is the mean flow depth, B_m and B_s are the effective width of the shear flow regions in the main channel and the vegetated zone, respectively, $\Omega \equiv C_D D_s d_0 / (2s^2)$, $\varepsilon_* \equiv \varepsilon / (u_{*0} d_0)$ where ε is the kinematic eddy viscosity, and the subscript 0 implies the values for flow unaffected by vegetation. If ε_* is specified, $\langle U(z) \rangle$ can be calculated. Although Ikeda *et al.* (1992) proposed a method to estimate ε_*, it is determined experimentally here for the following example of calculation of bedload lateral transport and sand ridge formation.

The lateral distribution of $\langle w_{rms} \rangle$ is approximated by the following according to the experimental results:

$$\langle w_{rms}(z) \rangle = \langle w_{rms\infty} \rangle + \left[\langle w_{rms0} \rangle - \langle w_{rms\infty} \rangle \right] \exp\left[-(z/B_m)^2 \right] \; (z \ge 0);$$

$$\langle w_{rms}(z) \rangle = \langle w_{rms}(0) \rangle \exp\left[-(z/B_s)^2 \right] \qquad (z \le 0) \tag{20.9}$$

where $\langle w_{rms\infty} \rangle$ is the value of $\langle w_{rms} \rangle$ in main channel (non-vegetated zone) unaffected by vegetation ($\langle w_{rms\infty} \rangle \cong \sqrt{C_f} \langle U_{m0} \rangle$). For the following example, the term $\langle w_{rms}(0) \rangle$ was determined experimentally.

Nakagawa & Tsujimoto (1980) proposed a formula to relate the sediment pick-up rate to the bed shear stress as follows:

$$\frac{p_s}{\sqrt{(\rho_s / \rho - 1) g / D}} = 0.03 \tau_* \left(1 - \frac{0.7 \tau_{*c}}{\tau_*} \right)^3 \tag{20.10}$$

where $\tau_* \equiv u_*^2 / \left[(\rho_s / \rho - 1) g / D \right]$; u_* is the shear velocity; ρ_s, and ρ are the mass densities of sand and water, respectively and τ_{*c} is the dimensionless critical tractive force. $u_*(z)$ is roughly estimated from $U(z)$ by assuming a constant value of C_f with respect to z. The mean step length is assumed to be a hundred times the sand diameter (Einstein, 1942): $\Lambda = 100d$.

The direction of bedload motion ϕ is governed by the flow direction β and the lateral bed-slope $\tan\theta \equiv \partial y_b / \partial z$, and the following equation (Nakagawa *et al.*, 1986) is employed:

$$\tan\phi \equiv \frac{\tan\theta}{\sqrt{\mu_d K_*}} + \tan\beta \tag{20.11}$$

where $K_* \equiv \mu_d \tau_* / \tau_{*c}$, μ_d is the friction coefficient of sand and $\tan\beta$ may be approximated as $\langle w \rangle / \langle U \rangle$. If the lateral bed-slope is neglected and $\langle w \rangle$ is regarded as repetitions of $\langle w_{rms} \rangle$ and $-\langle w_{rms} \rangle$, $\phi_j(z)$ is approximated as follows:

$$\phi_j(z) = \frac{(-1)^j \langle w_{rms}(z) \rangle}{\langle U(z) \rangle} \tag{20.12}$$

where $\langle U(z) \rangle$ and $\langle w_{rms}(z) \rangle$ are expressed by Equations 20.6 and 20.9, respectively.

Based on the models given above, the net flux of bedload lateral transport is calculated. The conditions for the calculation are as follows: the vegetation layer is characterized by the parameter Ω; the bed materials by d and ρ_s/ρ; and the hydraulic condition by the undisturbed Froude number, $Fr_0 \equiv U_{m0}/\sqrt{gd_0}$ and the relative roughness, D/d_0. When a flat bed is considered, the Manning-Stricklers equation is applied to estimate C_f from Fr_0 and D/d_0.

Figure 20.14 shows the calculated examples of net lateral transport rate of bedload with changing parameters, Fr_0, D/d_0 and Ω. At the position where q^{net}_{Bz*} becomes an inflection, $\partial q^{net}_{Bz}/\partial z$ reaches a minimum which corresponds to the peak of a sand ridge. Such an inflection point shifts toward the inside of the vegetated zone with an increase of the Froude number. The geometry of a sand ridge becomes less pronounced for a lower value of Ω (vegetation density).

MOBILE BED EXPERIMENTS OF SAND RIDGE FORMATION

Mobile bed experiments with uniform size sand were conducted in the laboratory flume shown in Figure 20.3, where $\Omega = 0.157$ ($R = 2.5$ mm, $s = 0.02$ m,

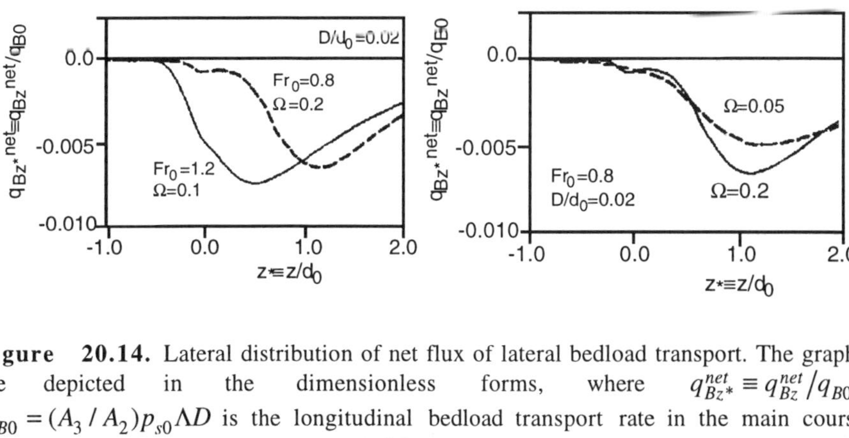

Figure 20.14. Lateral distribution of net flux of lateral bedload transport. The graphs are depicted in the dimensionless forms, where $q^{net}_{Bz*} \equiv q^{net}_{Bz}/q_{B0}$, $q_{B0} = (A_3/A_2)p_{s0}\Lambda D$ is the longitudinal bedload transport rate in the main course unaffected by the vegetation, and $z_* \equiv z/d_0$. See text for computational techniques

K_s = 1.6 m s^{-1}); Fr_0 = 1.44 and D/d_0 = 0.021 (D = 0.88 mm, d_0 = 0.041 m). The mobile bed was initially flattened, and the sediment measurement finished before bed undulation became appreciable.

A sediment trap consisting of several compartments (see Figure 20.15) was installed to measure the lateral flux of bedload for both directions separately by the compartments (a) and (b). The result of measured net lateral transport rate is plotted in Figure 20.16 together with the calculated results. The measured lateral bed profile, $y_b(z)/d_0$, is shown in Figure 20.17, where the calculated change of $\partial q_{Bz}^{net}/\partial z$ is also plotted, which is expected to be proportional to $(-\partial y_b/\partial t)$. The comparisons between the experiment and the calculation suggest that the present scenario of modelling provides a reasonable explanation of the phenomenon.

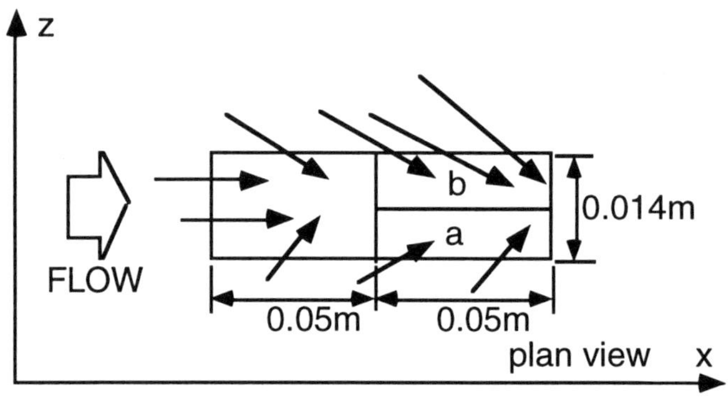

Figure 20.15. Plan view of the sand trap used to measure the lateral flux of bedload in the laboratory flume. Dimensions are not to scale. Arrows represent the direction of sand motion

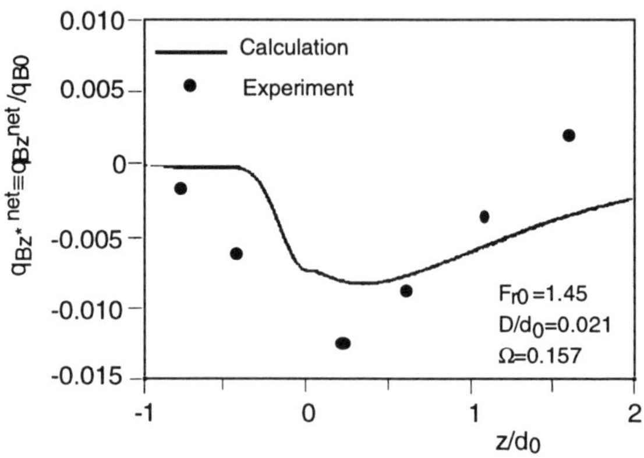

Figure 20.16. Comparison between measured and calculated net flux of bedload lateral transport

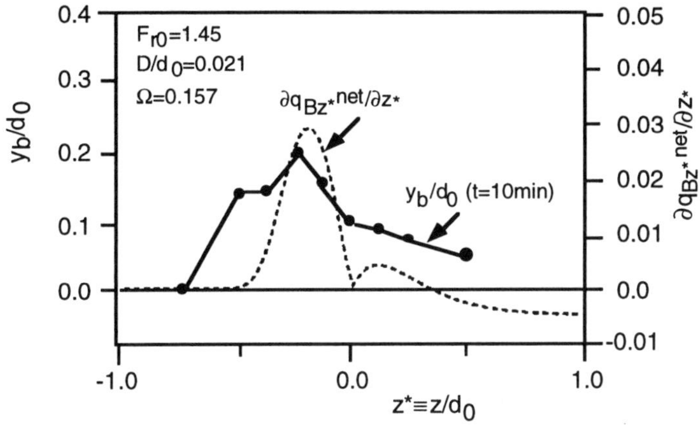

Figure 20.17. Comparison of lateral gradient of net lateral flux of bedload and geometry of sand ridge

SORTING OF HETEROGENEOUS BEDLOAD AND LONGITUDINAL STRIPE FORMATION

When a bed is composed of graded material, the selective fractional transport of different sediment sizes leads to sediment sorting near or inside the vegetation zone. In this chapter, a non-equilibrium transport model for each grain-size of graded bed material (Tsujimoto, 1991) is applied to explain the formation process of longitudinal sorting patterns near the vegetation zone.

FRACTIONAL TRANSPORT OF HETEROGENEOUS BED MATERIAL

If the volumetric ratio of the i-th grain-size class of sand due to grain-size (representative diameter, d_i) to the total volume of the surface layer of the bed is represented by p_i, the pick-up rate and the mean step length of this sand class represented by p_{si} and Λ_i, respectively. $p_i(t)$ represents the temporal change of bed-surface composition. Equation 20.4 is then modified for each grain-size of graded bed material as:

$$q_{Bzij}(z) = -(-1)^j \frac{A_3 D_i}{2A_2} \times \int_0^\infty \left\{ p_i(z+(-1)^j l_x) p_{si}(z+(-1)^j l_x) \exp\left[\frac{l_x}{\Lambda \sin|\phi_{ij}(z)|} \right] \right\} dl_x \tag{20.13}$$

where the subscript i indicates the class of sand according to grain-size (i = 1, 2, ... N where N is the number of classes), while j indicates the time duration when $\langle w' \rangle > 0$ ($j = 1$) or $\langle w' \rangle < 0$ ($j = 2$). ϕ_{ij} represents the bedload transport direction of the i-th sand class for each period ($\langle w' \rangle > 0$ or $\langle w' \rangle > 0$), and is identical with the flow direction so long as the bed is still flat. The net lateral transport for each grain-size is given as:

$$q_{Bzi}^{net}(z) = \frac{q_{Bzi1}(z) + q_{Bzi2}(z)}{2} \tag{20.14}$$

The apparent volume of the i-th sand class depositing during Δt due to lateral net flux of bedload is written per unit longitudinal distance as:

$$\Delta q_i = -\frac{1}{1-\rho_0}\frac{\partial q_{Bzi}^{net}}{\partial z}\Delta z \Delta t \tag{20.15}$$

where $\Delta t >> \Delta T$ and $\Delta q_i > 0$ results in deposition. If the thickness of the so-called exchange layer of mobile bed, E, is assumed to be constant and is of the order of the maximum size in the bed material, the change of bed elevation is given by applying the mass-continuity as:

$$\frac{\partial y_b}{\partial t} = \sum_{k=1}^{N} \frac{\Delta q_k}{\Delta z \Delta t} \tag{20.16}$$

The mixing process of eroded and/or depositing sediment with bed material is illustrated in Figure 20.18 for a simplified mixture composed of two fractions, where p_{i0} is the volumetric ratio of the i-th sand class in the total volume of the substratum (the initial surface layer). By referring to this figure, the bed-surface composition after Δt is expressed as:

$$p_i(t+\Delta t) = \frac{p_i(t)\left[E\Delta z - \sum_{k=1}^{N} \Delta qk(t)\right] + \Delta q_i(t)}{E\Delta z} \quad \text{(when } \partial y_b / \partial t < 0\text{);}$$

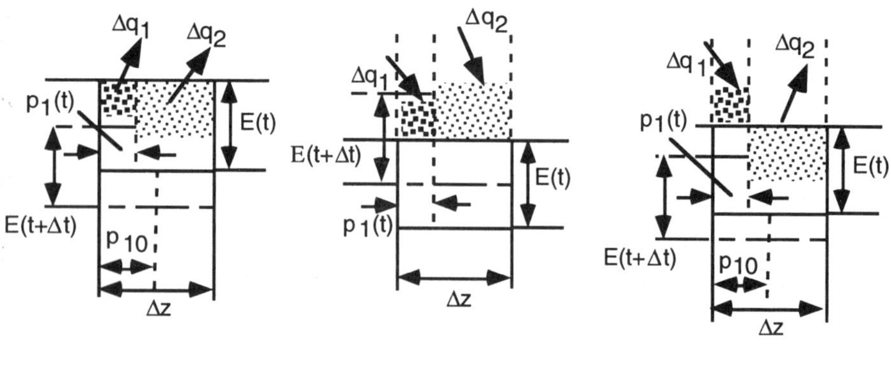

Figure 20.18. Exchange layer model

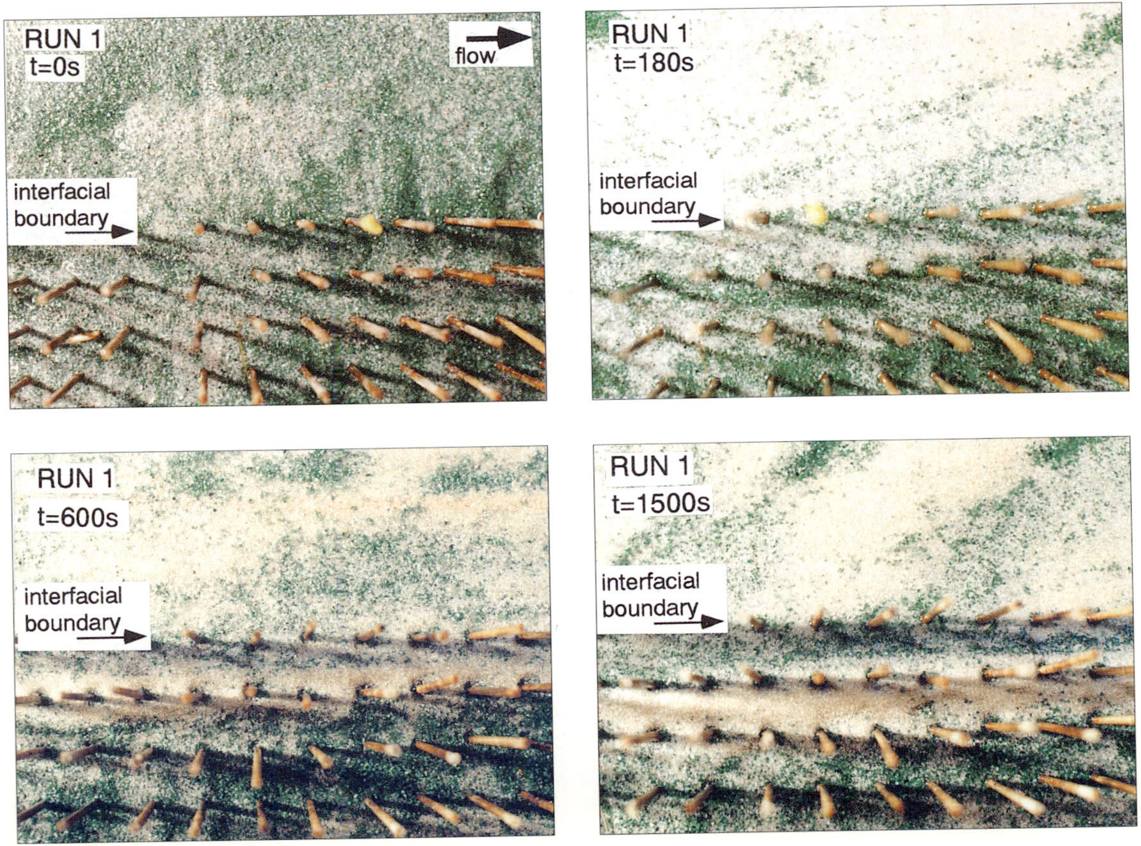

Figure 20.19. Sorting of sediment stripes in the mobile bed flume experiments. The rods are 2.5 mm in diameter

$$p_i(t+\Delta t) = \frac{p_i(t)E\Delta z - p_{i0}\sum_{k=1}^{N}\Delta q_k(t) + \Delta q_i(t)}{E\Delta z} \qquad (\text{when } \partial y_b / \partial t > 0) \quad (20.17)$$

A numerical calculation, based on the above, should select an appropriate Δt such that $|\Sigma\,\Delta q_k| < E\Delta z$, and $|\Delta q_i| < E\Delta z$ assuming that $\Delta q_i = -p_i(t)E\Delta z$ when $\Delta q_i < -p_i(t)E\Delta z$.

The pick-up rate for each grain-size, p_{si}, is found by modifying Equation 20.10 to give:

$$\frac{p_{si}}{\sqrt{(\rho_s/\rho - 1)g/D_i}} = 0.03\tau_{*i}\left(1 - \frac{0.7\tau_{*ci}}{\tau_{*i}}\right)^3 \qquad (20.18)$$

where $\tau_{*i} \equiv u_*^2/\{(\rho_s/\rho - 1)gD_i\}$, τ_{*ci} is the dimensionless critical tractive force for each grain-size and where the critical tractive force of sand with D_m $(=\Sigma[D_i p_i])$ in the mixture is represented by τ_{*cm},

$$\frac{\tau_{*ci}}{\tau_{*cm}} = \left(\frac{D_i}{D_m}\right)^b \qquad (20.19)$$

When $b \neq -1$, the difference between the critical shear stress for different particle sizes causes sorting and armour layer formation (Andrews & Parker, 1987). However, even with $b = -1$ when there is no difference between the critical tractive force for each grain-size, the difference between non-equilibrium transport rate for each grain-size brings about sorting such as diffuse gravel sheets and laterally alternate sorting (Tsujimoto, 1991). This is the case when a part of bed material moves $-1 < b < 0$ or when all fractions of bed material move i.e. $b = -1$. In the case of uniform size sand τ_{*cm} is assumed to be identical.

MOBILE BED EXPERIMENT: STRIPE FORMATION AND SORTING

Mobile bed experiments using a mixture composed of two grain-size distributions of coarser sand ($D_1 = 0.88$ mm) and finer sand ($D_2 = 0.34$ mm) with $p_{10} = p_{20} = 0.5$, were undertaken in the laboratory flume. The vegetation zone model was the same as used in the experiments with uniform sand. Quasi-uniform flows were imposed in the flume with a bed-slope of 6.67×10^{-3} and 3.33×10^{-3}. The coarser sand was coloured by bright green ink and the prepared sand mixture was put in the flume and screeded flat. During the experiment, a pronounced stripe of fine-grained sand was deposited inside the vegetated zone near the interface to the main talweg, as shown in Figure 20.19. Figure 20.20 illustrates the temporal change in the cross-stream bed-profile and the location of the finer-grained sand stripe. These figures suggest that sorting accompanies the formation of a longitudinal ridge, but the ridge is less developed compared with the case of uniform-size sand.

Under the conditions for the above experiments, numerical simulations were also conducted. Figure 20.21 shows the change of net lateral transport rate for each sand at $t = 0$, which has a maximum at a different location for each sand. This explains the appearance of sorting process. The bed-surface composition varies, and thus the spatial variation of the net transport rate varies with time. The sediment transport process was calculated with the present model by assuming the effect of the change of bed elevation on the flow is minimal. The calculation shows the temporal

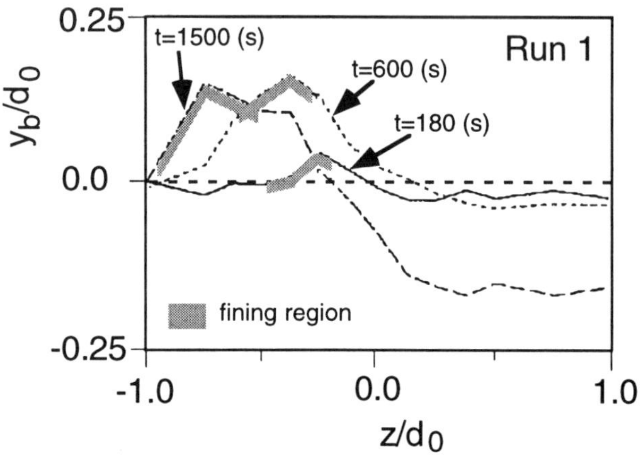

Figure 20.20. Temporal changes of the bed profile and location of the finer-grain stripe from the mobile bed flume experiments

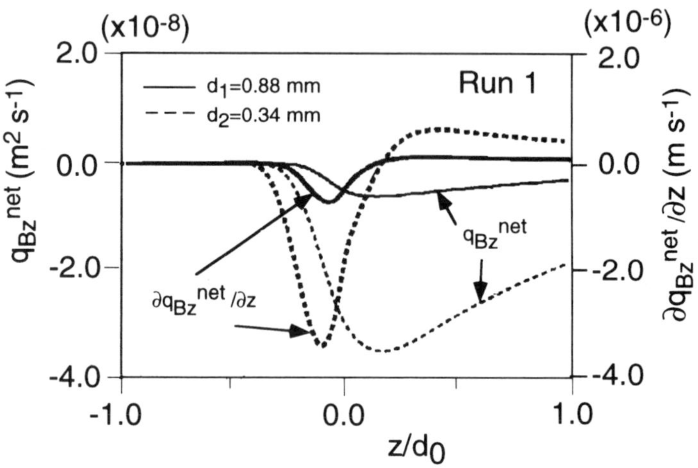

Figure 20.21. Changes in net lateral flux for each grain-size and their lateral gradient

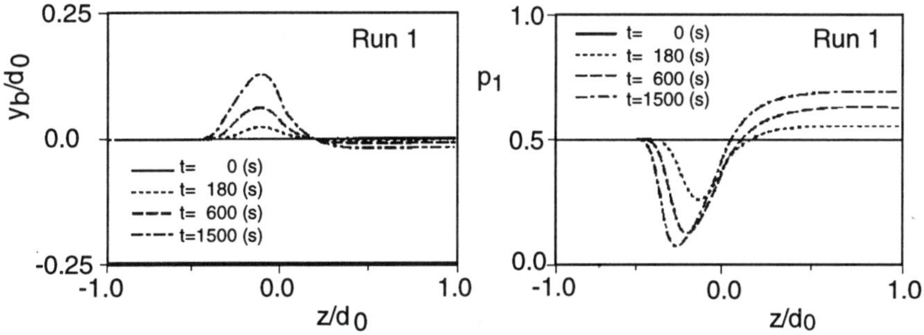

Figure 20.22. Temporal evolution of a sand ridge during sediment sorting

evolution of sand ridge formation and the temporal decrease in the coarser-grain component of the surface layer near the crest of the ridge (Figure 20.22). The location covered with the finest-grain mixture is not the same as the crest of the ridge. The model also describes the temporal shift of the stripes towards the inside of the vegetation zone.

CONCLUSION

The near-boundary flow of the vegetated zone is characterized by organized fluctuations of the water-surface and the longitudinal and lateral components of velocity. Such fluctuations contribute to the lateral exchanges of momentum and water mass. This is supported by measurements in a laboratory flume with simulated vegetation. The frequency of the water-surface and velocity fluctuations are so low and the lateral velocity fluctuation is so intensive that bedload motion is significantly affected. The observation of a net flux of bedload in the lateral direction can be predicted by applying a non-equilibrium transport model which also explains longitudinal sand ridge formation. In the case of heterogeneous bed material, sediment sorting accompanies sand ridge formation and a longitudinal stripe of sorted sediment will result. Such a process may be explained by applying non-equilibrium transport model for each grain-size.

REFERENCES

Andrews, E.D. & Parker, G. 1987 'Formation of a coarse surface layer as the response to gravel mobility', in Thorne, C.R. Bathurst, J.C. & Hey, R.D (Eds) *Sediment Transport in Gravel-bed Rivers*, John Wiley & Sons Ltd., Chichester, 269–300.

Chu, V.H., Wu, J.H. & Khayata, R.E. 1991. 'Stability of lateral shear flows in shallow open channel', *J. Hydraul. Eng.*, **117**, 1370–1388.

Einstein, H.A. 1942. 'Formulas for the transportation of bed load', *Trans. ASCE*, **117**, 561–597.

Fukuoka, S. & Fujita, K. 1988. 'Hydraulic effects of luxuriant vegetation on flood flow', *Rep. Public Works Res. Inst.*, Ministry of Construction, Japan, **180**, 129–192, (in Japanese).

Ikeda, S., Ohta, K. & Hasegawa, H. 1992. 'Periodic vortices at the boundary of vegetated area along river bank', *Proc. JSCE*, **443**, 47-54.

Nakagawa, H. & Tsujimoto, T. 1980. 'Sand bed instability due to bed load motion', *J. Hydraul. Div.*, ASCE, **106**, 2029–2051.

Nakagawa, H., Tsujimoto, T. & Murakami, S. 1986. 'Non-equilibrium bed load along side bank', *Proc. 3rd Int. Symp. River Sedimentation*, Jackson, Mississippi, USA, 1029–1065.

Tsujimoto, T. 1992. 'Fractional transport rate and fluvial sorting', in *Proc. Grain Sorting Seminar*, Mitteilungen **117**, ETH Zürich, 227–249.

Tsujimoto, T. & Kitamura, T. 1992. 'Experimental study on open-channel flow with vegetated zone along side wall - Correlative structure of fluctuations of velocity and free surface', *KHL Commun.*, Kanazawa Univ., **3**, 21–35.

Tsujimoto, T. & Kitamura, T. 1992b, 'Appearance of organized fluctuations in open-channel flow with vegetated zone', *KHL Commun.*, Kanazawa Univ., **3**, 37–45.

NOTATION

A_2 A_3 two- and three-dimensional geometrical coefficients for sand
b constant
B_m B_s effective width of shear flow regions in main course and that in vegetation zone
C_D drag coefficient of cylinder
C_f resistance coefficient of main channel
d_0 mean water depth
$d'(t)$ time series of the water-surface elevation
d_{rms} standard deviation of water-surface elevation
D diameter of sand
D_m mean diameter of fraction sand
E thickness of exchange layer
Fr_0 undisturbed Froude number
g gravitational acceleration
K_s proportionality constant
K_* dimensionless parameter
l_x longitudinal distance of pair of measuring points
p volumetric ratio of sand fraction to the total volume of exchange layer
p_s pick up rate
q_{B0} longitudinal bedload transport rate in the main channel unaffected by the vegetation zone
q_{Bz} lateral component of non-equilibrium bedload transport rate
q_{Bz}^{net} net lateral transport rate of bedload
q_{Bz*}^{net} dimensionless net lateral transport rate of bedload

R	diameter of cylinders
s	interval between the centre of cylinders
S	energy gradient
t	time
T	period of low frequency fluctuation
u_{rms}	standard deviation of longitudinal component of velocity
u_*	shear velocity
u_{*0}	shear velocity in main course unaffected by vegetation zone
$u'(t)$	longitudinal velocity fluctuation
U	mean primary velocity
U_{m0}	mean primary velocity in main channel unaffected by vegetation zone
U_{s0}	characteristic velocity in vegetation zone
U_*	dimensionless mean primary velocity
$-u'w'$	Reynolds stress in lateral direction
$w'(t)$	lateral velocity fluctuation
w_{rms}	standard deviation of longitudinal component of velocity
$w_{rms\infty}$	the value of $w_{\rm rms}$ in main channel unaffected by vegetation zone
x	longitudinal co-ordinates
y	vertical co-ordinates
y_b	bed elevation
z	lateral co-ordinates
z_*	z/h_0
β	flow direction
γ	ratio between characteristic velocity in vegetation zone and mean primary velocity in main channel unaffected by vegetation zone ($=U_{s0}/\langle U_{m0}\rangle$)
ε	kinematic eddy viscosity
ε_*	dimensionless kinematic eddy viscosity
θ	lateral bed-slope
Λ	mean step length of bedload motion
μ_d	friction coefficient of sand
ρ	mass density of water
ρ_s	mass density of sand
ρ_0	porosity of sand
$\tau_{\rm p}$	temporal shift of peak of cross-correlation coefficient
τ_*	dimensionless tractive force
τ_{*c}	dimensionless critical tractive force
τ_{*0}	dimensionless tractive force in main course unaffected by vegetation zone
τ_{*cm}	dimensionless critical tractive force of sand with $d_{\rm m}$ in mixture
ϕ	deflection angle of trajectory of bedload particle against x axis
ψ_{hw}	phase shift between $w'(t)$ and $h'(t)$
ψ_{wu}	phase shift between $u'(t)$ and $w'(t)$
ψ_{hu}	phase shift between $u'(t)$ and $h'(t)$
Ω	dimensionless vegetation density

$\langle\rangle$ depth-averaged value

Subscript

i value for i-th class sand
j value when $\langle w' \rangle > 0$: (j=1) or $\langle w' \rangle < 0$: (j=2)

21 The Structure of River Bed Relief

A. SIDORCHUK
Laboratory of Soil Erosion and Fluvial Processes, Geographical Faculty, Moscow State University, GSP-3, Moscow 119899, Russia

ABSTRACT

In a river flow the hierarchical system of macroturbulent vortex-like structures can cover a wide range of scales from the order of channel depth to meander length. These structures are formed due to the interaction between flow and the channel bed. The river bed relief, which is also in a hierarchical system ranging from ripples, dunes to bars, is a product of the 'imprint' of the macroturbulence on the movable channel bed. Morphological changes of the channel relief generate a transformation of the flow structure, which results in channel deformation.

The quantitative basis of the description of stream flow – channel bed interaction is the analysis of the initial instability of wave-like structures of the flow and channel bed relief. Three-dimensional equations of momentum and conservation in curvilinear orthogonal co-ordinates with depth-averaging following Kartvelishvily (1973) are used for this analysis. This approach permits the inclusion of helical flow effects and the vertical distribution of dynamic pressure. The equations can be linearized for perturbations with minimal loss of terms. A standard procedure of stability analysis of these equations with non-trivial boundary conditions leads to solutions which predict the continuous (both in longitudinal and lateral directions) spectrum of channel bed waves which are unstable in time.

The topography of this continuous spectrum is complicated. Five main types of the unstable bed undulations are defined: two- and three-dimensional ultramicroforms; symmetrical microforms; elongated mesoforms; elongated macroforms; and long and narrow megaforms. These unstable waves have their analogues in the relief of natural river channels. Megaripples correspond to ultramicroforms, dunes of first and second orders correlate with microforms, mesoforms are transitional between dunes and bars and the macroforms correspond to bars, islands and channel meanders. Parallel braided channels are the natural analogues for megaforms.

The main difference between the theoretical and empirical spectra of river channel forms is related to the non-linearity of the channel – flow interaction in a natural river. One such non-linear processes is the secondary flow circulation which leads to the transformation of the bed elevations due to channel meander formation.

INTRODUCTION

The basic principles of structural analysis of channel bed relief were formulated as a theoretical generalization of the empirical knowledge of river channel management. There are four main principles:

1. The principle of interaction between the flow and the movable channel bed

Coherent Flow Structures in Open Channels.
Edited by P.J. Ashworth, S.J. Bennett, J.L. Best and S.J. McLelland.

(Velikanov, 1954). According to Makkaveev (1955), this interaction takes the form of flow structures producing an 'imprint' on the movable bed, with associated feedback.

2. The principle of the discreteness of the fluvial processes (Kondrat'ev, 1953), which determines the existence of individual forms of river bed relief in the channel and of macroturbulent structures in the flow.
3. The principle of continuity of fluvial processes. This determines the continuity (at the macrolevel) of the water flow and that of the surface of the channel bed, assuming close interconnections and interrelations of the individual fluvial forms and their groups.
4. The principle of the structural organization of fluvial process (Kondrat'ev, 1953). This expresses the combination of fluvial forms into different organizational groups (different structural levels). Each level of organization is characterized by a specific morphologic and dynamic type of fluvial form.

The quantitative, and even qualitative description of these main principles in terms of bed relief structure, morphology and dynamics, still represents an unsolved problem. One of the most important problems is the understanding of the genesis of river bed relief and its complex structure. The interaction between the flow and the movable bed is a fundamental process that drives the self-organizing dynamic 'stream flow-channel bed' system. The complicated quasi-periodic changes of the flow velocity field result in regular changes in the rate of sediment transport (or of other material, for example, soluble matter) and leads to the creation of fluvial bedforms. These forms change the depth and resistance distribution in the channel and positively feedback to effect the flow field. For the case of long-term stability in averaged values of the main factors, this interaction produces a dynamic equilibrium of the 'stream flow-channel bed' system, and results in predictable morphological and dynamic patterns of the river bed relief.

INTERACTION BETWEEN THE STREAM FLOW AND CHANNEL BED

STREAM FLOW STRUCTURE

It is important to study stream flow-channel bed interaction in the whole range of flow structure and bedform sizes. For the analysis of large flow structures, velocity measurements over periods of several hours are needed. Following this guideline, measurements of longitudinal velocities (Sidorchuk, 1992) were carried out in the lower Terek River (Southern Russia) using a pygmy propeller-type current meter during 16 hours with a 600 s step, during 2 hours with a 10 s step and during 10 minutes with a 0.4 s step. The energy spectrum was calculated separately for each realization, and then the parts of the spectrum linked (Figure 21.1).

The energy spectrum is of the 'red noise' type. In general, the spectral density

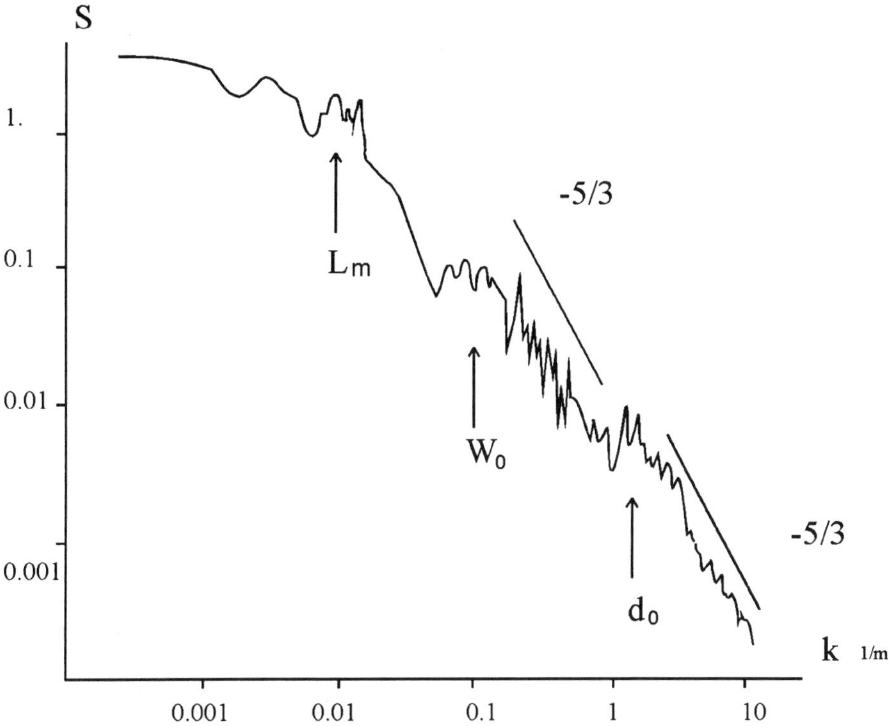

Figure 21.1. Macroturbulence spectrum of the Terek River flow, Grebenskaya station, August 1981

decreases with frequency increase. For the high frequency part of the spectrum this relation is typically described by the '-5/3' law, commonly found in high energy flows (Grinvald, 1974), with the local maximum of spectral energy in the part of spectrum with wavelength of the order $\sim d_0$. The part of the spectrum with the maximum spectral energy is usually interpreted as the region of energy inflow to the turbulent structures of boundary layer thickness size, and the portion with the '-5/3' law is interpreted as an inertial interval with energy transmission without dissipation (Monin & Yaglom, 1967). However, in the functions of spectral density for the Terek River the same two parts of the spectrum exist in the low frequency range. The local maxima are related to wavelengths of the order of width W_0 and of the meander length L_m,with corresponding inertial intervals. The same type of energy spectrum has been described for the lower Dnestr channel flow (Grinvald & Nikora, 1988) and for oceanic turbulence (Monin & Ozmidov, 1981). In these studies the interpretation of the low frequency region was the same as for the high frequency region of the spectrum. Following this interpretation, three main types of turbulent structures exist in the river flow: (i) three-dimensional structures with longitudinal dimension of the order of river depth; (ii) essentially two-dimensional structures with longitudinal dimension of the order of river width; and (iii) essentially one-dimensional structures with longitudinal dimension of the order of meander length.

There is one more type of flow structure in the stream flow – the quasi-periodic longitudinal and lateral change of the depth-averaged velocities above the waves of the bed relief. This is shown by the data from the Niger River (Figures 21.2 and 21.3). In the case of the dune field at the channel bed the mean velocity maximum is situated above the dune top and the minimum above the trough between dunes (Figure 21.2). In the case of large forms, such as bars, the maximum of mean velocity is situated above the trough between bars and the minimum above the bar top (Figure 21.3). This quasi-periodic change of the velocities only partly reflects the change of the channel depth and resistance. For the dune case the influence of the depth change on the velocity field can be estimated using the equation of water continuity and for the bar case using the Chézy-Manning formula. It is then found that the unperturbed velocity field is also quasi-periodic. In the case of dunes, the unperturbed velocity maximum moves to the dune trough and the minimum to the dune top (Figure 21.2). For the bar case, the positions of maximum and minimum mean velocity do not change, but the variability of velocity increases (Figure 21.3).

The velocity field associated with stationary flow structures on the bar scale can be calculated from channel morphology and discharge: $u = Q/(W_0 d_0)$. The velocity field of the meandering Pinega river (northern Russia) calculated from its morphology shows a stochastic nature, with a 'red-noise' type of energy spectrum (Figure 21.4). The spectrum consists of the same two main parts, i.e. the part with spectral density increase (energy inflow), and the part with spectral density decrease. The law of '-5/3' is evident for the last part of the Pinega River spectrum. The similarity of the spectra for the Terek and Pinega rivers, calculated with the use of quite different data, shows the high level of coherence between bedform relief and flow structures, either rapidly moving or stationary.

CHANNEL RELIEF STRUCTURE

The structure of channel relief can be also described in terms of the wave-number spectrum of the continuous parameters of channel morphology: bottom elevations, channel width, channel direction and their derivatives. The method of analysis of these data is the same as for time series of flow velocities. The spectrum over a wide range of wave-numbers is combined from 2–3 components, calculated from measurements with different steps for river reaches of various length. The spectra of the bottom elevations in the Amazon, Enisey, Lena and Niger rivers (Figure 21.5) were calculated from 100–500 km long profiles with a 100–500 m step-length (measured on the river maps), 3–5 km long profiles with a 5–10 m step-length, and 250–500 m long profiles with a 0.5 m step-length (measured from bathymetric profiles).

The spectra of the bed elevations of the river channels are continuous, i.e. a wide

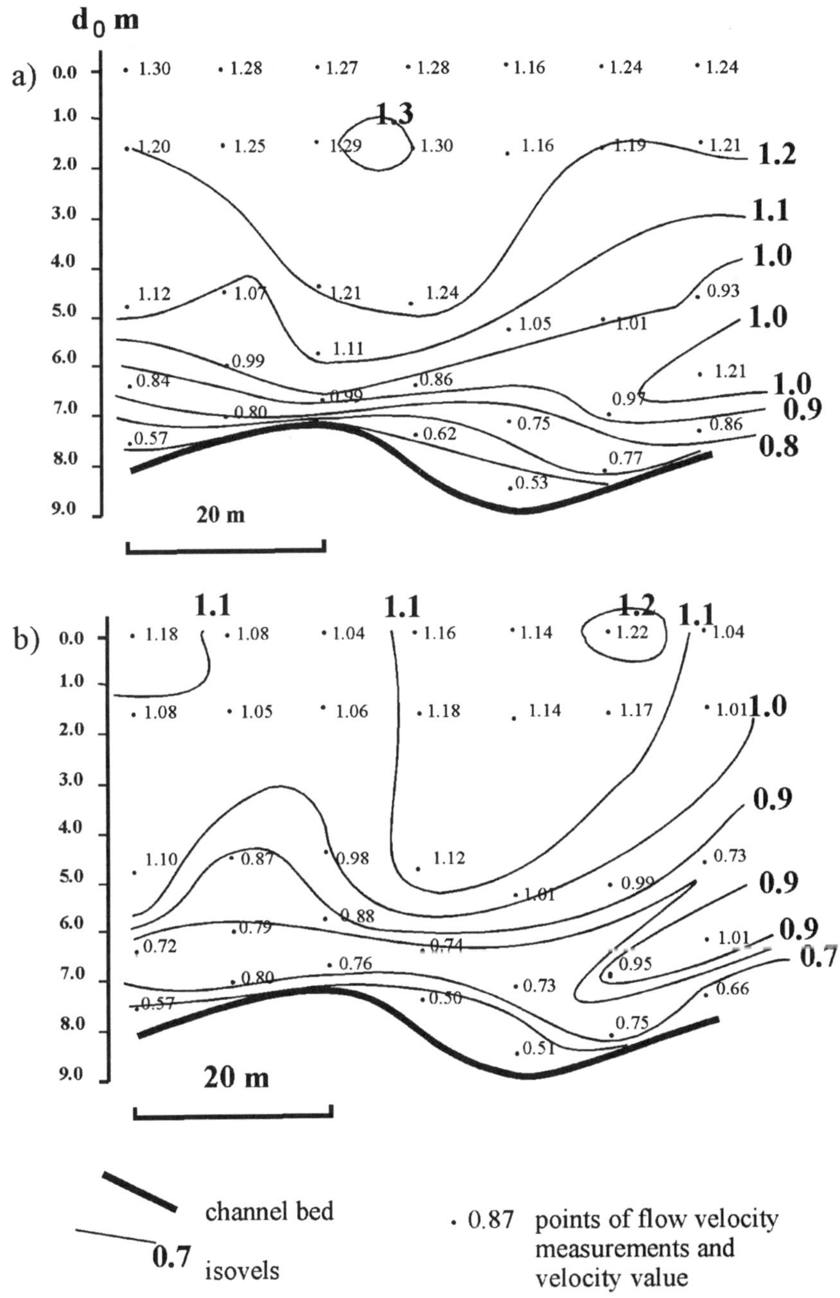

Figure 21.2. The flow field above dunes in the Niger River channel near Adjaokuta during the flood of 1978: (a) observed velocities; and (b) same without the influence of bed relief. All measurements are in m s^{-1}

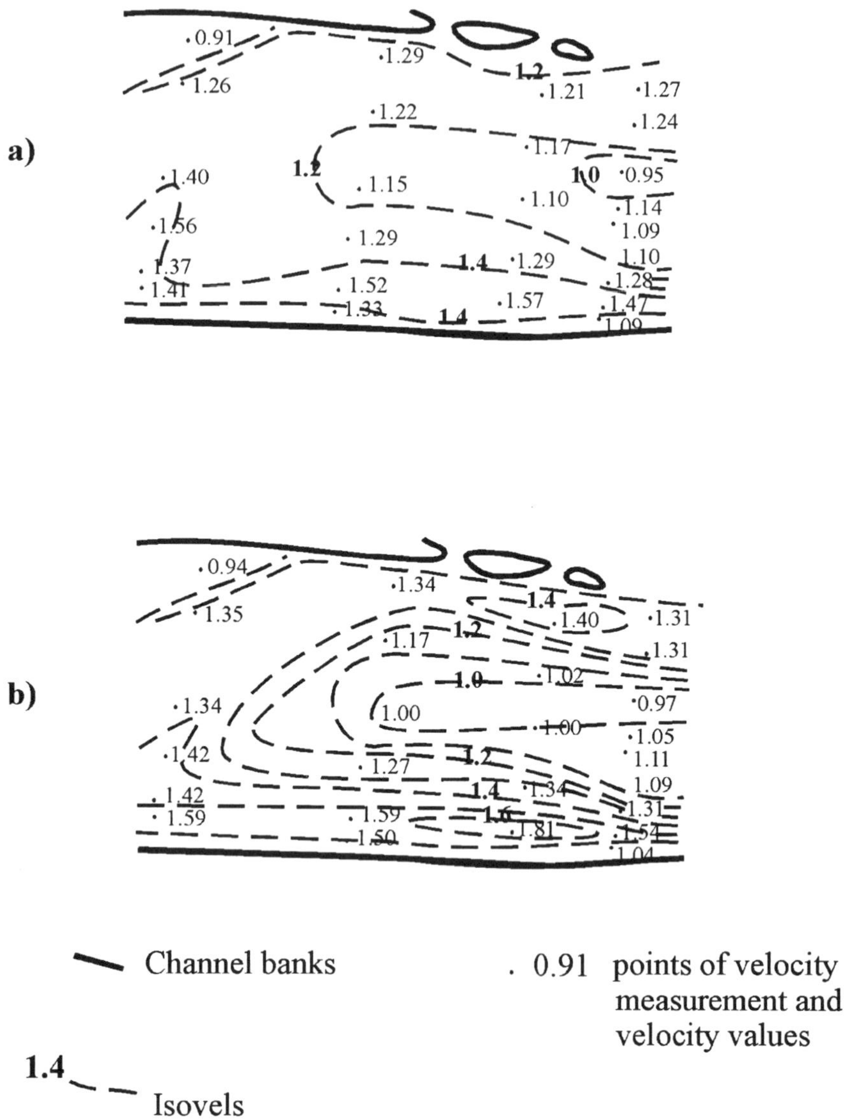

Figure 21.3. The flow field above a bar in the Niger River channel near Adjaokuta during the flood of 1978: (a) observed velocities; and (b) same without the influence of bed relief. All measurements are in m s^{-1}

range of wavelengths coexist in the bed relief. These waves are naturally combined into three main groups, which form the steps on the spectra: (i) long waves with a length more than the channel width (e.g. meanders, islands, bars and their combinations); (ii) medium waves with a length more than the channel depth

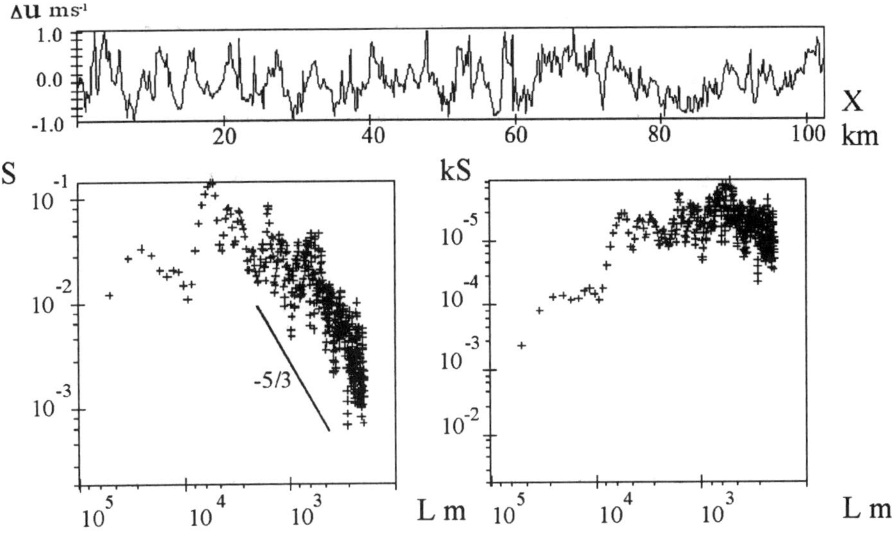

Figure 21.4. The energy spectrum of the Pinega River, calculated with the use of the longitudinal sequence of the depth-averaged velocities

(various dunes); and (iii) short waves of the order of the depth (megaripples). The spectral density of the bed elevations is a measure of the volume of the corresponding forms. The steps in the spectrum show the relative volumes of the main types of channel relief. They represent different levels of structural organization of bed relief. These levels are separated by narrow zones of the spectrum where the relative volume of channel relief is much less.

Within each structural level local maxima and minima of spectral density (or channel relief volume) are observed. This means that smaller groups of channel-relief forms with relatively similar characteristics are found within the main structural levels. These secondary groups may be called 'structural associations'. For example, within the 'dune' part of the Enisey River spectrum, two associations of dunes of the first and the second order can be defined.

CORRESPONDENCE OF FLOW AND CHANNEL RELIEF STRUCTURES

Comparison of the flow velocity spectra and channel bed elevations shows a close similarity between them. Both types of spectra are continuous over a wide range of wave-numbers, which represents the whole hierarchy of the flow and channel relief structures. Both of them contain regions of energy inflow and channel-relief-form volume increase together with other regions of rapidly decreasing spectral energy/volume of the channel forms. Presumably, this similarity is the result of the interaction between the relief and flow velocity field of the three main types of turbulent flow structure. The moving structures (as in the Terek River example)

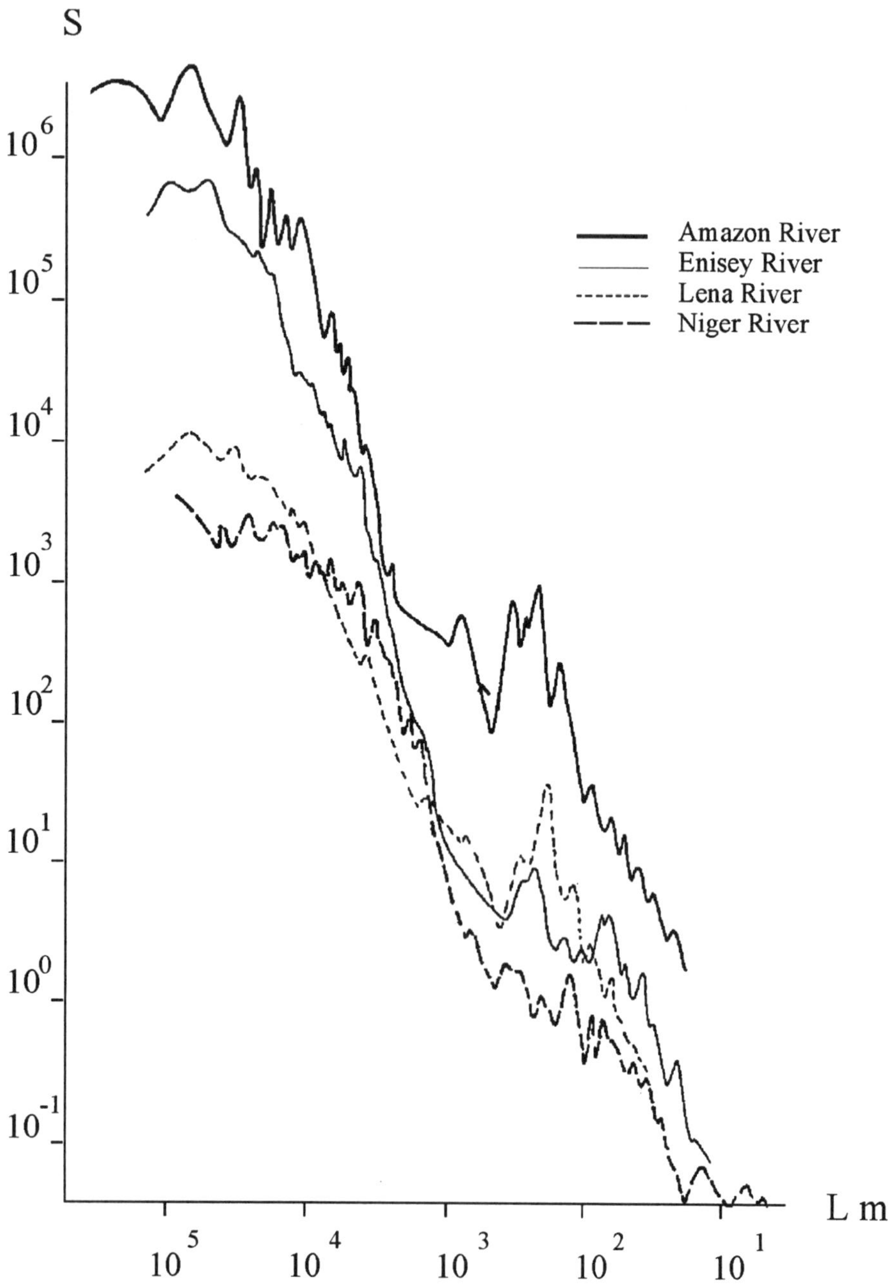

Figure 21.5. Spectra of the bottom elevations in river channels

cannot be 'imprinted' on the channel bed because the rate of their movement is of the order of flow velocity u, and the rate of the most rapidly moving bedforms is of the order 10^{-3}–$10^{-4}u$. Furthermore, it is suggested that the stationary quasi-periodical structures of the flow (as in the Pinega River example) may be the main cause of bed morphology development.

MECHANICS OF THE FLOW-CHANNEL BED RELATIONSHIP

BASIC PRINCIPLES OF THE FLOW-CHANNEL BED INTERACTION

The analysis of the initial instability of the quasi-periodic structures in the flow and on the channel bed gives a quantitative basis for the description of the interaction between stream flow and channel bed. The main cause of this instability is an excess of flow energy above the threshold value and an ability of the channel bed to be transformed. The application of the initial instability analysis to bed relief calculations has been previously reported by a number of authors (see reviews by Callander, 1978; Sidorchuk & Mikhinov, 1985). The works of Callander (1969), Parker (1976), Fredsøe (1978) and Mikhinov (1983) are most relevant to this study.

In all these works the two-dimensional equations of fluid motion and continuity were used with the equation of bed deformation. The flow resistance was described in terms of the Chézy formula. At the stage of depth integration of the three-dimensional equations of momentum and conservation, Callander (1969), Parker (1976) and Fredsøe (1978) used the assumption of a static vertical pressure distribution (Saint-Venant approach). Mikhinov (1983) took into account the dynamic vertical pressure distribution and curvature of the flow surface (Boussinesq approach). This more exact method of depth-averaging led to detection of a very important mode of instability in the region of short waves (megaripples) with the length of channel depth order. In all these solutions the helicoidal flow effects were lost.

At the stage of assignment of the initial perturbation geometry, Fredsøe (1978) used bi-sinusoidal (in longitudinal and lateral directions) moving waves. Callander (1969), Parker (1976) and Mikhinov (1983) used more general forms of moving sinusoidal waves and their shape in the lateral direction was estimated by solving linearized equations of perturbation. In all the foregoing works, the spectrum of perturbation was continuous in the longitudinal direction and discontinuous in the lateral direction.

At the stage of linearization of the perturbed equations Callander (1969), Parker (1976) and Mikhinov (1983) used the assumption of uniform velocity and depth distribution in the unperturbed flow. Fredsøe (1978) included the lateral slope of the unperturbed channel bed. This approach led to the appearance of the mode of instability at the region of long waves (bars). The solution of Fredsøe shows the importance of the lateral terms of equations for the instability of long waves.

However, field observations show that bars can appear in a channel with an initially flat bed (Lewin, 1976).

This brief analysis of the main studies of initial instability in the case of the two-dimensional equations of motion and conservation leads to the following conclusions:

1. All terms in the three-dimensional equation of motion and conservation are important for future instability analysis. The effects of the helical circulation and of vertical distribution of the dynamic pressure must be taken into account during the depth-averaging.
2. The form of the equations must allow the perturbation equations to be linearized with the minimal loss of terms.
3. The geometry of the initial perturbation must be assigned in a very general form in order to obtain a continuous spectrum of waves both in longitudinal and lateral directions.

GENERAL FORM OF THE TWO-DIMENSIONAL HYDRAULIC EQUATIONS

To satisfy these conditions, the three-dimensional equations of momentum and conservation were written in curvilinear orthogonal co-ordinates and the depth-averaging performed following the recommendations of Kartvelishvily (1973):

$$\begin{aligned}
&\frac{\partial}{\partial t}\int_y^h u\mathrm{d}Y+\frac{1}{H_x^2H_z}\frac{\partial}{\partial X}\left(H_xH_z\int_y^h u^2\mathrm{d}Y\right)+\frac{1}{H_x^2H_z}\frac{\partial}{\partial Z}\left(H_1^2\int_y^h uw\mathrm{d}Y\right)-\\
&\frac{1}{H_x^2}\frac{\partial H_x}{\partial X}\int_y^h u^2\mathrm{d}Y-\frac{1}{H_xH_z}\frac{\partial H_z}{\partial X}\int_y^h w^2\mathrm{d}Y+\frac{1}{H_x}\frac{\partial^2}{\partial t\partial X}\int_y^h\int_Y^h v\mathrm{d}\xi\ \mathrm{d}Y+\\
&\frac{1}{H_x}\frac{\partial}{\partial X}\left\{\frac{1}{H_xH_z}\left[\frac{\partial}{\partial X}\left(H_z\int_y^h\int_Y^h vu\mathrm{d}\xi\ \mathrm{d}Y\right)+\frac{\partial}{\partial Z}\left(H_x\int_y^h\int_Y^h vw\mathrm{d}\xi\ \mathrm{d}Y\right)\right]\right\}-\\
&\frac{1}{H_x}\frac{\partial}{\partial X}\int_y^h v^2\mathrm{d}Y+\frac{v^2}{H_x}\frac{\partial h}{\partial X}+\frac{1}{H_x}gd_0\frac{\partial h}{\partial X}+g\frac{\sqrt{(u^2+w^2)}u}{C_0^2}=0;\\
&\frac{1}{H_xH_z}\left(\frac{\partial}{\partial X}H_z\int_y^h u\mathrm{d}Y+\frac{\partial}{\partial Z}H_x\int_y^h w\mathrm{d}Y\right)+\frac{\partial d_0}{\partial t}=0;\\
&\frac{1}{H_xH_z}\left(\frac{\partial}{\partial X}H_z\int_y^h \rho u\mathrm{d}Y+\frac{\partial}{\partial Z}H_x\int_y^h \rho w\mathrm{d}Y\right)+\frac{\partial(h-d_0)}{\partial t}=0
\end{aligned}\tag{21.1}$$

Here H_x and H_z are co-ordinate scale factors and $\mathrm{d}\xi$ is the variable of integration for the Y-axis. The vertical distribution of the dynamic pressure is taken into account in terms with a vertical velocity component v. The equation of motion for the lateral velocity component can be derived from the first part of Equation 21.1 by substituting Z for X and w for u.

All terms, which contain the integrals with depth-averaging, can be replaced by expressions which include the depth d_0, components of surface velocities u^* and w^*, bottom sediment concentration ρ_b and kinematic coefficients α_i and β_i:

$$\int_y^h u\mathrm{d}Y = \alpha_0 u^* d_0; \quad \int_y^h u^2 \mathrm{d}Y = \alpha_1 u^{*2} d_0; \quad \int_y^h w\mathrm{d}Y = \alpha_2 w^* d_0;$$

$$\int_y^h uw\mathrm{d}Y = \alpha_3 u^* w^* d_0; \quad \int_y^h u\rho\mathrm{d}Y = \alpha_4 u^* \rho_b d_0; \quad \int_y^h w\rho\mathrm{d}Y = \alpha_5 w^* \rho_b d_0; \tag{21.2}$$

$$\int_y^h \int_Y^h v\mathrm{d}\xi\mathrm{d}Y = \beta_1 d_0^2 v^*; \quad \int_y^h \int_Y^h uw\mathrm{d}\xi\mathrm{d}Y = \beta_2 d_0^2 u^* v^*; \quad \int_y^h \int_Y^h u^2 v\mathrm{d}\xi\mathrm{d}Y = \beta_3 d_0^2 u^{*2} v^*$$

The kinematic coefficients can be calculated, if the vertical distribution of the velocity components and sediment concentration is known. A power-law profile is used for the longitudinal velocity

$$u = u^* \left(\frac{Y}{d_0} \right)^n \tag{21.3}$$

and a linear (as a first approximation according to Rozovsky, 1957) equation for the lateral velocity:

$$w = w^* \left[(2 + \alpha_2) \frac{Y}{d_0} - (1 + \alpha_2) \right] \tag{21.4}$$

Here α_2 varies in the range 0.1–0.2. A uniform vertical profile is assumed for the vertical velocity component distribution, and the surface vertical velocity is calculated from the boundary conditions on the flow surface:

$$v^* = \frac{\partial h}{\partial t} + \frac{u^*}{H_x} \frac{\partial h}{\partial X} + \frac{w^*}{H_z} \frac{\partial h}{\partial Z} \tag{21.5}$$

When the value of exponent n changes from 0.25 to 0.1, α_0 varies in the range 0.80–0.89, α_1 varies in the range 0.67–0.80, α_3 varies in the range 0.40–0.12, and $\beta_1 = 1.0$, $\beta_2 = \alpha_0$, $\beta_3 = \alpha_1$. The vertical distribution of sediment concentration is calculated using the Rouse (1938) formula:

$$\rho = \rho_b \left(\frac{d_0 - Y}{Y} \frac{\Delta}{d_0 - \Delta} \right)^{\frac{\omega}{\kappa u_*}} \tag{21.6}$$

where κ is the von Kármán constants. Mean sediment concentration calculated using Rossinskiy & Kuz'min (1964):

$$\rho_m = 1.5 \times 10^{-5} \frac{u^3}{\omega d_0} \tag{21.7}$$

When the flow velocity changes from 1.0 to 4.0 m s^{-1} and suspended sediments consist of particles with diameter 0.2–1.0 mm, α_4 varies in the range 0.012–1.05, and α_5 varies in the range from -0.018 to -1.35

INITIAL INSTABILITY ANALYSIS

According to the method of initial instability analysis, the values of surface velocities and surface and bottom elevations were presented as the sums of two components: the main unperturbed flow and perturbation:

$$u^* = U + u' ;\ w^* = w' ;\ d_0 = D_0 + d' ;\ h = h' \tag{21.8}$$

Expressions 21.2–21.8 were substituted into the parts of Equation 21.1. After subtraction of the equations for the main flow and linearization the following equations of momentum, conservation and deformation for perturbation were obtained:

$$\begin{aligned}
&\alpha_0 \frac{\partial u'}{\partial t} + (\alpha_0 - \alpha_1)\frac{U}{D_0}\frac{\partial d'}{\partial t} + \alpha_1 U \frac{\partial u'}{\partial s_x} + (\alpha_3 - \alpha_2)U\frac{\partial w'}{\partial s_y} + g\frac{\partial h'}{\partial s_x} + \\
&2\alpha_1 g \frac{u' U}{C_0^2 D_0} - \alpha_1 g \frac{U^2 d'}{C_0^2 D_0^2} + D_0\left(\beta_1 \frac{\partial^3 h'}{\partial^2 t \partial s_x} + 2\beta_2 U \frac{\partial^3 h'}{\partial t \partial s_x^2} + \beta_3 U^2 \frac{\partial^3 h'}{\partial s_x^3}\right) = 0; \\
&\alpha_2 \frac{\partial w'}{\partial t} + \alpha_3 U \frac{\partial w'}{\partial s_x} + \alpha_1 U^2 K_1^* + g\frac{\partial h'}{\partial s_z} + \alpha_3 g \frac{wU}{C_0^2 D_0} + \\
&+ D_0\left(\beta_1 \frac{\partial^3 h'}{\partial^2 t \partial s_z} + 2\beta_2 U_0 \frac{\partial^3 h'}{\partial t \partial s_x \partial s_z} + \beta_3 U_0^2 \frac{\partial^3 h'}{\partial s_x^2 \partial s_z}\right) = 0; \\
&\alpha_0 D_0 \frac{\partial u'}{\partial s_x} + \alpha_0 U \frac{\partial d'}{\partial s_x} - \alpha_0 U D_0 K_2^* + \alpha_2 D_0 \frac{\partial w'}{\partial s_z} + \frac{\partial d'}{\partial t} = 0; \\
&\alpha_4 \rho_b D_0 \frac{\partial u'}{\partial s_x} - \alpha_4 U \rho_b D_0 K_2^* + \alpha_5 \rho_b D_0 \frac{\partial w'}{\partial s_z} + (1 - \varepsilon)\frac{\partial (h' - d')}{\partial t} = 0;
\end{aligned} \tag{21.9}$$

Here $s_x = H_x X;\ s_z = H_z Z;\ K_x^* = -\frac{1}{H_x}\frac{\partial H_x}{\partial s_z};\ K_z^* = -\frac{1}{H_z}\frac{\partial H_z}{\partial s_x}$

The main unperturbed flow is assumed to be straight and have uniform velocities and depth distribution. The products and powers of the perturbation components were excluded due to linearization. The use of curvilinear orthogonal co-ordinates allows the terms with curvature of the longitudinal and lateral co-ordinates in the linearized equations, which describe the lateral movement, to be retained. In this case the geometry of curvilinear co-ordinates must be known. The Rozovsky-type geometry was used in this study (Figure 21.6) with the longitudinal co-ordinate (X) coincident with the flow line of the longitudinal component of perturbed velocity

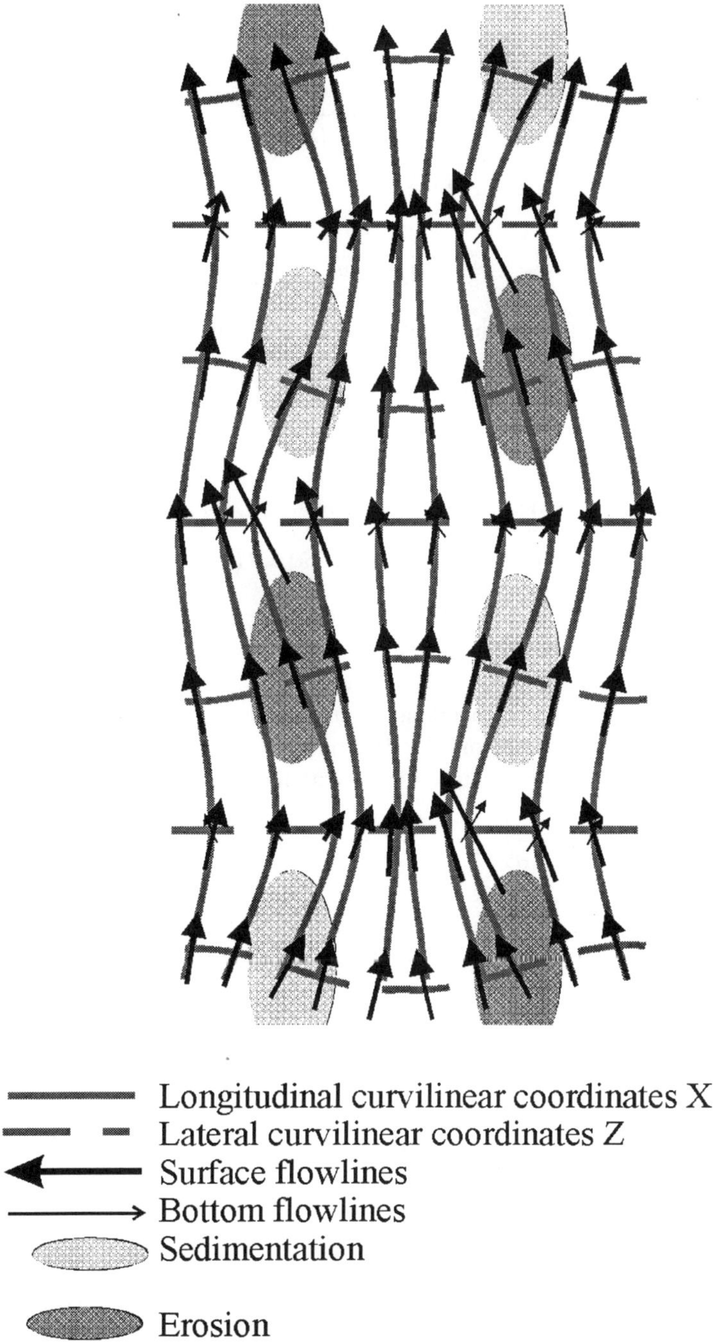

Figure 21.6. The Rozovsky-type geometry of curvilinear co-ordinates

u', and with the curvature K_x^* related to lateral velocity of the helicoidal flow (equal to the lateral component of the perturbed flow):

$$K_x^* = -\gamma_0 \frac{w'}{UD_0} \tag{21.10}$$

The coefficient γ_0 is a function of the Chézy coefficient C_0 and flow depth D_0:

$$\gamma_0 = \frac{1 - 7.2\sqrt{D_0}/C_0}{7.4 - 13.0\sqrt{g}/C_0}$$

Usually γ_0 values vary in the range 0.08–0.1. The curvature of the lateral co-ordinates for the case of initial perturbation is assumed to be small: $K_z^* \approx 0$.

The initial perturbations of the flow components and surface and bed elevations are presented in the form of longitudinal- moving sinusoidal waves with complex celerity $c = c_r + ic_i$:

$$u' = A(s_z)\exp(ik_x(s_x - ct)); \quad w' = B(s_z)\exp(ik_x(s_x - ct))$$

$$h' = P(s_z)\exp(ik_x(s_x - ct)); \quad (h' - d') = T(s_z)\exp(ik_x(s_x - ct)) \tag{21.11}$$

The functions A, B, P and T, which describe the perturbation shape in the lateral direction are also complex. This implies that the perturbation can be composed of two moving sinusoidal waves with different initial amplitude and phase (only the real part of Equation 21.11 is being taken in consideration).

After substitution of Equations 21.10 and 21.11 into Equation 21.9 the equations for momentum and conservation can be rewritten in the form:

$$a_1 A + b_1 \frac{\partial B}{\partial s_z} + c_1 P + d_1 T = 0; \quad b_2 B + c_2 \frac{\partial P}{\partial s_z} = 0;$$

$$a_3 A + b_3 \frac{\partial B}{\partial s_z} + c_3 P + d_3 T = 0; \quad a_4 A + b_4 \frac{\partial B}{\partial s_z} + d_4 T = 0 \tag{21.12}$$

Here:

$$a_1 = \alpha_1 Uik_x - \alpha_0 ik_x c + 2g\alpha_1 \frac{U}{C_0^2 D_0}; \; b_1 = (\alpha_3 - \alpha_2)U;$$

$$c_1 = gik_x - (\alpha_0 - \alpha_1)\frac{U}{D_0} ik_x c - g\alpha_1 \frac{U^2}{C_0^2 D_0^2} - D_0(\beta_1 ik_x^3 c^2 - 2U\beta_2 ik_x^3 c + \beta_3 ik_x^3);$$

$$d_1 = (\alpha_0 - \alpha_1)\frac{U}{D_0} ik_x c + g\alpha_1 \frac{U^2}{C_0^2 D_0^2};$$

$$b_2 = \alpha_3 Uik_x - \alpha_4 ik_x c + g\alpha_3 \frac{U}{C_0^2 D_0} - \alpha_0 \gamma_0 \frac{U}{D_0}$$

$$c_2 = g - D_0(\beta_1 k_x^2 c^2 - 2U\beta_2 k_x^2 c + \beta_3 k_x^2); \quad a_3 = \alpha_0 D_0 ik_x; \; b_3 = \alpha_2 D_0;$$

$$c_3 = \alpha_0 Uik_x - ik_x c; \; d_3 = -\alpha_0 Uik_x + ik_x c; \; a_4 = \alpha_4 \rho_b D_0 ik_x; \; b_4 = \alpha_5 \rho_b D_0;$$

$$d_4 = -(1-\varepsilon)ik_x c$$

When A, B and T are eliminated from Equation 21.12, the result gives a single ordinary second-order differential equation in terms of the lateral function of the surface perturbation P:

$$\frac{d^2 P}{ds_z^2} = \lambda \times P \tag{21.13}$$

The boundary conditions on the channel banks can be non-zero for this function. This leads to continuous eigenvalues of the Equation 21.13:

$$\lambda = k_z^2 = \left(\frac{\sigma\pi}{W_0}\right) = \frac{b_2(a_3c_1d_4 + a_4c_3d_1 - a_1c_3d_4 - a_4c_1d_3)}{c_2(a_1b_3d_4 + a_3b_4d_1 + a_4b_1d_3 - a_1b_4d_3 - a_3b_1d_4 - a_4b_3d_1)} \tag{21.14}$$

where σ runs through all positive, real values. The use of non-zero boundary conditions to be satisfied by Equation 21.13 leads to a continuous spectrum of perturbation waves both in lateral and longitudinal direction.

For the case of boundary conditions

$$P(0) = P(W_0); \; \frac{\partial P}{\partial s_z}(0) = -\frac{\partial P}{\partial s_z}(W_0) \tag{21.15}$$

the eigenfunctions of Equation 21.13 will have form of

$$P \sim \cos\left(\sqrt{\lambda}\left(s_z - \frac{W_0}{2}\right)\right) \tag{21.16}$$

which implies asymmetrical geometry of the perturbation.

For the case of boundary conditions

$$P(0) = -P(W_0); \; \frac{\partial P}{\partial s_z}(0) = \frac{\partial P}{\partial s_z}(W_0) \tag{21.17}$$

the eigenfunctions of Equation 21.13 will have form of

$$P \sim \sin\left(\sqrt{\lambda}\left(s_z - \frac{W_0}{2}\right)\right) \tag{21.18}$$

which implies symmetrical geometry of the perturbation.

From Equations 21.12 and 21.13 the dispersion equation for the complex celerity can be written in the form

$$A_1c^4 + A_2c^3 + A_3c^2 + A_4c + A_5 = 0 \tag{21.19}$$

The coefficients A_i are functions of the main flow parameters, kinematic coefficients, and longitudinal and lateral wave-numbers.

The dispersion represented in Equation 21.19 was analysed by Mikhinov (1983). He showed that the four complex roots of this equation represent waves of three types: (i) waves with the rate of motion $c_r >> U$; (ii) two waves (one moving downstream and the other upstream) with $c_r \sim U$; and (iii) slow waves with the rate of motion $c_r << U$. The first two types of waves belong to the flow and cannot represent channel relief forms due to their high rate of celerity. Only the third type of perturbation with a low rate of motion along the channel can be compared with the channel relief forms.

RESULTS OF THE INITIAL INSTABILITY ANALYSIS

The dispersion Equation 21.19 was simplified with the use of the condition $\partial / \partial t = 0$ in the momentum equations. Both waves of the second type are thereby eliminated. The equation was solved numerically for the wide range of flow parameters and wave-numbers. Only slow waves of the third type were analysed. The solutions were represented in the form of two-dimensional spectra of the increments of exponential growth of the waves amplitudes (Figure 21.7). This spectrum is continuous both in longitudinal and lateral directions according to boundary conditions. At the same time, the spectrum of the surface relief is complicated and several regions of specific relations between c_i, k_x and k_z can be defined.

Two main regions are defined by the sign of c_i In the region of positive c_i the amplitudes of initial perturbations increase in time. Such waves are unstable and can form the bed relief. In the region of negative c_i the amplitudes of perturbations decrease in time, such waves are stable and cannot form the bed relief.

Inside the region of unstable developing waves several special fields, lines and points were defined. In Figure 21.7b, the main line is the maxima of the product of k_x and c_i. In accordance with the study of channel bed instability by Kennedy (1963), the waves with the maximum increment of amplitude growth $k_x c_i$ are assumed to be responsible for bed relief formation. There are four special points on this line: (i) the local maximum of amplitude of short waves; (ii) the point at the beginning of rapid increase in the elongation of waves; (iii) the local minimum amplitude of long waves; and (iv) the local maximum amplitude of long waves. The lines which can be drawn through these points subdivide the region of unstable waves into five fields:

1. The field of very short waves (ultramicroforms). Two main types of them are defined: two-dimensional and three-dimensional ultramicroforms. The solid horizontal line on Figure 21.7b shows the maxima amplitudes of two-dimensional ultramicroforms in the spectrum. Mikhinov (1983) derived the formula for calculation of the length of these waves as:

$$L_x = 5.4 D_0 Fr \tag{21.20}$$

a)

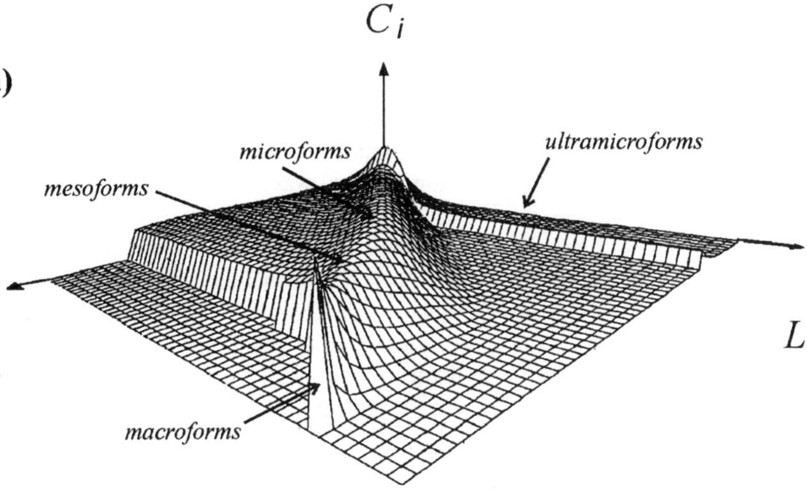

b)

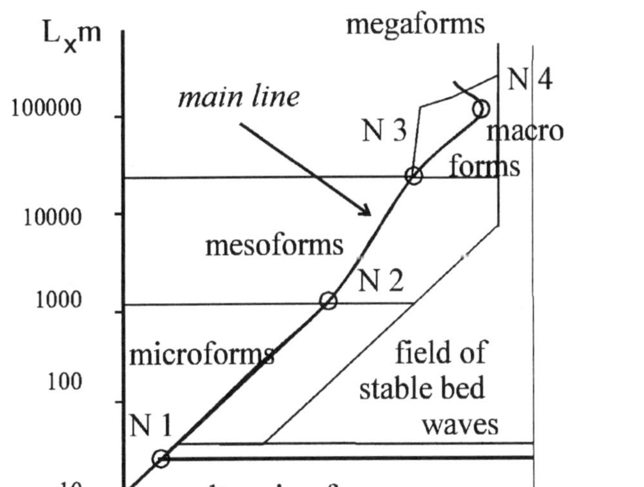

Figure 21.7. The two-dimensional spectrum of unstable waves amplitude increase (a) and the main theoretical types of river bed undulations (b) for the case of $U = 1.2$ m s^{-1}; $D_0 = 35$ m; $C_0 = 50$ (Amazon River parameters)

Point N1 on the main line of the spectrum represents the most probable location of three-dimensional ultramicroforms. Their lengths are 1.2–2.0 times larger than those of two-dimensional ones but the type of relation of depth to Froude number is the same for both types of ultramicroforms, if $Fr < 0.8$ (Figure 21.8).

2. The field of short waves (microforms). The most probable microforms are three-dimensional and isometric in the horizontal plane. There are no local maxima on the spectrum in the microform field; the rate of amplitude growth decreases steadily with longitudinal wave number.
3. The field of medium-size waves (mesoforms). The most probable mesoforms are also three-dimensional, but their shape in the horizontal plane is elongated. The rate of amplitude growth in this field also decreases steadily with the longitudinal wave number. The position of the point N2 between the fields of symmetrical microforms and elongated mesoforms changes with depth and resistance to flow (Figure 21.8):

$$L_x = 8.2 D_0 \exp(2.0 Fr) \qquad (21.21)$$

4. The field of long waves (macroforms). The most probable macroforms are three-dimensional, with an elongated shape on the horizontal plane. The

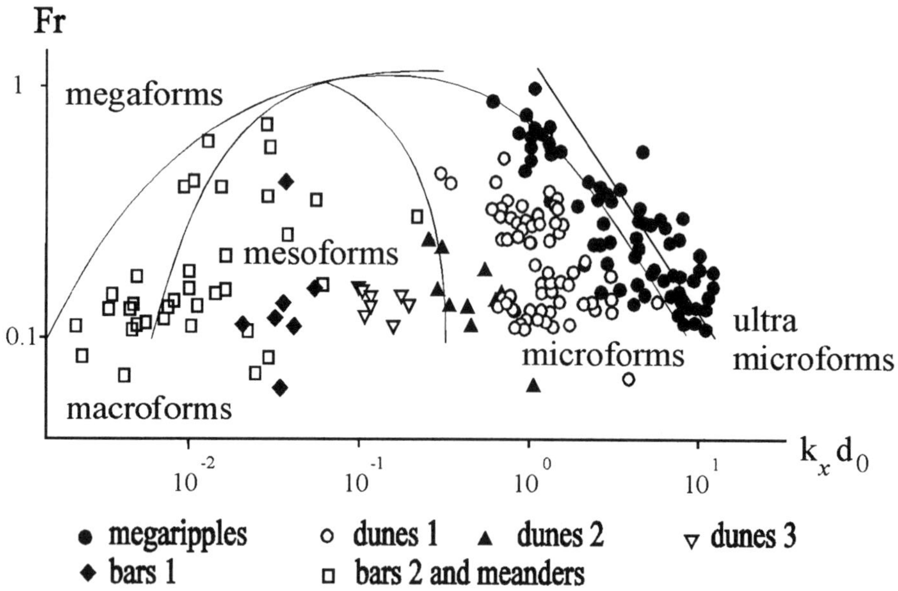

Figure 21.8. The correspondence of the theoretical types of river bed undulations to the bed forms and forms of channel relief, observed in flumes and natural rivers (based on the data from 21 publications of different scientists and from measurements of the author)

degree of elongation is related to the intensity of helicoidal circulation, Froude number and flow resistance:

$$L_x / L_z = \frac{0.2}{Fr}\sqrt{\frac{C_0^2\beta_0}{2\alpha_2 g} - 1} \qquad (21.22)$$

The rate of amplitude growth in this field increases steadily with longitudinal wave number and reaches its maximum at the point N4 (Figure 21.7b). The lengths of the long waves at this point increase with main channel depth D_0 and the magnitude of the Chézy coefficient, but decrease with growth of Froude number (Figure 21.8):

$$\begin{aligned} L_x &= 6.28D_0\left(\frac{C_0^2}{2g}\right)Fr^{-1} && 0.1 \le Fr \le 0.5 \\ L_x &= 62.8D_0\left(\frac{C_0^2}{2g}\right)\exp(-3.1Fr) && 0.5 < Fr < 1.0 \end{aligned} \qquad (21.23)$$

The point N3 (Figure 21.7b) is situated between the fields of mesoforms and macroforms and corresponds to the minimum of wave amplitude growth rate. Its position changes with depth, Froude number and resistance to flow (Figure 21.8):

$$L_x = 12.6D_0\left(\frac{C_0^2}{2g}\right)\exp(-1.8Fr) \qquad (21.24)$$

5. The field of very long waves (megaforms). The width of these forms is very small compare to their length. Typically, megaforms would take the shape of long and narrow strips in the channel. The rate of their amplitude growth is relatively low.

COMPARISON OF THEORETICAL AND OBSERVED CHANNEL FORMS

The hierarchy of channel bed relief forms in rivers and experimental flumes consists of four to ten levels, but the usual number of levels is from six to eight. For example, in the Vychegda River channel megaripples, dunes of three orders, bars of two orders and channel meanders were defined (Sidorchuk, 1989). The actual relief of river channels is more complicated than indicated by the theoretical spectrum. The main application of the theoretical spectrum is the use of the different fields of unstable channel bed undulations and the boundaries between them, for classification of the channel relief forms, and as a basis for comparison between the channel forms in different rivers.

The experimental data from different authors on ripples in flumes and river channels, and our measurements of megaripple parameters on the Terek, Niger, Ob' and Lena Rivers, satisfactorily fit the field of ultramicroforms in the theoretical spectrum (Figure 21.8). Two-dimensional and three-dimensional megaripples usually coexist in the channel and form wavy bands stretching across the channel. The antidunes in the experiments of Guy *et al.* (1967) also correspond to the field of ultramicroforms, so these bedforms belong to the megaripples type, existing in flows with a high Froude number.

First and second order dunes correspond to the field of microforms. These bedforms are most common in large river channels with sandy alluvium (e.g. Amazon, Jamuna, Mississippi, Niger, Enisey, Ob'). They form a well-defined maximum on the spectrum of natural channel bottom elevations (see Figure 21.5). This local maximum is absent on the theoretical spectrum in the field of microforms (see Figure 21.7). The discrepancy between the natural observations and theoretical results shows the limitations of the linear approach in initial instability analysis.

Third order dunes, bars of the second order and (partly) of the first order, correspond to field observations of mesoforms. The boundary between dunes (bedforms) and bars (which determine the channel morphology during the period of low flow) is rather ill-defined in natural channels. At the high-water stage the dominant channel relief can be submerged (becoming bedforms), and at the period of low water the same forms often constitute the bars. The boundary between dunes and bars on the theoretical spectrum is also unclear, and both large dunes and small bars can be observed in the field of mesoforms. In the case of wide and shallow channels (for example, upper Ob' River), the small bars and islands lie in the field of microforms. This raises some classification problems, because the length of small mesoforms and microforms (which can be both dunes and bars) increases with the Froude number, and the length of large mesoforms (which can also be dunes and bars) decreases with Froude number (Figure 21.8).

The field of macroforms usually corresponds to large bars, islands and channel meanders. As is the case with dunes, these forms are usually most common in river channels. Well-defined maxima on the spectrum of natural channel bed elevations correspond to these forms (see Figure 21.5). A local maximum of the rate of unstable amplitude growth also exists in the field of macroforms in the theoretical spectrum. This maximum represents the length and width of the most probable macroforms, which are described in Equations 21.22 and 21.23.

The maximum in the spectrum of observed data is always shifted to a smaller wavelength than the maximum on the theoretical spectrum. The main reason for this phenomenon is the non-linearity of the real processes of channel development. The rate of growth of the amplitudes of long bed waves is related not only to initial instability, which is described by the theoretical spectrum, but also to secondary effects, which become very important as the amplitude grows. When the width of a long unstable wave is less than two widths of the channel, the influence of this wave on the bank erosion is small. When the width of a long unstable wave (the initial bar) becomes more than two widths of the initial channel, bank erosion takes place around this bar. The alluvial form often stops its down-channel movement, being stabilized by vegetation and thin floodplain alluvium, and therefore forms a stable channel pattern, with an island or meander. The rate of growth of the deflection angle Θ of this planform decreases with an increase of its length. For the case of a sine-generated curve this rate can be calculated with the formula:

$$\frac{d\Theta}{dt} = \frac{4q_b}{L_m H_b} \frac{J_0^2(\Theta)}{J_0(dH_0/d\Theta) - H_0(dJ_0/d\Theta)} \qquad (21.25)$$

In Equation 21.25, J_0 and H_0 are Bessel and Struve functions of zeroth order, respectively. From the wide range of unstable undulations which could cause bank erosion, the shortest ones with the width $L_z \sim 2 \times W_0$ will increase the deflection angle at the maximum rate and thereby form the most developed meanders. As a result, two shallow riffles appear on each meander in the zones of channel curvature change, and these secondary waves of bottom elevation have the same length as the initial undulations that cause the meander formation. The most probable value of length/width ratio for macroforms is about four (Equation 21.22), so the length of developed meanders L_m must be about eight widths of the channel W_0. It is well known from numerous investigations, that a common value of meander length -to-channel width ratio is 8–10 (e.g. Leopold & Wolman, 1960). The most developed meanders form the most developed riffles, and the associated maximum of spectral density appears on the spectrum of the bottom elevations.

There has been relatively little investigation of very long and narrow forms of channel relief (megaforms). Coupled lateral vortices with the horizontal axis along the channel were studied by Losievsky (1934), and recently several papers have described the longitudinal narrow strips of ridges and troughs formed by such vortices in a flume (e.g. Nezu & Nakagawa, 1989). The phenomenon of parallel braiding in a natural river channel was described by Berkovich *et al.* (1989). It occurs quite often in wide and relatively shallow river channels, when two branches of the same river are following parallel courses for a long distance, usually near the opposite sides of the valley bottom. These branches are divided by a chain of islands and are usually connected only by short lateral channels. This phenomenon has been observed on the upper Ob', lower Lena and lower Northern Dvina (Sidorchuk, 1992). Parallel braiding may be the natural analogue of the theoretical megaforms.

CONCLUSIONS

1. The interaction between the flow and movable bed in a self-organizing dynamic system produces a quasi-periodic flow structure and different scales of channel bed.
2. Three main types of quasi-periodic structure are formed in the channel flow which are all supplied by energy from the mean velocity: (i) three-dimensional structures with a longitudinal dimension of the order of channel depth; (ii) essentially two-dimensional structures with a longitudinal

dimension of the order of river width; and (iii) essentially one-dimensional structures with a longitudinal dimension of the order of the river meander length. These structures in the flow can either move along the channel with the velocity of the main stream, or be quasi-stationary due to an interaction with the channel bed relief.

3. The forms of channel bed relief are classified into three main groups defined by the different levels of structural organization: (i) long waves with lengths greater than the channel width (this include meanders, islands and bars); (ii) medium waves with lengths exceeding the channel depth (different types of dunes); and (iii) short waves of the order of flow depth (megaripples). Despite this classification, the spectrum of the bottom elevation of the river channels is continuous: waves of various length coexist in the channel bed relief.
4. The analysis of initial instability of the wave-like structures of the flow and channel bed relief represents a quantitative basis of the description of stream flow-channel bed interaction. Three-dimensional equations of momentum and conservation in curvilinear orthogonal co-ordinates with depth-averaging following Kartvelishvily (1973) were used for this analysis. This approach allows the effects of the vertical distribution of dynamic pressure and helicoidal flow to be considered. It also permits the possibility of linearization of the equations of perturbation with minimal loss of terms.
5. The geometry of the initial perturbations and boundary conditions for them are described in the most general form to obtain the continuous spectrum of waves both in longitudinal and lateral directions.
6. The dispersion equation is derived from the analysis of the initial instability and solved numerically for a wide range of flow parameters and wave-numbers. Solutions for the slowest waves (with the rate of longitudinal motion $c_r << U$) were represented in the form of a two-dimensional spectrum of the increments of exponential growth in the waves' amplitude. In the region of unstable waves five main types of bed undulations were defined: two- and three-dimensional ultramicroforms; symmetrical microforms; elongated mesoforms; elongated macroforms and long and narrow megaforms.
7. All these unstable waves have their analogues in the bed relief of natural river channels. The ripples in flumes and megaripples in the river channels correspond to ultramicroforms. First and second order dunes are correlated with microforms. The transitional forms between dunes and bars belong to the field of mesoforms. The macroforms correspond to bars, islands, and meanders. Parallel braids are the natural analogues for megaforms.
8. The natural process of channel form development is non-linear. The secondary non-linear effects become very important in the case of the development of dunes and meanders.

ACKNOWLEDGEMENTS

The author is very grateful for the help given by the organizers of the conference at the University of Leeds, where this paper was presented, and to the International Science Foundation of G. Soros for the financial support, which allowed the author to attend the conference. The suggestions of Dr. H.M. Pantin and an anonymous referee have proved valuable and have been incorporated into the text.

REFERENCES

Berkovich, K.M., Ruleva, S.N. & Chalov, R.S. 1989. 'The channel behaviour of the upper Ob' River', *Geogr. and Nat. Resour.*, **4**, 54–61. (in Russian).

Callander, R.A. 1969. 'Instability and river channels', *J. Fluid Mech.*, **36**, 465–480.

Callander, R.A. 1978. 'River Meandering', *Ann. Rev. Fluid Mech.*, **10**, 129–158.

Fredsøe, J. 1978. 'Meandering and braiding of rivers', *J. Fluid Mech.*, **84**, 609–624.

Grinvald, D.I. 1974. *The Turbulence of Channel Flow,* Gidrometeoizdat, Leningrad, 165 pp. (in Russian).

Grinvald, D.I. & Nikora, V.I. 1988. *River Turbulence*, Gidrometeoizdat, Leningrad, 153 pp. (in Russian).

Guy, N.P., Simons, D.B. & Richardson, E.V. 1967. 'Summary of alluvial channel data from flume experiments', *US Geol. Surv. Prof. Pap.*, **462-I**, 1–96.

Kartvelishvily, N.A. 1973. *The flows in non-erodible channels*, Gidrometeoizdat, Leningrad, 279 pp. (in Russian).

Kennedy, J.F. 1963. 'The mechanics of dunes and antidunes in erodible-bed channels', *J. Fluid Mech.*, **16**, 521–544.

Kondrat'ev, N.E. 1953. 'About a discreteness of the fluvial processes', in *The Problems of Fluvial Processes*, Gidrometeoizdat, Leningrad, 16–20. (in Russian).

Leopold, L.B. & Wolman, M.G. 1960. 'River meanders', *Bull. Geol. Soc. Amer.*, **71**, 769–793.

Lewin, J. 1976. 'Initiation of bed form and meanders in coarse grained sediment', *Bull. Geol. Soc. Amer.*, **57**, 281–285.

Losievsky, A.I. 1934. 'Laboratory study of riffle formation process', *Trudy MOTSNIIVT*, **86**, Moscow, 1–156. (in Russian).

Makkaveev, N.I. 1955. *The channel of the river and erosion on its basin*, Akademizdat, Moscow, 346 pp. (in Russian).

Mikhinov, A.Ye. 1983. 'Instability of the bottom forms on alluvial channel', *Meteorology and Hydrology*, **11**, 84–91. (in Russian).

Monin, A.S. & Ozmidov, R.V. 1981. *Oceanic Turbulence*, Gidrometeoizdat, Leningrad, 320 pp. (in Russian).

Monin, A.S., Yaglom, A.M. 1967. *Statistical Hydromechanics,* Vol. **2**, Nauka, Moscow, 720 pp. (in Russian).

Nezu, I. & Nakagawa, H. 1989. 'Self forming mechanism of longitudinal sand ridges and troughs in fluvial open-channel flows', in *Hydraul. and Environ., Proc. 23rd Congr.*, IHAR, Ottawa, **B**, 65–72.

Parker, G. 1976. 'On the cause and characteristic scales of meandering and braiding in rivers', *J. Fluid Mech.*, **76**, 457–480.

Rossinskiy, K.I. & Kuz'min, I.A. 1964. 'Sediment budget approach to calculations of the channel bed deformations', *Trudy Gidroproekta*, **12**, 265–271. (in Russian).

Rouse, H. 1938. 'Experiments on the mechanics of the sediment suspension', *Proc. 5th Int. Congr. Appl. Mech.,* Cambridge, Mass., 47–53.

Rozovsky, I.L. 1957. *The Water Motion on the Alluvial Channel Bend*, AN UkrSSR, Kiev, 188 pp. (in Russian).
Sidorchuk, A.Yu. 1989. 'The hierarchy of the channel forms: structure and dynamics', in Logachev *et al.* (Eds) *Problems of the Methodology of Geomorphology*, Nauka, Novosibirsk, 93–96. (in Russian).
Sidorchuk, A.Yu. 1992. *The Structure of the River Channel Relief*, Gidrometeoizdat, Sankt-Petersburg, 125 pp. (in Russian).
Sidorchuk, A.Yu. & Mikhinov, A.Ye. 1985. *The morphology and dynamics of the channel relief*, VINITI Pub. House, Moscow, 166 pp. (in Russian).
Velikanov, M.A. 1954. *Dynamics of the Channel Flows*, Vol. **1**, Gostekhizdat, Moscow, 323 pp. (in Russian).

LIST OF NOTATION

$A\ B\ P\ T$	functions of lateral co-ordinate in Equation 21.11
A_i	coefficients in dispersion equation
C_0	Chézy coefficient
$c = c_r + ic_i$	complex celerity
D_0	depth of unperturbed channel
d_0	channel depth
d'	depth perturbation
Fr	Froude number $U/\sqrt{gD_0}$
g	acceleration due to gravity
H_b	river banks height
$H_x\ H_z$	scale factors
$H_0\ J_0$	functions of Struve and Bessel of zero order
h	flow surface elevations
h'	perturbations of the flow surface elevations
$K_x^*\ K_z^*$	curvature of longitudinal and lateral curvilinear orthogonal co-ordinates
$k_x\ k_z$	longitudinal and lateral wave numbers for perturbations
L_m	channel meanders length
$L_x\ L_z$	longitudinal and lateral length of the forms of channel relief
n	exponent in Equation 21.3
Q	discharge
q_b	lateral sediment flux
s_x, s_z	distances along longitudinal and lateral curvilinear orthogonal co-ordinates
t	time
U	longitudinal velocity of unperturbed flow
u_*	shear velocity
$u\ v\ w$	longitudinal, vertical and lateral velocity components at a point
$u^*\ v^*\ w^*$	longitudinal, vertical and lateral velocity components at the flow surface

u' v' w'	perturbations of the longitudinal, vertical and lateral velocity components at the flow surface
W_0	channel width
X Y Z	longitudinal, vertical and lateral curvilinear orthogonal co-ordinates
y	channel bed elevations
α_i β_i	kinematic coefficients
γ_0	coefficient in the formula of Rozovsky
Δ	roughness elements height
ε	bed sediments porosity
Θ	meanders angle of deflection
κ	von Kármán constant
λ	eigenvalues in Equation 21.13
ρ	sediment concentration at a point
ρ_b	bed sediment concentration
ρ_m	depth-averaged sediment concentration
ω	fall velocity of sediment particles

22 Some Speculations on the Relation Between Channel Morphology and Channel-scale Flow Structures

GARY PARKER
St. Anthony Falls Laboratory, University of Minnesota, Mississippi River at 3rd Avenue, Minneapolis, MN 55414, USA

ABSTRACT

A set of speculations concerning the relation between flow mechanics and river channel morphology is presented in the hope of encouraging research on the subject. In some cases the speculations are little more than plausible inferences based on limited information, whereas in other cases at least some direct evidence can be offered in their support. The speculations include the following:

- *Tributaries tend to enter meandering streams at the outside of bends. This configuration may represent a quasi-stable state.*
- *Concave-bank benches on meandering streams can evolve into downstream-migrating features of permanent form associated with flow separation at the outside of sharp bends.*
- *Scroll bars on meandering streams may evolve from highly elongated, partially beached oblique dunes.*
- *Under the correct conditions, channelization on submarine fans is spontaneous, is driven by the interaction between turbidity currents and the morphology they create and can occur even when the flows are everywhere net-depositional.*
- *When a channel is allowed to aggrade by the deposition of heterogeneous gravel, the deposit can be expected to display downstream fining. In the case of a mixture of sand and pea-sized gravel, however, downstream coarsening can occur. This may be due to the existence of a partial viscous sublayer, which shields the finer grains from the effects of turbulence.*
- *Many of the features of alluvial fans formed by braided rivers can be reproduced with constant rates of water and sediment inflow. The depth-slope product of the active anabranches appears to tend toward a specific value, a condition that allows for the formulation of a mechanistic theory of fan formation.*
- *The phenomena of discontinuous headcuts in cohesive material and chute-and-pool topography in fine non-cohesive materials appear to represent limiting cases of a single phenomenon, here termed cyclic steps. Cyclic steps are repetitive upstream-migrating bedforms characteristic of supercritical flow. They may in turn bear some relation with the step-pool topography of steep coarse-bed streams.*
- *Large rivers often display a relatively sharp transition from gravel to sand. Over time one might expect the associated gravel front to prograde out to the zone at which base level is maintained, e.g. the sea. The fact that gravel-sand transitions are so prevalent suggests, however, that some mechanism must act to stabilize their location. The two stabilizing mechanisms are hypothesized to be basin subsidence and abrasion.*

Coherent Flow Structures in Open Channels.
Edited by P.J. Ashworth, S.J. Bennett, J.L. Best and S.J. McLelland.

INTRODUCTION

The study of river mechanics and morphology has advanced rapidly in the last 20 years. This progress notwithstanding, the field remains blessed with an extensive list of fascinating and practically relevant phenomena that remain partially or completely unexplained in a mechanistic sense. Several such phenomena are considered in this paper with the goal of encouraging further research on the topics in question. All of the topics listed below are characterized by the interaction of morphology, sediment transport and large-scale flow structures. Some of them have yet to be the subject of detailed research. Others represent areas within which the author and colleagues are actively pursuing research, but where the results are as yet incomplete.

TRIBUTARY CONFLUENCES ON MEANDERING STREAMS

In the early 1980s the author had the opportunity to participate in a project involving a bridge on a meandering reach of the Minnesota River, USA. (Parker *et al.*, 1982). The Trunk Highway 169 bridge was built in 1960. The site before the construction of the bridge is shown in Figure 22.1a. The bridge was constructed so that its approaches would not be overtopped. The floodplain is 1400 m wide in this reach; during a severe flood all of the flow must be concentrated and passed under the bridge deck. Were the bridge deck only to span the width of the river itself, i.e. about 90 m, unacceptably severe constriction scour would result during floods. With this in mind the area outlined in white in Figure 22.1a was excavated to provide a flood-way, and the span of the bridge was increased proportionally.

Of interest here is the tributary, Le Sueur Creek, entering the Minnesota River from the east side (bottom of Figure 22.1a) just opposite the area outlined in white. It is seen that in 1960 the tributary entered on the outside of a gentle meander bend, slightly upstream of the bend apex. A small delta is visible at the mouth of the tributary. While the meander bends of the Minnesota River are known to migrate in a relatively orderly fashion (e.g. Johannesson & Parker, 1985; MacDonald *et al.*, 1991), the position of the tributary at the outside of the same bend has been preserved since at least the end of the last century.

Between 1960 and 1981 a series of events, including the record flood of 1969, resulted in the reversal of the sense of the bend, and the concomitant deposition of a large delta at the mouth of Le Sueur Creek. This reversal is documented in Figure 22.1b. The nature of the events leading to this reversal is discussed in Parker *et al.* (1982). It suffices to say that the reach of the Minnesota River near the bridge was destabilized by the delta, and rapid channel migration to the west threatened the western abutment of the bridge.

A study of the problem at St. Anthony Falls Laboratory suggested that the installation of two in-channel permeable dikes on the west side and a T-spur dike on

(a)

(b)

Figure 22.1. (a) View of the Minnesota River before the construction of the Trunk Highway 169 bridge. Flow is from left to right. The tributary at the bottom, Le Sueur Creek, enters from the east. (b) View of the same site in 1981. A large delta has deposited at the mouth of the tributary and the sense of the bend has been reversed. The channel downstream of the bridge is partially covered with ice. *Continued on next page*

(c)

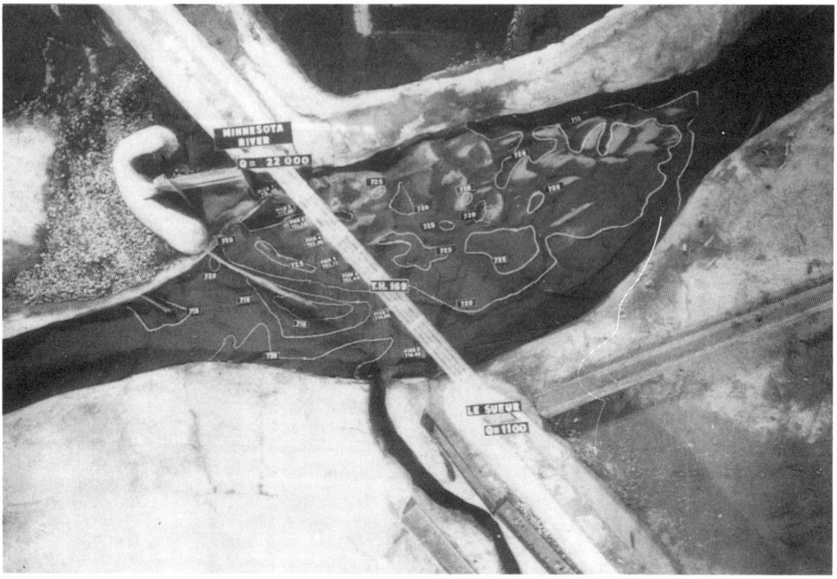

(d)

Figure 22.1. *continued* (c) Results of a model study to correct the bend reversal. The two in-channel permeable dikes and the T-spur dike on the floodplain are visible, as well as a large bar on the west bank further downstream. (d) The Minnesota River near the confluence of Le Sueur Creek in 1987. The river training works shown in Figure 22.1c have been installed. The sense of the bend at the confluence of Figure 22.1a has been restored

the western floodplain upstream of the bridge would encourage the channel to relocate in its older, more stable configuration, as shown in Figure 22.1c. These river training works were then installed. Four years after installation the river and the confluence had indeed recovered their previous configuration, as shown in Figure 22.1d. The training works have continued to function as designed, and survived the near-record flood of 1993 with neither serious damage nor degradation of river configuration.

The above engineering study suggests the following conundrum for basic research. Perhaps tributaries tend to enter meandering streams on the outside of a bend because the confluence represents a quasi-stable configuration. The outside of a bend typically represents a zone of increased flow velocity. In particular, the region just upstream of the apex typically represents a zone within which sediment transport capacity is high and increasing in the downstream direction, and sediment supply may be limited (Bathurst *et al.*, 1979). This would render the stream capable of picking up the extra sediment load from the tributary without major channel instability. The meander bend would migrate in time, perhaps eventually causing the tributary to avulse to the outside of another bend. The inside of a bend, on the other hand, would represent a highly unstable location for a confluence, due to its depositional nature.

An unpublished study of tributary confluences on meandering streams in Minnesota was conducted by R. Hills, a former graduate student of the author's, subsequent to the Minnesota River investigation. The study revealed that the outside of a bend upstream of the apex was the most common location for tributary confluences. An example of such a confluence from Papua New Guinea is shown in Figure 22.2. The example of Figure 22.3 from Siberia is from the study of meandering rivers by Alabyan (this volume). It is of interest to note that the stream confluence in Illinois studied by Rhoads (this volume) also has the same configuration. This topic would seem worthy of more study.

CONCAVE-BANK BENCHES

The conventional wisdom is that the outside banks of bends are subject to erosion and the inside subject to deposition. A river must, however, deposit sediment on the outside of a bend if the bend is to migrate downstream. If the bend is sufficiently sharp, part of this deposition can take the very specific morphology of a concave-bank bench (e.g. Hickin, 1979; Page & Nanson, 1982). Such a bench develops in an indentation of the outside of a bend just upstream of the apex within which flow separation occurs, creating the local depositional zone. The bend itself often has a characteristic 'hammer head' shape and commonly impinges upon a zone of reduced erodibility such as a valley side-wall or clay plug.

A concave-bank bench can be seen as the large grassed area in Figure 22.4a. The river in question is the Ok Tedi, Papua New Guinea. The fate of the bench between

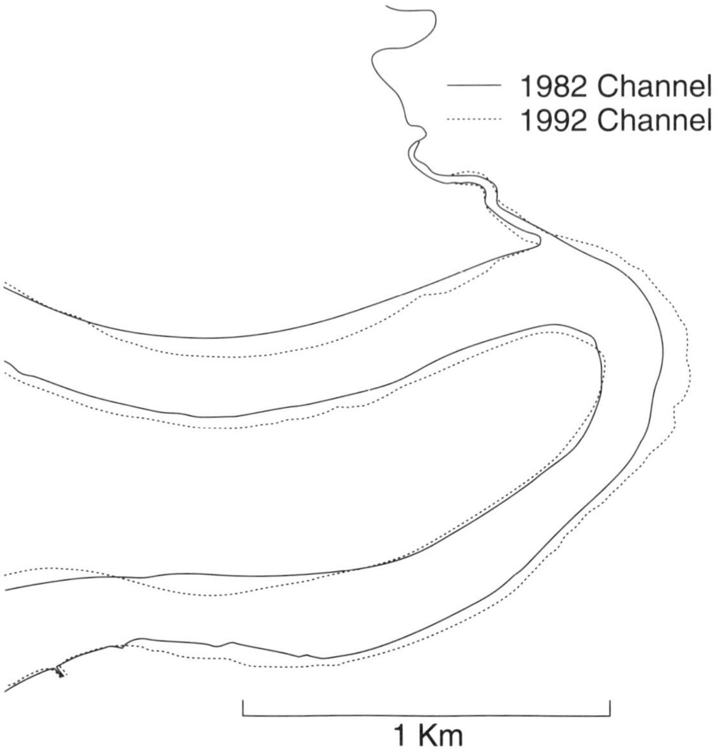

Figure 22.2. Tributary confluence on the Ok Tedi, Papua New Guinea, showing the bank lines of the main stream in 1982 and 1992

1982 and 1992 is shown in Figure 22.4b; it essentially preserves shape as the bend migrates downstream. Outward migration of the bend is prevented by the valley wall visible in Figure 22.4a, atop which lies the village of Atkamba.

In the course of perusing aerial photographs of the Fly River, Papua New Guinea, W.E. Dietrich and the author verified that under the right conditions the flow-sediment interaction responsible for the formation of concave-bank benches can be self-preserving over hundreds of years. The bench visible in Figure 22.4c on the Fly River, Papua New Guinea has maintained itself over many river widths of channel migration, as indicated by the crescentic features reminiscent of scroll bars on the floodplain just upstream. Indeed, the bench may persist until such time as the bend it occupies cuts-off to form an oxbow lake, as shown in Figure 22.4d, also from the Fly River. While many useful descriptions of concave-bank benches exist in the literature, their mechanistic description as a channel morphology of permanent form remains to be explored. It is of value to note in this regard that the three benches shown here are all associated with sharp bends which is a likely requirement for the formation of the flow structure that leads to their generation.

Figure 22.3. Tributary confluence on a meandering stream in Siberia. Photograph courtesy of A. Zaitsev

THE ORIGIN OF SCROLL BARS ON THE FLOODPLAINS OF MEANDERING STREAMS

One of the most beautiful and striking features of meandering rivers consists of the scroll bars that provide a record of a bend's previous migration. These become remarkably clear just as the flow stage declines below bankfull, as shown in Figure 22.5a for the Fly River, Papua New Guinea just downstream of the confluence with the Strickland River. What are scroll bars, and how are they formed?

Figures 22.5b and 22.5c are from the body of research reported in Parker (1976), in which self-formed meanders were studied in the laboratory. Figure 22.5b shows the bed configuration one hour after the commencement of an experiment, with the flow turned off. The two features that characterize the bed are bars and oblique dunes. The same channel is shown three hours later with the flow turned off in Figure 22.5c. The oblique dunes have evidently been rotated and beached successively downstream of the apex of each bend so as to form features that are strongly reminiscent of scroll bars. R.L. Gerard pointed this out to the author in the course of discussion in the 1970s.

(a)

(b)

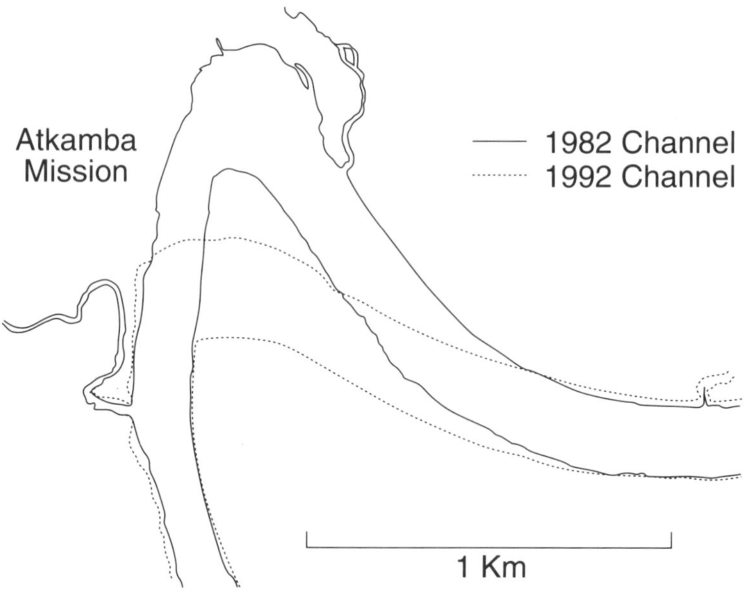

Figure 22.4. (a) View of a concave-bank bench on the Ok Tedi, Papua New Guinea. The village visible in the photograph, Atkamba, rests atop the valley wall. Flow is directed from bottom left to middle left of the photograph. Photograph courtesy of W.E. Dietrich. (b) Configuration of the concave-bank bench of Figure 22.4a, showing the bank lines in 1982 and 1992. *Continued on facing page*

(c) (d)

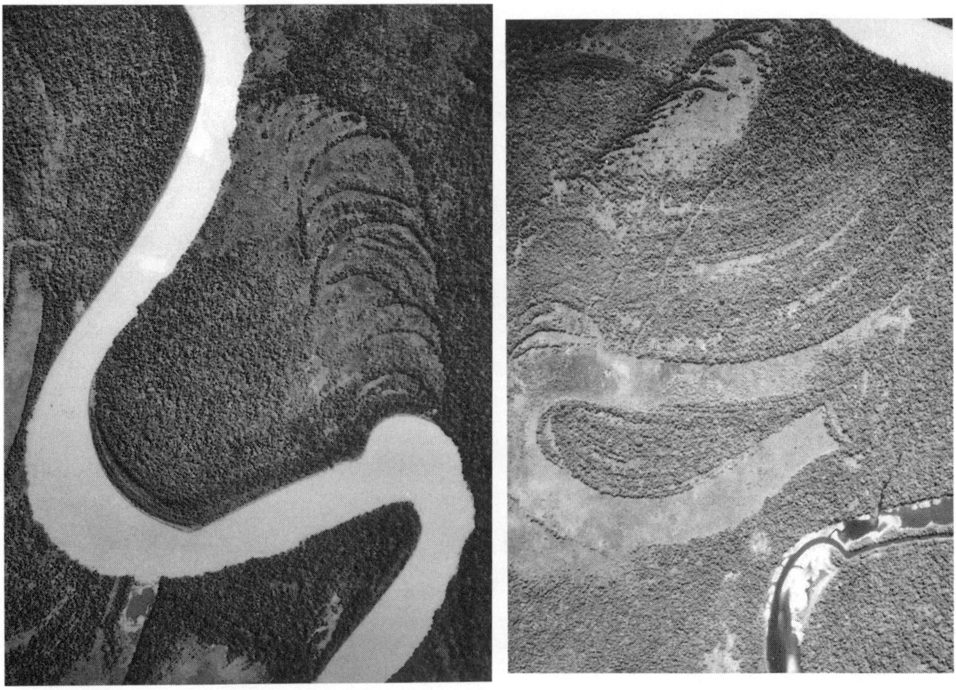

Figure 22.4. *continued* (c) An active concave-bank bench on the Fly River, Papua New Guinea. The crescentic features on the floodplain above it record its migration. River width is on the order of 250 m. (d) A relict concave-bank bench on a cut-off of the Fly River, Papua New Guinea, and its associated crescentic deposits

The author would like to suggest that scroll bars may be identifiable with highly elongated, partially beached oblique dunes. Jackson (1976) provides some field evidence from the Wabash River, USA suggesting that this might be the case, but the idea seems not to have been pursued further. A study of oblique dunes can be found in Hasbo *et al.* (1995). An exploration of their relation to scroll bars would be of value. In particular, a delineation of the conditions under which sediment ceases to accrete upon a given beached oblique dune, resulting in the creation of its successor, would seem quite interesting.

FLOODPLAIN FORMATION

Perhaps one of the most overlooked topics of rivers is floodplain formation. Schumm & Lichty (1963) documented the removal of the vegetated, fine-grained floodplain of a small meandering river by a prolonged drought followed by a sequence of severe floods. The narrow meandering river was converted to a much

(a)

Figure 22.5. (a) Scroll bars on the floodplain of the Fly River, Papua New Guinea, just downstream of the confluence with the Strickland River. Photograph courtesy Environment Department, Ok Tedi Mining Ltd. *Continued on facing page*

wider braided stream. Over the next several decades the stream narrowed substantially as it reconstructed its floodplain.

The mechanisms by which floodplains are formed and removed remain the subject of considerable speculation. James (1985) and Pizzuto (1987) have proposed diffusion-type models, and Ikeda & Izumi (1991) and Tsujimoto (1994) have documented large-scale eddies associated with channel-floodplain interaction (see Willetts & Rameshwaran, this volume; Naish & Sellin, this volume) that may be responsible for this process. Advective processes are probably also important, as suggested by the observed pattern of deposition on the Iowa River, Iowa, USA due to the record flood of 1993, as shown in Figure 22.6.

The topic of floodplain formation remains relatively under-investigated, particularly in light of the fact that floodplains tend to be repositories for contaminants adsorbed to fine-grained sediments. Simm (1993), however, is one of several authors who have recently studied the process in the field. A predictive mechanistic theory would be of great value.

(b) (c)

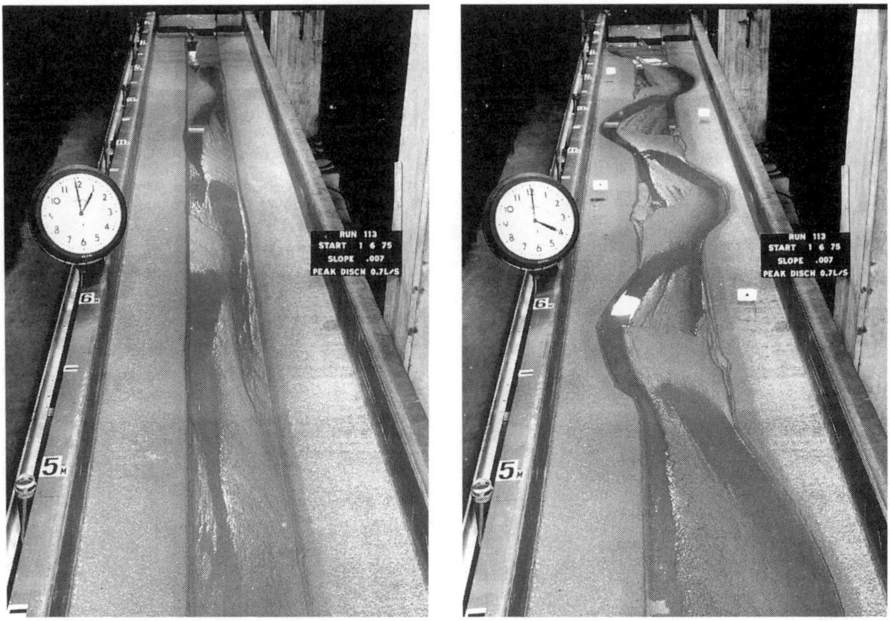

Figure 22.5. *continued* (b) Laboratory channel 1 hour after commencing the flow. Alternate bars and oblique dunes are visible in the channel. Flow is from bottom to top of the photograph. (c) The same laboratory channel 3 hours later. The oblique dunes on the downstream side of the bars have been rotated and beached in a pattern reminiscent of scroll bars

CHANNELIZED SUBMARINE FANS

Fine-grained sediment is easily mobilized. If fresh deposits of fine-grained material on the floodplain of a terrestrial meandering river are not to be removed by the next flood, then some mechanism must decrease their erodibility. One candidate for this is cohesion, either real due to grain-grain electrochemical effects or apparent due to, for example, biologically generated binding. A second candidate is vegetation.

The submarine fan constitutes an interesting environment within which at least one of these features, vegetation, is absent. The agent of flow, sediment transport and sediment deposition is the turbidity current. Turbidity currents are distinguished from their subaerial analogues, rivers, in that the water of a river drags the sediment down-slope, whereas the sediment in a turbidity current drags the water down-slope. Turbidity currents are capable of building magnificent submarine fans, many of which are coursed by large meandering channels bounded by high natural levees. This is shown for the case of the Amazon Fan in Figures 22.7a and 22.7c (Damuth

Figure 22.6. Floodplain deposition on the Iowa River, Iowa in response to the record flood of 1993. Photograph courtesy T. Nakato

et al., 1988); similar observations for the Mississippi Fan can be found in Twitchell *et al.* (1991).

A question of interest concerns the spontaneous self-formation of a well-defined channel bounded by natural levees in the environment of a fan, which by its nature must be depositional in the overall sense. More specifically, why does deposition not obliterate the channel? The author is conducting joint research on this topic with J. Imran, N. Katopodes and S. Bradford. A tentative first numerical calculation showing the initial stages of a channel bounded by natural levees formed by a depositional turbidity current is shown in Figure 22.7d. Mechanistic research on this subject, however, has only just begun. Although the submarine environment is different in many respects from the subaerial one, an explanation of self-channelization in the former case may lend understanding to the latter.

DOWNSTREAM FINING

Most gravel-bed rivers tend to become finer in the downstream direction (for example, Paola *et al.*, 1992 provide a list of relevant references). This tendency can be driven by abrasion or selective deposition of the coarser clasts. The latter effect in turn requires a depositional environment created by such factors as channel

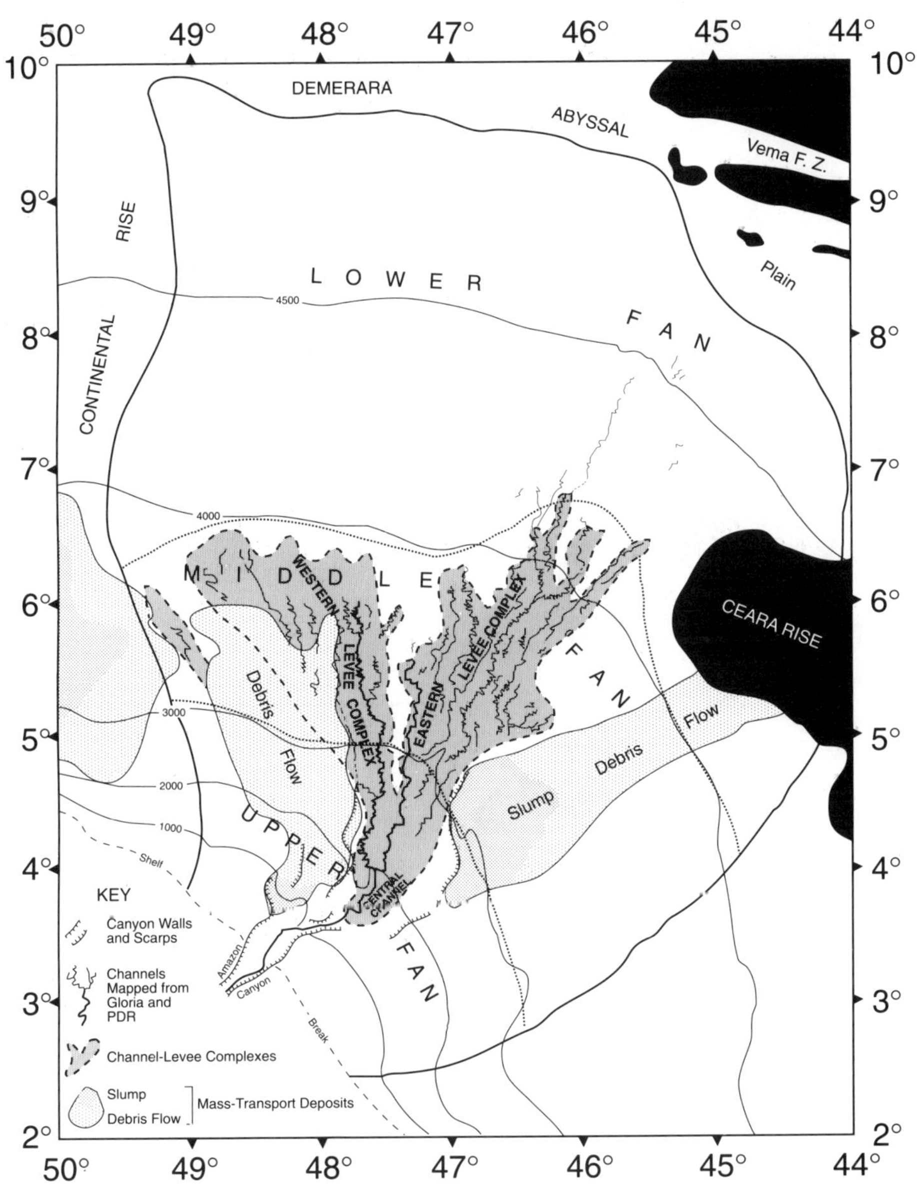

Figure 22.7. (a) Major morphologic and physiographic features of the Amazon Submarine Fan. From Damuth *et al.* (1988). *Continued on next page*

(b)

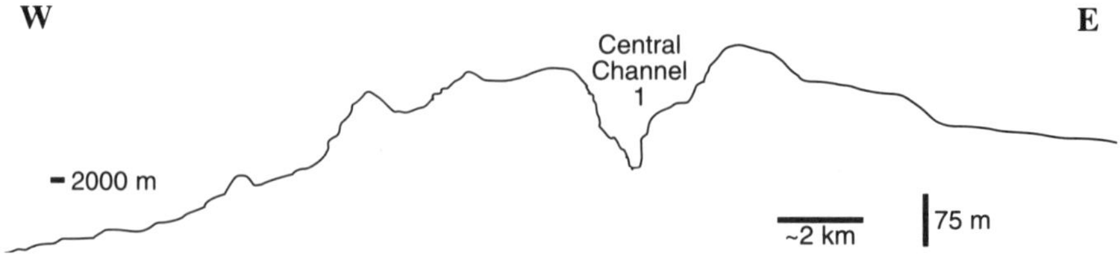

(c)

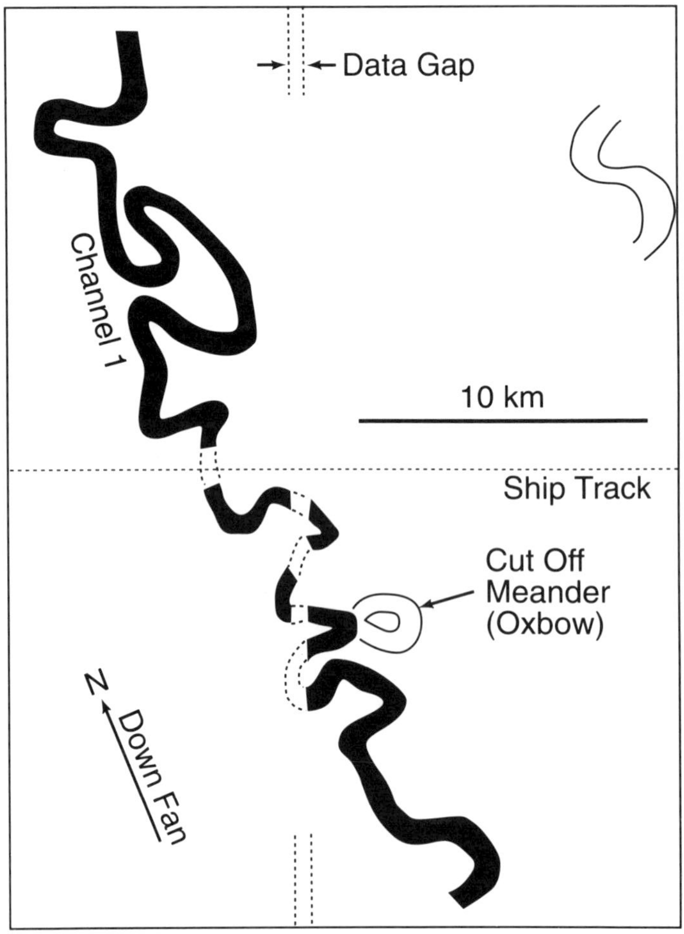

Figure 22.7. *continued* (b) Acoustic profile across the upper Amazon Fan showing the levee-bounded main channel. From Damuth *et al.* (1988). (c) Interpretation of an acoustic image of a meandering distributary on the Amazon Fan. From Damuth *et al.* (1988). *Continued on facing page*

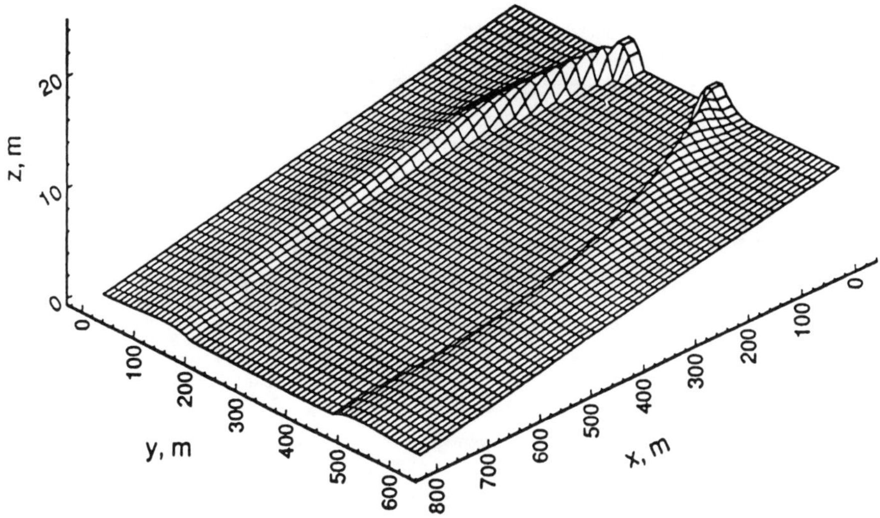

Figure 22.7. *continued* (d) Numerical prediction of the initial development of a channel bounded by natural levees produced by a depositional turbidity current. The initial condition is that of a flat, slightly tilted plane. The calculation was done by J. Imran

aggradation or basin subsidence. The author has been involved in an extensive research program on the subject with the following co-workers: C. Paola, P. Wilcock. J. Southard, R. Seal, S.K. Sinha, B. McArdell, Y. Cui and C.M. Toro-Escobar. The effort has yielded a number of publications, a complete list of which is available from the author. This work and the work of others (e.g. Mikos, 1993; Hoey & Ferguson, 1994; Kodama, 1994; Pizzuto, 1995) has resulted in an overall understanding of the mechanics of downstream fining, although a number of details remain to be understood.

The purpose here is not to suggest new research topics on the subject. Rather, Figures 22.8a–22.8e, which illustrate downstream fining as modelled in the laboratory in an aggradational environment, serve as a foil for two other interesting topics discussed below; (i) the counterintuitive phenomenon of downstream coarsening, and (ii) gravel-sand transitions. The experiment is Run 1 of Seal (1994). A mixture of gravel and sand ranging from under 0.1 mm to over 70 mm was fed into a flow of water in a channel with a width of 0.3 m and a length in excess of 60 m. The mixture was 70% gravel and coarser and 30% sand; it was weakly bimodal with the paucity near 2 mm, i.e. in the vicinity of the transition between sand and gravel sizes. The sediment was fed in at a rate of 11.3 kg min^{-1}; the water discharge was 49 l s^{-1}. The deposit was allowed to prograde freely over an antecedent, nearly horizontal bare bed. The first three photographs, Figures 22.8a–22.8c, show the surface of the deposit at the end of the run at distances of 13, 33 and 37 m downstream of an arbitrary origin near the sediment feed point.

(a)

(b)

Figure 22.8. Run 1, experiment on downstream fining, St. Anthony Falls Laboratory. The surface of deposit is shown for (a) 13 m and (b) 33 m downstream from the origin. Flow is from left to right. *Continued on facing page*

(c)

Figure 22.8. *continued* Run 1, experiment on downstream fining, St. Anthony Falls Laboratory. (c) The surface of the deposit is shown 37 m from the origin. *Continued on next page*

The progressive downstream fining is readily apparent. In this particular case, the mechanism for fining is the differential mobility of grains exposed on the surface due to imperfect hiding. That is, the effect of protrusion, which favours the mobility of coarse grains, does not quite balance out the effect of weight, which favours fine grains. As a result, the finer grains are rendered somewhat more mobile and, therefore, deposited less easily under aggrading conditions.

The last two photographs, Figures 22.8d and 22.8e, were taken at a different scale from the other three. They illustrate a feature of interest, i.e. a sharp transition from gravel-bed to sand-bed. The gravel deposit ended in a distinct prograding front with a slip face. All the gravel deposited before or at the slip face. The few gravel clasts visible beyond the slip face were placed there by reworking when the channel was drained. During the run itself only material in suspension could cross the zone of sharply dropping shear stress beyond the slip face, resulting in a deposit of pure sand.

(d)

(e)

Figure 22.8. *continued* Run 1, experiment on downstream fining, St. Anthony Falls Laboratory. (d) A view of the gravel-sand transition, located approximately 38 m from the origin. The gravel mixed in with the sand was placed there by reworking due to draining at the end of the run. (e) A view just downstream of the gravel-sand transition. Note the scale of photos 22.8d and 22.8e is different to that of Figures 22.8a–22.8c

DOWNSTREAM COARSENING

A number of authors have provided evidence for a reversal of mobility in the pea-sized gravel range (e.g. Straub, 1935; Kellerhals, 1967), according to which the mobility of a mixture of sand and pea-sized gravel is biased toward the coarser material. In late 1994 the author and a colleague, K. Taki accidentally reproduced this phenomenon in an attempt to model step-pool topography in a small channel with a length of 2 m and width of 0.05 m. The size distribution of the feed sediment is shown in Figure 22.9a; it consists of a unimodal mix of sand and gravel ranging from about 0.2 to 7 mm. A deposit was allowed to form by aggradation, and then the supply of sediment was cut off in the hope of inducing degradation to an armoured step-pool topography. The flow depth was shallow, i.e. on the order of 0.01 m and the sediment moved as bedload. In the course of degradation, however, neither armouring nor a step-pool topography evolved. When an aggradational phase was repeated, it was noticed that the coarser material was preferentially transported to the prograding front of the deposit.

Subsequent to this trial experiment, a more extensive set of experiments using the same sediment mixture as that of Figure 22.9a was conducted by G. Chang, then an undergraduate working under the supervision of the author. During one experiment, sediment was fed at the upstream at a rate of 8 g s^{-1} and water was supplied at the rate of 0.178 l s^{-1}. The initial bed was flat; a tailgate set a few centimetres above the bed ensured the formation of a progradational front.

As the deposit developed and the front prograded, most of the pea-sized gravel was observed to migrate to the vicinity of the front before depositing, with most of the sand depositing well behind the front. As opposed to the runs in the larger facility using coarser sediment discussed above, the aggrading profile showed virtually no upward concavity. The tendency for downstream coarsening during aggradation was also clearly displayed in bedload samples as well. When the bed finally reached equilibrium slope, downstream fining no longer appeared in the bedload. Rather, bedload transport took place as a series of alternating coarse and fine sheets. The deposit below, however, preserved the record of downstream coarsening during the aggradational phase, as shown in Figure 22.9b.

The photograph of the deposit given in Figure 22.9c illustrates the tendency for downstream coarsening. The very coarse layer at the bottom records the passage of the coarse progradational front, which was then buried in finer sediment. A second coarse layer higher up records a period of 'hesitation' after the progradational front had reached the tailgate but before rotational aggradation about the tailgate had commenced.

The author would like to propose the following hypothesis to explain downstream coarsening. Viscous effects are relatively unimportant in a consideration of the relative mobility of grain sizes in a mix of gravel. As noted above, the effect of relative protrusion is not quite sufficient to counterbalance the effect of weight, resulting in a tendency for preferential transport of the finer of

(a)

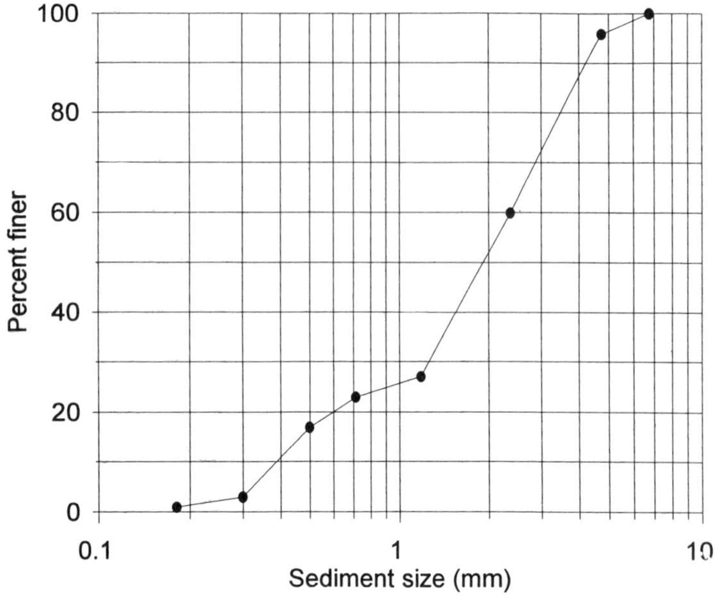

(b)

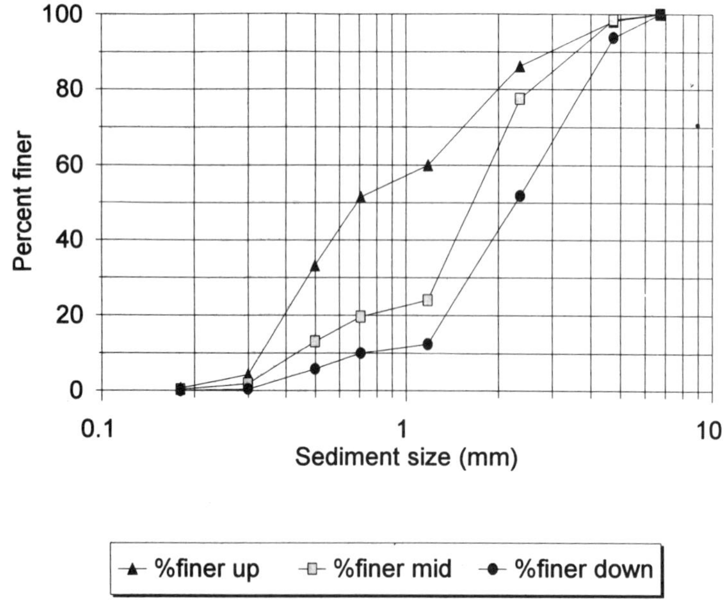

Figure 22.9. (a) Size distribution of the sediment used in the experimental study of downstream coarsening. (b) Size distributions of three vertical columns of excavated deposit, one located upstream, one a the middle of the flume and one located downstream. The tendency for downstream coarsening is readily evident. *Continued on facing page*

Figure 22.9. *continued* (c) Photograph of the deposit. The coarse bottom layer records the passage of the coarse progradational front. Above it, the quantity of coarser material is seen to increase from left to right, i.e. in the downstream direction

those grains exposed on the surface. In the case of a mixture of sand and pea-sized gravel, however, under the right conditions a partial viscous sublayer can exist. The finer grains may hide below the viscous sublayer, thus shielded from the turbulence and in particular the sweeps responsible for grain mobilization. The coarser grains feel the direct effect of boundary layer turbulence and are thus moved preferentially. Whilst this explanation is only a hypothesis, it would seem worthy of further investigation.

EVOLUTION OF SUBAERIAL ALLUVIAL FANS

Subaerial fans are no less interesting than their subaqueous counterpart. Subaerial fans can be divided into two types: (i) those dominated by debris or sheet flow; and (ii) those formed by the constant avulsion of braided or anastomosing channels that at any given time occupy only a small part of the fan surface. A field example of the latter type is provided by the fluvial fan of the Kosi River, India shown in Figure 22.10a (Gole & Chitale, 1966). Fans represent a fairly regular morphology created by fairly complex phenomena, the relative importance of which is not easily discerned. Occasionally, however, one finds a simple case that allows direct access to the heart of the matter.

The Hibbing Taconite Company operates a large iron mine in northern Minnesota. The tailings produced by the mine are delivered to a series of tailings basins. The specific gravity of the tailings is 3.0; sizes range from mechanically produced clay-size particles to clasts on the order of 10 mm, with the majority of material in the silt and sand sizes. The most proximal of the tailings basins in particular provides an excellent model of a fluvial fan dominated by braided channels, as shown in Figure 22.10b. Water and sediment discharge into the basin

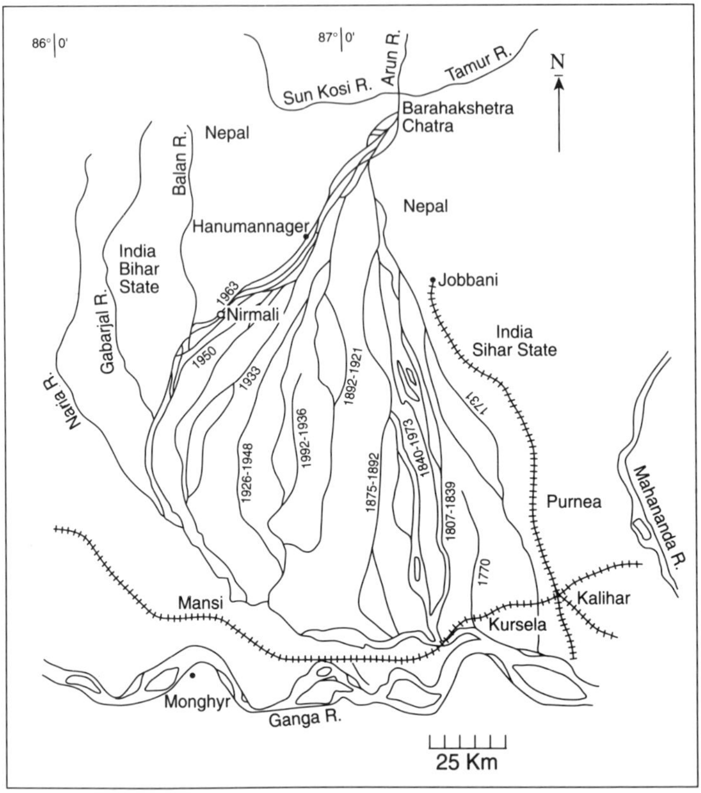

Figure 22.10. (a) The fluvial fan of the Kosi River, India. From Gole & Chitale (1966). *Continued on facing page*

are, however, held constant at 7.5 and 0.235 $m^3\ s^{-1}$, respectively. The entire process of fan building can thus be studied without the unpredictable vagaries of water and sediment delivery characteristic of natural fans.

The case possesses one more analogy to the natural case. In order to prevent overflow of the sediment, the dikes surrounding the basin are gradually built up in time. The result is the maintenance of a nearly constant vertical rate of aggradation, as shown in Figure 22.10c. This renders the case perfectly analogous to a natural fan the horizontal extent of which does not grow due to a perfect balance between deposition and subsidence. The implications of this balance have been studied in detail for the case of alluvial basins by Paola (1988).

The author is working with three co-workers, K.X. Whipple, C. Paola and D. Mohrig to develop a theoretical framework of fan evolution. At the time of writing the theory is incomplete, although an important part can be found in Paola (this

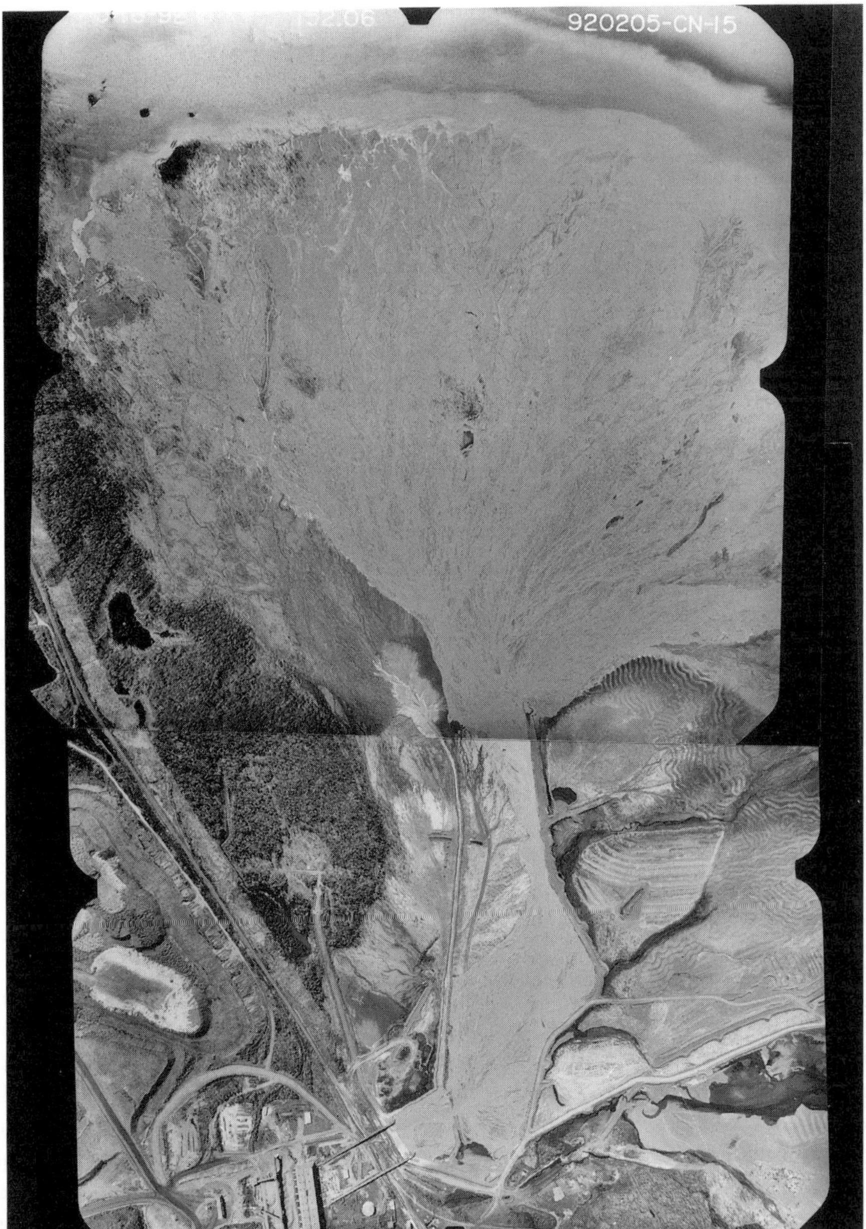

Figure 22.10.*continued* (b) View of the fan in the West Basin of the tailings pond of Hibbing Taconite Company, Hibbing, Minnesota, USA. Photograph courtesy of Hibbing Taconite Co. *Continued on next page*

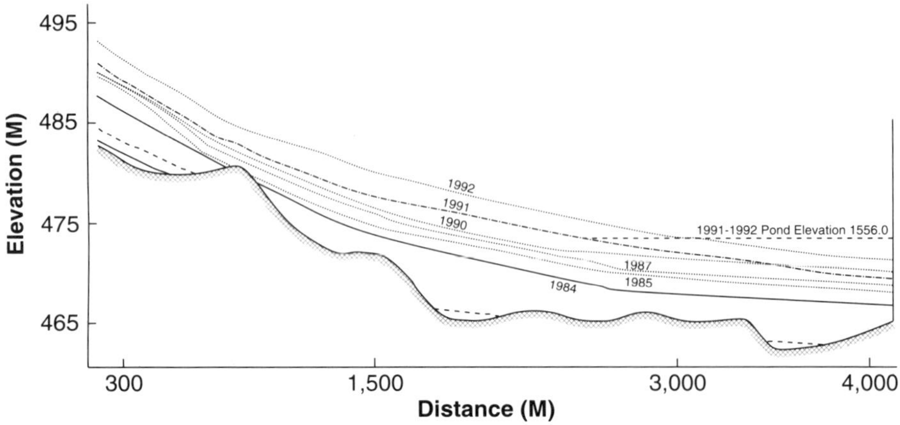

Figure 22.10. *continued* (c) Successive bed profiles of the West Basin of the tailings basin. Courtesy of Hibbing Taconite Co.

volume). In addition, some related experimental work is reported in Whipple *et al.* (1995). An analysis of field data suggests, however, that the composite active width of the braided channels can be determined in terms of a simple relation, according to which the quantity corresponding to the product of the depth and slope of an anabranch divided by characteristic grain size remains approximately constant. It is of some interest to note that the relation determining channel width does not directly involve width! Relations of this form were first postulated by Parker (1978a, 1978b). The study of the morphology of fluvial fans from a mechanistic point of view promises to be a most fascinating research area.

CYCLIC STEPS

As noted above, a number of physical model experiments were conducted in the course of the study of alluvial fans discussed above (Whipple *et al.*, 1995). The model basin was a quarter-circle with a radius of 2.1 m; typical bed slopes of the fan formed in the basin ranged from 0.05 to 0.10. An interesting periodic morphology was observed in the course of these experiments. When fine sand or silt was used, a series of slowly upstream-migrating steps often formed in the braid anabranches. Each step was manifested as a zone of steeply dropping flow bounded at the downstream end by a hydraulic jump. These steps were gentler for coarser material, appearing analogous to chute-and-pool topography. For finer material the steps formed sharp headcuts. The flow immediately upstream of the jump was supercritical, whereas that downstream of the jump was subcritical. These features are called cyclic steps here because of their spatially periodic manifestation.

It was found that these same steps could be observed in the field at the Hibbing Taconite tailings basin. They have also been observed in the field by Winterwerp (1992). None of these cases correspond to equilibrium overall conditions; in the model and field fans the bed was aggrading, whereas the bed was degrading in the case of the observations of Winterwerp. In order to improve the study of this phenomenon, these steps were reproduced under overall equilibrium conditions in the same small flume used to study downstream coarsening and described above. Three bed materials were used in the experiments, with characteristic sizes of 19, 45 and 120 μm. Water discharges ranged from 0.01 to 0.09 $l\ s^{-1}$, and sediment feed rates ranged from 0.5 to 8 $g\ s^{-1}$. Two such steps in succession are shown in Figure 22.11a. Close-up views of steps in the vicinity of the hydraulic jump are given in Figure 22.11b for a relatively gentle step and Figure 22.11c for a very sharp step. The mode of sediment transport was predominantly suspension, with erosion dominating in the supercritical zone and deposition in the subcritical zone.

The above steps may be described as transportational in nature, in that although they are formed by spatially alternating erosion and deposition, on average they must transport the entire load fed to them from the upstream end of the channel. It appears, however, that transportational cyclic steps have a purely erosional analogy associated with hillslope gullying. Under the correct conditions, an incising channel on a steep slope will devolve into a series of headcuts otherwise known as discontinuous gullying. Reid (1989) has observed such purely erosional cyclic steps in the field and Sawai (1977) has observed them in the laboratory. An example of erosional cyclic steps is provided in Figure 22.11d; a close-up view of a headcut is given in Figure 22.11e.

The author and N. Izumi have completed a theoretical description of erosional cyclic steps (Parker & Izumi, 1995). Joint research with K. Taki proceeds in regard to the transportational case. Both cases provide excellent examples of quasi-equilibrium erodible-bed behaviour for which supercritical flow plays a crucial role.

STEP-POOL TOPOGRAPHY

As outlined above, the study of downstream coarsening arose from a failed attempt by K. Taki and the author to reproduce step-pool topography in a small flume. When the sediment consisted of sand and pea-sized gravel step-pool topography would not form. When coarser material was added to the mix to increase the maximum size to about 15 mm, however, the bed again displayed downstream fining during aggradation and eroded back to a step-pool topography during degradation.

The purpose of the pilot experiments was a test of the hypothesis that step-pool topography is another manifestation of the same mechanism that gives rise to cyclic steps. To this end a water discharge of 0.18 $l\ s^{-1}$ and a sediment feed of 2.7 $g\ s^{-1}$ were introduced into the same small channel previously used for a study

(a)

(b)

Figure 22.11. (a) Transportational cyclic steps reproduced in a small laboratory flume. (b) Close-up of a relatively gentle step. *Continued on facing page*

(c)

(d)

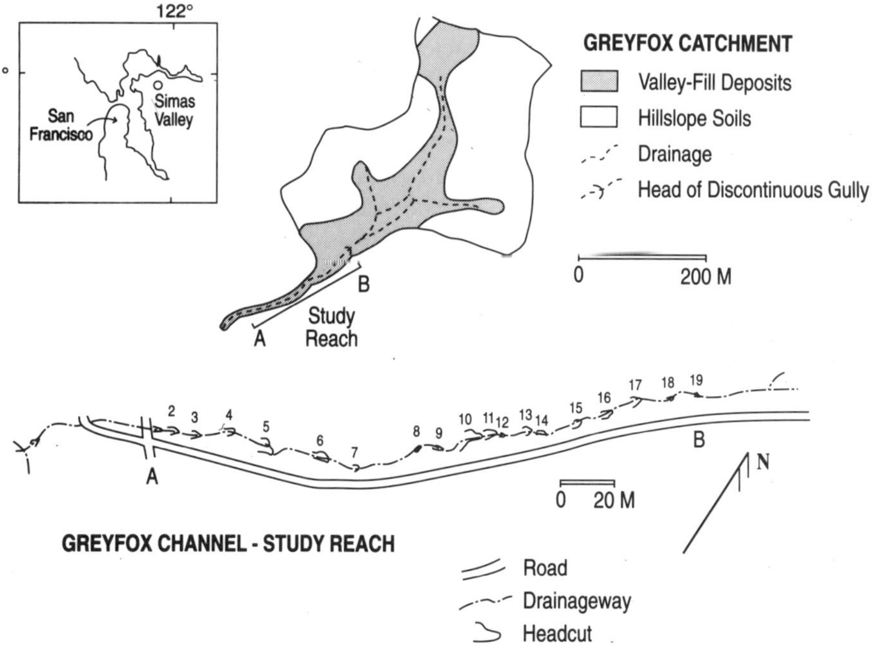

Figure 22.11. *continued* (c) Close-up of a steep step. (d) A series of erosional cyclic steps in an incising hillslope channel (from Reid, 1989). *Continued on next page*

(e)

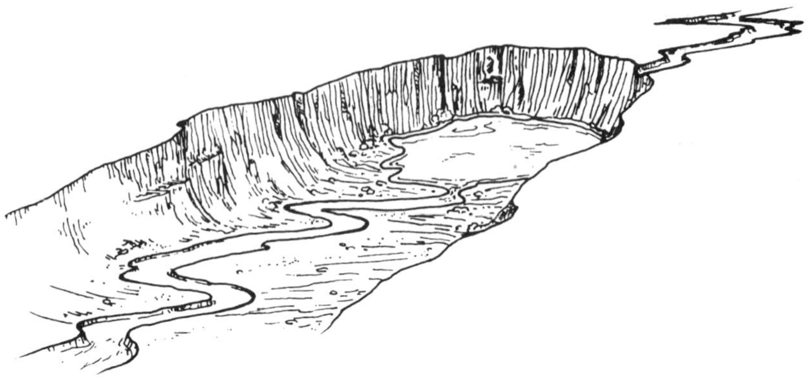

Figure 22.11. *continued* (e) Close-up of an erosional step. From Reid (1989)

of cyclic steps and later used for a study of downstream coarsening. The bed was allowed to aggrade until near-equilibrium conditions were reached. At these conditions the bed was characterized by a train of well-developed antidunes that showed little tendency to migrate either upstream or downstream. The coarsest grains tended to collect under the crests of the antidunes. The clusters of coarse clasts created in this way were not stable. The continuous transport of all sizes caused the clusters to break up and reform at random intervals.

After maintaining the equilibrium state for a period, the supply of sediment to the channel was cut off. The channel then degraded, so that the coarse material below the crest of each antidune stabilized into a step-pool topography. This topography is shown in Figure 22.12. The stabilized long profile of the bed was upward concave, with an upstream slope near 0.10 and a lower slope near 0.07.

The experiment indicated that while step-pool topography is a cousin of cyclic steps in that supercritical flow plays an important role, it is not a sibling. While the stabilized coarse steps themselves are often characterized by hydraulic jumps reminiscent of cyclic steps, the jumps are not present under the antecedent equilibrium conditions that cause the coarser grains to collect under antidunes. That is, whereas cyclic steps can be generated under conditions of dynamic quasi-equilibrium transport or erosion of sediment, step-pool topography is best described in terms of a static equilibrium. The explanation of step-pool topography as stabilized antidunes caused by a reduction in sediment supply originally offered by Whittaker & Jaeggi (1982), and confirmed by Ashida *et al.* (1986) and Grant & Mizuyama (1991), appears to be the correct one.

The above comments notwithstanding, the understanding of step-pool topography nevertheless remains somewhat incomplete. In particular, it should be possible to provide a complete explanation of the evolution of step-pool

Figure 22.12. Photograph of step-pool topography modelled in the laboratory

topography in terms of a numerical model of degradation of heterogeneous sediments. The armouring process might be treated using, for example, the method of Parker & Sutherland (1990); any description of the flow field must encompass the antidune instability itself.

GRAVEL-SAND TRANSITIONS

As noted above, the deposit produced by an experiment on downstream fining shown in Figures 22.8a–22.8e possessed a sharp transition from a bed in which gravel predominated to a bed of pure sand. Such transitions also occur commonly in nature, although not necessarily for the same reasons. In particular, it is common for a gravel-bed stream to undergo a rather sudden metamorphosis to a sand-bed stream downstream of a specific point. The point of transition is characterized by a sudden drop in characteristic grain size, and is often accompanied by a simultaneous sharp change in slope. In addition, morphology often changes from a gravel-bed braided or wandering state to a sand-bed meandering state. The classic study documenting this is Yatsu (1955); see also Sambrook Smith & Ferguson (1995). The case of the Kinu River, Japan is documented in Figures 22.13a and 22.13b. Of

(a)

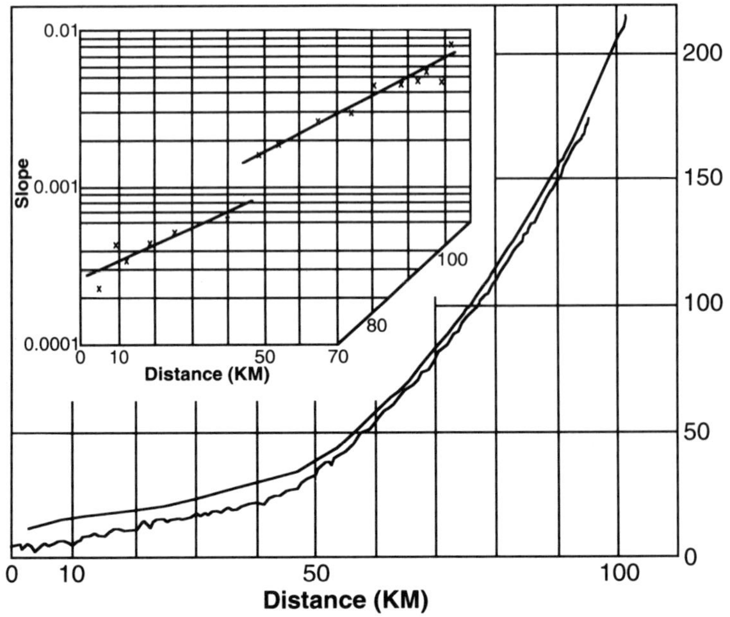

(b)

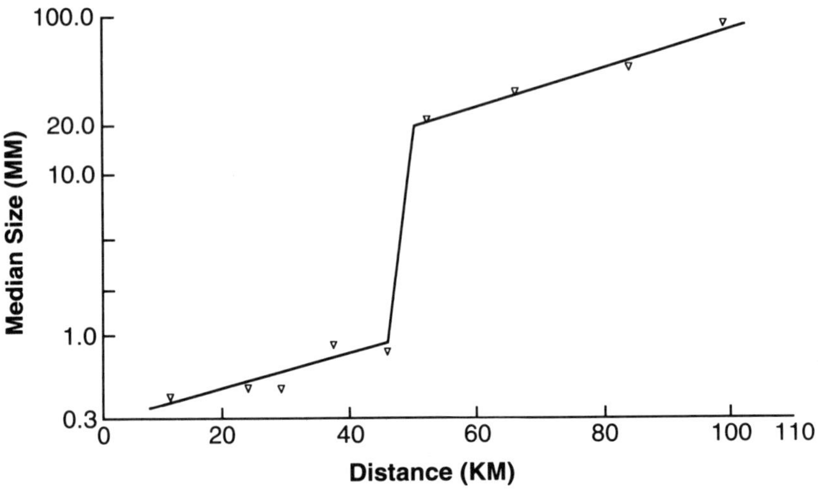

Figure 22.13. (a) Long profile of the Kinu River, Japan, showing a sudden change in bed slope. Based on an original from Yatsu (1955). (b) Downstream grain size variation of the Kinu River, Japan, showing the gravel-sand transition. Based on an original from Yatsu (1955)

interest here is the fact that gravel-sand transitions appear to be large-scale quasi-equilibrium rather than transient geomorphic forms.

It is reasonable to expect that the gravel-sand transition is associated with the tendency for bimodality in gravel-sand mixtures, such that the pea gravel size range is under-represented. At least two explanations have been offered for this paucity. As described above, the mobility of pea-sized gravel in a mixture with sand may be enhanced, increasing the frequency of saltation, collision with the bed and other particles and thus breakdown. In addition, in the case of crystalline rock, pea-sized clasts often consist of a relatively few crystals and thus may be subject to shattering. These issues are discussed in Shaw & Kellerhals (1982).

If the presence of bimodality is accepted, then at least two possible explanations for quasi-equilibrium gravel-sand transitions can be offered. The first of these is abrasion. In this case, the equilibrium bed profile is set by a balance between sediment deposition and abrasion. Gravel is assumed to abrade down to a critical size, e.g. 6 mm, beyond which it spontaneously (or quickly) shatters into its constituent sand-size crystals as described above. The result is a concave long profile with a sharp gravel-sand transition.

Another possibility is basin subsidence. Here it is assumed that the input sediment is already bimodal, and that no further abrasion is allowed. A balance between deposition and basin subsidence yields a concave long profile that fines in the downstream direction. The gravel-sand transition occurs at the point where sizes in the gap should dominate.

It is likely that both the mechanisms of abrasion and subsidence play a role in quasi-equilibrium gravel-sand transitions. Y. Cui, C. Paola and the author are presently pursuing a generalized theory of the phenomenon. The results obtained to date are promising.

TWO PARTING SHOTS AND A CONCLUSION

Two final phenomena are worthy of mention. Trains of antidunes are ubiquitous in the tailings basin of Hibbing Taconite Company, as shown in Figure 22.14. At lower slopes the trains deform gradually. At higher slopes, however, a given train often grows in amplitude until the onset of a sudden domino-like breaking of all the waves in the train and the restoration of a flat bed. This breaking releases stored water in the form of a roll wave. The pattern of antidune growth and breaking with the release of a roll wave was found to be periodic, with a period of the order of a minute. This same phenomenon has been reproduced in the laboratory by Foley (1975). It is a ripe plum for a mechanistic explanation in terms of non-linear wave interaction.

Figure 22.15 was supplied to the author by J. Fredsøe as a result of discussion that was initiated at the time of the conference which this volume summarizes. It shows two adjacent rivers in Iceland originating from the same glacial source. One

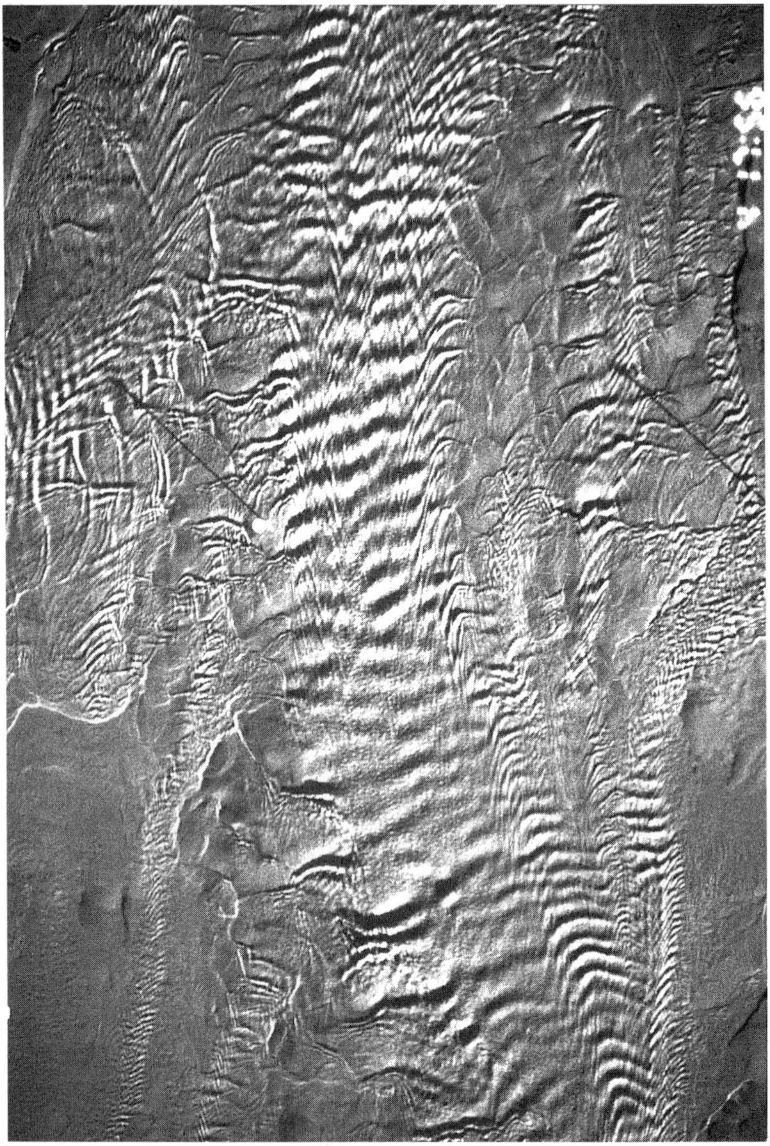

Figure 22.14. Antidune trains in braid anabranches of the West Basin, Hibbing Taconite Company, Hibbing, Minnesota, USA. The photograph was taken from a balloon operated by D. Mohrig. The flow is from top to bottom. The photograph encompasses a transverse length of about 14 m

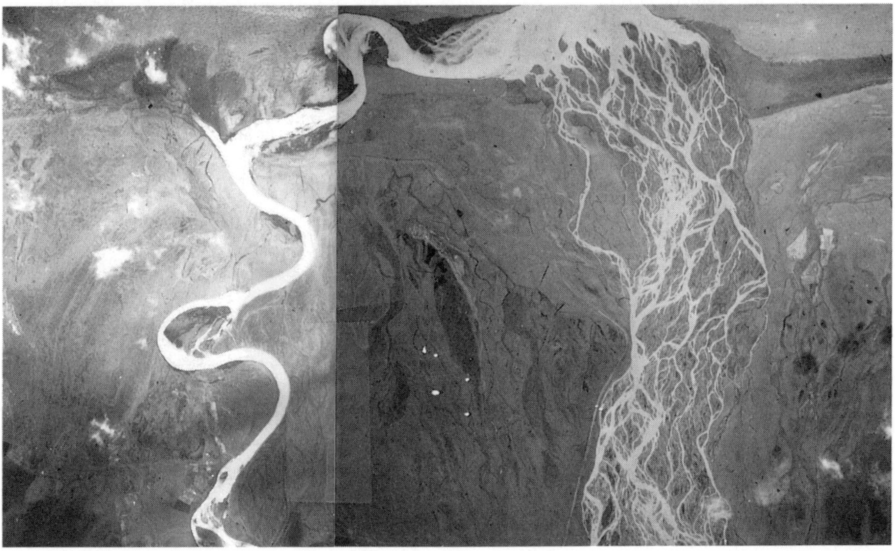

Figure 22.15. Two streams, one braided and one meandering, flowing from the same glacial source in Iceland. The flow is from top to bottom. The photograph was provided by J. Fredsøe

is strongly meandering and the other is strongly braided. Several explanations can be surmised to explain the difference. For example, Fredsøe (1978) suggests that suspended sediment enhances braiding. According to this argument the bulk of the fine sediment from underneath the glacier would be entering the channel to the right. Another possible explanation is that a disproportionate amount of all sizes, but particularly the coarse sizes, are entering the channel to the right. The right channel may thus be aggrading, or at least near equilibrium, whereas the sediment-starved channel to the left may be degrading. The relative paucity of sediment in the channel to the left might then promote a quasi-incised meandering morphology.

Are either of these explanations right? If not, then what is the explanation for the striking difference in morphology? Evidence such as that of Figure 22.15 forces the researcher to ask questions that cut to the heart of river mechanics and morphology.

The conclusion of this paper can be stated simply. There is a lot more work remaining to be done on the subject of water-sediment interaction and resulting morphology, and a lot more pleasure to be had in the doing of it.

ACKNOWLEDGEMENTS

The hypotheses and concepts presented here are not solely due to the author; instead they represent a distillation of discussions and joint research with a long list of co-workers and colleagues, and include influences from a similarly long list of researchers who have contributed to the published literature. A perforce incomplete attempt has been made to name the people in question in the text. Any misinterpretation of the work of others is, however inadvertent, the sole responsibility of the author. The research of the author and colleagues reported here was supported by the National Science Foundation, the Minnesota Department of Transportation, the Hibbing Taconite Company and Ok Tedi Mining Ltd.

REFERENCES

Ashida, K., Egashira, S. & Nishimoto, N. 1986. 'Sediment transport mechanism on step-pool bed form', *Annuals*, Disast. Prev. Res. Inst., Kyoto Univ., **29**, 1–14.

Bathurst, J.C., Thorne, C.R. & Hey, R.D. 1979. 'Secondary flow and shear stress at river bends', *J. Hydraul. Div., ASCE*, **105**, 1277–1295.

Damuth, J.A., Flood, R.D., Kowsmann, R.O., Belderson, R.H. & Gorini, M.A. 1988. 'Anatomy and growth pattern of Amazon deep-sea Fan as revealed by long-range side-scan sonar (GLORIA) and high-resolution seismic studies', *AAPG Bull.*, **71**, 885–911.

Foley, M.G. 1975. *Scour and Fill in Ephemeral Streams*, Report No. **KH-R-33**, W.M. Keck Laboratory of Hydraulics and Water Resources, California Institute of Technology, 189 pp.

Fredsøe, J. 1978. 'Meandering and braiding of rivers', *J. Fluid Mech.*, **84**, 609–624.

Gole, C.V. & Chitale, S.V. 1966. 'Inland delta building activity of the Kosi River', *J. Hydraul. Div., ASCE*, **92**, 111–126.

Grant, G.E. & Mizuyama, T. 1991. 'Origin of step-pool sequences in high-gradient streams: a flume experiment', *Proc. Japan-US Workshop on Snow Avalanche, Landslide, Debris Flow Prediction and Control*, 523–532.

Hasbo, P.B., Hansen, E.A. & Fredsøe, J. 1995. 'Mathematical modelling of flow over oblique bedforms', in Ashworth, P.J., Bennett, S.J., Best, J.L. & McLelland, S.J. (Eds) *Coherent Flow Structures in Open Channels: Abstract Volume*, Leeds, 25.

Hickin, E. J. 1979. 'Concave-bank benches on the Squamish River, British Columbia, Canada', *Can. J. Earth Sci.*, **16**, 200–203.

Hoey, T.B. & Ferguson, R.I. 1994. 'Numerical simulation of downstream fining by selective transport in gravel bed rivers: model development and illustration', *Water Resour. Res.*, **30**, 2251–2260.

Ikeda, S. & Izumi, N. 1991. 'On the coefficient of transverse diffusivity of suspended sediment', *Proc., Japan Soc. Civ. Engrg.*, **434**, 47–55.

Jackson, R.G. II. 1976. 'Large scale ripples of the lower Wabash River', *Sedimentology*, **23**, 593–623.

James, C.S. 1985. 'Sediment transfer to overbank sections', *J. Hydraul. Res.*, **23**, 435–452.

Johanesson, H. & Parker, G. 1985. 'Computer simulated migration of meandering rivers in Minnesota', *Proj. Report* **242**, St. Anthony Falls Laboratory, Univ. of Minnesota, 82 pp.

Kellerhals, R. 1967. 'Stable channels with gravel-paved beds', *J. Waterway, Port, Coastal and Ocean Eng., ASCE*, **93**, 63–84.

Kodama, Y. 1994. 'Experimental study of abrasion and its role in producing downstream fining in gravel-bed rivers', *J. Sedimentary Res.*, **A64**, 76–85.
MacDonald, T.E., Parker, G. & Leuthe, D.P. 1991. 'Inventory and analysis of stream meander problems in Minnesota', *Report*, St. Anthony Falls Laboratory, Univ. of Minnesota, 37 pp.
Mikos, M. 1993. 'Fluvial Abrasion of Gravel Sediments', *Mitteilungen Versuchsanstalt für Wasserbau, Hydrologie und Glaziologie*, **123**, ETH Zürich, 322 pp.
Page, K. & Nanson, G. 1982. 'Concave-bank benches and floodplain formation', *Earth Surf. Processes Landf.*, **7**, 527–543.
Paola, C. 1988. 'Subsidence and gravel transport in basins', in Kleinspehn, K.L. & Paola, C. (Eds) *New Perspectives in Basin Analysis*, Springer-Verlag, 231–243.
Paola, C., Parker, G., Seal, R., Sinha, S.K., Southard, J.B. & Wilcock, P.R. 1992. 'Downstream fining by selective deposition in a laboratory flume', *Science*, **258**, 1757–1760.
Parker, G. 1976. 'On the cause and characteristic scales of meandering and braiding in rivers', *J. Fluid Mech.*, **76**, 457–480.
Parker, G. 1978a. 'Self-formed straight rivers with equilibrium banks and mobile bed. Part 1. The sand-silt river', *J. Fluid Mech.*, **89**, 109–125.
Parker, G. 1978b. 'Self-formed straight rivers with equilibrium banks and mobile bed. Part 2. The gravel river', *J. Fluid Mech.*, **89**, 127–146.
Parker, G. & Izumi, N. 1995. 'Purely erosional cyclic and solitary steps in cohesive material. *J. Fluid Mech.*, submitted.
Parker, G., Martinez, I. & Hills, R. 1982. 'Model study of the Minnesota River near Trunk Highway No. 169 Bridge, Minnesota', *Proj. Report* **213**, St. Anthony Falls Laboratory, Univ. of Minnesota, 91 pp.
Parker, G. & Sutherland, A.J. 1990. 'Fluvial armor', *J. Hydraul. Res.*, **28**, 529–544.
Pizzuto, J.E. 1987. 'Sediment diffusion during overbank flows', *Sedimentology*, **34**, 310–317.
Pizzuto, J.E. 1995. 'Downstream fining in a network of gravel-bedded rivers', *Water Resour. Res.*, **31**, 753–759.
Reid, L.M. 1989. 'Channel incision by surface runoff in grassland catchments', Unpub. PhD Thesis, Univ. of Washington, 135 pp.
Sambrook Smith, G.H. & Ferguson, R.I. 1995. 'The gravel-sand transition along river channels', *J. Sediment. Res.*, **A65**, 423–430.
Sawai, K. 1977. *Sediment Hydraulics Research on the Mechanism of Variation of the Bed of Cohesive Channels*, Unpub. PhD Thesis, Kyoto Univ., 100 pp. (in Japanese).
Schumm, S.A. & Lichty, R.W. 1963. 'Channel widening and floodplain construction along the Cimarron River in southwestern Kansas', *US Geol. Surv. Prof. Paper*, **352-D**, 71–88.
Seal, R. 1994. *Mechanisms of Downstream Fining in Gravel Bed Rivers*, Unpub. PhD Thesis, University of Minnesota, 226 pp.
Shaw, J. & Kellerhals, R. 1982. 'The composition of recent alluvial gravels in Alberta river beds. *Bull.* **41**, Alberta Research Council, Calgary, 151 pp.
Simm, D.J. 1993. *The Deposition and Storage of Suspended Sediment in Contemporary Floodplain Systems: a Case Study of the River Culm, Devon.* Unpub. PhD Thesis, University of Exeter, 347 pp.
Straub, L.G. 1935. 'Some observations of sorting of river sediments', *Trans. AGU*, **16**, 463–467.
Tsujimoto, T. 1994. 'Study on turbulence and sediment transport in flow with vegetation', *Report*, Dept. of Civil Engineering, Kanazawa Univ., 135 pp.
Twitchell, D.C., Kenyon, N. H., Parson, L.M. & MacGregor, B.A. 1991. 'Depositional patterns of the Mississippi Fan surface: evidence from GLORIA II and high-resolution

seismic profiles', in Weimer, P. & Link, M.H. (Eds) *Seismic Facies and Sedimentary Processes of Submarine Fans and Turbidite Systems*, Springer Verlag, New York, 349–363.

Whipple, K.X., Parker, G. & Paola, C. 1995. 'Experimental study of alluvial fans', in Sonu, J.H., Lee, K.S., Seo, I.W. & Bhowmik, N.G. (Eds) *Proc. Intl. Joint Seminar on Reduction of Natural and Environmental Disasters in Water Environment*, Seoul National Univ., 282–295.

Whittaker, J.G. & Jaeggi, M.N.R. 1982. 'Origin of step-pool systems in mountain rivers', *J. Hydraul. Div., ASCE*, **108**, 758–773.

Winterwerp, J.C., Bakker, W.T., Mastbergen, D.R. & van Rossum, H. 1992. 'Hyperconcentrated sand-water mixture flows over erodible bed', *J. Hydraul. Engrg., ASCE*, **119**, 1508–1525.

Yatsu, E. 1955. 'On the longitudinal profile of the graded river', *Trans. AGU*, **36**, 655–663.

23 The Origin and Downstream Development of Coherent Flow Structures at Channel Junctions

STUART J. M^C^LELLAND
School of Geography and Department of Earth Sciences, University of Leeds, Leeds, LS2 9JT, UK

PHILIP J. ASHWORTH
School of Geography, University of Leeds, Leeds, LS2 9JT, UK

JAMES L. BEST
Department of Earth Sciences, University of Leeds, Leeds, LS2 9JT, UK

ABSTRACT

The three-dimensional structure of flow at river channel junctions controls local bed scour, sediment transport pathways and downstream bar growth. This study presents results from a combined laboratory flume and field research programme which provide the first three-dimensional flow data concerning the origin and transformation of secondary flows generated at confluences with appreciable bed scour. Flume and field measurements suggest that the origin of secondary flows is intimately associated with zones of flow separation generated on the lee-side of avalanche faces at the mouth of each confluent channel. Three-dimensional reconstructions of mean flow and turbulent kinetic energy illustrate the presence of flow separation associated with avalanche faces, lateral distortion of the mixing interface and downstream transformation of coherent flow structures in response to vertical and lateral gradients in the mean flow field. A descriptive model of the evolution of coherent flow structures at channel confluences is presented in which flow mixing and the redistribution of turbulent kinetic energy at channel junctions are shown to be closely connected to the downstream transformation of these coherent flow structures. These flow patterns lie at the heart of explanations of mixing between converging flows, the evolution of confluence bed morphology and downstream development of flow divergence and diffluence bar growth.

INTRODUCTION

River channel junctions are the nodes of drainage networks (Howard, 1971; Richards, 1980; Roy & Roy, 1988; De Serres & Roy, 1990) which control downstream flow mixing, sediment transfer, bar evolution and channel change (Southard *et al.*, 1984; Carson & Griffiths, 1987; Ashworth *et al.*, 1992a;

Coherent Flow Structures in Open Channels.
Edited by P.J. Ashworth, S.J. Bennett, J.L. Best and S.J. M^C^Lelland.

Ferguson *et al.*, 1992; Ashmore, 1993; Bridge, 1993). The basic planform morphology of channel junctions is either Y or ⊢-shaped (Mosley, 1976; Best, 1986; Bristow *et al.*, 1993) which often develops into an X-shape as a result of mid-channel bar growth immediately downstream from the confluence (Carson & Griffiths, 1987; Ferguson, 1993; Ashworth, 1996). This 'confluence-diffluence' unit has been shown to be one of the fundamental building blocks of braided rivers (Ashworth *et al.*, 1992a; Ferguson *et al.*, 1992; Thorne *et al.*, 1993; Goff & Ashmore, 1994) because it is characterized by rapid rates of scour and fill, bar aggradation and bank retreat which impinge on adjacent and downstream channel development.

To understand the dynamics of the confluence-diffluence unit, it is necessary to quantify the three-dimensional flow structure at the junction of the mixing tributary flows and immediately downstream as the flow diverges. Previous work has shown that the flow structure within the junction controls the rate of sediment erosion and bed scour (Mosley, 1976; Kjerfve *et al.*, 1979; Ashmore & Parker, 1983; Best, 1987; Roy & Bergeron, 1990), bedload transport pathways (Mosley & Schumm, 1977; Best, 1987, 1988; Roy & Bergeron, 1990) and the dispersal of suspended pollutants (Muskatirovic & Miloradov, 1980; Gaudet & Roy, 1995; Kenworthy & Rhoads, 1995). Experimental and field studies (e.g. Mosley 1976; Best & Reid, 1984; Best, 1987; Ashmore *et al.*, 1992; Biron *et al.*, 1993; Rhoads & Kenworthy, 1995; Biron *et al.*, in press) have shown that the key features of the flow field at a junction are: (i) flow stagnation at the upstream junction corner; (ii) a shear layer and mixing interface between the combining flows; (iii) flow separation at the downstream junction corner and in the lee of the tributary bar avalanche faces; and (iv) counter-rotating secondary flow cells at the zone of flow convergence. These fluid dynamic features have been allied to the characteristic confluence zone bed morphology (Mosley, 1976; Ashmore & Parker, 1983; Best, 1987; Ashmore, 1993; Biron *et al.*, 1993; Roy *et al.*, 1988; Roy & Bergeron, 1990; Ashworth, 1996) which has several elements: (i) a central scour; (ii) tributary mouth bars which may end abruptly at an avalanche face dipping into the scour; (iii) bars within separation zones at the downstream junction corners; and (iv) a mid-channel or lateral bar downstream from the junction. The size, location and magnitude of the flow features listed above and the form of the channel morphology are controlled by the junction angle, planform asymmetry and the water and sediment discharge ratio between tributary flows (Mosley, 1976; Kanda & Kitada, 1978; McGuirk & Rodi, 1978, Best & Reid 1984; Best, 1987, 1988; Rhoads & Kenworthy, 1995; Biron *et al.*, in press; Rhoads, this volume).

The majority of studies detailing junction flow dynamics have concentrated solely on the confluence zone whilst more recent work on the dynamics of the confluence-diffluence unit has focused on the sedimentation and kinetics of channel change (e.g. Ashmore, 1993; Bridge, 1993; Bristow *et al.*, 1993). Sediment dynamics at the confluence-diffluence unit are driven by the downstream transformation of the three-dimensional flow structure. Richardson *et al.* (this volume) show that surface-convergent secondary flow cells at the confluence are transformed into surface-divergent flow cells in each anabranch around the mid-channel bar, following the model proposed by Ashworth *et al.* (1992b). However, little is known about how mid-channel bar growth is initiated, what determines the spacing of scour-bar deposition and how the three-dimensional flow structure is modified by bar aggradation and subsequent bank erosion (Ashworth, 1996).

The presence of secondary flow cells at channel junctions has been suggested both in field studies (Ashmore *et al.*, 1992; Bridge & Gabel, 1992; Rhoads & Kenworthy, 1995; Rhoads, this volume) and laboratory experiments (Mosley, 1976; Ashmore, 1982; Best, 1987; Fujita & Komura, 1988; Weerakoon *et al.*, 1991). Several mechanisms have been suggested to account for the formation and subsequent downstream transformations of secondary flow cells at channel junctions: (i) streamline curvature produced by planform morphology (Mosley, 1976; Ashmore & Parker, 1983; Fujita & Komura, 1988; Weerakoon *et al.*, 1991; Bridge, 1993); (ii) flow separation in the lee-side of tributary bar avalanche faces (Best, 1988; Biron *et al.*, 1993); (iii) distortion of the shear layer leading to the entrainment of fluid towards the bed along the shear layer at the mixing interface and subsequent upwelling towards the flow surface (Best & Roy, 1991; Biron *et al.*, 1993; in press); and (iv) turbulence anisotropy resulting from a non-uniform distribution in turbulence, similar to the mechanism for secondary flow generation in straight channels (Nezu & Rodi, 1985; Tominaga *et al.*, 1989; Nezu *et al.*, 1993). Currently, it remains unclear which mechanism dominates the formation and transformation of secondary flow cells at junctions because of the lack of detailed three-dimensional flow data. However, the presence of such large-scale coherent flow structures is clearly important as they redistribute flow momentum (e.g. Tominaga *et al.*, 1989, 1993; Nezu *et al.*, 1993) and will influence sediment transport and the development of bed morphology in the confluence zone. Field and experimental studies have shown the high turbulence intensities at the mixing interface shear layer (Chu & Babarutsi, 1988; Biron *et al.*, 1993; Plesniak *et al.*, 1994; Weygandt & Mehta, 1995) as well as the shear layer associated with flow separation over tributary bar avalanche-faces which may be regarded as backward-facing steps (Best, 1988). The redistribution of turbulent energy associated with these shear layers may also be determined by downstream transformation of the three-dimensional flow structure which is linked to the transition from flow convergence and bed scour to flow divergence, sediment deposition and bank erosion.

The object of this study is to investigate the origin of coherent secondary flow cells at channel junctions and quantify their development as the tributary flows mix. In order to accomplish this objective, results from two complementary investigations are presented: (i) a laboratory flume experiment designed to investigate the mixing of two parallel streams with pronounced, equal height avalanche-face bars which enables study of flow separation behind the avalanche faces and mixing processes in the absence of pronounced streamline curvature; and

(ii) a field study that investigates the three-dimensional flow structure at an asymmetric junction in a dynamic mobile bed environment. In both studies the spatial organization of the turbulent flow field is quantified from detailed three-dimensional flow measurements. The differences in morphology between the two studies permits discussion of the sources of turbulence and the mechanisms responsible for the creation and transformation of secondary flows at and downstream from a channel junction. It is shown that flow mixing and the redistribution of turbulent kinetic energy at channel confluences may be explained by the development and transformation of coherent secondary flow structures.

DATA COLLECTION

LABORATORY MEASUREMENTS

Laboratory experiments were conducted in a glass sided, water-recirculating flume, 10 m long, 0.3 m wide, with a bed slope of 0.002. The test section consists of two narrow, parallel tributary channels each 0.138 m wide and 1.5 m long which merge at the end of a 5 mm wide blunt-ended splitter plate (Figure 23.1). The upstream section of the splitter plate can be adjusted to subdivide the incoming flow, allowing control of the relative discharge in each tributary channel. Logarithmic velocity profiles at the end of each tributary channel confirm that the turbulent boundary layer is fully developed for both flows. Bed height in both tributary channels is increased by inserting a tapered ramp 1.1 m in length which ends in a 0.05 m high, vertical negative step at the end of the splitter plate. Both

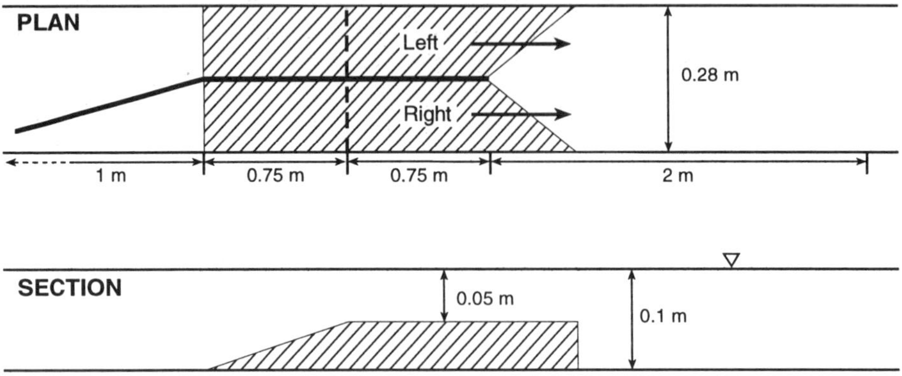

Figure 23.1. Schematic diagram showing plan and side views of the laboratory flume configuration. The raised tributary channels and avalanche face steps are shown as a shaded region. The bold line in the plan view diagram shows the splitter plate which separates the tributary flows and is used to control the flow rate entering each tributary channel. Diagram is not to scale

steps extend downstream along the channel wall such that the top of each step is inclined at 45° across the tributary channel flow (Figure 23.1). These backward-facing steps are designed to simulate the presence of steep avalanche faces prograding into the confluence as frequently found at natural river channel junctions (Best, 1988; Ashmore *et al.*, 1992; Biron *et al.*, 1993; Bristow *et al.*, 1993). The step height -to- tributary flow depth ratio (see below) is also characteristic of many confluences (Mosley, 1976; Kjerfve, 1979; Ashmore & Parker, 1983; Best, 1986). Flow depths are kept constant at 0.05 m in both the tributary channels. Downstream from the avalanche faces, the flow depth is twice that of the tributary channels (i.e. 0.1 m) since the tributary flows mix freely in the post-confluence channel which is 0.28 m wide and 2 m long.

Velocity measurements were conducted using laser Doppler anemometry (LDA). A DANTEC two-component, Argon-ion, fibre-optic LDA system was used in backscatter detection mode with frequency shift and a correlation-type burst processor. A 0.161 m focal length lens mounted on 0.038 m diameter transmission optics produces an effective measurement volume 0.081 mm long, 0.912 mm wide and 0.081 mm high with a beam intersection angle of 6.7°. The flow is seeded with 10 μm diameter water-based paint which diffuses fully and provides an excellent source of light scattering for the LDA.

The three-dimensional flow structure is quantified by taking velocity measurements in two perpendicular planes. In each plane, two orthogonal flow components are measured simultaneously. The first measurements (series A) are made in the downstream-vertical (x-y) plane which permits quantification of the streamwise (U_A) and vertical (V) velocity components. Since the focal length of the LDA optics is less than the channel width, measurements are taken through the glass sidewall from both sides of the channel. Measurements are repeated for 13 downstream cross-sections located at x/d_{max} = -0.5, 0, 0.5, 1, 1.5, 2, 3, 4, 6, 8, 11, 13, 17, where d_{max} is the maximum depth in the post-confluence channel and x is the downstream distance with $x = 0$ located at the end of the splitter plate. At each cross-section, measurements are spaced at 0.1 and $0.2d_{max}$ in the vertical (y) and cross-stream (z) directions respectively and to within $0.1d_{max}$ from the channel boundaries. This yields approximately 1350 individual measurements of the two-dimensional velocity. The second set of measurements (series B) was made in the streamwise-spanwise (x-z) plane, to quantify the streamwise (U_B) and cross-stream (W) velocities. These measurements are taken through the water surface using a small, streamlined, perspex fairing to prevent refraction of laser light due to water surface irregularities. The fairing has no measurable effect on flow velocities within the measurement volume of the LDA. The same sampling grid was used for series A and B measurements, but, due to the physical size of the fairing, series B measurements can only be made to within $z/d_{max} \approx 0.5$ of the sidewall, $y/d_{max} \approx 0.3$ of the water surface and $x/d_{max} \approx 0.5$ downstream of the splitter plate (approximately 700 measurements in total). At each measurement location the two perpendicular components of velocity are sampled simultaneously for 60 s at a sampling rate of at least 20 Hz. The actual sampling rate is determined by the number of tracer particles that pass through the LDA measuring volume, which is dependent on the local tracer concentration and flow velocity.

FIELD MEASUREMENTS

The field study was conducted in the Sunwapta River, Alberta, Canada, a proglacial outwash stream flowing from the Columbia Icefield to the Athabasca River (Figure 23.2). The study site is located within a 1 km wide braidplain of the Sunwapta River, approximately 10 km downstream of the Athabasca glacier. The meltwaters of the Athabasca, Dome and Stutfield glaciers produce a regular diurnal peak in discharge which dominate the smaller contribution from non-glaciated tributary valleys (Rice, 1982). The terminal moraines of these glaciers are the primary source of the river bed sediments which fine progressively downstream along Beauty Creek flats (Dawson, 1988; Sambrook Smith & Ferguson, 1995). Further background information on the Sunwapta River is given in Ashmore *et al.* (1992), Ashworth *et al.* (1992a), Ferguson *et al.* (1992) and Goff & Ashmore (1994).

The study reach consists of two tributary channels which join at an angle of approximately 100° with the left and right channels oriented at 60° and 40° respectively to the post-confluence channel. Average water-surface slope at the study site is 0.01. Figure 23.2 shows the planform morphology, bed topography and water surface features of the study reach. Low water surface waves mark the mixing interface between the tributary flows and a separate zone of 'boils' indicates flow upwelling near the left downstream junction corner. The mixing interface is deflected towards the right of the post-confluence channel since 70% of the total discharge is derived from the narrower and deeper left tributary. The confluence bed morphology is characteristic of that found at many asymmetric junctions (Best, 1987, 1988; Bristow *et al.*, 1993) with two avalanche faces dipping into a scour hole that is approximately four times deeper than the tributary channel depth. At the zone of maximum scour, the right-channel avalanche face is twice as steep (30°) as the left face (15°). This cross-sectional asymmetry persists downstream from the scour hole. Consequently, the channel talweg remains parallel to the right bank as the scour depth decreases and channel width increases. The grain size of talweg bed material decreases downstream from the zone of maximum scour: at cross-section 1 the median grain diameter (D_{50}) is 0.037 m whilst at cross-section 4 the D_{50} is 0.019 m.

Flow measurements were taken at four cross-sections located at $x/d_{max} \approx 3, 6, 8, 11$ where $x/d_{max} \approx 0$ is the confluence apex (labelled as 1 to 4 in Figure 23.2). Cross-sections 2 to 4 are approximately perpendicular to the channel banks and parallel to each other, whilst cross-section 1 spans the scour hole at an angle of 34° relative to the downstream cross-sections. Velocities were measured using seven two-component electromagnetic current meters (ECMs; see Table 23.1). The ECMs were arranged on two wading rods; instrument rack A consisted of one model 512

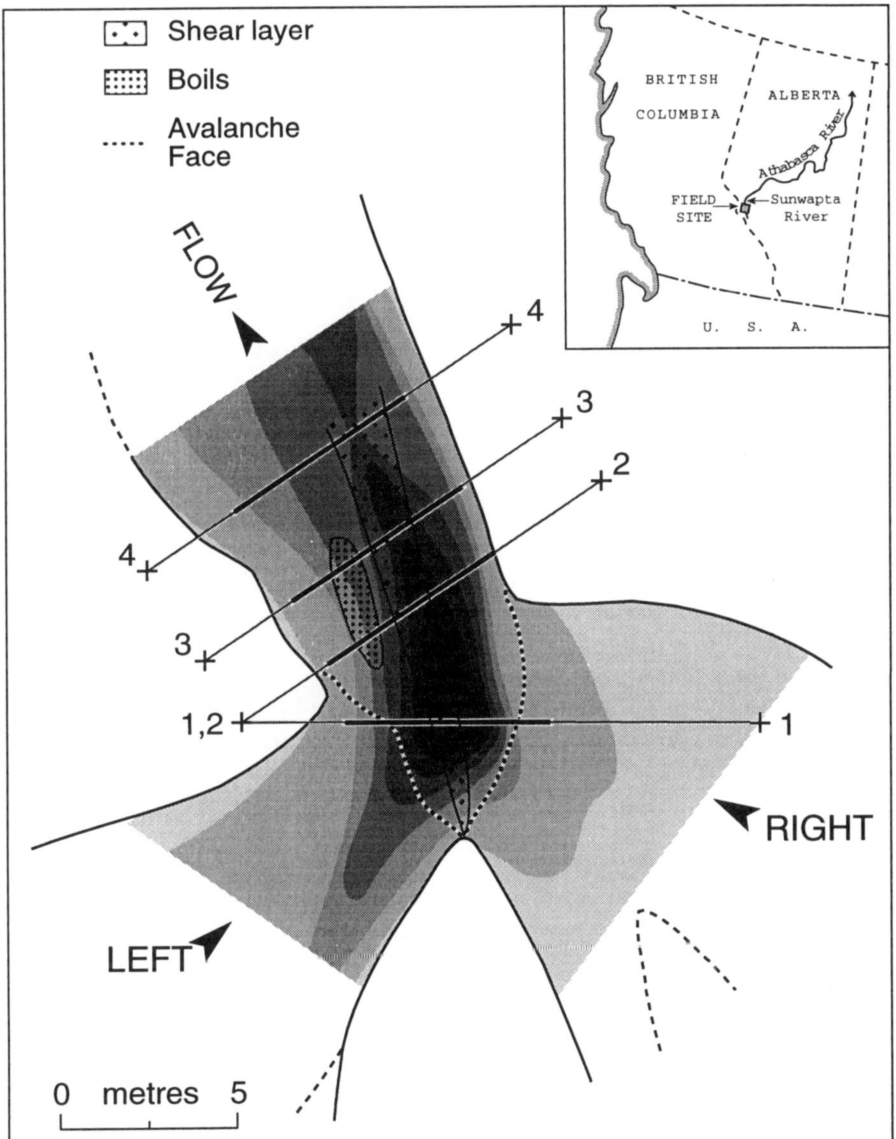

Figure 23.2. Morphology of the Sunwapta field site with inset showing the study area location. Bed elevation contours are at 0.1 m intervals and the surveyed position of the avalanche face crest is shown by a dotted line. The water margin is for bankfull conditions. Discharge of the left tributary is 70% of the total discharge. Flow measurements in Figures 23.7–23.10 are from the region shown as a thick line on each survey cross-section. The time-averaged positions of the mixing interface shear layer and the zone of fluid upwelling (boils), from visual observations, are superimposed on the bed morphology

Table 23.1. Description of ECMs used in field study. The differences in ECM characteristics are due to probe shape and the characteristics of the electronic filter which removes high frequency electronic noise from the output signal. A detailed discussion of these issues can be found in Lane *et al.* (1993).

Manufacturer	Model	Probe diameter m	Range m s^{-1}	Resolution m s^{-1}	Coil frequency Hz	Response time Hz	Noise level m s^{-1}
Colnbrook	Mark 2	0.027	±2.5	±0.002	80	5	0.03
Marsh-McBirney	511	0.038	±3.0	±0.014	30	5	†
Marsh-McBirney	512	0.038	±3.0	±0.014	30	16.7	0.01
Marsh-McBirney	523	0.012	±3.0	±0.014	30	16.7	0.01

† The manufacturers do not give a noise level for the Marsh-McBirney 511 model.

and three model 523 probes whilst rack B consisted of one Mark 2 and two model 511 probes (Table 23.1). In order to measure the full three-dimensional flow structure, each array was deployed separately but in close succession: array A simultaneously measuring the streamwise (U_A) and vertical (V) velocities; and array B simultaneously measuring the streamwise (U_B) and cross-stream (W) velocities. Measurements were made at verticals spaced 0.5 m apart and at four or five intervals on each vertical, beginning at $y/d_{max} \approx 0.09$ from the channel bed and ending at $y/d_{max} \approx 0.1$ from the water surface. At each measurement point, velocities were measured for 120 s at a frequency of 10 Hz for series A and 5 Hz for series B. The sampling frequencies were selected to suit the response time of the different ECMs (Table 23.1) and do not contribute to significant differences in the measured turbulence intensities.

The total discharge at cross-section 4 was measured as 3.6 m^3 s^{-1} with no appreciable input or output of flow from the post-confluence channel. Discharge changed by ±3% during the 4 hour measurement period.

HYDRAULIC CONDITIONS IN FLUME AND FIELD STUDIES

Table 23.2 summarizes the hydraulic conditions for the laboratory and field experiments. In both studies the left tributary dominates the total discharge such that the discharge ratio (Q_{LEFT}/Q_{RIGHT}) is approximately 2 in both situations. The channel bed in the laboratory experiment is smooth and the Reynolds and Froude numbers show that flow is fully turbulent and sub-critical throughout the test section.

The width-to-depth ratio of the post-confluence channel in the field increases by 55% and the relative depth (y_{max}/d_{max}) decreases by 40% due to sediment deposition and, to a lesser extent, bank erosion. The relative scour depth (step height -to-tributary flow depth ratio) is predominantly determined by the junction angle (Mosley, 1976; Kjerfve, 1979; Ashmore & Parker, 1983; Best, 1986) and therefore

Table 23.2. Hydraulic conditions in the laboratory flume experiment and field. Reynolds and Froude numbers are calculated from mean cross-sectional flow velocity and depth. Measurements at the junction are from $x/d_{max} = 1.5$ in the laboratory experiment and at $x/d_{max} = 3$ in the field study. In both cases, the downstream measurements are from the furthest downstream section.

		Laboratory experiment	Field study
Channel slope		0.002	0.01
Maximum flow depth (d_{max})		0.1 m	1.09 m
Maximum flow velocity in the left tributary (U_{max})		0.517 m s^{-1}	1.23 m s^{-1}
Discharge ratio		1.9	2.1
Step height -to- tributary flow depth ratio		1	4
Width-to-depth ratio ($width/y_{max}$)	At junction	3	6
	Downstream	3	11
Relative depth (y_{max}/d_{max})	At junction	1.0	1.0
	Downstream	1.0	0.6
Flow Reynolds number	At junction	1.8×10^4	2.2×10^5
	Downstream	0.8×10^4	2.9×10^5
Froude number	At junction	0.25	0.25
	Downstream	0.12	0.52

a smaller relative scour depth was reproduced in the laboratory experiment which has parallel tributary channels. The tributary flows in both studies decelerate as they enter the junction since the cross-sectional area expands and water depth increases. The change in channel morphology at the field site promotes flow acceleration as the depth decreases downstream from the maximum scour, with a resulting increase in the Reynolds and Froude numbers. Downstream changes in bed height and bank position were not replicated in the laboratory flume, hence the relative depth and aspect ratio of the experiment remain constant and the Reynolds and Froude numbers both decrease downstream. The advantage of the laboratory configuration is that the experimental results can be attributed to the processes of flow mixing in the absence of downstream morphological change.

RECONSTRUCTION OF THE THREE-DIMENSIONAL FLOW FIELD

The flow measurements in both the flume and field studies are obtained from two series of separate measurements in orthogonal planes. The mean and fluctuating streamwise velocities are then calculated from the average of series A and B such that:

$$U = (U_A + U_B)/2 \tag{23.1}$$

$$u' = (u'_A + u'_B)/2 \tag{23.2}$$

where U and u' are the mean and root-mean-square (RMS) of the streamwise velocity respectively and the subscript indicates the data series. The mean differences in U and u' between series A and B were 3% of U_{max} and 5% of u'_{max} for the laboratory experiment and 8% of U_{max} and 12% of u'_{max} for the field study. These errors are small given the presence of free shear layers and the natural variations in flow strength at the field site. For example, Tominaga *et al.* (1989) utilized a similar scheme to measure secondary currents in straight channels with uniform flow and found differences between data sets within 2% of U_{max} and 5% of u'_{max} which is comparable to those obtained here. In the present study, the differences between each measurement series may be attributed to either: (i) errors in positioning the probes at the same measurement point; (ii) velocity fluctuations at time scales too long to be characterized within the sampling period; or (iii) velocity fluctuations smaller than the sampling frequency. The ECM probes were orientated relative to each cross-section marker and alignment errors were within 5°. Errors due to long time-scale velocity fluctuations are likely to be reduced by averaging the measurements of two sampling periods.

This study sought to investigate the secondary flow structure in addition to the mean streamwise flow pattern. Secondary currents may be defined as a persistent component of velocity which is not parallel or tangential to the centreline of the channel and not parallel to the local bed slope (Bhowmik, 1982). In the laboratory experiment, the bed slope and channel centreline are fixed and invariant, thus the primary and secondary velocities are unambiguously defined. However, in the field study, primary and secondary velocities are not readily defined because each tributary flow is in a different direction and the channel centreline may be independent from the channel pattern due either to abrupt variations in channel curvature and width or zones of recirculating flow. In addition, there is no consistent trend in the local bed slope.

In order to compensate for the uncertainty in locating the primary and secondary flow directions, a number of methods for correcting cross-section orientation have been proposed (*cf.* Bathurst *et al.*, 1977; Dietrich & Smith, 1983; Markham & Thorne, 1992). However, these methods are not necessarily suited to the three-dimensional flow structure of braided river confluences because converging and diverging flow associated with rapid changes in bed morphology and channel width produce lateral and vertical discharge fluxes between cross-sections. In this situation, these correction techniques give a poor definition of the downstream flow direction as they assume minimal downstream flow acceleration. Furthermore, the interpretation of flow structures between sections is difficult because the correction angle does not necessarily follow an obvious frame of reference. To avoid the confusion of a changing cross-section orientation, a single, fixed frame of reference is most suitable for identifying coherent flow structures and tracking their downstream development. Since sections 2–4 in the field study are approximately perpendicular to the streamwise flow direction their orientation is chosen as the common frame of reference. Series B velocity measurements are then corrected at section 1 using:

$$U_B = \sin(\theta - \alpha)U_\theta \tag{23.3}$$

$$W = \cos(\theta - \alpha)U_\theta \tag{23.4}$$

where U_θ and θ are the magnitude and direction of the resultant horizontal (x-z plane) flow vector respectively and α is the angle by which section 1 deviates from the orientation of sections 2–4 (34°). The corrected velocity components are therefore parallel to those of the downstream cross-sections. Series A measurements cannot be re-oriented to this new frame of reference because the resultant horizontal vector is not known. Hence, at section1, Equations 23.1 and 23.2 are not applied to the streamwise velocity component i.e. $U = U_B$ and $u' = u'_B$.

RESULTS

FLUME EXPERIMENT

The pattern of streamwise velocity and secondary flow vectors at five representative cross-sections in the laboratory flume are shown in Figure 23.3 with an interpretation of the secondary flow pattern and shear layer position summarized in Figure 23.4. Four features of the flow field are described below: (i) flow separation in the lee of the step and along the right sidewall; (ii) mixing between the tributary flows; (iii) the secondary flow structure; and (iv) the spatial redistribution of turbulent kinetic energy.

Flow separation

Two zones of flow separation occur at the junction: (i) in the lee of the backward-facing steps; and (ii) alongside the right sidewall downstream from the step separation zone. The step separation zone has a complex three-dimensional shape (Figures 23.3 and 23.4) which is controlled by both the different velocities of the tributary flows and proximity to the channel sidewall. Under the slower flow on the right of the channel, the step separation zone bulges upwards whilst below the faster left tributary flow the separation zone is thinner (see section $x/d_{max} = 1.5$). Additionally, the step separation zone is higher at the channel edges, reaching a maximum along the left sidewall and producing a 'peak and trough' profile to the surface of the separation zone (Figures 23.3 and 23.4). At the end of the steps, negative streamwise velocities (deep purple to dark blue colours in Figure 23.3) occur between the channel bed and $y/d_{max} = 0.4$ and a thin shear layer forms above the region of flow reversal. Flow reattachment occurs between 6 and 9 step heights downstream from the end of the splitter plate with a mean reattachment point (at $y/d_{max} = 0.1$ above the bed) of 7 step heights. Similar reattachment lengths for separated flow over backward-facing steps have been measured in two-dimensional flows (De Brederode & Bradshaw, 1972; Papadopoulos & Ötügen, 1995; Keogh & Addison, this volume). The region of flow reattachment is also skewed across the channel, being at 6 step heights below the faster left tributary flow and 8 step heights below the slower right tributary flow. At both sidewalls, the separation zone extends 9 step heights downstream from the splitter plate.

A second, more extensive, zone of flow separation occurs adjacent to the right sidewall (Figures 23.3 and 23.4) and can be divided into two regions: near-surface and near-bed. At the surface, negative flow velocities occur near the sidewall in a region from $x/d_{max} = 4$ to 12. This zone is widest at the surface, extending over $z/d_{max} = 0.5$ into the post-confluence channel and reaches down to $y/d_{max} = 0.35$ below the water surface. Near the channel bed, negative velocities occur at $x/d_{max} = 9$ to 13.5 in a zone that is widest at the bed and up to $y/d_{max} = 0.35$ high at the sidewall. These two zones of flow reversal are incorporated within an extensive region of flow stagnation which occupies 25% of the total flow volume (Figure 23.3). Within this region of flow stagnation, streamwise flow velocities are decelerated to $0.1U_{max}$. Dye injection suggests that there is limited entrainment into this separation zone and the residence time is greater than that in the step separation zone.

Flow mixing between tributary flows

From the splitter plate to the furthest downstream section there is both a deceleration of the mean velocity and redistribution of flow momentum as the tributary flows mix. Figure 23.5 shows the mean vertical profile of the resultant velocity ($U_r = \sqrt{(U^2 + V^2 + W^2)}$) at the channel centreline and on the left and right side of the post-confluence channel at the same downstream locations as Figure 23.3. In the region of the step separation zone ($x/d_{max} = 1.5$ and 3), the upper section of the velocity profile has a steep gradient associated with the step shear layer, whilst the negative gradient near the channel bed is related to flow recirculation within the step separation zone. The combining tributary flows can be identified as two separate flows with limited mixing before $x/d_{max} = 3$ and a core of maximum velocity on the left at the water surface (Figures 23.3 and 23.5). Downstream from the step separation zone, at $x/d_{max} = 6$, there is a reduction in the velocity gradient as flow accelerates in the lower region and decelerates in the upper region (Figure 23.5). The vertical velocity gradient on the right of the channel reverses as a lobe of faster fluid moves outwards from the centreline towards the right of the channel, with the maximum velocity occurring near the channel bed (Figure 23.3). The decrease in size of the lateral separation zone ($x/d_{max} = 11$ to 17) is accompanied by a deceleration of flow on the left of the channel and an acceleration on the right of the channel. At the centreline, near-bed velocities decelerate whilst near-surface velocities accelerate suggesting that the faster left tributary flow continues to move beneath the slower flow from the right-

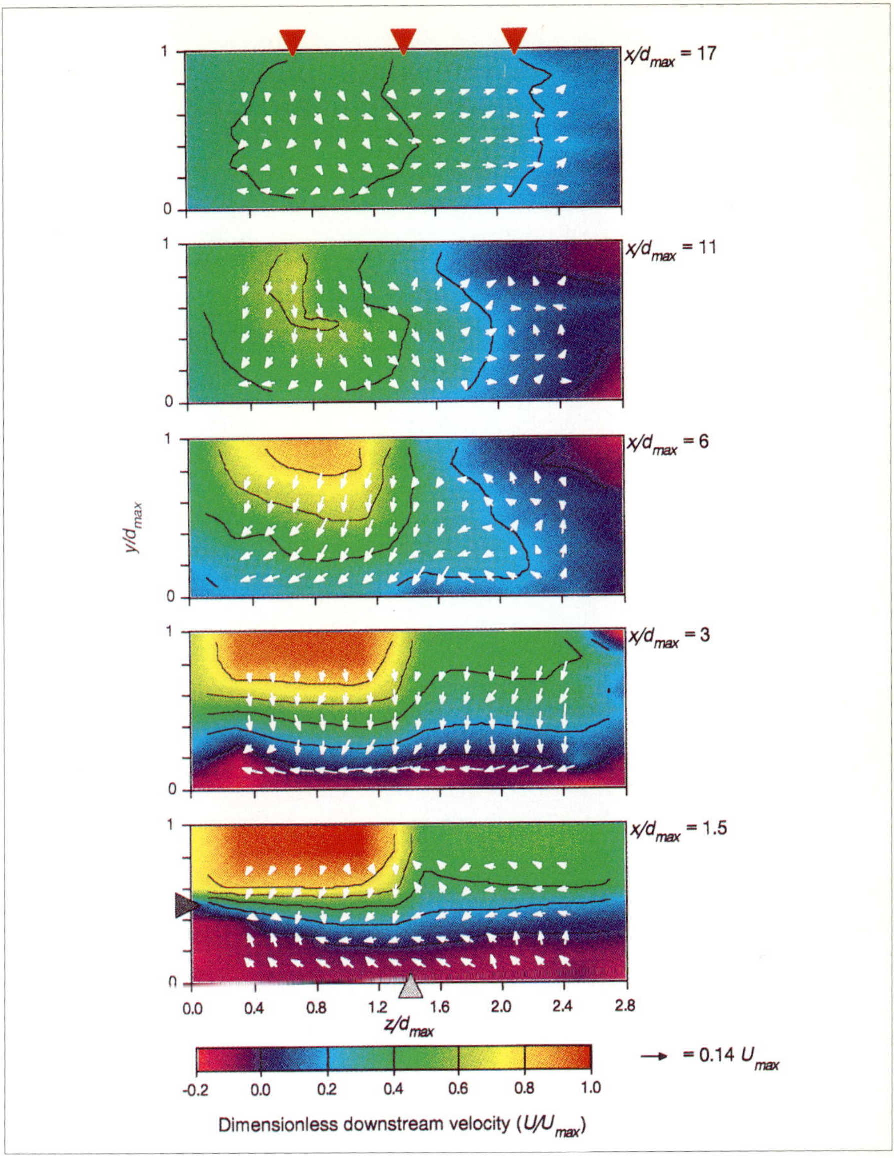

Figure 23.3. Three-dimensional flow field in the flume channel at five cross-sections located at x/d_{max} = 1.5, 3, 6, 11, 17. Cross-stream and vertical distances are expressed in relation to the maximum flow depth. Downstream flow direction is into the page. Vectors show the secondary flow velocities and are superimposed on contours of the downstream flow strength. Contour lines of downstream velocity correspond to the divisions on the colour bar at 0, 0.2, 0.4, 0.6, and $0.8U_{max}$. Contours are interpolated from a regular grid of 14x9 measurements and vectors from a 10x7 grid. The light grey arrow shows the splitter plate position and the dark grey arrow shows the avalanche face height which ends immediately upstream of this section (x/d_{max} = 1.5). Red arrows on the downstream cross-section indicate the position of profiles shown in Figure 23.5

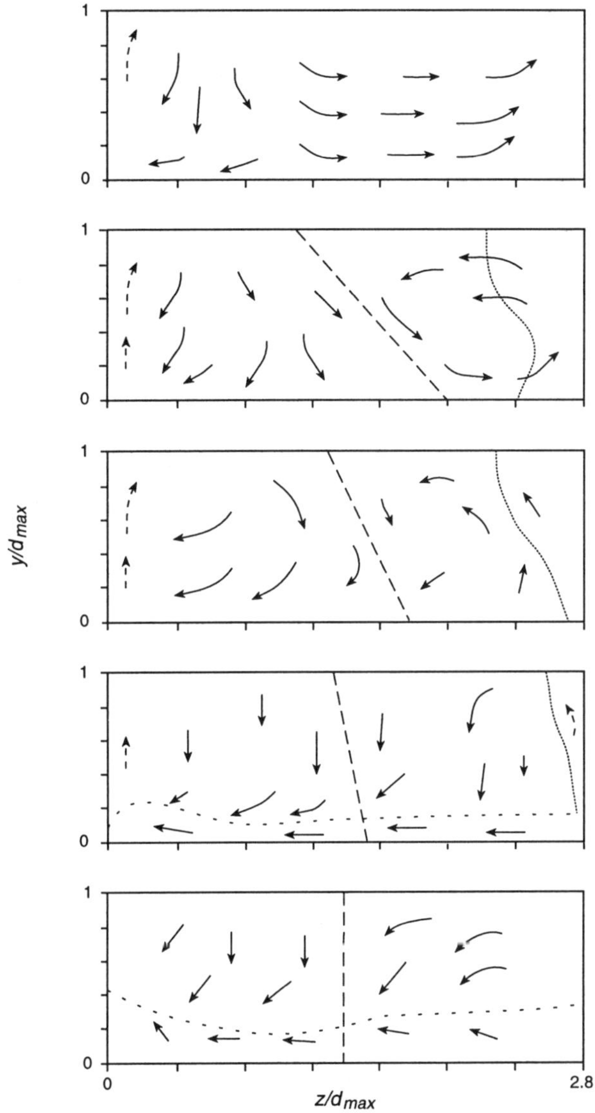

Figure 23.4. Summary sketch of the secondary flow pattern for the sections shown in Figure 23.3. Dashed vectors represent flow circulation assuming conservation of mass. The vertical long-dash lines indicate the approximate position of the mixing interface shear layer estimated from three-dimensional flow pattern and turbulent kinetic energy distribution. For the furthest downstream section ($x/d_{max} = 17$), this line is not shown because the shear layer is indistinct. The approximate boundary of the step separation zone is shown by a transverse line of short dashes and the vertical dotted lines indicate the interface between the lateral separation zone and the main flow (see text for further explanation)

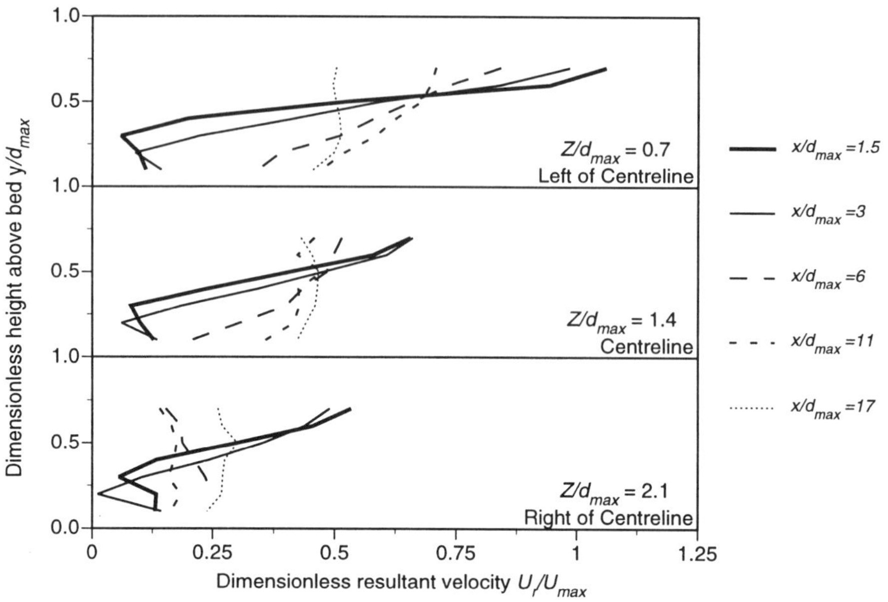

Figure 23.5. Mean vertical profiles of the resultant velocity $U_r = \sqrt{(U^2 + V^2 + W^2)}$ at the channel centreline ($z/d_{max} = 1.4$) and on the left ($z/d_{max} = 0.7$) and right ($z/d_{max} = 2.1$) sides of the post-confluence channel at the same downstream location as Figure 23.3 ($x/d_{max} = 1.5$, 3, 6, 11, 17)

channel. In conjunction with the distortion of the mixing interface, the core of maximum velocity moves below the water surface and towards the right channel wall (Figures 23.3 and 23.5).

Secondary flow structure

The secondary flow vectors are both shown in Figure 23.3 superimposed on the streamwise velocity and are summarized in Figure 23.4 by an explanatory sketch which also includes vectors inferred from continuity. Above the step separation zone, flow is directed downwards towards the channel bed as the total flow area expands. Within the separation zone, flow from the right-channel moves to the left and then upwards and outwards towards the left sidewall. This reflects the strong lateral pressure gradient within the step separation zone, with lower pressure on the left side being created by the differential tributary flow velocities and variation in separation zone thickness. Initially, there are no strong cross-stream flows in the upper region of the flow since there is limited mixing between the two flows ($x/d_{max} < 3$). However, beyond the step separation zone ($x/d_{max} = 6$), as the mixing rate between the flows increases, the vectors on the left of the channel indicate clockwise flow rotation whilst those on the right of the channel indicate an anticlockwise flow rotation. At $x/d_{max} = 11$, this anticlockwise rotation expands to occupy the entire region of $z/d_{max} > 0.9$, consistent with the distortion of the streamwise velocity contours which indicate deformation of the shear layer between the flows (Figures 23.3 and 23.4). At the final cross-section, a region of downflow persists in the zone of the maximum streamwise velocity with secondary flow diverging outwards near the channel bed. On the right of the channel, flow rotation diminishes and the vectors diverge outwards towards the sidewall.

Redistribution of turbulent kinetic energy

Turbulent kinetic energy (k) is defined as

$$k = 0.5\left(\overline{u'^2} + \overline{v'^2} + \overline{w'^2}\right) \quad (23.5)$$

and $\overline{u'^2}$, $\overline{v'^2}$ and $\overline{w'^2}$ are the mean square values of the velocity fluctuations in the streamwise, vertical and cross-stream directions respectively. Turbulent kinetic energy represents the energy extracted from the mean flow by motion of the turbulent eddies (Bradshaw *et al.*, 1967; Kline *et al.*, 1967) and its production involves interaction of the Reynolds stresses with the mean velocity gradients. Turbulent kinetic energy is advected by the interactions of velocity and pressure fluctuations and dissipated by the action of viscous forces.

Turbulent kinetic energy is calculated directly from the measured flow data for $x/d_{max} = 1.5$ to 17 in the region where the full three-dimensional flow field is known, i.e. between $z/d_{max} = 0.5$ to 2.3 and from the bed to a height of $y/d_{max} = 0.7$. Cross-stream velocities are not measured beyond these widths and heights. In previous studies where cross-stream turbulence has not been measured, the magnitude of $\overline{w'}$ has been estimated as either equal to $\overline{v'}$ (Harsha & Lee, 1970), $0.58\,\overline{v'}$ (Bowden, 1956) or $0.5\,\overline{v'}$ (Gordon & Dohne, 1973; Gordon, 1975; Prinos *et al.*, 1988). For this study, multiple regression between the RMS of the measured three-dimensional flow data yields

$$\overline{w'} = 2.04 \times 10^{-3} + 0.312\overline{u'}_A + 0.560\overline{v'} \qquad n = 669;\ r^2 = 0.860 \quad (23.6)$$

This relationship is applied where $\overline{w'}$ is not measured to give the complete distribution of turbulent kinetic energy throughout the measurement section, as displayed in Figure 23.6 (non-dimensionalized with respect to U^2_{max}).

The magnitude of the dimensionless turbulent kinetic energy varies from 0 to 0.038. Intense production of turbulent kinetic energy is associated with: (i) flow separation in the lee of the steps; and (ii) the mixing interface between the tributary flows where Kelvin-Helmholtz instabilities are created at the shear layer between flows of different velocity (Winant & Browand, 1974; Jimenez, 1983; Browand & Troutt, 1985; Biron *et al.*, 1993). The maximum turbulent kinetic energy occurs downstream from the left-channel step where the vertical velocity gradient is

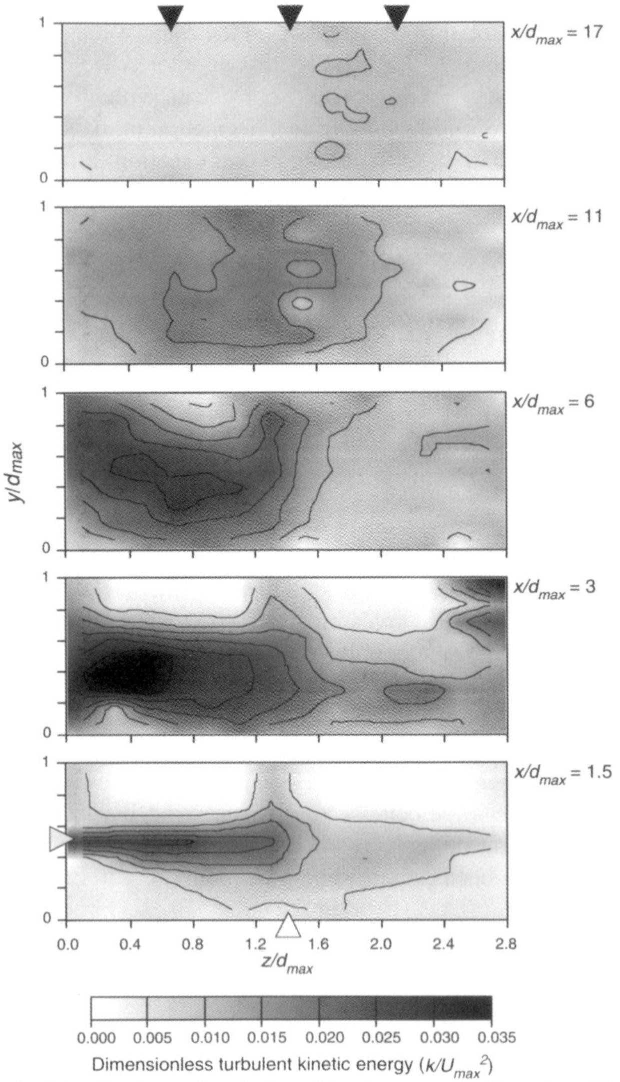

Figure 23.6. Distribution of turbulent kinetic energy (Equation 23.5) in the test section of the flume channel at the same cross-sections shown in Figure 23.3 (x/d_{max} = 1.5, 3, 6, 11, 17). Downstream flow direction is into the page and axes are the same as Figure 23.3. Contour lines of turbulent kinetic energy correspond to divisions on the colour bar at 0.005, 0.01, 0.015, 0.02, 0.025, 0.03, $0.035k/U_{max}^2$. Contours are interpolated from a grid of 14×9 measurements. The white arrow head on the upstream section (x/d_{max} = 1.5), indicates the mid-channel position of the splitter plate and the grey arrow head shows the height of the avalanche face step. The black arrow heads on the downstream cross-section (x/d_{max} = 17) indicate the lateral position of the velocity profiles shown in Figure 23.5

steepest (Figure 23.5). Less turbulent kinetic energy is generated on the right of the channel because the velocity gradient across the step separation zone is 25% of that on the left of the channel. Near the point of flow reattachment ($x/d_{max} = 3$), two cores of maximum kinetic energy are associated with bulges in the step separation zone on both sides of the channel. Beyond the reattachment point ($x/d_{max} = 6$), the kinetic energy core on the right of the channel is rapidly dissipated whilst the core on the left is transported downstream. This core then merges with the region of kinetic energy associated with the mixing interface shear layer. At sections $x/d_{max} = 11$ and 17, the maximum turbulent kinetic energy tends to follow the maximum lateral velocity gradient associated with the streamwise velocity distribution across the mixing interface. In addition, turbulent kinetic energy is redistributed following the pattern of secondary flow rotation and divergence (Figures 23.3 and 23.4). A region of low turbulent kinetic energy exists from $x/d_{max} = 6$ to 11 near the right sidewall and is associated with the lateral flow separation zone.

FIELD STUDY

In order to facilitate comparison between laboratory flume and field, the three-dimensional flow structure at the field study site is described in the same format as the flume study. The bed topography and three-dimensional flow field at each cross-section through the field confluence are shown in Figure 23.7 and the secondary flow pattern is summarized in Figure 23.8.

Flow separation

Reorientation of the velocity measurements at section 1 (see Figure 23.2 for location), results in the cross-stream velocity vectors (Figures 23.7 and 23.8) on both sides of the channel being oriented approximately perpendicular to the corresponding avalanche-face crest. Hence, the secondary flow vectors show pronounced flow recirculation in the lee of both avalanche faces as would be expected in the presence of flow separation. Figure 23.9 shows vertical profiles of the resultant velocity (U_r) at the talweg and at $z/d_{max} \approx 1$ on either side of the talweg. At section 1, the velocity profile has a marked inflection which suggests flow recirculation within the avalanche-face separation zone with the minimum resultant velocity occurring within the separation zone shear layer.

A region of separated flow also occurs at the downstream junction corner on the left of the post-confluence channel. It is not possible to measure the flow structure in this zone because of the shallow depths, but the presence of sand-size sediment and upstream ripple orientation suggest this is an area of low flow velocity and flow reversal consistent with flow separation at the downstream junction corner (McGuirk & Rodi, 1978; Best 1986, 1987; Petts & Thoms, 1987; Rhoads & Kenworthy, 1995).

Flow mixing

An acceleration of the mean flow velocity and decrease in channel area exists from section 1 to 4. At sections 1 and 2, the core of maximum streamwise velocity is located to the left of the talweg near the channel bed. At section 3, the region of maximum streamwise velocity spreads both outwards to the right of the channel talweg and upwards towards the surface on the left of the channel. As flow accelerates towards section 4, the core of maximum streamwise velocity reduces in size and becomes concentrated near the bed on the right of the talweg.

Although downstream acceleration of flow differs between the laboratory and field studies, changes in the resultant velocity gradient are similar (Figures 23.5 and 23.9). At section 1, the maximum resultant velocities occur near the water surface and the upper part of the velocity profile is steep with the gradient being greatest on the left of the channel where flow separates in the lee of the left avalanche face. At section 2, acceleration of flow occurs near the bed whilst there is a deceleration near the surface on the left of the channel. As maximum flow acceleration occurs near the channel bed, there is a reduction in the streamwise velocity gradient across the channel width. At section 3, all the velocity gradients show further near-bed flow acceleration maintaining the negative streamwise velocity gradient. Although this pattern persists at the talweg of section 4, greater flow acceleration at the water surface leads to a more uniform streamwise velocity distribution on either side of the talweg.

Secondary flow structure

The secondary flow vectors (Figures 23.7 and 23.8) illustrate flow rotation in the plane perpendicular to the streamwise flow direction. At section 1, flow from the tributary channels converges at the water surface, turns downwards at the mixing interface ($z/h_{max} \approx 3.25$) and then diverges outwards away from the centre of maximum bed scour. This pattern of two counter-rotating flow cells has been proposed as characteristic of channel junctions in many studies (Mosley, 1976; Fujita & Komura, 1988; Weerakoon *et al.*, 1991; Ashmore *et al.*, 1992; Rhoads & Kenworthy, 1995; Rhoads, this volume) and these are the first three-dimensional field measurements to confirm this flow structure. At sections 1 and 2, the strong lateral surface flow of the clockwise rotating cell on the left of the channel impinges on the near-bed flow of the right cell. Downstream from the zone of maximum scour (section 2), surface convergence of the anticlockwise rotating cell on the right of the channel has diminished. At section 3, there is also a marked anticlockwise rotation of the flow vectors in the lower part of the right cell, but the lateral flow from the right bank to the channel centre is confined to a narrow near-surface zone. To the left of the clockwise rotating cell the flow vectors are directed towards the water surface and flow diverges laterally away from the cell towards the channel bank. This flow divergence is marked by a region of lower streamwise velocity. The counter-rotating flow pattern evident in section 3 transforms into a

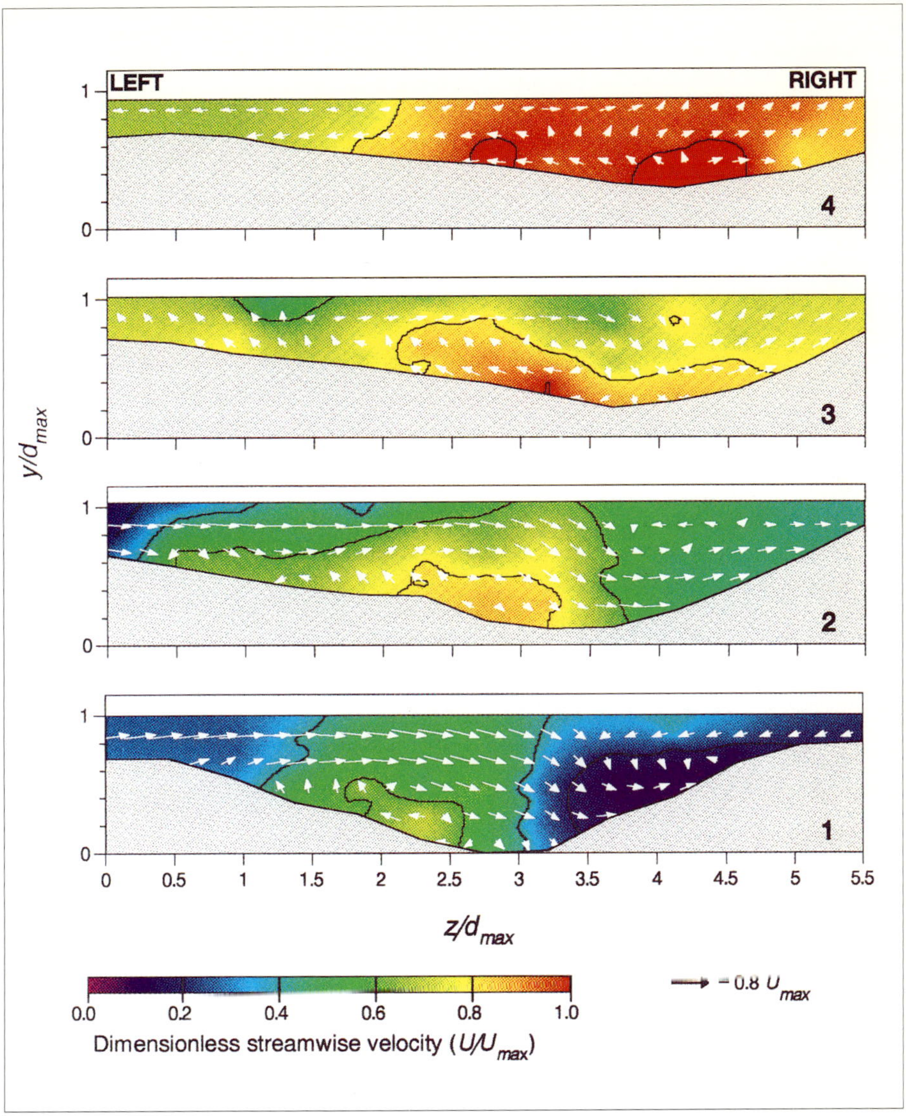

Figure 23.7. Bed topography and three-dimensional flow structure at the field confluence study site for the four measurement cross-sections shown in Figure 23.2 located at x/d_{max} = 3, 6, 8, 11. Downstream flow direction is into the page. Distances across the channel and above the bed are expressed in relation to the maximum flow depth (d_{max}). Results are shown from z/d_{max} = 0 to 5.5 (as shown by the thicker line in Figure 23.2) and from the channel bed (shaded grey) to the water surface. Secondary flow vectors are superimposed on contours of the streamwise flow strength. Contour lines of streamwise velocity correspond to divisions on the colour bar at 0.2, 0.4, 0.6, 0.8 and 1.0U_{max}. Vectors and contours are interpolated from an average of 40 measurements at each cross-section

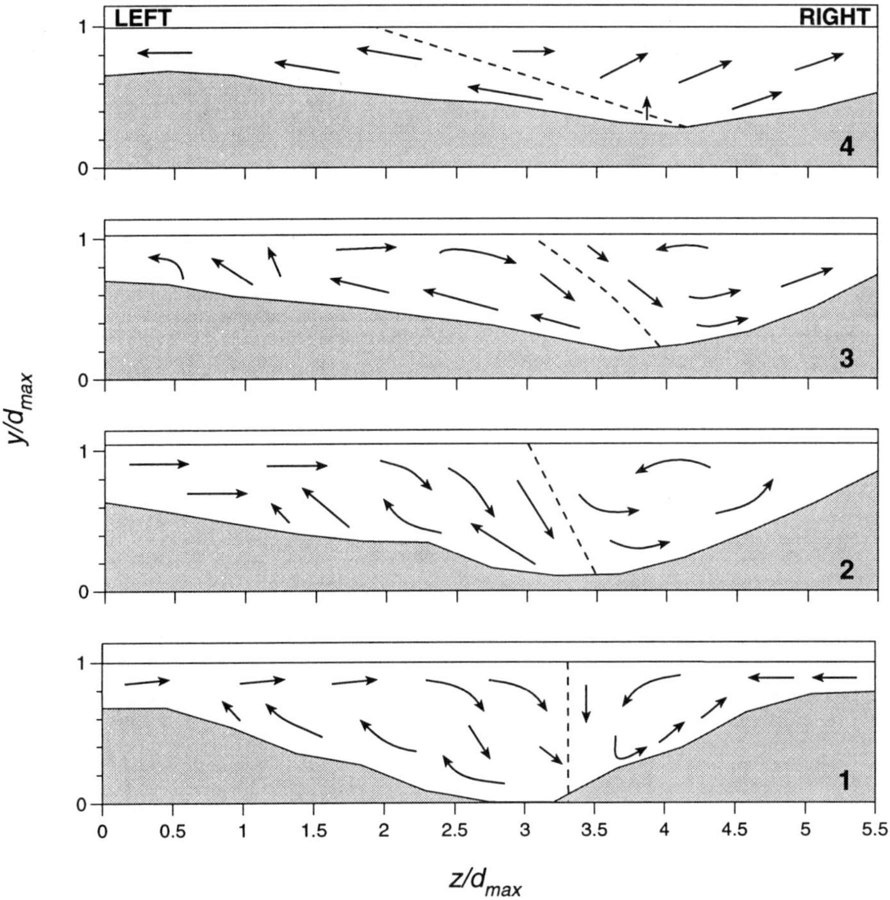

Figure 23.8. Summary sketch of the secondary flow pattern in the field confluence for the sections shown in Figure 23.7. The vertical long-dash lines indicate the approximate position of the mixing interface shear layer as estimated from secondary flow vectors and the distribution of turbulent kinetic energy

simpler diverging flow at section 4. Flow vectors diverge at $z/d_{max} = 4.25$ near the bed and $z/d_{max} = 2$ near the surface suggesting a laterally inclined shear layer between the diverging bodies of flow (Figures 23.7 and 23.8). Most vectors at section 4 are directed towards the water surface and indicate topographic forcing of flow as the scour hole shallows and the average flow depth decreases.

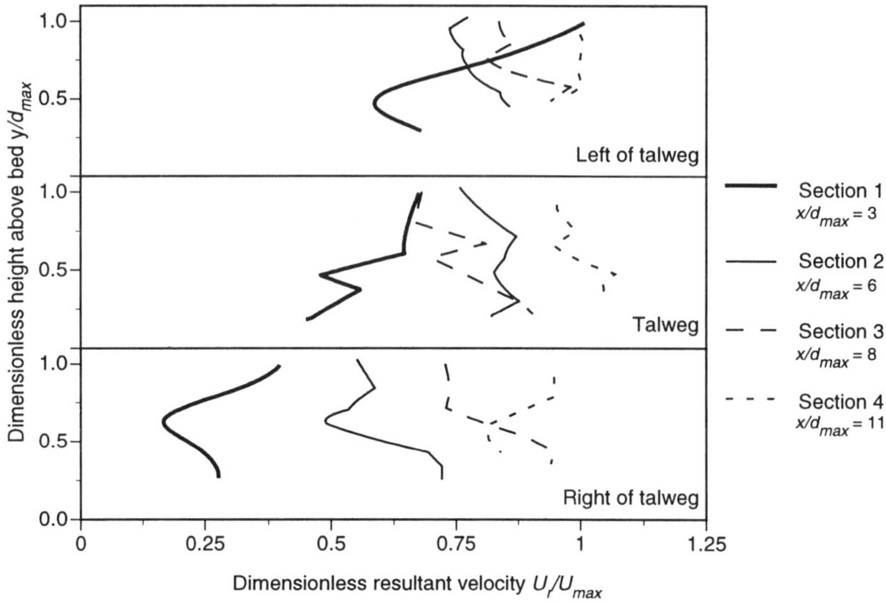

Figure 23.9. Mean vertical profile of the resultant velocity $U_r = \sqrt{(U^2 + V^2 + W^2)}$ at the channel talweg, and at $z/d_{max} = 1$ to the left and right sides of the talweg, at the four measurement cross-sections shown in Figure 23.2 (located at $x/d_{max} \approx 3, 6, 8, 11$)

Redistribution of turbulent kinetic energy

The magnitude of the dimensionless turbulent kinetic energy through the confluence varies from 0 to 0.04 (Figure 23.10), similar to the range present in the flume experiment. At section 1, the maximum turbulent kinetic energy (Figure 23.10) occurs in the lee of the avalanche face on the left side of the channel suggesting high turbulence intensities associated with separation of flow from the bed. A core of high turbulent kinetic energy also occurs at $z/d_{max} = 3$ and coincides with the talweg and region of downflow illustrated in Figures 23.7 and 23.8. This narrow region of high turbulent kinetic energy production is generated at the shear layer between the two combining flows. Downstream at section 2, the core of turbulent kinetic energy associated with the mixing interface shear layer increases in both magnitude and width, suggesting an expansion of this interface between the two flows and greater shear in this region. The core of maximum turbulent kinetic energy remains above the talweg but is suppressed further below the water surface than in section 1. This is due to the action of the secondary flows which transport momentum from the surface and then towards the bed in this region. In addition, a zone of high turbulent kinetic energy extending from the mixing interface towards the right bank is associated with strong outwards near-bed flow velocity (Figures

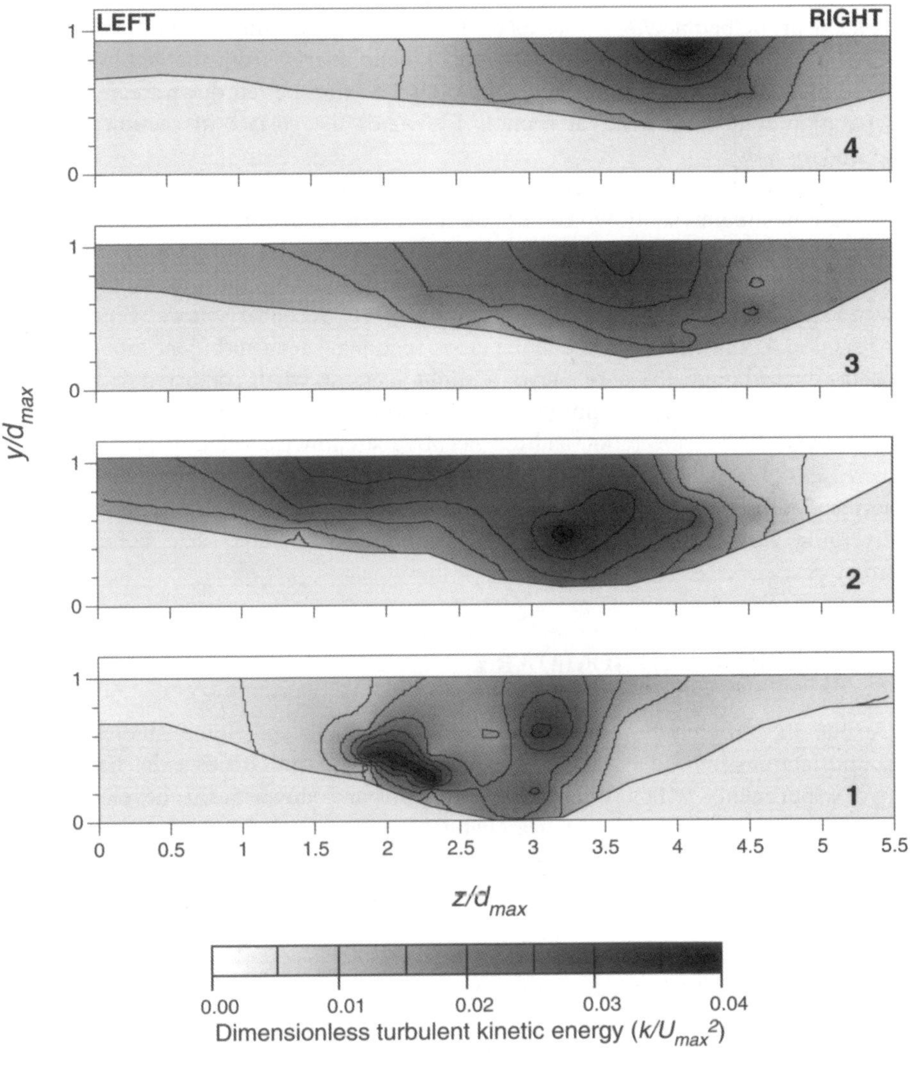

Figure 23.10. Distribution of turbulent kinetic energy (Equation 23.5) at the field study site for the four measurement cross-sections shown in Figure 23.2 ($x/d_{max} \approx 3$, 6, 8, 11). Flow direction and axes are the same as Figure 23.7. Contour lines of turbulent kinetic energy correspond to divisions on the colour bar at 0.005, 0.01, 0.015, 0.02, 0.025, 0.03, 0.035, $0.04 k/U_{max}^2$. The highest values of turbulent kinetic energy occur in the avalanche face separation zone (section 1; $z/d_{max} = 1.75$, $y/d_{max} = 0.5$). Contours are interpolated from an average of 40 measurements at each cross-section.

23.7 and 23.8). A region of high turbulent kinetic energy also extends to the left of the mixing interface in proximity to a zone of low turbulent kinetic energy which exists near the bed at $z/d_{max} = 1.5$. This reflects the upwelling of fluid and negative vertical velocity gradient which transport kinetic energy from the bed towards the surface. Additionally, kinetic energy may also be transported downstream from the separation zone shear layer at section 1 towards the surface at section 2 by the secondary flow.

At section 3, a general decrease in turbulent kinetic energy occurs despite an increase in streamwise flow velocity, suggesting a reduction in the efficiency with which turbulence extracts energy from the mean flow. The region of maximum turbulent kinetic energy continues to follow the talweg and moves towards the water surface, mirroring the reduction in strength of secondary flows (Figure 23.8). At section 4, a further decrease in the cross-sectional mean turbulent kinetic energy occurs even though the core shows a slight increase when compared to section 3. This increase in the maximum turbulent kinetic energy may be attributed to streamwise flow acceleration and the negative streamwise velocity gradient which increase the production and diffusion of turbulence towards the surface. The corresponding reduction of cross-sectional mean turbulent kinetic energy reflects the diverging flow pattern and decreasing importance of large-scale coherent flow structures.

DISCUSSION AND SUMMARY

Despite the differences in morphology between field and flume studies and the simplifications introduced in the flume model, many similarities exist between the two experiments which illustrate the origin and downstream development of coherent flow structures through a channel junction. It is important to stress that coherent secondary flows are produced in the flume experiment in the absence of streamline curvature and both studies reveal that the downstream transformation of the three-dimensional flow structure produces a simpler, diverging flow. The presence of streamline curvature and increased relative scour depth in the field study appear to intensify these secondary flows, but they may not *cause* their generation or subsequent transformation. Several features of the three-dimensional flow structure are common both to the flume and field studies and may be used to propose a generalized model of flow at channel junctions with pronounced bed scour.

CHARACTERSITICS OF FLOW SEPARATION

In the region near the backward-facing steps or avalanche faces, flow separation over the step produces a zone of recirculating fluid and a bounding free shear layer. Dye tracing in the flume experiment shows a rapid exchange of fluid across the steep velocity gradient (Figures 23.3 and 23.5) and that this region is characterized by high turbulence intensities associated with Kelvin-Helmholtz instabilities. Hence, there is intense production of turbulent kinetic energy along this interface. Flow vectors reveal a strong ($> 0.1U_{max}$) downflow of fluid into the separation zone in both flume and field and lateral transfer of fluid towards the channel edges. Such lateral transfer of fluid within the avalanche-face separation zone in the field may be shown by the strong fluid upwellings on the left of the channel near section 2 (Figures 23.2 and 23.7) which possibly owe their origin to fluid entrained into the avalanche-face separation zone which is then advected downstream by the secondary flow (Figure 23.3). This is similar to the upwelling documented by Weerakoon *et al.* (1991) which was associated with flow separation at the downstream junction corner and secondary flows generated in the absence of avalanche-face steps. The step separation zone in the flume is higher near the channel edges than the centreline due to wall-flow interactions similar to those observed by Papadopoulos & Ötügen (1995). Their study showed that two streamwise eddies form in the lee of a straight backward-facing step which bring fluid upwards at the wall and downwards near the channel centre. The effect of this circulation is to lift the reattachment zone at the channel edges and reduce the reattachment length at the channel centre. In the flume, the strong down flow in the lee of both negative steps illustrates the transfer of high momentum fluid towards the channel bed. The near-bed flow in the lee of the steps shows a strong lateral movement of flow towards the left channel wall. It appears that fluid from the left and right flows is entrained into the left side of the step separation zone, transported further leftwards and then ejected towards the left channel wall. This asymmetric flow pattern is produced by the higher velocities in the free-stream of the left-channel which generate lower pressures in the left of the step separation zone and entrain fluid into this region. Similarly, Figure 23.9 shows that the resultant near-bed flows are greater on the left side of section 1 in the field, below the faster left tributary flow. Furthermore, the height of the inflection point on the resultant velocity profile suggests that the step separation zone is larger below the slower moving right-channel flow. Immediately downstream from the step separation zone in the flume, there is very little recirculation of fluid at the channel wall and therefore the region of maximum velocity remains near the surface since insufficient near-bed and near-wall low momentum fluid is moved towards the surface. However, in the field, much stronger secondary circulations are present which may be attributed to a combination of factors including negative steps that are skewed with respect to the mean tributary flow directions, higher lateral velocity gradients, streamline curvature due to planform shape, larger step height -to- tributary flow depth ratios, differential bed roughness and acceleration of flow downstream from the scour. Stronger secondary flows in the field transfer low momentum fluid from the channel bed and banks towards the channel centre, suppressing the core of maximum velocity below the water surface towards the channel bed.

FLOW MIXING

Downstream of the reattachment region of the step separation zone in the flume, the mixing interface shear layer becomes distorted (Figures 23.3 and 23.4; $x/d_{max} \approx 6$). A similar distortion is also observed in the field even where the cross-sectional mean streamwise velocity increases downstream (Figures 23.7 and 23.8). In the flume, the reduced streamwise velocity gradient at $x/d_{max} = 6$ promotes clockwise flow rotation on the left of the channel, anticlockwise flow rotation on the right and strong downwelling at the centreline. This suggests that momentum is transferred inwards at the surface, down towards the bed and then outwards. The strongest outwards movement is towards the left channel wall, although anticlockwise rotation on the right side of the channel moves low momentum fluid from the bed towards the surface and then inwards towards the channel centreline (Figures 23.3 and 23.4). This causes the faster left-channel flow to penetrate beneath the slower flow on the right and hence the slower moving fluid overrides the faster flow. This general flow circulation pattern is also present in the field, as shown by deformation of the shear layer in Figure 23.8. The secondary flow appears to promote this mixing process in the flume since the kinetic energy distribution indicates limited lateral distortion of the shear layer at the mixing interface.

Entrainment of fluid from the right to the left of the flume test section is caused by turbulent entrainment across the mixing interface shear layer and lateral transfer of fluid associated with the step separation zone. This movement of fluid generates flow separation at the right wall whilst, in the field, flow separation occurs at the left downstream junction corner. Consequently, the secondary flow pattern on the right of the flume channel grows at the expense of that on the left, opposite to that in the field where the left tributary cell develops preferentially. These differences are an important indication of how flow cells develop and suggest that lateral separation zones enhance secondary flow development due to the lateral streamwise velocity gradient between the main and separated flow regions. However, despite these differences, there is a downstream transformation of coherent flow structures into a simpler diverging flow in both the flume and field. In both cases there is a change in the streamwise transport of vorticity. As a consequence, vortex stretching in the field occurs due to accelerating flow which distorts the secondary flow. Such streamwise flow acceleration and vortex stretching are absent in the flume. However, the redistribution of flow momentum by the secondary flow reduces the streamwise velocity gradient and creates strong lateral velocity gradients thereby reinforcing the diverging flow pattern.

The mixing zone shear layer between the two confluent flows, like that over the lee-side separation zones, is characterized by Kelvin-Helmholtz instabilities that rotate around the main axis of the shear layer (Biron *et al.*, 1993). This mixing layer is also a region of high kinetic energy production in both flume and field and can be seen to become laterally distorted through the junction (Figures 23.5 and 23.10), as postulated in other studies (Best & Roy, 1991; Gaudet & Roy, 1995;

Rhoads & Kenworthy, 1995; Biron *et al.*, in press; Rhoads, this volume). In the field, the core of maximum turbulent kinetic energy moves towards the surface due to the negative vertical streamwise velocity gradient which controls the transport of turbulent kinetic energy. This is in marked contrast to measurements obtained in straight channel flow where the peak turbulent kinetic energy occurs in the lower region of the flow (e.g. Parthasarathy & Muste, 1994). Furthermore, the log-law velocity distribution cannot be applied at sites of flow convergence and divergence where the vertical transfer of momentum is greatly different from that of two-dimensional flows.

DEVELOPMENT OF COHERENT FLOW STRUCTURES

These diagnostic features of the three-dimensional flow structure in the field and flume studies may be used to propose a schematic model of coherent flow structure development at channel junctions with pronounced bed scour (Figure 23.11). The strength of flow rotation and the dimensions of the coherent structures will depend on many factors including: (i) the height and orientation of the avalanche faces; (ii) flow velocity in the tributary channels; (iii) streamline curvature induced by the gross channel planform; (iv) lateral velocity gradients produced by flow separation from the channel walls; and (v) differential bed roughness. Flow recirculation in the lee of the avalanche faces develops downstream into coherent structures which may then be greatly influenced by streamline curvature, turbulence anisotropy and distortion of the shear layer. In the present study, flow separation over the avalanche faces is intrinsic to the generation of these coherent flow structures. Such a mechanism may also explain the origin of secondary flow cells previously documented for junctions with distinct avalanche faces both in the flume (Mosley, 1976; Ashmore, 1982; Best, 1987, 1988) and field (Ashmore *et al.*, 1992). In the absence of flow separation in the lee of avalanche faces, for example at very low junction angles where avalanche face bars are absent, factors (ii) to (v) listed above may give rise to secondary flows. However, with the bed morphology studied here, which is characteristic of many confluences (i.e. the presence of appreciable scour), the generation of channel-scale coherent secondary flow structures is strongly influenced by avalanche-face flow separation. Downstream distortion of these coherent flow structures occurs in response to both vorticity transport by the streamwise or lateral velocity gradients and distortion of the mixing zone shear layer between the flows. These cause deformation of the coherent flow structures and may amplify or suppress the development of secondary flows.

IMPLICATIONS FOR SEDIMENT TRANSPORT AND CHANNEL CONFLUENCE DYNAMICS

It has been shown in two-dimensional flows that turbulent kinetic energy is directly related to the Reynolds stresses (Bradshaw, 1967; Harsha & Lee, 1970; Gordon &

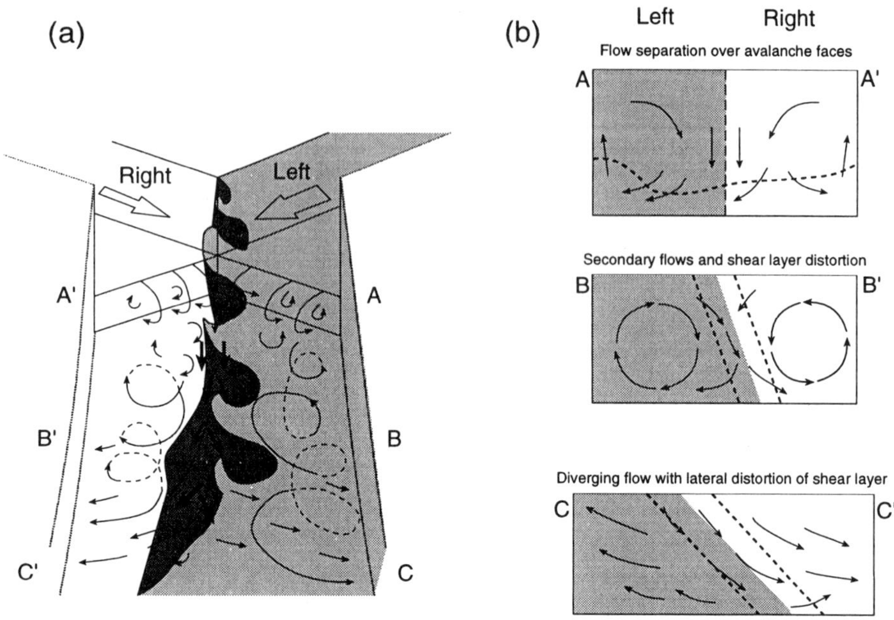

Figure 23.11. A schematic model of coherent flow structures at channel junctions with pronounced bed scour (see text for explanation): (a) three-dimensional view illustrating shear layer between the two flows and development of coherent flow structures downstream from the avalanche faces; and (b) the secondary flow pattern at three representative cross-sections shown in (a) as A-A', B-B' and C-C'. The flow direction in (a) is towards the observer whilst in (b) flow is directed into the page, consistent with the flume and field sections. In both (a) and (b) the shaded zone indicates flow originating from the left tributary channel. The long dashed lines in (b) illustrate the approximate width of the mixing interface which may vary between the channel bed and water surface dependent on local differences in lateral gradients of both turbulence intensity and velocity

Dohne, 1973; McCave, 1979) which are also of prime importance in the transport of sediment. Flow at junctions is strongly three-dimensional and the turbulent kinetic energy will be related to all the components of Reynolds stress. In the present study $u'v'$ and $u'w'$ are measured and their relationship with turbulent kinetic energy is

$$|u'v'| + |u'w'| = -1.80 \times 10^{-4} + 0.380k \qquad n = 669;\ r^2 = 0.875 \qquad (23.7)$$

$$|u'v'| + |u'w'| = -9.28 \times 10^{-3} + 0.530k \qquad n = 99;\ r^2 = 0.688 \qquad (23.8)$$

for the flume (Equation 23.7) and field (Equation 23.8). Part of the unexplained relationship between the Reynolds stresses and turbulent kinetic energy can be attributed to the $v'w'$ and $u'v'w'$ components of Reynolds stress. The $v'w'$ component may be particularly important in the field where secondary flows are stronger. The difference between the gradients of these relationships suggests a reduced efficiency in extracting energy from the Reynolds stresses in the laboratory flume as compared to the field situation. This may be attributed to both the smaller strength of secondary flow rotation and the absence of surface roughness in the laboratory experiment, since large-scale coherent flow structures are the most efficient mechanism for extracting energy and surface roughness will result in an increase in the intensity of turbulent fluctuations. Equations 23.7 and 23.8 imply that zones of high turbulent kinetic energy reflect strong Reynolds stresses and therefore may be influential in promoting sediment transport. Such zones of high kinetic energy associated with the step separation zones and mixing interface shear layer may be influential in generating bed scour by positive feedback processes. Additionally, the counter-rotating secondary flow cells documented here will serve to segregate sediment loads from each tributary as they move through the scour and towards the diffluence (Best, 1988; Ashworth *et al.*, 1992b; Kenworthy & Rhoads, 1995).

The downstream transformation of coherent flow structures from secondary flow rotation into a simpler diverging flow results in the movement of the core of high turbulent kinetic energy away from the bed, reducing both the near-bed Reynolds stresses and turbulent kinetic energy gradient. This process provides a mechanism for deposition of sediment and the subsequent growth of a mid-channel bar downstream from the junction. The field example shows that the redistribution of turbulent kinetic energy can occur together with a general acceleration of the streamwise velocity. Scour morphology may also be influenced by the concentration of high turbulent kinetic energy associated with the mixing interface and separation zone shear layers whilst the rate at which the coherent flow structures transform downstream controls the onset of flow divergence and the development of the diffluence bed morphology. Future work should investigate whether the distance between the junction scour hole and mid-channel bar is a function of the initial growth and strength of secondary circulation and determine the rate at which these three-dimensional flow structures transform downstream.

ACKNOWLEDGEMENTS

The field results presented here are part of a large collaborative data set collected jointly with Peter Ashmore, Pascale Biron, Rob Ferguson, Alistair Kirkbride, André Roy and Greg Sambrook Smith. We are grateful for their permission to use these data. Fieldwork for PJA and JLB was funded by a Royal Society Overseas Fieldwork grant co-ordinated by Rob Ferguson with additional financial support from the Foundation for Canadian Studies. SJM was supported by a NERC studentship (GT4/92/13/P). The Environment Canada Parks Service kindly provided a research permit for Jasper National Park and the Canadian Hostelling Association, Alberta District, provided welcoming accommodation at Beauty Creek. JLB is grateful for the award of a Nuffield Foundation Science Research Fellowship which aided completion of this work. We would like to thank Willi Hager and Bruce Rhoads for their detailed and thoughtful reviews of the paper.

REFERENCES

Ashmore, P.E. 1982. 'Laboratory modelling of gravel braided stream morphology', *Earth Surf. Processes Landf.*, **7**, 201–225.

Ashmore, P.E. 1993. 'Anabranch confluence kinetics and sedimentation processes in gravel braided streams', in Best, J.L. & Bristow, C.S. (Eds) *Braided Rivers: Form, Process and Economic Applications*, Sp. Pub. Geol. Soc., **75**, London, 129–146.

Ashmore, P.E., Ferguson, R.I., Prestegaard, K.L., Ashworth, P.J. & Paola, C. 1992. 'Secondary flow in anabranch confluences of a braided, gravel-bed stream', *Earth Surf. Processes Landf.*, **17**, 299–311.

Ashmore, P.E. & Parker, G. 1983. 'Confluence scour in coarse braided streams', *Water Resour. Res.*, **19**, 392–402.

Ashworth, P.J. 1996. 'Mid-channel bar growth and its relationship to local flow strength and direction', *Earth Surf. Processes Landf.*, **21**, 103–123.

Ashworth, P.J., Ferguson, R.I., Ashmore, P.E., Paola, C., Powell, D.M. & Prestegaard, K.L. 1992a. 'Measurements in a braided river chute and lobe: 2. sorting of bedload during entrainment, transport and deposition', *Water Resour. Res.*, **28**, 1887–1896.

Ashworth, P.J., Ferguson, R.I. & Powell, D.M. 1992b. 'Bedload transport and sorting in braided channels', in Billi, P., Hey, R.D., Thorne, C.R. & Tacconi, P. (Eds) *Dynamics of Gravel Bed Rivers*, John Wiley & Sons Ltd., Chichester, 497–513.

Bathurst, J.C., Thorne, C.R. & Hey, R.D. 1977. 'Direct measurements of secondary currents in river bends', *Nature*, **269**, 504–506.

Best, J.L. 1986. 'The morphology of river channel confluences', *Prog. Phys. Geog.*, **10**, 157–174.

Best, J.L. 1987. 'Flow dynamics at river channel confluences: implications for sediment transport and bed morphology', in Ethridge, F.G., Flores, R.M. & Harvey, M.D. (Eds) *Recent Developments in Fluvial Sedimentology*, Soc. Econ. Palaeon. and Mineralogists, Sp. Pub. **39**, 27–35.

Best, J.L. 1988. 'Sediment transport and bed morphology at river channel confluences', *Sedimentology*, **35**, 481–498.

Best, J.L. & Reid, I. 1984. 'Separation zone at open-channel junctions', *J. Hydraul. Engrg.*, **110**, 1588–1594.

Best, J.L. & Roy, A.G. 1991. 'Mixing layer distortion at the confluence of channels of different depth', *Nature*, **350**, 411–413.

Bhowmik, N.G. 1982. 'Shear stress distribution and secondary currents in straight open channels', in Hey, R.D., Bathurst, J.C. & Thorne, C.R. (Eds) *Gravel-bed Rivers*, John Wiley & Sons Ltd., Chichester, 31–55.

Biron, P., De Serres, B., Roy, A.G. & Best, J.L. 1993. 'Shear layer turbulence at an unequal depth channel confluence', in Clifford, N.J., French, J.R. & Hardisty, J. (Eds) *Turbulence: Perspectives on Flow and Sediment Transport*, John Wiley & Sons Ltd., Chichester, 197–214.

Biron, P., Best, J.L. & Roy, A.G. in press. 'The fluid dynamics effects of bed height discordance at river channel confluences', *J. Hydraul. Engrg.*

Bowden, K.F. & Fairbairn, L.A. 1956. 'Measurements of turbulent fluctuations and Reynolds stresses in a tidal current', *Proc. Roy. Soc. London*, **A**, **237**, 422–438.

Bradshaw, P., Ferriss, D.H. & Atwell, N.P. 1967. 'Calculation of boundary layer development using the turbulent energy equation', *J. Fluid Mech.*, **28**, 593–616.

Bridge, J.S. 1993. 'The interaction between channel geometry, water flow, sediment transport and deposition in braided rivers', in Best, J.L. & Bristow, C.S. (Eds) *Braided Rivers: Form, Process and Economic Applications*, Sp. Pub. Geol. Soc., **75**, London, 13–71.

Bridge, J.S. & Gabel, S.L. 1992. 'Flow and sediment dynamics in a low sinuosity braided river: Calamus River, Nebraska Sandhills', *Sedimentology*, **39**, 125–142.

Bristow, C.S., Best, J.L. & Roy, A.G. 1993. 'Morphology and facies models of channel confluences', in Marzo, M. & Puigdefábregas, C. (Eds) *Alluvial Sedimentation*, Sp. Pub. of the International Association of Sedimentologists, **17**, 91–100.

Browand, F.K. & Troutt, T.R. 1985. 'The turbulent mixing layer: geometry of large vortices', *J. Fluid Mech.*, **158**, 489–509.

Carson, M.A. & Griffiths, G.A. 1987. 'Bedload transport in gravel channels', *J. Hydrol. (New Zealand)*, **26**, 1–151.

Chu, V.H. & Babarutsi, S. 1989. 'Confinement and bed-friction in shallow turbulent mixing layers', *J. Hydraul. Engrg.*, **114**, 1257–1274.

Dawson, M. 1988. 'Sediment size variation in a braided reach of the Sunwapta River, Alberta, Canada', *Earth Surf. Processes Landf.*, **13**, 599–618.

De Brederode, V. & Bradshaw, P. 1972. 'Three-dimensional in nominally two-dimensional separation bubbles. I. Flow behind a rearward-facing step', *Imperial College, Aeronautics Department, London*, Report **72-19**.

De Serres, B. & Roy, A.G. 1990. 'Flow direction and branching geometry at junctions in dendritic networks', *Prof. Geographer*, **42**, 194–201.

Dietrich, W.E. & Smith, J.D. 1983. 'Influence of the point bar on flow through curved channels', *Water Resour. Res.*, **19**, 1173–1192.

Ferguson, R.I. 1993. ' Understanding braided processes in gravel-bed rivers: progress and unsolved problems', in Best, J.L. & Bristow, C.S. (Eds) *Braided Rivers: Form, Process and Economic Applications*, Sp. Pub. of Geol. Soc., **75**, London, 73–87.

Ferguson, R.I., Ashmore, P.E., Ashworth, P.J., Paola, C. & Prestegaard, K.L. 1992. 'Measurements in a braided chute and lobe 1. flow pattern, sediment transport and channel change', *Water Resour. Res.*, **28**, 1877–1886.

Fujita, I. & Komura, S. 1988. 'Visualization of the flow at a confluence', *Proc. of the 3rd Int. Symp. on Refined Flow Modelling and Turbulence Measurements*, 611–618.

Gaudet, J.M. & Roy, A.G. 1995. 'Effect of bed morphology on flow mixing length at river confluences', *Nature*, **373**, 138–139.

Goff, J.J. & Ashmore, P. 1994. 'Gravel transport and morphological change in braided Sunwapta River, Alberta, Canada', *Earth Surf. Processes Landf.*, **19**, 195–212.

Gordon, C.M. 1975. 'Sediment entrainment and suspension in a turbulent tidal flow', *Marine Geol.*, **18**, M57–M64.

Gordon, C.M. & Dohne, C.F. 1973. 'Some observations of turbulent flow in a tidal estuary', *J. Geophys. Res.*, **78**, 1971–1978.

Harsha, P.T. & Lee, S.C. 1970. 'Correlation between turbulent shear stress and turbulent kinetic energy', *AIAA J.*, **8**, 1508–1510.

Howard, A.D. 1971. 'Optimal angles of stream junction: Geometric stability to capture and minimum power criteria', *Water Resour. Res.*, **7**, 863–873.

Jimenez, J. 1983. 'A spanwise structure in the plane shear layer', *J. Fluid Mech.*, **132**, 319–336.

Kanda, T. & Kitada, T. 1977. 'An implicit method for unsteady flows with lateral inflows in urban rivers', *Proc. of 17th Congr. IAHR*, Baden-Baden, Paper **A102**, 213–220.

Kenworthy, S.T. & Rhoads, B.L. 1995. 'Hydrologic control of spatial patterns of suspended sediment concentration at a stream confluence', *J. Hydrol.*, **168**, 251–263.

Kjerfve, B., Shao, C.-C. & Stapor, F.W. 1979. 'Formation of deep scour holes at the junction of tidal creeks: an hypothesis', *Marine Geol.*, **33**, M9–M14.

Kline, S.J., Reynolds, W.C., Schraub, F.A. & Runstadler, P.W. 1967. 'The structure of turbulent boundary layers', *J. Fluid Mech.*, **30**, 741–773.

Lane, S.N., Richards, K.S. & Warburton, J. 1993. 'Comparison between high frequency velocity records obtained with spherical and discoidal electromagnetic current meters', in Clifford, N.J., French, J.R. & Hardisty, J. (Eds) *Turbulence: Perspectives on Flow and Sediment Transport*, John Wiley & Sons Ltd., Chichester, 121–163.

Markham, A.J. & Thorne, A.C.R. 1992. 'Geomorphology of gravel-bed river bends', in Billi, P., Hey, R.D., Thorne, C.R. & Tacconi, P. (Eds) *Dynamics of Gravel Bed Rivers*, John Wiley & Sons Ltd., Chichester, 433–450.

McCave, I.N. 1979. 'Tidal currents at the North Hinder lightship, southern North Sea: Flow directions and turbulence in relation to maintenance of sand banks', *Marine Geol.*, **31**, 101–114.

McGuirk, J.J. & Rodi, W. 1978. 'A depth-averaged mathematical model for near field of side discharges into open-channel flow', *J. Fluid Mech.*, **86**, 761–781.

Mosley, M.P. 1976.' An experimental study of channel confluences', *J. Geol.*, **84**, 535–562.

Mosley, M.P. & Schumm, S. 1977. 'Stream junctions – a probable location for bedrock placers', *Econ. Geol.*, **72**, 691–94.

Muskatirovic, J. & Miloradov, M. 1980. 'Some experiences in improving river confluences', *IAHR Symp. on River Engineering and its Interaction with Hydrological and Hydraulic Research*, Belgrade, Yugoslavia, D1–D11.

Nezu, I. & Rodi, W. 1985. 'Experimental study on secondary currents in open channel flow', *21st IAHR Congress*, Melbourne, Australia, **2**, 115–119.

Nezu, I., Tominaga, A. & Nakagawa, H. 1993. 'Field measurements of secondary currents in straight rivers', *J. Hydraul. Engrg.*, **119**, 598–614.

Papadopoulos, G. & Ötügen, M.V. 1995. 'Separating and reattaching flow structure in a suddenly expanding rectangular duct', *J. Fluids Eng.*, **117**, 17–23.

Parthasarathy, R.N. & Muste, M. 1994. 'Velocity measurements in asymmetric turbulent channel flows', *J. Hydraul. Engrg.*, **120**, 1000–1020.

Petts, G.E. & Thoms, M.C. 1987. 'Morphology and sedimentology of a tributary confluence bar in a regulated river: North Tyne, UK', *Earth Surf. Processes Landf.*, **12**, 433–440.

Plesniak, M.W., Mehta, R.D. & Johnston, J.P. 1994. 'Curved two-stream turbulent mixing layers: three-dimensional structure and streamwise evolution', *J. Fluid Mech.*, **270**, 1–50.

Prinos, P., Tavoularis S. & Townsend, R. 1988. 'Turbulence measurements in smooth and rough-walled trapezoidal ducts', *J. Hydraul. Engrg.*, **114**, 43–53.

Rhoads, B.L. & Kenworthy, S.T. 1995. 'Flow structure at an asymmetrical stream confluence', *Geomorphology*, **11**, 273–293.

Rice, R.J. 1982. 'The hydraulic geometry of the lower portion of the Sunwapta valley train, Jasper National Park, Alberta', in Davidson-Arnott, R., Nickling, W. & Fahey, D.B. (Eds) *Research in Glacial, Glacio-fluvial and Glacio-lacustrine Systems*, Geobooks, Norwich, 151–174.

Richards, K. 1980. 'A note on changes in channel geometry at tributary junctions', *Water Resour. Res.*, **16**, 241–244.

Roy, A.G. & Bergeron, N. 1990. 'Flow and particle paths at a natural river confluence with coarse bed material', *Geomorphology*, **3**, 99–112.

Roy, A.G. & Roy, R. 1988. 'Changes in channel size at river confluences with coarse bed material', *Earth Surf. Processes Landf.*, **13**, 77–84.

Roy, A.G., Roy, R. & Bergeron, N. 1988. 'Hydraulic geometry and changes in flow velocity at a river confluence with coarse bed material', *Earth Surf. Processes Landf.*, **13**, 583–598.

Sambrook Smith, G.H. & Ferguson, R.I. 1995. 'The gravel sand transition along river channels', *J. Sed. Res.*, **A65**, 423–430.

Southard, J.B. Smith, N.D. & Kuhnle, R.A. 1984. 'Chutes and lobes: newly identified elements of braiding in shallow gravelly streams', in Koster, E.H. & Steel, R.J. (Eds)

Sedimentology of Gravels and Conglomerates, Can. Soc. Petroleum Geologists, Mem. **10**, 51–59.

Thorne, C.R., Russell, A.P.G. & Alam, M.K. 1993. 'Planform pattern and channel evolution of the Brahmaputra River, Bangladesh', in Best, J.L. & Bristow, C.S. (Eds) *Braided Rivers: Form, Process and Economic Applications*, Sp. Pub. Geol. Soc., **75**, London, 257–276.

Tominaga, A., Nezu, I., Ezaki, K. & Nakagawa, H. 1989. 'Three-dimensional turbulent structure in straight open-channel flows', *J. Hydraul. Res.*, **127**, 149–173.

Tominaga, A., Nezu, I. & Nagao, M. 1993. 'Turbulence control of channel flow by forced secondary currents over longitudinal ridge', *9th Symp. on Turbulent Shear Flows*, Kyoto, Japan, 14.4.1–14.4.6.

Weerakoon, S.B., Kawahara, Y. & Tamai, N. 1991. 'Three-dimensional flow structure in channel confluences of rectangular section', in *24th Congr. IAHR, Proc. of the Technical Session on Study of Streams and Watersheds of High Irregularity*, Madrid, **A**, 373–380.

Weygandt, J.H. & Mehta, R.D. 1995. 'Three-dimensional structure of straight and curved plane wakes', *J. Fluid Mech.*, **282**, 279–311.

Winant, C.D. & Browand, F.K. 1974. 'Vortex pairing: the mechanism of turbulent mixing-layer growth at moderate Reynolds number', *J. Fluid Mech.*, **63**, 237–255.

NOTATION

d_{max}	maximum measured flow depth (m)
n	sample size
k	turbulent kinetic energy (J kg^{-1})
r	correlation coefficient
Q	discharge (m^3 s^{-1})
u'	streamwise velocity fluctuations (m s^{-1})
U	mean streamwise velocity (m s^{-1})
U_r	mean resultant velocity (m s^{-1})
U_{max}	maximum streamwise velocity (m s^{-1})
U_θ	mean magnitude of horizontal flow vector (m s^{-1})
v'	vertical velocity fluctuations (m s^{-1})
V	mean vertical velocity (m s^{-1})
w'	cross-stream velocity fluctuations (m s^{-1})
W	mean cross-stream velocity (m s^{-1})
x	downstream distance (m)
y	vertical distance (m)
y_{max}	maximum flow depth at a cross-section (m)
z	spanwise distance (m)
α	cross-section offset angle for velocity correction (°)
θ	mean orientation of horizontal flow vector (°)

SUBSCRIPTS

A	denotes series A measurements

B	denotes series B measurements
LEFT	denotes left tributary
RIGHT	denotes right tributary

24 Mean Structure of Transport-effective Flows at an Asymmetrical Confluence when the Main Stream is Dominant

BRUCE L. RHOADS
Department of Geography, University of Illinois at Urbana-Champaign, Urbana, IL 61801, USA

ABSTRACT

A field experiment has been conducted at an asymmetrical stream confluence in east-central Illinois, USA to examine how mean flow structure, sediment transport and bed morphology vary with momentum ratio and flow stage. Previous work showed that for transport-ineffective flows with momentum ratios less than one (i.e. when the main stream is dominant), mean flow structure at the entrance to the downstream channel consists of twin surface-convergent helical cells embedded within a converging flow field. This study confirms that this flow structure also occurs at transport-effective stages. Similar to low-stage events, the twin cells are separated by the mixing interface between the two confluent streams. As flow moves through the downstream channel the helical cell on the inner (tributary) side of the mixing interface remains strong, whereas the counter-rotating cell on the opposite side of the mixing interface decreases in strength. The strong helical motion over the inner portion of the downstream channel may be the result of greater curvature of the tributary flow relative to the main stream as flow in the tributary turns to become aligned with the downstream channel. Patterns of bedload transport reflect the general structure of the flow field. Distinct segregation of the incoming sediment loads occurs at the entrance to the downstream channel, but further downstream sediments from the main stream are deflected inward, leading to mixing of the two loads over the inner portion of the channel. The evolution of the bed morphology as the momentum ratio changes from greater-than-one to less-than-one also is consistent with the observed mean flow structure and patterns of sediment transport.

INTRODUCTION

Over the last 20 years scientific interest in the fluvial dynamics of stream confluences has increased considerably. Recent work has demonstrated that hydrodynamic conditions at confluences are characterized by a variety of coherent flow structures, including vortices with vertical axes along the shear layer between converging flows (e.g. Biron *et al.*, 1993a), vortices with longitudinal axes that

Coherent Flow Structures in Open Channels.
Edited by P.J. Ashworth, S.J. Bennett, J.L. Best and S.J. M^c^Lelland.

develop in the lee of steep avalanche faces (Best, 1987, 1988; McLelland *et al.*, this volume) that can also distort the shear layer (Best & Roy, 1991), large-scale secondary circulation associated with channel curvature (Ashmore & Parker, 1983; Ashmore *et al.*, 1992; Rhoads & Kenworthy, 1995), and one or more zones of recirculation caused by separation of flow from the channel boundary (Mosley, 1976; Best & Reid, 1984; Rhoads & Kenworthy, 1995). Each of these hydrodynamic features potentially can generate distinct spatial patterns of sediment mobilization or deposition and therefore characteristic sedimentologic and morphologic features (Biron *et al.*, 1993b; Mayer, 1995). However, not all of these types of fluid motion occur at every confluence. At present, the scales, interactions and morphologic importance of various coherent flow structures have yet to be determined. Laboratory studies and mathematical simulations provide a starting point for the development of a general model of confluence dynamics (e.g. Best, 1987, 1988; Weerakoon *et al.*, 1991), but the morphologic variety and hydrodynamic complexity of confluences necessitate field evaluations of experimental and theoretical findings over a broad range of natural conditions. In particular, the need exists to document more precisely interactions between flow structure, sediment transport and channel morphology at confluences of various planform types, ranging from symmetrical (Y-shaped) to asymmetrical (those in which the downstream channel is a linear extension of one of the tributaries), as hydrologic conditions vary (Bridge, 1993). Most studies to date have focused on characteristics of flow structure during events that are not transport-effective (e.g. Ashmore *et al.*, 1992; Rhoads & Kenworthy, 1995) or on the movement of individual particles through confluences during events that do not substantially alter channel form (Best, 1988; Roy & Bergeron, 1990).

The study described in this Chapter documents mean flow structure, patterns of sediment transport and changes in bed morphology at an asymmetrical stream confluence when flow is at a transport-effective stage and the main channel upstream from the confluence is dominant relative to the lateral tributary. It is part of a larger research programme examining the response of this confluence to variations in hydrologic conditions. The results confirm that the flow structure found at transport-ineffective stages (Rhoads & Kenworthy, 1995) persists and strengthens as stage increases and that this structure influences spatial patterns of sediment movement and changes in bed topography. The study contributes to knowledge of the role of time-averaged coherent structures in confluence dynamics and complements other contributions to this volume that focus on confluences (e.g. McLelland *et al.*, this volume; Richardson *et al.*, this volume).

FIELD METHODOLOGY

The field data for this study were collected during the summer 1992 at the confluence of the Kaskaskia River (KR) and Copper Slough (CS) in east-central

Illinois, USA, the site of a long-term investigation of confluence dynamics (Figure 24.1). The planform of this confluence is asymmetrical with a junction angle of 60°. The Kaskaskia River has a slightly larger drainage area than the Copper Slough (54 km^2 versus 41 km^2) and is therefore considered the main stream at the confluence. Both streams are contained within straight, trapezoidal channels constructed by local landowners to improve the drainage of adjacent agricultural land. The channels have bottom widths of 7 to 9 m, top widths of 15 to 17 m, and are 3 to 4 m deep. Bed material in the Kaskaskia River consists predominantly of sand, whereas the Copper Slough contains mixed sand and gravel (Figure 24.2).

Hydrologic variability at the confluence is monitored using four continuous water-stage recorders consisting of pressure transducers linked to data loggers. Two recorders are located on each stream at distances of 15 to 20 m and 0.9 to 1.7 km upstream from the confluence (Figure 24.1). No tributaries enter either stream between the two recorders. Water levels at each location are recorded on an hourly basis with the frequency of readings increasing to 5 minute intervals during storm events. Changes in bed morphology produced by flow variability are documented through repeat topographic surveys of the confluence.

To characterize the structure of the mean flow field at the confluence, time-

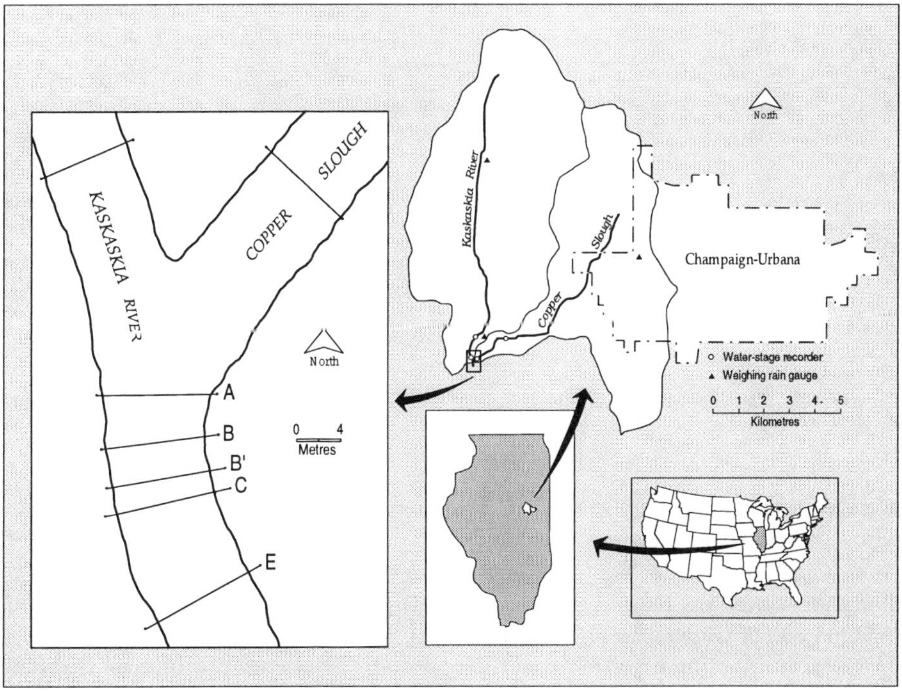

Figure 24.1. Map of field site showing locations of flow-measurement cross-sections

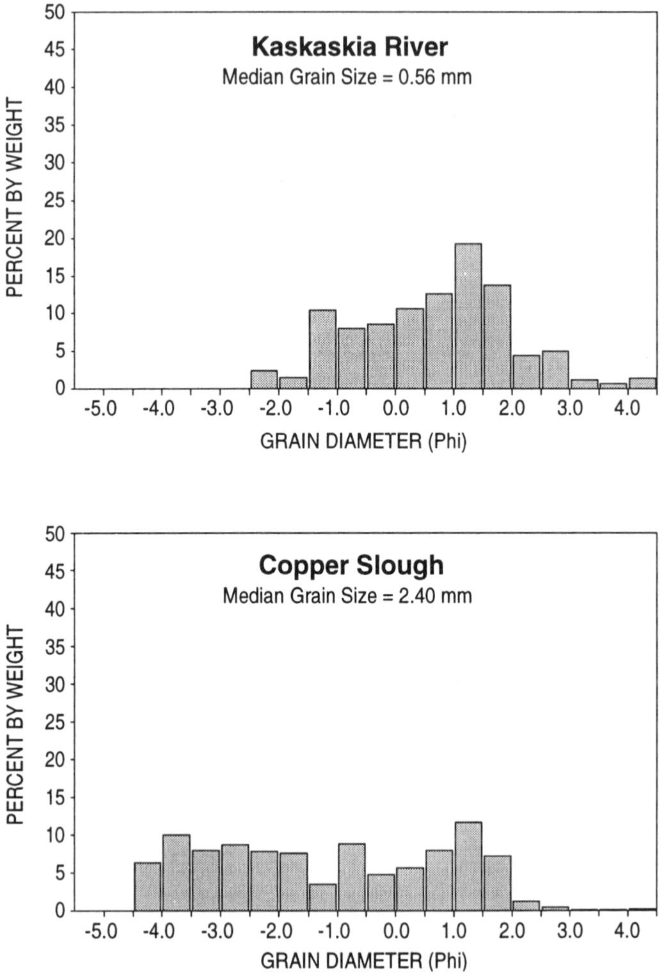

Figure 24.2. Textural characteristics of bed material in the Kaskaskia River and Copper Slough

averaged readings of downstream (U) and cross-stream (W) velocities are obtained with a Marsh-McBirney Model 511 electromagnetic current meter (time constant = 0.2 s, sampling frequency = 1 Hz, sampling interval = 60 s) along cross-sections located at and immediately downstream from the downstream junction corner (Figure 24.1). A temperature probe mounted behind the flow sensor provides information on mean water temperature (T) at each position in the flow field. This information is useful for defining the position of the mixing interface between the confluent streams, which usually have slightly different (2–3° C) water temperatures. The values of W are decomposed into discrete components, W_{py} and

W_{sy}, associated with primary flow (U_p) and secondary flow (W_s), respectively, where U_p and W_s are defined on the basis of zero net lateral discharge in a plane normal to the depth-averaged flow direction at each vertical (Bathurst *et al.*, 1977). Details of this procedure, as well as the technique used to ensure that velocity measurements are obtained within a fixed plane of reference at each cross-section, are discussed by Rhoads & Kenworthy (1995). The purpose of the decomposition is to illustrate the separate contributions of flow convergence/divergence and secondary circulation to the cross-stream flow field. Values of W_{py} show how skewing of the depth-averaged flow in relation to the orientation of the channel cross-section contributes to cross-stream flow, whereas values of W_{sy} represent the component of cross-stream flow produced by differences in orientations of individual velocity vectors at a measurement vertical. Mean velocities measured in each stream immediately upstream from the confluence are used to compute the momentum flux ratio (M_r) of the incoming flows:

$$M_r = \frac{\rho_2 Q_2 \overline{U_2}}{\rho_1 Q_1 \overline{U_1}} \qquad (24.1)$$

where Q is discharge (m^3 s^{-1}), ρ is the water density (kg m^{-3}), U is mean velocity over the cross-section (m s^{-1}), and subscripts 1 and 2 refer to the Kaskaskia River and Copper Slough, respectively.

A 76.2×76.2 mm Helley-Smith bedload sampler oriented in the direction of the near-bed velocity vector (*cf.* Gaudet *et al.*, 1994) is used to collect samples of bedload at several locations along one or more of the flow measurement cross-sections. Each location is sampled for 120 s during separate traverses of each cross-section to produce composite 240 s samples. Although this procedure may not provide accurate estimates of average rates of bedload transport, especially for sandy bedload entering the confluence from the Kaskaskia River (see Gomez *et al.*, 1990, 1991), it does indicate which portions of the channel bed are more active than others. The main purpose of the bedload sampling programme is to determine the pattern of bedload movement through the confluence for different types of hydrologic conditions. Because the texture of bed material in each stream is distinctly different (Figure 24.2), the bedload samples provide valuable information on the pattern of mixing of separate sediment loads from the Kaskaskia River and Copper Slough within and downstream from the confluence.

RESULTS

HYDROLOGY AND BED MORPHOLOGY

The temporal pattern of runoff in the Kaskaskia River and Copper Slough during the summer of 1992 was characterized by persistent low flow in June and frequent high-stage events in July (Figure 24.3). The extant bed morphology in late June

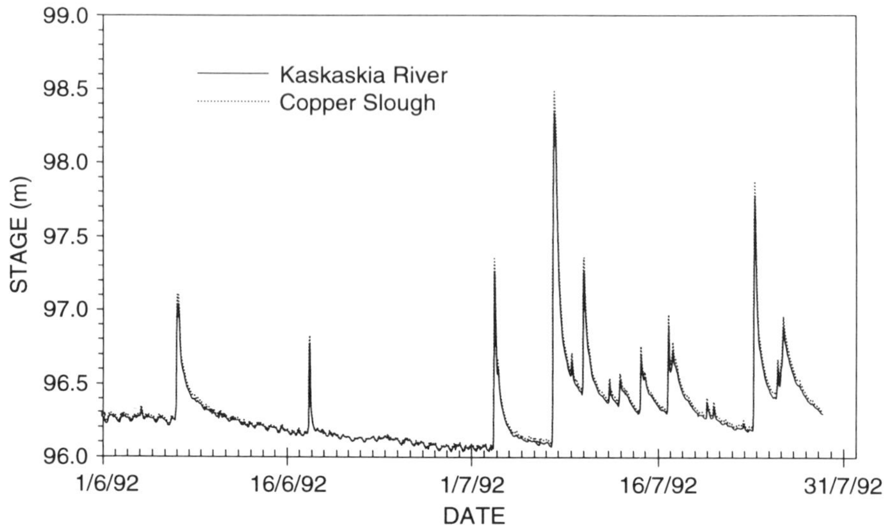

Figure 24.3. Variation in water stage (arbitrary datum) during June and July 1992 for two stage recorders within 20 m of the confluence

1992 consisted of a large main bar at the downstream junction corner, a secondary bar in the lee of the main bar and a narrow scour hole along the west bank of the downstream channel (Figure 24.4). This morphology was produced primarily by two high-stage, high momentum-ratio ($M_r >> 1.0$) events in July and August 1991 and by sustained high momentum-ratio, low-stage conditions between August 1991 and June 1992 (Rhoads & Kenworthy, 1995).

The period of high flow in July 1992 marked the end of a ten-month interval during which no major channel-shaping events occurred in either stream and bed morphology at the confluence did not change appreciably. During July 1992, over 350 mm of precipitation fell on Champaign-Urbana, Illinois (Figure 24.1) – the largest monthly precipitation total recorded over a 105 year period. Precipitation occurred on 18 days during the month, producing considerable fluctuation in flow stage on the Kaskaskia River and Copper Slough (Figure 24.3).

An initial storm on 2–3 July affected mainly the Copper Slough basin, producing a high momentum-ratio ($M_r >> 1.0$) at the confluence. On 7 July over 80 mm of precipitation occurred in the upper Kaskaskia basin. An initial surge of water from the partly urbanized drainage basin of the Copper Slough generated a momentum ratio greater than one on the rising limb of the runoff hydrograph, however, the momentum ratio dropped below one on the receding limb of this event and remained below one for the following 2 weeks (Table 24.1). The channel bed at the confluence adjusted rapidly to the low momentum-ratio conditions. The scour hole enlarged in depth and breadth, resulting in removal of sediment over the

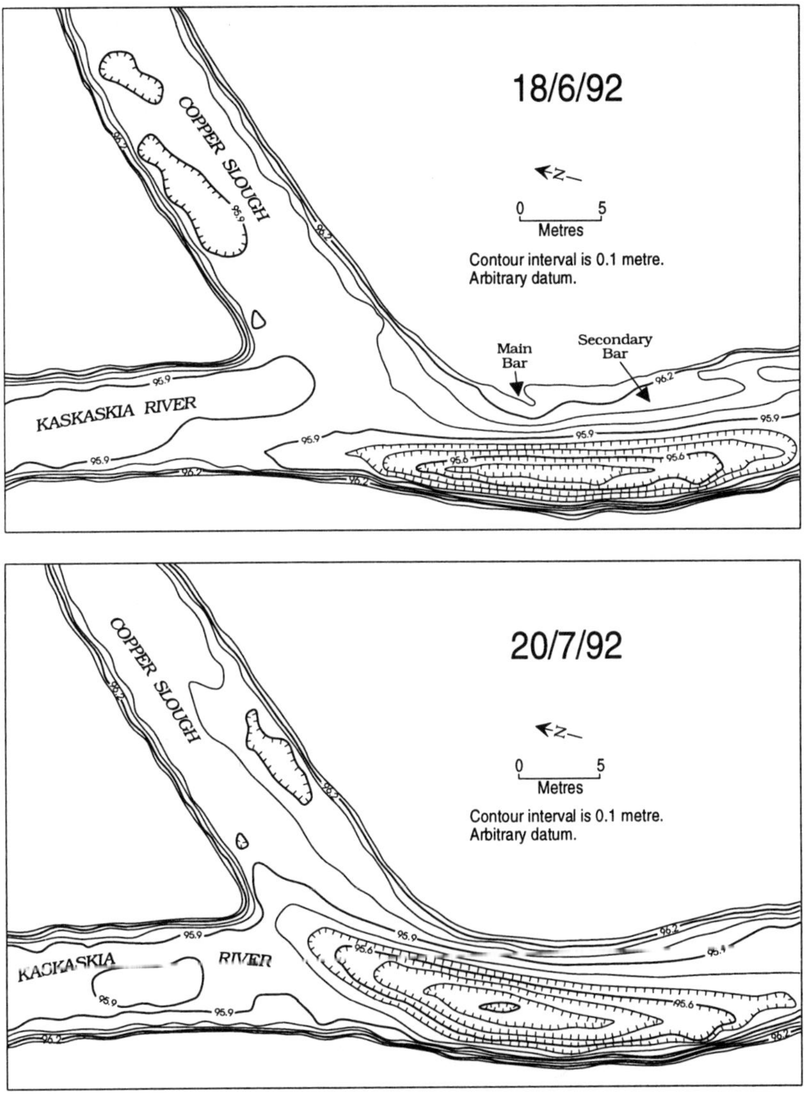

Figure 24.4. Bed morphology in late-June and mid-July 1992

inner half of the downstream channel where the main and secondary bars had been located (Figures 24.4 and 24.5). The zone of scour also expanded into the central region of the confluence (Figure 24.4). Toward the end of July 1992 the momentum ratio generally exceeded 1.0 (Table 24.1) and the transverse profile of the downstream channel became more asymmetrical as a wedge of sediment advanced into the confluence from the Copper Slough (Figure 24.5).

Table 24.1. Field measurements of momentum ratio during July 1992

Date	Time	Momentum ratio CSC/KRC	Momentum ratio CSU/KRU
7 July 1992	p.m.		3.81
8 July 1992	a.m.	0.50	
	p.m.	0.47	
10 July 1992	a.m.		0.97
11 July 1992	p.m.		0.80
13 July 1992	a.m./p.m.	0.68	0.90
15 July 1992	a.m.	0.35	
	p.m.	0.43	
16 July 1992	a.m.	0.46	
23 July 1992	p.m.	3.54	
24 July 1992	a.m.	1.00	
	p.m.	0.82	
26 July 1992	a.m.		1.21
31 July 1992	a.m.	1.51	
	p.m.	1.51	

CSC = Copper Slough at confluence, KRC = Kaskaskia River at confluence, CSU = Copper Slough 1675 m upstream of confluence, KRU = Kaskaskia River 940 m upstream of confluence

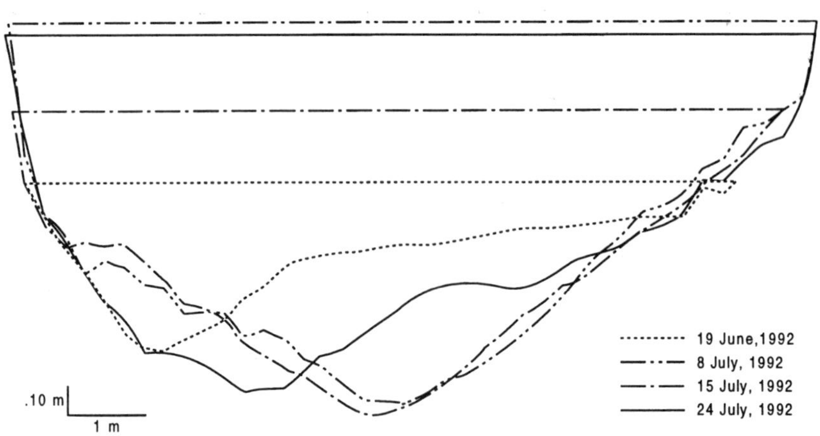

Figure 24.5. Change in the channel profile at cross-section A between late-June and late-July 1992

FLOW STRUCTURE

Data on *U* and *W* were collected at cross-sections A and C on 8 July and at cross-sections A, B, C, and E on 15 July 1992. Both of these events had momentum ratios substantially less than one (Figure 24.6). On each date, measurements were obtained on the receding limb of a hydrograph, thus total discharge and flow stage declined over the period of data collection (Figure 24.6). The drop in flow stage at individual cross-sections was approximately 5% of the maximum flow depth on 8 July and only 1% of the maximum flow depth on 15 July. The momentum ratio changed slightly throughout each day, but the ratio on 15 July was consistently lower than the ratio on 8 July.

Depth-averaged velocity vectors

The vectors of depth-averaged velocity for each date show that the pattern of flow through the downstream channel is roughly parallel to the channel direction (solid lines) and bed contours (Figures 24.4 and 24.6). This pattern of flow is similar to that observed at this confluence at low stage when $M_r < 1.0$ and a large bar exists at the downstream junction corner (Rhoads & Kenworthy, 1995). Some flow convergence is evident over the inner portion of cross-section A, especially on 15 July, but this convergence is less pronounced than that which occurs for low-stage flows with $M_r > 1.0$ in the presence of a main bar (Rhoads & Kenworthy, 1995). Further downstream the vectors are roughly parallel to one another and the zone of flow separation that develops for low-stage conditions in the presence of a main bar (Rhoads & Kenworthy, 1995) is not evident on these two dates. Instead, the measurement locations closest to the inner bank, which are within 0.5 m of a zone of partially submerged vegetation along this bank, generally have strong downstream velocities.

Water temperatures

On 8 July the water temperature of the Copper Slough was about 3° C warmer than that of the Kaskaskia River, whereas on 15 July this difference in temperature was approximately 2° C. The temperature of each stream varied slightly throughout each day, but the difference in temperature between the tributaries remained fairly constant. This temperature contrast should not have a major influence on mixing dynamics at the confluence because it corresponds to a density difference of only 0.05–0.07%, which is small in comparison to the disparity in inertia (velocity) between the flows (40–60%). The difference in temperature, however, is large enough to reveal the general location of the mixing interface between the flows, which is defined by a zone of closely spaced isotherms within each cross-section (Figure 24.7). The temperature difference across this zone is about 3° C on 8 July and 2° C on 15 July, values which correspond to the difference in temperature between the two confluent streams on these dates. The transverse gradient in temperature within the zone of clustered isotherms defining the mixing interface is not uniform, but at each cross-section a central region can be identified where the spacing between isotherms is smallest. For example, on 8 July the centre of the

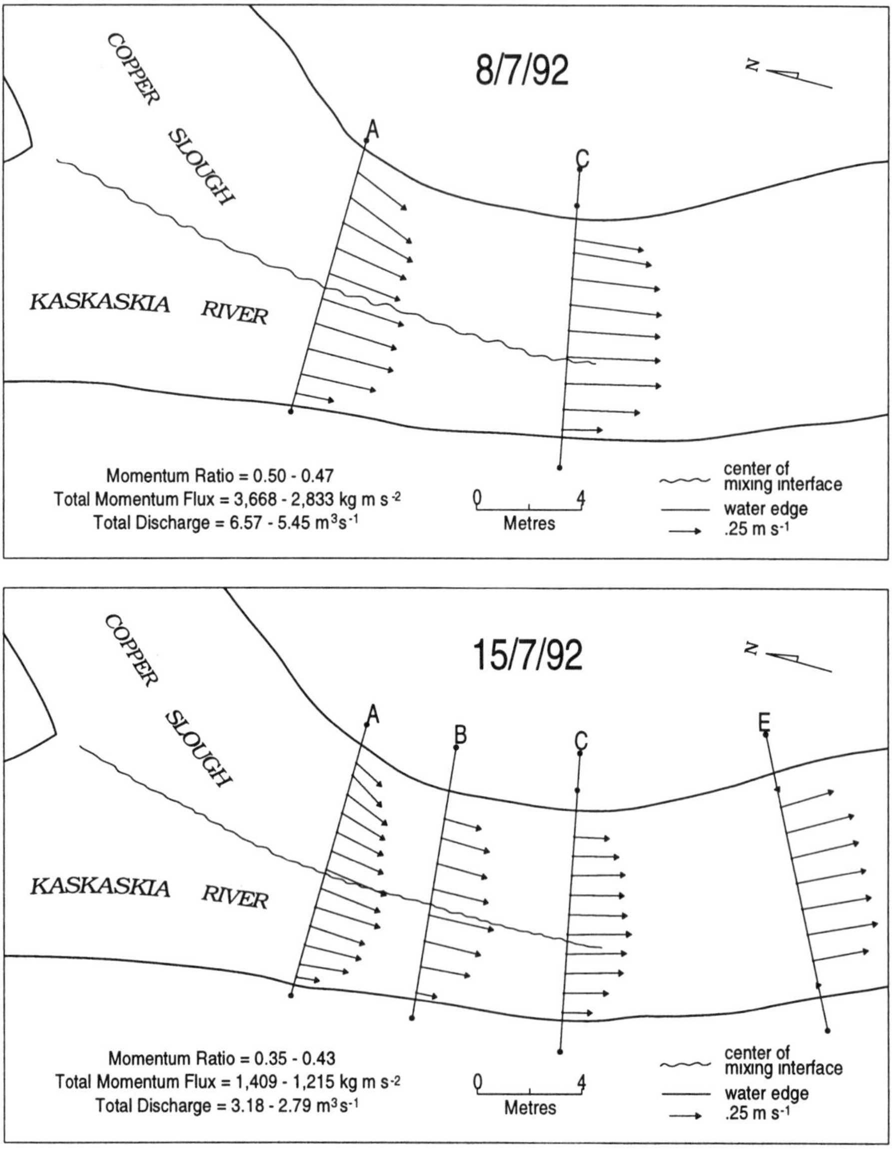

Figure 24.6. Depth-averaged velocity vectors and surficial position of the centre of the mixing interface on 8 and 15 July 1992. Values of momentum ratio, total momentum flux and total discharge are for the beginning and end of the measurement period

mixing interface at cross-section A is located between the fifth and sixth measurement verticals from the west bank, whereas on 15 July it is located between the sixth and seventh measurement verticals (Figures 24.6 and 24.7). Although

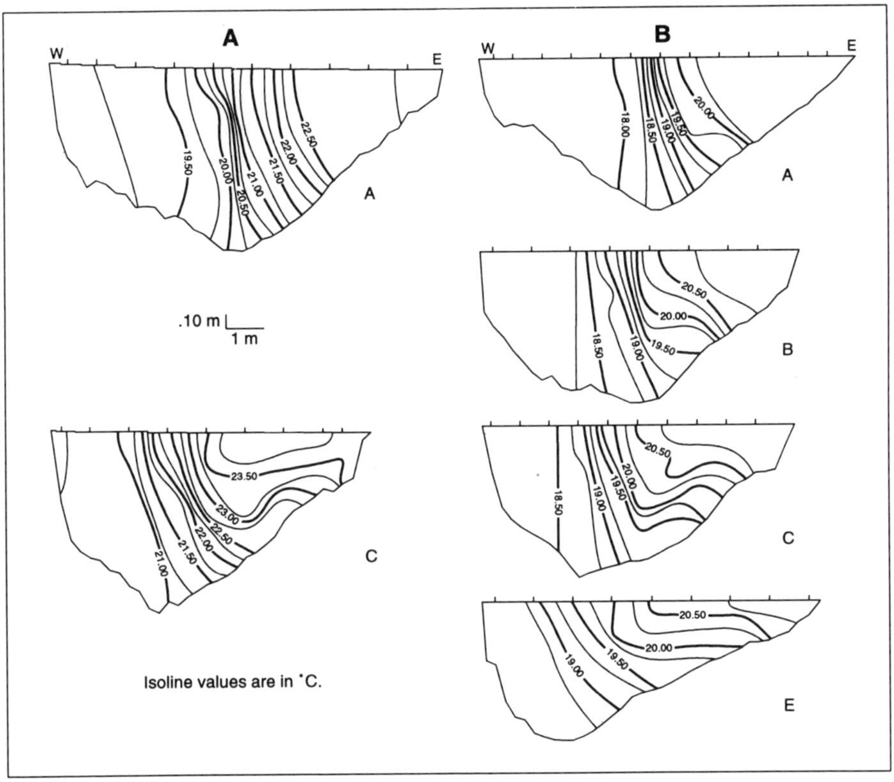

Figure 24.7. Water temperatures for (A) 8 July and (B) 15 July 1992. (looking upstream, locations of measurement verticals indicated by ticks along water-surface)

diurnal trends resulted in some systematic change in the temperatures of the incoming flows between the times that measurements were obtained on the west side versus the east side of each cross-section, such trends could not be responsible for clustering of the isotherms at each cross-section because: (i) the maximum total increase or decrease in temperature during the measurement period at any cross-section was considerably less (|0.50|°C on 8 July; |0.25|°C on 15 July) than the difference in temperature between the two streams; and (ii) the rate of change in temperature of the incoming flows was fairly uniform throughout the period of measurement at each cross-section.

The spatial evolution of the mixing interface downstream from the confluence is similar for 8 and 15 July and is consistent with the pattern observed at this confluence during a low-stage event with a momentum ratio less than one and when a large bar is present at the downstream junction corner (Rhoads & Kenworthy, 1995). At cross-section A the mixing interface is located near the centre of the channel (Figure 24.7). The mixing interface has a vertical orientation, but near the bed the isotherms curve toward the opposing channel banks with inflection toward the east (inner) bank exceeding curvature toward the west (outer) bank (Figure 24.7). The interface is broad compared to its width during shallow flows at this confluence (Rhoads & Kenworthy, 1995), and the width of the interface on 8 July exceeds the width on 15 July. Together these relationships indicate that the width of the mixing interface scales with flow depth. This increase in width with flow depth probably is related to two factors. First, as flow depth increases, the diameter of coherent vertical vortices within the mixing interface should increase (Chu & Babarutsi, 1988). Second, increases in flow depth may enhance lateral oscillation of the shear layer between the flows (e.g. Biron *et al.*, 1993a), in which case the mixing interface as defined on the basis of the temperature field will in part represent the spatially averaged location of an oscillating free shear-layer.

Downstream of cross-section A, the mixing interface becomes increasingly distorted and disrupted. Isotherms at the bottom part of the interface spread apart and curve progressively inward along the bed and upward toward the inner channel bank, whereas those defining the upper part of the interface remain closely spaced and shift toward the outer channel bank (Figures 24.6 and 24.7). The width of the upper part of the interface remains remarkably constant between cross-sections A and C, suggesting that lateral mixing in this region of the flow is limited. On 15 July, displacement of the isotherms toward the outer bank at the surface and toward the inner bank near the bed continues downstream of cross-section C, and the mixing interface becomes increasingly indistinct as the isotherms near the surface spread apart.

Downstream velocity

The downstream velocity field at cross-section A on both dates is characterized by local velocity maxima on each side of the channel – a pattern similar to that observed for low-stage events with $M_r < 1.0$ (Rhoads & Kenworthy, 1995). On 8 July the two zones of maximum velocity are located beneath the surface and are separated by an intervening region of relatively low-velocity fluid near the centre of the channel (Figure 24.8). The low-velocity region is located in the same position as the core of the mixing interface between the two flows (Figures 24.7 and 24.8), indicating that the two maximum-velocity zones represent the separate talwegs of the Kaskaskia River and the Copper Slough. Along the margins of the low-velocity region the isovels extend downward, suggesting that high-velocity fluid near the surface is being transported toward the channel bed.

Two velocity maxima also straddle the mixing interface between the two flows on 15 July; the inner maximum has velocities between 0.48 and 0.50 m s^{-1} and velocities in the outer maximum are between 0.65 and 0.70 m s^{-1} (Figures 24.7 and 24.8). In contrast to conditions on 8 July, however, the high-velocity zones are not submerged, but instead are located at the water surface. Moreover, the pattern of downstream velocity between the high-velocity zones is more complicated and

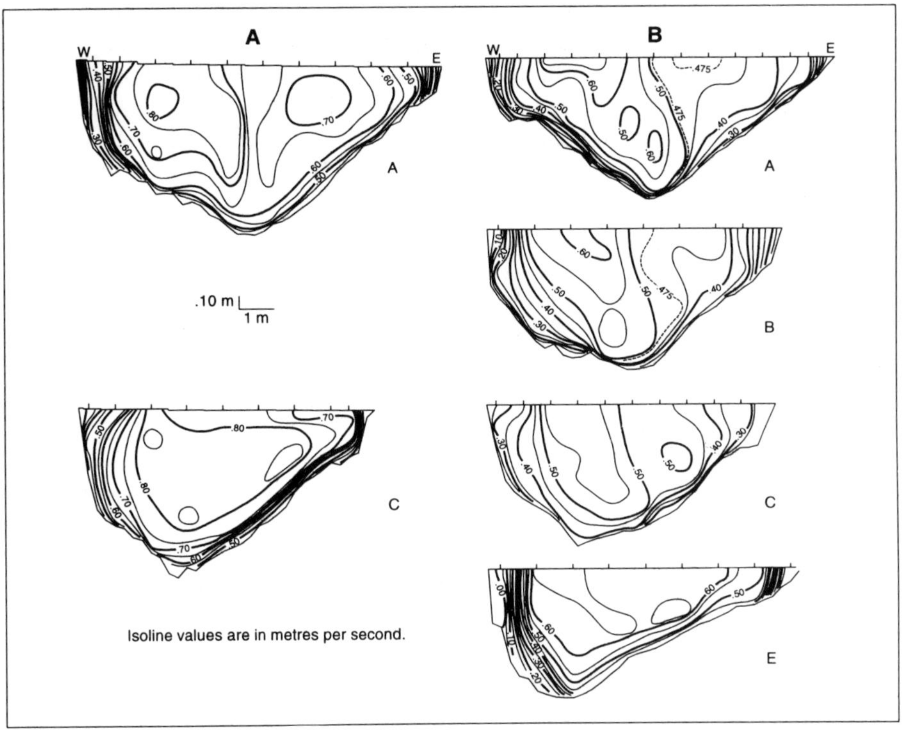

Figure 24.8. Downstream velocities (U) for (A) 8 July and (B) 15 July 1992. (looking upstream, locations of measurement verticals indicated by ticks along water surface)

discontinuous than on 8 July. Near the bed in the centre part of the channel, a core of high-velocity fluid exists along the outer margin of the mixing interface. This near-bed high-velocity core is located within a broad region of high-velocity flow that extends from the surface toward the base of the scour hole over the outer portion of the cross-section. The velocities near the bed in the scour hole are similar to those near the surface over the outer part of the cross-section and are about 25% greater than those in the surficial high-velocity zone over the inner portion of the cross-section.

On 8 July, the two cores of maximum velocity merge between cross-sections A and C to form a single, expansive region of maximum velocity extending from the centre of the channel toward the inner bank (Figure 24.8). The two surficial high-velocity zones also converge downstream of cross-section A on 15 July. At cross-section B, the inner zone has merged with adjacent flow of the same velocity and is represented as a local increase in the near-surface spacing between the 0.50 m s^{-1} and 0.48 m s^{-1} isovels. Isovels near the bed over the inner portion of this cross-section bulge inward and upward, suggesting that high-velocity fluid is being transported into this region. At cross-sections C and E, this inward transport of momentum produces a distinct core of maximum velocity near the bed over the inner portion of the channel. The flow accelerates between cross-sections C and E as cross-sectional area decreases due to a reduction in channel depth at the downstream end of the scour hole and the presence of a recirculation zone in the lee of failed bank material on the west side of the channel.

Cross-stream velocity

On 8 and 15 July the cross-stream velocity field at cross-section A is characterized by strong outward movement of flow over the inner portion of the channel (Figures 24.9 and 24.10). Close to the bed outward movement decreases or reverses direction, suggesting that helical motion is occurring within the flow. Over the outer portion of the cross-section near-bed fluid generally is moving toward the outer bank, although some cross-stream velocities measured within 0.07 m of the bed deviate from this trend. On both dates actively migrating dunes and irregularities in exposed glacial till produced local relief of 0.10 to 0.12 m over the outer portion of the downstream channel. Thus, some of the velocities obtained close to the bed may reflect flow conditions within an internal boundary layer associated with these bedforms.

Decomposition of the cross-stream velocity field into W_{py} and W_{sy} components shows how flow convergence/divergence and secondary circulation contribute to the cross-stream movement of water (Figures 24.9 and 24.10). The outward movement of flow over the inner portion of cross-section A is the result mainly of the flow direction over this part of the channel being skewed in relation to the orientation of the cross-section (Figure 24.6). The decrease in values of W_{py} toward the outer bank confirms that flow convergence occurs at cross-section A. On both dates, the general pattern of W_{sy} consists of outward motion near the surface and inward motion near the bed over the inner portion of the cross-section, and inward motion near the surface and outward motion near the bed over the outer portion of the cross-section. Previous work has suggested that a similar pattern of weak secondary circulation occurs during low-stage, low momentum-ratio flows at this confluence (Rhoads & Kenworthy, 1995). The data for 8 and 15 July confirm that twin counter-rotating helical cells are embedded within the converging flow at this confluence when $M_r < 1.0$. The transition from anticlockwise to clockwise motion (looking upstream) lies between verticals five and six on 8 July and verticals six and seven on 15 July, locations that correspond to the position of the core of the mixing interface on each date. Thus, the two cells are located on either side of the mixing interface and exist within separate flows entering the confluence from the two upstream channels. On 15 July a small anticlockwise-rotating cell is also present near the outer bank. This cell may be similar to the outer-bank cells observed in meander bends, which have been attributed either to interaction between the adjacent main circulation cell and the bank geometry or to anisotropy of

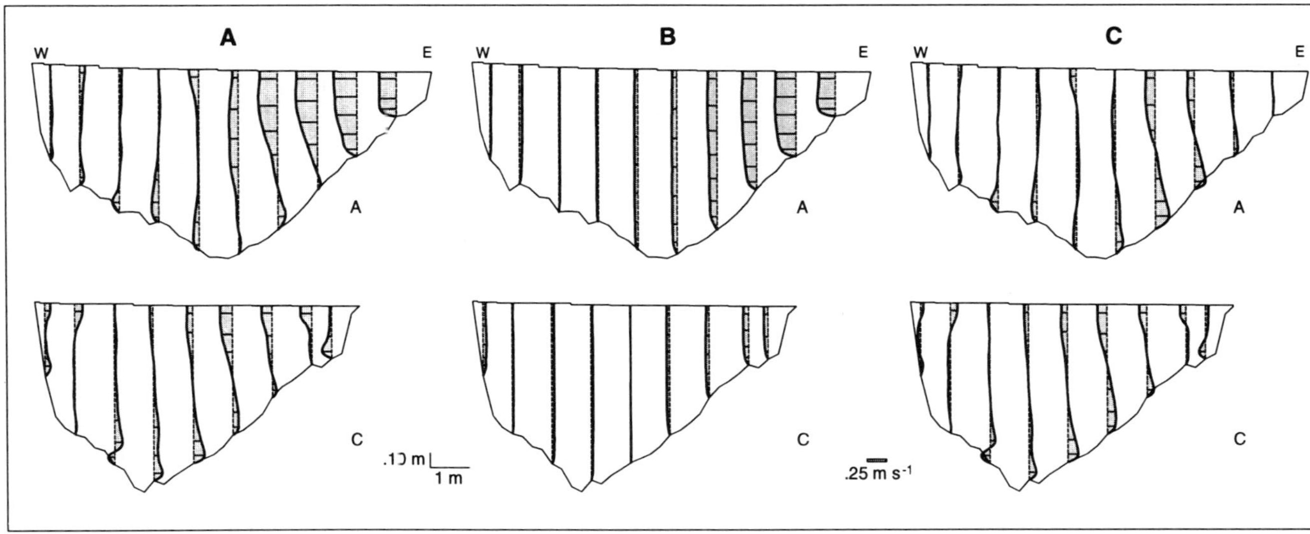

Figure **24.9.** Mean flow structure on 8 July 1992: (A) cross-stream velocities (W); (B) cross-stream component of primary velocities (W_{py}); and (C) cross-stream component of secondary velocities (W_{sy})

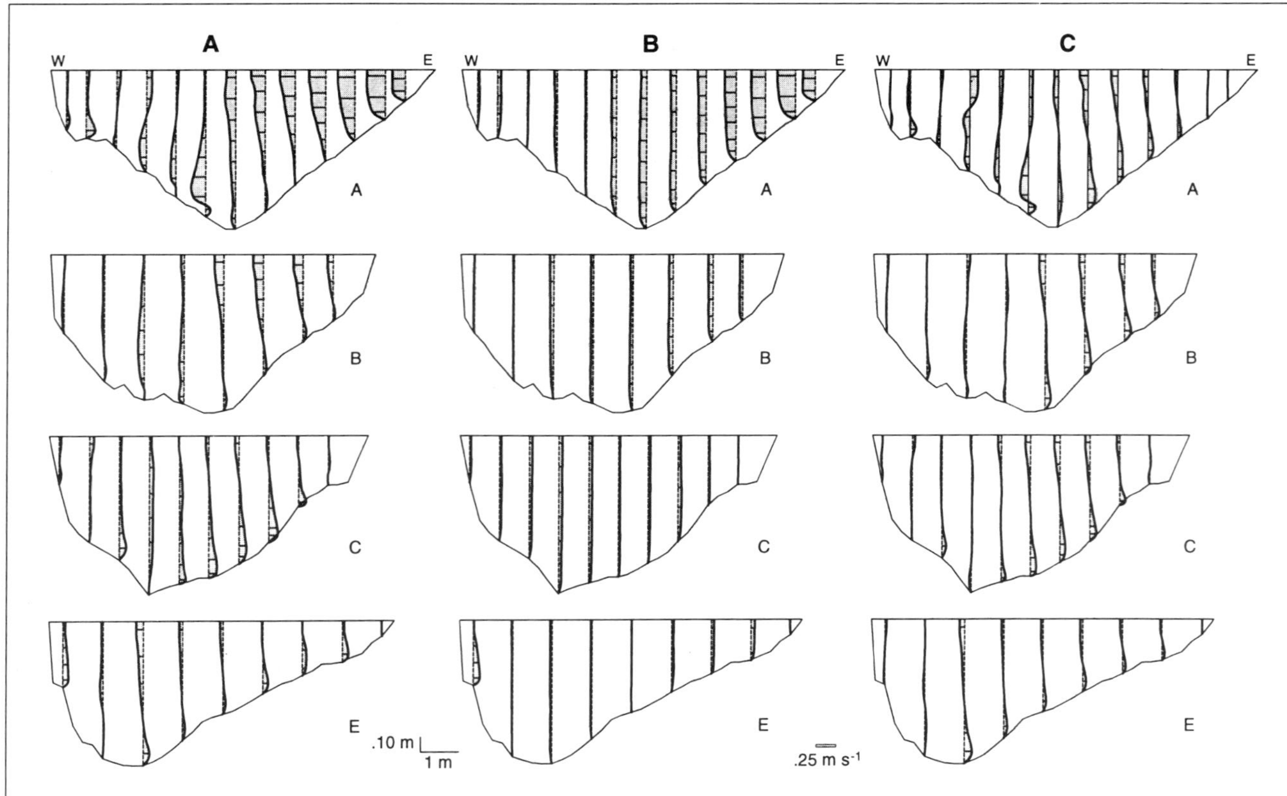

Figure 24.10. Mean flow structure on 15 July 1992: (A) cross-stream velocities (W); (B) cross-stream component of primary velocities (W_{py}); and (C) cross-stream component of secondary velocities (W_{sy})

turbulence near the bank-bed interface (Bathurst *et al.*, 1979). The combination of strong circulation in the main clockwise cell, the abrupt break in slope in the channel bed near the outer bank, and the sharp interface between the outer bank and the bed on 15 July should enhance one or both of these two generative mechanisms.

Downstream of cross-section A, values of W_{py} diminish, indicating that the flow path at each measurement location is not strongly skewed in relation to the orientation of the cross-section (Figures 24.9 and 24.10). Moreover, the small differences in the magnitudes of W_{py} among verticals at a particular cross-section show that the flow is not strongly converging or diverging. The data for W_{sy} reveal that the inner anticlockwise-rotating cell at cross-section A maintains its strength and expands towards the outer bank in the downstream direction. Meanwhile, the secondary velocities in the adjacent clockwise-rotating cell diminish in intensity to such an extent that this cell becomes indistinct. The expansion of the inner helical cell corresponds with the outward shift in the surficial position of the core of the mixing interface. On 8 July, the core of the mixing interface at cross-section C is centred on the fourth measurement vertical from the west bank (Figures 24.6 and 24.7); this vertical also marks the outer limit of pronounced anticlockwise helical motion at this cross-section (Figure 24.9). On 15 July, the core of the mixing interface is located between the fourth and fifth verticals at cross-sections B and C, and, although it is less distinct, between roughly the second and third verticals at cross-section E. At each of these cross-sections, the vertical on the inner side of the mixing-interface core corresponds to the outer limit of anticlockwise helical motion (Figure 24.10).

PATTERNS OF BEDLOAD TRANSPORT

Samples of bedload were collected at cross-section A on 8 July and at cross-sections A, C, and E on 24 July (Figures 24.11–24.14). On the latter date, M_r at the confluence varied from 1.0 early in the day to 0.82 when sampling was completed in mid-afternoon (Figures 24.12–24.14). Total discharge at the confluence declined throughout the day on 24 July, following the occurrence of a high momentum-ratio, high-stage event late in the day on 23 July (Figure 24.3). This event shifted the zone of maximum channel depth toward the outer bank, increasing the asymmetry of cross-section A (Figure 24.5).

Both on 8 and 24 July, the separate loads from the Kaskaskia River and Copper Slough are distinctly segregated at cross-section A (Figures 24.11 and 24.12). On 8 July the channel bottom between the edge of water and a distance of 1.25 m from the west end (zero point) of the cross-section (Z = 1.25 m) consisted of exposed glacial till. Immobile sand covered the bed at Z = 1.25 m. Large amounts of sand associated with the migration of low dunes into the confluence from the Kaskaskia River were sampled at distances of 2.5, 3.5 and 4.5 m from the west endpoint of the cross-section (Figure 24.11). The latter location (Z = 4.5 m) corresponds to

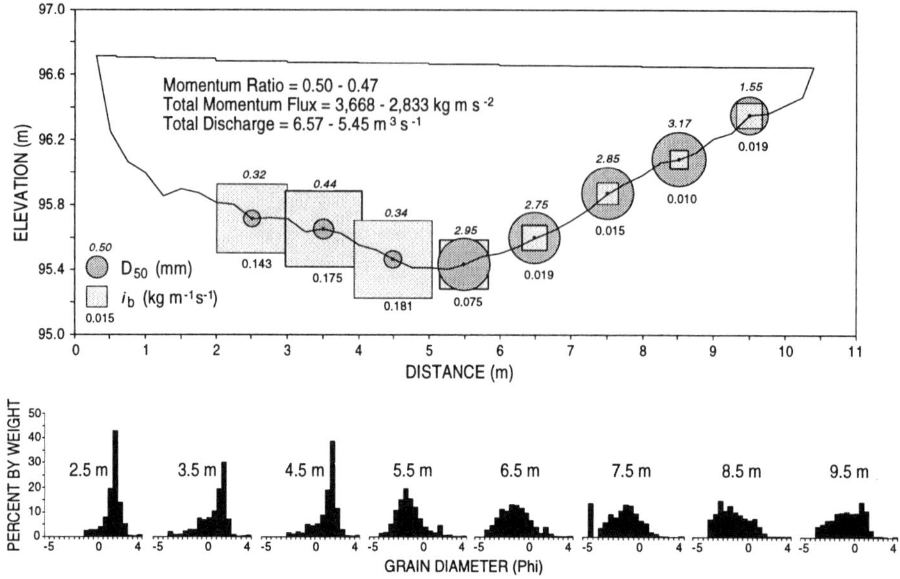

Figure 24.11. Cross-channel variation in i_b, D_{50}, and grain-size distribution of bedload at cross-section A on 8 July 1992

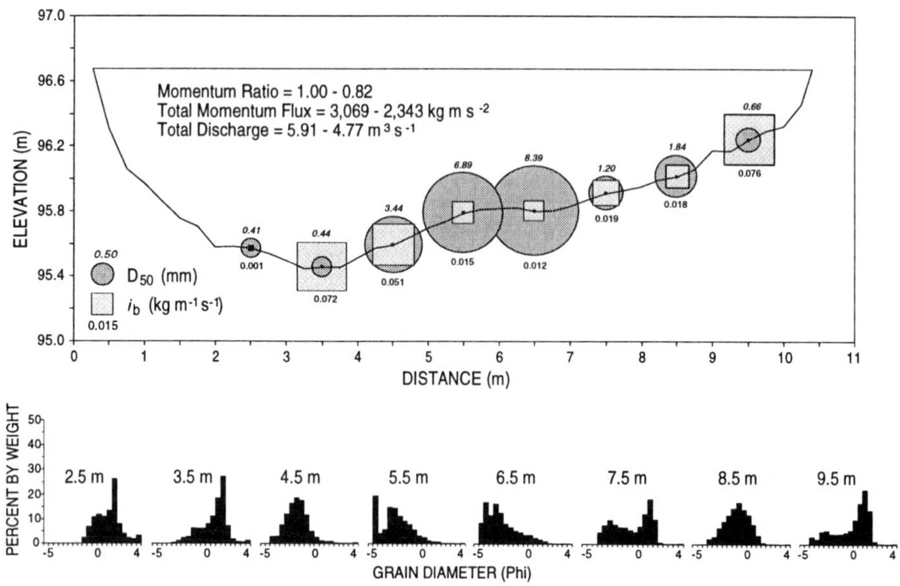

Figure 24.12. Cross-channel variation in i_b, D_{50}, and grain-size distribution of bedload at cross-section A on 24 July 1992

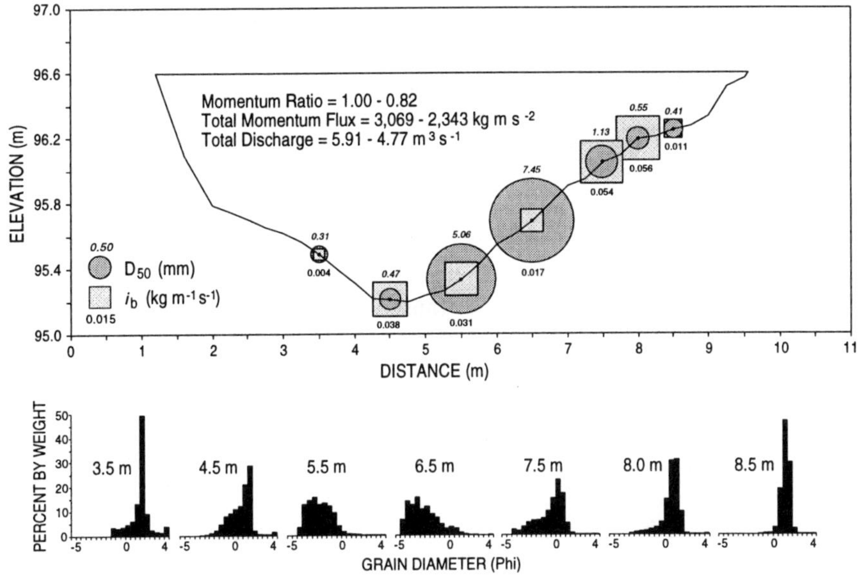

Figure 24.13. Cross-channel variation in i_b, D_{50}, and grain size-distribution of bedload at cross-section C on 24 July 1992

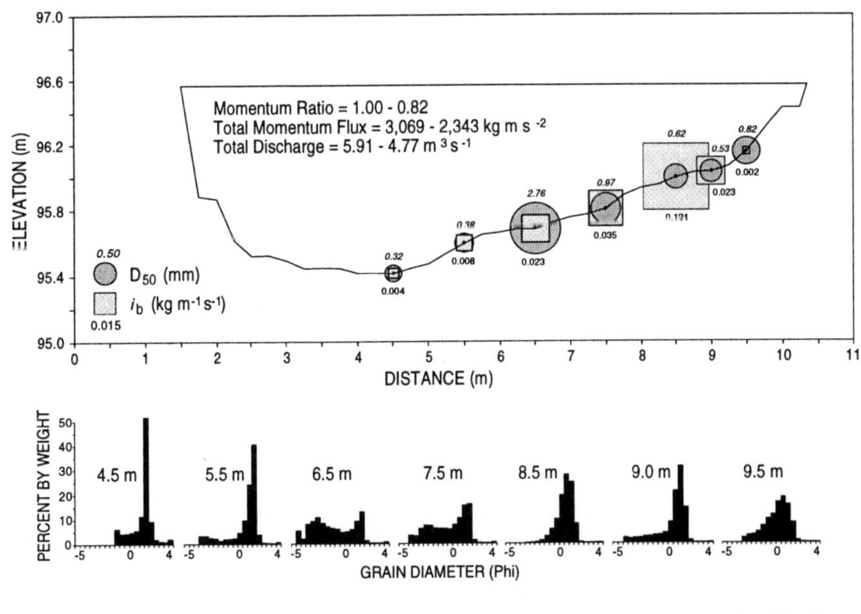

Figure 24.14. Cross-channel variation in i_b, D_{50}, and grain size-distribution of bedload at cross-section E on 24 July 1992

the fifth vertical from the west bank for this date (see Figures 24.7–24.10). At and inward of $Z = 5.5$ m, which corresponds to the sixth vertical from the west bank, the bedload consisted predominately of gravel or mixed sand-and-gravel from the Copper Slough. This transition zone between sandy bedload from the Kaskaskia River and mixed sand-and-gravel bedload from the Copper Slough is located in the same position as the core of the mixing interface between the confluent streams. This location also marks the transition from anticlockwise helical motion over the inner portion of the cross-section to clockwise helical motion over the outer part of the channel. Although the measured transport rates may be prone to large errors, it is interesting to note that the highest transport rates for the segregated sand and mixed sand-and-gravel bedloads are found immediately adjacent to the deepest part of the channel, suggesting that the highest bed shear stresses occur in this part of the cross-section.

On 24 July the zone of exposed till extended almost to a distance of 2.5 m from the west endpoint of cross-section A (Figure 24.11). A small amount of sandy bedload was detected at $Z = 2.5$ m and migrating dunes entering the confluence from the Kaskaskia River produced a sharp increase in the sand transport rate at $X = 3.5$ m. At and inward of $Z = 3.5$ m, except for a small area adjacent to the inner bank, the bedload is comprised of mixed sand-and-gravel from the Copper Slough. The zone of sandy bedload along the inner bank represents part of a continuous, narrow sheet of rippled sand that extended upstream along the south bank of the Copper Slough and downstream along the inner bank of the downstream channel. This sand sheet was clearly visible in the shallow water adjacent to the inner bank. Because measurements of flow structure and temperature were not obtained on 24 July, the position of the mixing-interface core on this date is uncertain. However, the location of the transition from sand to mixed sand-and-gravel bedload suggests that the interface probably was located between $Z = 3.5$ and 4.5 m. This location generally corresponds to the observed position of Kelvin-Helmholtz vortices on the water-surface on this date. The difference between the inferred location of the mixing-interface core on 24 July ($Z \approx 4.0$ m) and the actual position of this core on 8 July ($Z \approx 5.0$ m) is consistent with the notion that the higher momentum-ratio on 24 July should shift the mixing interface toward the outer bank. As on 8 July, the highest transport rates for sandy bedload from the Kaskaskia River and mixed sand-and-gravel bedload from the Copper Slough (excluding the sand sheet along the inner bank) occur immediately adjacent to the transition zone separating the two types of bedload.

The bedload data for cross-sections C and E on 24 July show that the zone of active transport is confined to the inner portion of the downstream channel. Little or no bedload was moving over the outer part of cross-sections C and E at the time of sampling and the bed in this portion of the channel consisted of exposed glacial till. Narrow zones marking a transition from sandy bedload near the centre of the channel to mixed sand-and-gravel bedload over the inner portion of the channel can be identified both at cross-section C ($Z = 4.5$–5.5 m) and cross-section E

($Z = 5.5$–6.5 m); however, the transition is less well-defined at cross-section E than at cross-section C. The abrupt change from sand to gravel near the centre of the channel at cross-section C indicates that the separate loads from each stream are still segregated at this location. The relatively uniform values of median grain size (D_{50}) at cross-section E suggest that the loads from each stream have mixed to some extent between cross-sections C and E. The highest transport rates at both cross-sections are associated with the sand sheet along the inner portion of the downstream channel.

DISCUSSION

The combined results on flow structure, water temperature, and sediment transport provide a detailed picture of the dynamics of this confluence for transport-effective events with $M_r < 1.0$. Under these conditions, the mean flow structure is characterized by two counter-rotating surface-convergent helical cells embedded within a converging flow field. Similar twin-cell patterns have been observed at symmetrical confluences in the laboratory (Mosley, 1976) and in the field (Ashmore *et al.*, 1992; Bridge & Gabel, 1992), and at this confluence during low-stage conditions (Rhoads & Kenworthy, 1995). The helical cells are located on either side of the mixing interface, which in turn lies between two cores of maximum downstream velocity associated with the separate talwegs of the confluent streams. During high-stage flows, helical motion can become sufficiently well-developed to shift each maximum of downstream velocity below the water surface, a phenomenon that commonly occurs in meander bends (e.g. Hey & Thorne, 1975; Hickin, 1978; Dietrich & Smith, 1983; Thorne *et al.*, 1985; Markham & Thorne, 1992), but that was not observed at this confluence during a low-stage event with small values of W_{sy} (Rhoads & Kenworthy, 1995). The well-developed twin helical cells substantially influence the downstream velocity field by transporting high-velocity, near-surface fluid downward toward the bed and inward along the bed toward each bank. The downward advection of high-momentum fluid near the centre of the channel extends the isovels of U toward the bed and can generate a near-bed core of maximum velocity over the zone of maximum channel depth, a feature that has been documented in some meander bends (Thorne *et al.*, 1985). It is well-established that areas of downwelling fluid produce a local increase in mean bed shear stress, thereby increasing rates of sediment transport and the potential for scour (e.g. Bathurst *et al.*, 1979). Although the data on bedload transport rates must be interpreted with caution, the highest rates of transport at the entrance to the downstream channel are associated with the region of downwelling fluid between the two surface-convergent cells. The opposing movement of fluid toward each bank along the bed enhances near-bed mixing of the separate flows as evidenced by divergence of the near-bed isotherms defining the mixing interface in the centre part of the channel (Figure 24.15). This pattern of near-bed fluid motion

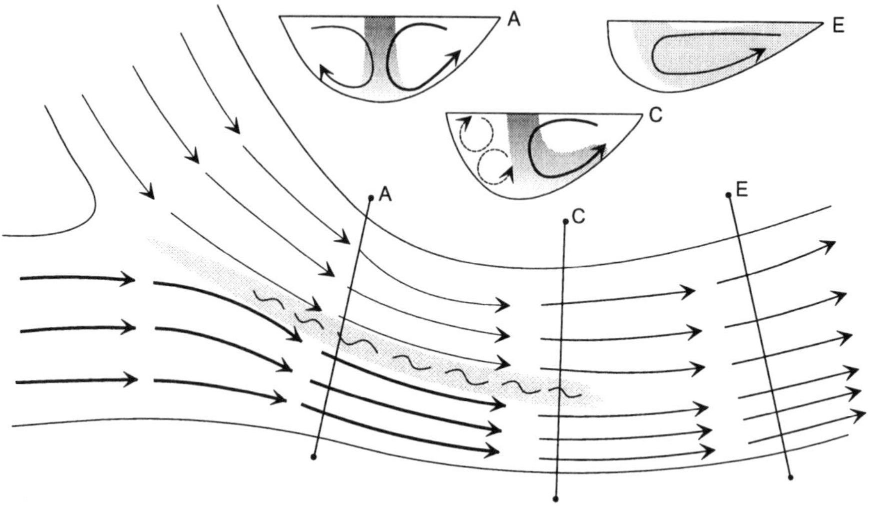

Figure 24.15. Conceptual model illustrating the pattern of depth-averaged streamlines, helical flow, and distortion of the mixing interface (shaded) at an asymmetrical confluence when the mainstream is dominant. Boldness of arrows depicting streamlines and secondary currents is directly proportional to flow strength. Dashed arrows at cross-section C indicate absence of well-defined helical motion. Variation in the shade of mixing interface reflects the relative magnitude of the lateral temperature gradient defining this interface (dark – strong gradient; light – weak gradient)

also confines the incoming sediment from each stream to opposite sides of the confluence, producing segregation of bedload at the entrance to the downstream channel. Near the steep outer bank of the downstream channel a small helical cell sometimes develops, producing a region of upwelling along the outer bank. This local upwelling of fluid reduces the mean bed shear stress, inhibiting sediment movement near the outer bank, a region where no transport could be detected in this study.

Downstream from cross-section A, the structures of high-stage and low-stage flows are quite similar. The combined effects of flow convergence and surface-convergent helical motion cause the twin maximum-velocity zones to merge into an extensive, near-surface maximum-velocity zone that occupies a substantial portion of the total cross-sectional area of the flow. The helical cell associated with the main stream decays, whereas the inner cell persists and possibly increases in intensity. As the surficial portion of the mixing interface shifts toward the outer bank, the anticlockwise-rotating inner cell expands, whereas the decaying clockwise-rotating outer cell is reduced in size. The strong anticlockwise helical flow over the inner part of the flow entrains cool fluid from the main stream along the mixing interface and transports it inward along the bed. As a result, the bottom part of the mixing interface becomes increasingly distorted and disrupted as the isotherms that define this feature spread apart and curve inward and upward toward the inner channel bank (Figure 24.15) – a phenomenon observed both in this study and for low-stage flows (Rhoads & Kenworthy, 1995).

As the clockwise cell over the outer portion of the channel decays, cross-stream flow near the bed becomes directed inward over the entire channel width. The inward directed flow sweeps the outer portion of the downstream channel clean of sediment, exposing the underlying glacial till. Over the inner portion of the channel, bedload from the main stream and the tributary converge and mix. Bedload from the lateral tributary is sorted over the inner portion of the downstream channel in a manner similar to that which occurs in the downstream portion of meander bends with mixed sand-and-gravel loads (Bridge, 1992, 1993); sand moves along the inner bank, whereas coarse gravel is located toward the deepest part of the channel.

This general depiction of the connection between flow structure and sediment transport is consistent with the documented change in bed morphology between late June and early July 1992. The shift in the position of the talweg from near the outer bank to the centre of the confluence as the flow became transport-effective and M_r decreased below one can be explained by the response of the flow field, mixing interface, and patterns of sediment transport to changes in M_r. The transition from high momentum-ratio conditions, which tend to promote accumulate of sediment along the inner bank of the downstream channel (Rhoads & Kenworthy, 1995), to low momentum-ratio conditions, shifts the mixing interface toward the centre of the confluence. The combined effects of large turbulent stresses within the mixing interface (e.g. Biron *et al.*, 1993a) and descending flow associated with the surface-convergent helical motion increase the bed shear stress, resulting in enhanced mobilization of sediment. Divergence of flow near the bed produces a diverging sediment flux, which forms and maintains the region of scour near the centre of the confluence.

Although other studies have documented helical motion at confluences, the mechanism responsible for this property of the flow is somewhat controversial. Two competing hypotheses have been proposed to account for this phenomenon. The first holds that helical motion is the result of curvature of flow streamlines as the confluent flows collide and become realigned to enter the downstream channel. The resulting super-elevation of the water surface within the confluence generates twin helical cells with near-bed flow directed toward the channel banks (Mosley, 1976; Ashmore *et al.*, 1992; Bridge, 1993). The second hypothesis attributes helical motion to separation of the flow over steep avalanche faces along the margins of the scour hole within the confluence (Best, 1987, 1988; Best & Roy, 1991; Biron *et al.*, 1993b). According to this hypothesis, as flow crosses the step-like faces at oblique angles, open separation vortices are generated along the margins of these faces (*cf.* Allen, 1982, p. 105).

The results of any individual field study cannot conclusively resolve this debate, but the data presented here generally are consistent with the flow-curvature hypothesis. The pattern of helical motion at the field site is quite similar to that predicted by a three-dimensional numerical model of flow through an asymmetrical open-channel confluence with a junction angle of 60° and a momentum ratio less than 1.0 (Weerakoon *et al.*, 1991). The main flow features predicted by this model include: (i) twin surface-convergent helical cells within the flow at the entrance to the downstream channel; (ii) stronger circulation in the inner cell than the outer cell and an increase in the lateral extent of the inner cell downstream of the confluence; and (iii) a region of separated flow along the inner bank of the downstream channel. The mechanism generating the two counter-rotating cells in the model is curvature of the confluent flows within the confluence. The tributary flow curves as it turns to become aligned with the downstream channel, whereas the main-stream flow curves slightly as it is deflected toward the outer bank by the tributary flow (Figure 24.15). The stronger circulation of the inner cell compared to the outer cell is the result of greater curvature of flow from the tributary compared to flow from the mainstream. The model also predicts zones of super-elevated water at the upstream junction corner and along the outer bank of the downstream channel, a finding that is consistent with conditions at the field site. Field surveys of water-surface elevations on each side of the downstream channel on 8 and 15 July (using an automatic level and surveying rod) showed that elevations over the outer part of the downstream channel were generally about 0.01 to 0.02 m higher than elevations along the inner bank. The lack of flow separation in this investigation probably is related to differences in the mechanics of flow around the curved downstream junction corner at the field site (Figure 24.6) compared to flow around the angular junction corner used in the numerical model.

Given the pattern of flow through the confluence on 8 and 15 July, it seems unlikely that helical motion is the result of flow separation from the channel bed. Although a well-defined scour hole existed within the confluence in July 1992, the sides of this feature had maximum transverse slopes of only 12–13° and thus cannot be considered gravitationally controlled avalanche faces. Moreover, the flow vectors at the entrance to the downstream channel as well as farther downstream are aligned closely with the bed contours on 8 and 15 July (Figures 24.4 and 24.6). The rate of decrease in bed elevation along the path of near-surface flow vectors, which will be least affected by any potential flow separation and thus provide the best indicator of the orientation of the incoming flow in relation to features on the channel bed, is less than 4°, a bed gradient that is unlikely to produce substantial flow separation (Allen, 1982, p. 113). Although it is possible that flow separation could be generated farther upstream within the centre part of the confluence, field measurements by Kenworthy (1994) indicate that near-surface velocity vectors in this region also are oriented along the bed contours.

CONCLUSION

This study has contributed to our understanding of the fluvial dynamics of stream confluences by documenting mean flow structure, patterns of sediment movement and bed morphology at an asymmetrical confluence when the main stream is dominant and the flow is at a transport-effective stage. Previous work at this site provided initial support for an analogy between the fluvial dynamics of this confluence and the fluvial dynamics of meander bends (Rhoads & Kenworthy, 1995). The findings of this investigation lend further support to such a comparison. Additional studies that include measurements of water surface topography are needed to verify this analogy. Another key issue requiring further research is the interaction between turbulent structures in the mixing interface and coherent motion within the mean flow field. Unravelling this interaction will lead to a better understanding of the relative importance of various scales of coherent flow features for flow mixing and sediment dynamics at confluences. Given the diverse array of confluence planforms and bed morphologies, as well as the hydrologic variability associated with these sites, further field studies are needed to document flow and sediment dynamics at different types of confluences. Ideally, such studies should focus on interactions among flow structure and sediment transport during formative events that actively reshape the channel bed. In a larger context, the role of confluences in network-scale fluvial dynamics has yet to be determined. The results of this investigation suggest that confluences can influence network-scale dynamics by locally regulating the downstream movement of sediment from conjoining links.

ACKNOWLEDGEMENTS

Steve Kenworthy, Dan Mayer, and Mark Welford helped to collect and compile the data for this study. Special thanks to Jane Domier for assistance in designing and producing the figures. This research was supported by grants from the National Science Foundation (SES-9024225 and SES 9242848), the Illinois Water Resources Centre, and the University of Illinois Research Board. Helpful comments by Pascale Biron and an anonymous reviewer are greatly appreciated.

REFERENCES

Allen, J.R.L. 1982. *Sedimentary Structures: Their Character and Physical Basis,* Vol. **II**, Elsevier, Amsterdam, 663 pp.

Ashmore, P.E. & Parker, G. 1983. 'Confluence scour in coarse braided streams', *Water Resour. Res.*, **19**, 392–402.

Ashmore, P.E., Ferguson, R.I., Prestegaard, K.L., Ashworth, P.J. & Paola, C. 1992. 'Secondary flow in anabranch confluences of a braided, gravel-bed stream', *Earth Surf. Processes Landf.*, **17**, 299–311.

Bathurst, J.C., Thorne, C.R. & Hey, R.D. 1977. 'Direct measurements of secondary currents in river bends', *Nature,* **269**, 504–506.

Bathurst, J.C., Thorne, C.R. & Hey, R.D. 1979. 'Secondary flow and shear stress at river bends', *J. Hydraul. Div., ASCE*, **105**, 1277–1295.

Best, J.L. 1987. 'Flow dynamics at river channel confluences: implications for sediment transport and bed morphology', in Ethridge, F.G., Flores, R.M. & Harvey, M.D. (Eds) *Recent Developments in Fluvial Sedimentology*, Soc. Econ. Paleon. and Mineralogists, Sp. Pub. **39**, Tulsa, Oklahoma, 27–35.

Best, J.L. 1988. 'Sediment transport and bed morphology at river channel confluences', *Sedimentology*, **35**, 481–498.

Best, J.L. & Reid, I. 1984. 'Separation zone at open channel junctions', *J. Hydraul. Engrg.*, **110**, 1588–1594.

Best, J.L. & Roy, A.G. 1991. 'Mixing layer distortion at the confluence of channels of different depth', *Nature*, **350**, 411–413.

Biron, P., De Serres, B., Roy, A.G. & Best, J.L. 1993a. 'Shear layer turbulence at an unequal depth confluence', in Clifford, N.J., French, J.R. & Hardisty, J. (Eds) *Turbulence: Perspectives on Flow and Sediment Transport,* John Wiley & Sons Ltd., Chichester, 197–213.

Biron, P., Roy, A.G., Best, J.L., & Boyer, C.J. 1993b. 'Bed morphology and sedimentology at the confluence of unequal depth channels', *Geomorphology*, **8**, 115–129.

Bridge, J.S. 1992. 'A revised model for water flow, sediment transport, bed topography and grain size sorting in natural river bends', *Water Resour. Res.,* **28**, 999–1013.

Bridge, J.S. 1993. 'The interaction between channel geometry, water flow, sediment transport and deposition in braided rivers', in Best, J.L. & Bristow, C.S. (Eds) *Braided Rivers,* Geol. Soc. Lond., Sp. Pub. **75**, 13–71.

Bridge, J.S. & Gabel, S.L. 1992. 'Flow and sediment dynamics in a low-sinuosity river: Calamus River, Nebraska Sandhills', *Sedimentology*, **39**, 125–42.

Chu, V.H. & Babarutsi, S. 1988. 'Confinement and bed-friction effects in shallow turbulent mixing layers', *J. Hydraul. Engrg.,* **114**, 1257–1274.

Dietrich, W.E. & Smith, J.D. 1983. 'Influence of the point bar on flow through curved channels', *Water Resour. Res.*, **19**, 1173–1192.

Gaudet, J.M., Roy, A.G., & Best, J.L. 1994. 'Effect of orientation and size of Helley-Smith sampler on its efficiency', *J. Hydraul. Engrg.*, **120**, 758–766.

Gomez, B., Hubbell, D.W. & Stevens, H.H. 1990. 'At-a-point bedload sampling in the presence of dunes', *Water Resour. Res.,* **26**, 2717–2731.

Gomez, B., Emmett, W.W. & Hubbell, D.W. 1991. 'Comments on sampling bedload in small rivers', *Proc. of the Fifth Federal Inter-agency Sedimentation Conf.*, Vol. **1**, Las Vegas, 2.65–2.72.

Hey, R.D. & Thorne, C.R. 1975. 'Secondary flows in river channels', *Area,* **7**, 191–195.

Hickin, E.J. 1978. 'Mean flow structure in meanders of the Squamish River, British Columbia', *Can. J. Earth Sci.*, **15**, 1833–1849.

Kenworthy, S.T. 1994. *Hydrologic and Morphologic Influences on Confluence Flow Structure,* Unpub. MSc thesis, Dept. of Geography, University of Illinois. 105 pp.

Markham, A.J. & Thorne, C.R. 1992. 'Geomorphology of gravel-bed river bends', in Billi, P., Hey, R.D., Thorne, C.R. & Tacconi, P. (Eds) *Dynamics of Gravel Bed Rivers*, John Wiley & Sons Ltd., Chichester, 433–456.

Mayer, D.R. 1995. *Hydrologic Control of Spatial Patterns of Surficial Bed-material Characteristics at a Stream Confluence,* Unpub. MSc thesis, Dept. of Geography, Univ. of Illinois. 120 pp.

Mosley, M.P. 1976. 'An experimental study of channel confluences', *J. Geol.*, **84**, 535–562.

Rhoads, B.L. & Kenworthy, S.T. 1995. 'Flow structure at an asymmetrical stream confluence', *Geomorphology*, **11**, 273–293.

Roy, A.G. & Bergeron, N. 1990. 'Flow and particle paths at a natural river confluence with coarse bed material', *Geomorphology*, **3**, 99–112.

Thorne, C.R., Zevenbergen, L.W., Pitlick, J.C., Rais, S., Bradley, J.B. & Julien, P.Y. 1985. 'Direct measurement of secondary currents in a meandering sand-bed river', *Nature*, **315**, 746–747.

Weerakoon, S.B., Kawahara, Y. & Tamai, N. 1991. 'Three-dimensional flow structure in channel confluences of rectangular section', *Proc. XXIV Congr., IAHR*, A, 373–380.

NOTATION

D_{50}	median grain size (mm)
i_b	bedload transport rate (kg m^{-1} s^{-1})
M_r	momentum flux ratio (kg m s^{-2})
Q	discharge (m^3 s^{-1})
T	water temperature (°C)
U	time-averaged downstream velocity (m s^{-1})
$\overline{U}$	time-averaged velocity at a cross-section (m s^{-1})
U_p	primary velocity (m s^{-1})
W	time-averaged cross-stream velocity (m s^{-1})
W_s	secondary velocity (m s^{-1})
W_{py}	cross-stream component of primary velocity (m s^{-1})
W_{sy}	cross-stream component of secondary velocity (m s^{-1})
Z	distance from west endpoint of cross-section (m)
ρ	fluid density (kg m^{-3})

Subscripts

1	Kaskaskia River
2	Copper Slough

25 Secondary Flow and Channel Changes around a Bar in the Brahmaputra River, Bangladesh

W.R. ROY RICHARDSON, COLIN R. THORNE
Department of Geography, The University of Nottingham, University Park, Nottingham, NG7 2RD, UK

SALEEM MAHMOOD
FAP-24, House 96, Road 23, Banani, Dhaka, Bangladesh

ABSTRACT

This paper reports preliminary results and interpretations based on direct measurements of primary velocities, secondary currents, sediment concentrations and channel changes in the braided Brahmaputra (Jamuna) River in Bangladesh. Measurements were made around a braid bar using an Acoustic Doppler Current Profiler (ADCP) to measure velocity profiles and relative variations in suspended sediment concentration, with position fixing via a Differential Global Positioning System. A total of 16 transects were investigated during the 1994 summer monsoon runoff. In this preliminary review, data for two sections are examined to characterize the flow patterns in a curved anabranch alongside the bar and at the bifurcation just upstream of the bar.

Results demonstrate linkages between large-scale secondary flow cells, primary isovel distributions and pathways of concentrated suspended sediment transport. Along the flanks of the braid bar curved anabranch flow exhibits helical flow in the talweg, outward flow over the bar at the inner margin and, possibly, a counter-rotating secondary cell close to the outer bank. Plunging flow near the outer bank brings fast, relatively sediment-free water close to the bank toe, promoting scour, instability and bank retreat. Near-bed convergence and upwelling at the crest of the bar at the inner bank concentrates sediment, promoting deposition and bar growth. These patterns are similar to those observed in meander bends of single-thread rivers. At the flow bifurcation, flow exhibits an incremental shifting of helical flow in both halves of the channel driven by the changing distribution of flow approaching the study section. This process, together with the stage-related scour and fill patterns observed, gave rise to the growth of a medial bar along the channel centreline, where secondary flow converged.

This study demonstrates the utility of ADCP for investigations of flow and sediment transport distributions in rivers. An understanding of the distributions of flow and sediment movement may be used to explain observed patterns of bed scour and fill that are, in turn, responsible for short-term morphological developments. Examination of satellite images demonstrates that such short-term developments can only be properly understood within the context of channel planform evolution at larger spatial and temporal scales. It is further shown how developments at the bifurcation produce changes in the relative dominance of the anabranches, leading to evolution of the braided planform.

Coherent Flow Structures in Open Channels.
Edited by P.J. Ashworth, S.J. Bennett, J.L. Best and S.J. McLelland.

INTRODUCTION

During the 1994 summer monsoon in Bangladesh a joint study was undertaken between the University of Nottingham and Flood Action Plan 24 – The River Survey Project (FAP-24). This was one of a number of special studies performed under FAP-24 with the aim of improving our understanding of river morphology. The specific objectives of the study are to measure primary and secondary flow distributions around a bifurcation-bar-confluence unit and to investigate the extent to which the pattern of primary isovels, pathways of sediment transport and changes in channel morphology are influenced by secondary flow structures. While good progress is being made towards these objectives, data processing is on-going and, therefore at present, only preliminary results are available for publication.

SECONDARY FLOWS IN MEANDERING RIVERS

Secondary flows may be defined as velocities in the plane perpendicular to the axis of primary flow (Prandtl, 1952). Two types of secondary flow are generally recognized, stress-induced or weak secondary currents, caused by non-uniform distribution of boundary shear stress, and skew-induced or strong secondary currents, caused by skewing of cross-stream vorticity into a streamwise direction due to channel curvature or bed topography (Perkins, 1970). Generally, in rivers it is found that secondary flow patterns are dominated by skew-induced currents (Bathurst, 1979; Bathurst *et al.*, 1979; see Rhoads, this volume), however, planform morphology, channel shape and the distribution of bed roughness are important factors in determining the strength and form of secondary flow.

Research in meandering rivers has established that the pattern of primary isovels and pathways of bed material transport (both bedload and suspended load) are strongly affected by skew-induced secondary currents (Engelund, 1974; Hey & Thorne, 1975; Bathurst *et al.*, 1977; Bridge, 1977, 1984; Bathurst *et al.*, 1979; Dietrich *et al.*, 1979; Thorne & Rias, 1984; Thorne *et al.*, 1985; Odgaard & Berg, 1988; Markham & Thorne, 1992). Surface convergence and plunging flows convect fast, near-surface water towards the solid boundary, promoting a steep near-bed primary velocity gradient, high local boundary shear stresses and bed scour. Conversely, convergent secondary flow near the bed causes upwelling of slow, near-bed flow, promoting deposition. Sediment concentrations are also affected by secondary flows. Plunging flow brings near-surface water that is relatively sediment free into proximity with the boundary, where it can pick-up sediment to cause scour. Upwelling carries sediment laden, near-bed water higher into the water column, where primary velocities sweep it along as a sediment plume. Such plumes have been observed to link sediment sources to sediment deposition zones in alluvial rivers (Peters, 1977; Peters & Goldberg, 1989). Secondary currents are known to occur at all flow stages in meander bends (Markham & Thorne, 1992) and at high flow they may persist through the inflexion point between meanders

(Thorne & Hey, 1979).

In a study of the Fall River, USA (Thorne *et al.*, 1985) measurements were made in meander bends at low, intermediate and bankfull stages. Figure 25.1 shows results for a section close to a bend apex at bankfull flow that are similar to those of several other studies (Hey & Thorne, 1975; Bathurst *et al.*, 1977; Bridge, 1977, 1984; Bathurst *et al.*, 1979; Thorne & Rias, 1984).

The main feature of the secondary flow is a helical, skew-induced cell which carries fast surface-water towards the outer bank and slower near-bed water towards the inner bank. However, a small cell of reverse rotation is also present at the outer bank. This cell occupies the channel between one and two times the bank height away from the outer bank. It is caused by local interaction of the skew-induced cell with the steep outer bank (Bathurst *et al.*, 1979). The skew-induced and outer-bank cells combine to depress the maximum primary velocity core below the water surface and compress primary isovel spacing close to the bank toe (Figure 25.1). This leads to toe scour that erodes and destabilizes the outer bank, and which quickly removes bank-failure debris to promote rapid bank retreat (Thorne & Lewin, 1979).

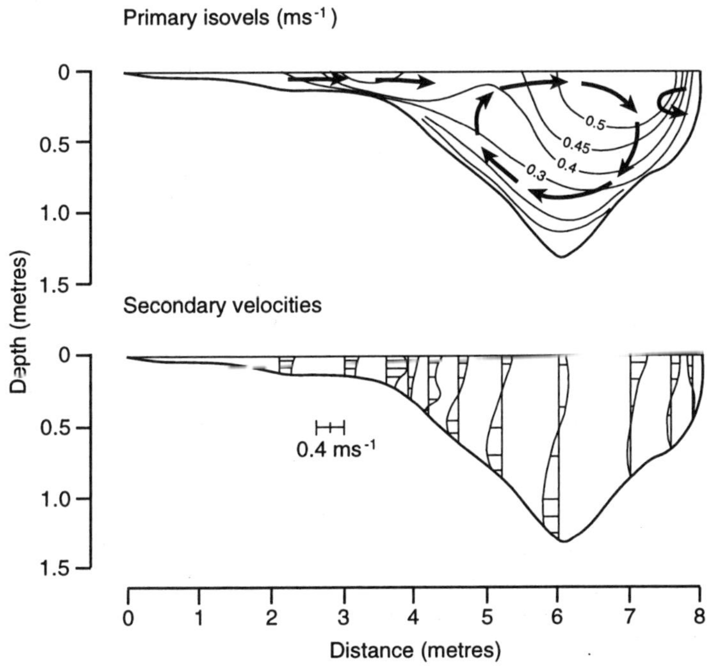

Figure 25.1. Primary isovels and secondary velocity profiles for bankfull flow at a meander apex on the Fall River, USA. Note that secondary arrows superimposed on primary isovels are inferred from isovel curvature and measured horizontal secondary velocities. (after Thorne *et al.* 1985)

Flow over the point bar in Figure 25.1 is directed radially outwards over the whole flow depth, corroborating the work of Dietrich *et al.* (1979). There are two reasons for outward flow. First, the flow area is decreasing in the downstream direction over the point bar, causing water to be directed outwards (topographically steered) to maintain continuity. Second, at the inner bank the outward acting centrifugal force, due to the channel curvature, overcomes the inward acting pressure gradient force, due to super-elevation, carrying water outwards over the point bar (Dietrich, 1982; Dietrich & Whiting, 1989). This pattern of flow leads to convergence of near-bed flow and strong upwelling where the skew-induced cell meets the zone of outward flow. The morphology of the bar is moulded by the secondary pattern to display a flat upper point bar platform dominated by outwards flow, a steep point bar face dominated by inwards flow and a sharp point bar crest beneath the zone of upwelling at the junction of the two secondary flow features. Growth of the point bar occurs through advance of the bar crest towards the retreating outer bank. This drives bend growth and migration that is typical of meandering rivers (Friedkin, 1945; Thorne & Lewin, 1979; Lapointe & Carson, 1986; Thorne, 1991).

Qualitative descriptions and explanations of secondary current patterns in single-thread meandering channels have proved to be a crucial prerequisite to successful mathematical modelling, because they have clarified which terms in the governing equations of fluid motion may, and may not, be neglected (Dietrich & Whiting, 1989; Smith & McLean, 1984, Odgaard & Berg, 1988).

SECONDARY FLOWS IN BRAIDED RIVERS

There have been few field measurements of velocity fields around bars in braided rivers, with those of Bridge & Gabel (1992) being an exception. To date there has been no systematic, field-based study of secondary currents in a large, sand-bed braided river. Bristow & Best (1993) have recently presented a comprehensive review of past studies. They found that field work has tended to emphasize flow and morphological dynamics at confluences, while the link between converging flow at a confluence and subsequent divergence, leading to bifurcation of the channel, has been relatively neglected. Consequently, rather little is known about flow patterns in large braided rivers (Peters, 1977) and even less about any role secondary currents may play in confluence-diffluence mechanics (Best & Bristow, 1993). Similarities between meandering and braided rivers do exist, however, (Thorne *et al.*, 1993) suggesting that secondary currents could be in part responsible for significant morphological forms in braided rivers.

Ashworth *et al.* (1992) suggested that a bifurcated channel could be viewed as consisting of back-to-back meander bends (Figure 25.2a), the left channel being a mirror image of the right. On this basis, they suggested that the secondary flow pattern would consist of twin skew-induced cells, diverging at the surface and

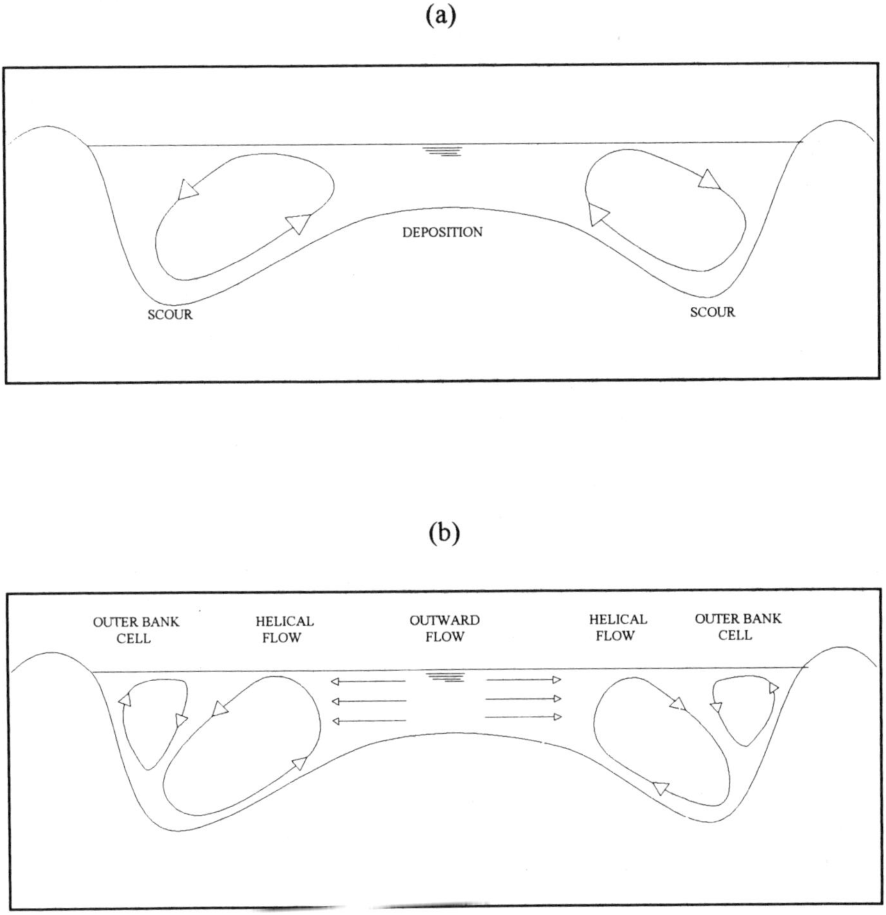

Figure 25.2. (a) Concept of secondary flow in a bifurcated channel as a mirror image of two meanders (after Ashworth *et al.*, 1992); (b) Concept of secondary flow in a bifurcated channel proposed in this study

converging at the bed (Figure 25.2a). The resulting circulation drives fast surface water outwards to the banks where it plunges, and brings slow, near-bed water inwards, where it upwells at the channel centre.

While the analogy of a bifurcated channel in a braided river with back-to-back meanders has merit, examination of the results reported in the previous section on flow in meander bends of single-thread channels suggests that the concept of secondary current cells in Ashworth *et al.* (1992) possibly over-simplifies the situation found in nature.

A mirror image of the current understanding of secondary flow at a meander bend (Figure 25.1) would indicate that the likely pattern of secondary currents in a bifurcated channel might be similar to that shown in Figure 25.2b. The salient features are:

- helical skew-induced flow confined to the deep anabranch talweg channels, with outwards flow driving fast surface water towards the outer, eroding banks and inwards flow bringing sediment laden near-bed flow towards the flanks of the braid bar;
- outer-bank cells generating surface convergence, with flow plunging a distance one or two times the bank height away from the outer banks and promoting basal scour and undercutting around the bank toe; and
- outward flow over the upper braid bar at the channel centre, concentrating bed material transport and deposition at the near-bed flow convergence/upwelling zone on the flanks of the braid bar and promoting deposition of fine sediment on the bar top through reducing velocity and unit discharge over the bar in the downstream direction.

THE BRAHMAPUTRA RIVER, BANGLADESH

The Brahmaputra is one of the world's largest rivers with a mean daily discharge of 19 600 m^3 s^{-1}, a mean annual peak discharge of 65 500 m^3 s^{-1} and a total sediment load of the order of 500×10^6 tonnes per year (Coleman, 1969; Barua, 1994). The flow regime is strongly seasonal, with the great majority of flow occurring during the summer due to Himalayan snowmelt and monsoonal rainfall. The river discharge usually peaks in late July or early August. In 1988 the monsoon flood had a peak discharge of 98 600 m^3 s^{-1} and an estimated return period of 65 years (Brammer, 1990). Measurements of discharge over the last 37 years show no discernible change in the annual maximum discharge (Sir William Halcrow & Partners Ltd., 1993) suggesting that the hydrological regime is steady, at least over engineering time-scales.

The main tributaries of the Brahmaputra River are the Teesta, Dudhkumar and Dharla rivers, each of which contributes a large volume of sediment derived from the southern flanks of the Himalayas. There are two important distributaries, both on the left bank (Figure 25.3). One of these, the Old Brahmaputra, is actually a former course of the main river, abandoned in the late 18th Century. The river downstream of the Old Brahmaputra offtake is called the Jamuna in Bangladesh.

The channel of the Brahmaputra is braided and its dominant morphological features are multiple anabranches separated by bars and islands. Coleman (1969), Bristow (1987) and Thorne *et al.* (1993) have established that there are two distinct braid bar levels. First order bars (islands) are of the order of 30 km in length and are scaled on the width of the primary channel. These islands have surface elevations close to floodplain level and are only partially submerged at bankfull flow. Second order bars (braid bars) are of the order of 6 km long and are scaled on the width of the major anabranches. These bars are completely submerged at bankfull flow.

The islands are vegetated, inhabited and persist in about the same location for decades. They grow laterally by sand accretion when braid bars are attached to their flanks, and vertically by silt deposition during over-bank flows, giving them a characteristically layered stratigraphy (Bristow, 1987). They erode when attacked by curving anabranches on one or both sides (Bristow, 1987; Thorne *et al.*, 1993). The second order braid bars are much more mobile and are continually being re-worked by the river. Bars grow and migrate downstream through erosion around the bar head coupled to sand deposition on the flanks and bar tail and their dynamics are related to seasonal stage changes in the main anabranches (Bristow, 1987; Bristow & Best, 1993). On the basis of their high mobility, rapidity of adjustment to imposed flow conditions and close association with the hydraulics of the major anabranches, the second-order braid bars were selected as the appropriate scale for the field measurements of this study.

FIELD STUDY

To investigate the patterns of secondary flow, sediment concentration and channel change, measurements around a bifurcation-bar-confluence unit were made during the summer monsoon in 1994. The study site was selected following a reconnaissance survey in May 1994. It encompasses a bar in the left-bank anabranch of the Brahmaputra, 10 km south of Bahadurabad (Figures 25.3 and 25.4). At the beginning of the field study, in May 1994, the left sub-channel flanking the bar was dominant, carrying approximately three times the flow of the right sub-channel. However, during 1994 the right sub-channel grew at the expense of the left, so that at the end of the survey period in November 1994, the sub-channels were of about equal size.

In the survey 16 transect lines were established, starting 5 km upstream of the study bar and continuing downstream of the bar (Figure 25.4). In this preliminary paper consideration is limited to results from Sections 3 and 6. These transects were selected to represent flow in two key areas: the curved sub-channel alongside the bar, and; the point of bifurcation upstream of the bar head. Measurements were made from the main vessel of the River Survey Project, using a 300 kHz Acoustic Doppler Current Profiler (ADCP) and dual frequency echo-sounder. Measurements were made in May, August and September 1994 corresponding to the rising, peak and falling stages of the monsoon event. Use of a differential global positioning system (DGPS) for position fixing meant that errors in relocating transects were of the order of ±5 m, which is negligible in an anabranch over 500 m wide.

An ADCP uses the Doppler effect to measure three-dimensional velocity profiles. It should be noted that the ADCP cannot measure velocities at the bed. The Doppler effect is the change in apparent sound pitch that results from relative

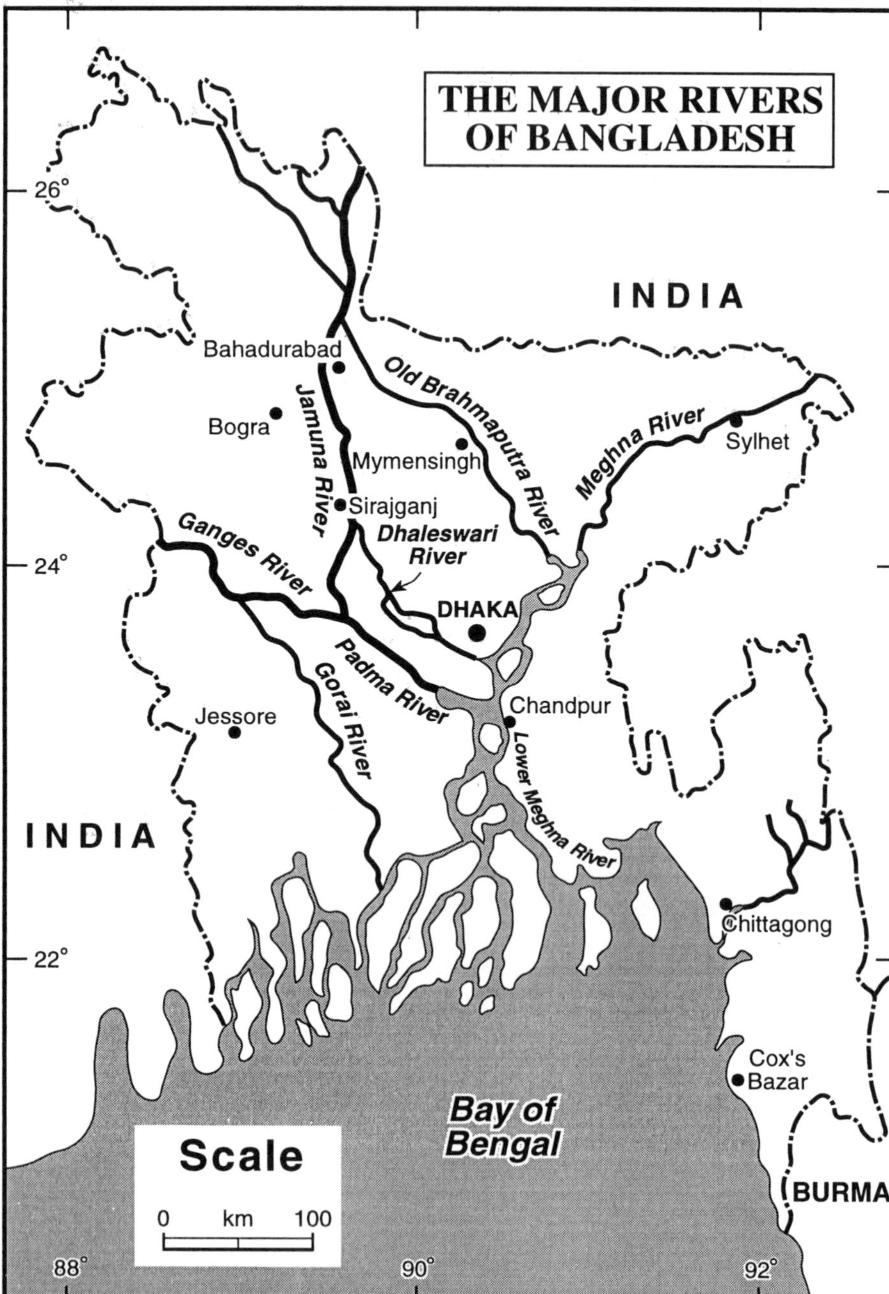

Figure 25.3. The major rivers of Bangladesh. Study site is just south of Bahadurabad on the Jamuna River

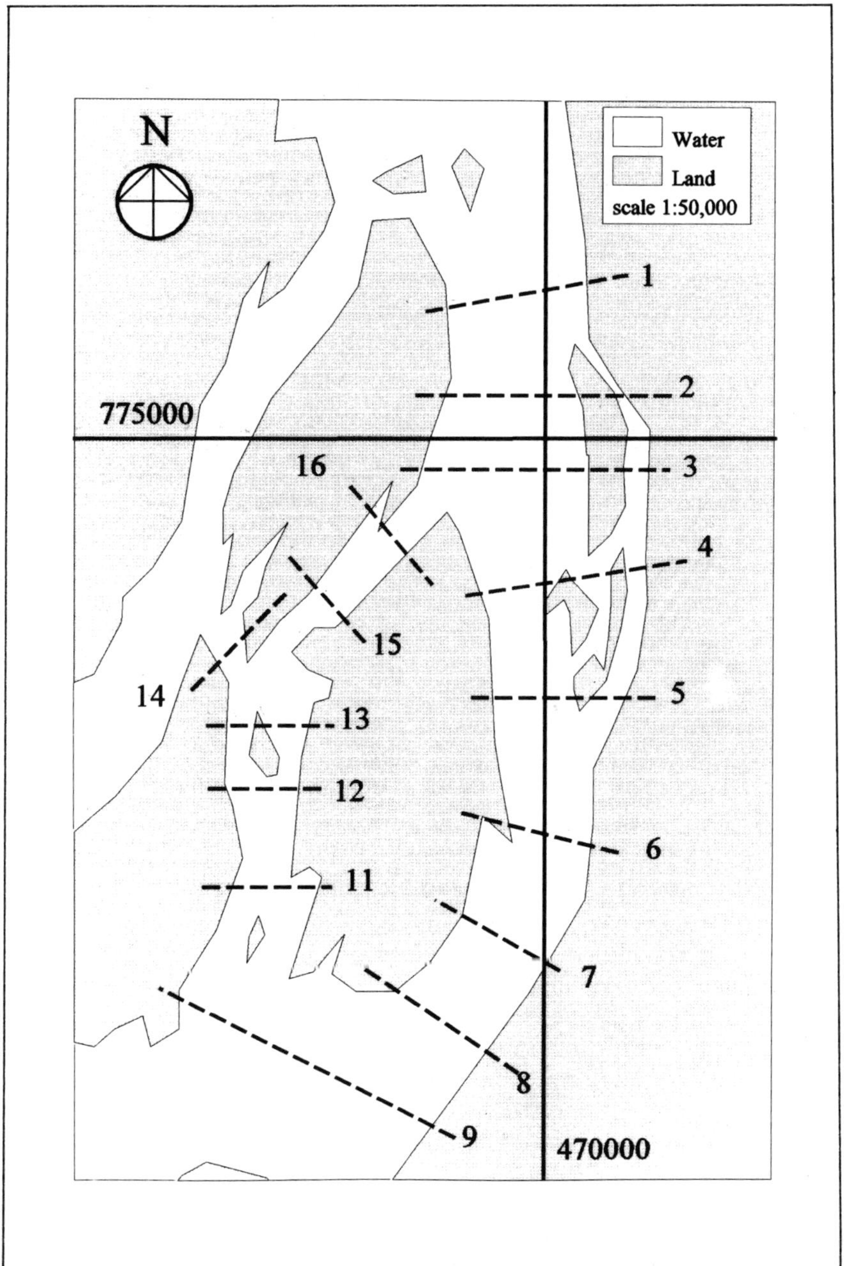

Figure 25.4. Map of study bar and transect lines used in the field survey based on LANDSAT image for 1994. Co-ordinates refer to transverse Mercator system for Bangladesh. Only results for Sections 3 (bifurcation) and 6 (curved sub-channel) are discussed here

motion between the observer and the source. The magnitude of Doppler shift in sound pitch measured by the ADCP is directly proportional to the relative radial velocity between the source and sound scatterers such as suspended sediment in the flowing water. The ADCP measures velocity profiles by transmitting acoustic pulses along four diverging beams in the water column every 0.5 s and then recording the shift in frequency of the returned signals. The ADCP then resolves the four measured radial velocities into north-south, east-west and vertical components. As there are four measured velocities but only three resolved components, the vertical component can be calculated twice using different pairs of beams. The difference between the two calculated vertical velocities is due to non-uniformity of the velocity field across the sampled area and is displayed by the ADCP as an 'error velocity'. The ADCP uses the time delay between transmission and return of the signal to sample the velocity profile in measurement ranges (termed bins). The ADCP was configured to sample the data into bins of 0.5 m depth, extending from 2.7 m below the water surface to near the bed, and to average velocity profile data in measurement ensembles every 10 pings. Using this method of successive measurement onboard a moving vessel, a continuous data set is obtained for each cross-section. With an average vessel speed of 4 knots, the velocity profile represents a flow field approximately 10 m wide.

The intensity of the returned signal in decibels is also logged. This is approximately proportional to the amount of suspended material scattering the acoustic signal and, therefore, can be used to indicate relative suspended sediment concentration (Thorne *et al.*, 1994).

DGPS is used to calculate the vessel velocity. This enables correction of the velocities measured by the ADCP for boat motion. The DGPS has a positioning accuracy of ±1 m in static mode, which in practice corresponds to an error in the measured water velocities of approximately ±0.15 m s^{-1}. However, within each velocity profile the same DGPS correction is applied over the whole flow depth, so that no bias is introduced in plots of secondary currents.

Measured north-south and east-west velocities do not necessarily correspond to the primary and secondary components. There are a number of methods that may be used to define the primary and secondary planes and these were reviewed recently by McLelland *et al.* (1994). As outlined by McLelland *et al.* (1994) none of the traditional methods of velocity correction in open channels is suited to studies on large braided rivers such as the Brahmaputra.

Braided rivers are characterized by a multi-threaded planform. In any one cross-section in a braided channel there may be several talwegs separated by various bed features, which may contain a core of local maximum velocity acting as a semi-independent stream. It is proposed that each of these streams of water will have their own primary and hence secondary orthogonal velocity directions. Therefore, any technique for correcting velocity measurements in a braided river should be able to define separate secondary flow planes for each stream of water in a multi-threaded channel cross-section.

In order to achieve this goal a secondary flow function was derived to search for secondary flow activity for various alignments in a river channel. The proposed secondary flow function measures helical rotation for various cross-sectional alignments and highlights peaks at certain orientations. Taking the measured flow velocity components as velocity magnitude (V_H) and velocity direction (φ), and referring to the corrected velocity components as primary (U_P) and secondary (U_S), the secondary flow function is defined as:

$$\left(\frac{U_P}{|U_S|}\right)(y - Y)W$$

where $U_S = V_H \cos(\varphi - \alpha)$, α is the flow orientation, y is the depth of the measurement 'bin' being considered, Y is the average depth and W is the width of the measurement ensemble. For each velocity profile an average depth is calculated and then the above function is summed for all individual velocity measurements. The process is then repeated for all velocity profiles in a given cross-section and a value of the secondary flow function is obtained. All possible orientations ($\alpha = 0$ to 180°) are then applied and a graph of the secondary flow function versus plane orientation is obtained (Figure 25.5a). A graph of the secondary flow function versus distance along the cross-section is then plotted for each value of plane orientation associated with a peak value of the secondary flow function (Figures 25.5b and 25.5c). Figures 25.5b and 25.5c can be used to determine which streams within the braided cross-section correspond to which plane orientation. The channel can then be sub-divided by locating the position of minimum depth average velocity between cores of local maximum velocity. Each stream of water can then be given a separate value of primary and secondary flow direction.

HISTORICAL PLANFORM CHANGES

The results of field measurements made during a single monsoon event must be interpreted in the context of longer term channel development at the study site. Historical LANDSAT satellite images covering the study area were supplied by Flood Action Plan 19 – Geographical Information Systems. Images were obtained from 1990, 1992, and 1994. The historical record of planform change was used to document and establish morphological changes around the study site over the medium term.

The recent record of planform developments in the left-bank anabranch of the Brahmaputra River (Figure 25.6) confirms that flow and channel change processes observed around a single bar during a single flow season can only be understood within the context of morphological changes at larger time and spatial scales. Figure 25.6 shows that the left-bank anabranch of the Brahmaputra River in 1990 is predominantly meandering with a sinuous alignment. There is a west-bank concave curve north of Bahadurabad and the future site of the study bar is at the

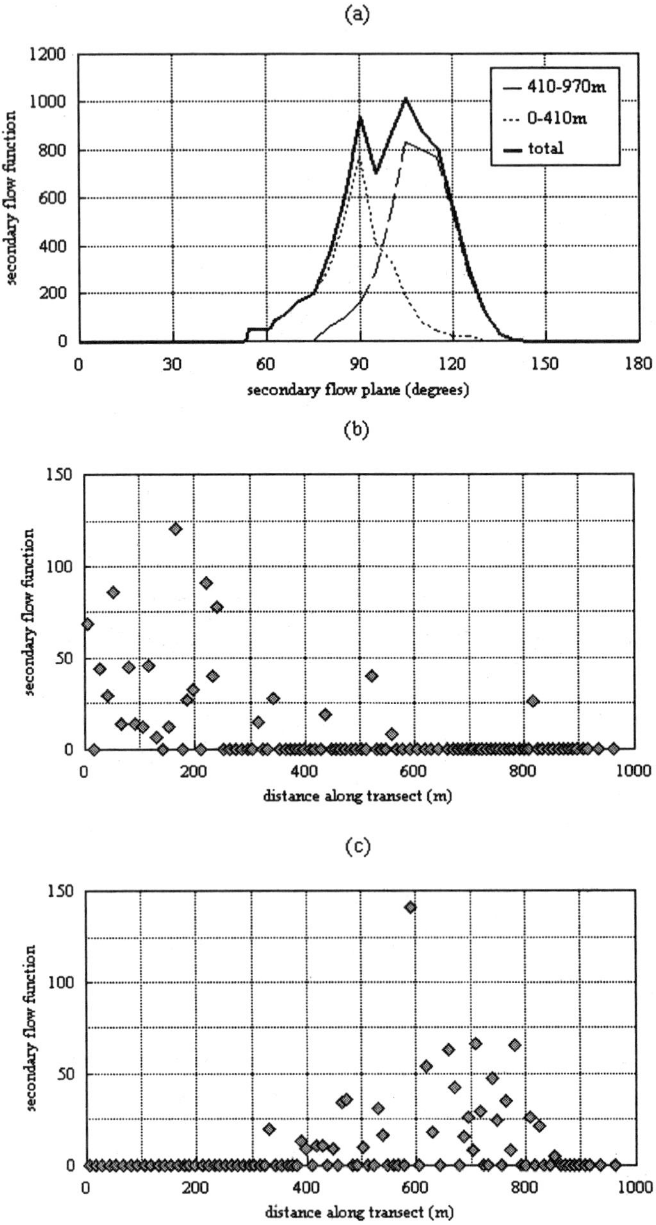

Figure 25.5. Velocity correction method applied to September data at section 3, (a) graph showing secondary flow function plotted against flow orientation; (b) and (c) graphs showing secondary flow function plotted against distance along transect for peak values of flow orientation shown in (a)

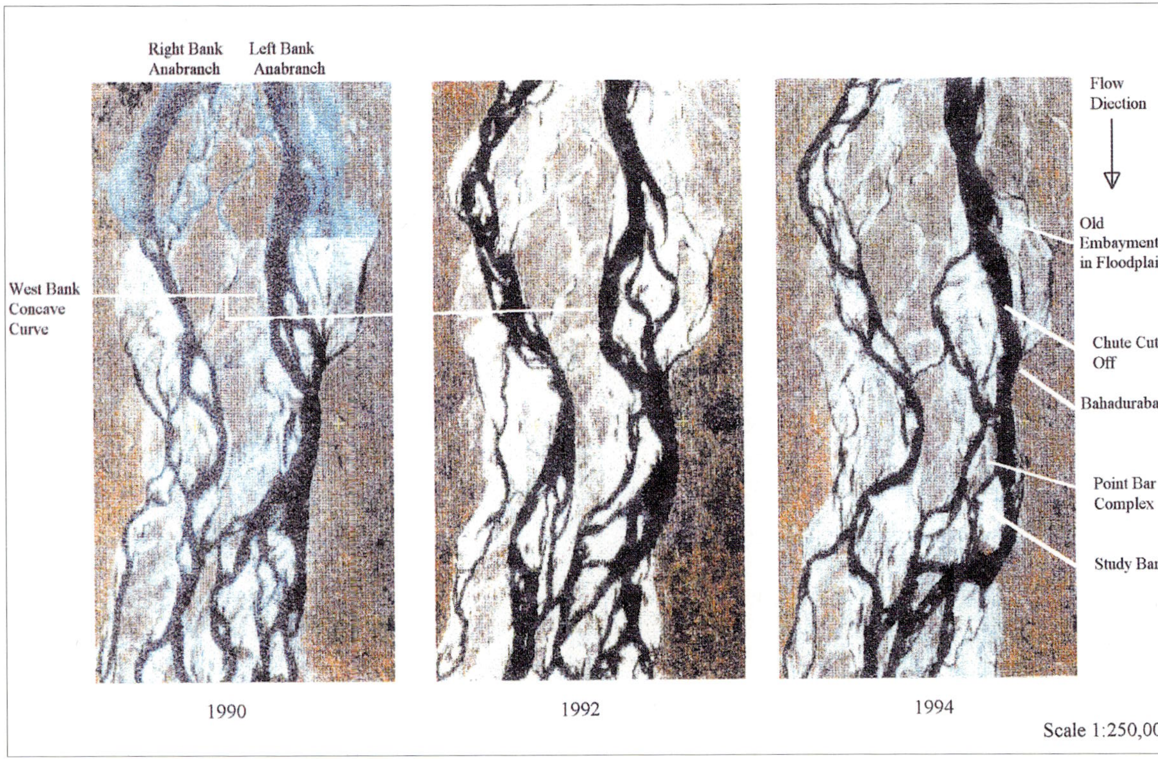

Figure 25.6. LANDSAT images for the Brahmaputra River around the study reach in 1990, 1992 and 1994. Yellow crosses may be used for cross-reference between images. Note increase in sinuosity of left-bank anabranch between 1990 and 1992 and cut-off north of Bahadurabad between 1992 and 1994. Flow orientation and curvature approaching the study bar are dependent on morphological changes immediately upstream

trailing edge of a growing point bar complex in an east-bank concave bend that is attacking the floodplain along the east-bank, south of Bahadurabad.

By 1992, the curves in the left-bank anabranch had grown in amplitude to give the left-bank anabranch a more sinuous course. The west-bank concave bend, north of Bahadurabad is strongly curved. The east-bank concave bend containing the study site has also grown and has shifted downstream and to the east, with the study bar beginning to emerge just downstream of the point bar apex.

Between 1992 and 1994 a chute cut-off of the west-bank concave bend north of Bahadurabad occurred. This is a common phenomenon for sharply curved anabranch bends in the Brahmaputra (Klaassen & van Zanten, 1989). The bend containing the study site has continued to grow and migrate downstream and the study bar is identifiable as the apex element of the point bar complex in this bend. However, the upstream cut-off has changed both the heading and curvature of the flow approaching the study reach. In 1992 the approach heading was about 140°, with a clockwise curvature which favoured the left sub-channel at the study bar. By 1994 the heading had changed to about 180°, and the planform curvature is anticlockwise, which starts to favour the right sub-channel. Historically, the Brahmaputra has shown a tendency to abandon and then re-occupy embayments cut into the floodplain by aggressive anabranch bends (Thorne *et al.*, 1993). If the left-bank anabranch re-occupies the old meander scar visible in the bank-line north of Bahadurabad, its heading will swing further to the south-west and its anticlockwise curvature will increase as it approaches the study area. This would further promote flow down the right sub-channel at the expense of the left sub-channel.

Between 1990 and 1994, planform evolution altered the heading of the flow approaching the study area from a situation that favoured the left sub-channel at the bar to a situation which favoured the right sub-channel. This demonstrates how planform evolution sets the scene for the changes observed in the flow structures, bed topography and planform change during the 1994 summer monsoon.

RESULTS

Data processing in this study is at an early stage and the analysis presented here is restricted to data for two transects. Sections 3 and 6 were selected specifically to address the issues raised in the review of secondary flows in meandering and braided channels presented earlier.

FLOW IN A CURVED SUB-CHANNEL

To establish whether there is any field evidence for the secondary flow pattern proposed in Figure 25.2b, results observed for section 6 during May 1994 are first considered. section 6 is located mid-way along the divided flow reach near the apex of the curve in the left sub-channel (Figure 25.4). Figure 25.7 shows plots of

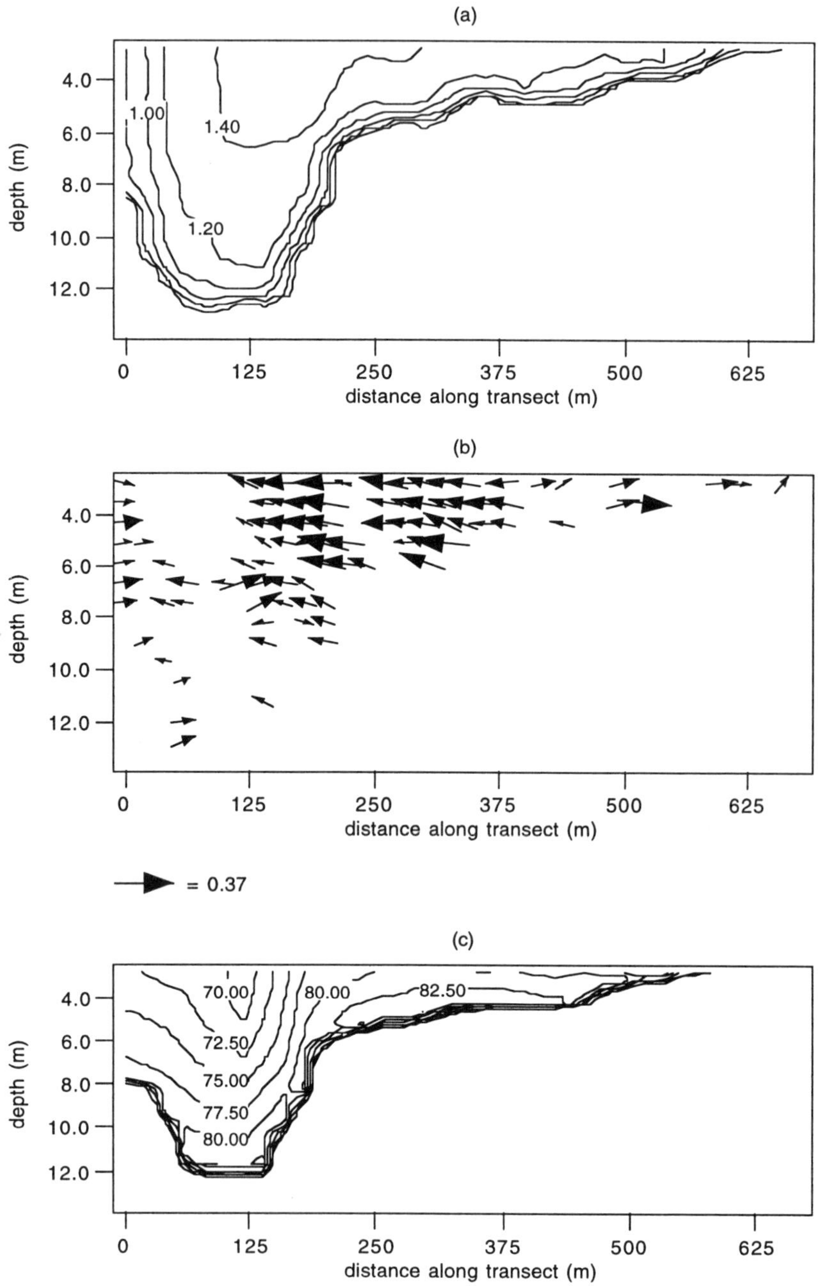

Figure 25.7. Results from section 6 for (a) primary isovels; (b) secondary velocities; and (c) ADCP backscatter from the May 1994 survey. Only significant secondary velocities are plotted (> 0.15 m s^{-1}) in cross-section (b)

primary isovels, secondary velocities and relative backscatter intensity.

Examination of the cross-sectional geometry shows that it displays morphological features similar to those found in a single-thread channel bend. The section is highly asymmetrical with deep scour close to the outer (east) bank and a well defined talweg channel, separated from a prominent bar at the inner bank by a sharp bar crest. The outer bank is known to be retreating from comparison of recent satellite images (Figure 25.6).

The primary isovels (Figure 25.7a) display a single core of maximum velocity above the talweg, with tightly packed isovels close to the bed at the bank toe and on the point bar. Primary isovels close to the outer bank near the water surface are more widely spaced. Above the point bar crest, upward bulging of the isovels suggests that there may be a zone of upwelling flow.

The plot of secondary velocities is consistent with this interpretation of the primary isovels. Helical circulation is evident in the talweg, with outward flow in the upper half of the profile and inward flow at depth. At the outer bank there is some evidence of an outer bank cell of reverse rotation in that secondary flow near the surface is directed inwards. The position and dimensions of the outer bank cell suggested by the secondary velocities coincide with the low velocity zone in the primary isovels and the zone of plunging flow is consistent with the area of closely packed isovels at the bank toe. At the point bar crest, secondary flow is directed radially outwards over the whole flow depth and a strong secondary velocity component can be seen to flow over the point bar crest to join the outwards surface flow of the helical cell in the talweg. Bed convergence and upwelling are evident on the point bar face. At the inner bank there is a zone of inward flow that may be indicative of a zone of flow separation (Markham & Thorne, 1992).

In Figure 25.7c the relative backscatter intensity plot at section 6 shows high sediment concentrations over the point bar platform, on the point bar face and close to the outer bank. These are areas of upwelling in the velocity measurements. There is an area of water with very low sediment concentration above the talweg, associated with the core of maximum primary velocity. Bulges in the isolines in Figure 25.7c are also consistent with the effects of upwelling evident in the primary velocity plot above the point bar crest and the secondary velocity plot at the point-bar face.

Taken together, the results for section 6 are consistent with patterns of primary velocity and secondary flow observed in bends of single-thread rivers (Figure 25.1) and display the elements hypothesized in Figure 25.2b. The backscatter results are also consistent with suspended sediment load measurements made in single-thread meander bends by Dietrich & Smith (1983).

FLOW AT A BIFURCATION

Figure 25.8 shows results for section 3 in May, August and September 1994, corresponding to rising, peak and falling monsoon flood stages. In May this section was located near the flow bifurcation, immediately upstream of the study bar. During the course of the summer monsoon the nose of the study bar migrated approximately 0.5 km downstream while another, smaller medial bar formed in the left channel to the east. When applied to the data for May, the procedure to locate the primary and secondary planes for section 3 revealed two primary directions in the flow field, with headings of 80° for the flow between 0 and 787 m across the section and 98° between 787 and 1044 m across (Figure 25.8a). That is, the velocity streams are diverging by 18°. It should be noted that the results plotted in Figure 25.8a display primary and secondary velocities resolved according to the primary direction for each segment of the cross-section.

The plot of primary isovels for May shows two cores of maximum velocity, separated by a band of low velocity water between 700 and 900 m (Figure 25.8a). The larger core is associated with the flow direction in the channel between 0 and 787 m, which enters the left sub-channel. The smaller core is associated with the flow between 787 and 1044 m, which enters the right sub-channel. These results illustrate that although the cross-section is roughly parabolic, with no topographic divide in the channel bathymetry, the flow has already begun to bifurcate hydraulically.

The secondary velocities confirm that a hydraulic bifurcation has occurred, with markedly different secondary patterns in the two velocity streams. On the left side of the channel (250 to 600 m) the secondary velocities display clockwise helical circulation. This is a skew-induced cell associated with curvature of the flow as it approaches the study reach through a large, west-bank concave bend (see Figure 25.6, 1994 image). Close to the left-bank (0 to 250 m) secondary velocities are complex and difficult to interpret. In the right side of the channel (850 to 1044 m) secondary velocities are directed strongly to the right (west) over the whole flow depth. This is associated with topographic steering of the flow towards the right sub-channel. At this stage and location there is, apparently, insufficient curvature of flow entering the right sub-channel to generate a strong skew-induced secondary cell. Around the flow bifurcation (600 to 830 m) secondary velocities are generally weak and their directions are diverse, including significant vertical components. This suggests an area of complex, highly turbulent flow. In the field the water surface in this area displayed lines of boils and vortices typical of a turbulent shear layer.

The backscatter intensity plot shows a filament of water with a relatively low sediment concentration associated with area of flow bifurcation (700 to 830 m). In the field flow in the left sub-channel was observed to be eroding the flank of the bar close to the bar head. An area of high sediment concentration is displayed in the left channel associated with the zone of upwelling in the helical secondary flow of the left flow stream (300 to 450 m).

The results for high-stage flow in August are shown in Figure 25.8b. The procedure to locate the primary and secondary planes for this section again showed two primary directions in the flow field, with headings of 78° for the flow between

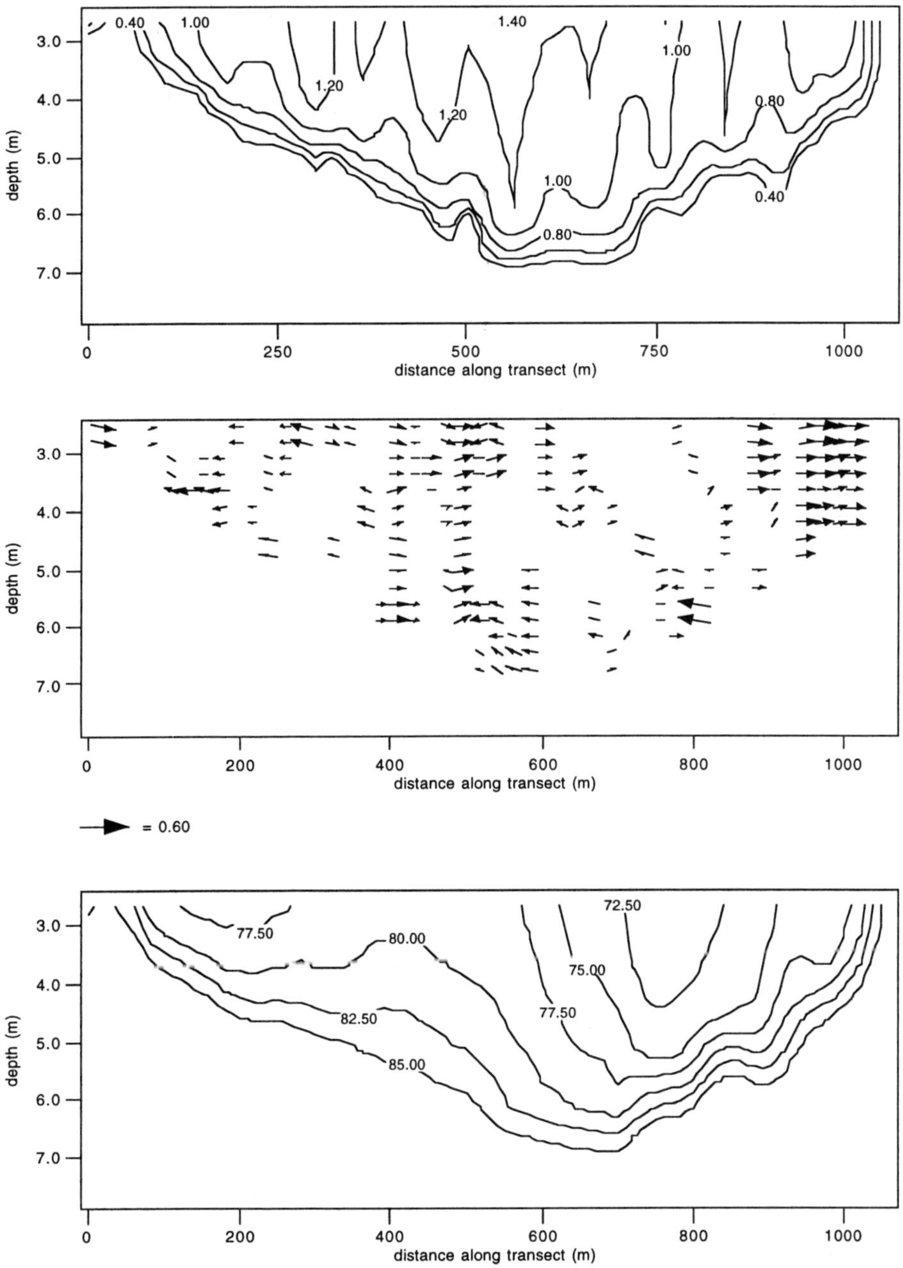

Figure 25.8. (a) Results from section 3 for primary isovels, secondary velocities and ADCP backscatter from May 1994. Only significant secondary velocities (> 0.15 m s^{-1}) are plotted. *Continued on next page*

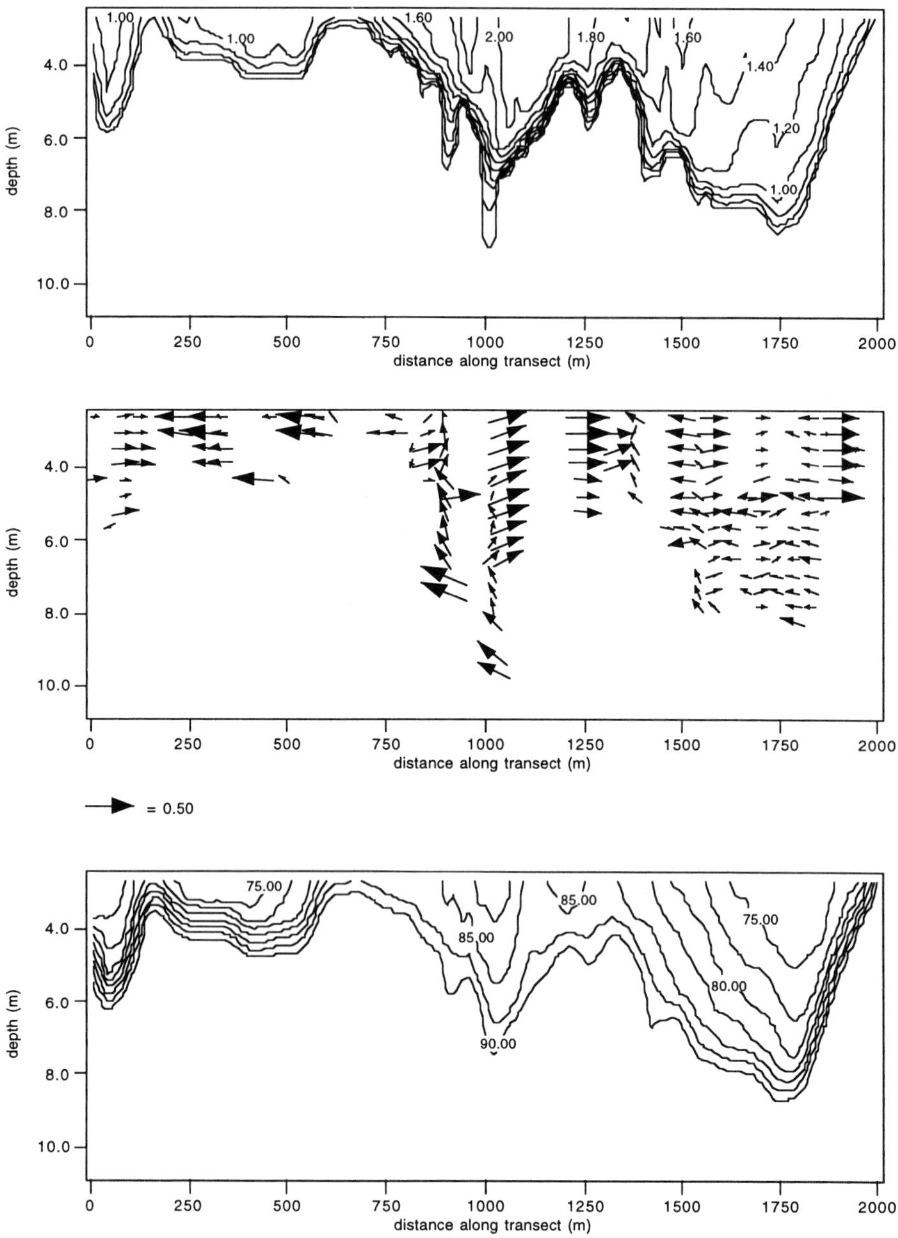

Figure 25.8. *continued* (b) Results from section 3 for primary isovels, secondary velocities and ADCP backscatter from August 1994. Only significant secondary velocities (> 0.15 m s^{-1}) are plotted. *Continued on facing page*

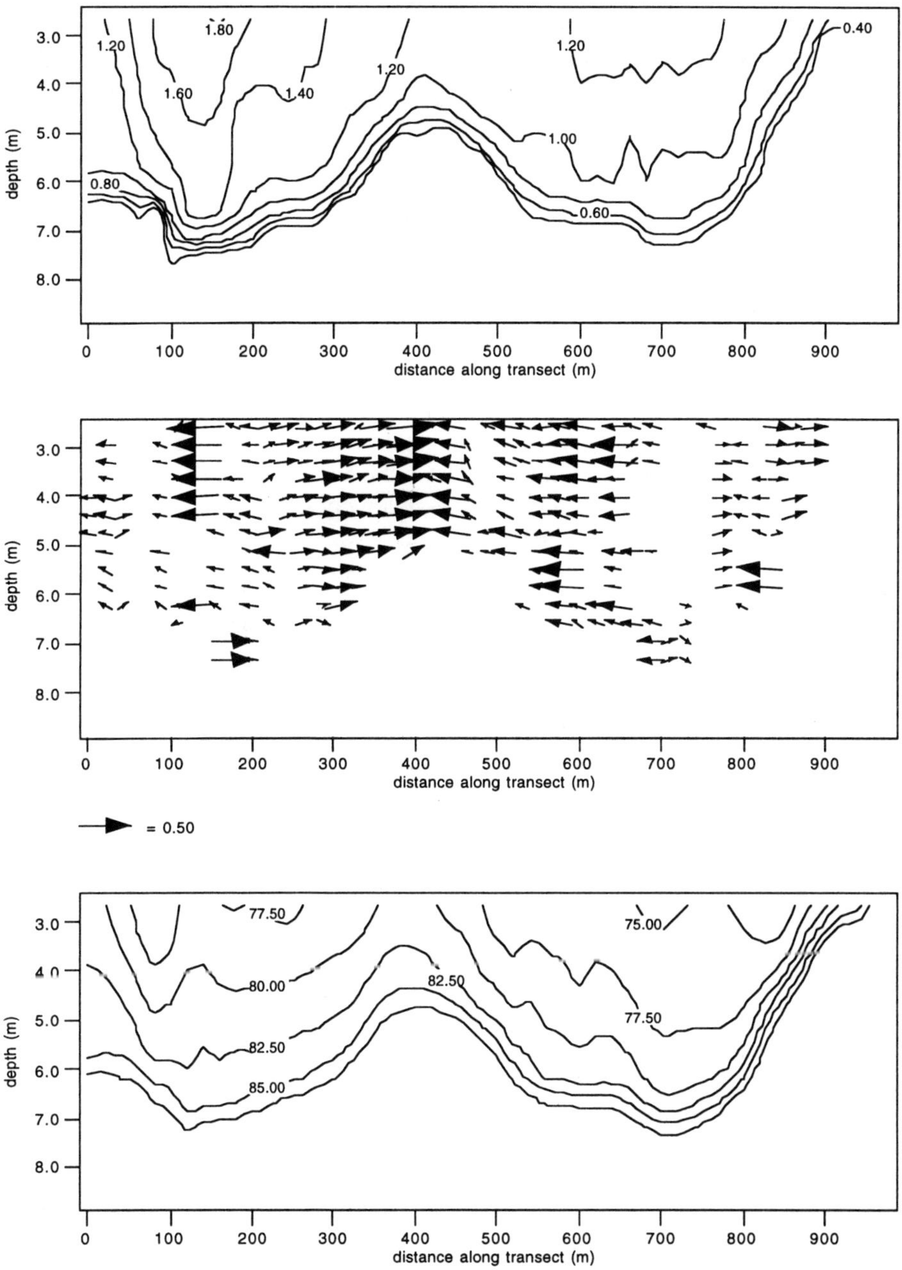

Figure 25.8. *continued* (c) Results from section 3 for primary isovels, secondary velocities and ADCP backscatter from September 1994. Only significant secondary velocities (> 0.15 m s^{-1}) are plotted

0 and 1440 m across the section and 103° between 1440 and 2000 m (Figure 25.8b). Compared to results for May, the area of flow bifurcation has remained in about the same location, but the angle of divergence between the flow streams has increased from 18 to 25°. The cross-section is no longer parabolic, but exhibits a pronounced medial bar (1100 to 1400 m) separating talwegs at 1000 and 1750 m across the transect.

The primary velocity isovels again show two high-velocity filaments, associated with flows heading towards the left and right sub-channels, but the high-velocity stream in the right sub-channel has expanded noticeably compared to May. The secondary flow pattern has changed radically. Clockwise helical flow is now present in the right flow stream, while flow in the left channel is directed to the west over in the upper portion of the section with strong vertical velocities associated with the complex bed features (800 to 1440 m in Figure 25.8b). It appears that the skew induced cell from the bend upstream has migrated westwards and that water is being drawn across the line of flow bifurcation from the diminishing left channel into the growing right channel. At this high stage, a relatively small back channel along the left bank (0 to 250 m) contains significant flow, which is following a path that is hydraulically separated from the rest of the cross-section by a wide, shallow shelf (250 to 750 m).

The backscatter plot for August shows low sediment concentration associated with the clockwise helical cell in the right sub-channel (1600 to 1800 m) and an area of high sediment concentration above the medial bar (1100 to 1400 m) where near-bed secondary components converge. The shape of the bed on the medial bar suggests that climbing dunes may be present on the gently sloping stoss (east) flank due to strong secondary flow from left to right over the bar.

The bed level change plot for the period May to August at section 3 (Figure 25.9a) illustrates the building of a medial bar with climbing dunes. Erosion is evident in the sub-channels on either side, but is most pronounced in the right sub-channel which exhibits both bed scour and bank retreat. This pattern of morphological development is consistent with the observed westward (rightwards) migration of the clockwise-rotating secondary cell, acceleration of primary flow in the right sub-channel and bed sedimentation where water crossing the bar from the diminishing left sub-channel meets an area of upwelling due to helical flow in the right sub-channel, to produce near-bed convergence.

The results for September are shown in Figure 25.8c. The procedure to locate the primary and secondary planes for this section again reveals two primary directions in the flow field, with headings of 88° for the flow between 0 and 417 m across the section and 114° between 417 and 963 m across (Figure 25.8c). That is, the flow in the left sub-channel has ceased to swing to the east, but has swung back towards the west, while flow in the right sub-channel has continued to veer west. As a result the angle of divergence has remained approximately constant at 26°. Falling stages on the recession of the monsoon flood restricts flow to the main channel, making it impossible to navigate over the shelf that was covered in August (0 to

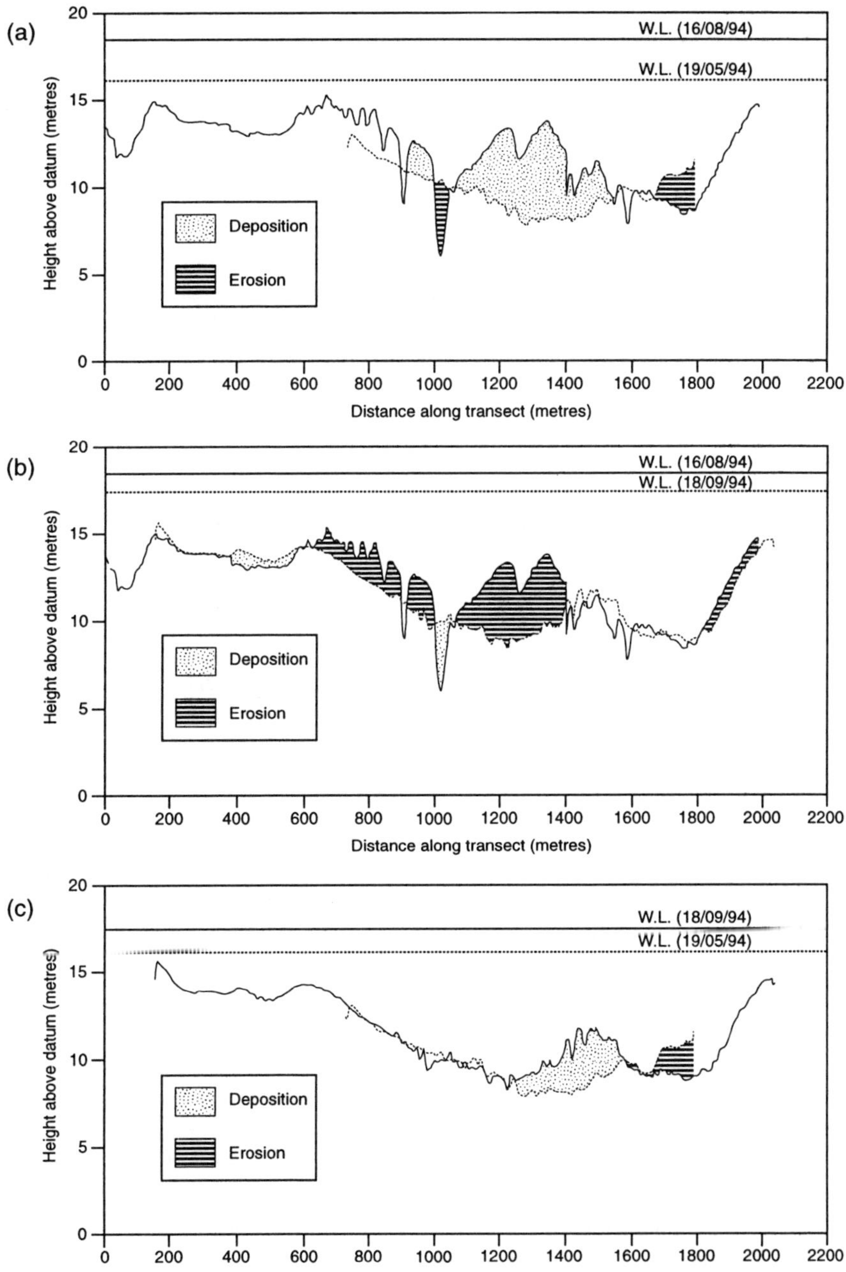

Figure 25.9. Bed level changes at section 3 between (a) May and August; (b) August and September; and (c) May and September 1994

800 m in Figure 25.8b). The bed in the main channel cross-section now clearly displays two sub-channels, with a pronounced medial bar (250 to 500 m) separating talwegs at 100 and 700 m across the transect in Figure 25.8c.

Primary isovels show two well-defined velocity maxima, separated by an area of slower flow over the medial bar. Secondary velocities indicate persistence of the clockwise helical flow in the right sub-channel that was identified in the results for August. Results for the left sub-channel are clearer and display an anticlockwise helix, to produce an almost symmetrical pattern with surface flow divergence in each sub-channel and secondary flow convergence over the medial bar at the centreline.

The backscatter plot in Figure 25.8c reveals filaments of relatively low sediment concentration in both sub-channels, associated with the two mirrored helical cells producing plunging flow near the banks. Sediment concentrations are high around the channel centreline, above the medial bar, where secondary flows converge and upwell.

The bed-level change plot for the period August to September at section 3 (Figure 25.9b) illustrates that during the falling stage general scour has occurred, smoothing the bed profile. The large climbing dune bedforms built on the rising limb have been destroyed and the area of deep scour channel in the left sub-channel (at 1000 m in Figure 25.9a) has been filled. Significant erosion of the right bank, first evident in Figure 25.9a, has continued. The pattern of morphological development is consistent with the observation that the height of bedforms scales on the flow depth and intensity, both of which have diminished. Continued westward (rightwards) migration of the main helical secondary cell is evident and there appears to have been development of a clockwise, skew-induced cell in the left sub-channel due to flow curvature around the growing medial bar at about the channel centreline. Erosion on the right margin drives enlargement of the right sub-channel that is associated directly with the change in dominance between the two sub-channels discussed earlier.

PRELIMINARY CONCLUSIONS

The ADCP measurements taken at section 6 indicate that cross-sectional morphology, primary isovels, secondary flows and suspended sediment distributions in bends of braided rivers share some common features with patterns observed in meander bends of single-thread rivers. The results support the hypothetical flow and cross-sectional distributions shown in Figure 25.2b. These findings suggest that elements of our understanding of process-form interactions in meandering rivers may be transferable to braided rivers.

The results for section 3 illustrate how hydraulic bifurcation of the flow affects the distributions of primary velocity, secondary flow and sediment concentration and drives morphological changes by controlling the distribution of sediment transport, erosion and deposition around the wetted perimeter. Measurements repeated during the rising, peak and falling limbs of the monsoon flood reveal that morphological change occurs as a result of the combination of stage-related scour and fill coupled with incremental shifting in the pattern of flow driven by evolution in the distribution and circulation of flow approaching the study Section.

Zones of relatively low sediment concentration in the plots of ADCP backscatter intensity correlate well with areas of erosion, while areas of relatively high concentration are associated with upwelling flow. The similarity between the sediment distributions recorded at section 6 in May 1994 and those measured using sediment samplers in meandering rivers suggests that the ADCP backscatter is capable of producing reliable, qualitative pictures of relative suspended sediment concentrations.

The main response of the channel to the imposed flow field during the 1994 monsoon flood was to enlarge the right sub-channel at the expense of the left sub-channel (Figure 25.9c). A medial bar grew along the channel centreline, where secondary flow converged. These changes may be explained in terms of the local flow patterns, but those patterns are themselves a product of planform evolution that has affected the orientation and curvature of flow approaching the study site. Historical satellite images (Figure 25.6) show that a bend cut-off upstream led to a change in flow heading and a reversal of planform curvature in the flow approaching the bifurcation. Strong, clockwise helical flow generated by a right-bank concave bend upstream of the bifurcation at the study site dominated flow patterns at section 3 and led directly to erosion at the right margin and deposition at the channel centre. Hence, it must be concluded that channel changes in the study area can only be explained when observed flow and sediment processes are viewed within the context of morphological changes at larger spatial and temporal scales.

ACKNOWLEDGEMENTS

Financial support for this study was provided by the Engineering and Physical Sciences Research Council (UK) through a CASE studentship (B/93561631) with Sir William Halcrow & Partners Ltd. The study was performed as part of a joint project between the University of Nottingham, Delft Hydraulics and the Danish Hydraulics Institute. Technical and logistical support were provided by all the staff and advisers at FAP-24, River Survey Project in Bangladesh. Particular thanks go to Hans Høyer, Pieter van Groen, Johan Grijsen, and Project Adviser J.J. Peters.

REFERENCES

Ashworth, P.J., Ferguson R.I. & Powell, M.D. 1992. 'Bedload transport and sorting in braided channels', in Billi, P., Hey, R.D., Thorne, C.R. & Tacconi, P. (Eds) *Dynamics of Gravel Bed Rivers*, John Wiley & Sons Ltd., Chichester, 497–513.

Barua, D.K. 1994. 'On the environmental controls of Bangladesh river systems', *Asia Pacific J. Environment Development*, Bangladesh Unnayan Parishad (BUP), **1**, 81–98.

Bathurst, J.C. 1979. 'Distribution of boundary shear stress in rivers', in Rhodes, D.D. & Williams, G.P. (Eds) *Adjustments of the Fluvial System*, Kendall/Hunt Publishing Co, Dubuque, Iowa, 95–116.

Bathurst, J.C., Thorne, C.R. & Hey, R.D. 1977. 'Direct measurements of secondary currents in river bends', *Nature*, **269**, 504–506.

Bathurst, J.C., Thorne, C.R. & Hey, R.D. 1979. 'Secondary flow and shear stress at bends', *J. Hydraul. Div., ASCE*, **105**, 1277–1295.

Brammer, H. 1990. 'Floods in Bangladesh, geographical background to the 1987 and 1988 floods', *Geogr. Journal,* **156**, 12–22.

Bridge, J.S. 1977. 'Flow, bed topography, grain size and sedimentary structure in open-channel bends: a three-dimensional model', *Earth Surf. Processes*, **2**, 401–416.

Bridge, J.S. 1984. 'Flow and sedimentary process in river bends; comparison of field observations and theory', in Elliot, C.M. (Ed) *River Meandering*, ASCE, New York, 857–872.

Bridge, J.S. & Gabel, S.L. 1992. 'Flow and sediment dynamics in a low sinuosity river: Calamus River, Nebraska Sandhills', *Sedimentology*, **39**, 125–142.

Bristow, C.S. 1987. 'Brahmaputra River: channel migration and deposition', in Ethridge, F.G., Flores, R.M. and Harvey, M.D. (Eds) *Recent Developments in Fluvial Sedimentology*, Soc. Econ. Paleon. and Mineralogists, Sp. Pub. **39**, 63–74.

Bristow, C.S. & Best, J.L. 1993. 'Braided rivers: perspectives and problems', in Best, J.L. & Bristow, C.S. (Eds) *Braided Rivers*, Geol. Soc. London Sp. Pub., **75**, 1–11.

Coleman, J.M. 1969. 'Brahmaputra River: channel processes and sedimentation', *Sediment. Geol.*, **3**, 129–239.

Dietrich, W.E. 1982. *Flow, Boundary Shear Stress and Sediment Transport in a River Meander*, Unpub. PhD Diss., Univ. Washington at Seattle, 261 pp.

Dietrich, W.E. & Smith, J.D. 1983. 'Influence of the point bar on flow through curved channels', *Water Resour. Res.*, **19**, 1173–1192.

Dietrich, W.E., Smith, J.D. & Dunne, T. 1979. 'Flow and sediment transport in a sand-bedded meander', *J. Geol.*, **87**, 305–315.

Dietrich, W.E. & Whiting, P. 1989. 'Boundary shear stress and sediment transport in river meanders of sand and gravel', in Ikeda, S. & Parker, G. (Eds) *River Meandering,* AGU, Water Resources Monograph **12**, Washington, 1–50.

Engelund, F. 1974. 'Flow and bed topography in channel bends', *J. Hydraul. Div., ASCE*, **100**, 1631–1648.

Friedkin, J.F. 1945. *A Laboratory Study of the Meandering of Alluvial Rivers*, USACE Waterways Experiment Station, 40 pp.

Hey, R.D. & Thorne, C.R. 1975. 'Secondary Flows in River Channels', *Area*, **7**, 191–196.

Hey, R.D. & Thorne, C.R. 1975. 'Secondary Flows in River Channels – discussion', *Area*, **8**, 235–236

Klaassen, G.J. & van Zanten, B.H.J. 1989. 'On cut off ratios of curved channels', *Proc. 23rd Congr. IAHR*, Ottawa, 111–122. (also Delft Hydraul. Pub. No.444).

Lapointe, M.F. & Carson, M.A. 1986. 'Migration patterns of an asymmetrical meandering river: The Rouge River, Quebec', *Water Resour. Res.*, **22**, 731–743.

Leopold & Wolman, 1957. 'River channel patterns: braided, meandering and straight', *US Geol. Surv. Prof. Paper*, **282b**, 39–85.

Markham, A.J. & Thorne, C.R. 1992. 'Geomorphology of gravel bed river bends', in Billi, P., Hey, R.D., Thorne, C.R. & Tacconi, P. (Eds) *Dynamics of Grave-bed Rivers*, John Wiley & Sons Ltd., Chichester, 433–450.

M[c]Lelland, S.J., Ashworth, P.J. & Best, J.L. 1994. *Velocity Correction Methods in Open Channels: Description, Critique & Recommendations*, FAP-24 Working Paper, August 1994, 16 pp.

Odgaard, A.J. 1986. 'Meander flow model. II: Applications', *J. Hydraul. Engrg.*, **112**, 1137–1150.

Odgaard, A.J. & Bergs, M.A. 1988 'Flow processes in a curved alluvial channel', *Water Resour. Res.*, **24**, 45–56.

Perkins, H.J. 1970. 'The formation of streamwise vorticity in turbulent flow', *J. Fluid Mech.*, **44**, 721–740.

Peters, J.J. 1977. 'Sediment transport in the Zaire River', in Nihoul, J.C.J. (Ed) *Bottom Turbulence, Proc. 8th Int. Liege Colloq. on Ocean Dynamics*, **19**, 221–236.

Peters, J.J. & Goldberg, A. 1989. 'Flow data in large alluvial channels', in Maksimovic, C. & Radojkovic, M. (Eds) *Computational Modelling and Experimental Methods in Hydraulics,* (Hydrocomp 1989), Elsevier, London, 77–86.

Prandtl, L. 1952. *Essentials of Fluid Dynamics,* Blackie, London, 452 pp.

Smith, J.D. & McLean, S.R. 1984. 'A model for meandering streams', *Water Resour. Res.*, **20**, 1301–1315.

Thorne, C.R. & Hey, R.D. 1979. 'Direct measurements of secondary currents at a river inflexion point', *Nature*, **280**, 226–228.

Thorne, C.R. & Lewin, J. 1979. 'Bank processes, bed material movement and planform development in a meandering river', in Rhodes, D.D. & Williams, G.P. (Eds) *Adjustments of the Fluvial System*, Kendall/Hunt Publishing Co, Dubuque, Iowa, 117–137.

Thorne, C.R & Rias, S. 1984. 'Secondary current measurements in a meandering river', in *River Meanders*, ASCE Sp. Pub., Proc. Rivers '83, ASCE Publications, 675–686.

Thorne, C.R. & Richardson, W.R.R. 1995. *Secondary Current Measurements Around a Bar in the Brahmaputra River, Bangladesh*, Draft Final Report to FAP-24, Univ. of Nottingham, 65 pp.

Thorne, C.R., Russell, A.P.G. & Alam, M.K. 1993. 'Planform pattern and channel evolution of the Brahmaputra River, Bangladesh', in Best, J.L. & Bristow, C.S. (Eds) *Braided Rivers*, Geol. Soc. London Sp. Pub., **75**, 257–276.

Thorne, C.R., Zevenbergen, L.W., Pitlick, J.C., Rais, S., Bradley, J.B. & Julien, P.Y. 1985. 'Direct measurements of secondary currents in a meandering sand-bed river', *Nature*, **316**, 746–747.

Thorne, P.D., Hardcastle, P.J., Flatt, D. & Humphery, J.D. 1994. 'On the use of acoustics for measuring shallow water suspended sediment processes', *IEEE J. Oceanic Engrg.*, **19**, 48–57.

NOTATION

U_P	primary velocity component
U_S	secondary velocity component
V_H	resultant velocity in the horizontal plane
W	width of the measurement ensemble
y	depth of measurement 'bin'
Y	average depth
α	flow orientation
φ	velocity direction

26 Morphology and Stage-dependent Flow Structure in a Gravel-bed River

NICHOLAS J. CLIFFORD
Department of Geography and Jackson Environment Institute, University College London, 26 Bedford Way, London, WC1H 0AP, UK

ABSTRACT

Data are presented from 170 time-series of high-frequency velocity measurements obtained in a braid-bar reach of a small gravel-bed river. Investigations of velocity variation and of the behaviour of coherent flow structures reveal systematic associations with individual grains, bed microtopography, channel bedforms and channel planform. These associations are also influenced by changes in flow stage and by the height at which measurements are made in the boundary layer. For a restricted range of depth-limited conditions, the results suggest that significant aspects of flow variation in the turbulent range can be indexed by time-averaged properties of the flow. Outside these conditions, however, caution is required when generalizing from point-specific measurement and when comparing results from simpler laboratory boundary layers.

INTRODUCTION

Over the last 20 years, with the advent of more reliable flow monitoring equipment, studies of small-scale flow processes in fluvial environments have become increasingly popular. Fundamental interest exists in demonstrating the behaviour and possible origin of structures in near-boundary flows which can be compared with laboratory-derived models of flow organization, but field studies are of potential value in a variety of other respects. The basic characteristics of flow velocity and stress variation are key determinants to the initiation of particle motion, while the existence, size and behaviour of flow structures reflects boundary flow resistance and controls local patterns of dispersive mixing. Despite this potential, however, field studies are still at a relatively early stage of development and lack the sophistication of their laboratory counterparts. Relatively crude comparisons between the laboratory 'burst-ejection' model and field measurements continue to dominate research, despite a need for application of small-scale process studies to more general questions of flow and sediment transport dynamics.

If the full potential of small-scale process studies is to be realized, however, two

Coherent Flow Structures in Open Channels.
Edited by P.J. Ashworth, S.J. Bennett, J.L. Best and S.J. McLelland.

outstanding and related issues must be addressed. First, the feasibility of linking point-instantaneous measurements with time- and space-averaged flow measures such as those required in flow models must be demonstrated. Second, an assessment of the nature and role of flow structures and their relationship to channel form at all scales from grain to bend must be undertaken. Essentially, both issues centre on questions of research design, as emphasis extends from the *physics* of flow processes operating at a point, to the geomorphological *context* of flow processes operating at the reach and sub-reach scales. The purpose of this paper is to present data which illustrate some basic scale-dependence in point flow measures and which may be related to contrasting form and flow dynamics as stage varies. Emphasis is focused upon possible interactions between small-scale (grain and microtopography-related) and larger-scale (bar and channel form related) structuring of the boundary layer. Although the study is preliminary in nature, patterns in the data are reasonably clear and offer some insight for improved research designs.

FIELD LOCATION, RESEARCH DESIGN AND INSTRUMENTATION

Data were obtained from a 200 m braid-bar section of the Langden Brook, NW Lancashire, a third order tributary of the River Hodder. The location of the study site and position of monitored sections is shown in Figure 26.1 together with information concerning the local channel centreline bed profile. Two series of measurements were undertaken: a principal series of measurements directed towards possible bar-scale effects on flow characteristics; and a second series (the 'beam experiments') directed towards assessment of much smaller-scale grain and microform effects on flow structure and sampling design.

For the principal series, measurements were undertaken at channel centreline positions. In order to ensure adequate incorporation of bar-scale effects, monitoring was extended to a smaller, undivided bar channel in addition to the main channel. The large channel consists of a riffle-pool-bar unit and an X-type bar pool confluence-diffluence (Ferguson, 1993), while the small channel consists of a simpler riffle-pool unit. Centreline depths for the large channel varied between 0.12 and 0.72 m, with corresponding depths for the smaller channel of 0.08–0.45 m. Maximum channel widths were 9.5 and 2.5 m respectively. Particle size characteristics for the study reaches (based on a 0.5 m grid and transect sampling following Wolman, 1954) revealed that maximum particle sizes were identical, but the D_{50} of the intermediate axis was 0.05 m in the small channel, compared to 0.07 m in the larger channel. The two channel sections thus cover contrasting width-to-depth ratio, sub-reach morphology, and relative roughness conditions. Measurements were made at four flow stages corresponding to discharge durations of 85, 75, 50 and 10%. At the highest recorded discharge, movement of finer material was observed 'over-passing' the coarser armoured bed. It is thought that

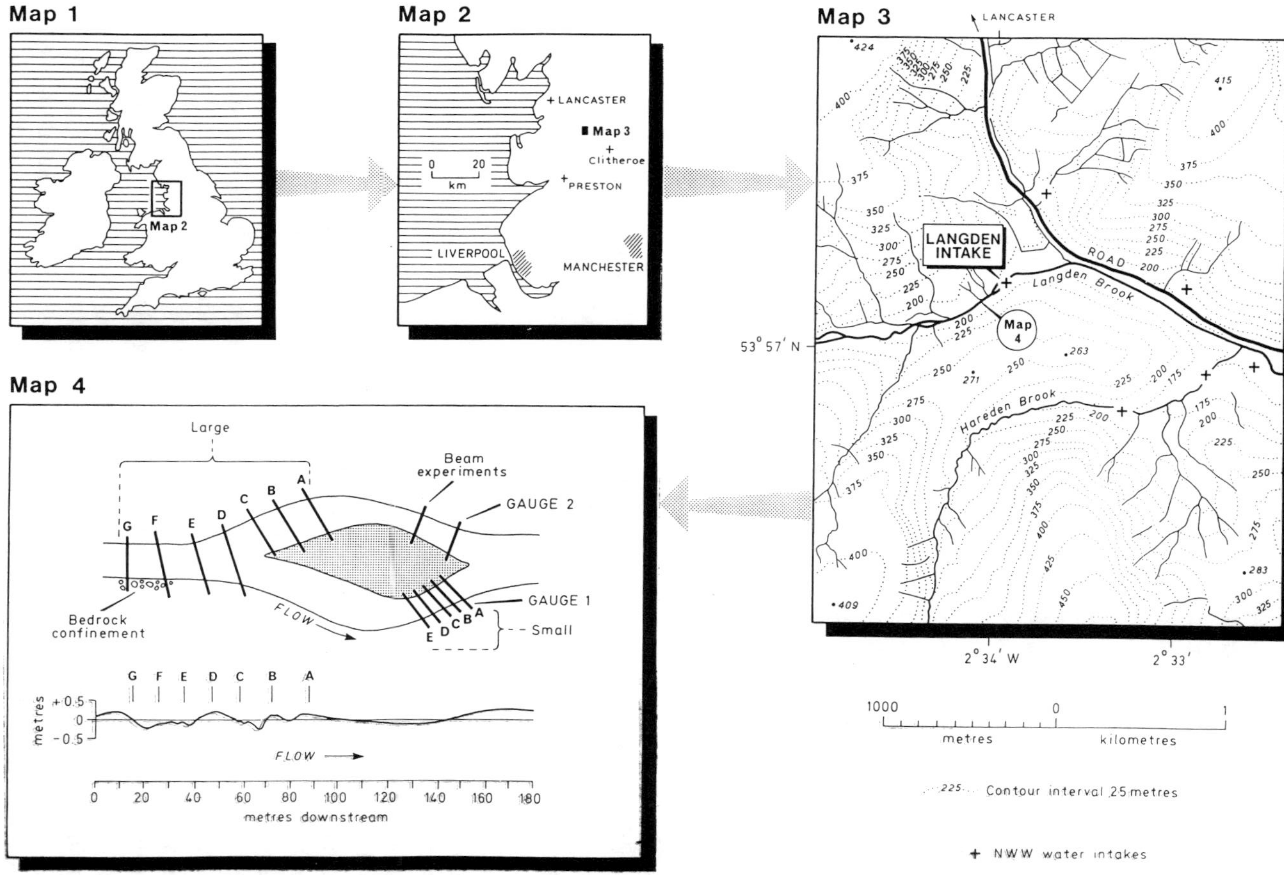

Figure 26.1. Location of study site (Maps 1–3) and details of research design (Map 4). Note that the planform shown in Map 4 is not to scale, although scale has been preserved on the accompanying long profile

this represents a stage below, but reasonably close to, the momentum extraction conditions discussed by Whiting & Dietrich (1990) as most relevant for process studies under 'active' conditions.

Instrumentation consisted of discoidal (0.055 m head) Valeport Series 800 electromagnetic current meters (ECMs) interfaced via a Campbell 21X data logger to a Toshiba laptop computer. Owing to restricted depth and high relative roughness in the majority of locations, the favoured sampling strategy used single ECM units, although some data series were obtained with ECM arrays. The ECMs were mounted on a tripod rig with a maximum leg separation of 3 m. A geared central column allowed accurate height adjustment and also enabled rotation in the vertical and horizontal planes where necessary. Units were mounted upstream from the rig legs to avoid interference, and an additional support strut could be fixed to minimize vibration during higher flow measurements. Data collection concentrated on vertical and longitudinal components of the flow, obtained at 0.2, 0.6 and 0.8 local depth. Current meters were located as accurately as possible at the same position for each measurement run, in 'open plane-bed' situations (*sensu* Brayshaw, 1985). Sampling rates and durations were selected according to criteria discussed extensively by Soulsby (1980). Comparison of spectra obtained with an ultrasonic current meter at varying frequencies and durations in the same study reach (Clifford *et al.*, 1992), and spectra derived from the ECM output for data series at similar mean velocities, confirmed that 10 Hz and 3 minute duration was acceptable in all cases with very similar spectral response and no cut-off.

For the beam experiments, an ECM was positioned at a single height (0.6 depth) and moved along a transect in conjunction with a point gauge assembly. Flow measurements were repeated every 0.25 m over 3 m transects upstream and cross-stream from a fixed origin centred on the large riffle bar cross-section A. Microprofile data accurate to within 5 mm were also taken at 0.01 m intervals in conjunction with the flow measurements, although detailed discussion of these is not presented here.

The beam experiments had two objectives. First, to assess the degree to which local flow characteristics vary over very short distances. This information is essential when assessing the ability to generalize point-specific measurements over larger spatial scales. Second, information was gathered as a preliminary attempt to explain observed variation in flow characteristics. The experiments were undertaken at the second lowest discharge in the riffle portion of the larger channel. Microtopographic bedforms could clearly be seen, comprising individual obstacle clasts and nucleated structures (Laronne & Carson, 1976; Billi, 1988) as well as series of lineated transverse rib features (McDonald & Bannerjee, 1971; Boothroyd & Ashley, 1975). Ribs were composed of individual clasts and clast groupings extending over 10–20% of the local channel width in a shallow staircase-like sequence. The low-flow water surface pattern was clearly related to the microtopography, comprising vortex streets from the obstacle clasts and clusters, jets through the gaps between clusters and in ribs, and small standing waves between series of ribs. The flow microtopography is thus a mixture of turbulent, near-turbulent and anchored flow structures typical of depth-limited environments with large roughness elements (Herbich & Schulits, 1964).

In total, the flow data set comprises 170 velocity time-series, falling into three groups: (i) ECM records from the small and large channel sections, at four discharge values; (ii) 30 ECM series relating to the 'beam experiments'; and (iii) 20 series obtained with an ultrasonic current meter used in calibration and sampling tests. The total number of series reported is approximately two thirds of that actually collected in the field: on many occasions, replicate runs were obtained to check the integrity of the data series and other series were rejected in view of possible corruption. The ECM output produces two component velocity series, longitudinal, u, and vertical, v. In this study, co-ordinates for u and v were chosen with respect to the local mean flow (parallel for u, with the cross-stream time-average = 0) and at 90° to the local channel bed. In practice, because of the use of open-bed sites, both components were close approximations to flows parallel to the channel centreline and in the vertical. No attempt was made to rotate output to give a time-averaged value of $v = 0$ (see Lapointe, 1992), since this was considered likely to obscure fundamental aspects of the spatial variation in first-order characteristics of the near-bed flow. Velocity components were subdivided into time-average and instantaneous fluctuating parts ($u = U + u'$ and $v = V + v'$) prior to subsequent analysis.

METHODOLOGY

Post-field analysis proceeded via the following stages: (i) visual inspection of time-series plots (this is important, at an early stage, to ensure data quality control); (ii) computation of series-averaged measures describing the distributional characteristics of the detrended velocity and stress series (enabling the comparison of turbulence data between different series and environments, and to relate turbulence characteristics to time-averaged properties of the flow); and (iii) determination of length scales of turbulent eddy motion, using pseudo-periodic autoregressive models where appropriate. A variety of procedures may be used to identify and parameterize coherent flow structures. Use of autoregressive modelling is a complement to more familiar spectral techniques and results derived from vortex-shedding relationships. Essentially, in analogy with regression modelling, a statistical model of the form

$$v_t = a_1 v_{t-1} + a_2 v_{t-2} + \ldots\ldots + a_n v_{t-n} + e_t \tag{26.1}$$

is fitted to the fluctuating part of each velocity series, where $a_1 - a_n$ are coefficients of the velocity, v_t, at a given lag of time, t, and e_t is a random component. Thus, the instantaneous value of velocity is a lagged regression of itself. The order of the best-fit model is an indicator of the complexity of the flow series, and is determined from inspection of the autocorrelation (ACF) and partial autocorrelation (PACF) functions. Where the ACF of a series has an initial exponential decay followed by a sinusoidal oscillation, *and* its PACF 'cuts-off' (becomes statistically insignificant) after only two lags, then it is possible to characterize the series by a second order autoregressive (AR2) process where

$$v_t = a_1 v_{t-1} + a_2 v_{t-2} + e_t \qquad (26.2)$$

Provided that $a_1^2 < -4a_2$, the AR(2) model describes a strictly pseudo-periodic process: that is, the outcome of a cyclic process whose period varies randomly thorough time (Jenkins & Watts, 1968; p. 166). For this kind of model, it is also possible to determine an 'average' frequency, f_0, since

$$\cos 2\pi f_0 = {}^{a_1}\!\Big/_{2\sqrt{-a_2}} \qquad (26.3)$$

The periodicity represented by such a model is the inverse of the frequency.

There is a long history of application of this kind of model in the environmental sciences where an output series is thought to represent a physically-determined process operating in conditions of unknown complexity and/or measurement error, which can be considered as 'random disturbances'. The methodology is extensively discussed in Clifford & French (1992a, 1992b) and Clifford (1993). Applications are also reviewed by Robert (1993) and discussed in Clifford *et al.* (1992) and Robert *et al.* (1993), with specific reference to the identification of flow structure in poorly-controlled and complex boundaries. More general discussions of the comparison between autoregressive models and spectral analysis are given in texts such as Harvey (1993).

RESULTS AND DISCUSSION

MEASUREMENTS AT VARIABLE STAGE AND RELATIVE DEPTH

The principal results are summarized in Figures 26.2–26.7. It should be noted that two kinds of graphical plots have been employed: standard bivariate graphs, and 'composite plots'. The composite plots show one or more variables (ordinate axis) plotted against a categorical axis, in which each category represents one measurement section. The categories are labelled as large channel sections, A–G and small channel, a–e as shown on Figure 26.1. Movement from left to right across these graphs is, therefore, equivalent to moving upstream through the measurement reaches. Results from the small channel are shown beyond a separating line to the right of each composite plot. In the composite plots, data have also been plotted in order of increasing height above the bed, and for some, a separate plot is also drawn for lower and higher discharges. Throughout, the 'lower discharge' figures relate to the combined data from 10 and 50% durations, and the 'higher discharge' figures to data for discharges of 75 and 85% durations. This format was chosen, therefore,

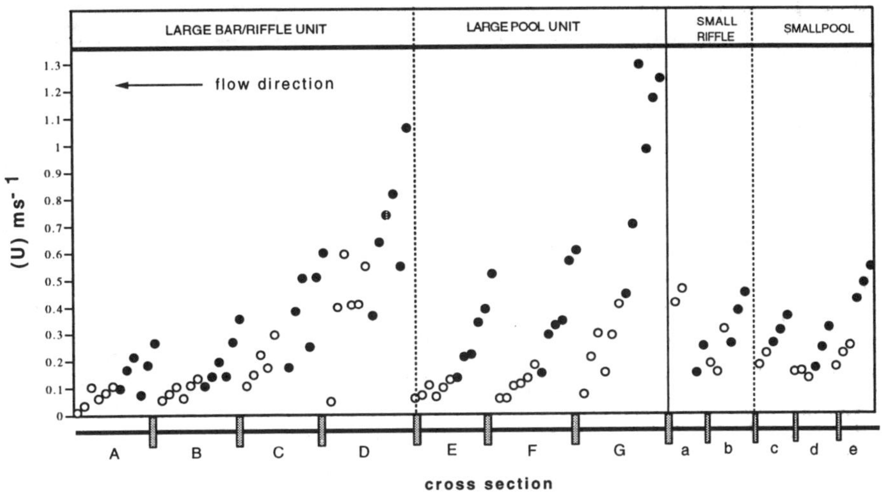

Figure 26.2. Variation in mean longitudinal velocity with changing location, discharge and relative depth. Open symbols represent data for low discharges; closed symbols represent data for high discharges

as a simple attempt to display the relative importance of grain-, form- and stage-dependency in the data. Any systematic relationship dependent upon these parameters will be apparent as 'runs' of data points of up to three data values at each cross-section.

Basic velocity and turbulence intensity

Mean longitudinal velocity (Figure 26.2) follows the expected pattern, rising with distance from the bed and with discharge in most cases, illustrative of a well-developed boundary profile. However, it is clear that the degree of variation is dependent on the scale of the channel and measurement location. Both the greatest velocities and most variation occurs at section D (a separation zone at the bar head) and G (the section influenced by a jet from the upstream bar), so that, effectively, two morpho-dynamic *sub-reaches* are present. Patterns of velocity variation about the mean (i.e. root-mean-square (RMS) turbulence intensities; Figures 26.3a and 26.3b) closely resemble those of the mean longitudinal velocity. It is again clear that the strength of any relationships is site dependent. Zones of high relative roughness (A and B) show least evidence of changes until the highest discharge. This may be indicative of a consistent and dominant *microtopography* effect over most of the low to intermediate discharge range in the riffle-like section of the main bar and the riffle portion of the small channel. The head of the large pool zone is again the site of greatest turbulence intensity, as is the pool head in the small channel, this time indicating a *bedform*-dependent control on flow behaviour. In

(a)

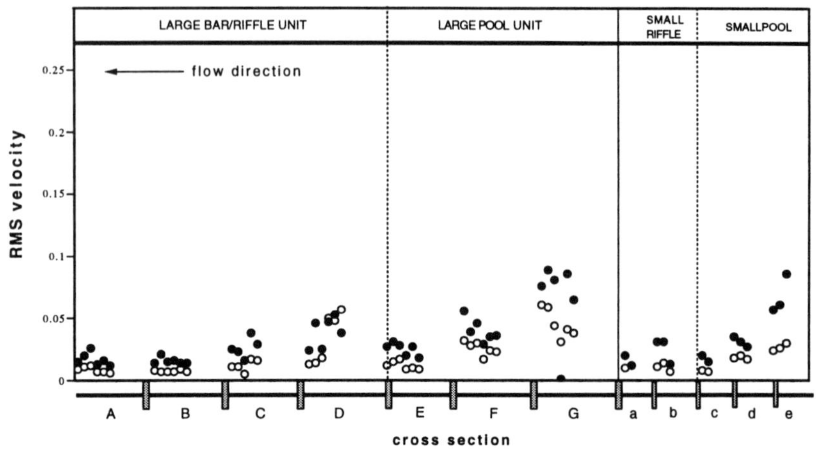

(b)

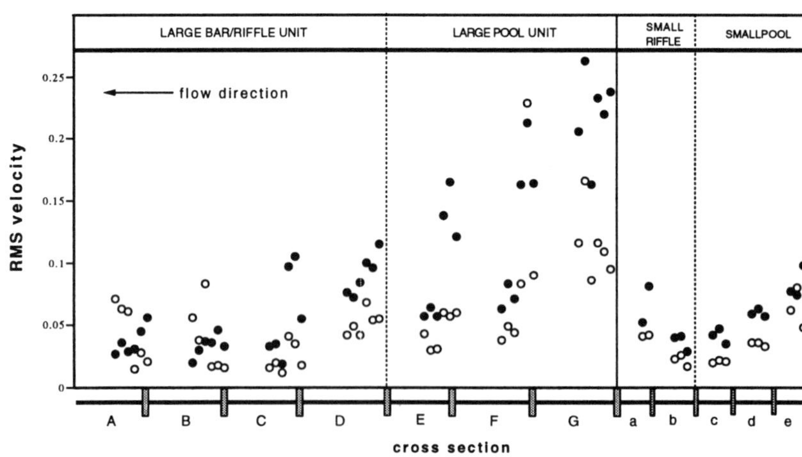

Figure 26.3. Variation in absolute turbulence intensity (RMS value, m s^{-1}) with changing location, discharge and relative depth. Open symbols refer to vertical velocity component; closed symbols refer to longitudinal velocity component. (a) Shows data obtained at low discharges; and (b) shows data obtained at high discharges

many instances, variation in the vertical velocity component follows the longitudinal pattern, but with lower absolute values. In a small number of cases, particularly over the shallowest riffle portion of the bar, vertical velocity variation is dominant.

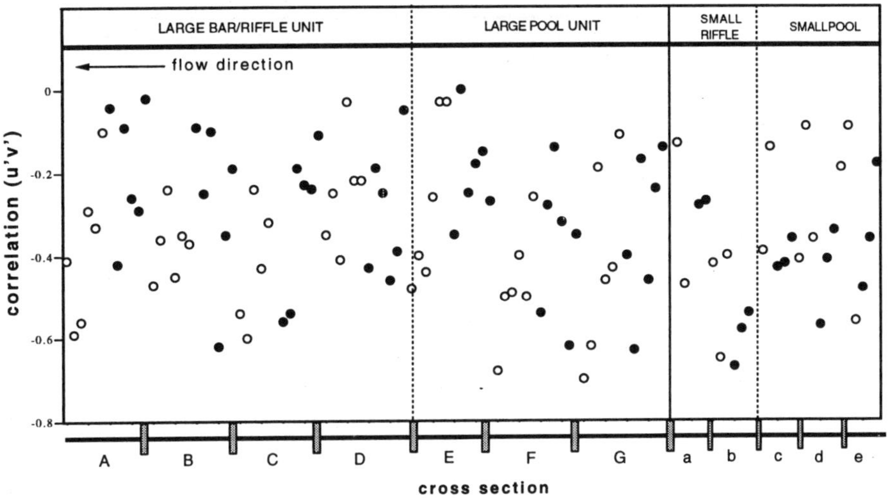

Figure 26.4. Variation in velocity correlation with changing location, discharge and relative depth. Open symbols represent data for low discharges; closed symbols represent data for high discharges

Velocity correlations

Correlation between fluctuating velocity components (Figure 26.4) are generally closer to 0 than the -0.4 reported in laboratory flows, with the highest (more negative) values frequently occurring closest to the bed. The pattern of decreasing correlation towards the water surface is consistent with other field measurements in gravel-bed rivers (Roy, pers. comm.), indicative of a tendency towards isotropy. Greater correlation near to the bed is also consistent with expectations of the generation of flow structures in this region as a result of shear and possible vortex-shedding phenomena. The greater strength of the near-bed relationship might prove useful in future modelling and is important with respect to characteristics of near-bed stress. There is some suggestion of a higher (if more variable) correlation in the large pool zone from F–G, but higher correlations also occur in the riffle-like zones of both the large and small channels closer to the bed. As with intensity measurements, therefore, the correlation results reflect at least two different scales of flow structuring: a bed/grain-dependent effect and a bar profile/planform effect. Interestingly, the existence of the systematic association of longitudinal and vertical turbulence intensity with behaviour of the mean velocity (illustrated by comparing Figures 26.2 and 26.3) is not clearly seen in Figure 26.4. This may be because so many individual observations are being correlated, thus reducing overall values. A more physically meaningful measure would be to correlate the 'event structure' in the separate velocity series. On the basis of the data presented here, there are no clear associations with stage or discharge. Clearly, given the -0.4 correlation

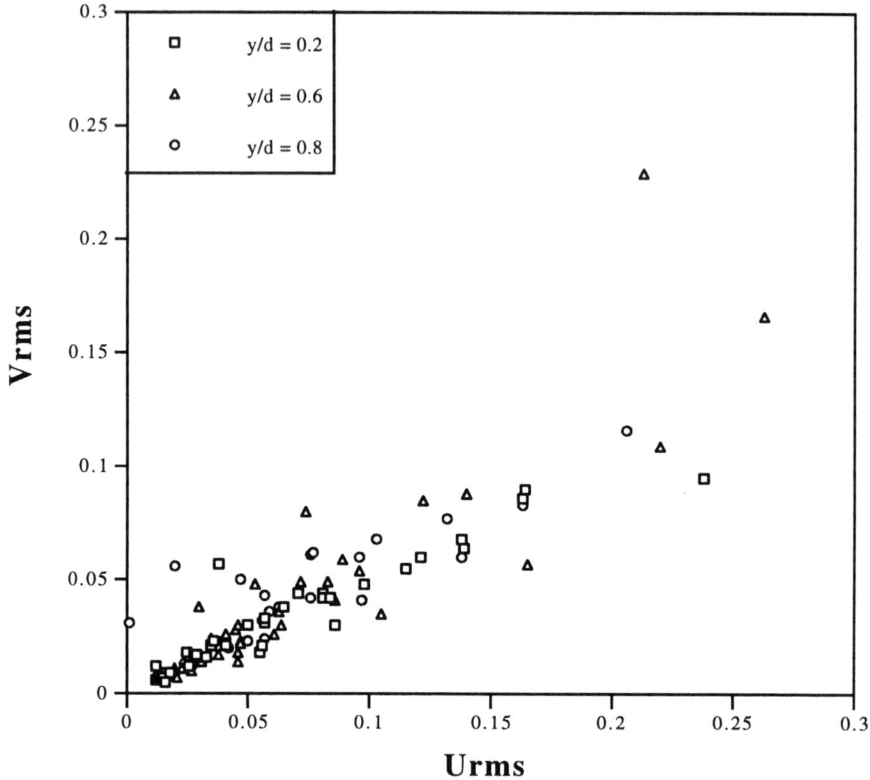

Figure 26.5. Comparison between turbulence intensity in longitudinal (U_{rms}) and vertical (V_{rms}) velocity (m s^{-1}) components for varying relative depth

reported in a wide variety of other natural and laboratory boundaries where elements of turbulent structure have also been identified, a simple correlation coefficient may not be the most indicative measure to interpret all significant aspects of any co-relationships. Figure 26.5, for example, presents a direct comparison of vertical and longitudinal turbulence levels. A very strong relationship between longitudinal (U_{rms}) and (V_{rms}) vertical intensities (RMS levels) can be seen. Regression results for observations at individual relative depths give slightly different linear regression coefficients, but there is no statistically significant difference between them. The relationship for the combined data is:

$$V_{rms} = 0.574 U_{rms} \qquad (26.4)$$

This result may be compared with Naden's (1981) analysis of McQuivey's (1973) data which yielded a coefficient of 0.77 for deeper sand-bed rivers, and with studies from a riffle-pool reach in a single thread gravel stream (Clifford, 1990) where an *average* coefficient of 0.38 was recorded. Significantly, in the latter case, higher values were recorded in riffle sections as discharge fell, being 0.36 at a stage of 0.44 m (two thirds bankfull), and 0.75 for a stage of 0.08 m (approximately baseflow). It is apparent again, therefore, that measurements are influenced by sub-reach morphology and flow dynamics which are themselves stage-dependent. Given the marked contrast in correlations between near-boundary and near-surface measurements in natural environments, interpretations of overall indicators of velocity correlation such as the -0.4 value must be similarly qualified. Clifford (in review) demonstrates that for deeper rivers of lower relative roughness, the effects of local mean velocity and local shear must be considered jointly. To some extent, these also cancel-out. Intensities do increase between profiles as mean velocity increases, but at different rates depending upon the relative depth of measurement. Within data sets from common profiles, intensities fall with increasing local mean velocity as shear is reduced away from the boundary. A simple classification of 'gravel-bed' or 'sand-bed' is inappropriate. Generalizing from the data from Langden Brook, in *some* environments of low relative depth and reasonably well-differentiated sub-reach flow processes, it may be possible to formulate useful regression relationships such as:

$$U_{rms} = 0.182U \tag{26.5}$$

$$V_{rms} = 0.104V \tag{26.6}$$

This mean velocity dependence might have practical significance where the capacity to make field measurements at more than one height in the profile is lacking, or where use is made of published sources which quote mean velocities obtained at differing relative depth, but where some indication of relative turbulence levels is also required. However, in addition to the problem of transferring results between contrasting reaches and rivers, the potential for modelling near bed turbulent stress, which is necessary in order to test relationships for sediment transport (e.g. Bridge & Bennett, 1992) is likely to be more difficult. Even where relationships such as Equations 26.5 and 26.6 do obtain, there will be problems in reconstructing stress behaviour because of its sensitive dependence on *two* distributions (Lu & Willmarth, 1973; Clifford *et al.*, 1991) which are frequently non-normal (Figure 26.8).

Flow structures

Figure 26.6 summarizes the combined percentage of the variance in the original velocity series (U and V) explained by autoregressive models of first and second order. Overall, the models fitted to V are a poorer fit (but still generally account for $>60\%$ of the variance) than for U (generally $>75\%$). In the large pool zone, models fitted to low discharge series show more consistency and greater success than for corresponding high discharges. As might be expected, where the overall

(a)

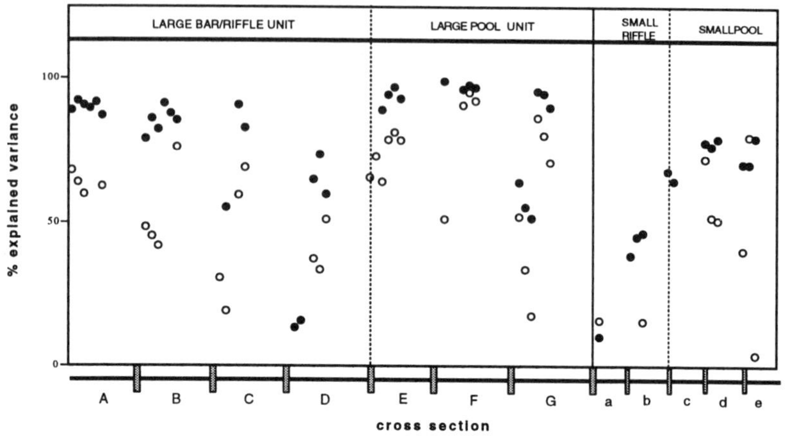

(b)

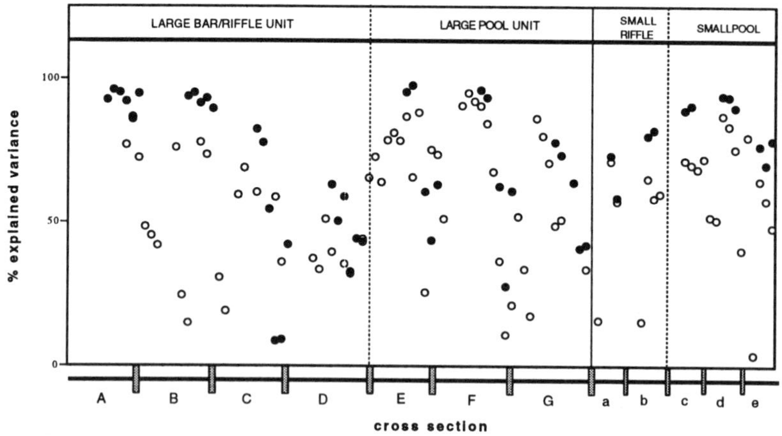

Figure 26.6. Combined goodness-of-fit of first and second-order autoregressive models to longitudinal (closed symbol) and vertical (open symbol) velocity time series at varying location, discharge and relative depth. (a) Shows data obtained at low discharges; and (b) shows data obtained at high discharges

model fit is poorest, the additional variance explained by the second order model is greater (Figure 26.7): for series with 'more complex' flows, any additional parameter is likely to add a greater proportion of explained variance. In total, 64% of the series are most appropriately fitted by a second order model, with 23% fitted by first order models, the remainder being more complex. There is some suggestion that the improvement in fit is greater with increasing discharge and at zones of

(a)

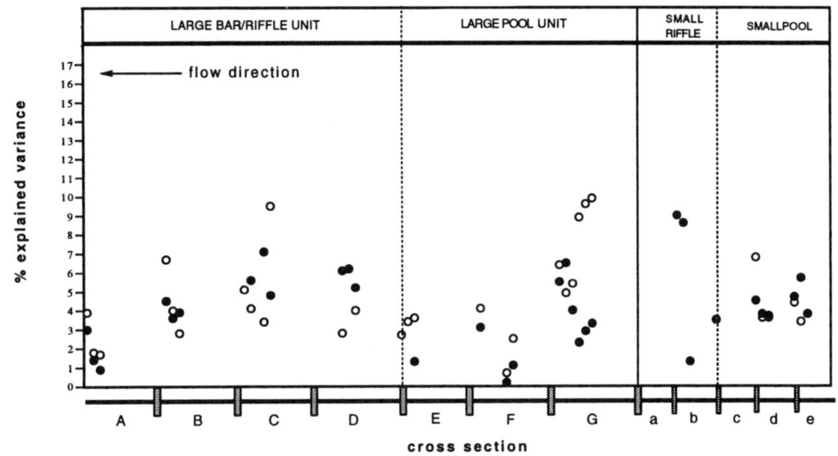

(b)

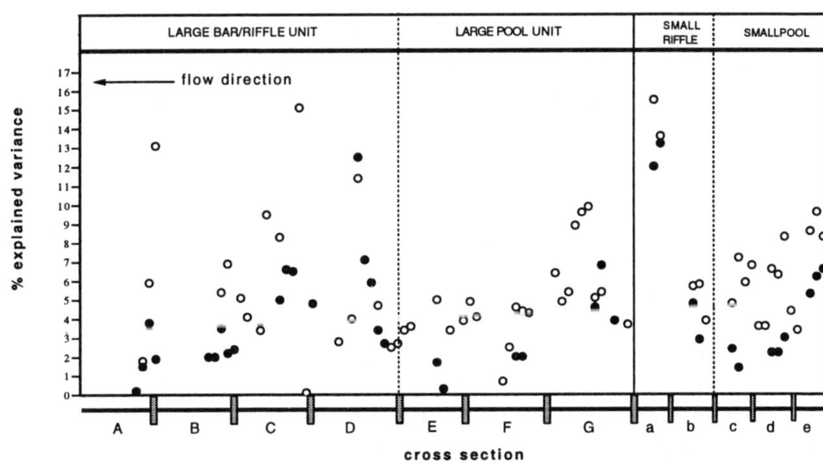

Figure 26.7. Additional percentage of variance explained by second order autoregressive models fitted to longitudinal (closed symbol) and vertical (open symbol) velocity time series at varying location, discharge and relative depth. (a) Shows data obtained at low discharges and (b) shows data obtained at high discharges

morphological transition where total fits are poorest. Also, away from the highest relative roughness sections, improvement of fit shows some tendency to increase with distance from the bed, particularly in *V*.

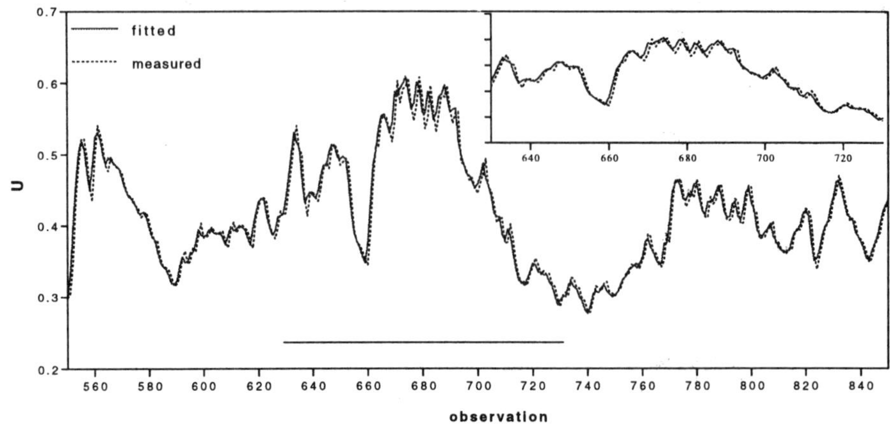

Figure 26.8. An example of a longitudinal velocity series (measurement units: m s^{-1}) obtained from Langden Brook and the results of fitting a second order autoregressive model to this. The main plot represents 320 observations taken over 32 s. Inset shows detail for a 12 s period

Just over 40% of the *U* series are fitted by pseudo periodic AR(2) models (i.e. two thirds of those described best by AR(2) models), which is slightly greater than the percentage reported in Clifford *et al.* (1992) but less than the 100% reported in Robert *et al.* (1993). Figure 26.8 is an illustration of the performance of a second order model in 'tracking' a measured longitudinal time series. In this case, just over 93% of the original variance can be reproduced by the model. Importantly, the fitted values very closely reproduce both the larger-scale, longer-duration and smaller-scale, shorter-term fluctuations of the measured series. As expected, the fitted values produce a slightly smoothed version of the original data, and this may also occasionally be out-of-phase with the measured series. Thus, while the figure indicates that autoregressive models of this kind are very useful in the analysis and characterization of individual velocity components, similar success when applied to the modelling of Reynolds stress (see Robert, 1993) is more doubtful because of its double velocity dependence. Clearly, in any attempt at model fitting, the sub-reach location at which monitoring is undertaken is important, as is relative depth and discharge. These effects should be acknowledged and explored further in future investigations (see discussion of beam experiments, below). A simple relationship between periodicity and mean velocity is not observed (Figure 26.9a), possibly because of stage and morphological effects, but if 'outlier' observations are ignored, both correlation between the velocity components and variations in period seem to be most consistent closest to the bed (Figure 26.9b). Again this may be encouraging for future attempts to model near-bed flow and sediment transport processes.

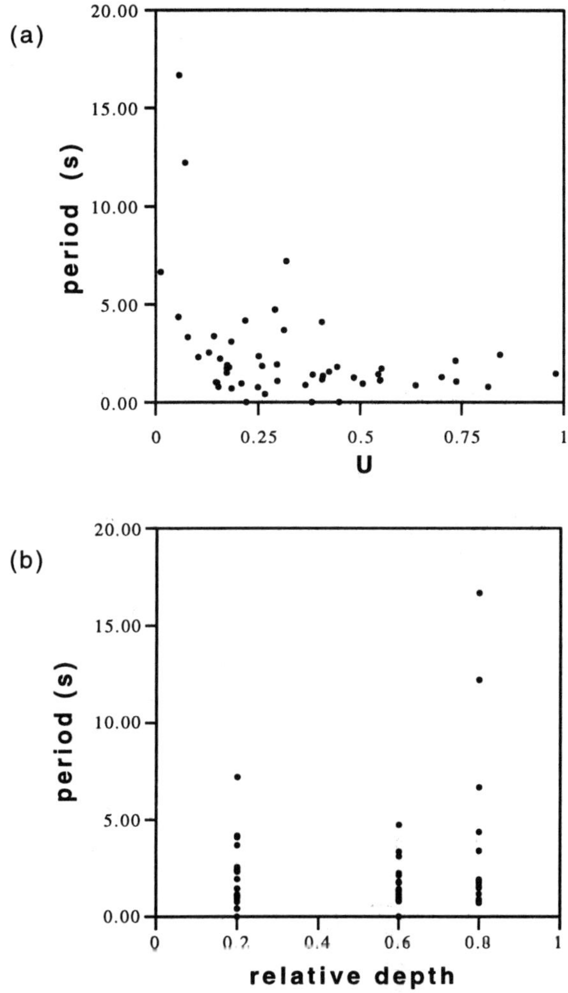

Figure 26.9. Comparison of periodicities obtained from pseudo-periodic second order autoregressive models of velocity time series with: (a) contrasting local mean (*U*) velocity (measurement units: m s^{-1}); and (b) contrasting relative depth

MEASUREMENTS AT CONSTANT STAGE AND RELATIVE DEPTH

The results of the beam experiments are summarized in Figures 26.10–26.12. Two aspects of the figures are of interest: variation over a short distance given fixed measurement height and discharge (potentially important for the practicalities of measurement), and the existence of 'runs' of values which indicate the efficacy of this kind of instrumentation and research design in resolving the presence of flow structure at reasonable levels of detail.

Velocity and turbulence intensity

With respect to simple velocity behaviour (Figure 26.10a), the general view is one of consistency, with individual outliers which may be related to jets-and-wakes at a microtopographic scale. Interestingly, however, outliers with respect to one measure are not necessarily reflected in another measure. This is especially striking when comparing mean velocity and turbulent intensities (Figures 26.10a and 26.10b). Runs of data points are also present within and between series of measurements. For example, between 1.75 and 3.5 m upstream, there is an increase and then a decline in mean velocity and in the correlation between u' and v' (Figures 26.10b and 26.10c). This corresponds to an upstream traverse through a vortex street, suggesting that eddy streets exhibit formation/maxima and damping characteristics which might be parameterized in future studies of this kind. Again, from 2.25–4.25 m in the cross-stream, correlation varies as several vortex streets are crossed and as jets between and standing waves within in the lee of transverse ribs are encountered. Results between 0.75–1.5 m upstream are highly variable, with some marked and unusual positive correlations. At the smallest scale, therefore, conventional boundary layer theory is not always appropriate, then, and relative variation in all measures over very short distances is disturbingly high. For example, there is a range in velocity from approximately 0–0.7 m s^{-1} and a correlation from +0.5 to -0.5 in only 0.5 m. Laboratory results over large roughness elements show that relative submergence is a major factor in the distribution of longitudinal turbulent intensity. Results are also dependent upon relative depth at a constant discharge (Wang *et al.*, 1993). The expectation of local variation must, therefore, be incorporated in future experimental designs. On a more positive note, it is encouraging to note that this kind of instrumentation can resolve very small-scale differences. By choosing sufficiently numerous measurements in 'open plane-bed' conditions, extreme local variations can be avoided in most cases, as the main experiments show. At the very small-scale, further research, possibly employing smaller measuring instruments, might also yield better relations to couple the detail of flow and bed microtopography. It may also be possible to suggest more appropriate small-scale co-ordinate schemes, as a prelude to identifying more details of turbulent 'event structure'. Measurements of this kind over a range of flow stages are also vital, since the microtopography itself may change as flow increases (Billi, 1988). The relative importance of anchored and turbulent structures with variations in stage also remains to be assessed.

Flow structure

Measures of flow dimensions derived from the autoregressive models (Figure 26.11) are similarly marked by very local-scale variation. Values of the percentage of explained variance may change from as low as 27% and as high as 85% within distances of only 0.25 m, which is consistent with well-marked microtopographic effects. Despite the variations, all but four of the 22 series are best fitted by AR1 or

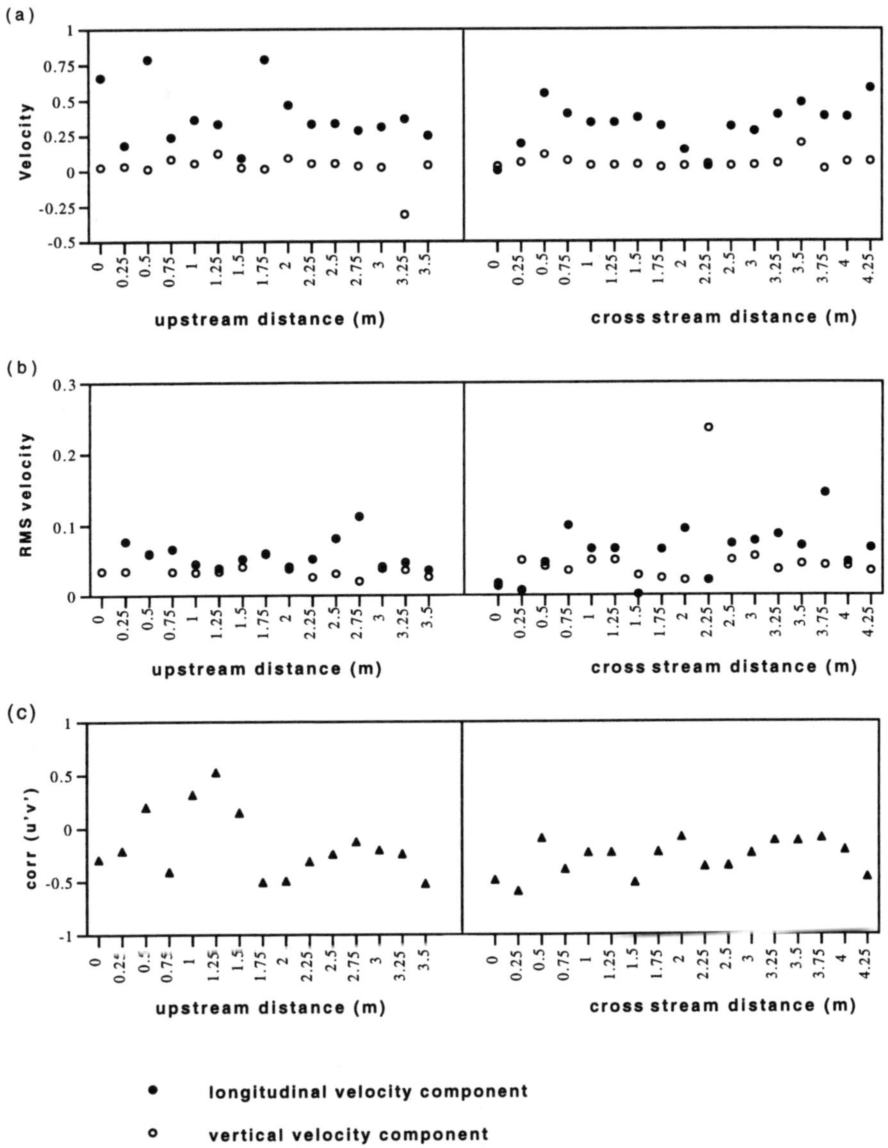

Figure 26.10. (a) Variation in local mean velocity (m s^{-1}); (b) turbulence intensity (m s^{-1}); and (c) velocity correlation, over short distances at constant relative depth and discharge

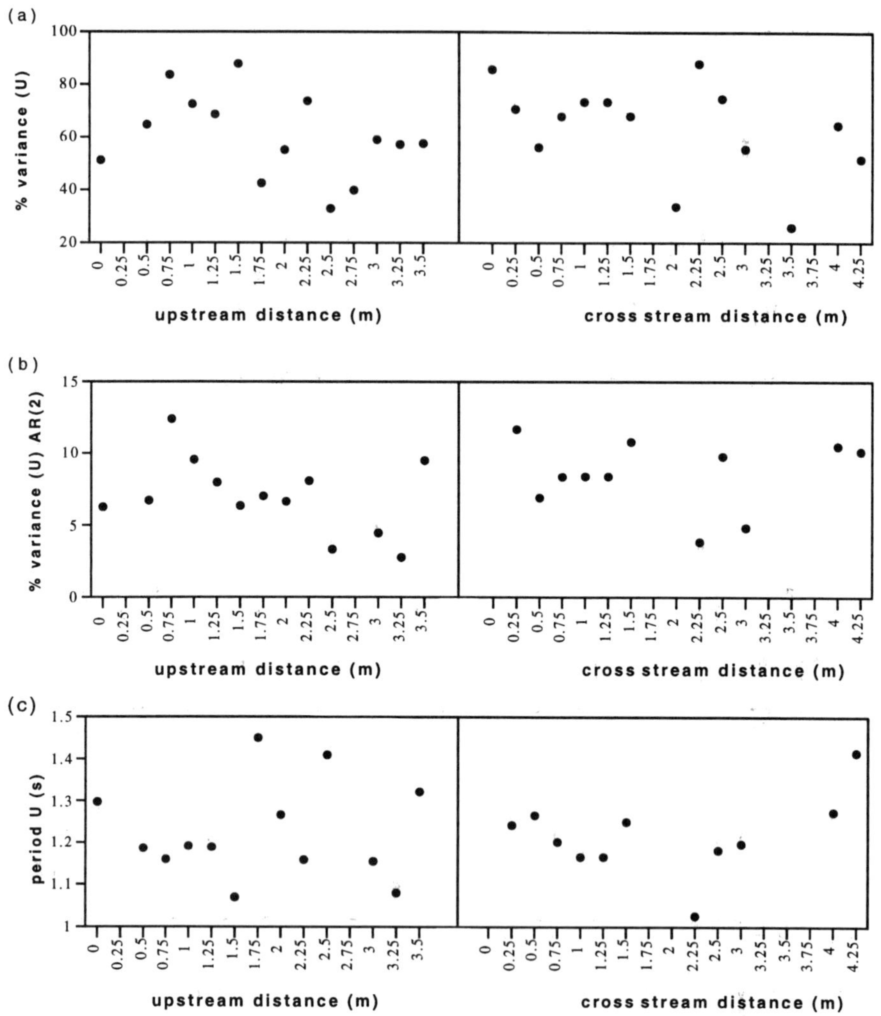

Figure 26.11. (a) Contrasting goodness-of-fit; (b) additional percentage of variance explained by a second order autoregressive model; and (c) periods obtained from a pseudo-periodic second order autoregressive model, for velocity time-series obtained over short distances at constant relative depth and discharge

AR2 models, and 13 of the 22 series are pseudo-periodic, which is a slightly higher percentage than seen in the main data sets. Significantly, there is a strong association between runs of data values in the mean velocity and in the periodicities of the AR2 models. This relationship is examined more closely in Figure 26.12, and appears to hold irrespective of whether the AR(2) model is fitted to the *U* or *V* series. Both the nature and strength of this relationship are interesting and might be

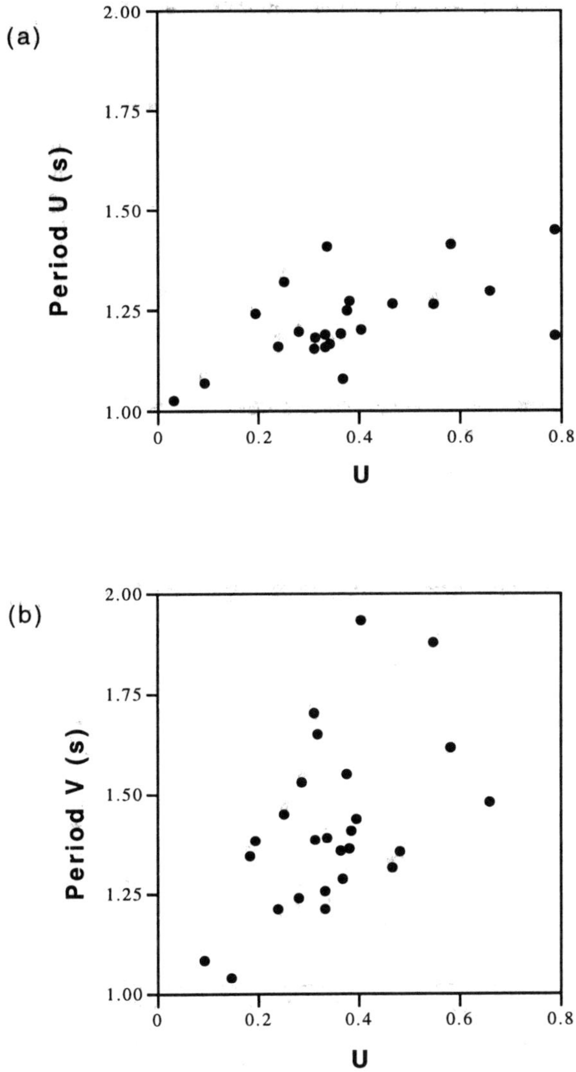

Figure 26.12. Comparison of periods obtained from a pseudo-periodic second order autoregressive model fit to velocity series obtained over short distances at constant relative depth and discharge. (a) Longitudinal and (b) vertical velocity components. Velocities measured in m s^{-1}

compared with the more complex associations in the principal data set. For both velocity components in the beam data series, period increases with mean longitudinal velocity. Although the marked contrast in the goodness-of-fit again suggests that it is worth controlling more carefully for location and discharge in future measurements, this positive relationship is of note, since it may lead to some important avenues of further research with respect to the *cause* of coherence in the flow. In particular, were the Strouhal vortex-shedding relationship to apply, then a negative, rather than a positive relationship between period and velocity would be expected: as velocity increases, eddy shedding becomes more intense, and periods decrease. Therefore, these results appear consistent with Best's (1993) comments regarding eddy detachment/coalescence and cast doubt on the general utility of the approach, despite some successful applications in similar environments (Clifford *et al.*, 1992; Robert *et al.*, 1993). As discharge rises, or in higher velocity, more 'open' waters at constant discharge, point observations are influenced by a wide range of eddy sizes and eddy coalescence, together with jet-and-wake effects (that is, *anchored* flow structures), so that the 'average' eddy frequency is reduced.

CONCLUSIONS

Data from Langden Brook illustrate points of some practical and fundamental significance with respect to studies of flow variation and flow structure in natural coarse-grained channels, and allow comparison between these and other environments. Most generally, the results demonstrate the viability of this kind of research design and instrumentation when used to resolve significant aspects of velocity variation at the small scale. These in turn, highlight spatial, boundary-profile, flow-stage and form-related variations in turbulence characteristics. Areas closest to the boundary and/or in zones of high relative roughness may behave in different ways to deeper zones which frequently reflect upstream jetting from riffle or bar confluences. For a limited range of flows and morphologies (i.e. measurements closest to the bed over riffle-like portions of the channel) the marked association between turbulence characteristics and mean velocity is especially encouraging for attempts to index turbulence characteristics from mean flow measurements. Measurements made over short distances at high spatial resolution may be valuable in disclosing the range and kind of turbulence phenomenon and their causes which are frequently 'lost' in other research designs. These variations suggest the need to direct careful attention to research design in future studies, and for caution in applying laboratory-derived models of flow structure to very different and spatially-variable natural boundary layers. More extensive and elaborate efforts to monitor small-scale, high frequency flow variation are likely, therefore, to pay considerable dividends.

ACKNOWLEDGEMENTS

This research was funded by NERC Grant GR9/918 held by NJC at the University of Hull. Mr Rob Brown and Mr Phil Whelan provided invaluable help with field data collection and analysis. In addition, Mr Peter Guy (Recreation and Conservation Officer) and Mike Conway and Ken Longshore (White Bull Treatment Works) of North West Water arranged for site access and for the provision of flow and rainfall records. Their co-operation is gratefully acknowledged.

REFERENCES

Best, J.L. 1993. 'On the interactions between turbulent flow structure, sediment transport and bedform development: some considerations from recent experimental research', in Clifford, N.J., French, J.R. & Hardisty, J. (Eds) *Turbulence: perspectives on flow and sediment transport,* John Wiley & Sons Ltd., Chichester, 61–92

Billi, P. 1988. 'A note on cluster bedform behaviour in a gravel-bed river', *Catena,* **15**, 473–481.

Boothroyd, J.C. & Ashley, G.M. 1975. 'Process, bar morphology and sedimentary structures on braided outwash fans, northwestern Gulf of Alaska', in Jopling, V.A. & McDonald, B. (Eds) *Glaciofluvial and Glaciolacustrine, Sedimentation,* SEPM Sp. Pub. **23**, 193–222.

Brayshaw, A.C. 1985. 'Bed microtopography and entrainment thresholds in gravel-bed rivers', *Geol. Soc. Am. Bull.*, **96**, 218–223.

Bridge, J.S. & Bennett, S.J. 1992. 'A model for the entrainment and transport of sediment grains of mixed sizes, shapes, and densities', *Water Resour. Res.*, **28**, 337–363.

Clifford, N.J. in review. 'A comparison of turbulent intensities from alluvial rivers: characteristics and implications for modelling flow processes', *Earth Surf. Processes Landf.*

Clifford, N.J. & French, J.R. 1993a. 'Monitoring and modelling turbulent flows: historical and contemporary perspectives', in Clifford, N.J., French, J.R. & Hardisty, J. (Eds) *Turbulence: Perspectives on Flow and Sediment Transport,* John Wiley & Sons Ltd., Chichester, 1–34.

Clifford, N.J. & French, J.R. 1993b. 'Monitoring and analysis of turbulence in geophysical boundaries: some analytical and conceptual issues', in Clifford, N.J., French, J.R. & Hardisty, J. (Eds) *Turbulence: Perspectives on Flow and Sediment Transport,* John Wiley & Sons Ltd., Chichester, 93–120.

Clifford, N.J., Hardisty, J., French, J.R. & Hart, S. 1993. 'Downstream variation in bed material characteristics: a turbulence-controlled form-process feedback mechanism', in Best, J.L. & Bristow, C.S. (Eds) *Braided Rivers*, Geol. Soc. Lond. Sp. Pub., **75**, 89–104.

Clifford, N.J., McClatchey, J. & French, J.R. 1991. 'Measurements of turbulence in the benthic boundary-layer over a gravel bed: Discussion', *Sedimentology*, **37**, 161–171.

Clifford, N.J., Robert, A. & Richards, K.S. 1992. 'Estimation of flow resistance in gravel-bedded rivers: a physical explanation of the multiplier of roughness length', *Earth Surf. Processes Landf.,* **17**, 111–126.

Ferguson, R.I. 1993. 'Understanding braiding processes in gravel-bed rivers: progress and unsolved problems', in Best, J.L. & Bristow, C.S. (Eds) *Braided Rivers*, Geol. Soc. Lond. Sp. Pub., **75**, 73–88.

Harvey, A.C. 1993. *Time Series Models*, 2nd edn. Harvester Wheatsheaf, 308 pp.

Herbich, J.B. & Schulits, S. 1964. 'Large-scale roughness in open channel flow', *J. Hydraul. Div., ASCE,* **90**, 202–230.

Jenkins, G.M. & Watts, D.G. 1968. *Spectral Analysis and its Applications*, San Francisco, Holden-Day, 525 pp.

Lapointe, M. 1992. 'Burst-like sediment suspension events in a sand bed river', *Earth Surf. Processes Landf.*, **17**, 253–270.

Laronne, J.B. & Carson, M.A. 1976. 'Interrelationships between bed morphology and bed material transport for a small gravel bed channel', *Sedimentology*, **23**, 67–85.

Lu, S.S. & Willmarth, W.W. 1973. 'Measurements of the structure of the Reynolds stress in a turbulent boundary layer', *J. Fluid Mech.*, **60**, 481–511.

McDonald, B.C. & Bannerjee, I. 1971. 'Sediments and bedforms on a braided outwash plain', *Can. J. Earth Sci.*, **8**, 1282–1301.

McQuivey, R.S. 1973b. 'Summary of turbulence data from rivers, conveyence channels and laboratory flumes', *US Geol. Surv. Prof. Paper*, **802B**, 66 pp.

Naden, P.S. 1981. 'Discussion of thresholds for initiation of sediment movement', *Univ. Leeds Working Paper*, **308**, 19 pp.

Robert, A. 1993. 'Bed configuration and microscale processes in alluvial channels', *Prog. Phys. Geog.*, **17**, 123–136.

Robert, A., Roy, A.G. & De Serres, B. 1993. 'Space-time correlations of velocity measurements at a roughness transition in a gravel-bed river', in Clifford, N.J., French, J.R. & Hardisty, J. (Eds) *Turbulence: Perspectives on Flow and Sediment Transport,* John Wiley & Sons Ltd., Chichester, 165–183.

Soulsby, R.L. 1980. 'Selecting record length and digitisation rate for near-bed measurements', *J. Phys. Oceanography*, **10**, 208–219.

Wang, J., Dong, Z., Chen, C. & Xia, Z. 1993. 'The effects of bed roughness on the distribution of turbulence intensities in open channel flow', *J. Hydraul. Res.*, **31**, 89–98.

Whiting, P.J. & Dietrich, W.E. 1990. 'Boundary shear stress and roughness over mobile alluvial beds', *J. Hydraul. Engrg.*, **116**, 1495–1511.

Wolman, M.G. 1954. 'A method of sampling coarse river-bed material', *Trans. AGU*, **35**, 951–956.

NOTATION

U	mean longitudinal velocity
U_{rms}	root-mean-square of mean longitudinal velocity
V	mean vertical velocity
V_{rms}	root-mean-square of mean vertical velocity
a_n	velocity coefficient
e_t	random component of velocity series
f_0	frequency of velocity fluctuations
t	lag time
u	instantaneous longitudinal velocity
u'	fluctuating part of longitudinal velocity
v	instantaneous vertical velocity
v'	fluctuating part of vertical velocity
v_t	fluctuating part of velocity at time t

27 A Computer Model of Bank Erosion Based on Secondary Flow Simulation

ANDREI ALABYAN
Geographical Faculty, Moscow State University, GSP-3, Moscow, 119899, Russia

ABSTRACT

A numerical model is described that predicts meander bend development based on a sediment budget with a transverse sediment flux component. The one-dimensional model relies on computing interlinked blocks of channel geometry, secondary flow and sediment load. Realistic simulations of a meandering section of the Seyakha River (Northern Siberia) are presented.

INTRODUCTION

The process of bank erosion can be considered as part of the broader problem of meander development and migration and is still one of the most interesting and mysterious puzzles in the Earth Sciences. Prediction of bend development is of great practical importance because it affects many human activities using both the meandering channel and adjacent floodplain. There are three main approaches to modelling bank erosion and bend development:

1. Qualitative and quantitative description of the mechanism of bank gravitational failure and fluvial erosion (e.g. Thorne, 1982; Osman & Thorne, 1988; Darby & Thorne, 1994). Currently, the computer realization of this approach is applicable mainly to straight sand-bed streams with cohesive bank materials (Darby, pers. comm., 1994).
2. Kinematic modelling of meander migration based on correlation between bank retreat rates and parameters of channel plan geometry (e.g. Makkaveyev, 1955; Kondrat'yev, 1968; Brice, 1974; Hickin & Nanson, 1975; Hooke, 1977; Hooke & Harvey, 1983). A computer model based on a relation between the rate of erosion and plan curvature of the channel was proposed by Ferguson (1984) but although interesting from the theoretical viewpoint cannot be applied to short-term analyses of the morphometric

Coherent Flow Structures in Open Channels.
Edited by P.J. Ashworth, S.J. Bennett, J.L. Best and S.J. McLelland.

and hydraulic characteristics in a natural channel.

3. Dynamic modelling of flow in a curved channel and its interaction with erodible banks (e.g. Begin, 1981; Ikeda *et al.*, 1981; Falcon & Kennedy, 1983; Zamyshlyaev, 1983; Hasegawa, 1989; Johannesson & Parker, 1989; Odgaard, 1989; Howard, 1992). Although some of the empirical equations are only two-dimensional, after linearization and further simplification, most of them are suitable for computer simulation. By focusing on different aspects of the flow dynamics these models have demonstrated that it is possible to produce a realistic sequence of meander deformation. However, from a practical viewpoint, some of these models have problems because a number of empirical coefficients are required as input and despite a complex computing procedure, the simulation is limited by the accuracy of model parameters.

The purpose of this paper is therefore to overcome some of these deficiencies and present a simple, physically-based model with minimum input data for quick short-term calculations of bank erosion within long reaches of meandering rivers.

FLOW PROCESSES IN RIVER BENDS

FLOW-BOUNDARY INTERACTION

The morphological development of a meandering channel pattern is dependent on the interaction between the flow and channel boundary. All within-channel processes including bank retreat are affected by this relationship (Figure 27.1). Channel bank properties such as the cohesion, structure, grain-size and vegetation cover govern both the rate of bank collapse and therefore channel cross-sectional and planform change. The behaviour of the bank-flow interaction can be considered within the framework of 'basal endpoint control', proposed by Carson & Kirkby (1972) which was later developed for eroding river banks by Thorne (1982). According to this concept the amount of basal accumulation depends on the rate of supply of material from bank failure or cut-bank erosion and the rate of removal of basal material by fluvial entrainment and transport downstream, away from the bank toe. If the flow can remove all incoming material and is able to scour the basal area, further bank erosion will be promoted. Alternatively, if the flow is unable to remove all the debris, near-bank accumulation occurs, and the deposited material protects the bank from further erosion. In natural rivers the cross-section shape and near-bank flow depth can be considered as an indirect index of bank erodibility, permitting the omission of bank cohesion effects. For the same reason, the process of cut-bank erosion by the streamwise flow can be disregarded, although the role of primary flow is still important for determining sediment load capacity. The model proposed below places the main emphasis on the role of secondary flow which is responsible for sediment removal from the bank toe.

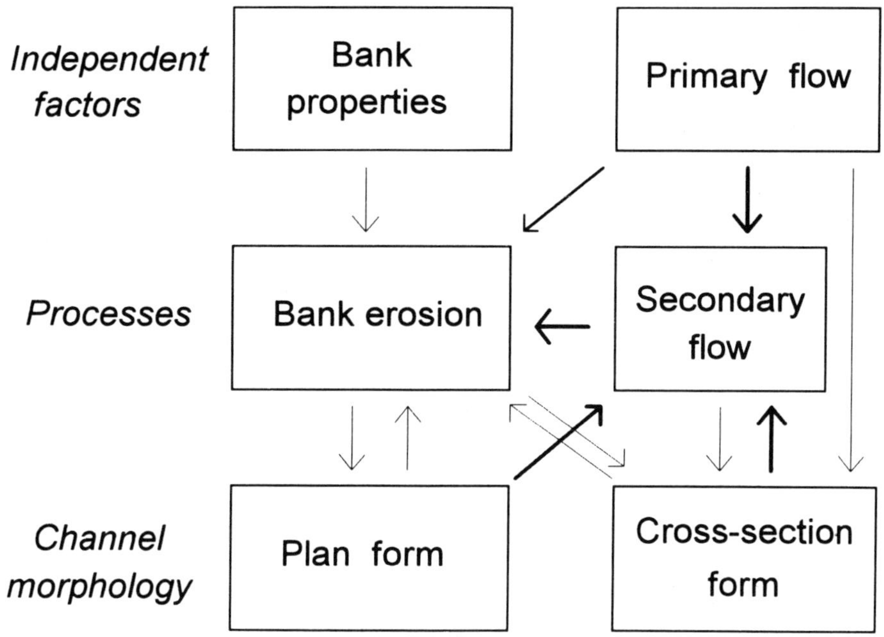

Figure 27.1. Schematic diagram of the flow-bank interaction. Bold arrows indicate relationships considered in the numerical model described in this paper

SECONDARY FLOW

More than a century ago Thomson (1876) suggested that the flow structure in a curved section is represented by the combination of primary downstream flow and secondary cross-stream currents which in-turn drive helical circulation (Figure 27.2a). A reason for this phenomenon is the centripetal force driving fast surface-flow jets towards the concave bank and inducing a transverse gradient of the water surface. As a result of the extra hydrostatic pressure, near-bed jets deflect towards the convex bank.

This scheme was confirmed in principle by Milovich (1914) who made detailed flume tests using ink for flow visualization (Figure 27.2b), and subsequently by numerous direct measurements in small and large rivers (Figure 27.2c and 27.2 d). Invention of the electromagnetic current meter allowed discoveries of other details of the secondary flow pattern but basically confirmed the existence of the main helical cell (e.g. Hooke & Harvey, 1983; Thorne *et al.*, 1985; see also Richardson *et al.*, this volume).

According to both theoretical investigations and direct measurements a velocity maximum of the helical circulation is located in the near-bed zone towards the convex bank. Its magnitude is directly proportional to mean downstream flow

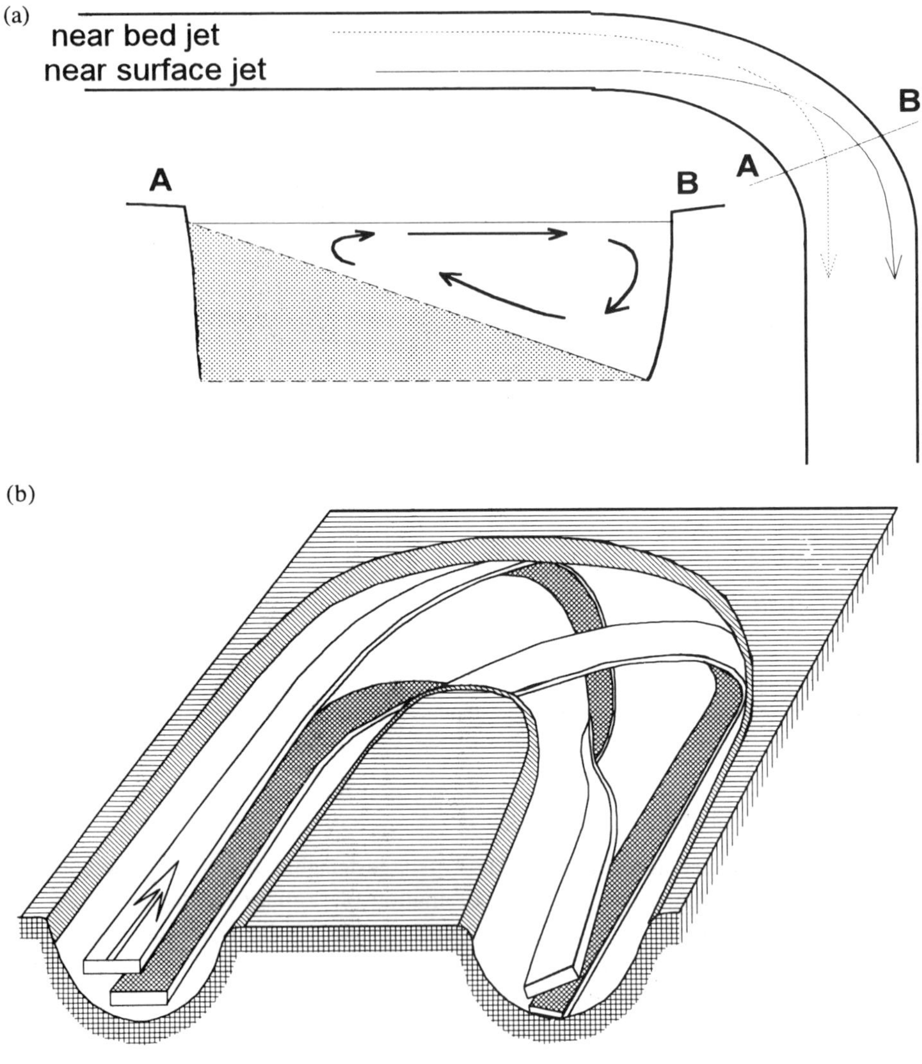

Figure 27.2. Secondary flow in river bends: (a) proposed by Thomson (1876); (b) shown visually in the flume test of Milovich (1914); (c) measured by propeller current meter in the Vilui river, Russia (Zaitsev, pers. comm., 1979); and (d) measured by a two-dimensional electromagnetic current meter in the Fall river by Thorne *et al.* (1985). *Continued on facing page*

velocity and depth and is inversely proportional to the radius of meander curvature depending on the friction factor (Goncharov, 1954; Rozovskiy, 1957; Falcon & Kennedy, 1983; Odgaard, 1989).

(c)

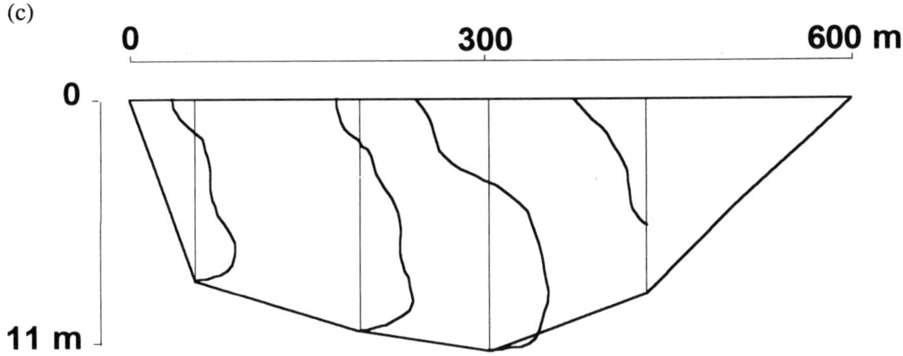

(d)

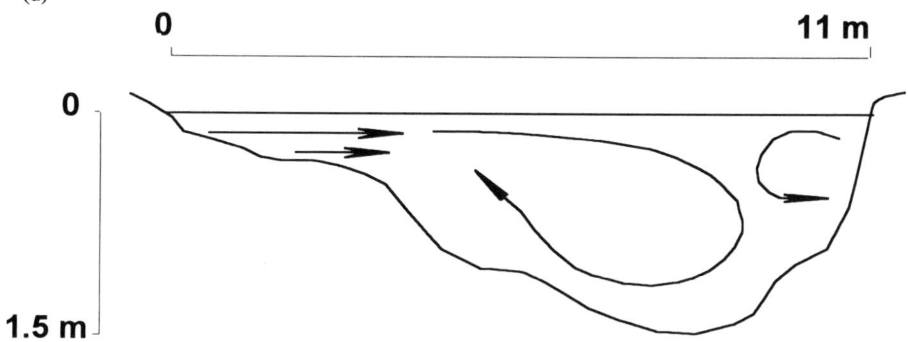

Figure 27.2. *continued*

ANALYTICAL STRATEGY

MAIN ASSUMPTIONS

The annual volume of bank retreat, c, can be derived from a balance of material supply and removal in the outer-bank zone (Figure 27.3)

$$cl(d+h) = G_Z t \qquad (27.1)$$

where l is length of the bank section, d is the mean flow depth, h the bank height above the water surface, G_z is the transverse sediment flux and t is time. Since the sediment load concentrates in the near-bed zone and the maximum transverse velocity also occurs there, it is reasonable to assume that as a first approximation the components of suspended flux are in direct proportion to the components of flow velocity in the near-bed zone.

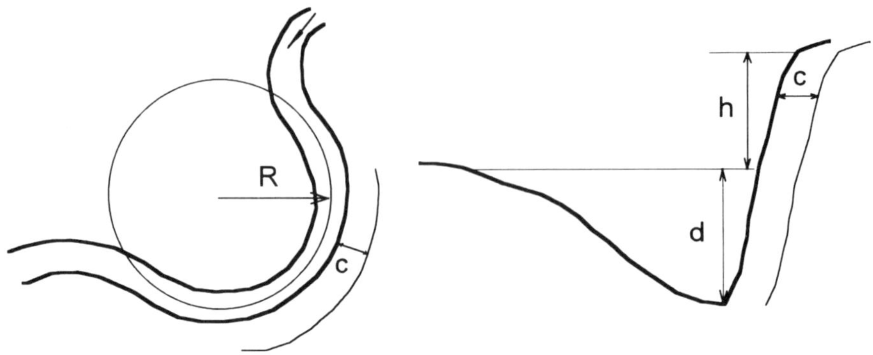

Figure 27.3. Morphometric characteristics and descriptive terms used in the bend model

$$\frac{G_Z}{G} = \frac{w}{\sqrt{(u^2 + w^2)}} \tag{27.2}$$

where G is the volumetric sediment load, and u and w the streamwise and transverse velocities near the bed, respectively. The streamwise velocity in the near-bed zone is assumed to be proportional to the mean downstream velocity, U

$$u = k_1 U \tag{27.3}$$

It should be noted that only suspended sediment transport is considered within the model framework. Trajectories of coarser sediment moving as bedload are more difficult to describe numerically, mainly because they are dependent on the local bed topography and lateral bed gradients (Seminara & Tubino, 1989). However, since bank material is usually much finer than the adjacent bed sediment it is reasonable to assume that the bulk of debris feeding the flow from the bank is transported as suspended load.

The mean, d, and maximum depth, d_{max}, in a cross-section are related as:

$$d = k_2 d_{\max} \tag{27.4}$$

as is appropriate for parabolic or trapezoidal channel shapes.

TRANSVERSE FLOW VELOCITY

The basics of the model are taken from the results of Rozovskiy (1957) who derived relatively simple expressions for the transverse flow velocity from two-dimensional flow equations. According to his model a bend and adjacent section can be subdivided into zones of developed circulation, rising circulation, and falling circulation (Figure 27.4). The rate of developed circulation is expressed as:

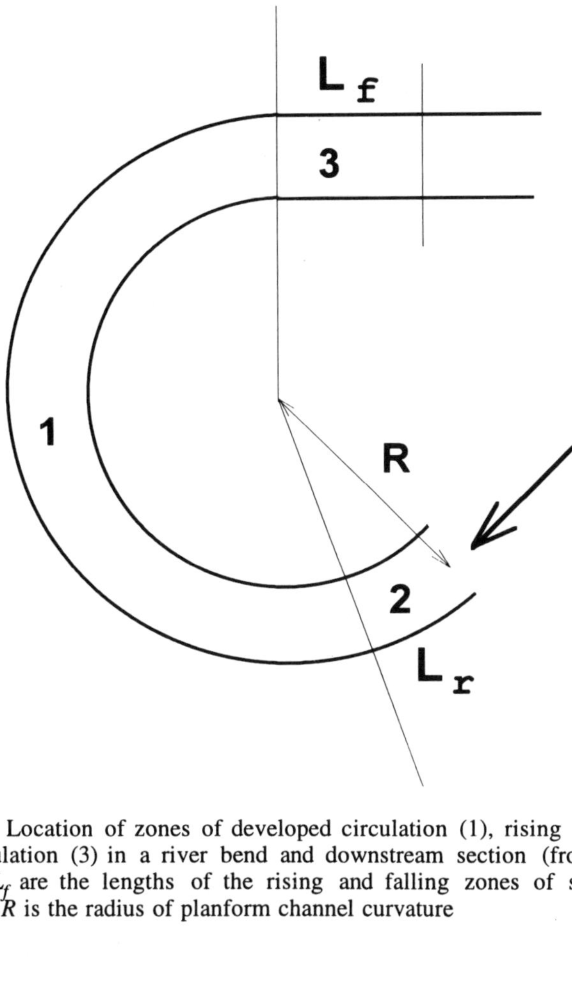

Figure 27.4. Location of zones of developed circulation (1), rising circulation (2), and falling circulation (3) in a river bend and downstream section (from Rozovskiy, 1957). L_r and L_f are the lengths of the rising and falling zones of secondary flow respectively and R is the radius of planform channel curvature

$$w = k_3 U \frac{d}{R}\left(1 - \frac{2\sqrt{g}}{C}\right) \tag{27.5}$$

where R is the radius of plan curvature, g is the gravity acceleration, C the Chézy coefficient and k_3 a constant. The length of a section of rising circulation, L_r, is given by:

$$L_r = \frac{k_4 C}{\sqrt{gd}} \tag{27.6}$$

In such a section of flow the circulation increases as:

$$w = w_0\left(1 - \exp\frac{\sqrt{g}}{C}\frac{X_0 - X}{d}\right) \tag{27.7}$$

where the subscript, 0, defines the value of the streamwise co-ordinate and transverse velocity at the start of the developed circulation zone and X is the streamwise co-ordinate. The section of falling circulation, L_f, has a length

$$L_f = \frac{k_5 C}{\sqrt{gd}} \tag{27.8}$$

and the decay of circulation is also expressed by an exponential law

$$w = w_0 \exp\frac{\sqrt{g}}{C}\frac{X - X_0}{d} \tag{27.9}$$

SEDIMENT DISCHARGE

There are a number of theoretical and empirical equations for the prediction of sediment discharge which vary considerably in the level of sophistication and complexity. However, almost all of them can be approximated by a power-law relation with flow velocity

$$G = k_6\left(\sqrt{u^2 + w^2}\right)^4 \tag{27.10}$$

The coefficient k_6 depends on the erosion and deposition conditions and is determined empirically on the basis of a sediment load rating curve.

NUMERICAL STRATEGY

The equations and assumptions discussed above are used to drive a one-dimensional numerical model. River sections that have a constant radius of plan curvature are approximated by circular arcs. Each of them must be represented by three base points: two points located on segment boundaries and one within a segment dividing it into two hemi-segments. Care must be taken to ensure a smooth transition of the main morphometric parameters between the base points. For each base point, plan co-ordinates of talweg, channel width, maximum depth and bank height above the water surface for bankfull stage are determined.

Because of the non-linear behaviour of the main computational formulae a number of calculation points must be chosen within each hemi-segment. Values of channel morphometry at calculation points are determined by linear interpolation between base points. Therefore, at any calculation point the channel is characterized by bankfull width, maximum depth and bank height as well as planform centreline curvature. Furthermore, the complete calculation requires input of values for water discharge (usually bankfull), channel slope or a friction factor parameter and a time interval. The sequence of calculation is as follows:

1. Approximation of the channel centreline by a chain of arcs (segments). For each segment calculation of the radius of curvature R. Right curvature is taken as positive and left as negative. This step is visualized on-screen for control purposes.
2. Interpolation of the channel morphometric parameters width, B, maximum depth, d_{max} and bank height h, and input into the calculation points.
3. Calculation of the mean flow velocity U

$$U = \frac{Q}{Wk_2 d_{max}} \qquad (27.11)$$

   at each point.
4. Calculation of the first approximation of transverse velocity w in the near-bed zone using Equation 27.5.
5. For each segment, determination of the zones of rising and decaying circulation using Equations 27.6 and 27.8 respectively, and selection of calculation points located within these zones.
6. For selected points, correction of w using Equations 27.7 or 27.9, or both if zones of falling and rising circulation overlap.
7. Estimation of flow velocity u in the near-bank zone by using Equation 27.3 and the sediment load, G using Equation 27.10.
8. Calculation of the transverse sediment flux, G_Z using Equation 27.2.
9. Determination of the volume of bank erosion, c from Equation 27.1 and on-screen visualization.

The constants k_1, k_2, k_3, k_4, k_5 and k_6 allow the model to be flexible during calibration. The computer program has a structure suitable for easy incorporation of other formulae for transverse velocity and sediment load instead of Equations 27.4 and 27.10 if required.

TESTING OF THE MODEL USING THE SEYAKHA RIVER

The Seyakha river is situated on the Yamal Peninsula in the north of Western Siberia. This region has a typical tundra landscape with low relief and gently meandering rivers. Both the river banks and bed contain mostly fine sand. Because of the very low elevation of the surrounding floodplain appreciable velocity in the Seyakha is only reached during the high flows caused by summer snowmelt which occur for 1–2 weeks per year. Channel morphological changes occur during this period but the river is very stable and inactive for the rest of the year.

A combination of aerial photograph analysis and radiocarbon dating in a study reach allowed determination of the previous channel locations (Figure 27.5). Rates of horizontal channel migration through the last few decades were determined for

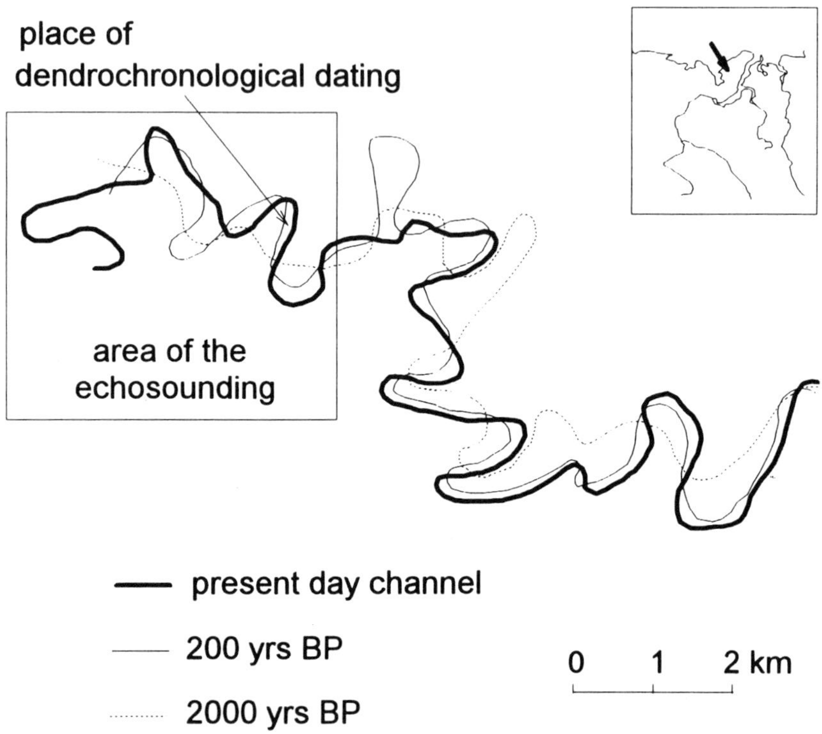

Figure 27.5. Reach of the Seyakha river, Siberia used as a test of the bend development computer model

one bend by using the dendrochronology of dwarf willows growing at the edge of convex meander banks (Figure 27.5). An echo-sounder profile along the talweg was available for the upper section of the study reach and is used as input to the model.

The numerical simulation consists of 16 segments of the study reach (Figure 27.6). Radius of curvature ranges from 235–1390 m. The number of calculation points between base points varied from 5 to 15 depending on the hemi-segment length. The simulation was undertaken for 7 days of bankfull flow. Water discharge is 600 m^3 s^{-1}, the Chézy coefficient C and constants k_1, k_2, k_3, k_4, k_5 and k_6 were taken as 70, 0.5, 0.65, 4.0, 2.3, 2.3 and 0.004 respectively.

According to aerial photographic and radiocarbon analyses, the average rate of channel migration for the last two centuries within steady developing meanders is 1.0–1.5 m $year^{-1}$. However, the rate of bank erosion can vary significantly over time and along the bank edge. Simulated rates of bank erosion (Table 27.1) are of the same order as those observed in the field and the spatial pattern of zones

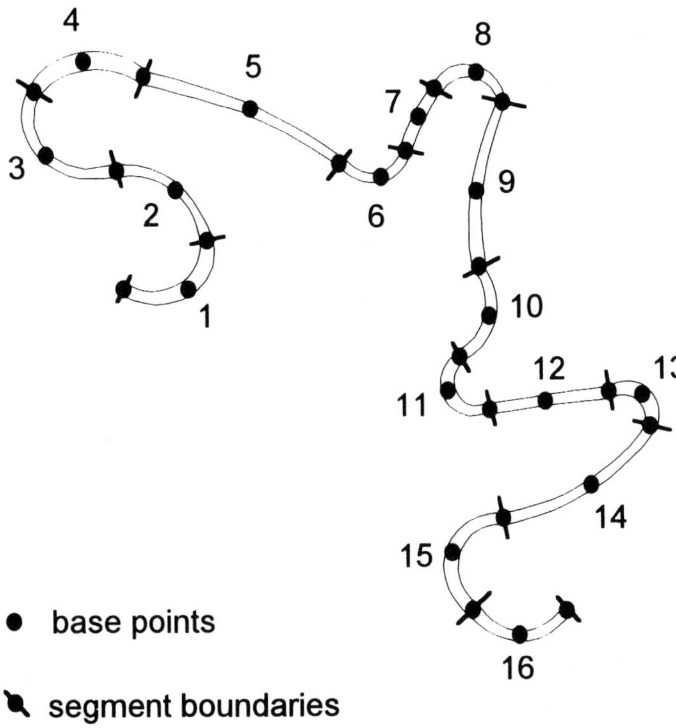

Figure 27.6. Schematic diagram of the modelled section of the Seyakha River showing the number and location of segments. Flow direction is from 1 to 16

susceptible to bank erosion correlate well with the zones of maximum erosion. Detailed comparison of the field and computed values was possible for segment 14, where dendrochronological data of the convex bank accretion are available (Table 27.2). Although this section is of relatively low curvature and is not distinguished by high values of bank retreat the changes along the channel demonstrate the reasonable agreement between the model and the field.

CONCLUSION

A numerical model is presented that is physically based, user friendly, requires minimal input data and allows the numerical solution of bend development in segments of one-dimension. The aim of this paper is not to develop a model from the basic equations of flow motion and flow-bank interaction but instead to have a

Table 27.1. Numerical simulation results for a reach of the Seyakha river

Segment number	Radius of curvature R (m)	Max. erosion c (m)	Location of max. erosion	Flow parameters in the location of maximum erosion B (m)	d_{max} (m)	u (m s^{-1})	w (m s^{-1})
1	390	7.6	middle	120	6.5	1.15	0.13
2	620	2.5	end	130	7.0	0.98	0.08
3	500	8.4	end	110	6.5	1.25	0.11
4	580	2.6	beginning	115	8.0	0.97	0.09
5	3890	0.5	end	130	6.5	1.06	0.01
6	280	4.3	end	130	7.5	0.92	0.16
7	3330	0.6	middle	130	6.1	1.06	0.01
8	260	3.9	end	120	8.5	0.88	0.18
9	2890	0.6	middle	135	6.5	1.02	0.02
10	700	1.2	end	140	7.5	0.85	0.06
11	240	3.7	beginning	135	7.9	0.84	0.18
12	13880	0.06	end	160	6.5	0.86	0.003
13	240	1.4	beginning	155	8.6	0.66	0.15
14	1850	0.7	middle	125	7.5	0.96	0.03
15	450	1.8	beginning	145	7.4	0.84	0.09
16	400	0.6	end	130	11.0	0.63	0.10

Table 27.2. Amount of calculated bank erosion using the bend model described here compared to the amount of convex bend accretion (estimated using dendrochronology) from segment 14 of the Seyakha river

Point number	Computed concave bank erosion for 7 days of bankfull flow (m)	Mean annual amount of convex bank accretion (m) in the Seyakha River
1	0.26	0.44
2	0.38	0.68
3	0.73	0.90
4	0.68	0.87
5	0.65	0.88

program structure that allows easy incorporation of different existing theoretical or empirical formulae which describe the flow structure and sediment transport processes in meander bends. This model can be used both for calculation of bank erosion rates under different hydrological conditions and as a tool for testing various formulae relevant to the bank erosion process.

ACKNOWLEDGEMENTS

I wish to thank Drs. Aleksei Sidorchuk and Boris Matveev for valuable advice and the use of unpublished field data for model testing. The International Science Foundation and The University of Leeds are acknowledged for providing funds to enable participation in the conference. My thanks to Dr. Janet Hooke and Prof. Chris Paola for constructive comments on the initial draft and the editors for improving the English text.

REFERENCES

Begin, Z.B. 1981. 'Stream curvature and bank erosion, a model based on the momentum equation', *J. Geol.*, **89**, 407–504.

Brice, J.C. 1974. 'Evolution of meander loops', *Bull. Geol. Soc. Am.*, **85**, 581–586.

Carson, M.A. & Kirkby, M.J. 1972. *Hillslope Form and Process*, Cambridge University Press, Cambridge, 475 pp.

Darby, S.E. & Thorne, C.R. 1994. 'Prediction of tension crack location and riverbank erosion hazards along destabilized channels', *Earth Surf. Processes Landf.*, **19**, 233–245.

Falcon, M.A. & Kennedy, J.F. 1983. 'Flow in alluvial-rivers curves', *J. Fluid Mech.*, **133**, 1–16.

Ferguson, R.I. 1984. 'Kinematic model of meander migration', in Elliot, C.M. (Ed) *River Meandering*, ASCE, New York, 942–951.

Goncharov, V.N. 1954. *Basics of Channel Flow Dynamics*, Gidrometeoizdat, Leningrad, 452 pp. (in Russian).

Hasegawa K. 1989. 'Studies on qualitative and quantitative prediction of meander channel shift', in Ikeda, S. & Parker, G. (Eds) *River Meandering*, Water Resources Monograph 12, American Geophysical Union, Washington, 215–236.

Hickin, E.J. & Nanson, G. 1975. 'The character of channel migration on the Beatton River, north east British Columbia, Canada', *Bull. Geol. Soc. Am.*, **86**, 487–494.

Hooke, J.M. 1977. 'The distribution and nature of changes in river channel pattern', in Gregory, K.J. (Ed) *River Channel Changes*, John Wiley & Sons Ltd., Chichester, 591–609.

Hooke, J.M. & Harvey, A.M. 1983. 'Meander changes in relation to bend morphology and secondary flows', in Collinson, J.D. & Lewin, J. (Eds) *Modern and Ancient Fluvial Systems*, Blackwell, Oxford, 121–132.

Howard A. 1992. 'Modelling channel migration and floodplain sedimentation', in Carling, P.A. & Petts, G.E. (Eds) *Lowland Floodplain Rivers: Geomorphological Perspectives*, John Wiley & Sons Ltd., Chichester, 1–40.

Ikeda, S., Parker, G. & Sawai, K. 1981. 'Bend theory of river meanders. I. Linear development', *J. Fluid Mech.*, **112**, 363–377.

Johannesson, J. & Parker G. 1989. 'Linear theory of river meanders', in Ikeda, S. & Parker, G. (Eds) *River Meandering*, Water Resources Monograph **12**, American Geophysical Union, Washington, 181–214.

Kondrat'yev, N.Ye. 1968. 'Hydromorphic principles of computations of free meandering, 1. Signs and indexes of free meandering', *Soviet Hydrol.*, **4**, 309–335.

Makkaveyev, N.I. 1955. *A River Channel and an Erosion in the Basin*, The USSR Academy of Science, Moscow, 247 pp. (in Russian).

Milovich, A.Ya. 1914. *Passive Flow Bend*, Moscow, 98 pp. (in Russian).

Odgaard, A.J. 1989. 'River-meander model. I: Development', *J. Hydraul. Engrg.*, **155**, 1433–1464.

Osman, M.A., & Thorn, C.R. 1988. 'Riverbank stability analysis I: Theory', *J. Hydraul. Engrg., ASCE*, **114**, 134–150.
Rozovskiy, I.L. (1957) *Flow of Water in Bends of Open Channels*, Ukrainian Academy of Science, Kiev, 233 pp. (in Russian, English translation, Israel Program for Scientific Translation, Jerusalem, 1961).
Seminara, G. & Tubino, M. 1989. 'Alternate bars and meandering: free, forced and mixed interactions', in Ikeda S. & Parker, G. (Eds) *River Meandering*, Water Resources Monograph **12**, American Geophysical Union, Washington, 267–320.
Thomson, J.J. (1876) On the windings of rivers in alluvial plains, *Proc. Roy. Soc. London*, **25**, 5–8.
Thorne, C.R. 1982. 'Processes and mechanisms of river bank erosion', in Hey R.D., Bathurst J.C. & Thorne C.R. (Eds) *Gravel Bed Rivers*, John Wiley & Sons Ltd., Chichester, 227–270.
Thorne, C.R., Zevenbergen, L.W., Pitlick, J.C., Raist S., Bradley, J.B. & Julien, P.Y. 1985. 'Direct measurements of secondary currents in a meandering sand-bed river', *Nature*, **316**, 746–747.
Zamyshlyaev, V.I. 1983. 'A mathematical model for plan deformations of a river channel', *Proc. Russian State Hydrological Institute*, **288**, 56–72. (in Russian).

NOTATION

B	channel width (m)
c	annual bank retreat (m)
C	Chézy coefficient ($m^{0.5}$ s^{-1})
d	mean flow depth (m)
d_{max}	maximum flow depth (m)
g	gravity acceleration (m s^{-2})
G	volumetric sediment load (m^3 s^{-1})
G_Z	transverse sediment flux (m^3 s^{-1})
h	bank height above water surface (m)
k_1–k_6	constants
l	length of the bank section (m)
L_f L_r	length of zones of falling and rising secondary flow (m)
Q	water discharge (m^3 s^{-1})
R	radius of channel plan curvature (m)
t	time (s)
u	streamwise flow velocity in near-bed zone (m s^{-1})
U	mean streamwise flow velocity (m s^{-1})
w	transverse flow velocity in near-bed zone (m s^{-1})
w_0	transverse flow velocity in near-bed zone at the start of the developed circulation zone (m s^{-1})
X	streamwise co-ordinate
X_0	streamwise co-ordinate at the start of the developed circulation zone
Z	transverse co-ordinate

28 Floodplain Secondary Circulation as a Mechanism for Flow and Shear Stress Redistribution in Straight Compound Channels

PETER R. WORMLEATON
Department of Civil Engineering, Queen Mary and Westfield College, University of London, Mile End Road, London, E1 4NS, UK

ABSTRACT

Experimental data are presented on the discharge and boundary shear stresses in straight compound trapezoidal channels with varying floodplain widths. It is shown that the effect of the channel floodplain interface extends across the full floodplain width in all cases. Evidence is discussed concerning the existence of a single secondary flow cell on the floodplain which has been thought to be responsible for the lateral transfer of momentum.

INTRODUCTION

Flow in compound channels consisting of a deep main-channel bounded by shallower floodplains is of considerable interest to river engineers. The interaction between the faster fluid velocities in the main channel and the slower moving flow on the floodplains causes shear stresses at their interface which significantly distort flow and boundary shear stress patterns. The nature of these distortions needs to be understood for the design of river engineering schemes that rely upon realistic flow or sediment routing models.

The UK Flood Channel Facility (FCF) was built at HR Wallingford Ltd., to obtain accurate data on flow in compound channels. The FCF is a 60 m long and 10 m wide flume, designed to operate in a range of Reynolds numbers sufficiently high to reduce scaling effects as far as possible. The overall research programme is divided into three series: *series A* – straight, fixed-bed channels, *series B* – meandering fixed-bed channels (e.g. Naish & Sellin, this volume) and *series C* – straight and meandering mobile-bed channels. *Series A* and *B* are complete, whilst *series C* is currently being conducted.

This paper concentrates on some of the results from the *series A* experiments, in

Coherent Flow Structures in Open Channels.
Edited by P.J. Ashworth, S.J. Bennett, J.L. Best and S.J. McLelland.

particular the effect of floodplain width on the flow and shear stress distributions in straight compound channels. It serves to highlight some of the unexpected results that were obtained from the initial series. It also underlines our comparative lack of understanding of the behaviour of even this simplest of compound cross-sections.

EXPERIMENTAL DATA

The three geometries chosen as part of the *series A* experiments on the FCF had straight, smooth trapezoidal main-channels of identical dimensions. On either side of the main channel were parallel symmetrical floodplains which were also smooth but of three different widths. The geometries, including the notation used in this paper, are shown in Figure 28.1. In each case the bed width, $2b$, of the main channel was 1.5 m and the bankfull depth was 0.15 m. This gave an aspect ratio of 10 which is typical of many British rivers. The size of the main channel was such that bankfull Reynolds number was approximately 4×10^5. The floodplain -to- main channel width ratios (B/b) are 6.67, 4.2 and 2.2 in the three cases respectively. Measurements of discharge, velocity and boundary shear stress distributions were made at each section over a range of relative depths, $(H-h)/H$, up to around 0.5. In addition, detailed laser Doppler anemometry (LDA) readings of secondary flows and turbulence parameters were made for section 2 (B/b = 4.2) at four relative depths between 0.1 and 0.3.

The boundary shear stress was measured with a Preston tube of either 2.7 or 4 mm diameter, and calculated using Patel's (1965) formula. There is some uncertainty surrounding the use of the Patel's calibration at low Reynolds numbers, since his readings were taken in a smooth pipe with Reynolds number greater than

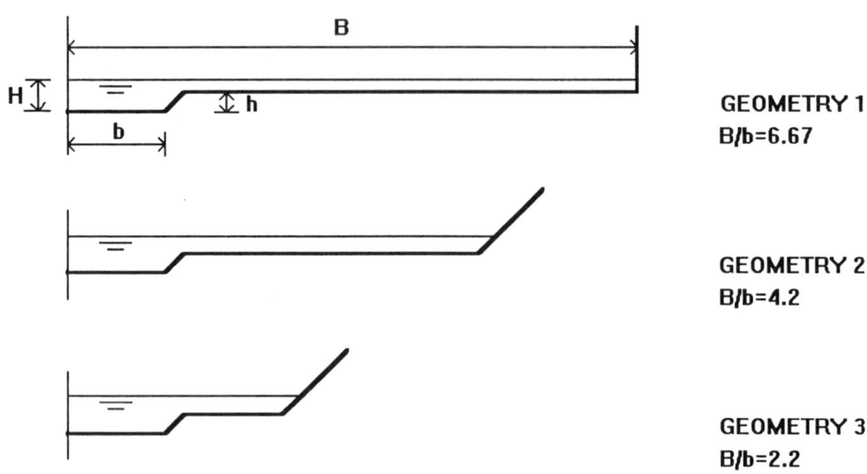

Figure 28.1. The three cross-sectional geometries showing notation

8.9×10^3, based upon centreline velocity and pipe diameter. The Reynolds number on the floodplain varied from 1.16×10^3 for run 1/1 to 60.7×10^3 for run 1/7, based upon depth of flow and mean velocity. Thus, at lower depths, the Preston tube was operating below the lower end of Patel's calibration range. All boundary shear stresses were adjusted *pro rata* to give zero error in their overall integrated value compared with the weight component of the flow in the downstream direction. It could be argued that this simple correction is inappropriate, bearing in mind that the Preston tube essentially reads a pressure which is converted to boundary shear stress through Patel's highly non-linear calibration equations. Nevertheless, the mean error in the integrated shear stresses for geometries 1–3 was only 2.77% with a standard deviation of 4.44%.

Figure 28.2 shows the distribution of depth-averaged velocity for the three sections and Figure 28.3 the distribution of boundary shear stress which was measured using a Preston tube. Looking particularly at the widest floodplain in section 1, it is clear that velocity and boundary shear stress are uniform over most of its width. This region of the floodplain, where velocity and shear stress are constant, is referred to here as the 'laterally uniform' zone. Figures 28.2 and 28.3 indicate that a 'laterally uniform' zone is identifiable for floodplain aspect ratios (floodplain width -to- floodplain depth) greater that about 40, which includes all geometry 1 data and the lower flow depths of the other two geometries. Variations due to the channel-floodplain interface and the floodplain side extend over a very limited distance from these boundaries. This suggests the existence of two-dimensional flow on the floodplain in the 'laterally uniform' zone, and consequently no lateral momentum transfer. If this were so then further widening of the floodplain, once a 'laterally uniform' zone exists, should have no effect on main-channel flow.

The channel cross-section can be divided into sub-sections along any plane within the fluid. Within each sub-section, if the flow is steady, the weight component of the fluid in the flow direction must be balanced by the sum of the shear forces acting around its solid boundary and on the fluid in the division plane. Thus, the magnitude of the shear force on the fluid interface will be given by the difference between the weight component of the fluid and the solid boundary shear forces. This interface shear force is referred to as the apparent shear force from which an associated apparent shear stress can be calculated. If z is taken as lateral distance across the channel from the floodplain edge, the apparent shear stress, τ_a, on a vertical plane at a distance $z = Z$ across the channel is thus given by:

$$\tau_a = \frac{1}{Y} \int_0^Z \left(\tau_b \left\{ 1 + \frac{1}{s^2} \right\}^{1/2} - \rho g Y(z) S_0 \right) dz \qquad (28.1)$$

where $Y(z)$ is the depth of flow at any lateral distance, s is channel lateral side slope, S_0 is longitudinal bed slope and $\tau_b(z)$ is bed shear stress. Figure 28.4 shows the apparent shear stress, τ_a, acting on a series of vertical planes within the fluid across geometry 1. The apparent shear stress is zero on the channel centreline as

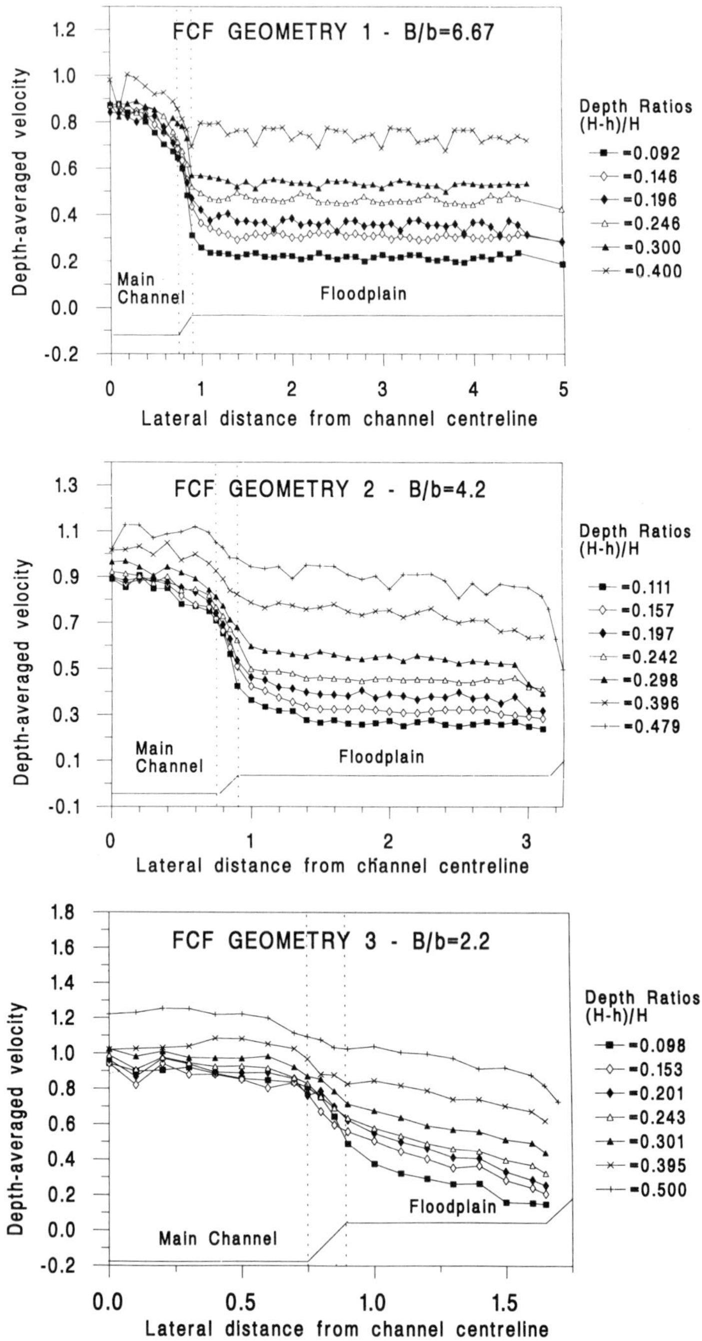

Figure 28.2. Observed depth-averaged velocity distributions for test geometries 1–3 (velocity measured in m s^{-1} and lateral distance measured in m)

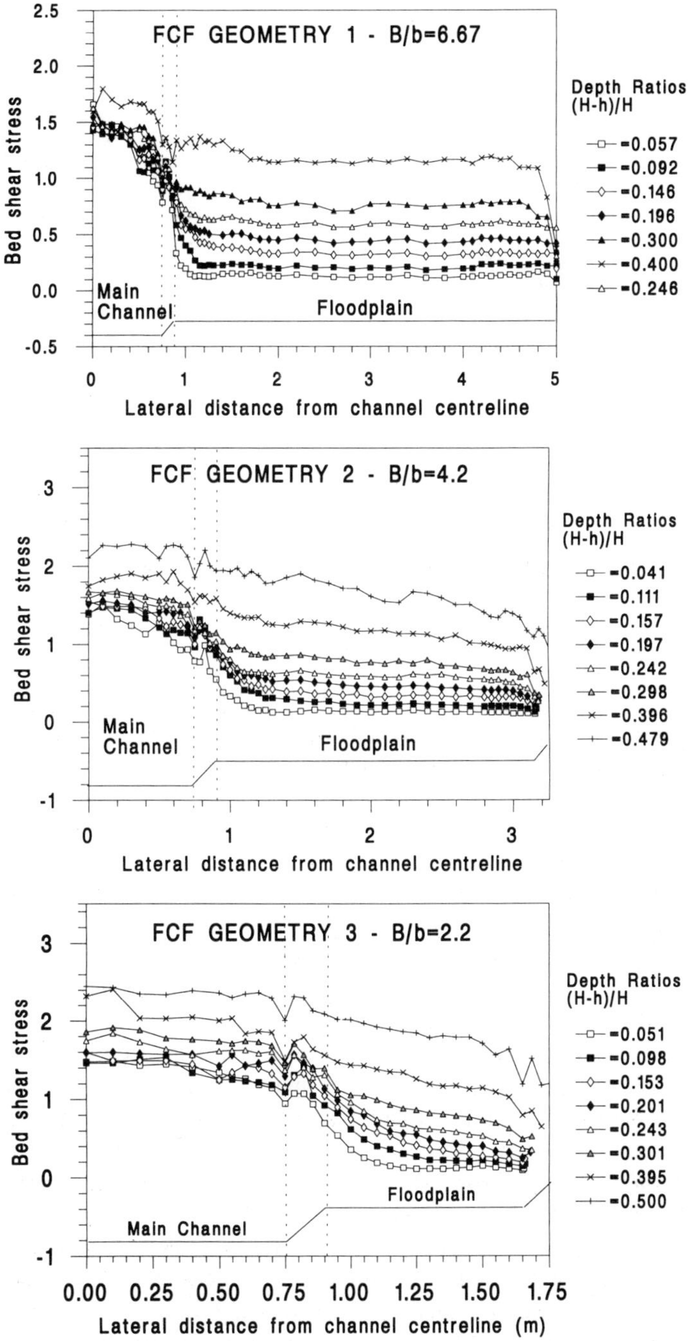

Figure 28.3. Observed bed shear stress distributions for test geometries 1–3 (bed shear stress measured in N m^{-2} and lateral distance measured in m)

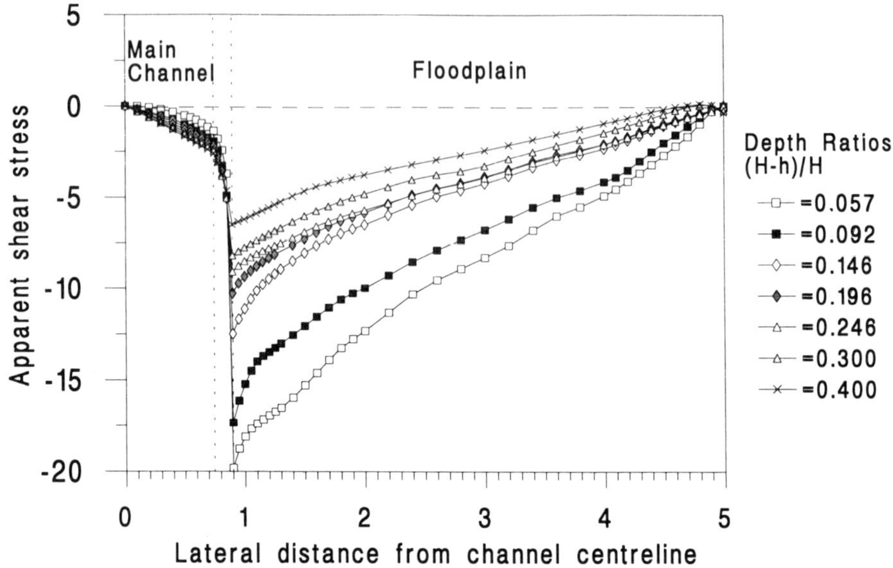

Figure 28.4. Apparent shear stress distribution for geometry 1 (apparent shear stress measured in N m^{-2} and lateral distance measured in m)

would be expected in a symmetrical channel, increases slightly across the main channel and then rapidly reaches a maximum at the main channel-floodplain interface. On the floodplain the apparent shear stress decreases quite rapidly near the interface and then in a fairly linear manner across the entire floodplain width to become zero again at the floodplain edge. Where there is an apparent shear stress there is transfer of momentum, hence the flow across the whole floodplain is not two-dimensional, as assumed.

This implies that the flow in the main channel will be a function of floodplain width, even when a 'laterally uniform' zone exists. A comparison of flow in the main-channel for the three floodplain widths, as shown in Figure 28.5, confirms this; an increase in floodplain width results in a greater apparent shear force at the interface which in turn reduces flow in the main channel. The decrease in main-channel flow with increasing floodplain width ratio is approximately linear over the range of depths shown. Mean results over a range of depth ratios, $0.05 < (H-h)/H < 0.5$, show that the main-channel flow decreases by 4.6% between floodplain width ratios of 2.2 and 4.2, and a further 5.4% between floodplain width ratios of 4.2 and 6.67. From these results it is clear that floodplain width is a significant factor in determining main-channel velocity and boundary shear distributions. Whilst primary velocities might indicate two-dimensional flow, there is a momentum transfer mechanism operating across the entire floodplain even at very high aspect ratios.

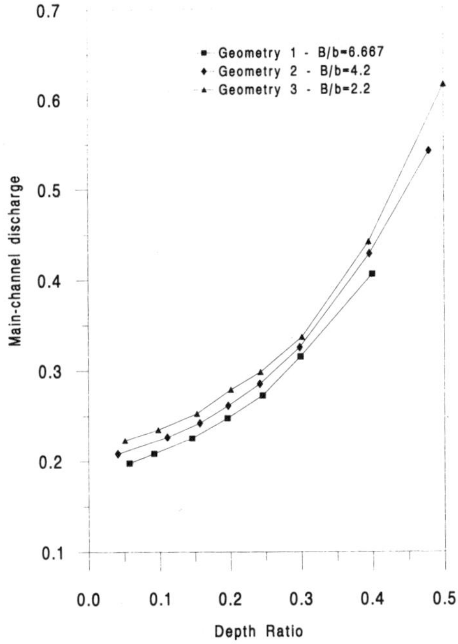

Figure 28.5. Variation of main-channel discharge with relative flow depth and floodplain width (discharge measured in $m^{-3}\,s^{-1}$)

THEORETICAL ANALYSIS

The equation for the streamwise component of momentum at a point in steady uniform flow is:

$$\rho\left(V\frac{\partial U}{\partial y}+W\frac{\partial U}{\partial z}\right)=\rho g S_o+\frac{\partial \tau_{yx}}{\partial y}+\frac{\partial \tau_{zx}}{\partial z} \qquad (28.2)$$

where U, V, and W are the mean velocity components in the x(streamwise), y(vertical) and z(lateral) directions respectively, τ_{yx} and τ_{zx} are Reynolds shear stresses, ρ is the fluid density and S_0 the bed slope. By combining the continuity equation with Equation 28.2, Shiono & Knight (1988, 1991) obtained the relationship:

$$\rho\left(\frac{\partial (UV)}{\partial y}+\frac{\partial (UW)}{\partial z}\right)=\rho g S_o+\frac{\partial \tau_{yx}}{\partial y}+\frac{\partial \tau_{zx}}{\partial z} \qquad (28.3)$$

This can be averaged by integrating over water depth, Y, to give:

$$\frac{\partial (Y\langle \rho UW\rangle)}{\partial z}-\frac{\partial y_s}{\partial z}\left(\rho U_s W_s-\{\tau_{zx}\}_s\right)=\rho g Y S_o+\frac{\partial (Y\langle \tau_{zx}\rangle)}{\partial z}-\tau_b\left(1+\frac{1}{s^2}\right)^{1/2}=\Gamma \qquad (28.4)$$

where Γ represents the sum of the forces acting on a vertical fluid element of unit width due to the weight component of the fluid, internal fluid shear and boundary shear. y_s is the elevation of the water surface and U_s, W_s, $(\tau_{zx})_s$ are values at the water surface. In this analysis it is assumed that $V_b = V_s = (\tau_{zx})_s = 0$ and depth-averaged terms are defined as:

$$\langle \rho UW \rangle = \frac{1}{Y}\int_0^Y (\rho UW)\mathrm{d}y \quad \text{and} \quad \langle \tau_{zx} \rangle = \frac{1}{Y}\int_0^Y \tau_{zx}\mathrm{d}y \tag{28.5}$$

Substituting for τ_a, from Equation 28.1 into Equation 28.4 gives:

$$\Gamma = \frac{\partial(Y\langle \tau_{zx} \rangle)}{\partial z} - \frac{\partial(Y\tau_a)}{\partial z} = \frac{\partial(Y\Psi)}{\partial z} \tag{28.6}$$

where

$$\frac{\partial(Y\tau_a)}{\partial z} = \tau_b\left\{1 + \frac{1}{s^2}\right\}^{\frac{1}{2}} - \rho g Y S_0 \tag{28.7}$$

and

$$\Psi = \frac{1}{Y}\int_0^z \Gamma \mathrm{d}z = \langle \tau_{zx} \rangle - \tau_a \tag{28.8}$$

Ψ represents the sum of the depth-averaged shear stresses acting on a plane in the fluid due to weight and boundary shear stress balance, i.e. the apparent shear stress, τ_a, and internal turbulent shear stress, τ_{zx}. Equation 28.8 enables calculation of Ψ, if the Reynolds shear stress, τ_{zx}, and apparent shear stress, τ_a, are known. τ_{zx} can be obtained from LDA measurements made at four depths in geometry 2 ($B/b = 4.2$). The resulting Ψ values were calculated by Shiono & Knight (1992) and are shown here as Figure 28.6. Since these were the only LDA readings under consideration, no further measurements of τ_{zx} are available.

Apparent shear stress can be calculated using Equation 28.7 from boundary shear stress readings for geometries 1–3 at all depths. Moreover, if there is no lateral variation in velocity, then $\partial U/\partial z = 0$ and so $\tau_{zx} = 0$ which was confirmed by the LDA readings on the floodplain. This enables Ψ, and hence Γ, to be evaluated on the floodplains in the 'laterally uniform' zone and the results of this are given in Table 28.1. As can be seen from Figure 28.2 this zone extends over most of the wider floodplains of geometry 1. The variation in Γ with depth ratio over the 'laterally uniform' zone is shown in Figure 28.7. The geometry 1 data have a linear relationship of $\Gamma = -(0.038+1.17\{H\text{-}h\})$ with correlation coefficient of 0.982 and standard error of estimate for Γ of 0.0071. The data from geometries 2 and 3 are incomplete since, in these cases, the interface and side effects spread across the entire floodplain, i.e. the floodplain aspect ratio is less than 40. The geometry 2 data show a wide deviation from the linear relationship, however, the relative magnitude of the differences is small. Thus, a 1% error in the bed shear will lead to an error in Γ of around 8%. Bearing this in mind, the values of Γ for geometries

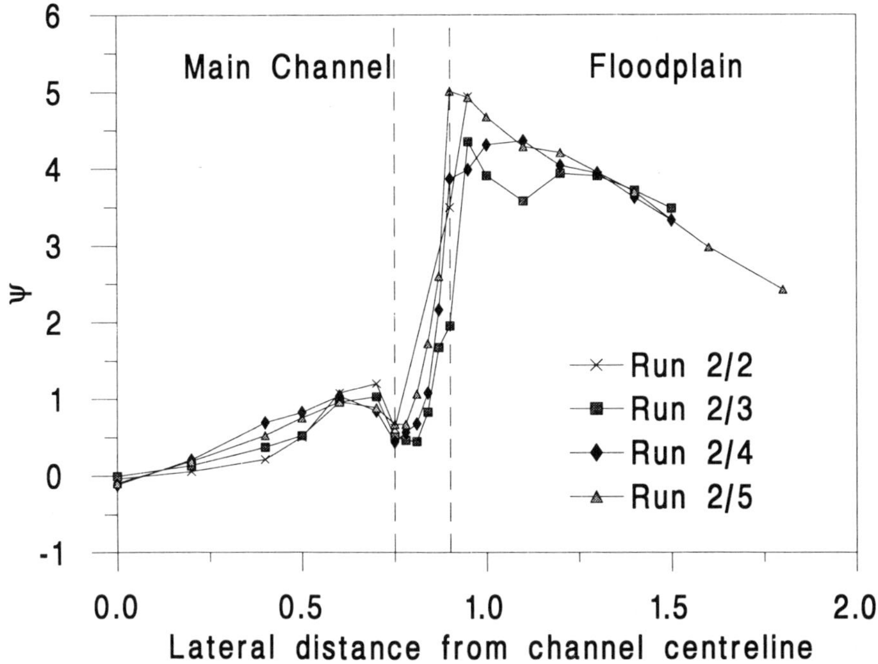

Figure 28.6. Depth-average Ψ values (measured in N m^{-2}) for geometry 2 (lateral distance measured in m)

1–3 in the 'laterally uniform' zone show a remarkably similar dependence upon depth ratio, with overall standard error of estimate for Γ of 0.0163.

THE 'LATERALLY UNIFORM' FLOODPLAIN ZONE

Equation 28.4 shows that Γ is comprised of two terms, the first $\partial(Y\langle\rho UW\rangle)/\partial z$ is a secondary flow term, whilst the second only exists if the water surface is not horizontal i.e. $\partial y_s/\partial z \neq 0$. Moreover in the 'laterally uniform' zone, it has been shown that the Reynolds stress, τ_{zx}, is zero. Thus, assuming a horizontal water surface and bed for the 'laterally uniform' zone, Equations 28.4 and 28.6 give:

$$Y\frac{\partial\langle\rho UW\rangle}{\partial z} = \Gamma = -Y\frac{\partial\tau_a}{\partial z} \qquad (28.9)$$

Since the observed values of Γ appear to be constant across the 'laterally uniform' zone, this clearly has implications for the floodplain flow structure and, in

Table 28.1. Observed values in 'laterally uniform' floodplain zones

Run number	Width ratio B/b	Depth ratio $(H-h)/H$	Floodplain depth (mm)	Mean bed shear stress (N m^{-2})	$\Gamma=-\partial Y\tau_a/\partial z$ (Equation 28.6) (N m^{-2} m^{-1})
1/1	6.67	0.057	8.99	0.131	0.042
1/2	6.67	0.092	15.19	0.214	0.060
1/3	6.67	0.146	25.63	0.326	0.063
1/4	6.67	0.196	36.56	0.443	0.074
1/5	6.67	0.246	48.81	0.586	0.093
1/6	6.67	0.300	64.43	0.767	0.125
1/7	6.67	0.400	100.12	1.157	0.147
2/1	4.2	0.041	6.49	0.135	0.071
2/2	4.2	0.109	18.73	0.215	0.026
2/3	4.2	0.157	27.84	0.330	0.048
2/4	4.2	0.198	36.95	0.432	0.060
2/5	4.2	0.242	47.96	0.596	0.112
3/1	2.2	0.051	8.00	0.121	0.041
3/2	2.2	0.098	16.27	0.203	0.041

particular, on the magnitude of the cross-stream velocity. With a laterally uniform primary velocity it suggests that momentum transfer results from a large secondary circulation cell extending across the floodplain with increasing values of cross-stream velocity nearer to the main-channel interface (Figure 28.8). If this is the case, it is possible to estimate the required increase in the cross-stream velocity by assuming vertical profiles for U and W, which enable the depth-average secondary flow term $\langle\rho UW\rangle$ to be calculated. The primary velocity profile can be taken as:

$$U = \frac{U_*}{\kappa}\ln\left(\frac{y_b U_*}{\upsilon}\right) + K = A \ln B y_b^* \tag{28.10}$$

where y_b is the distance from the bed, $y_b^* = y_b / Y$, $\kappa = 0.415$ is the von Kármán constant, K is a constant with an experimental value of 5.29 for hydraulically smooth open-channels (Nezu & Rodi, 1986), U_* is the shear velocity which can be calculated from the Preston tube measurements of boundary shear stress and A and B are empirical constants. Figure 28.9 shows the comparison between mean observed primary velocity values and those calculated from Equation 28.10 over a range of depths in the 'laterally uniform' zone (geometry 1). The comparison is generally satisfactory at the higher depths, although at the lower depths the calculated velocities are greater than those observed. This suggests that the shear velocity and, hence the boundary shear stress may be less than those determined by the Preston tube. Nevertheless, Equation 28.10 provides a reasonable estimate of primary velocities over the range of depths considered in this analysis.

Assigning a vertical profile to the cross-stream velocity is somewhat more

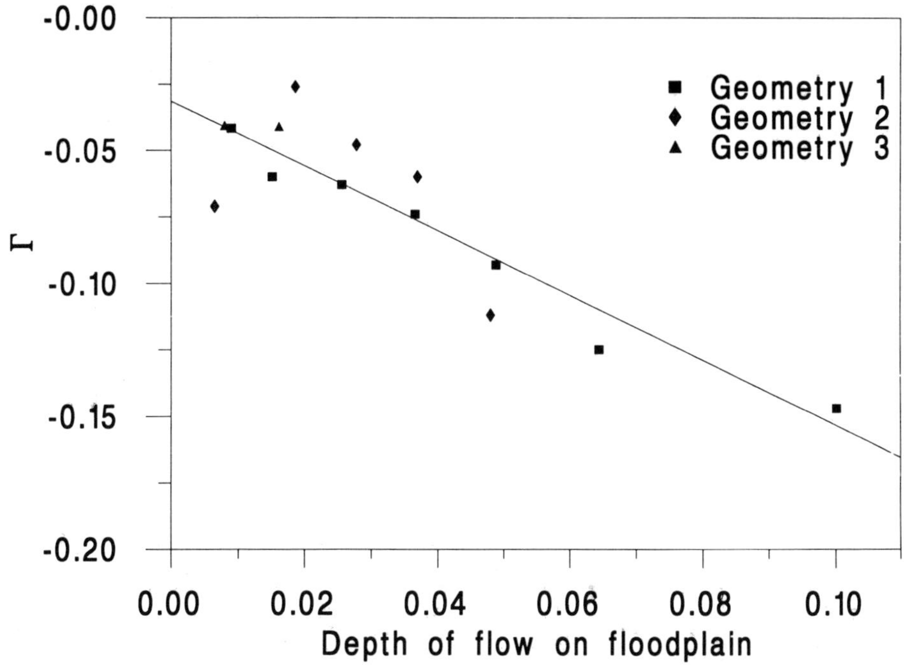

Figure 28.7. Variation in Γ (measured in N m^{-2}) over the 'laterally uniform' floodplain zone for geometries 1–3 (lateral distance measured in m)

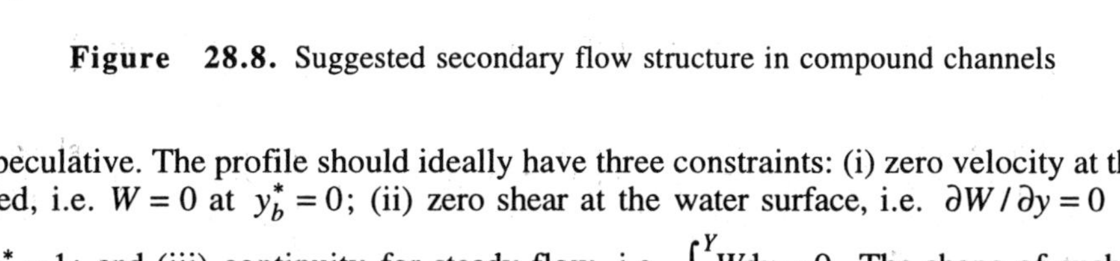

Figure 28.8. Suggested secondary flow structure in compound channels

speculative. The profile should ideally have three constraints: (i) zero velocity at the bed, i.e. $W=0$ at $y_b^*=0$; (ii) zero shear at the water surface, i.e. $\partial W/\partial y=0$ at $y_b^*=1$; and (iii) continuity for steady flow, i.e. $\int_0^Y W\mathrm{d}y=0$. The shape of such a profile is shown in Figure 28.10a. Two simple curves that can approximately meet these requirements yet retain sufficient degrees of freedom to have a variable 'cross-over' depth, y_c, at which $W=0$, are the logarithmic and quartic polynomial curves given as Equations 28.11 and 28.12 respectively.

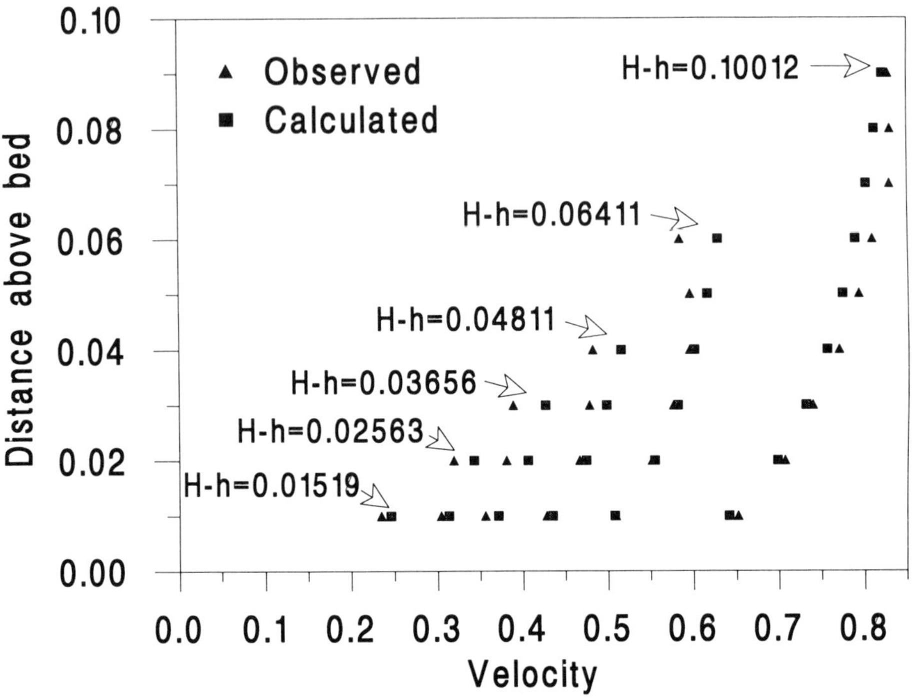

Figure 28.9. Observed and modelled primary velocity profiles for the 'laterally uniform' floodplain region of geometry 1 (velocity measured in m s^{-1} and lateral distance measured in m). Observed values averaged over a 1 m width between 2.5 and 3.5 m from the main-channel centreline

$$W = \alpha\left\{(1 - y_b^*)^2 + \gamma\right\} \ln \beta y_b^* \qquad (28.11)$$

$$W = a\left(y_b^{*4} + b y_b^{*3} + c y_b^{*2} + d y_b^*\right) \qquad (28.12)$$

The values of the parameters β and γ in Equation 28.11, and b, c and d in Equation 28.12, can be adjusted to meet the three constraints given above as far as is possible and also to provide a range of cross-over depths. The parameters α and a change the magnitude of the cross-stream velocity but not the geometric similarity of the profiles. The drawback of both of these profiles is that they only apply over limited ranges of cross-over depths. Outside these ranges the profiles either take on physically unrealistic shapes or mathematical singularities occur. It was therefore decided to consider the use a simple linear profile with variable cross-over depth as shown in Figure 28.10b. Clearly this profile does not comply with the bed and surface boundary conditions, but continuity gives the following equations in terms

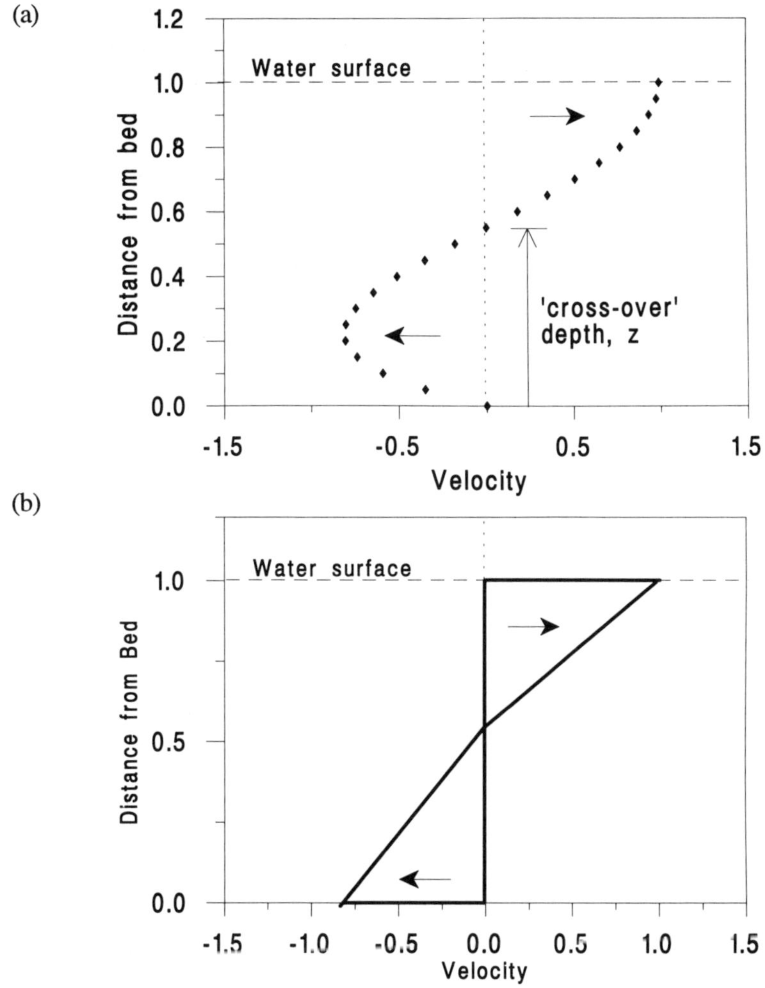

Figure 28.10. Idealized lateral velocity profiles (dimensionless variables) of the velocity, W_s, at the water surface.

$$W = W_B\left(1-\frac{y_b^*}{y_c^*}\right) = W_S\left(\frac{1}{y_c^*}-1\right)\left(1-\frac{y_b^*}{y_c^*}\right) \quad \text{for } y_b^* \le y_c^*$$

$$W = W_S\left(\frac{y_b^* - y_c^*}{1-y_c^*}\right) \quad \text{for } y_b^* \ge y_c^* \qquad (28.13)$$

where $y_c^* = y_c / Z$.

One purpose of this paper is to estimate the cross-stream velocities required for the secondary flow term $\partial\langle\rho UW\rangle/\partial z$ to fully account for the observed lateral variation in the apparent shear stress, τ_a, across the floodplain, as in Equation 28.9. The values of $\langle\rho UW\rangle$ are given below, for the three types of profile in combination with the logarithmic primary velocity profile (Equation 28.10). Constant terms have been evaluated to meet continuity and surface conditions where appropriate.

Logarithmic
$$\langle\rho UW\rangle = \frac{\rho A\alpha}{54}\left\langle 19 - \frac{15\ln\beta\{\ln\beta - 2\}}{\{\ln\beta - 1\}}\right\rangle \tag{28.14}$$

Polynomial
$$\langle\rho UW\rangle = -\rho Aa\left\langle\frac{c}{24} + 0.09\right\rangle \tag{28.15}$$

Linear
$$\langle\rho UW\rangle = \frac{\rho AW_s}{2}\left\langle 1 - \left\{\frac{2y_c^* - 1}{y_c^* - 1}\right\}\ln y_c^*\right\rangle \tag{28.16}$$

It is interesting to note that none of the above relationships for $\langle\rho UW\rangle$ are dependent upon the parameter B in the primary velocity profile. The linear relationship contains the cross-over depth, y_c^*, explicitly, whilst Equations 28.14 and 28.15 contain it implicitly as a function of β and c respectively. Maximum lateral velocities are relatively insensitive to the chosen vertical profile, hence, the linear profile has been used for the most part in this paper due to its mathematical simplicity and wide operating range in terms of y_c. However comparisons with the other profiles are made where appropriate.

In Equation 28.16, the secondary flow term is dependent upon three parameters, A, W_s and y_c^*. In the 'laterally uniform' zone, the primary velocity and, hence A are laterally constant. Therefore, the derivative of $\langle\rho UW\rangle$ in the lateral direction is:

$$\frac{\partial(Y\langle\rho UW\rangle)}{\partial z} = \frac{A}{2}\left\langle 1 - \frac{(2y_c^* - 1)}{(y_c^* - 1)}\right\rangle\frac{\partial W_s}{\partial z} + \frac{AW_s}{2(y_c^* - 1)}\left\langle\frac{1}{y_c^*} - 2 - \frac{\ln y_c^*}{(1 - y_c^*)}\right\rangle\frac{\partial y_c^*}{\partial z} \tag{28.17}$$

The vertical velocity (V) profile can be derived from the continuity equation for steady uniform flow $((\partial V/\partial y) + (\partial W/\partial z) = 0)$. This gives zero V values at the water surface and bed which increase parabolically to a maximum value, V_{max}, at the cross-over depth. The maximum vertical velocity is given by:

$$V_{max} = \frac{Y}{2}\left\langle(1 - y_c^*)\frac{\partial W_s}{\partial z} - W_s\frac{\partial y_c^*}{\partial z}\right\rangle \tag{28.18}$$

If it is assumed that the profile of the cross-stream velocity remains geometrically similar across the 'laterally uniform' zone, then the cross-over depth does not change and so $\partial y_c^*/\partial y = 0$. This enables $\partial W_s/\partial y$ to be calculated from Equations 28.9 and 28.17. V_{max} can then be found from Equation 28.18. These values for a cross-over at $y_c^* = 0.5$ are given in Table 28.2. Equivalent values of $\partial W_s/\partial y$ and V_{max} for the logarithmic and polynomial profiles are also shown for comparison and these profile parameters are similar over the range shown for all

Table 28.2. Calculated values in 'laterally uniform' floodplain zone

Run number	Linear profile $\partial W_s/\partial z$ (m s^{-1} m^{-1})	Linear profile V_{max} (m s^{-1})	Logarithmic profile $\partial W_s/\partial z$ (m s^{-1} m^{-1})	Logarithmic profile V_{max} (m s^{-1})	Polynomial profile $\partial W_s/\partial z$ (m s^{-1} m^{-1})	Polynomial profile V_{max} (m s^{-1})
1/1	0.3361	0.00076	0.3710	0.00105	0.3153	0.00106
1/2	0.2219	0.00084	0.2449	0.00118	0.2081	0.00119
1/3	0.1115	0.00071	0.1230	0.00100	0.1046	0.00101
1/4	0.0791	0.00072	0.0873	0.00100	0.0742	0.00102
1/5	0.0646	0.00079	0.0713	0.00110	0.0606	0.00111
1/6	0.0574	0.00092	0.0633	0.00129	0.0538	0.00130
1/7	0.0354	0.00089	0.0391	0.00124	0.0332	0.00125
2/1	0.7731	0.00125	0.8534	0.00175	0.7253	0.00177
2/2	0.0773	0.00036	0.0854	0.00051	0.0725	0.00051
2/3	0.0777	0.00054	0.0857	0.00076	0.0728	0.00076
2/4	0.0641	0.00059	0.0707	0.00083	0.0601	0.00083
2/5	0.0785	0.00094	0.0866	0.00131	0.0736	0.00132
3/1	0.3825	0.00077	0.4222	0.00107	0.3588	0.00108
3/2	0.1457	0.00059	0.1608	0.00083	0.1366	0.00083

three profiles. Values of $\partial W_s / \partial z$ for the linear profile lie between those of the logarithmic and polynomial. This suggests that conclusions drawn in terms of cross-stream velocity gradients are relatively insensitive to the vertical profile chosen.

When $\partial y_c^* / \partial z = 0$, the fact that Γ is constant across the 'laterally uniform' zone implies that $\partial W_s/\partial z$ is also constant, as can be seen from Equations 28.9 and 28.17. Thus, W_s decreases linearly with lateral distance from the main-channel centreline. Moreover, Equation 28.18 shows that V_{max}, does not change laterally. This is in line with the secondary circulation pattern suggested in Figure 28.8. However, Table 28.2 shows that the values of $\partial W_s/\partial z$ for all three profiles are very large indeed. For example, if the $\partial W_s/\partial z$ remains constant across the 4.1 m wide floodplain of run 1/1 then W_s, at the interface, would be $4.1\times0.336 = 1.344$ m s^{-1} for a depth ratio of 0.196. This would reduce to 0.145 m s^{-1} at the highest depth ratio of 0.4. Cross-stream velocities of the size calculated above would have been visually apparent in the experiments on the FCF, yet they were not seen. Furthermore, the LDA measurements of cross-stream velocities for geometry 2 are only of the order of 0.01 m s^{-1}.

Table 28.2 shows that the logarithmic and polynomial curves give almost equal values of V_{max}, whilst the linear values are some 28% less. They do not vary in a systematic fashion with depth as does the cross-stream velocity. The values are lower than the measured mean vertical velocities, at 1.5 m from the main-channel centreline, which were -0.005, -0.0045 and -0.007 m s^{-1} for runs 2/3, 2/4 and 2/5 respectively. LDA readings were not taken at a greater lateral distance on the floodplain and these are on the very inner limit of the observable 'laterally uniform' zone. However, the LDA readings for the vertical components show no evidence of decreasing further with greater lateral distance onto the floodplain.

These data indicate that a single circulation cell would need to be much stronger than that observed to explain the high values of bed shear in the 'laterally uniform' zone. Reasons for this disparity need to be sought. A particular problem is posed by the fact that the model over-estimates the $\partial W_s / \partial z$ values and under-estimates the V_{max} values. However, these are related through the continuity equation, thus, an increase in the absolute value of one would require a corresponding increase in the other and *vice versa*.

An implicit assumption in the above analysis is that the cross-stream velocity profile remains geometrically similar across the floodplain. Substituting Equation 28.17 into Equation 28.9 results in a first order differential equation in W_s, which can be solved to give values of cross-stream velocity across the floodplain, assuming that $W_s = 0$ at the outer edge. To test the effect of changing cross-over depth ratio, y_c^*, was allowed to vary linearly across the floodplain. Typical results are given in Table 28.3 for run 1/4 at a point 2 m from the channel centreline, i.e. towards the inner edge of the 'laterally uniform' zone. This shows that an increase in y_c^* from the main-channel interface to the edge of the floodplain depresses W_s and W_b, although still not to the low values recorded in the FCF. For the larger values of $\partial y_c^* / \partial z$, the cross-stream velocity reaches a peak value on the floodplain and then decreases towards the main-channel interface whilst for smaller values it increases steadily as it approaches the interface. In the latter case the vertical velocity increases with distance from the main-channel interface since $\partial W / \partial z$ varies exponentially in the solution. However, in general the V_{max} values decrease in line with the reduced variation of cross-stream velocity.

A second implicit assumption in the solution is that of a horizontal water surface. Measurements indicate that the water surface was horizontal, however small variations of the order of ±1 mm may well have been undetected. To assess the effect of lateral water-surface slope, an increase in level of 1 mm ($s = 2.44 \times 10^{-4}$

Table 28.3. Effect of changing cross-over depth at 2.0 m from channel centreline. Run 1/4: depth ratio 0.196

Values of y_c^* on floodplain		Cross-stream velocity (m s^{-1})		Maximum vertical velocity (m s^{-1})
inner	outer	W_s	W_b	
0.5	0.5	0.237	-0.237	0.00072
0.4	0.6	0.176	-0.208	0.00063
0.3	0.7	0.133	-0.188	0.00055
0.2	0.8	0.102	-0.173	0.00048
0.1	0.9	0.076	-0.162	0.00042

for geometry 1) between the main-channel interface and the floodplain outer edge was imposed. The additional terms relating to the water surface on the left-hand side of Equation 28.4 are now activated. Determination of $\partial\langle\rho UW\rangle/\partial z$ from Equation 28.16 for varying water-surface level involves depth variation in the velocity profile parameter A (i.e. $\partial A/\partial y$) which can be evaluated from best-fit curves. This analysis leads to a differential equation in W_s, similar to that above, which can be solved by letting $W_s = 0$ at the floodplain edge. The imposition of this lateral water-surface slope gives a value of W_s = 0.206 m s^{-1} for run 1/4 at 2 m from the channel centreline. This is a reduction of 12.9% from the value for a horizontal water surface. The reduction in velocity, for the same change in lateral water level of 1 mm across the floodplain is larger for the lower depths, for example, there is reduction of 32.8% for run 1/1 at the lowest depth ratio (0.057), yet only a reduction of 6.0% for run 1/7 at the highest depth ratio (0.400). It is apparent that even a relatively modest lateral water surface slope can significantly reduce the cross-stream velocity which is required to account for the increased boundary shear stress on the floodplain. However, a much larger lateral slope would be needed to reduce calculated cross-stream velocity values to the observed values – and such a large lateral slope was not observed on the FCF.

Clearly then, if the increase in apparent shear stress across the floodplain is to be balanced by the secondary flow term $\partial\langle\rho UW\rangle/\partial z$ alone, the variation required in the cross-stream velocity is considerably greater than that observed. This remains true when realistic adjustments due to change in velocity profile shape or non-horizontal water surface are considered. Having highlighted this anomaly in the cross-stream velocities, it may be worthwhile briefly re-considering the experimental data upon which it is based. The calculations hinge upon three measurements on the floodplain, namely the depth of flow, Y, the boundary shear stress, τ_b, and the Reynolds stress, τ_{zx}. Determination of the flow depth offers little scope for error. Reservations expressed earlier concerning the applicability of the Preston tube at low Reynolds numbers and the *pro rata* boundary shear stress corrections do not adequately explain the consistency and magnitude of the differences between the boundary shear stress and the weight component of flow in the 'laterally uniform' zone. Hence, the apparent shear stress, which arises from this difference, cannot sensibly be ascribed to experimental error.

The Reynolds stress term $\partial(Y\langle\tau_{zx}\rangle/\partial z$ is assumed to be zero in the analysis of the 'laterally uniform' zone. LDA readings of τ_{zx} for geometry 2 are plotted in Figure 28.6 and these show that it rapidly falls to zero on the floodplain beyond the main-channel/floodplain interface. Moreover Figure 28.6 illustrates that if the $\partial(Y\tau_{zx})/\partial z$ term were to balance the apparent shear stress term $\partial(Y\tau_a)/\partial z$, then not only must τ_{zx} be very much greater at the interface than measured by the LDA, it must also vary right across the floodplain. A systematic error of this magnitude in the LDA readings is extremely unlikely.

The evidence suggests that the high Γ values on the floodplain cannot be totally accounted for by a secondary circulation cell, as shown in Figure 28.8, since the cross-stream velocities needed to drive it would be much larger than those observed. Nevertheless, if experimental errors are discounted then the magnitude of Γ is correct. It is therefore of interest in practical applications to determine the effect that these observed Γ values have on calculated velocity, boundary shear stress and discharge in compound channels.

SOLUTION OF THE TWO-DIMENSIONAL FLOW EQUATION

The depth-averaged flow Equation (33.4) can be calculated either with or without the Γ term. As shown in Figure 28.6 $\Omega = 1/Y\int_0^Z \Gamma \mathrm{d}z$ increases in the main channel from zero at the centreline to a point near the bottom of the side slope at $z = 0.75$ m. It then reduces slightly, before rising rapidly to a maximum on the floodplain adjacent to the top re-entrant corner of the main channel at $z = 0.9$ m. From this it decreases across the full width of the floodplain, the decrease being linear in the 'laterally uniform' zone where $\partial Y\Psi/\partial z = \Gamma = -\tau_a$. A simplified model for the variation in Ψ is given in Figure 28.11a. Here the small reduction in the Ψ value near the bottom of the channel side slope at $z \approx 0.75$ m is ignored. The Ψ values in the main channel appear to depend little upon depth, therefore, a linear variation from zero at the centreline to 1.06 N m^2 at $z = 0.75$ m is taken as a reasonable fit for all depths. The value of Ψ on the floodplain is modelled from the relationship $\Gamma = -(0.038 + 1.17(H - h))$ which was obtained earlier for the 'laterally uniform' zone. Hence, if Ψ is assumed to be zero at the floodplain edge, then it will be equal to $(0.038/(H - h) + 1.17)(B - b - sh)$ at the re-entrant corner $z = 0.9$ m and Ψ varies linearly at the interface zone between $0.75 \text{ m} < y < 0.9 \text{ m}$. These three zones of linear variation in Ψ translate into zones of constant Γ (see Figure 28.11b).

The Reynolds shear stress, τ_{zx}, may be related to the lateral gradient of the depth mean velocity as:

$$\langle\tau_{zx}\rangle = \rho\langle\varepsilon_{zx}\rangle\frac{\partial U}{\partial z} \tag{28.19}$$

where ε_{zx} is the eddy viscosity. A common basis for many turbulence models (e.g. Lean & Weare, 1979; Wormleaton, 1988; Shiono & Knight, 1988) is that ε_{zx} can be expressed in terms of the depth, Y, and shear velocity, U_*, as:

$$\langle\varepsilon_{zx}\rangle = \lambda Y U_* \tag{28.20}$$

where λ is a parameter of the flow. The shear velocity and boundary shear stress are related to the mean velocity, U, using the Darcy-Weisbach friction factor f. Thus:

$$U_* = \left(\frac{f}{8}\right)^{1/2} U \quad \text{and} \quad \tau_b = \frac{\rho f U^2}{8} \tag{28.21}$$

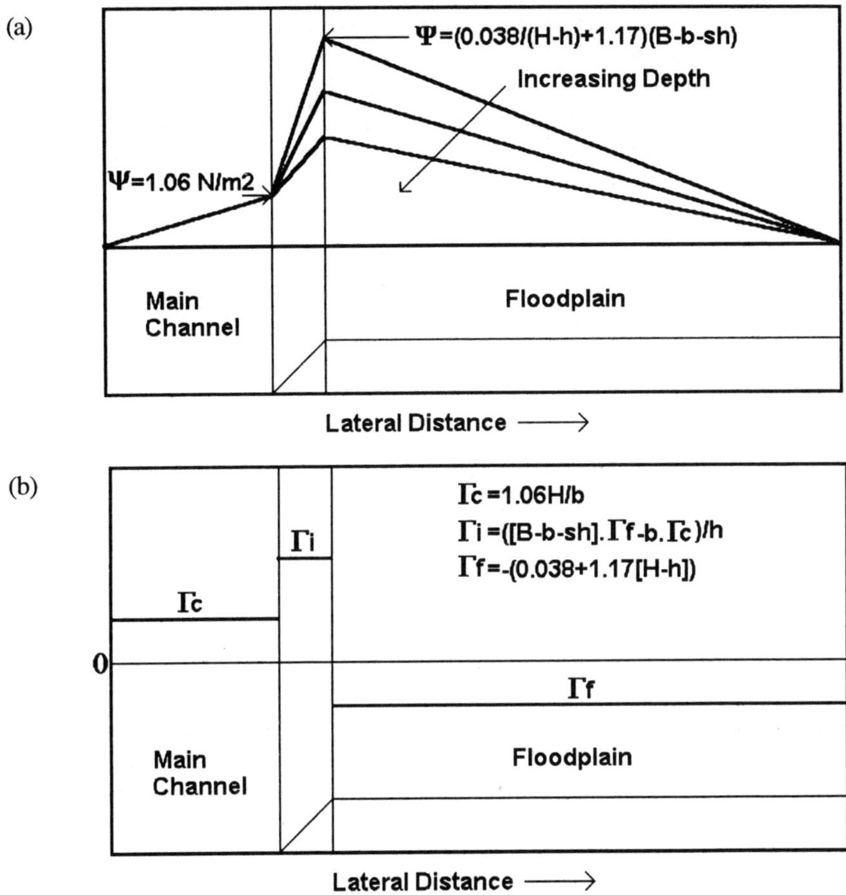

Figure 28.11. (a) Ψ and (b) Γ variations used in the two-dimensional mathematical model for velocity and boundary shear stress distribution

Substitution of Equations 28.19–21 into Equation 28.4 results in a differential equation in terms of velocity U. This has been solved analytically both for constant λ (Shiono & Knight, 1988, 1992) as well as numerically for variable λ by modifying the friction factor f across the section (Wormleaton, 1988). Wormleaton (1988) used an ε_{zx} value comprising two components, one similar to Equation 28.20 representing bed generated turbulence and a second, based on lateral distance and velocity parameters, representing shear generated turbulence. Since the purpose of this exercise is to isolate the improvement in accuracy brought about by the inclusion of Γ, the simplest model was chosen to highlight the comparison. Thus, Equation 28.20 was used for ε_{zx}, with a single λ value of 0.067 across the entire section. This is a generally accepted value for depth-averaged two-dimensional open-channel flow. The main channel was taken as hydraulically smooth, whilst the floodplain was assumed to have a roughness height of $k_s = 0.237$ mm, as inferred by the numerical analysis of Thomas & Williams (1995). The standard Colebrook-White equation was used to determine local values of f across the section.

The depth-averaged flow equation was used to analyse the flows measured in the FCF for all depth ratios of geometries 1–3. It should be recalled that depth ratios of greater than 0.242 (run 5) for geometry 2 and 0.098 (run 2) for geometry 3 did not have a recognizable 'laterally uniform' zone.

The importance of Γ for calculating the depth-averaged velocity and boundary shear stress distributions is shown in Figures 28.12 and 28.13 which compare experimental data with numerical solutions. Depth-averaged velocity distributions for geometries 1–3 at depth ratios of approximately 0.1 and 0.3 are shown in Figures 28.12a and 28.12b respectively. Boundary shear stress results at the same two depth ratios are similarly shown in Figures 28.13a and 28.13b. Figure 28.12 shows that the model for Γ reduces velocity in the main channel where Γ is positive and increases it on the floodplain where Γ is negative. This generally improves the agreement with observed data in the main channel for geometries 1 and 2. For geometry 3, particularly at the lower depth ratio of 0.1, the inclusion of Γ reduces the main-channel velocity below that observed. The main channel-floodplain interaction is reduced as the floodplain becomes narrower, as indicated by the smaller apparent shear stress at the interface. This leads to increased main-channel discharge as shown in Figure 28.5. It is probable, therefore, that the value of Γ, which is related to the main channel-floodplain interaction, is less for geometry 3 than the model value obtained from geometry 2. On the floodplain the inclusion of Γ improves the agreement with the observed data, particularly in the 'laterally uniform' zone. Γ serves to transfer momentum outwards across the floodplain from the main-channel interface. It is particularly encouraging to see the close agreement between the observed and calculated velocity results on the 'laterally uniform' zone, since the former were not used in any way in the development of the model for Γ. The prediction of floodplain velocity in the shear zone, near the interface, can be improved by using different values of λ on the floodplain, as described by Knight & Shiono (1988, 1991) or by using a modified formula for ε_{zx}, as described by Wormleaton (1988).

The boundary shear stress results in Figure 28.13 show a similar pattern of variation to the depth-averaged velocities. This is not surprising since the two values are related in the sense that $\tau_b \propto U^2$. It is noticeable however, that the analytical boundary shear stresses inclusive of Γ generally agree more closely with the observed values in the main channel than do their velocity counterparts. This is more pronounced for lower depth ratios and the narrower floodplain of geometry 3. It might be argued that errors in boundary shear stress should be twice those in the velocity, as suggested by the relationship given above. However, in this study the boundary shear stress readings are used to formulate the model for Γ and the velocity values are not. The solution is therefore weighted in favour of correct boundary shear stress readings. The fact that it also gives improved agreement in the velocities is not only a check on the method itself, but on the independent observations of boundary shear stress and velocity from the FCF.

Bulk discharge in the whole channel or in the individual main channel and floodplain subsections can be determined by integrating the velocity distributions. The errors between calculated and observed discharge rates for geometries 1–3 are shown in Figure 28.14. Results from calculations when Γ is omitted give reasonable values of the whole discharge, however, this includes a large over-estimation of main-channel discharge which cancels-out a large under-estimation of floodplain discharge. This is a common feature of most conventional discharge calculation methods when they are applied to compound channels. These large errors in the constituent main channel and floodplain discharges are important for numerical modelling and also in the calculation sediment transport. The solution including Γ gives much reduced errors for flows in the individual channel sub-sections and also significant improvement in the total discharge over the range of depths studied. It is encouraging to note the good level of agreement that can be obtained for these bulk values using such a simple model for the eddy viscosity. The results do, however, underline the importance of including Γ in the calculation.

CONCLUSIONS

Results were presented from experiments on flow over a range of depths in three smooth compound channel geometries with different floodplain widths. For floodplain aspect ratios greater than about 40, a region of the floodplain was identifiable in which the boundary shear stresses and velocities did not vary laterally. Readings of boundary shear stress in the 'laterally uniform' zone indicated the presence of an apparent shear stress on vertical planes within the fluid. This increased linearly from the floodplain edge to the main-channel interface. It is shown that a wider floodplain results in greater apparent shear stress at the interface which reduces main-channel discharge. Furthermore, even with a floodplain aspect ratio of over 400, the floodplain width still affects main-channel discharge.

The variation in apparent shear stress requires a momentum transfer across the 'laterally uniform' zone. The mechanism for this was sought, and analysis of the depth-averaged flow equations indicates that it may be due to secondary circulation. The variation would suggest a single large cell extending across the full width of the floodplain which gradually reduces in strength towards the floodplain edge.

To drive a circulation of the required strength, the cross-stream velocities would need to be very much larger than those observed in the experiments on the FCF. Extending the analysis to lateral profiles of varying geometry and including the possibility of a non-horizontal water surface would go some way towards explaining this discrepancy. However, there still remained a large difference between observed cross-stream velocities and those required by a circulation cell strong

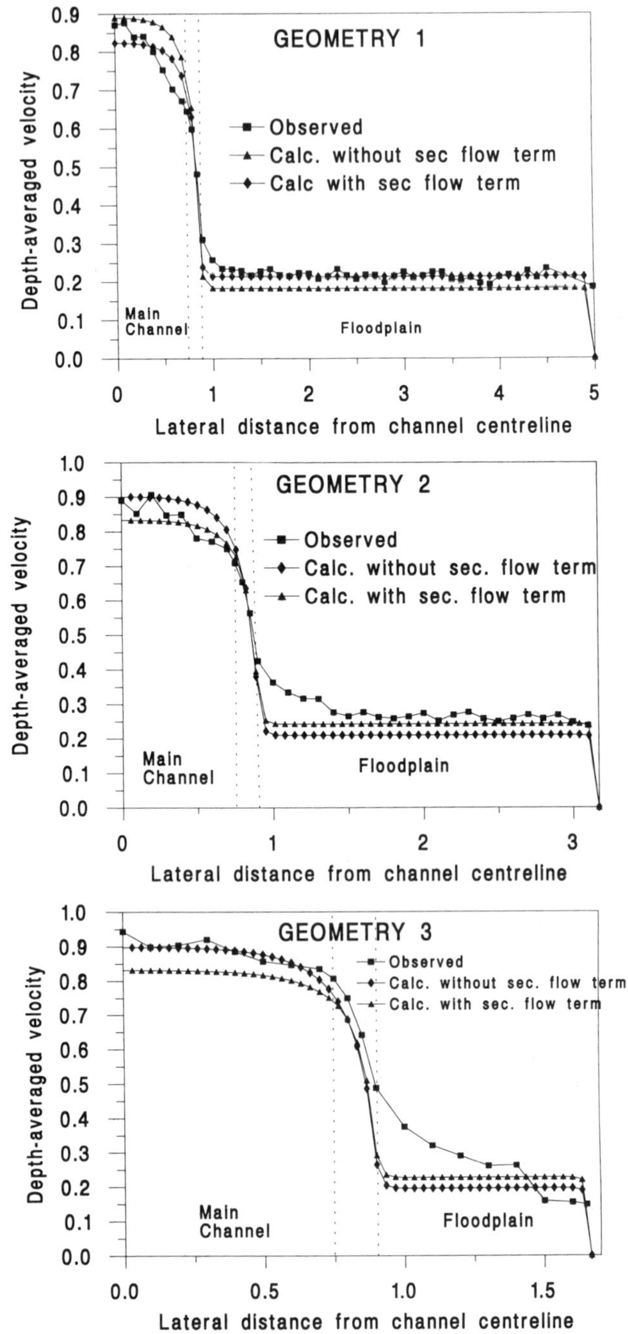

Figure 28.12a. Observed and calculated depth-averaged velocity distributions for depth ratio of 0.1 (velocity measured in m s^{-1} and lateral distance measured in m)

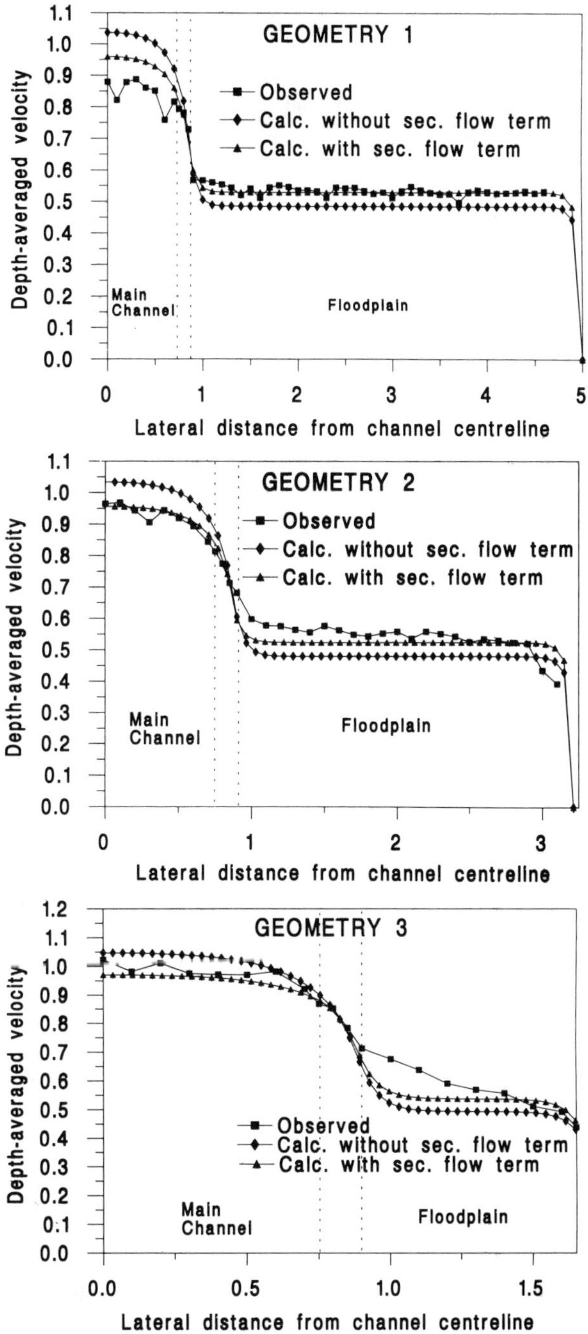

Figure 28.12b. Observed and calculated depth-averaged velocity distributions for depth ratio of 0.3 (velocity measured in m s^{-1} and lateral distance measured in m)

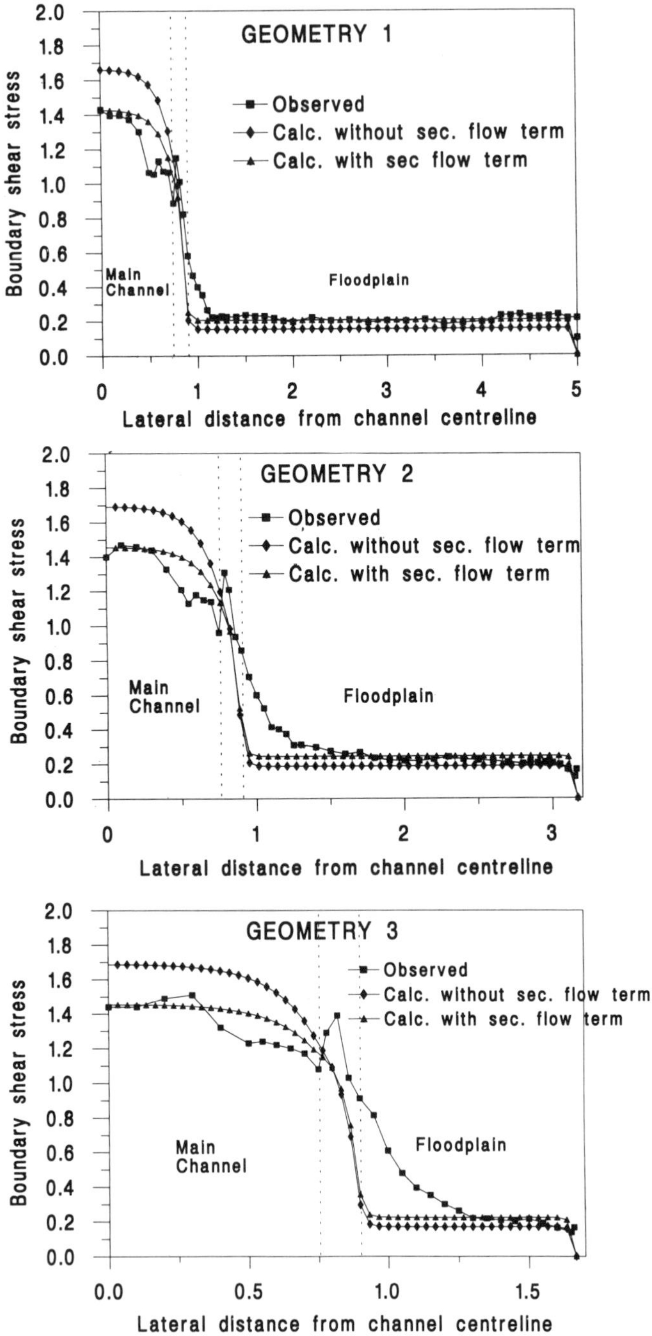

Figure 28.13a. Observed and calculated boundary shear stress distributions for depth ratio of 0.1 (bed shear stress measured in N m^{-2} and lateral distance measured in m)

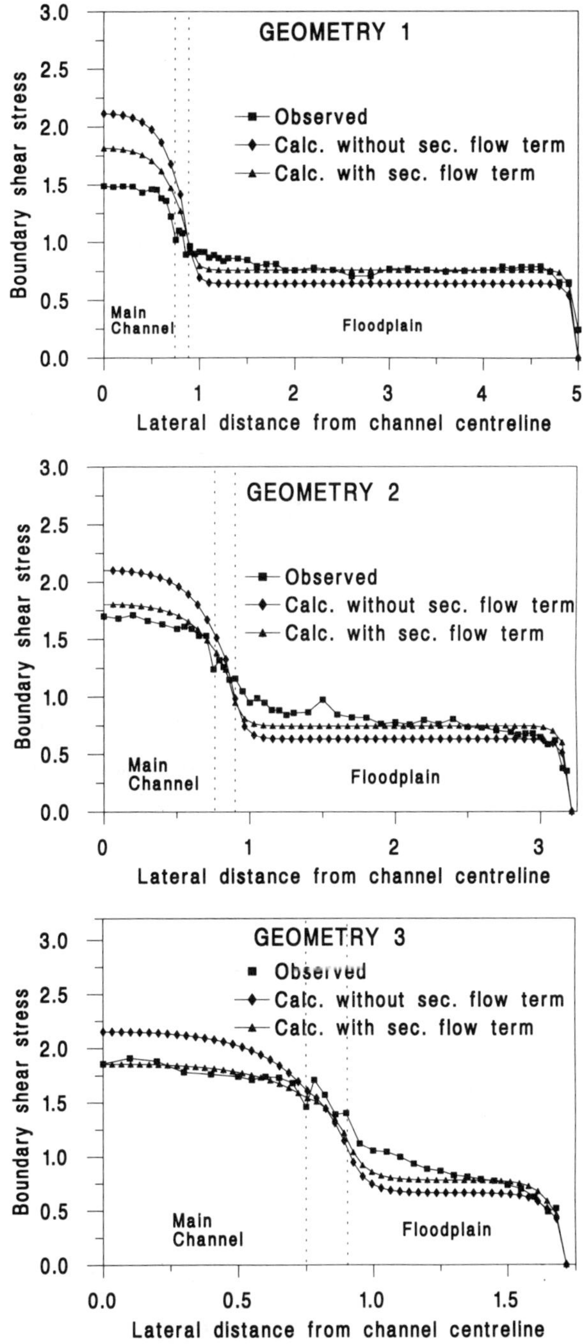

Figure 28.13b. Observed and calculated boundary shear stress distributions for depth ratio of 0.3 (bed shear stress measured in N m^{-2} and lateral distance measured in m)

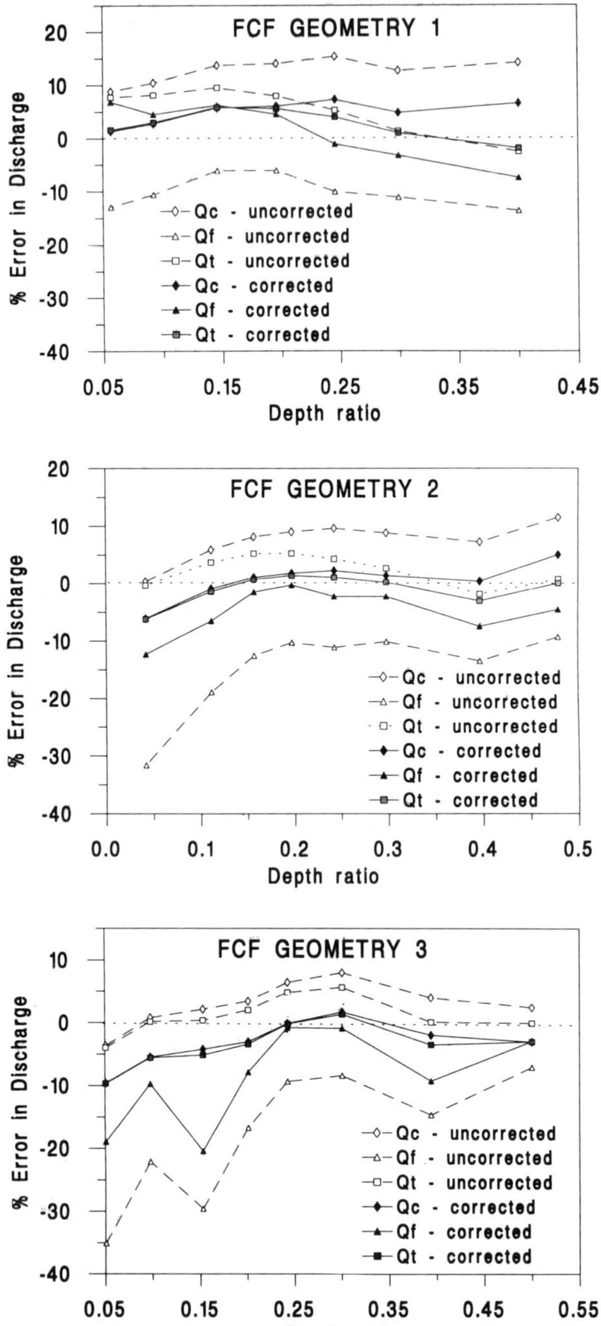

Figure 28.14. Errors in calculated discharge values

enough to increase the bed shear stresses on the floodplain to the experimental values. This leads to the conclusion that the increased bed shear stress on the floodplain are not entirely explained in terms of a single secondary circulation cell spanning the floodplain, as shown in Figure 28.8.

A numerical solution of the depth-averaged flow equations showed the importance of including the 'circulation' term, Γ, in the analysis. This improved the agreement between the observed and calculated velocity and boundary shear stress distributions. The discharge errors in the overall channel and particularly in the individual main channel and floodplain sub-areas were also significantly reduced by the inclusion of Γ in the solution.

Finally, it is worth noting that the three sets of experiments discussed in this paper were carried out with a hydraulically smooth floodplain. This is often the case in artificial compound channels built as flood protection works, however, in natural rivers the floodplain is invariably rougher. Whilst experiments have been carried out with roughened floodplains (Wormleaton *et al*, 1982; Knight, 1992), secondary velocity measurements and boundary shear stresses on the floodplain have not been recorded. A roughened floodplain will retard the floodplain flow (e.g. Naish & Sellin, this volume), thus increasing the lateral shear in the region of the main channel-floodplain interface (Wormleaton *et al.* 1982) which drives the secondary circulation cell on the floodplain. However, this will be offset by the increased dissipative effect of the roughened floodplain bed. There is clearly a need for detailed experimental data before the net result of these opposing effects can be quantified.

REFERENCES

Knight, D.W. (Ed) 1992. *SERC Flood Channel Facility*, HR Report **SR314**, HR Wallingford Ltd.

Lean, G.H. & Weare, T.J. 1979, 'Modelling two-dimensional circulating flow', *J. Hydraul. Div., ASCE*, **105**, 78–92.

Nezu, I. & Rodi, W 1986. 'Open channel flow measurements with a laser Doppler anemometer', *J. Hydraul. Div., ASCE*, **112**, 335–354.

Patel, V.C. 1965. 'Calibration of the Preston tube and limitations on its use in pressure gradients', *J. Fluid Mech.*, **23**, 185–208.

Shiono, K. & Knight, D.W. 1988. 'Two dimensional analytical solution for a compound channel', in. Isawa, Y., Tamai, N. & Wada, A. (Eds) *Proc. 3rd. Int. Symp. on Refined Flow Modelling and Turbulence Measurements*, Tokyo, Japan, 503–510.

Shiono, K. & Knight, D.W. 1991. 'Turbulent open channel flows with variable depth across the channel', *J .Fluid Mech.*, **222**, 617–646.

Thomas, T.G. & Williams, J.J.R. 1995. 'Large eddy simulation of a symmetric trapezoidal channel at a Reynolds number of 430,000', *J. Hydraul. Res.*, **33**, 825–842.

Wormleaton, P.R. 1988. 'Determination of discharge in compound channels using the dynamic equation for lateral velocity distribution', *Proc. Int. Conf. on Fluvial Hydraulics*, Budapest, Hungary, 98–103.

Wormleaton, P.R., Allen, J. & Hadjipanos, P. 1982. 'Discharge assessment in compound channel flow', *J. Hydraul. Div., ASCE*, **108**, 975–993.

NOTATION

a	constant
A	constant
b	main-channel half bed width
B	floodplain half-width
$b \quad c \quad d$	constants in Equation 28.12
f	Darcy-Weisbach friction factor
g	acceleration due to gravity
h	undisturbed main-channel bankfull depth
H	undisturbed main-channel flow depth
K	constant
s	main-channel side slope
S_0	downstream bed slope
$U \quad V \quad W$	time-averaged velocity at-a-point in the downstream, vertical and lateral directions respectively
U_*	shear velocity in the flow direction
V_{max}	maximum vertical velocity
$x \quad y \quad z$	distance in the downstream, vertical and lateral directions respectively
y_c	vertical distance of point of zero cross-stream velocity from the bed
y_c^*	vertical distance of point of zero cross-stream velocity from the bed, non-dimensionalized with respect to Z
y_b	vertical distance from the bed
y_b^*	vertical distance from the bed, non-dimensionalized with respect to Y
Y	depth of flow at a point
$\alpha \quad \beta \quad \gamma$	constants
Γ	sum of the forces acting on a vertical fluid element of uniform width
ε_{zx}	eddy viscosity
κ	von Kármán constant
λ	flow parameter
ν	kinematic viscosity
ρ	fluid density
τ_a	apparent shear stress
τ_b	bed shear stress
$\tau_{yx}, \quad \tau_{zx}$	Reynolds shear stresses
Ψ	sum of depth-averaged shear stresses acting on a plane in the fluid
Ω	$1/Y\int_0^Z \Gamma dz$

SUBSCRIPT

b	Value at the bed
s	Value at the water surface

29 Meandering Overbank Flow Structures

B.B. WILLETTS & P. RAMESHWARAN
Department of Engineering, King's College, University of Aberdeen, Aberdeen, AB9 2UE, UK

ABSTRACT

Small-scale experiments are reported on overbank flow in systems in which the inner-channel meanders in a straight, uniform floodplain. Of the several factors that influence flow in such systems, attention is focused on channel bed relief, floodplain roughness and floodplain longitudinal slope. Experiments were performed in systems of regularized geometry, with flat floodplains of uniform roughness into which were incised inner channels of uniform sinuosity, consisting of a number of regular meanders. Each meander was composed of circular arcs, joined to its upstream and downstream neighbours by common tangents.

Meandering channels are shown to have quite different internal flow structures when flow is overbank, compared with inbank flow. Significant differences in overbank flow resistance between cases in which the independent variables take different values are shown to be associated with changes in the flow structures in and near the inner channel. In particular, the introduction of quasi-natural bed relief in the inner channel clearly influences the transverse and streamwise flow distributions and also the flow resistance.

INTRODUCTION

The flow resistance of river systems when banks are over-topped is of great interest because it dictates the elevation of the water surface for a particular flood discharge. It is believed to be strongly influenced by coherent flow structures associated with the interaction between flow over the floodplain and flow in the inner-channel cross-section (e.g. Sellin, 1964; Wormleaton, this volume; Naish & Sellin, this volume). The energy dissipation produced by mixing between these flows, and the associated vorticity, has an important influence on the stage-discharge behaviour and its practical consequences. Figure 29.1 outlines the view of flow structure that has emerged in the case of floodplain flows with a meandering inner channel from a number of experimental studies undertaken predominantly in laboratory channels with a sequence of regular meanders cut into a plane floodplain of uniform width (Toebes & Sooky, 1967; James & Brown, 1977; Liu, 1991; Willetts & Hardwick, 1993; Ervine *et al.*, 1993). While, in both this case and that involving a straight inner channel (e.g. Ackers, 1991; Wormleaton, this volume) interaction between

Coherent Flow Structures in Open Channels.
Edited by P.J. Ashworth, S.J. Bennett, J.L. Best and S.J. McLelland.

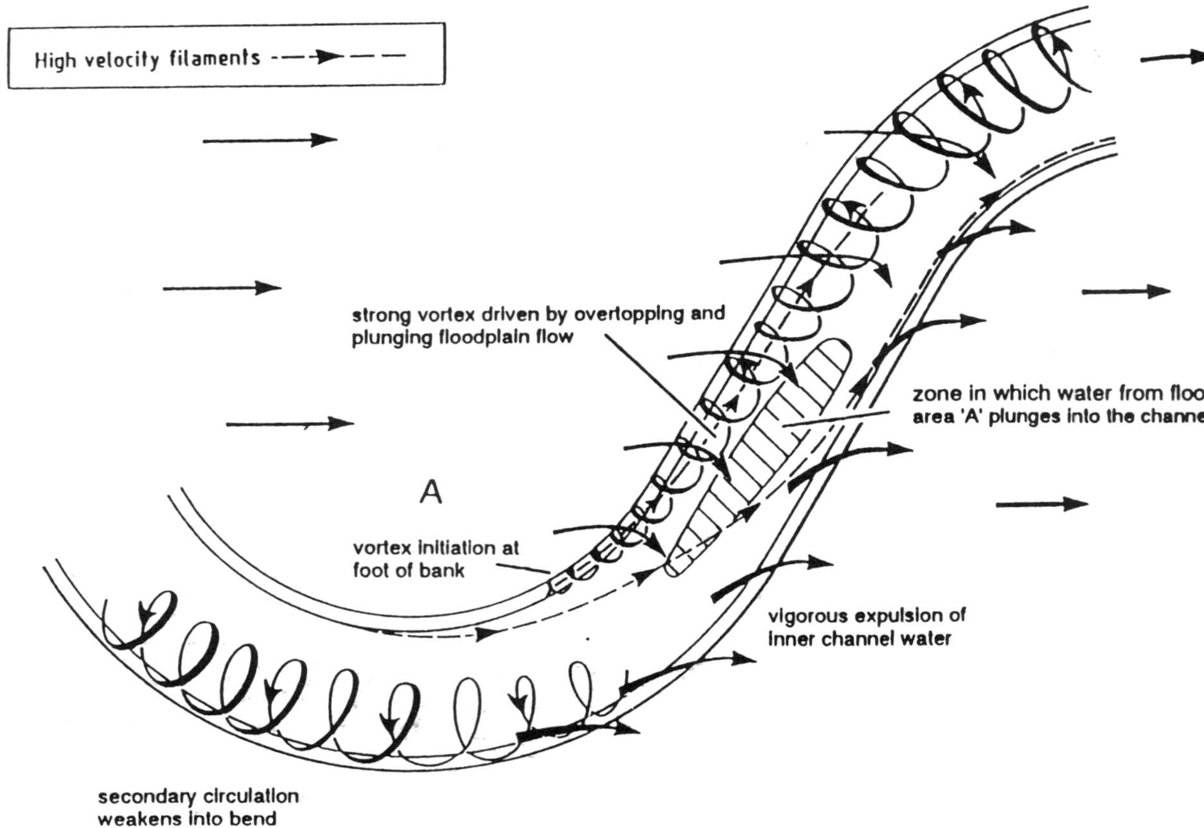

Figure 29.1. Flow features within a flooded meandering channel (after Hardwick, 1992)

'channel' and 'floodplain' flows is important, the flow structures involved in the interaction are quite different.

This paper is concerned with the flow structures responsible for mixing and energy dissipation. The objective of the reported study was to test the dependency of the features of flow shown in Figure 29.1, and therefore the energy dissipation rate and stage-discharge behaviour, on some of the variables which characterize the river-floodplain system. In the work reported here the variables studied are floodplain longitudinal slope and roughness and inner channel bed relief, but this work is part of a larger programme involving research teams at the Universities of Glasgow and Bristol which will also study the influence of sinuosity, channel aspect ratio and bank slope, floodplain side slope, meander amplitude -to- floodplain width, and system scale.

The paper is based on observations made predominantly in a model river-floodplain system of small-scale and having an inner channel of low aspect (width-to-depth) ratio. While reference is also made to work at larger-scale in the UK Flood Channel Facility (FCF) at HR Wallingford Ltd. (Wormleaton, this volume; Naish & Sellin, this volume), dependence on small-scale results implies some restriction of the general applicability of the findings and such limitations are noted. Results will be presented in terms of the stage-discharge behaviour and of the distribution of streamwise and spanwise flow components in the inner-channel for different values of the studied variables.

EXPERIMENTS

EQUIPMENT AND PROCEDURE

Observations were made in a system consisting of an inner channel of sinuosity approximately 1.4, cut in a flat, 1.2 m wide floodplain in a tilting flume. The inner-channel top width was always 200 mm and the bank slope always 60°. In some experiments the channel bed was flat, 52.5 mm below the floodplain, so that the inner channel was trapezoidal; in others, the channel bed was allowed to form naturally by placing mobile material with an initially flat surface 52.5 mm below the floodplain and exposing it to bankfull flow, after which the flow-moulded surface was fixed to prevent further evolution of the bed topography. During formation of this 'natural bed', material was fed at the upstream end of the channel at the equilibrium transport rate. The resulting topography showed the characteristic features of natural channel beds, namely shoaling near the inner bank and deepening near the outer bank immediately downstream of bend apices, with fairly symmetrical cross-sections in the inter-bend regions. The topography of the 'natural bed' is detailed in Figure 29.2a and illustrated in Figure 29.2b. This channel form is termed the 'natural channel' throughout the text. It does not accurately mimic any natural channel in the field because the plan geometry is fixed independently of

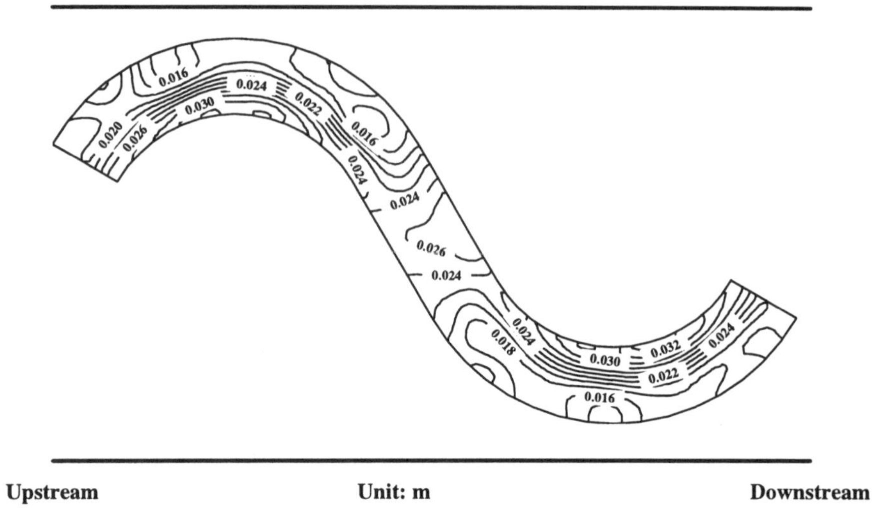

Figure 29.2a. Contours of 'natural' channel bed topography

Figure 29.2b. Photograph showing 'natural' channel bed

the sediment transport processes and the influence of sorting, while present, is not scaled to that in a full-size river.

In some experiments the floodplain roughness was equal to that of the inner-channel, whilst in other experiments the floodplain surface was uniformly roughened using a regular array of surface mounted 5 mm spheres. Figure 29.3 shows the variation with depth of the roughness height on the unroughened and roughened surfaces. The roughness is calculated for two array spacings, however, flow response will be reported only for the 'smooth' case and the rougher of the two 'rough' cases (hereafter called 'rough B'). Thus, flow behaviour is reported for: two trapezoidal inner-channel cross-sections with different floodplain roughness conditions; and two quasi-natural inner-channel cross-sections with different floodplain longitudinal slopes (see Table 29.1).

Discharge was measured by means of a flow meter in the delivery pipe to the headwater tank. The horizontal velocity component was measured using a miniature propeller meter (15 mm diameter) with a rotating vane to orientate the meter in the mean flow-direction. Horizontal velocity was measured at several levels in the flow, as detailed in Figure 29.4. Water-surface elevation was measured using a point gauge position-registered in the vertical by a rack and pinion lift connected to a

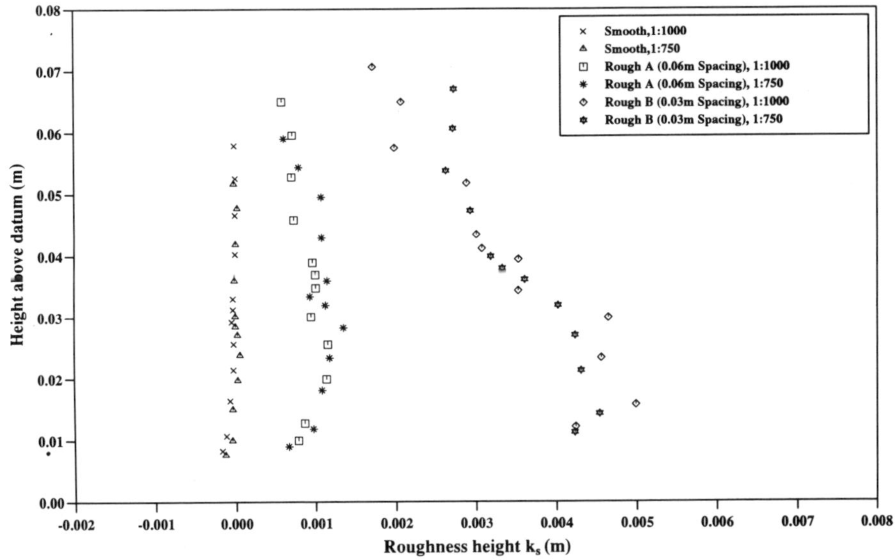

Figure 29.3. The variation of roughness height with water height for different fixed-bed experimental configurations. Roughness height was calculated using the Colebrooke-White equation (Ackers, 1991). Calibration experiments were performed in a 0.3 m wide flume and a side-wall correction has been applied to the roughness height calculation

Table 29.1. Summary of the experimental programme

Test No.	Sinuosity	Cross-section	Floodplain longitudinal slope	Roughness	Reported
1	1.374	Trapezoidal	1:1000	Smooth	YES
2	1.374	Trapezoidal	1:750	Smooth	
3	1.374	Trapezoidal	1:750	Rough A (0.06 m Spacing)	
4	1.374	Trapezoidal	1:1000	Rough A (0.06 m Spacing)	
5	1.374	Trapezoidal	1:1000	Rough B (0.03 m Spacing)	YES
6	1.374	Trapezoidal	1:750	Rough B (0.03 m Spacing)	
7	1.374	Natural	1:1000	Rough B (0.03 m Spacing)	YES
8	1.374	Natural	1:750	Rough B (0.03 m Spacing)	YES

rotary potentiometer. The instruments were mounted on a gantry running on rails parallel to the floodplain.

Systematic flow visualization was undertaken using dye injection at the bend apex and the cross-over region to assist the interpretation of the measured data. Three colours were injected simultaneously at the same level, but at different spanwise positions (Figure 29.5). The process was repeated at several depths, and video records made of each case. These sequences provide valuable non-quantitative information about vertical transpositions within the flow and, hence, the geometry of some of the persistent flow recirculation cells.

RESULTS

The overall resistance to flow is best characterized by means of stage-discharge relationships. Figure 29.6 shows the results in this form for four experiments which together reflect the resistance changes accompanying changes in floodplain roughness, channel bed relief and floodplain longitudinal slope. The terms 'smooth' and 'rough B' in the legend of this and later figures refer to friction behaviour described in Figure 29.3.

It was a hypothesis of the study that flow resistance is governed, to a large extent, by the flow circulations and redistributions associated with the mixing of 'channel' and 'floodplain' flow. In order to validate this hypothesis, it is necessary to investigate whether, and in what respect, flows with different stage-discharge characteristics exhibit different secondary flow behaviour. To this end, the distribution of spanwise and longitudinal velocity components, relative to the local channel direction, have been plotted for the inner channel of all the cases being compared at two different cross-sections (the bend apex and the cross-over, which is midway between adjacent bend apices). These distributions are shown in Figures 29.7a–29.7d and 29.8a–29.8d.

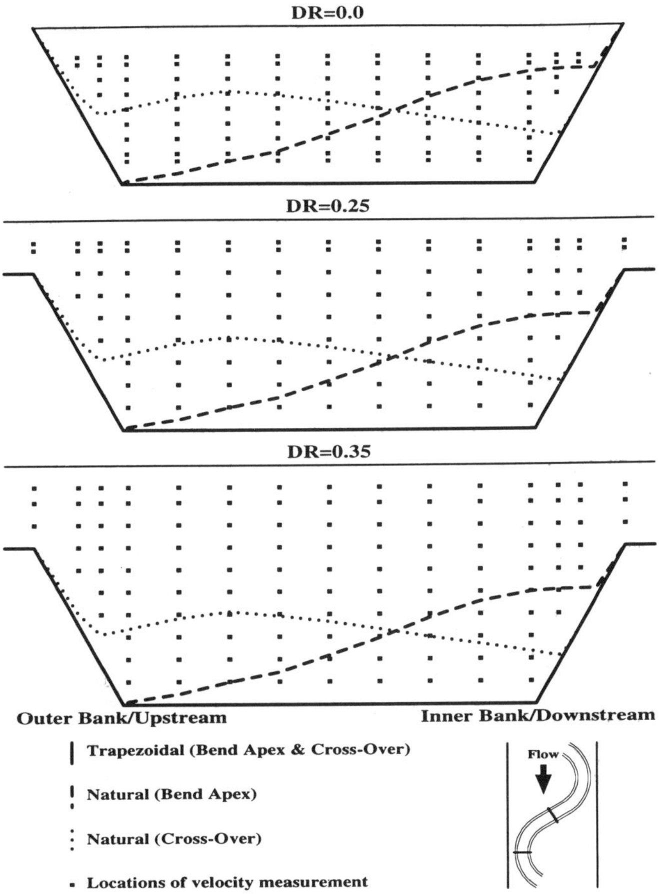

Figure 29.4. Locations of velocity measurement points. Solid baseline shows trapezoidal channel bed and broken lines show 'natural' channel bed (see key). Inset shows location of bend apex and cross-over measurement sections

DISCUSSION

The transverse velocities for the inbank case shown in Figure 29.7c accord well with field observation. In addition to the well-known primary circulation cell associated with bend flow, the small counter-rotating cell near the outer bank noted by Thorne *et al.*(1985) and the net outward flow near the inner-bank observed by Dietrich *et al.*(1979) are both present in Figure 29.7c. This degree of conformity with field observation reduces concern that the scale of the experiments and the artificiality of the procedure for evolving the 'natural bed' may impair the validity of the results.

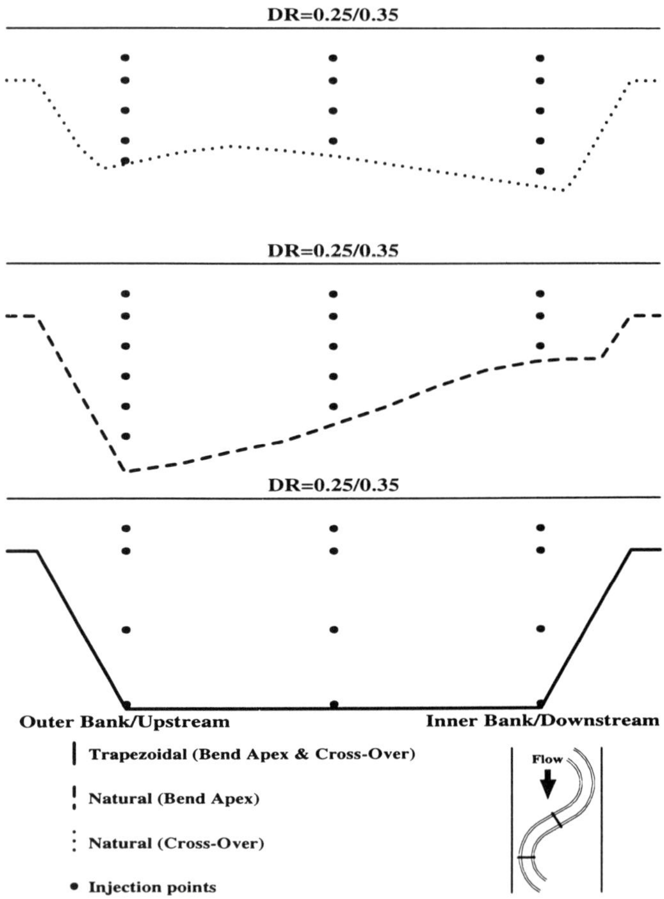

Figure 29.5. Locations of dye injection points. Solid baseline shows trapezoidal channel bed and broken lines show 'natural' channel bed (see key). Inset shows location of bend apex and cross-over measurement sections

The stage-discharge curve of the trapezoidal channel with a smooth floodplain (Figure 29.6) exhibits the features that have been seen in several reports of earlier studies (e.g. Sellin *et al.*, 1993). Discharge increases steadily with depth up to bankfull. The increase ceases for a range of small, overbank flow depths and then, at a depth ratio (floodplain depth -to- inner channel depth), DR, of approximately 0.15, resumes at a greater and accelerating rate with increasing flow depth. When the flow first goes overbank, the cross-sectional area of flow increases suddenly. However, this is not accompanied by an increase in discharge because the retarding effect of the boundary on the flow is also greatly increased. Indeed, the flow in the incised channel is reduced (compared with the bankfull condition) by the restraining

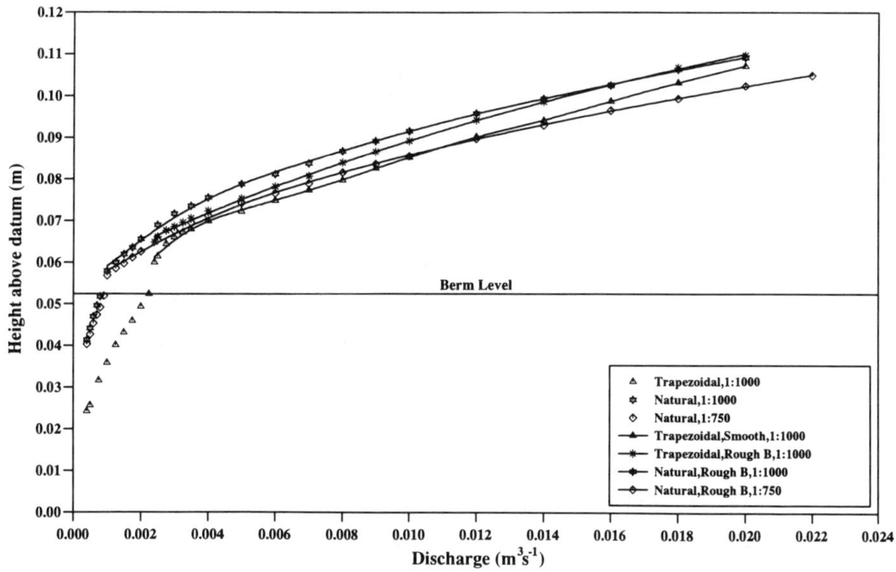

Figure 29.6. Stage-discharge curves showing relationship between flow height and discharge

influence of the floodplain flow which, being shallow, is more sluggish and which interacts with the channel flow through turbulent flow structures at the interface. These flow structures dissipate sufficient energy at shallow overbank depths to offset completely any increase in system discharge above the bankfull discharge.

The effect of roughening the floodplain without changing the channel cross-section, shown in a second curve in Figure 29.6 (Trapezoidal channel, Rough B, 1:1000), was to reduce discharge at all measurable overbank flow depth. The influence is greatest for high depth ratios, at which the system discharge can be expected to be increasingly dominated by the floodplain flow. Indeed, it has been suggested that at high floodplain depths, the channel may be regarded as a form of gross roughness in the base of the floodplain, resisting rather than supplementing the system discharge. For small overbank depths the floodplain roughness has negligible influence on system discharge.

A pronounced difference can be noted between the stage-discharge curve at high overbank flows in these experiments (which is very like that for the smooth floodplain case) and that reported in experiments with roughened floodplains in the FCF (Sellin *et al.*, 1993; Naish & Sellin, this volume). In the FCF experiments the reduction of discharge compared with the smooth floodplain case increased very rapidly and continuously with depth at large aspect ratios (see, e.g. Sellin *et al.*, 1993, Figures 6 and 7). The contrast between the FCF rough floodplain curve and that reported here is attributable to the different type of roughness elements used in

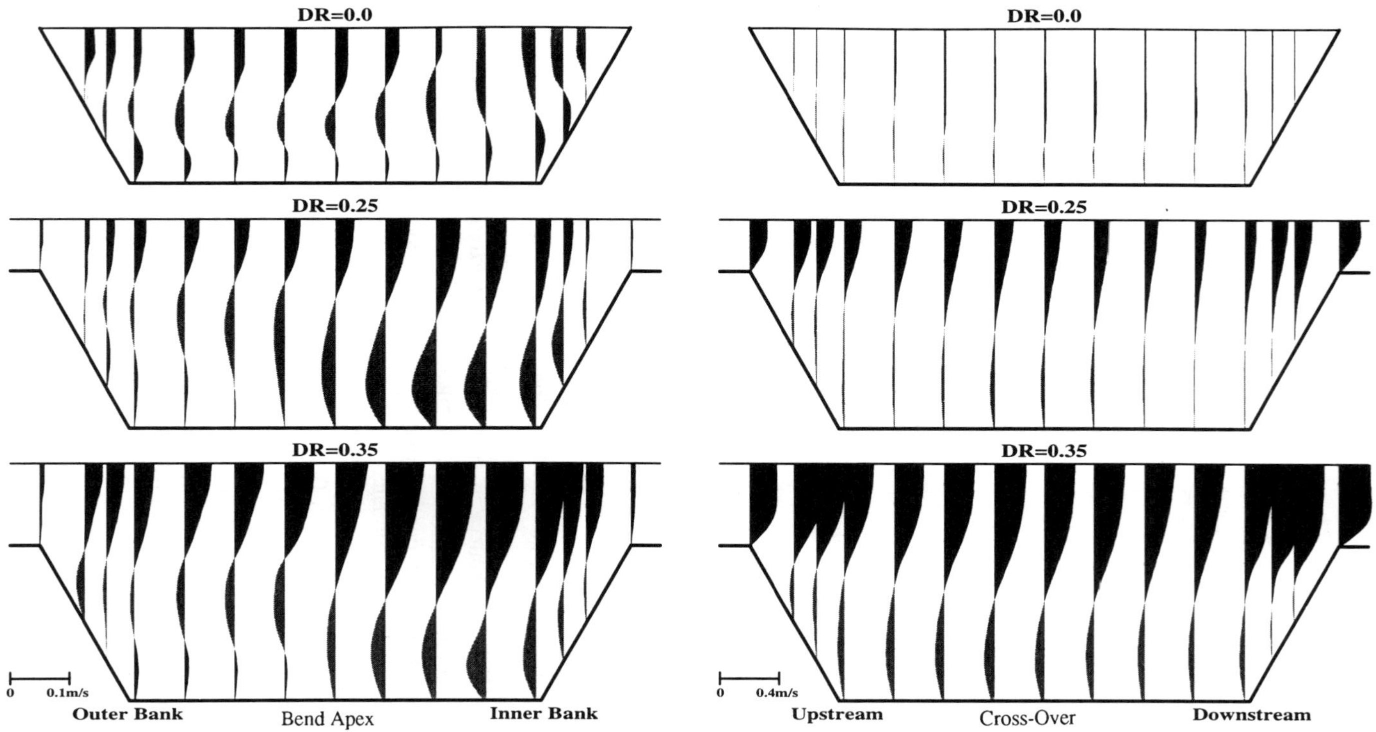

Figure 29.7a. Cross-sections showing transverse velocity structure at the bend apex (left) and cross-over (right) regions for *test number 1*. At each location, velocity profiles are shown for three depth ratios (DR = 0.0, 0.25 and 0.35). Test conditions are: trapezoidal channel shape, smooth floodplain and longitudinal floodplain slope of 1:1000 (see Table 29.1)

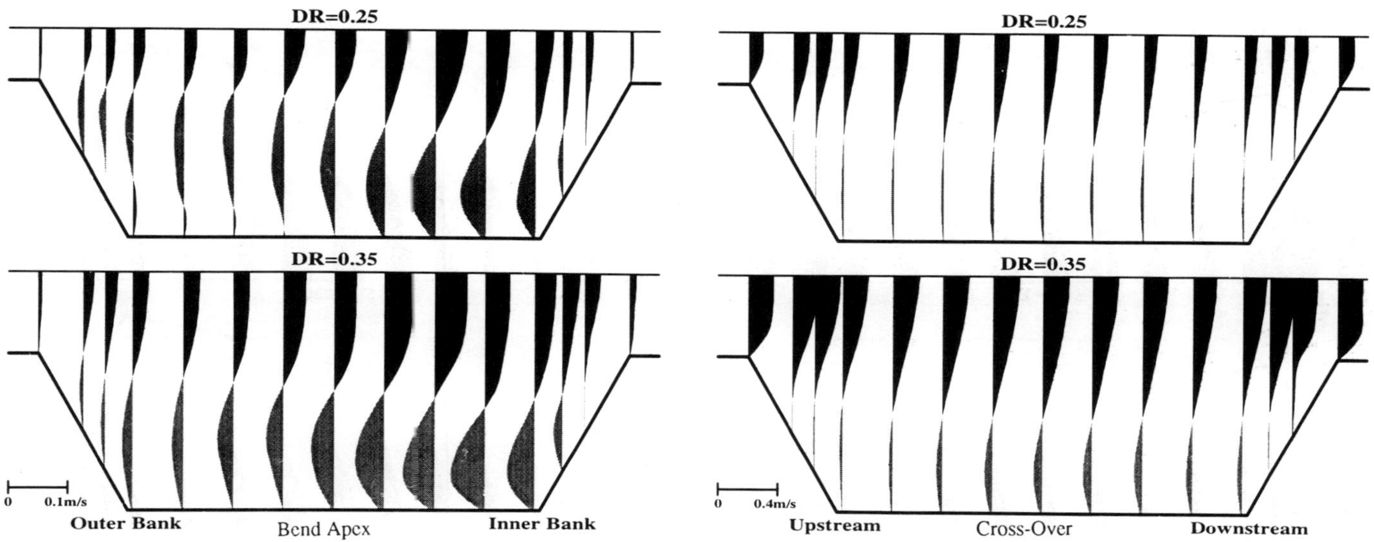

Figure 29.7b. Cross-sections showing transverse velocity structure at the bend apex (left) and cross-over (right) regions for *test number 5*. At each location, velocity profiles are shown for two depth ratios (DR = 0.25 and 0.35). Test conditions are: trapezoidal channel shape, 'rough B'-type floodplain roughness and longitudinal floodplain slope of 1:1000 (see Table 29.1)

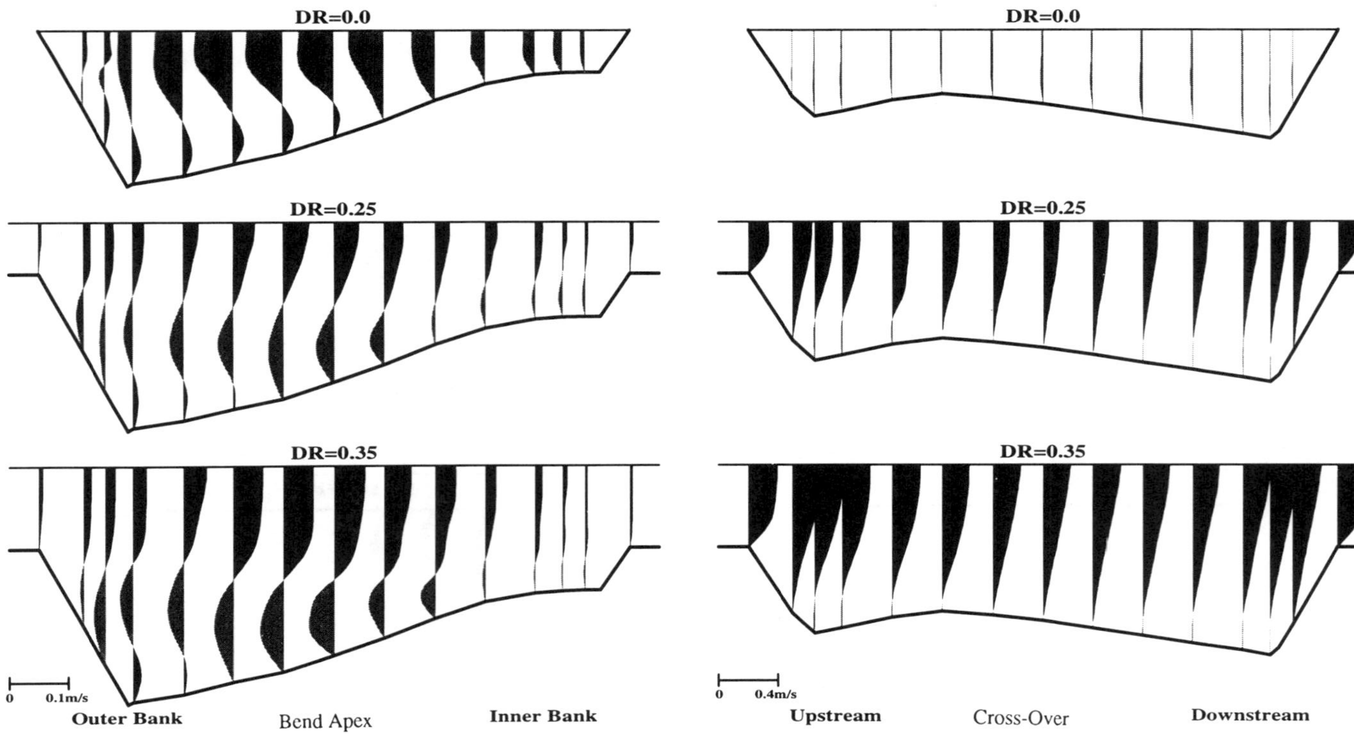

Figure 29.7c. Cross-sections showing transverse velocity structure at the bend apex (left) and cross-over (right) regions for *test number 7*. At each location, velocity profiles are shown for three depth ratios (DR = 0.0, 0.25 and 0.35). Test conditions are: quasi-natural channel shape, 'rough B'-type floodplain roughness and longitudinal floodplain slope of 1:1000 (see Table 29.1)

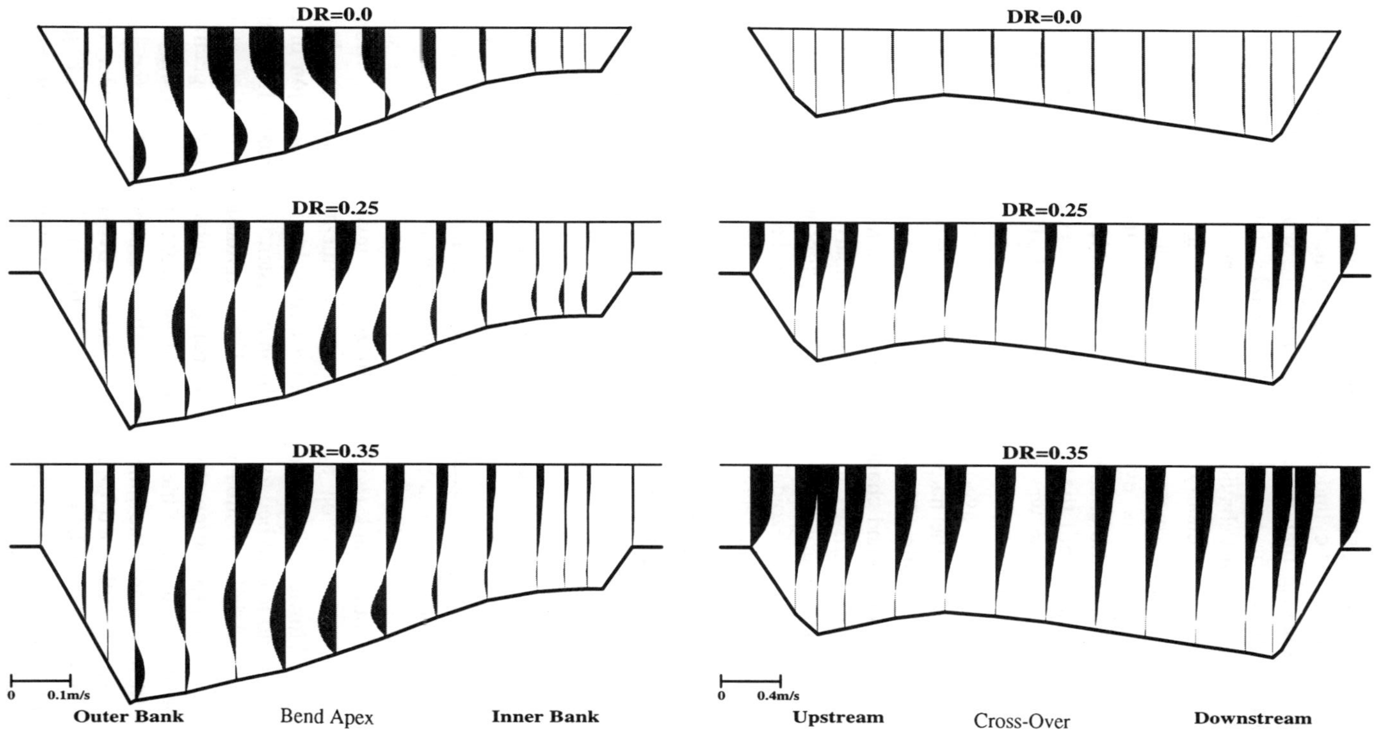

Figure 29.7d. Cross-sections showing transverse velocity structure at the bend apex (left) and cross-over (right) regions for *test number 8*. At each location, velocity profiles are shown for three depth ratios (DR = 0.0, 0.25 and 0.35). Test conditions are: quasi-natural channel shape, 'rough B'-type floodplain roughness and longitudinal floodplain slope of 1:750 (see Table 29.1)

the two studies. In the FCF experiments, floodplain roughening was accomplished by means of a pattern of vertical cylinders which always penetrated the water surface and produced flow resistance that increased continuously with depth, unlike the resistance behaviour shown in Figure 29.3. At depths for which flood-plain discharge dominates the system behaviour, it is not unexpected that this difference in floodplain roughness is so clearly reflected in the stage-discharge curve. In comparing the discharge performance of the trapezoidal and natural channel configurations (Figure 29.6), it is evident first that the inbank conveyance is considerably lower for the natural channel. This is attributable partly to reduction of cross-sectional area compared with the trapezoidal section, because the aspect ratio chosen for the trapezoidal section was unstable once the bed became mobile, and partly to an increased friction factor.

However, once the flow goes overbank the rate of increase of discharge with depth is greater in the case of the natural than the trapezoidal inner channel, and resumes from a lower depth ratio. This suggests that less energy is dissipated in the interaction between floodplain and channel flows in the former than in the latter case. Some support for this surmise can be found by comparing Figures 29.7b and 29.7c, which show that secondary circulations are more vigorous, both at the apex and at the cross-over in the trapezoidal as compared with the natural cross-section when flow is overbank. Because secondary currents promote mixing between the channel and floodplain flows, this difference is consistent with a greater energy dissipation rate through such mixing in the trapezoidal case.

Major differences are apparent in Figure 29.7 between the inbank and overbank secondary circulations at the bend apex. For the right-hand bend shown, the primary circulation cell is anticlockwise for inbank flow, but clockwise for overbank flow. This accords with findings in experiments at larger scale and has an important bearing on the direction of boundary shear stress vectors. For inbank flow, these vectors have a component that is directed towards the inner bank when its magnitude is high, and this directs sediment transport towards the inner bank over much of the bed width. The effect is particularly pronounced in the natural cross-section. However, for overbank flow, the large components of lateral shear stress are directed towards the outer bank (with a small zone of reversal near the deepest part of the cross-section). Thus, on a mobile bed, sediment transport might be expected to be inclined towards the outer bank over a considerable proportion of the bed width and this would lead to bed topography in persistently overbank flow quite different to that characteristic of inbank flow. The longitudinal velocities at the bend apex are also distributed differently in overbank and inbank flow respectively (Figure 29.8). The strongest part of the current relocates from an area near the outer bank (for inbank flow) towards the inner bank (in the overbank case). In the latter case, sediment would be expected to be much more active near the inner bank because these longitudinal velocity gradients are the main agents of grain disturbance, being far stronger than the lateral velocity gradients shown in Figure 29.7. Figure 29.8 also demonstrates that overbank flow generally has slower longitudinal components in the cross-over section, midway between adjacent bends, than does inbank flow and, therefore might be expected to reduce sediment transporting capability at this point. This implies that longitudinal sediment transport gradients are different in the two cases reinforcing the expectation of different topography.

The reversal of the secondary circulation cell direction in the overbank case is clearly driven by the momentum exchange with the flow coming off the floodplain in the cross-over region where the channel and floodplain flows are oblique in plan. The velocity profile information in Figure 29.7 indicates mixing between these two flows which is highly dissipative of energy and is an energy sink not present when flow is inbank. The overbank secondary cells involve a greater volume of water than the inbank flow cells and redistribute flow momentum such that high longitudinal velocities occur in shallower water, to a greater extent than for the inbank case. Both of these changes suggest greater rates of energy dissipation in the overbank case. Thus, the pattern of processes observed to be characteristic of overbank flow might be expected to be highly dissipative of energy and to be dependent on the momentum exchange associated with mixing between channel and floodplain flows. Further study is needed to establish quantitatively the component dissipation rates, and to determine their dependency on plan and cross-section geometry and boundary roughness.

Comparison of Figures 29.7c and 29.7d reveals very little difference in secondary current pattern or strength between cases in which the floodplain longitudinal slope is 1:1000 and 1:750. This is illustrated for experiments with roughened floodplains and natural channel cross-section. Figures 29.8c and 29.8d show the greater streamwise flow velocities that might be expected when the floodplain longitudinal slope is higher and indicate some minor redistribution of velocity in the cross-section, most noticeable at the cross-over when flow is overbank. The greater floodplain longitudinal slope, naturally, increases the discharge for a given stage (Figure 29.6).

The influence of floodplain roughness on the flow field can be seen by comparing Figures 29.7a and 29.8a with Figures 29.7b and 29.8b. While the longitudinal velocity component is lower when the floodplain is rough (as intuition might suggest) the overbank secondary flow cells are stronger for the rough floodplain case in comparison to the smooth floodplain condition.

CONCLUSIONS

- Momentum is exchanged between 'channel flow' and 'floodplain flow' because of the different influence of the solid boundary in these two zones.
- This momentum exchange depends on mixing between the two flows which is closely related to secondary current activity in the inner-channel cross-section (i.e. to large coherent flow structures).

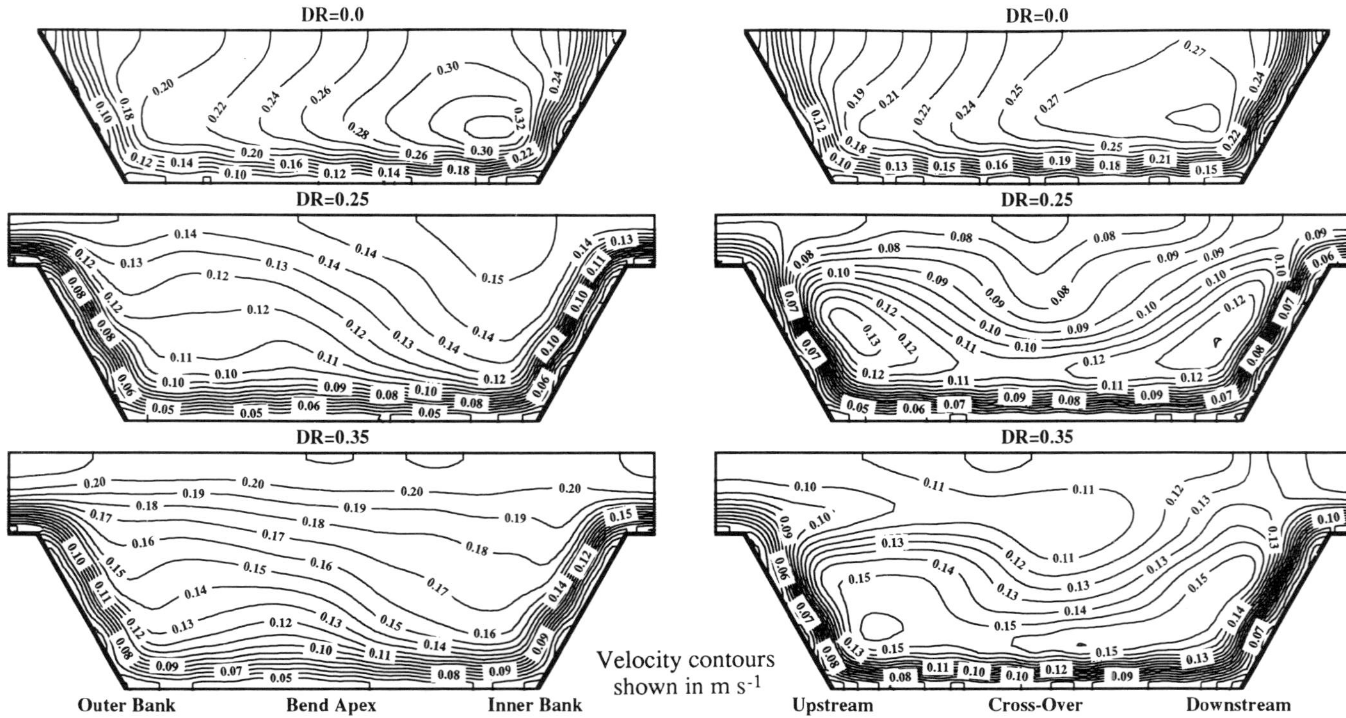

Figure 29.8a. Cross-sections showing longitudinal velocity structure at the bend apex (left) and cross-over (right) regions for *test number 1*. At each location, velocity contours are shown for three depth ratios (DR = 0.0, 0.25 and 0.35). Test conditions are: trapezoidal channel shape, smooth floodplain and longitudinal floodplain slope of 1:1000 (see Table 29.1)

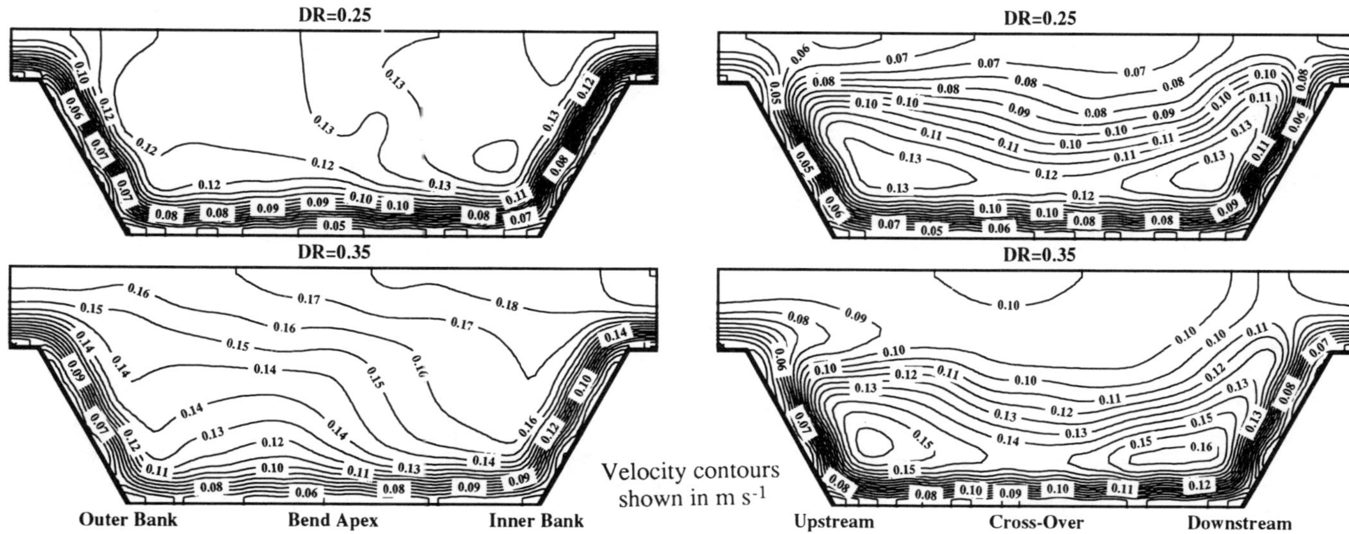

Figure 29.8b. Cross-sections showing longitudinal velocity structure at the bend apex (left) and cross-over (right) regions for *test number 5*. At each location, velocity contours are shown for two depth ratios (DR = 0.25 and 0.35). Test conditions are: trapezoidal channel shape, 'rough B'-type floodplain roughness and longitudinal floodplain slope of 1:1000 (see Table 29.1)

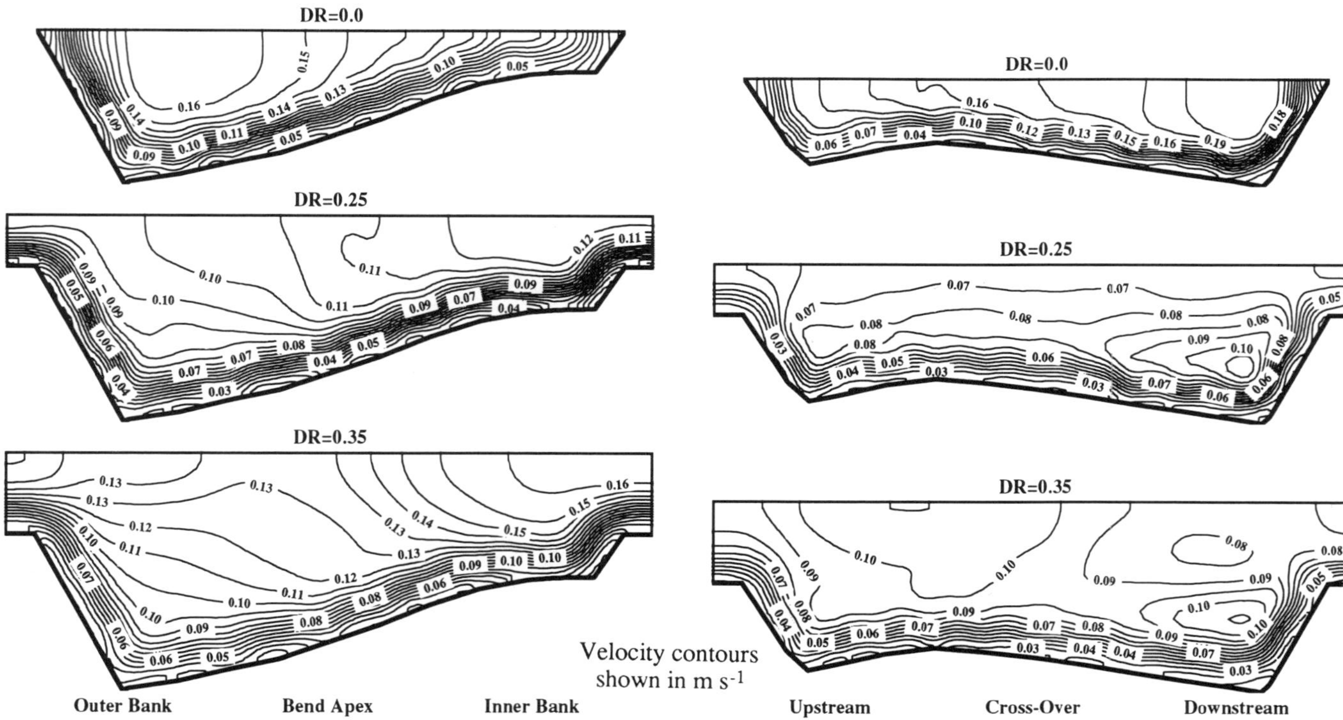

Figure 29.8c. Cross-sections showing longitudinal velocity structure at the bend apex (left) and cross-over (right) regions for *test number 7*. At each location, velocity contours are shown for three depth ratios (DR = 0.0, 0.25 and 0.35). Test conditions are: quasi-natural channel shape, 'rough B'-type floodplain roughness and longitudinal floodplain slope of 1:1000 (see Table 29.1)

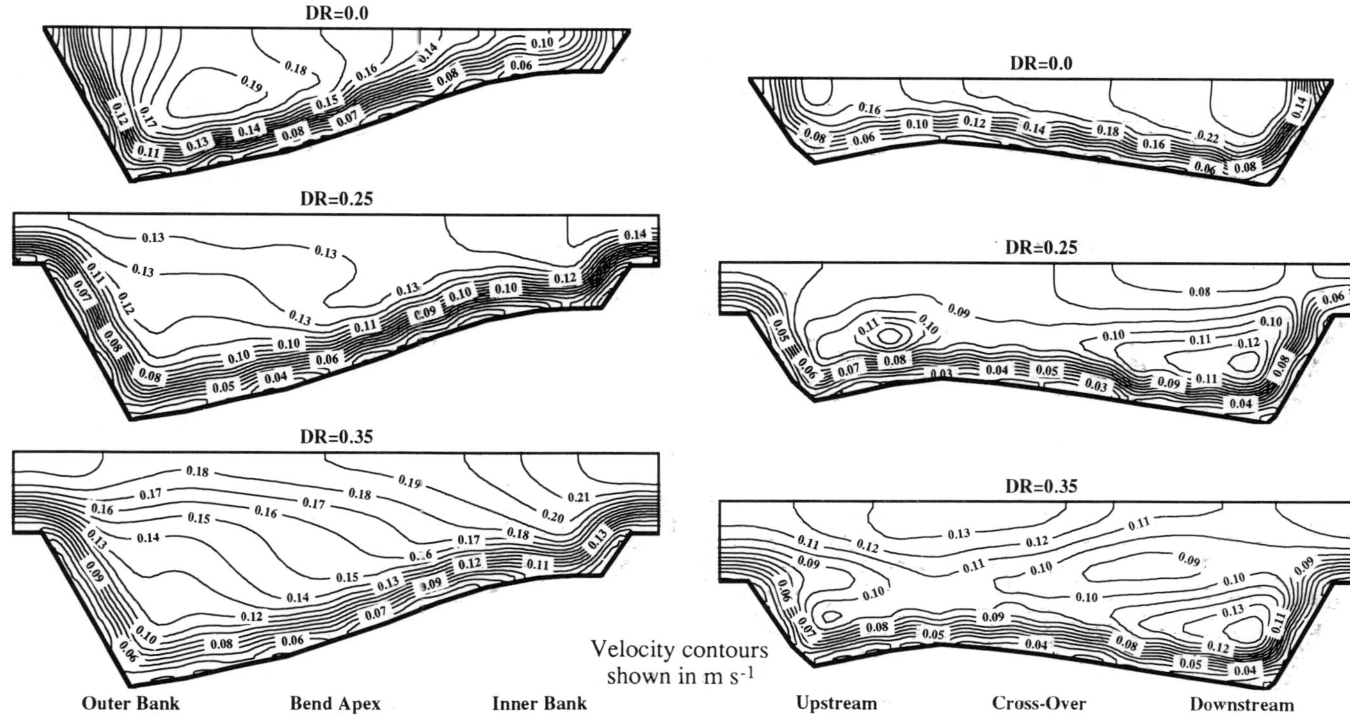

Figure 29.8d. Cross-sections showing longitudinal velocity structure at the bend apex (left) and cross-over (right) regions for *test number 8*. At each location, velocity contours are shown for three depth ratios (DR = 0.0, 0.25 and 0.35). Test conditions are: quasi-natural channel shape, 'rough B'-type floodplain roughness and longitudinal floodplain slope of 1:750 (see Table 29.1)

- These secondary flow cells influence energy dissipation directly and also indirectly by promoting dissipative exchanges between channel and floodplain.
- The momentum transfer and energy dissipation processes are quite different when the inner channel meanders from those when it is straight.
- Boundary shear stress and therefore sediment transport, is differently distributed in overbank as compared with inbank flow. Thus, the influence of overbank flow on channel morphology can also be expected to be distinct.
- Because it is quite clear that inner-channel shape profoundly influences the exchange processes, the results of laboratory studies that employ trapezoidal or rectangular cross-sections have limited significance in the context of real system behaviour.
- In the range studied, floodplain longitudinal slope has little influence on the flow structures, and therefore the character of the dissipation processes. However, an increase of floodplain roughness was observed to diminish channel longitudinal flows, but slightly increase the vigour of secondary currents in a trapezoidal section.

ACKNOWLEDGEMENTS

The Engineering and Physical Science Research Council provide the resource for this study. We are grateful for the benefit of helpful discussions with Dr D.A. Ervine, Professor R.H.J. Sellin and their colleagues.

REFERENCES

Ackers, P. 1991. 'Hydraulic design of compound channels', *Report* **SR281**, 2 vols., HR Wallingford Ltd.

Dietrich, W.E., Smith, J.D. & Dunne, T. 1979. 'Flow and sediment transport in a sand bedded meander', *J. Geol.*, **87**, 305–315.

Ervine, D.A., Willetts, B.B., Sellin, R.H.J. & Lorena, M. 1993. 'Factors affecting conveyance in meandering channels', *J. Hydraul. Engrg., ASCE,* **119**, 1383–1399.

Hardwick, R.I. 1992. *The Behaviour of Meandering Channels in Flood*, Unpub. PhD Thesis, Univ. of Aberdeen, 192 pp.

James, M & Brown, B.J. 1977. 'Geometric parameters that affect floodplain flows', US Army Corps of Engineers, Waterways Experiment Station, Report **H-77-1**, 141 pp.

Liu, B-Y. 1991. *Study on Sediment Transport and Bed Evolution in Compound Channels*, Unpub. PhD Diss., Disaster Prevention Research Institute, Kyoto Univ., 227 pp.

Sellin, R.H.J. 1964. 'A laboratory investigation into the interaction between the flow in the channel of a river and that over its floodplain', *La Houille Blanche,* **7**, 793–802.

Sellin, R.H.J., Ervine, D.A. & Willetts, B.B. 1993. 'Behaviour of meandering two-stage channels', *Proc. Instn. Civ. Engnrs., Wat., Marit. and Energy*, **101**, 99–111.

Thorne, C.R., Zevenbergen, L.W., Pitlick, J.C., Rais, S., Bradley, J.B. & Julien, P.Y. 1985. 'Direct measurements of secondary currents in a meandering sand-bed river', *Nature,* **315**, 746–747.
Toebes, G.H. & Sooky, A.A. 1967. 'Hydraulics of meandering rivers with floodplains', *J. Waterways and Harbours Div., ASCE, 93,* 213–236.
Willetts, B.B. & Hardwick, R.I. 1993. 'Stage dependency for overbank flow in meandering channels', *Proc. Instn. Civ. Engnrs, Wat., Marit. and Energy*, **101**, 45–54.

NOTATION

DR depth ratio

30 Flow Structure in a Large-scale Model of a Doubly Meandering Compound River Channel

C. NAISH & R.H.J. SELLIN
Department of Civil Engineering, University of Bristol, Queen's Building, University Walk, Bristol, BS8 1TR, UK

ABSTRACT

This paper presents the findings of current research carried out by the University of Bristol in association with the NRA into the flow structure in a doubly meandering compound river channel. The construction of a major trunk road along the Blackwater Valley in Hampshire has resulted in the need to relocate a length of the River Blackwater. The relocated river's hydraulic performance is being investigated at prototype, 1:5 and 1:25 scale.

Transverse velocity profiles and longitudinal velocity contours have been constructed from flow direction and velocity measurements taken from a 1:5 scale model constructed in the UK Flood Channel Facility at HR Wallingford Ltd. Flow visualization experiments have also been undertaken. Longitudinal velocities were found to increase with greater inbank flow depths. However, with low overbank flow the inner-channel velocities reduced to below those noted at bankfull flow, whilst higher overbank flow depths further reduced the inner-channel longitudinal velocities, but increased floodplain longitudinal velocities. Secondary flow cells were detected at all flow depths examined. The rotational direction of inner-channel cells was found to be unaffected by overbank flow and plunging floodplain flows were not detected in the inner-channel cross-over region.

Dye injections revealed localized floodplain flow travelling upstream or trapped in stationary vortices. These flow structures appeared to be triggered by flow separation at the floodplain bend apices. Incorporating a lateral cross-fall of 1/30 into the floodplain slope expanded the size and strength of these vortices, intensifying the inner-channel longitudinal velocities and reducing the floodplain longitudinal velocities.

INTRODUCTION

The construction of a major trunk road along the Blackwater valley in Hampshire has resulted in the need to relocate a length of the River Blackwater. This gave the National Rivers Authority (Thames Region) an opportunity to design and study the hydraulic performance of a doubly meandering, compound channel designed to meet specific environmental standards. There has been growing public concern in recent years that previous artificial river changes, such as this relocation, have decreased

Coherent Flow Structures in Open Channels.
Edited by P.J. Ashworth, S.J. Bennett, J.L. Best and S.J. McLelland.

the richness of wildlife habitats and the quality of the landscape. A potential conflict of interest exists in the design of flood alleviation/protection schemes, as often the most efficient hydraulic solution may not be the most acceptable on environmental grounds. As a result, the use of straight, uniform-section channels is being replaced in many cases by either singly or doubly meandering compound channels.

The use of compound channels is an effective and aesthetically pleasing method of providing flood capacity in rivers. The inner channel carries the flow resulting from extended periods of light rain, as well as the base flow, while the upper (floodplain) channel carries the flood waters resulting from periods of heavy rain or severe storms. The environmental benefits are that this method of design can often incorporate an existing lower channel, thus retaining the old channels' habitat and form, and during low summer flows the floodplain berm can provide a marginal marsh habitat.

The majority of research into the performance of compound meandering channels has been carried out in small laboratory flumes (e.g. Lipscomb, 1956; Toebes & Sooky, 1967; James & Brown, 1977; Smith, 1978; Myers, 1990; Willetts & Hardwick. 1993). However, it is not yet clear how reliably these small-scale results can be incorporated in prototype design methods. The Flood Channel Facility (FCF), constructed at HR Wallingford Ltd. in 1985–1986, was intended to provide a large scale experimental channel in which research teams from Universities and industry could study the mechanics of compound or two-stage channel flow. The conception and planning of the facility is described by Knight & Sellin (1987). Several series of experiments have already been carried out which were concerned with more generalized compound channel geometries. Results from straight compound channel are reported by Myers & Brennan (1990) and Wormleaton (this volume) and the skewed compound channel results by Elliott & Sellin (1990), Knight & Shiono (1990), Wormleaton & Merritt (1990), and Sellin (1995). The meandering compound channel results are reported by Ervine *et al.* (1993, 1994), Sellin *et al.* (1993) and Willetts & Rameshwaran (this volume).

Because of the expense and difficulties involved in collecting field data, especially at times of flood, very few prototype-scale studies of compound river channels have been undertaken. Myers & Martin (1989) reported their study of a 1 km reach of the River Maine, Northern Ireland and Sellin *et al.* (1990) a 3 km length of the River Roding, Essex. The latter concluded that conventional design methods may result in errors up to 30%.

The River Blackwater Project was planned as a logical extension to the earlier series of experiments carried out in the FCF. The test reach of the relocated River Blackwater channel has a meandering inner-channel of medium sinuosity ($s = 1.18$) and a wider upper channel (the berm) of low sinuosity ($s = 1.05$). The prototype has a longitudinal slope of 0.001 and was constructed with berms inclined at 1/30 (cross-fall), but scale model tests were carried out with both inclined and horizontal berms. Figure 30.1 shows the planform shape of the test reach while Figure 30.2

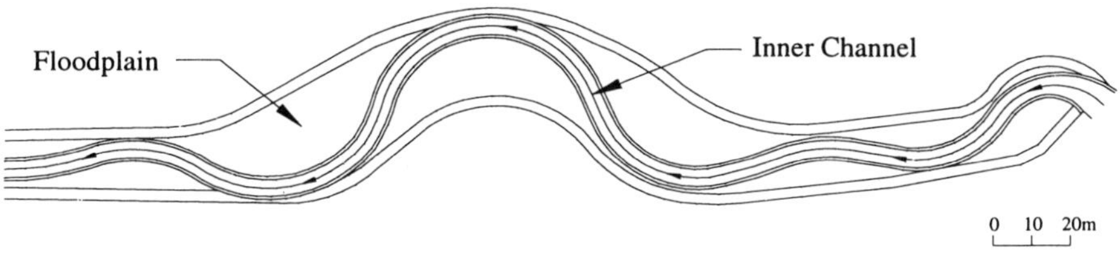

Figure 30.1. Planform of the inner channel and floodplains (prototype scale)

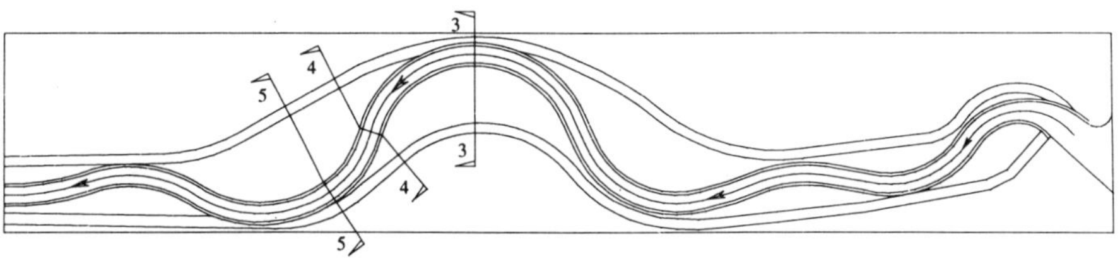

Figure 30.2. Typical channel cross-section (prototype scale)

shows a typical cross-section morphology.

Sellin *et al.* (1993) carried out the 1:5 scale model study and found that the use of a berm cross-fall increases conveyance at lower overbank flow depths since the flow on the berm is guided to follow more closely the inner-channel flow direction. At high overbank depths this effect becomes insignificant and the discharge, compared to that for floodplains with no transverse inclination, is less because the cross-sectional area is smaller.

The small cross-fall on the floodplain was adopted for the following reasons:

- The resulting cross-section will give a low lying berm area immediately adjacent to the inner channel which will encourage the development of a marsh habitat.
- The portion remote from the inner channel will be correspondingly higher and better drained and will provided better conditions for maintenance purposes.

As a result of hydrological studies of the 50 km^2 catchment the inner channel has been designed to carry 1.5 $m^3\ s^{-1}$, and the total channel 9.1 $m^3\ s^{-1}$. Stage and discharge data are currently being collected on the river. A large 1:5 scale model has been constructed and tested in the FCF, and both distorted and undistorted 1:25 scale models have been constructed and tested in the Hydraulics Laboratory of the University of Bristol.

To the authors' knowledge, this project will make available for the first time data from a doubly meandering compound channel at prototype, 1:5 and 1:25 distorted and undistorted scales and will give the best opportunity to date to establish scaling effects in hydraulic modelling. The use of a large scale model (FCF) allowed an extensive series of flow direction and velocity measurements to be undertaken which are the subject of this paper.

EXPERIMENTAL CONFIGURATION

The undistorted 1:5 scale model of the Blackwater was constructed in the 56 m long, 10 m wide FCF. Figure 30.3 shows the model planform and the location of the three cross-sections at which velocity measurements were made. The model surface is constructed of carefully trowelled cement mortar laid over a gravel base. The inner channel is trapezoidal in cross-section, 150 mm deep, 850 mm wide at the base and 1150 mm wide at the top (bankfull level). The floodplains were initially horizontal but were later remodelled to incorporate a 1/30 cross-fall. Water is supplied from a sump to the inlet by seven pumps connected in parallel. Flow enters through a stilling tank and leaves over a line of variable-level weirs. Discharge is measured by orifice plate meters fitted to each of the pump delivery lines. Water levels are measured in the model at eight tapping points situated along the centreline of the inner channel and connected to external observation wells which are read to 0.01 mm using digital point gauges.

To investigate the structure of the flow in the model an extensive series of point velocity measurements (both direction and magnitude) were made at heights of 132, 162, 187 and 237 mm above the inner-channel bed and at sections 3, 4 and 5. Throughout the velocity measurement programme, the inner channel was roughened with 8 mm gravel. Three Floodplain roughnesses ere used: mortar finish and gravel of 8 and 13 mm. Under these conditions the 1:5 model operated with flow Reynolds numbers in excess of 6×10^4. Figure 30.4 details the mean flow Reynolds number over the total depth range for all 1:5 scale tests considered here. Measurements were made at each section over a basic rectangular grid, 50 mm (cross-stream) by 25 mm (vertical). Measurements of local flow direction in the horizontal plane were made using a balanced vane connected to a rotary potentiometer by a light vertical shaft. All data from the vane measurements were stored in a computer which also controlled the positioning of the instrument carriage on its automated bridge. These directional data are then used by the computer in the subsequent propeller meter traverse to align the propeller meter (disc diameter 10 mm with and accuracy of ±4%) at each measurement position

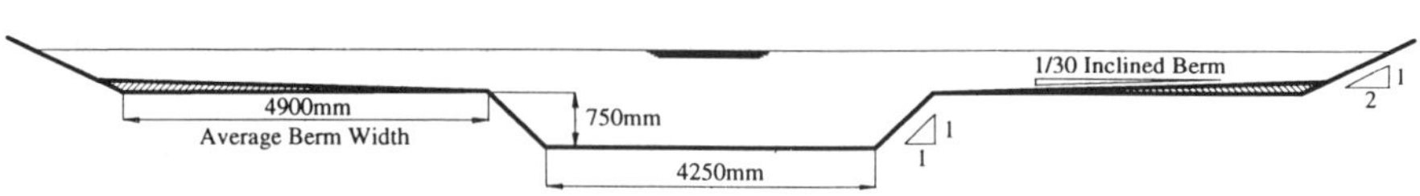

Figure 30.3. Location of cross section positions on 1:5 scale model

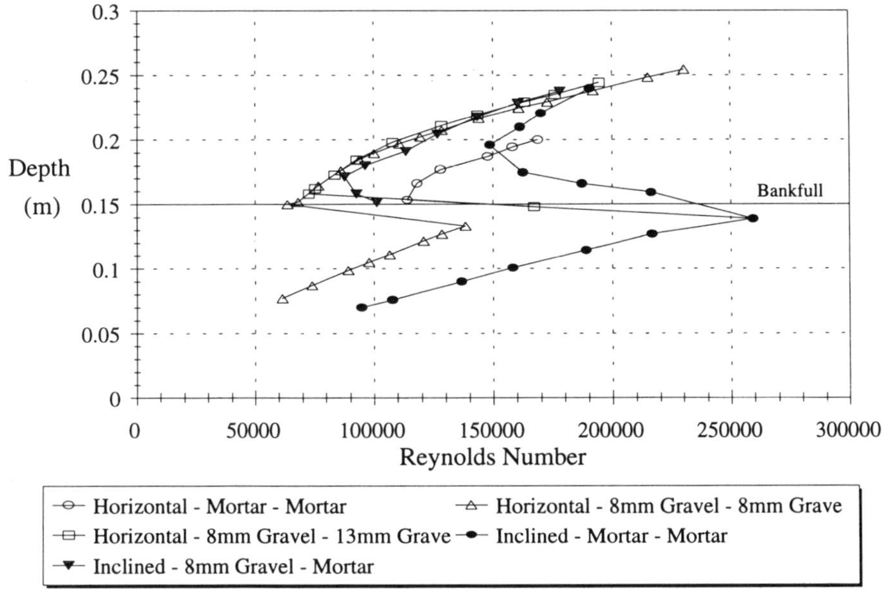

Figure 30.4. 1:5 scale model Reynolds numbers

with the disc perpendicular to the predetermined local flow direction.

The velocity distribution diagrams described in the following section are constructed for the 1:5 scale model by resolving the measured velocity vectors parallel with the local inner-channel direction, for the longitudinal velocity contour diagrams, and in the perpendicular direction for the transverse velocity profiles. All diagrams are drawn looking downstream.

OBSERVATIONS ON THE DISTRIBUTION OF LONGITUDINAL AND TRANSVERSE VELOCITY COMPONENTS

SECTION 3 – HORIZONTAL BERM

With a flow depth of 132 mm (inbank flow) and the inner channel roughened with 8 mm gravel, Figure 30.5a shows a longitudinal velocity range of 0.3 to 0.4 m s^{-1} across the inner channel with the maximum velocity core on the outside of the bend as expected. Figure 30.5b shows a strong clockwise-rotating cell extending across the greater part of the cross-section.

At bankfull depth (Figure 30.6a), the increased longitudinal velocity ranges from 0.35 to 0.45 m s^{-1}, the core of which has moved from the inside bend to the centre of the inner channel. Figure 30.6b shows a clockwise-rotating cell across the whole

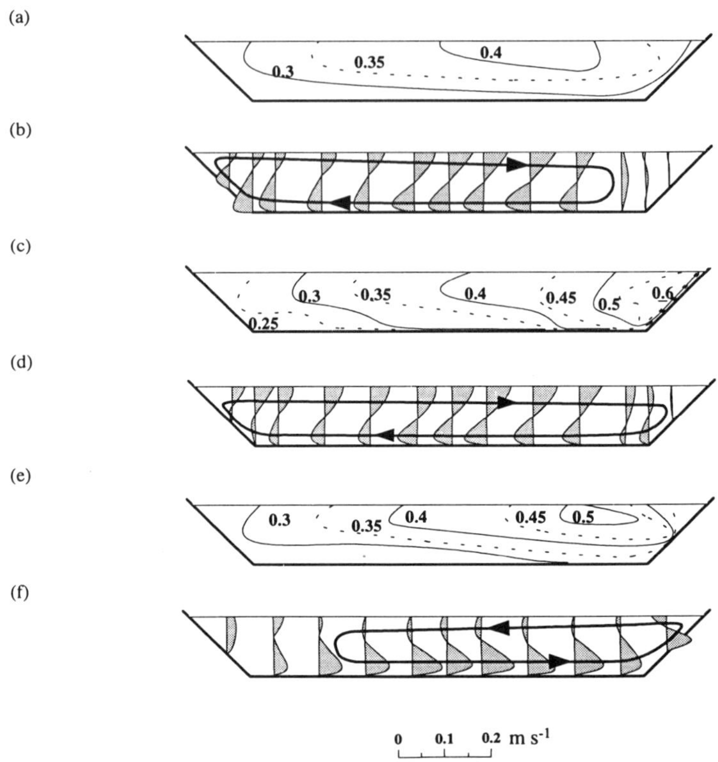

Figure 30.5. Longitudinal and transverse velocity patterns at Sections 3, 4 and 5 for inbank flow conditions. The flow depth is 132 mm above the inner channel mortar base and the inner channel bed is roughened with 8 mm gravel. Cross-sections shown are: (a) longitudinal and (b) transverse velocities for section 3; (c) longitudinal and (d) transverse velocities for section 4; and (e) longitudinal and (f) transverse velocities for section 5. Longitudinal velocity contours are measured in m s^{-1} and the scale for transverse velocities is shown below the last cross-section

width of the cross-section, producing larger transverse velocities near the inside of the bend and leaving an area of low transverse velocity near the surface, at the outside of the bend.

With a low overbank flow depth of 187 mm and a mortar finish to the horizontal floodplains, Figure 30.7b indicates that the longitudinal velocity range has decreased to between 0.3 and 0.4 m s^{-1}, which is similar to the inbank values shown in Figure 30.5a. The high velocity core has also moved closer to the inside bend of the inner channel. The maximum velocity in the shallow flow over the floodplain is the same locally as the maximum velocity found in the inner channel. Figure 30.7b shows that a weak anticlockwise cell has appeared near the outside of the bend. This replaces the weak inner-channel transverse flow area described in the

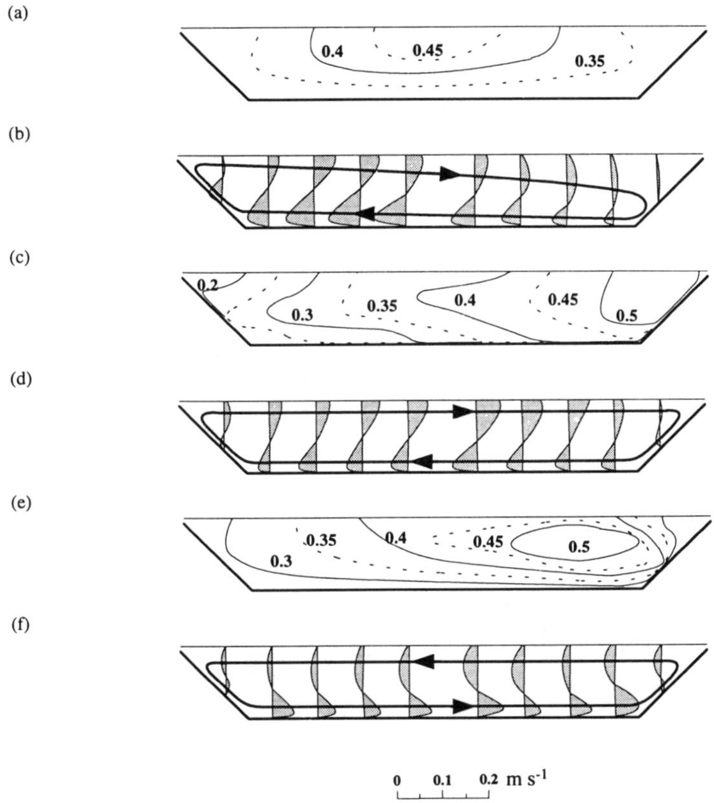

Figure 30.6. Longitudinal and transverse velocity patterns at Sections 3, 4 and 5 for bankfull flow conditions. The flow depth is 162 mm above the inner channel mortar base and the inner channel bed is roughened with 8 mm gravel. Cross-sections and scales are the same as Figure 30.5

bankfull flow case (Figure 30.6b). The inner clockwise cell persists and is now linked to flow discharging off the floodplain.

At this low overbank flow depth of 187 mm, increasing the floodplain roughness, using 8 mm gravel reduces the floodplain longitudinal flow velocity from 0.4 to 0.2 m s^{-1} (Figure 30.8a). The inner-channel longitudinal velocity distribution is relatively unaffected, however, Figure 30.8b reveals that the anticlockwise cell shown in Figure 30.7b has grown in size, but not velocity, and is positioned above the original clockwise cell. Both cells appear to be driven by a strong lateral flow from the floodplain. As the flow depth increases to 237 mm, Figure 30.9a shows that the floodplain longitudinal velocity has increased to match the inner-channel longitudinal velocity. The inner-channel velocity range has decreased slightly from 0.3–0.4 m s^{-1} to 0.25–0.4 m s^{-1}, but the flow structure

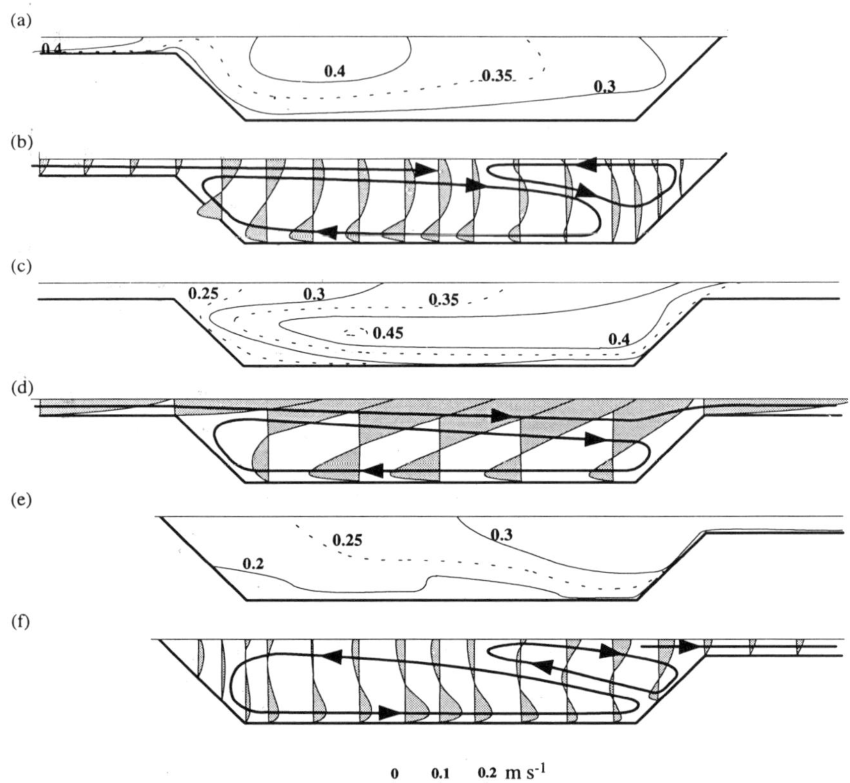

Figure 30.7. Longitudinal and transverse velocity patterns at Sections 3, 4 and 5 for low overbank flow conditions with inner channel roughness greater than floodplain roughness and horizontal floodplain berms. The flow depth is 187 mm above the inner channel mortar base and the inner channel bed is roughened with 8 mm gravel. Cross-sections and scales are the same as Figure 30.5

remains unchanged with the longitudinal velocity core positioned near the inside bank of the inner channel. Figure 30.9b shows that the increase in flow depth has not altered the secondary flow pattern, with an anticlockwise cell, driven by the plunging flow from the floodplain, positioned above the clockwise cell.

Increasing the floodplain surface roughness with 13 mm gravel at low overbank flow depths (Figure 30.10a) reduces the longitudinal floodplain velocities. However, the distribution of longitudinal velocity within the inner channel has been unaffected. Figure 30.10b shows that the weak anticlockwise cell, first noted in Figure 30.7b, has dramatically reduced in size and the clockwise cell now dominates the secondary flow structure. Figure 30.11a shows that with an overbank flow depth of 237 mm the increased floodplain surface roughness reduces the inner-channel velocity range from 0.25–0.4 to 0.25–0.35 m s^{-1}, but the velocity

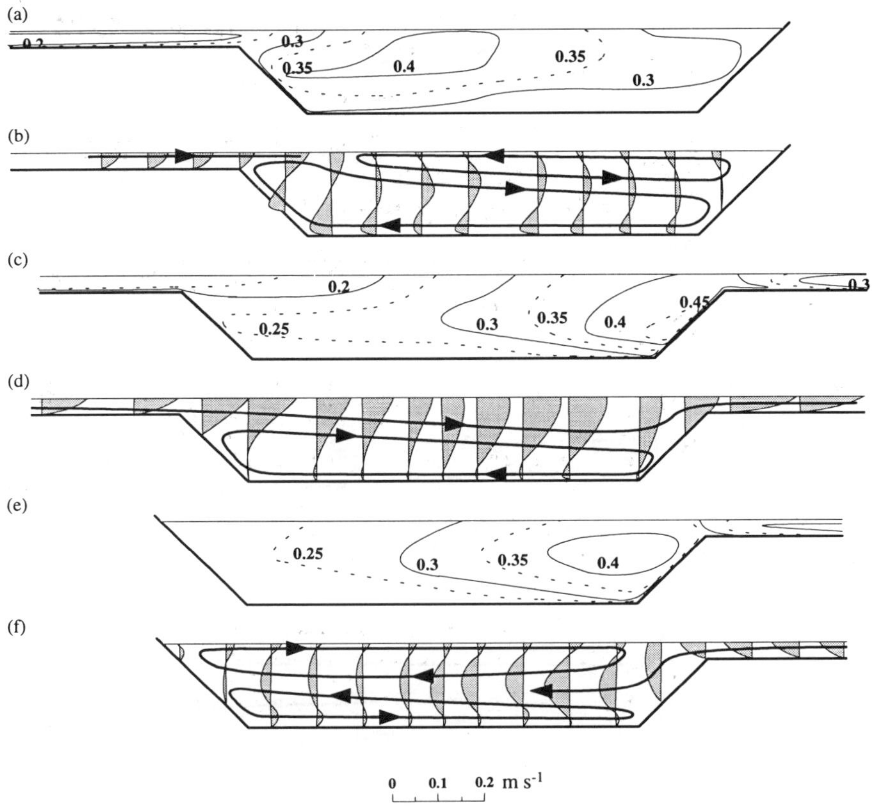

Figure 30.8. Longitudinal and transverse velocity patterns at Sections 3, 4 and 5 for low overbank flow conditions with inner channel roughness equal to floodplain roughness and horizontal floodplain berms. The flow depth is 187 mm above the inner channel mortar base. The inner channel bed and the floodplain region are both roughened with 8 mm gravel. Cross-sections and scales are the same as Figure 30.5

distribution remains the same. The longitudinal floodplain velocity is unaffected by the increase in surface roughness at this flow depth. Figure 30.11b reveals that the anticlockwise cell seen for the low overbank flow condition (Figure 30.10b) has merged with the large clockwise cell driven by flow from the floodplain. The increase in floodplain roughness has not reduced the flow velocities discharging off the floodplain, but this small change has, in effect, suppressed the anticlockwise cell just detectable in Figure 30.10b.

SECTION 3 – INCLINED BERM

With the floodplain roughened using 13 mm gravel, Figure 30.12a shows that the

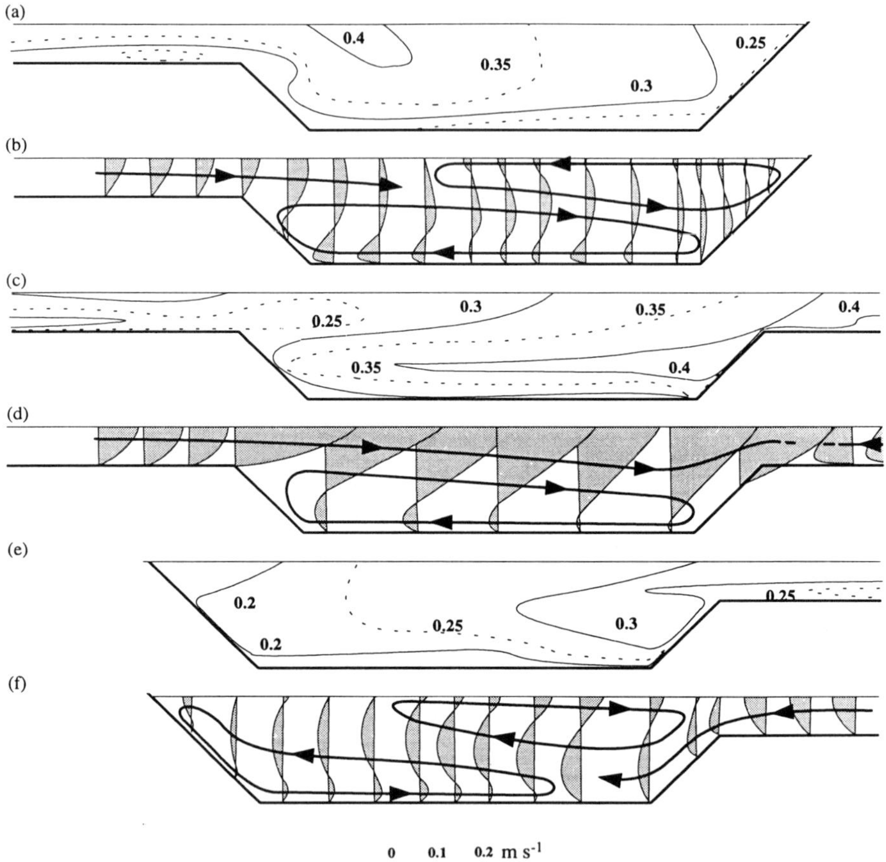

Figure 30.9. Longitudinal and transverse velocity patterns at Sections 3, 4 and 5 for high overbank flow conditions with inner channel roughness equal to floodplain roughness and horizontal floodplain berms. The flow depth is 237 mm above the inner channel mortar base. The inner channel bed and the floodplain region are both roughened with 8 mm gravel. Cross-sections and scales are the same as Figure 30.5

1/30 berm inclination has increased the low overbank velocity range from 0.3–0.4 m s^{-1} to 0.35–0.5 m s^{-1} in the inner channel and that the shallow flow velocities on the inclined floodplain are very low. The inner-channel core of high velocity, increased from 0.4 to 0.5 m s^{-1} and has moved from the inside bend to the water surface in the centre of the inner channel. Figure 30.12b shows that the inclination of the berm leaves the structure of the secondary flow cells unchanged. anticlockwise cell at the outside of the bend has expanded compared with Figure 30.10b. Figure 30.13a shows that with the greater flow depth of 237 mm above the inner channel mortar base, the longitudinal velocities there are 0.1 m s^{-1} higher

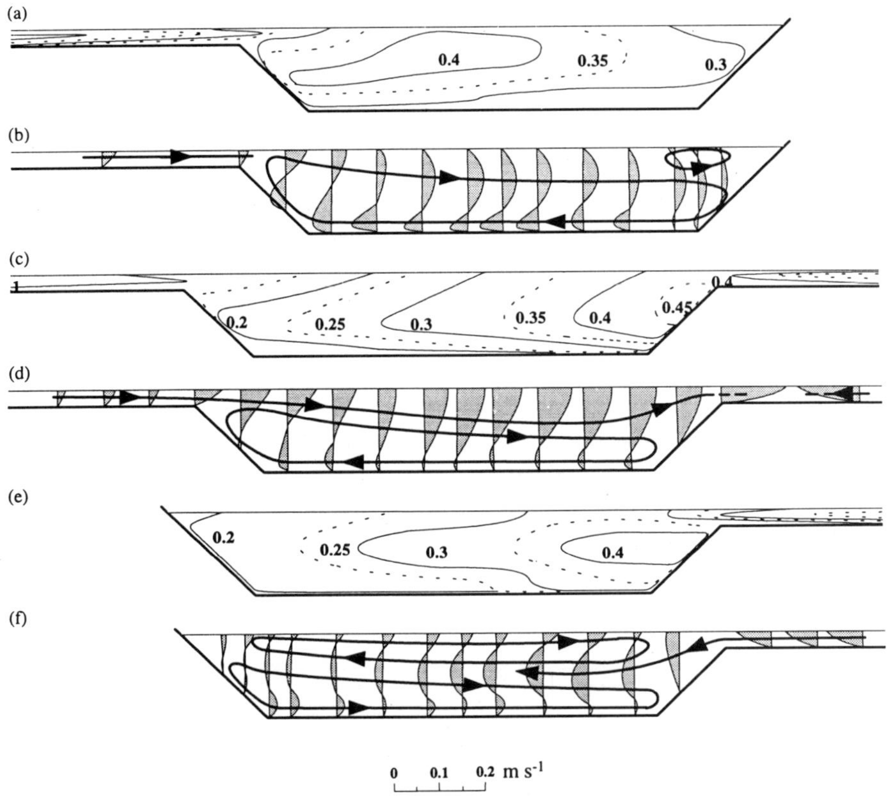

Figure 30.10. Longitudinal and transverse velocity patterns at Sections 3, 4 and 5 for low overbank flow conditions with inner channel roughness less than floodplain roughness and horizontal floodplain berms. The flow depth is 187 mm above the inner channel mortar base. The inner channel bed is roughened with 8 mm gravel and the floodplain region is roughened with 13 mm gravel. Cross-sections and scales are the same as Figure 30.5

when compared with the 'horizontal' case in Figure 30.11a. The floodplain longitudinal velocities have increased slightly with the higher overbank flow depth. Figure 30.13b shows that with this greater flow depth, the two cells apparent in Figure 30.12b (low overbank flow) have merged to form a large clockwise cell. The same behaviour was noted in Figure 30.11b, the equivalent horizontal berm case.

SECTION 4 – HORIZONTAL BERM

At section 4, with an (inbank) flow depth of 132 mm and the inner channel roughened with 8 mm gravel, Figure 30.5c shows a longitudinal velocity range of 0.25 to 0.6 m s^{-1}. Here the maximum velocity core has been forced to the

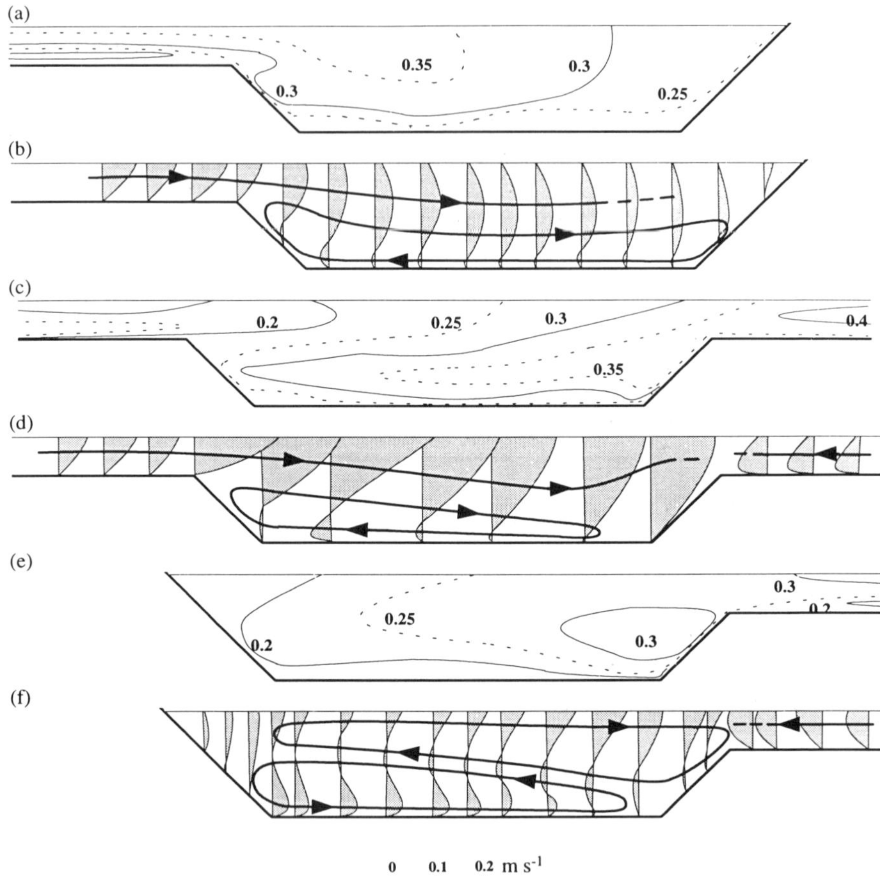

Figure 30.11. Longitudinal and transverse velocity patterns at Sections 3, 4 and 5 for high overbank flow conditions with inner channel roughness less than floodplain roughness and horizontal floodplain berms. The flow depth is 237 mm above the inner channel mortar base. The inner channel bed is roughened with 8 mm gravel and the floodplain region is roughened with 13 mm gravel. Cross-sections and scales are the same as Figure 30.5

downstream bend of the inner channel. Figure 30.5d shows that the clockwise cell noted at section 3 has expanded to cover the entire cross-section.

With a bankfull flow depth Figure 30.6c shows that the longitudinal velocity has decreased with the increase in inner-channel flow depth, having a velocity range of 0.2 to 0.5 m s^{-1}. The velocity core is still situated near the downstream bank of the inner channel. Figure 30.6d shows that the clockwise cell has expanded to cover the area of low transverse velocities detected in section 3.

With a low overbank flow depth of 187 mm and a mortar finish to the

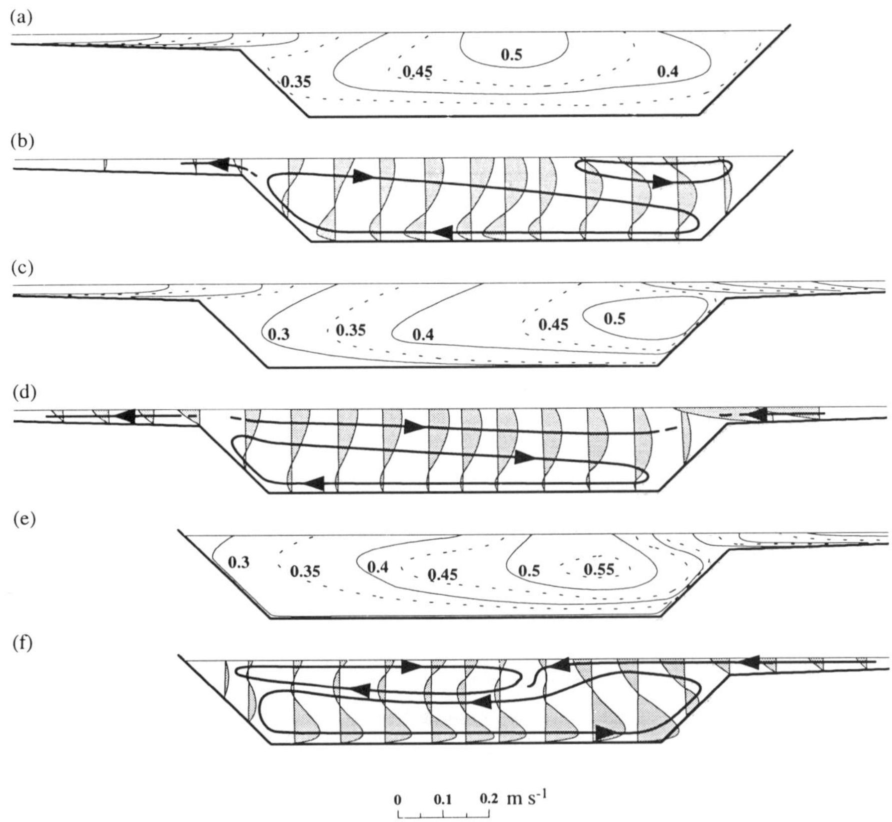

Figure 30.12. Longitudinal and transverse velocity pattern at Sections 3, 4 and 5 for low overbank flow conditions with inner channel roughness less than floodplain roughness and floodplain berms inclined at a slope of 1/30. The flow depth is 187 mm above the inner channel mortar base. The inner channel bed is roughened with 8 mm gravel and the floodplain region is roughened with 13 mm gravel. Cross-sections and scales are the same as Figure 30.5

horizontal floodplains, Figure 30.7c shows that the longitudinal velocity range in the inner channel has decreased further to between 0.25 and 0.45 m s^{-1} with the maximum velocity core now spread across the whole inner channel. The longitudinal velocity of the shallow floodplain flow has reduced to less than 0.25 m s^{-1} which is accounted for by the high transverse floodplain velocities shown in Figure 30.7d which cut across the inner-channel flow, enhancing the existing clockwise cell.

At this low overbank flow depth with floodplain roughness increased using 8 mm gravel, Figure 30.8c shows that the longitudinal velocity range in the inner channel is unaffected but the maximum velocity core has returned to the right-hand

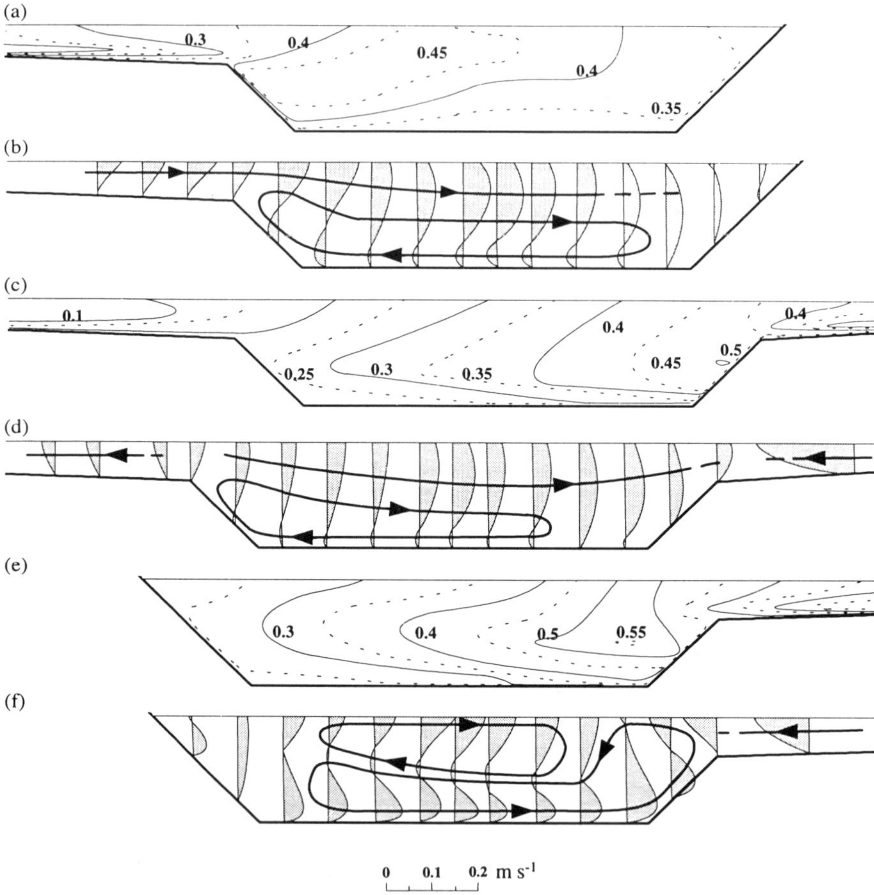

Figure 30.13. Longitudinal and transverse velocity patterns at Sections 3, 4 and 5 for high overbank flow conditions with inner channel roughness less than floodplain roughness and floodplain berms inclined at a slope of 1/30. The flow depth is 237 mm above the inner channel mortar base. The inner channel bed is roughened with 8 mm gravel and the floodplain region is roughened with 13 mm gravel. Cross-sections and scales are the same as Figure 30.5

inner-channel bank. The upstream floodplain velocity has been reduced from 0.25 to 0.15 m s^{-1} due to the increased floodplain roughness, however, the downstream floodplain velocity has been less affected, being reduced from 0.35 to 0.3 m s^{-1}. Figure 30.8d shows that roughening the floodplain has reduced all transverse velocities and the slower flow crossing the inner channel has flattened the clockwise cell apparent in Figure 30.7d. As the flow depth increases to 237 mm, Figure 30.9c shows that the inner-channel longitudinal velocity core has reduced from 0.45

to 0.4 m s^{-1} and has expanded across the width of the inner channel but reduced in thickness. This also occurred in Figure 30.7c under lower flow depth and floodplain roughness conditions. In addition, the longitudinal floodplain velocity has increased to match the local inner-channel velocity, with the upstream floodplain velocity still slower than that on the downstream floodplain. Figure 30.9d shows that as the depth of flow increases, the transverse velocities also increase, with the clockwise cell becoming less compressed at the downstream bank. This flow structure is also visible in Figure 30.7d, where it occurs at shallower overbank flow because of the smoother floodplain surface. Under the present roughness conditions the downstream floodplain is attempting to discharge back into the inner channel, but the strong transverse currents prevent this and the flow is forced downstream parallel with the inner channel.

With the 13 mm gravel on the floodplain surface, Figure 30.10c shows that again the inner-channel longitudinal velocity flow structure has not been affected. The inner-channel longitudinal floodplain velocity has decreased from 0.15 to 0.1 m s^{-1} with the longitudinal velocity over the upstream floodplain increasing from 0.3 to 0.35 m s^{-1}. From Figure 30.10d it is apparent that the increased floodplain roughness has reduced the strength of the transverse flow discharging off the floodplain, with the weaker clockwise cell still being confined by the slower moving but expanding floodplain flow crossing the inner channel. The confluence of the upstream floodplain flow crossing the inner channel with the flow from the downstream floodplain, first noted in Figure 30.9d, can also be seen here. Figure 30.11c shows that with the greater flow depth of 237 mm, the inner-channel longitudinal velocity range has been further reduced from that noted in Figure 30.9c, to 0.2–0.35 m s^{-1}, but the flow structure is unaltered. Again the longitudinal floodplain velocity is unaffected by the increase in roughness at this flow depth. Figure 30.11d shows that the transverse velocity flow structure is not affected by the slight increase in floodplain roughness and remains unchanged from Figure 30.9d.

SECTION 4 – INCLINED BERM

With the floodplain still roughened with 13 mm gravel, Figure 30.12c shows that the 1/30 berm inclination has again increased the inner-channel longitudinal velocities by 0.1 m s^{-1} without affecting the inner-channel velocity distribution. The longitudinal velocity core has maintained its position at the downstream bank. Figure 30.12d shows clearly the effect of the berm inclination on the flow cell at this cross-over region. The strong clockwise cell developed in the bend is still apparent but now weak flows enter or leave the floodplain in the opposite direction to that observed with the horizontal floodplain (Figure 30.10d). Figure 30.13c shows that with the greater flow depth of 237 mm, the inner-channel velocities have decreased slightly with increasing flow depth, but they are still higher than those experienced under horizontal berm conditions. Conversely, the floodplain velocities have increased slightly but remain lower than experienced with the horizontal berm. The inner-channel longitudinal flow structure remains the same with the velocity core still positioned near the downstream inner-channel bank. Figure 30.13d shows that the secondary flow cells remain unchanged compared with the low overbank case shown in Figure 30.12d. However, comparison with Figure 30.11d shows the greatly reduced transverse velocities resulting from the sloping berms.

SECTION 5 – HORIZONTAL BERM

At section 5, with an (inbank) flow depth of 132 mm and the inner channel roughened with 8 mm gravel, Figure 30.5e shows that the longitudinal velocity range has been reduced to between 0.3 and 0.5 m s^{-1}. The maximum longitudinal velocity core is now on the inside (right bank) of the bend, not on the outside as normally expected. This suggests that the bends are not independent of each other, conditions noted at section 3 continuing to influence the flow at Sections 4 and 5. Figure 30.5f reveals that a strongly developed anticlockwise circulation is established close to the right-hand bank. It should be noted that the strong undercurrent and weak surface current apparent at this cross-section suggests that the flow is accelerating in the downstream direction near the right-hand bank, a condition which accords with the influence of free vortex flow near the inside of this bend.

With bankfull flow depth, Figure 30.6e shows that the longitudinal velocity range has remained constant despite the increase in inner-channel flow depth. A longitudinal velocity range of 0.3 to 0.5 m s^{-1} is apparent with the core remaining near the inside of the bend but having moved from the free surface to mid-depth. Figure 30.6f reveals that the anticlockwise cell noted in the inbank case (Figure 30.5f), now has a more balanced near-bed and surface current strength.

With a low overbank flow depth (187 mm) and a mortar finish to the horizontal floodplains, Figure 30.7e shows that the shallow overbank flow has caused a dramatic reduction in the inner-channel longitudinal velocity. The longitudinal velocity core, reduced from 0.5 to 0.3 m s^{-1}, is still positioned at the inside of the inner-channel bend and now extends onto the floodplain. Figure 30.7f shows that the anticlockwise cell previously detailed in Figure 30.6f is still in existence, however a clockwise cell has also developed at the inside bend and this feeds water out onto the floodplain.

Increasing the floodplain roughness at this low overbank flow depth of 187 mm from a mortar finish to a 8 mm gravel, further reduces the floodplain longitudinal velocity from 0.3 to 0.2 m s^{-1} but the longitudinal inner-channel velocity range has increased to between 0.25 and 0.4 m s^{-1} (Figure 30.8e). Figure 30.8f shows that although the clockwise cell, first detected in Figure 30.7f, has not changed its rotational sense, the transverse flow direction on the floodplain has changed direction and is now contributing water to the inner channel. This upper cell has expanded to fill the full inner-channel width, correspondingly reducing the size of the anticlockwise cell below it. As the flow depth increases to 237 mm, Figure 30.9e shows that the inner-channel longitudinal velocity range has been reduced from 0.25–0.4 to 0.2–0.3 m s^{-1}, but the position of the inner-channel longitudinal velocity core is unchanged. This inner-channel longitudinal flow structure is similar to that seen in Figure 30.7e but the higher floodplain surface-roughness has reduced the flow velocities. Figure 30.9f shows that the two cells which were previously above each other have separated laterally, with the inside (right) bend cell being driven by the plunging floodplain discharge and the original anticlockwise cell being pushed towards the outside of the bend.

With the floodplain roughness further increased (13 mm gravel), Figure 30.10e shows that the inner-channel longitudinal flow structure at the low overbank flow depth of 187 mm is unaffected. The inner-channel maximum velocity region remains at the inside of the bend but the range of values has increased slightly and the maximum floodplain velocities have decreased from 0.3 to 0.2 m s^{-1}. The secondary flow structure shown in Figure 30.10f is unchanged by the increased floodplain roughness and is broadly similar to that shown in Figure 30.8f.

Figure 30.11e shows that with a high overbank flow depth (237 mm), the inner-channel longitudinal velocities remain unchanged compared with Figure 30.9e but the floodplain velocities have been reduced slightly. Figure 30.11f reveals that as the floodplain roughness and flow depth have increased, the mid-depth transverse velocities have decreased while the bed and water surface velocities have increased. This progression can best be seen by comparing Figures 30.9f, 30.10f and 30.11f.

SECTION 5 – INCLINED BERM

With the floodplains still roughened using 13 mm gravel, Figure 30.12e shows that the 1/30 berm inclination has increased the inner-channel longitudinal velocity range from 0.2–0.4 m s^{-1} to 0.3–0.55 m s^{-1} and the longitudinal velocity core has moved away from the inside bend of the inner channel. Figure 30.12f shows that the upper clockwise cell in the inner channel has been reduced in size, due to the reduced cross-boundary flows, and has moved over towards the outside of the bend Simultaneously, the lower anticlockwise cell has grown in size under the influence of the flow in the bend. Figure 30.13e shows that with the greater flow depth of 237 mm, the inner-channel longitudinal velocity has decreased slightly, and the floodplain velocities have increased and both are significantly higher than those experienced under the horizontal berm condition (Figure 30.11e). Figure 30.13f shows that the secondary flow cells are unchanged from the low overbank condition (Figure 30.12f) with the exception that the greater flow depth has caused the surface transverse velocities to increase dramatically.

SUMMARY OF VELOCITY DISTRIBUTION TRENDS

Longitudinal velocity contours

- For inbank flow, the left-hand bend influences the longitudinal flow structure at the succeeding right-hand bend.
- Longitudinal velocities increase with higher inbank flow depths.
- As flow depth increases from bankfull to low overbank flow depths, the inner-channel longitudinal velocity is reduced by the slow moving water on the floodplain.
- Increasing the floodplain roughness reduces both inner-channel and floodplain longitudinal velocities.
- With inclined berms, cross-boundary flows are reduced and therefore inner-channel longitudinal velocities increase whilst floodplain velocities reduce.
- Increasing the overbank flow depth for all floodplain roughness conditions further reduces the inner-channel longitudinal velocities and increases the floodplain longitudinal velocities.

Transverse velocity profiles

- For both inbank and bankfull flows, a single cell is apparent, filling the whole inner channel and being driven by centripetal force at the bends.
- With low overbank flow and the inner channel roughness greater than the horizontal floodplain roughness, a second circulation cell develops at the bend apex. This appears to be driven by the flow leaving the floodplain and is rotating in the opposite direction to the original cell. At section 4, the cross-over region, the original inner-channel cell is flattened by the overlying transverse flow.
- Low overbank flow combined with identical horizontal floodplain and inner channel roughness reduces both the inner channel and the floodplain transverse velocities. At the bend apex, the secondary (upper) cell grows laterally and at the cross-over region, the original (lower) cell is more obviously confined by the slower moving flow passing from the floodplain.
- With low overbank flow and the horizontal floodplain roughness increased so that it exceeds that of the inner channel, the inner channel and floodplain transverse velocities are further reduced but with no effect on the secondary flow structure.
- Higher overbank flow depths increase the size and strength of the secondary flow cells but their structure remains unaltered.
- With the berms inclined the principle flow cell now dominates at the bend apex and the secondary (floodplain driven) cell is reduced in size. At the cross-over region, (section 4) the strong surface cross-currents are absent.

FLOW VISUALIZATION EXPERIMENTS

Flow visualization experiments were undertaken by injecting a vegetable dye into the flow 20 mm above the bed at Sections 3, 4 and 5. The resulting flow patterns shown by the dye stream were recorded by a video camera mounted above the flume.

HORIZONTAL BERMS

Figure 30.14 shows the results of dye injections on the 1:5 scale model with a smooth mortar finish for both the inner channel and the floodplain and an inner-channel flow depth of 200 mm. Dye injection at the inner channel at section 3 revealed the expected flow patterns, with the dye stream following the line of the inner channel. Dye injections in the inner channel at section 5 gave similar results. Injections into the floodplain flow at section 5 also produced the predicted result, with the dye stream crossing the floodplain and then becoming entrained with the inner-channel flow. However, dye injected at the floodplain boundary at section 3 (apex) revealed large vortex structures being generated and moving downstream across the floodplain. A further injection in the left-hand side of the inner channel at section 4 revealed flow being expelled from the inner channel up onto the floodplain. A further injection on the floodplain at section 4 highlighted the presence of a large stationary vortex in that region.

INCLINED BERMS

Figure 30.15 shows that inclining the berms has had the effect of increasing the size and strength of the vortices when compared to those observed on the horizontal berms. Dye injections on the floodplain at section 3, with an inner-channel flow depth of 250 mm, suggest that the vortices are now being shed slightly further upstream relative to the bend apex. In addition, a stationary vortex also occurs on the floodplain upstream from section 3 which confirms that the flow structures found at section 4 are not confined to the specific geometry of the floodplain at section 3.

DISCUSSION

The flow mechanisms that have been postulated from the velocity data collected on the 1:5 scale River Blackwater model differ from those detected in published accounts of large- and small-scale flume studies of sinusoidally meandering compound channels.

Kiely (1990) reported on the flow mechanisms observed during a small-scale singly meandering compound model study. The model was built in a flume 14.4 m long by 1.2 m wide with an inner-channel sinuosity of 1.25. He noted that for

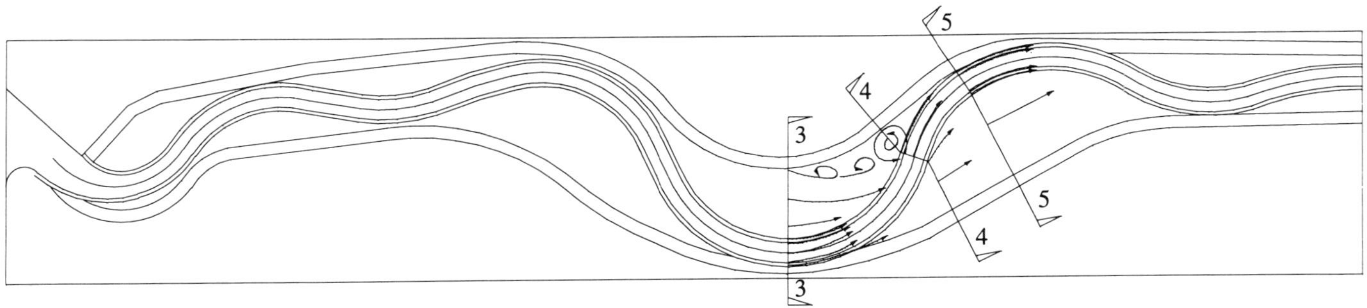

Figure 30.14. Flow patterns detected in the 1:5 scale model with 200 mm inner-channel flow depth and mortar finish to the inner channel and horizontal floodplain berms

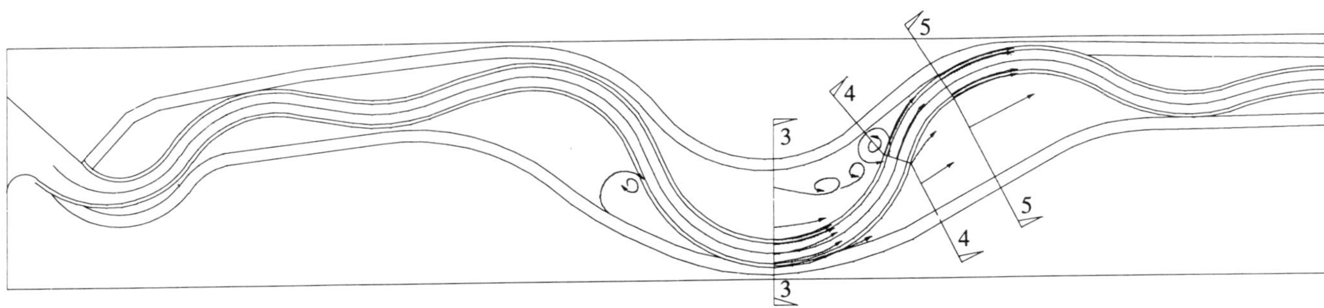

Figure 30.15. Flow patterns detected in the 1:5 scale model with 250 mm inner-channel flow depth and mortar finish to the innerchannel and inclined floodplain berms

overbank flow, the secondary flow directions are opposite to those observed with inbank flow. Kiely stated that the direction of the secondary cells identified at the bends was not due to the inner-channel curvature, as for the case of inbank flow, and suggested that it was due to the intense shear layer across the interface of the outer bend. This was not found to be the case for the Blackwater results. For inbank flow conditions the secondary currents flow towards the inner bank in the lower part of the inner channel and a return flow from the inner to the outer bank occurs in the upper part of the deep channel cross-section, as described by Henderson (1966). The direction of these secondary flow cells are unaffected by overbank flow in the case of the 1:5 Blackwater model. The floodplain flow entering the inner channel at section 3 is linked to the existing inner-channel cell, whilst the cross channel flow at section 4 drives the existing clockwise inner-channel cell. At section 5, the inbank anticlockwise cell remains, but a second cell has developed on the inside of the bend, rotating in the opposite direction.

Willetts & Hardwick (1993) reported on singly meandering compound channel experiments carried out on a small-scale laboratory model 11 m long by 1.2 m wide. Inner-channel sinuosities tested were 1.20, 1.40 and 2.04. They described the vigorous exchange of water between inner channel and floodplain in and beyond the downstream half of each bend (see Willetts & Rameshwaran, this volume; Figure 30.1). They stated that the floodplain water entering the channel penetrates to its full depth in the cross-over region. Throughout the Blackwater tests, under all different floodplain roughness conditions, the inner-channel cell in the cross-over region (section 4) extended across the whole width of the channel and the floodplain flow passed above it.

Sellin *et al.* (1993) reported on large-scale, meandering compound channel experiments with inner-channel sinuosities of 1.37 and 2.04 conducted in the FCF. They also found that, as a result of the momentum exchange with the floodplain flow in the cross-over region, the channel flow enters the bend with a secondary circulation counter to that which the bend flow would itself induce. As a result, water from the channel cross-section at the bend apex (including water from the channel bed) escapes onto the downstream floodplain immediately after the bend. This process induces water from the upstream floodplain to plunge vigorously into the channel, not at the upstream bank, but near the channel centreline. These results compare well with those of Willetts & Hardwick (1993).

The above three model studies all used meander belt widths which were narrower than the overall floodplain width, therefore the expulsion of the inner-channel water was not restricted by floodplain boundaries. With the doubly meandering channel geometry of the Blackwater, the floodplain boundaries formed an extension of the inner-channel side walls at the bend apices, thus preventing cross-channel flow in these locations (section 3 and section 5). The cross-channel flow experienced at section 4 drives the cell in a clockwise direction, but the effect of the meandering floodplain boundary meeting the inner channel further downstream prevents this driving action and the anticlockwise cell induced by the channel bend develops at section 5. A similar process also occurs at section 3.

The effect of a sinuous floodplain boundary was detected in the flow visualization experiments. Willetts & Hardwick (1993) found that the inner-channel water is expelled downstream of the bend apex (see Willetts & Rameshwaran, this volume; Figure 30.1), but a dye injection in the 1:5 scale Blackwater model at the base of the inner channel between Sections 4 and 5 (where the berm is reducing in width) revealed water being expelled onto the upstream berm and feeding the berm vortex there. This is caused by the flow separation at the floodplain bend apex preventing cross-channel flow in this region, thereby enabling water to move up onto the upstream berm. This results shows that a relatively low floodplain boundary curvature can trigger unexpected flow structures downstream. The presence of this vortex does not contribute to the longitudinal conveyance of flood flows and effectively reduces the floodplain width at this point.

The coherent flow structures identified here will provide a more reliable physical basis for the development of numerical models for untested compound channel designs and therefore provide a more accurate prediction of their stage-discharge relationships.

CONCLUSIONS

Longitudinal velocity contours and transverse velocity profiles were constructed from the data collected at three sections located around a large meander on the River Blackwater 1:5 scale model by resolving the measured velocity vector data both parallel and perpendicular to the local inner-channel direction. Secondary flow cells were detected at all flow depths at the three cross-sections examined.

Longitudinal velocities increase with higher inbank flow depths, but with low overbank flow, the inner-channel velocities decrease below those recorded at the bankfull condition. A higher overbank flow depth further reduces the inner-channel longitudinal velocities but increases the velocities on the floodplain. The direction of rotation of the inner-channel secondary flow cells is unaffected by overbank flow. In the cross-over region, the inner-channel cell extends across the whole width of the channel, preventing flow plunging from the floodplain into the inner-channel.

Flow visualization experiments on the 1:25 and 1:5 scale models of the River Blackwater reveal floodplain flow travelling upstream or trapped in stationary vortices in certain locations associated with a shrinking berm area. These trapped flow structures appear to be triggered by flow separation on the floodplain bend apices and reduce the conveyance of the channel. An Increase in flow depth tends to decrease the size and strength of these 'trapped' vortices.

Inclining the berms has the effect of increasing the size and strength of the stationary vortices when compared with those found on the horizontal berm. In addition, inner-channel longitudinal velocities are increased while the floodplain velocities are reduced.

ACKNOWLEDGEMENTS

The authors acknowledge the financial support of the UK SERC and the NRA in both this research and the work carried out on the 1:5 scale model by Dr M Lambert of the University of Adelaide, Australia.

REFERENCES

Elliott, S.C.A & Sellin, R.H.J. 1990. 'SERC flood channel facility: skewed flow experiments', *J. Hydraul. Res.*, **28**, 197–214.

Ervine, D.A., Sellin, R.H.J. & Willetts, B.B. 1994. 'Large flow structures in meandering compound channels', in White, W.R. & Watts, J. (Eds) *Proc. 2nd Int. Conf. on River Flood Hydraulics*, John Wiley & Sons Ltd., Chichester, 459–469.

Ervine, D.A., Willetts, B.B., Sellin, R.H.J. & Lorena, M. 1993. 'Factors affecting conveyance in meandering compound flows', *J. Hydraul. Engrg.*, **119**, 1383–1399.

Henderson, F.M. 1966. *Open Channel Flow*, MacMillan Publishing Co., New York, 252 pp.

James, M. & Brown, B.J. 1977. 'Geometric parameters that influence flood plain flows', *US Army Corps of Engineers, Waterways Experiment Section, Vicksburg, Mississippi, Report* **H-77-1**.

Kiely, G.K. 1990. 'Overbank flow in meandering compound channels – the important mechanisms', in White, W.R. (Ed) *Proc. Int. Conf. on River Flood Hydraulics*, John Wiley & Sons Ltd., Chichester, 207–217.

Knight, D.W. & Sellin, R.H.J. 1987. 'The SERC Flood Channel Facility', *J. IWEM*, **1**, 198–204.

Knight, D.W. & Shiono, K. 1990. 'Turbulence measurements in a shear layer of a compound channel', *J. Hydraul. Res.*, **28**, 175–196.

Lipscomb, E.B. 1956. 'Hydraulic capacity of meandering channels in straight floodways', *US Army Corps of Engineers, Waterways Experiment Section, Vicksburg, Mississippi*, Tech. Memorandum No. 2.

Myers, W.R.C. 1990. 'Physical modelling of a compound river channel', in White, W.R. (Ed) *Proc. Int. Conf. on River Flood Hydraulics, Wallingford*, 381–390.

Myers, W.R.C. & Brennan, E.K. 1990. 'Flow resistance in compound channels', *J. Hydraul. Res.*, **28**, 141–155.

Myers, W.R.C. & Martin, L.A. 1989. 'River Maine Project', *Proc. MAFF Conf. of River and Coastal Engineers*, Loughborough University.

Sellin, R.H.J. 1995. 'Hydraulic performance of a skewed two-stage channel', *J. Hydraul. Res.*, **33**, 43–64.

Sellin, R.H.J., Ervine, D.A. & Willetts, B.B. 1993. 'Behaviour of meandering two-stage channels', *J. Water, Maritime and Energy, ICE*, **101**, 99–111.

Sellin, R.H.J., Giles, A. & van Beesten, D.P. 1990. 'Post-implementation appraisal of a two-stage channel in the River Roding, Essex', *J. IWEM*, **4**, 119–130.

Sellin, R.H.J., Lambert, M.F. & Werner, M.G.F. 1993. 'Large scale model investigation of a two stage channel', *NRA R&D Note* **366**.

Smith, C.D. 1978. 'Effect of channel meander on flood stage in valley', *J. Hydraul. Div., ASCE*, **104**, 49–58.

Toebes, G.H. & Sooky, A.A. 1967. 'Hydraulics of meandering rivers with floodplains', *J. Waterways and Harbours Div., ASCE*, **93**, 213–236.

Willetts, B.B. & Hardwick, R.I. 1993. 'Stage dependency for overbank flow in meandering channels', *J. Water, Marit. Energy, ICE*, **101**, 45–54.

Wormleaton, P.R. & Merrett, DJ 1990. 'An improved method of calculation for steady uniform flow in prismatic main channel/floodplain sections', *J. Hydraul. Res., IAHR*, **28**, 157–174.

31 Linking Hydraulics, Bedload Transport and River Bed Adjustment with the Conceptual FAST Model

PETER ERGENZINGER & CARMEN de JONG
Institut für Geographische, Wissenschaften Freie Universität Berlin, Grunewaldstraße 35, Berlin 12165, Germany

ABSTRACT

Studies of the interactions between hydraulic conditions, sediment transport, bed roughness and geometry in straight, natural gravel-bed river reaches are relatively sparse. Observations at Squaw Creek, Montana show that there are strong interrelationships between hydraulic conditions and the mode of coarse bed-material transport. Each mode of bedload transport (random, pulsed and carpet-like) is associated with specific hydraulic conditions and results in different bed roughness and geometry. Random transport occurred at low discharge with a predominance of unstructured turbulent flow. Pulsed transport developed at flow depths of more than 8–10 times the roughness height and was associated with the presence of secondary currents. These secondary currents are organized into helical cells which generated a sculptured water surface of well-developed, regularly-sized waves. At the highest transport rates, carpet-like transport is associated with intensive interactions between particles and water in the mobile bed layer and weak interrelations between the lower and upper flow layers which leads to a generally flat water surface with irregular upwellings, boils and small-scale waves.

The FAST (Fluid And Sediment Transfer) conceptual model links flow structure, bedload transport, bed roughness and channel geometry during these different phases. FAST also incorporates the threshold between different phases, for example, the depth required for the development of helical flows or the type of roughness and bedload transport for the switch between pulsed transport and sheet transport.

INTRODUCTION

Investigations on the interrelationships between bedload transport, river-bed adjustment and the related hydraulic conditions are rare and the feedback system is poorly understood. Since the characteristic nature of the vertical variation of velocity with depth was determined by Prandtl (1926) most measurements and interpretations have concentrated on two-dimensional studies of flow, often

Coherent Flow Structures in Open Channels.
Edited by P.J. Ashworth, S.J. Bennett, J.L. Best and S.J. McLelland.

restricted to steady flow conditions. It is only with recent development of measurement techniques that a growing number of studies have focused upon turbulent as well as structured flow. Knowledge of the three-dimensional turbulent flow field may allow the determination of more useful parameters for sediment transport calculations.

Field data on the response of river beds to the passage of unsteady flood waves are uncommon. Several attempts at data collection were made by the Lainbach research team (Ergenzinger, 1992), but normally the river bed changes are only determined between flood flows. A convincing example for investigations of long-duration events is presented by Laronne & Duncan (1992). Nevertheless, most published data comes from flumes. Extensively tested laser techniques are available for measurement of bed profiles in flumes, whereas similar measurements in the natural environment still lack the appropriate tools. In shallow flow conditions, typical of many gravel-bed rivers, the use of echo-sounders may be limited due to short signal distances (0.2–0.6 m) and the large diversity of reflecting surfaces. Without non-trapping and non-invasive measurements of bedload transport within flood waves, it is difficult to consider the interrelationships between discharge, bedload and river bed properties.

Since 1980, the electromagnetic detector system developed by Ergenzinger & Custer (1983) has been measuring the passage of naturally magnetic particles at Squaw Creek, Montana, USA. During 1991 and 1992 a new electronic system was installed by Christaller (Ergenzinger *et al.*, 1994a), enabling studies on the transfer of magnetic pebbles and cobbles with a time resolution as small as 10 s. With this instrumentation it is possible to shed light on the dynamics of gravel sheets (e.g. Whiting *et al.*, 1988; Dinehart, 1992) and other modes of bedload transport.

Following research presented in Ergenzinger *et al.* (1994a) the aim of this paper is to define the conditions that control the transition between the different phases of bedload transport as defined by the FAST model. In short, we are dealing with feedback cycles between hydraulic conditions, sediment transport, bed roughness and channel geometry. We emphasize that these processes are determined both by hydraulic conditions and the effects of bed material transport.

FIELD SITE

Squaw Creek is a gravel-bed, mountain stream in the headwaters of the Missouri. Its source is in the Gallatin Range, Bozeman County, Montana. The river drains a catchment of 106 km^2 with a maximum elevation of 3154 m. The predominance of Tertiary volcanic rocks, mainly andesites, with an average of 7% magnetite is fundamental to the bedload measuring technique. The lower quarter of the basin is occupied by Palaeozoic sedimentary rocks and older metamorphic rocks. Important sources of coarse sediments are glacial and fluvial deposits within the middle reach of the basin. Mean annual precipitation is about 810 mm (Bunte, 1992) and floods normally occur during snowmelt from May to mid-June. Snowmelt floods peak with a mean discharge of about 6 $m^3 s^{-1}$. Data were gathered during the snowmelt induced flood of the 5–6 July 1991.

The study site (Figure 31.1) lies within the last straight reach situated just above the confluence with the Gallatin river (1634 m) next to the Bible Camp bridge and below a bend. The active gravel-bed is on average 10 m wide with a slope of about 0.02. The armoured surface layer consists of grain-sizes ranging from pebbles to cobbles with a D_{50} of 34 mm and a D_{84} of 80 mm (where D represents grain diameter and the subscript indicates the percentage of material finer than that grain diameter). Subsurface material has a D_{50} of 20 mm and a D_{84} of 50 mm.

MEASUREMENT TECHNIQUES

ELECTRONIC MEASUREMENT OF BEDLOAD TRANSFER

This technique is described in detail by Ergenzinger *et al.* (1994a). At the measuring site (Figure 31.1) there are two detector sills with sensitive copper coils which sense the weak electric current generated whenever a magnetic particle passes the sill with sufficient velocity, according to Faraday's law of electromagnetic induction. The system is powerful enough to detect the movement of all magnetic particles with *b*-axes greater than 30 mm (at least two-thirds of the coarse bedload particles).

The upper and lower sills are separated by 30 m in the downstream direction and each spill spans the active width of the river bed. To determine the lateral variation in bedload transport rates the upper sill comprises six coil sections and the lower sill five sections. Each section covers 1.5 m of river bed. The time resolution of the system is 10 s which allowed the passage of single, coarse particles to be detected under all conditions, during the 1991 season.

MEASUREMENT OF THE RIVER BED AND WATER SURFACE

The 'Tausendfüssler' was used to determine the water surface and river bed elevations and their changes through time (see Ergenzinger & Stüve (1989) for further details). The device was attached to the measuring bridge (Figure 31.1) allowing the distances between the river bed and bridge to be surveyed every 0.1 m and between the water surface and bridge every 0.5 m, both with an accuracy of ±0.01 m. The spacing of the roughness interval was chosen in order to parameterize dynamic grain roughness as well as bedform roughness. From fractal studies of roughness in comparison to the grain size distribution (de Jong, 1995) it was found that the smallest interval necessary to determine grain roughness lies at $K_3(10)$ for Squaw Creek, where $K_3(10)$ is the maximum vertical difference between three adjacent bed elevations, measured at 0.1 m intervals. Water surface heights were

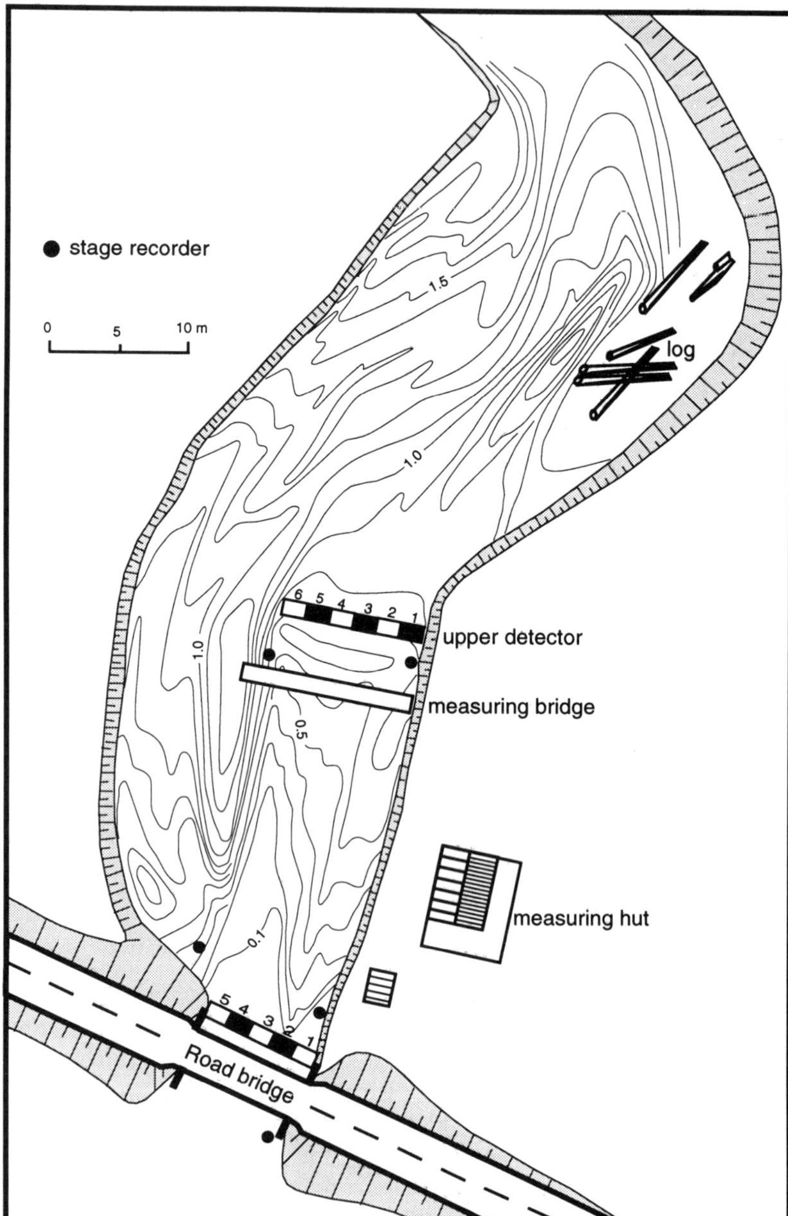

Figure 31.1. The Squaw Creek measuring site indicating the measuring bridge with the Tausendfüssler, two bedload monitoring detector sills and the locations of five stage recorders

determined at wider intervals due to the less dynamic nature of water surface fluctuations. The total time required for a complete survey of the section was 30–35 minutes. Data obtained from the Tausendfüssler were utilised to calculate the roughness created by the protrusion of particles. The mean K_3 coefficient for the cross-section is calculated as the mean of the moving vertical difference.

MEASUREMENT OF FLOW VELOCITY AND WATER SLOPE

A multi-rotor system with five rotors was used to measure water velocities every 0.5 m. The results were used to calculate discharge and local shear velocities (Table 22.1). Due to a number of difficulties there are no measurements after 00:30, 6 June (P. Carling, pers. comm., 1992).

In 1992 an attempt was made to determine surface velocities within the reach by tracking fluorescent ping-pong balls using two photogrammetric cameras combined with stroboscopic illumination. Floating balls were released instantaneously from a specially constructed bar attached to the upper end of the measuring bridge spaced at 0.2 m intervals. As the balls drifted down the reach, stroboscopic light illuminated them at 1 s intervals. At the same time, two Zeiss cameras, raised above the road bridge at the lower end of the reach allowed overlapping pictures to be taken of the tracers simultaneously at 1 s intervals. This enabled three-dimensional pictures of the water surface to be constructed with a special program developed at the Geodetic Institute of the Technical University of Berlin. The streamwise (x), vertical (y) and cross-stream (z) positions of the balls were obtained with an accuracy of ±0.01 m. This technique results not only in three-dimensional surface water velocities but also can be used to calculate local water surface topography.

In addition, some two-dimensional velocity measurements were obtained using an electromagnetic current meter (ECM) with 0.06 m diameter probe. The instruments were attached to a rail below the measurement bridge and the primary axis oriented towards the main flow direction perpendicular to the bridge. Velocities were measured at a frequency of 10 Hz and averaged over 60 s at each position.

The water-surface slope was determined with pressure hoses between the two stage recorders on each side of the measuring sites (Figure 31.1). The intake of the hose was 20 m above the intake of the lower hose. Water surface gradients were calculated from the difference in water level between the two hoses, measured with an accuracy of ±1 mm.

RESULTS

DISCHARGE CONDITIONS

The selected snowmelt induced flood occurred during 5 and 6 June 1991 (Figure 31.2). The flood began to rise at 15:00, peaked at about 19:00 and then maintained

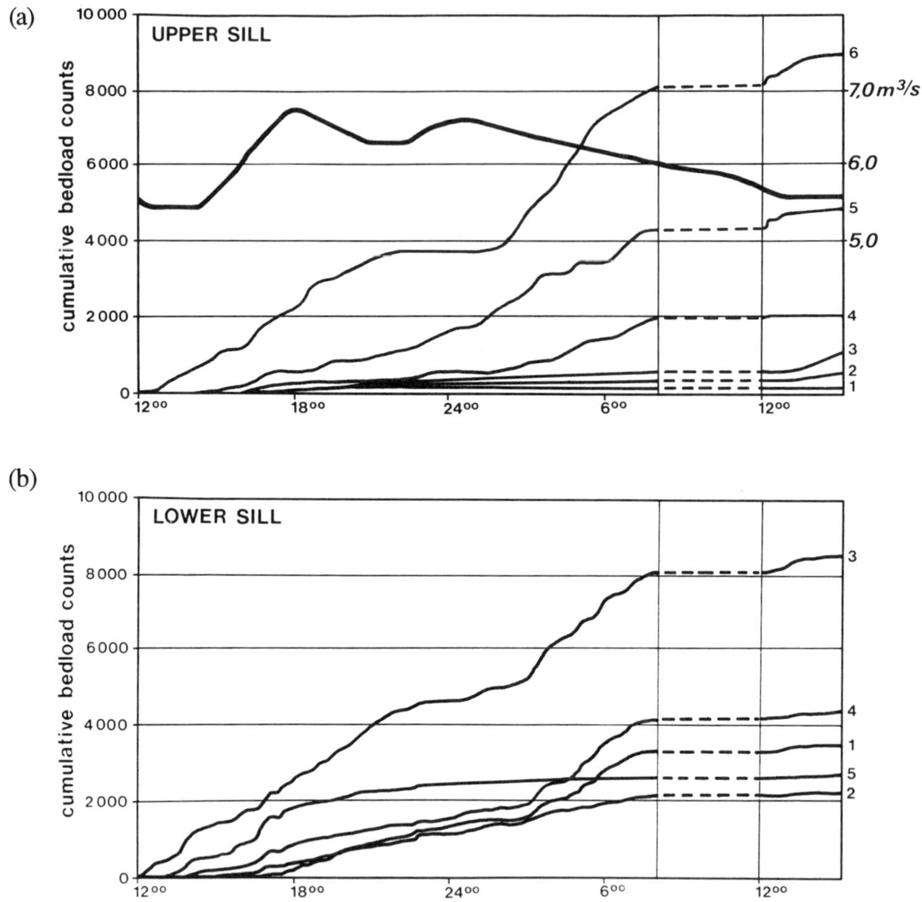

Figure 31.2. Cumulative plots of the number of magnetic particles detected during 5–6 June flood, 1991; (a) is for the upper sill and (b) for the lower sill. The bold line in (a) shows the discharge hydrograph

a steady discharge due to the impact of prolonged rainfall. A second peak developed at 00:00. The recession lasted until 14:00 next day. Absolute changes in discharge magnitude were remarkably low with a minimum discharge of 5.6 $m^3 s^{-1}$ and peak discharge of 6.8 $m^3 s^{-1}$.

BEDLOAD TRANSPORT

In contrast to the slow changes of water discharge the bedload transport rates were more variable. In total, 17 000 particles were recorded moving over the upper sill whilst 20 000 moved over the lower sill. Figure 31.2 illustrates the activity in different sectors, but excludes the non-magnetic material which comprises 33% of the total bedload. A failure of the electronic system explains the horizontal dashed line between 07:25 and 12:00 on 6 June. Sectors 4–6 in the inner channel are the most active areas of the bed, however, sectors 1–3 at the bar only carried magnetic bedload during distinctive phases. The temporal distribution of transport rates can be differentiated from the gradient of the curves. Dominant phases have gradients of greater than 50° during the times of the rising flood and during the recession. Such phases of intense transport, i.e. with transport rates above 17 particles per 10 minutes (including the non-magnetic proportion), normally only last for 20–40 minutes and record the transit of gravel sheets or carpet-like transport events. The pulsed transport involved longer periods of up to 60 minutes. These have typical slopes of 30–45° on the cumulative curves, corresponding to typical transport rates of 8–13 particles per 10 minutes, including non-magnetic material. Phases of 'random transport' have gentler inclinations on the cumulative curves (less than 15°) and represent the most common mode of sediment transport. Comparisons between the bedload sums of the upper and lower sills allow isolation of the beginning and end of different transport phases (Figure 31.3). The time lag is often about 20 minutes for 30 m distance but there are some events with almost no time-delay.

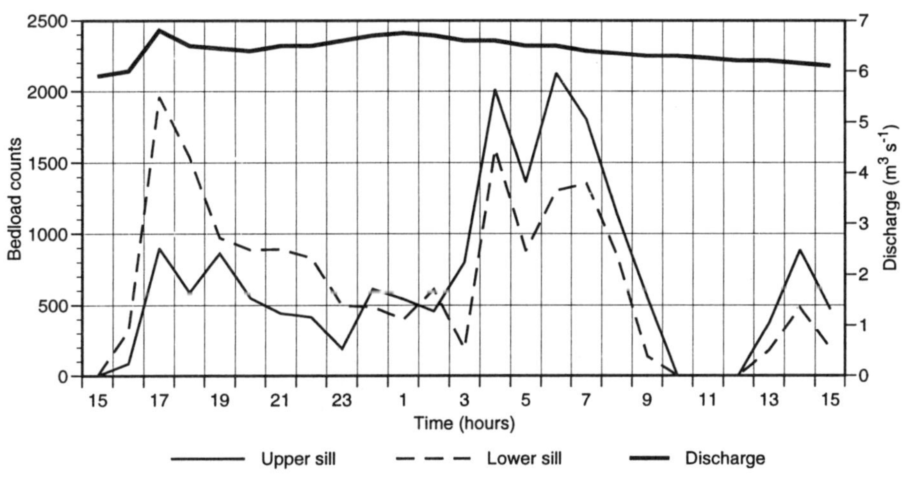

Figure 31.3. Bedload counts for the upper and lower sill, calculated over hourly intervals during the flood of 5–6 June for 10 minute bedload sums

In Figure 31.2 the cumulative curves show the combined bedload counts for all the sectors of both sills. In both data sets the slopes of the curves are discontinuous. Slopes of more than 50° correspond to transport rates of greater than 9 particles per minute per 1.5 m channel width. These are the phases of carpet-like or sheet-like transport. The onset of such movements occurred twice simultaneously only in sectors 5 and 6: (i) at 16:37; and (ii) at 02:30. Sectors 5 and 6 cover the first 3 m of the inner channel. The next 1.5 m is adjacent to the interface where gravel sheet transport did not develop. Slope angles are gentler, with 30–45° and this corresponds to about 5–7 particles per minute per 1.5 m width and is typical of pulsed transport. The more gentle slopes below 15° connect the steeper parts of the slopes and are the result of periods with very low transport rates of about 1 particle per minute per 1.5 m width. These represent random transport and are typical for the sectors feeding the bar (1–3). If the most active sector of the upper sill (sector 6) is compared with the most active sector of the lower sill (sector 3), it is striking that a similar number of about 8500 magnetic particles across each 1.5 m wide sector of the sill. The horizontal slope occurring between 07:25–12:00 was caused by a failure of the electronic system and is not interpreted.

The high temporal resolution of bedload transport allows a detailed examination of the dynamics of single particle transport. For most of the time the transport events do not consist of randomly-distributed, single particles, but of 'families'. A more detailed picture of such a family was published by Ergenzinger *et al.* (1994b). Figure 31.4 shows such a situation with a 160 s duration. In the main channel 23 magnetic particles crossed sector 6 of the sill. The first pulse contained 6, the second 4 and the third 12 magnetic particles. The transit time of a cluster lasts approximately 60 s and all transit times were in the range 20–60 s. These sedimentological units can be seen when the water is clear and are normally composed of one or two bigger cobbles and an entourage of several smaller particles.

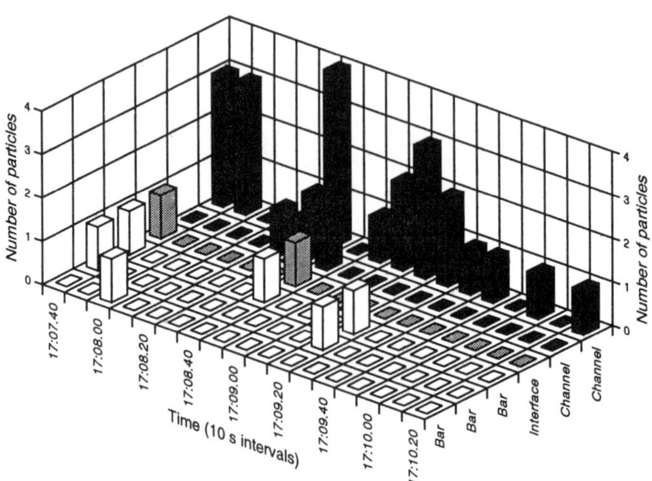

Figure 31.4. Transit of magnetic particles across the different sections of the upper sill. The signals are summed across 10 s. The most active section is the first 1.5 m of the inner channel

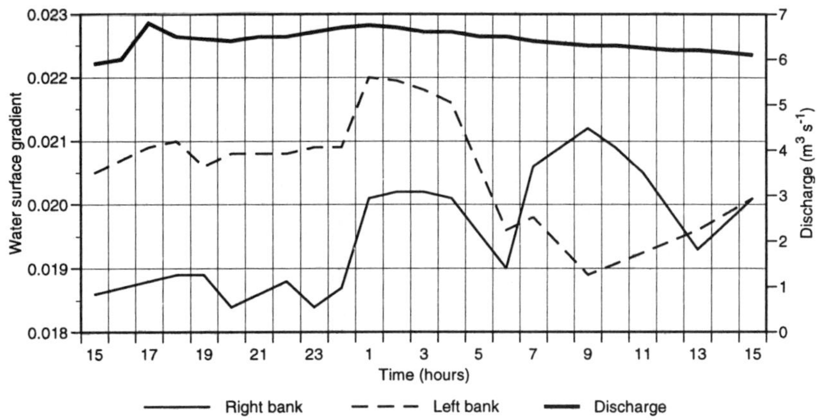

Figure 31.5. Variation in water-surface slope near the right and left channel banks during the flood (bold line shows the flood hydrograph)

RIVER BED AND HYDRAULIC CONDITIONS

The changing properties of slope, roughness and geometry of the cross-section below the measuring bridge reflect both the impact of changing hydraulic conditions and of bedload transport modes and rates. The slope of the water surface was measured next to both banks by four stage recorders (Figure 31.5; Ergenzinger *et al.*, 1992a). Until 00:00 there was an increasing difference between the water slopes parallel to the left and right banks. The slope along the left bank stagnated at about 0.0185 whereas the slope along the right bank increased from 0.0210 to 0.0215. Considerable changes occurred during the second peak and the recession after 00:00: both slopes increased between 00:30–01:30 to 0.0220 and 0.0200 for the left and right banks respectively. After this, both slopes decreased dramatically. By 06:30 the situation changed and both slopes approached each other with values of 0.0190 and 0.0197, but then the left bank rose sharply to greater than 0.0210 whereas the right bank declined further to 0.0190. Just before the rise of the meltwater flood on 6 June, both slopes returned to a common level of 0.0200. Such changes in water surface gradient between the two banks are quite common. At Squaw Creek they were first described by Bunte (1991, pp. 108–109) who related them to changing bed conditions. The sudden increases in slopes after 00:00 were not caused primarily by rising water but by aggradation in the area around the upper sill. When this area reached threshold conditions, with water surface slopes of about 0.0220, erosion began. During the first phase of intensive bedload transport the water surface slopes declined rapidly. Whilst during the bedload transport recession (06:00–09:00), the divergence between the two river banks resumed. This time the left bank reached its maximum when the opposite bank was at a minimum. The water-surface slope is a more effective predictor of bed changes and bedload transport than discharge.

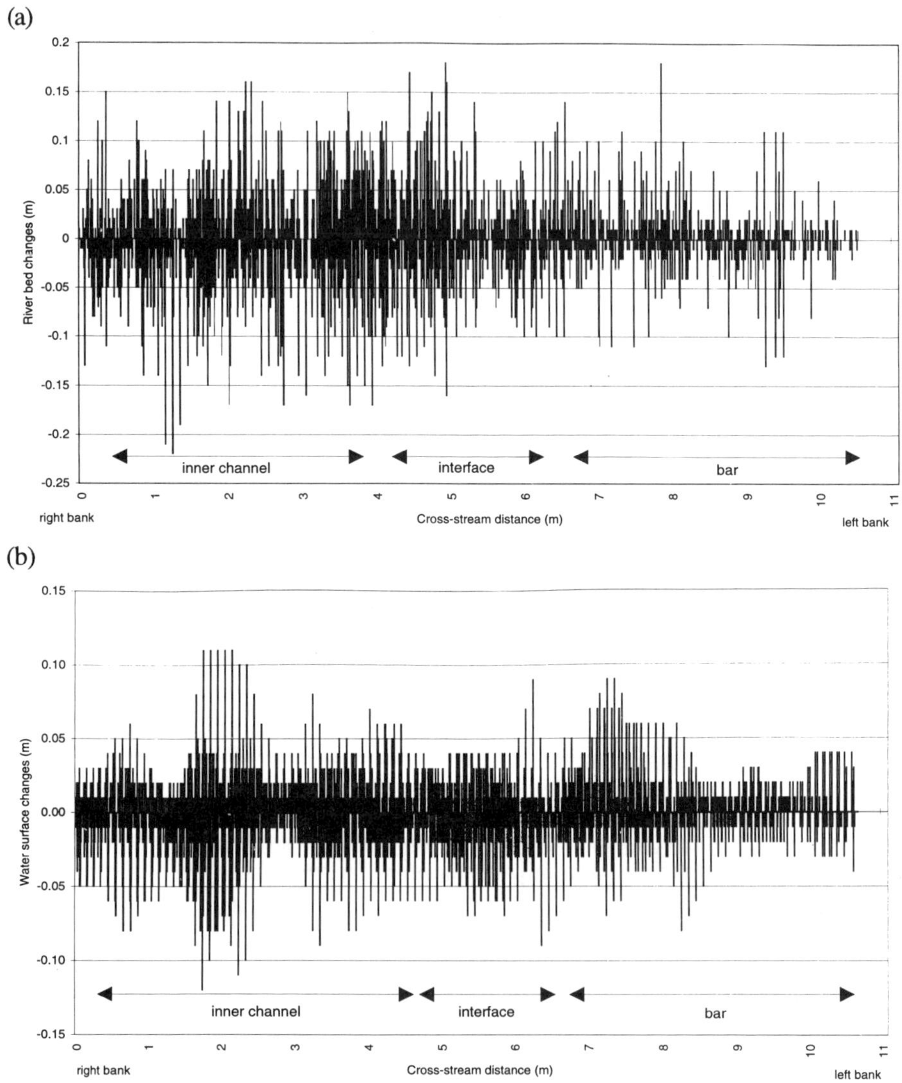

Figure 31.6. (a) River bed height and (b) water surface elevation changes for 5–6 June 1991, calculated temporally between successive Tausendfüssler measurements

The Tausendfüssler measurements below the measuring bridge suggest that there was more erosion and aggradation during the passage of the meltwater flood than just one layer of particles. Figure 31.6a shows the composite changes of 21 measured cross-sections between 5 and 6 June. Within the inner channel (0.5–5 m) erosion and deposition depths were within the range ±0.15 m. On the slope of the interface (the bank between the bar and main channel, 5–8 m), erosion and deposition depths are smaller. Since average water depth is only about 0.4 m, it is not surprising that the adjustment of the river bed influences the topography of the water surface (Figure 31.6b). The maximum deflections from the mean water-surface elevation occur near to the banks of the inner channel.

Erosion and deposition patterns are not directly related to discharge variation. During the first bedload pulse between 15:00–23:00 on the 5 June, erosion was predominant (with an exception at 20:30) and during the passage of the second bedload wave between 23:00 on 5 June and 10:00 on 6 June accretion was dominant (with the exception of 05:30). From the Du Boys equation, reach-average bed shear stress is equal to the product of unit weight of water, hydraulic radius and the spatially-averaged energy slope. In a situation where the hydraulic radius is proportional to discharge (*cf.* Petit, 1990), bed adjustments are not related to the reach-average bed shear stress, but may be attributed to the bedload balance between the upper and lower sill (Figure 31.7). So long as sediment output exceeds input, erosion is predominant and phases of intensive accretion only occur during recession, when there was greater sediment input than output.

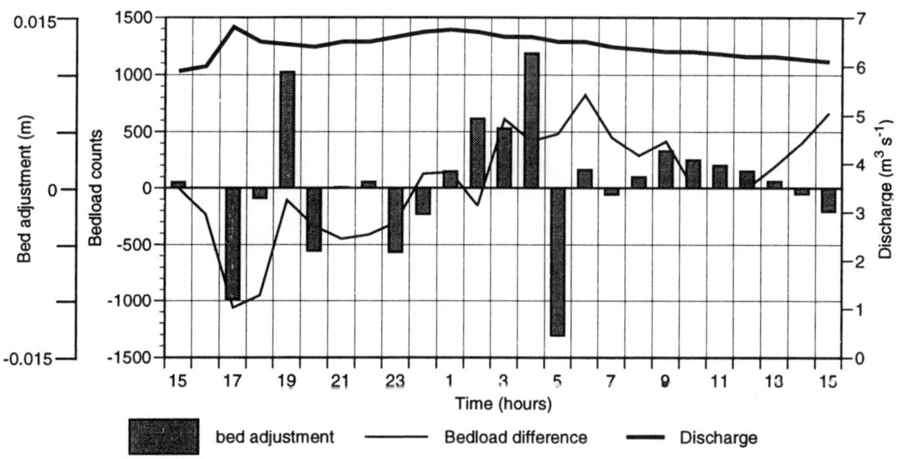

Figure 31.7. Bed adjustment in relation to discharge and the difference in bedload input and output to and from the test section during the flood of 5–6 June 1991

Roughness variation is illustrated by the measured changes in the K_3 parameter within the inner channel which fluctuated during the passage of the flood between 0.045 and 0.08 m (Figure 31.8). Again the biggest change occurred during early morning at the recession of the most intensive bedload transport. In general, whenever an intensive wave of bedload transport begins, the roughness values are low (Lambert & Paris, 1992; Tait & Willetts, 1992). During the recession of the bedload wave roughness increases rapidly. The reason for the low roughness values

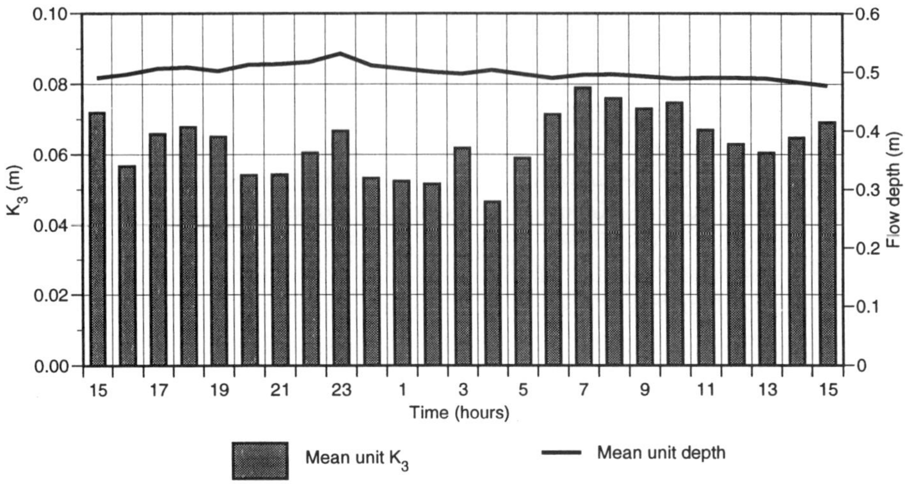

Figure 31.8. Average flow depth and K_3 roughness of the inner channel during the flood 5–6 June 1991

whilst bedload is moving has to be considered in terms of the river bed. During intense bedload transit, the entire bed is mobile and the bed acts like a conveyor belt which transmits as much bedload as possible. Large obstacles which obstruct the flow are removed. If only the inner channel is considered, then the roughness may increase locally and temporarily due to the transfer of large particles, however over the entire river bed roughness decreases (Ergenzinger *et al.*, 1994a; de Jong, 1995). On the other hand, as soon as bedload transport decreases, bed roughness rebuilds again as large particles are stranded. This concentrates bedload into narrow zones and transport becomes size selective (Bunte, 1992). The same is true for other flood events studied at Squaw Creek during the same year and for studies undertaken in the comparable Lainbach river, southern Germany (de Jong, 1995).

Since K_3 only describes the changes in geometry of the inner channel, it is suggested that river bed adjustment is related both to a smoothing by accretion of finer particles and to coarsening when fines are selectively transported. Nevertheless, these changes cannot be accurately supported by measurements since the time required for one cross-sectional measurements is too long for such rapid adjustments.

The hydraulic conditions for the first bedload wave are summarized in Table 31.1. Mean velocities and friction velocities are uniform, hence, there are minimal changes in Froude and Reynolds numbers. This example illustrates that the adjustment and changes of water-surface slope, channel geometry and bed roughness correspond strongly to the mode of bedload transport. Whenever discharge is above the critical threshold for bedload transport the resulting changes depend on the

Table 31.1. Values of mean velocity, shear velocity, discharge, Reynolds and Froude numbers for the flood of 5 June 1991

Time	Mean velocity $m\ s^{-1}$	Shear velocity $m\ s^{-1}$	Discharge $m^3\ s^{-1}$	Reynolds number	Froude number
16:20	1.52	0.32	5.7	2.88×10^5	0.82
17:20	1.60	0.28	6.8	3.19×10^5	0.82
18:20	1.55	0.24	6.5	3.02×10^5	0.81
20:20	1.52	0.38	6.4	3.06×10^5	0.77
22:20	1.52	0.35	6.5	3.03×10^5	0.76
00:20	1.53	0.34	6.7	3.08×10^5	0.78

supply of coarse particles. The interrelationship between bedload transport rate and hydraulic conditions will be illustrated below for several specific conditions.

HYDRAULIC CONDITIONS AND BEDLOAD TRANSPORT

Observations of the water surface during flood discharges suggest that for most of the time there are both waves and troughs present in the measurement cross-section. Einstein & Li (1958) observed that drifting material gathered at troughs in the water surface. Figure 31.9 shows positions of tracked markers at 1/30 s intervals in the inner channel as a result of the experiments undertaken in 1992 on the distribution of surface velocity. After about 6 m downstream (Figure 31. 9) most of the markers had become grouped into troughs and this occurred with a discharge below 5 $m^3\ s^{-1}$. The troughs are comparable to the general dynamics of the water surface (Figure 31.6b) shown in relation to the cross-section for the duration of the flood wave of 5–6 June 1991. Whereas the bar was covered by crossing shear waves, the inner channel was marked by wave crests along the bank and the upper limit of the interface. During the experiments two further wave crests developed so that the waves now had amplitudes of about 1 m. The non-uniform water surface was maintained for several hours during the experiments.

Results of two-dimensional flow velocities are shown in Figure 31.10. Stage reached 0.42 m in Figure 31.10a and 0.37 m in Figure 31.10b. Systems of well-developed inverse secondary velocities developed between the water surface and river bed when the stage was 0.42 m (similar to Bhowmik, 1982), whereas at a lower stage (0.37 m) the secondary velocities were more erratic and less-structured. In the spring of 1992, discharge seldom became critical for bedload transport, hence, these measurements only support a general pattern. With rising stage the secondary flow system becomes increasingly important. Cross-over points in secondary flow direction occur at 9.7, 8.4, 7.4 and 6.8 m, whilst at 9.1 m a potential cross-over point was disturbed by a downstream obstacle. These distance are about twice the water depth and reflect the development of troughs at the water surface.

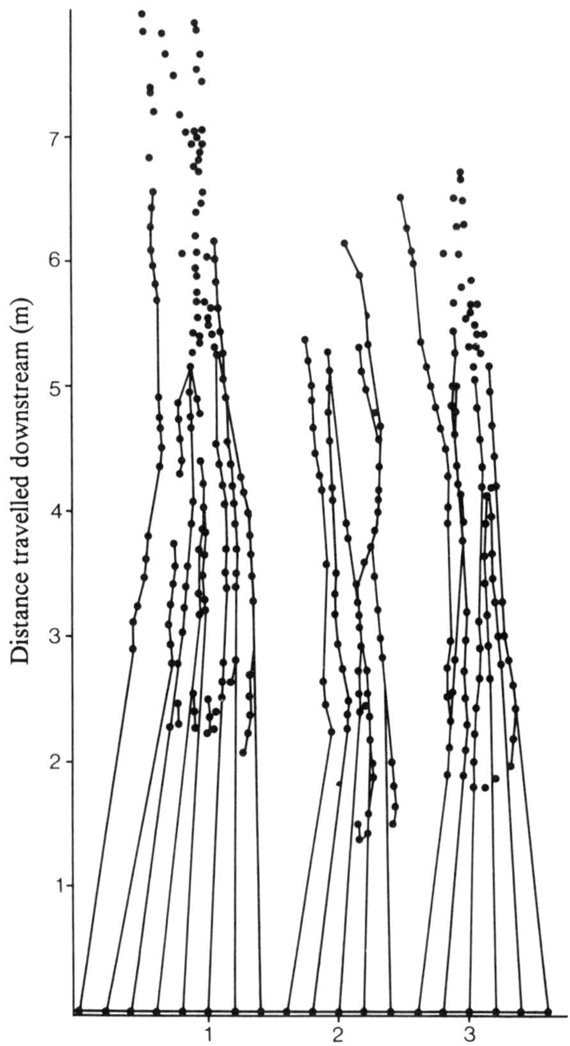

Figure 31.9. Results of the experiments determining the structure and velocity of the water surface by tracked markers

Changes in water surface elevation not only reflect the hydraulic structure but are maintained and developed by local roughness conditions. This is explained by some cross-stream profiles measured during the flood of 5–6 June 1991. The first three measured cross-stream profiles are shown at the onset of the flood (Figure 31.10) when unit depth (local depth/mean depth) fluctuates by ±20% in every case. If the

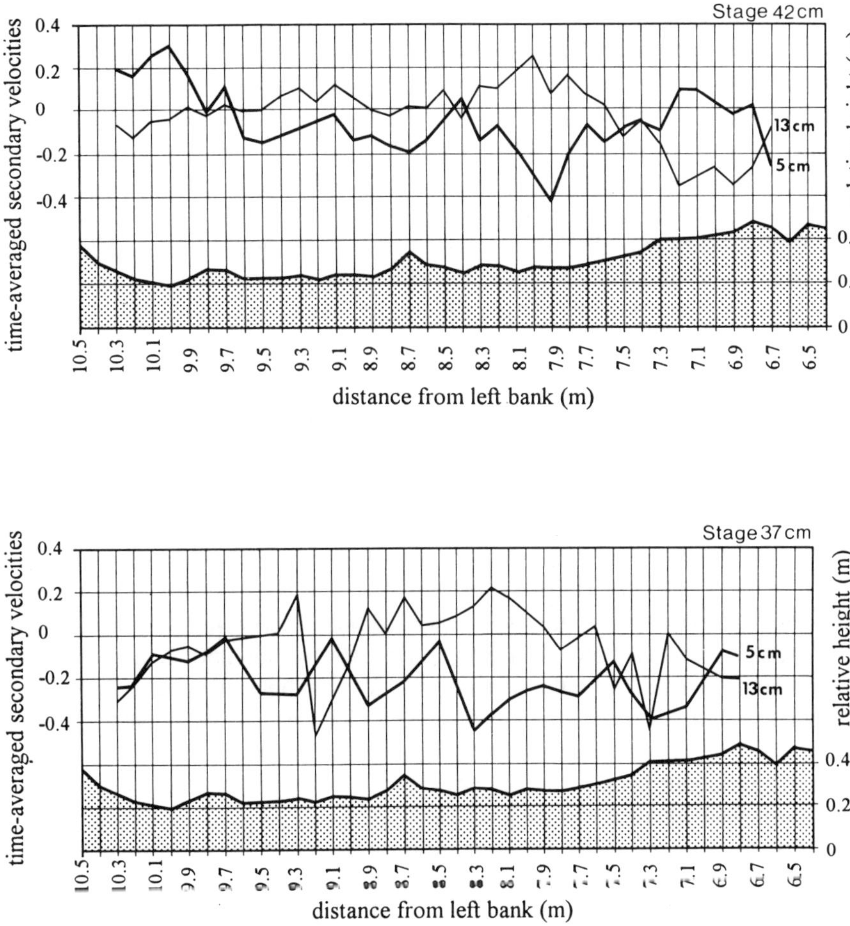

Figure 31.10. Secondary velocities in the inner channel of Squaw Creek 20 May 1992. The velocities are measured simultaneously 0.05 m below the water surface and 0.13 m above the river bed. positive velocities are deflections towards the left banks, negative velocities are due to deflection towards the right bank. (a) stage 0.42 m; and (b) stage 0.34 m

areas of unit depth greater than unity are compared with those below unity, the crests and the troughs of the inner channel cross profile can be separated. At 15:30 there were four lateral crests, at 16:00 three lateral crests and at 16:45 four similar sized crests. At 15:30, five troughs developed corresponding to these crests. Comparably, four troughs developed at 16:00 and five troughs at 16:45. Whenever there was a trough, an obstacle with a unit K_3 roughness (local K_3/mean K_3) above unity coexisted. At the beginning there was only one 'pile' of large roughness but between 1.5 and 3.3 m at the end there were five peaks of roughness developed across the section at 0.3, 2, 3.2, 4.2 and 4.7 m.

These results are combined in Figure 31.12, where for every available cross-section the first or second maximum of unit K_3 roughness was related with the corresponding minimum unit depth. There is a relatively good correlation between these parameters ($r^2 = 0.4$, 95% level of significance). When the roughness elements are greater than 1.5 times the average roughness, the water surface is pulled down further than the corresponding decrease in water depth. According to Studerus (1982) this effect is due to the development of secondary flows and the transport of particles into zones of greater roughness. The development of the wave crests is a direct replica of this situation. The cross-sectional waves correspond to vortices with lengths twice the flow depth and thus, the zones of maximum roughness are again twice the flow depth. The areas of maximum erosion always correspond to upwelling near to the bed (Figures 31.11a–31.11c).

As stated before, the interaction between bed dynamics and the hydraulic conditions starts when discharge is critical for bedload movement (compare the rapid changes between 16:00 and 16:45). When the transport rate of bedload per hour rises above 200 magnetic particles (or a total of 270 particles) roughness characteristics change. This includes reduction of K_3 values at the beginning of the bedload wave so that the vortex flow described above is no longer present (Figure 31.11). This happened between 16:45 and 17:10. About 250 magnetic particles were counted in sector 6 of the bedload sill. K_3 roughness values decreased and a good correlation between bed roughness and surface waves is maintained only to the right of the 3 m point marked on the cross-section. The wave at 1 m is not caused by local bed conditions but should be related to the stable bank conditions just above the measured cross-section.

The most uniform roughness distribution (Figure 31.13b) with rather high roughness values (Figure 31.8) occurred at 07:10. This situation happened after a slug of intensive transport (06:45 to 07:10) and at the same time marked a phase of maximum concurrent erosion and accretion within the measuring section. Since the inner channel became enlarged and smoothed during that phase (Figure 31.13) within only a few minutes the geometry and roughness of the bed changed from a situation with wavy water (well-developed vortex flow) to a new hydraulic situation. According to observations of the water surface, the flow structures are characterized by a lower layer with intensive interaction of particles and water and it is assumed that there is an upper layer with a more-or-less vertical logarithmic velocity profile during the times of the transit of gravel sheets. Coherent secondary flows are destroyed and replaced by several boils with vertical axes creating an irregular, choppy pattern of small surface-waves. During certain phases, very steep irregular waves occur at the surface, whilst the water exhibits surface boils at other sites.

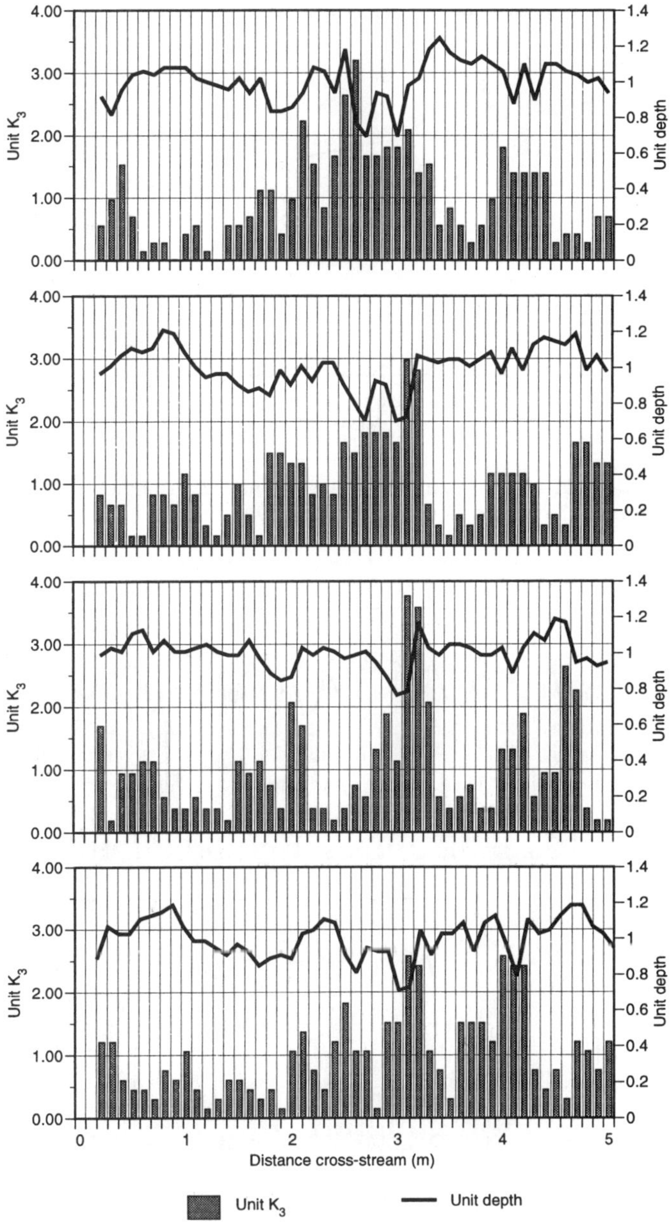

Figure 31.11. A sequence of cross profiles of Squaw Creek during 5 June 1991. Unit K_3 roughness (local K_3 /average K_3) is compared with the unit depth (local depth/average depth). (a) measurements at 15:30 (begin of flood); (b) measurements at 16:00 (rising hydrograph); (c) measurements at 16:45 (begin of bedload transport); and (d) measurements at 17:10 (during bedload transport)

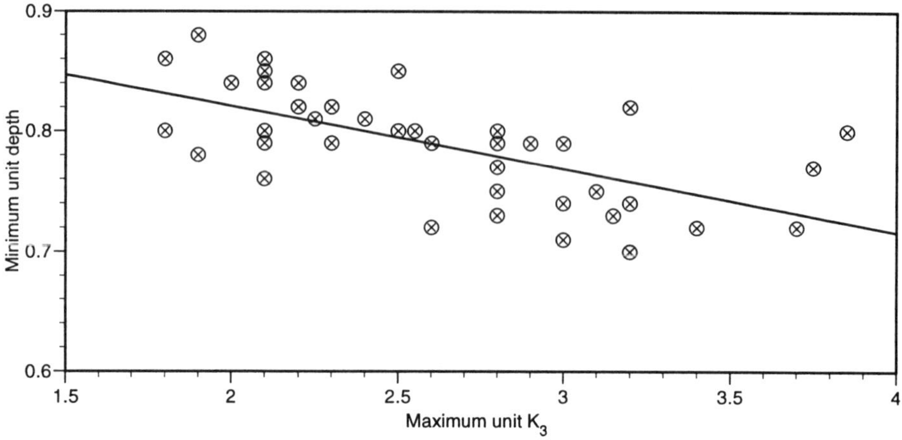

Figure 31.12. Plot of the maximum values of unit K_3 roughness with the induced minimum unit depth during the flood of 5–6 June 1991

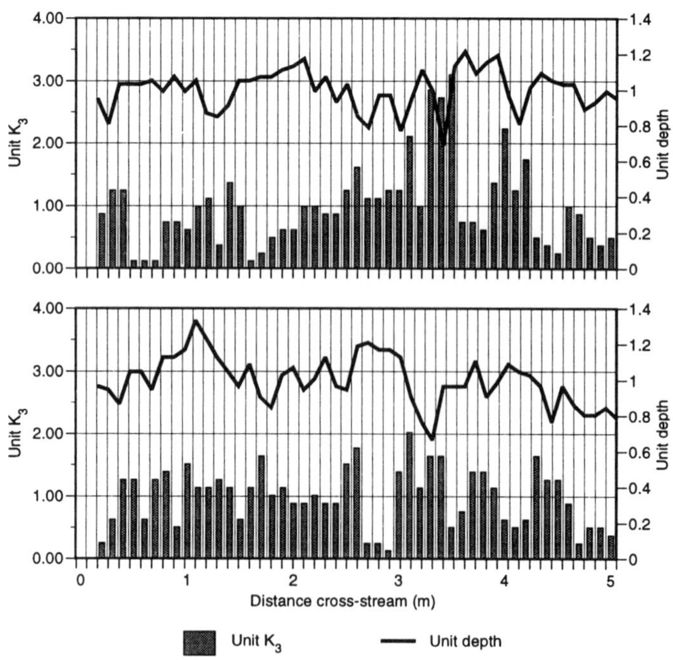

Figure 31.13. Comparison of the development of unit K_3 roughness and unit depth during a phase with high bedload transport. Transit of bedload at the different segments of the upper and lower sill. The measurements of: (a) 06:45; and (b) 07:10 relate only to the first three sections along the upper sill

DISCUSSION AND INTERPRETATION

HELICAL FLOW AND PULSED BEDLOAD TRANSPORT 'FAMILIES'

During the snowmelt floods, vortex or helical flows were predominant in the inner channel. This was illustrated by channel geometry, bed roughness and the occurrence of pulsed bedload transport.

Stage measurement and the electronic bedload counts are the only variables measured continually. Intensity and time-periods for five different transport phases and their related hydraulic conditions can be derived for the flood of 5–6 June. Table 31.2 includes some results on the duration and intensity of bedload transport. According to these results the most transport is associated with the transit of families of stones (*c.* 60% of flood time). This mode of quasi-continuous transport of small amounts of bedload is caused by an unstable spatial arrangement of roughness elements and a related shift of vortex cells. So long as the flood is powerful enough to maintain an unstable river bed, the hydraulic conditions are affected by the distribution of roughness elements. Similarly, vortex cells can cause a particular arrangement and development of the river-bed structure. This interplay causes about 5 to 8 magnetic particles to be in transport per minute for 1.5 m width and they move to a new location on the river bed involving a short transport distance of 10–20 m. Locally, this causes new flow conditions and is the reason for the next adjustment of the hydraulic system in a spatially-distributed form-process feedback mechanism.

Table 31.2. Duration and intensity of counted magnetic bedload particles at sector 6 (upper sill: 7.7 to 9.2 m from the left bank) and sector 3 (lower sill: 3 to 4.5 m from the left bank)

	Mode of transport	Duration (hours minutes)	Duration (%)	Intensity particles min^{-1} (% of total transport)
Upper sill:	gravel sheet	03 48	17	16
	pulsed (families)	13 56	62	06
	random (singles)	04 33	20	01
Lower sill:	gravel sheet	02 50	13	14
	pulsed (families)	12 53	58	07
	random (singles)	06 38	29	01

TWO LAYERED FLOW AND TRANSPORT OF PARTICLE SHEETS

The occurrence of phases of intensive transport associated with gravel sheets is both temporally and spatially unstable and difficult to predict. According to the measurements at Squaw Creek this transport stage takes place during parts of both the rising and receding limbs of the hydrograph. The onset of gravel sheet transport during the recession is caused by the steepening of the slope of surface water due to sedimentation above the upper sill (Figures 31.3 and 31.5). As shown in Figure 31.2 the onset of gravel movement can happen at the same time in two or even three neighbouring sectors of the sill but only at 16:40 was there intensive transport in all five sectors over the lower sill. Sometimes there is frequent alteration between gravel sheet and pulsed transport (sector 5, upper sill, 02:30 to 07:25). The longest period of gravel-sheet transport occurred in sector 6 of the upper sill lasting for 1 hour and 46 minutes.

The transport of a layer of coarse material causes abrupt and intense changes in bed roughness and geometry. The K_3 values are lowered by local accretion and/or erosion at the beginning of carpet-like transport. Simultaneously, the water surface changes and in response to the newly developed hydraulic situation. Velocities close to the river bed could not be measured but it is proposed that this situation resembles the 'two-layered' flow system, measured by Wang *et al.* (1993).

THE *FAST* MODEL: SOME CONCLUSIONS

A first attempt to combine these observations on the interrelationship between bedload transfer and river-bed adjustment in geometry and roughness was published by Ergenzinger *et al.*. (1994a), Ergenzinger & de Jong (1994) and de Jong (1995). The proposed conceptual model, FAST, attempts to discriminate between different unsteady hydraulic situations during the transit of a flood wave associated with the feedback with different modes of bedload transport. Even when the calculated shear stresses are similar the resulting hydraulic system can be modified intensively due to bedload transport. The ideas described above concern the characterization of flood flows in coarse-grained rivers with high width-to-depth ratios. Intensive and non-intensive phases of bedload transport are related to the development or destruction of helical flow cells. In the case of low intensity bedload transport the dominant flow structure is helical or vortex flow causing transport of coarse particle families. In the case of high-intensity bedload transport, two-layered flow structures are related to the occurrence of particle sheets.

These phases are important for determining whether roughness values are large or small and whether the river bed is adjusting in terms of erosion or deposition (de Jong & Ergenzinger, 1994). The model demonstrates the importance of flow and river bed adjustment in controlling the nature of bedload transport and *vice versa.* Since bedload does not relate directly to the flood discharge, river bed adjustment and flow hydraulics form an important interactive process in determining: (i) whether the next phase of bedload transport is intensive or non-intensive; (ii) whether the formation of flow cells is expected; and (iii) how the river bed will respond in terms of erosion and deposition.

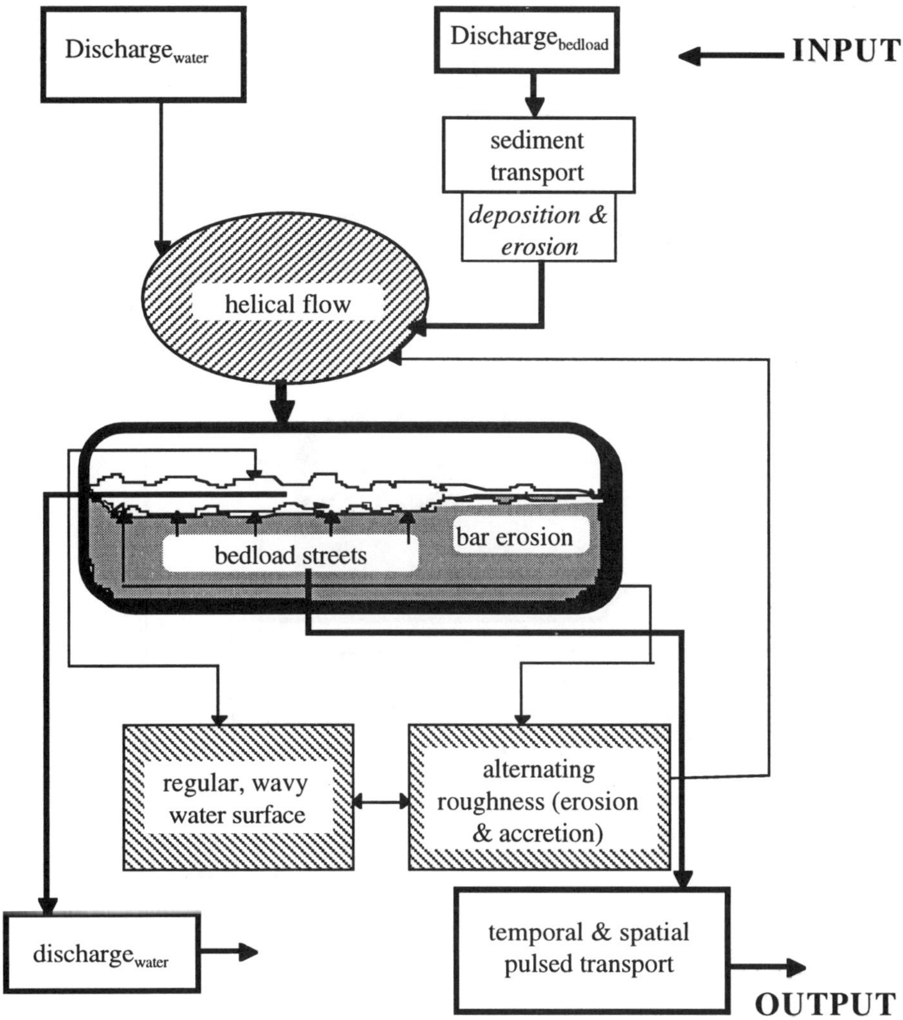

Figure 31.14. Flow chart illustrating interactions between discharge, bedload transport and hydraulics during helical flow (FAST model)

Figure 31.14 displays the case of helical flow schematically. As mentioned above this type of flow and bedload transport occurred during more than half of the flood wave. The prerequisite for this development is a discharge such that the river bed can be eroded locally. When the water level in the inner channel rises to about 0.5 m (about 8–10 times the roughness height), there is vigorous development of the river bed and water surface. Rough zones draw the water surface down and there

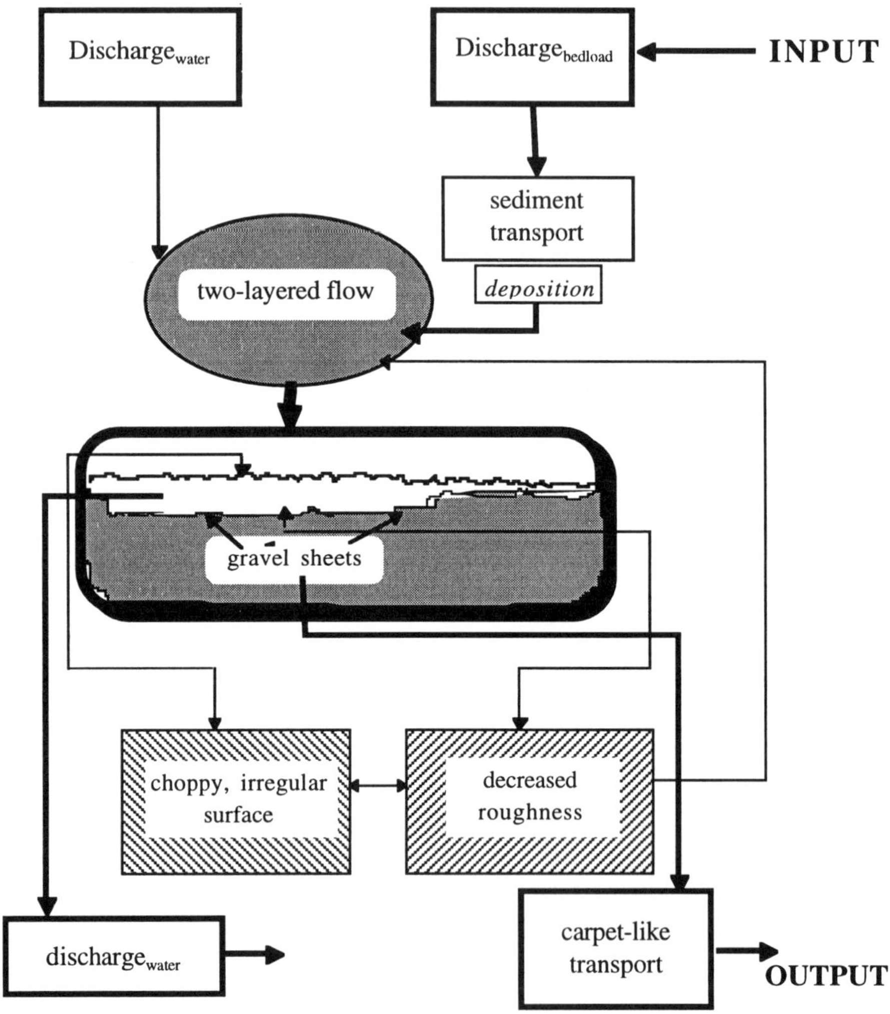

Figure 31.15. Flow chart illustrating interactions between discharge, bedload transport and hydraulics during two-layered flow (FAST model)

is a concurrent flow upwelling above the low roughness zones. As simulated by Studerus (1982) the distance between the rough zones is about twice the water depth. This indicates that helical flow cells are well-developed across the channel.

Water depth is not very variable during the flood. Instead, roughness is far more variable than the water depth, although the distribution and size of water surface depressions vary substantially. In studies of downstream variations in roughness as compared to cross-stream, a close correlation was obtained for the different phases (de Jong, 1995). During intense bedload transport the bed smoothed in both directions, and during less intense transport, when flow cells developed, the bed again roughened in both directions, due to deposition of obstacles and winnowing of fines. This suggest that the river bed must be considered as a complete system for different transport phases.

Cross-sectional and longitudinal measurements of roughness in the Lainbach River show that the may be differences in the magnitude of K_3 roughness in each direction, but, the relative changes in each direction are the same (de Jong, 1995). The similarity between cross-sectional and longitudinal roughness is a prerequisite for the establishment of helical flow cells. Continuous electronic counts of magnetic bedload particles demonstrate that this hydraulic situation is connected with the transport of families of bed material. Bedload transport occurs in narrow zones characterized by low roughness and upwelling water. With the exception of the channel banks and larger bed particles, the geometry and especially the roughness changes due to the erosion and accretion of small numbers of particles associated with these narrow bedload transport zones. The dynamics of these zones could not be measured due to the large sector width, however, the system is unstable and shifts slowly in both space and time.

The two-layered flow system (Figure 31.15) develops due to the impact of intensive bedload transport. The vertical distribution of velocities of such hydraulic systems are still poorly understood. Due to the large number of particles in the ground layer the interaction between water and particles must be far more intensive. The importance of this situation for the development of turbulence is not known in detail. According to the observations at Squaw Creek the water surface is either more or less horizontally smooth with many vertical boils or asymmetrical, small and steep waves. At the beginning of the onset of a bedload sheet there is a strong tendency to develop a smooth bed surface, to decrease roughness and to enlarge the inner channel laterally.

Even during the flood there are phases with very low bedload transport, when the dynamics of river bed adjustment are reduced. The hydraulic system will change back from helical flow to an unstructured turbulent flow only when the water level drops below six times the roughness heights. The change from random bedload transport to helical flow and *vice versa* can occur as a continuous process. The change from helical flow to gravel sheets is sudden and the hydraulics change abruptly at this threshold.

ACKNOWLEDGEMENTS

This work was funded by the Deutsche Forschungsgemeinschaft (German Research Council). Thanks are due to Kristin Bunte, Steve Custer, Karin Möhrke, the Christaller research group and the Carling research group. CDJ's work was supported by a scholarship of the Studienstiftung.

REFERENCES

Bhowmik, N.G. 1982. 'Shear stress distribution and secondary currents in straight open channels', in Hey, R.D., Bathurst, J.C. & Thorne, C.R. (Eds) *Gravel-bed Rivers*, John Wiley & Sons Ltd., Chichester, 31–61,

Bunte, K. 1992. *Untersuchungen der Zeitlichen Variation des Grobgeschiebetransportes und Seiner Korngroßenzusammensetzung* (*Squaw Creek, Montana, USA*), Unpub. PhD Thesis, Freie Univ. Berlin, 200 pp.

de Jong, C. 1992. 'Micro-profiling of mobile gravel beds', in Bogen, J., Walling, D.E. & Day, T.J. (Eds) *International Symposium on Erosion and Sediment Transport Monitoring Programmes in River Basins*, IAHS **210**, Oslo, 63–67.

de Jong, C. & Ergenzinger, P. 1994. 'Interpreting the temporal and spatial dynamics of bedload transport phases according to F.A.S.T. (Fluid And Sediment Transfer Model)', in Olive, L.J., Loughran, R.J. & Kesby J.A. (Eds) *Variability in Stream Erosion and Sediment Transport*, IAHS **224**, Canberra, 25–31.

de Jong, C. 1995. 'Temporal and spatial interactions between river bed roughness, geometry, bedload transport and flow hydraulics in mountain streams – examples from Squaw Creek, Montana (USA) and Schmiedlaine/Lainbach (S. Germany)', *Berliner Geographische Abhandlungen*, **59**, 229.

Dinehart, R.L. 1992. 'Evolution of coarse gravel bed forms: field measurements at flood stage', *Water Resour. Res.*, **28**, 2667–2689.

Einstein, H.A. & Li, H. 1958. 'Secondary currents in straight channels', *Trans. AGU*, **39**, 1058–1088.

Ergenzinger, P. 1992. 'River bed adjustments in a step-pool system: Lainbach, Upper Bavaria', in Billi, P., Hey, R.D., Thorne, C.R. & Tacconi, P. (Eds) *Dynamics of Gravel-bed Rivers*, John Wiley & Sons Ltd., Chichester, 415–430.

Ergenzinger, P. & Custer, S.G. 1983. 'Determination of bedload transport using naturally magnetic tracers. First experiences at Squaw Creek, Gallatin County, Montana', *Water Resour. Res.*, **19**, 187–193.

Ergenzinger, P. & de Jong, C. 1994. 'Monitoring and modelling the transport of coarse single particles in mountain streams', in Cotreneo, G.V. & Rumer, R.R. (Eds) *Hydraulic Engineering*, **1**, ASCE, Buffalo, New York. 634.

Ergenzinger, P., de Jong, C. & Christaller, G. 1994a. 'Interrelationships between bedload transfer and river bed adjustment in mountain rivers', in Kirkby, M. (Ed) *Process Models and Theoretical Geomorphology*, John Wiley & Sons Ltd., Chichester, 141–158.

Ergenzinger, P., de Jong, C., Laronne, J.B. & Reid, I. 1994b. 'Short term temporal variations in bedload transport rates: Squaw Creek, Montana, USA and Nahal Yatir and Nahal Eshtemoa, Israel', in Ergenzinger, P. & Schmidt, K.-H. (Eds) *Dynamics and Geomorphology of Mountain Rivers*, Springer, Berlin, 251–264.

Ergenzinger, P., Reid, I., Laronne, J.B. & de Jong, C. 1992. 'Short term temporal variations in the spatial pattern of bedload transport rate: Squaw Creek, Montana, USA and Nahal Yatir & Eshtemoa, Israel', in Bogen, J., Walling, D.E. & Day, T.J. (Eds) *International Symposium on Erosion and Sediment Transport Monitoring Programmes in River Basins*, IAHS **210**, Oslo, 77–81.

Lamberti, A. & Paris, E. 1992. 'Analysis of armouring processes through laboratory experiments', in Billi, P. Hey, R.D. Thorne, C.R. & Tacconi, P. (Eds) *Dynamics of Gravel-bed Rivers*, John Wiley & Sons Ltd., Chichester, 227–250.

Laronne, B., Outhet, D.N., Duckham, J.L. & McCabe, T.J. 1992. 'Determining event bedload volumes for evaluation of potential degradation sites due to gravel extraction', in Bogen, J., Walling, D.E. & Day, T.J. (Eds) *International Symposium on Erosion and Sediment Transport Monitoring Programmes in River Basins*, IAHS **210**, Oslo, 87–94.

Laronne, J.B. & Duncan, M.J. 1992. 'Bedload transport paths and gravel bar formation', in Billi, P. Hey, R.D. Thorne, C.R. & Tacconi P. (Eds) *Dynamics of Gravel-bed Rivers*, John Wiley & Sons Ltd., Chichester, 177–202.

Laronne, J.B., Reid, I., Yitschak, Y. & Frostick, L.E. 1992. 'Recording bedload discharge in a semiarid channel, Nahal Yatir, Israel', in Bogen, J., Walling, D.E. & Day, T.J. (Eds) *International Symposium on Erosion and Sediment Transport Monitoring Programmes in River Basins*, IAHS **210**, Oslo, 79–86.

Müller, A. & Studerus, X. 1979. 'Secondary flow in an open channel', in *Proc. 18th Congr. IAHR*, Cagliari, Italy, **B.a.3**, 142–144.

Petit, F. 1989. 'The evaluation of grain shear stress from experiments in a pebble-bedded flume', *Earth Surf. Processes Landf.*, **14**, 499–508.

Prandtl, L. 1926. 'Über die ausgebildete turbulenz', in *Verhandlungen des II Intl. Kongr. für Angewandte Mechanic*, Zürich, 62–75.

Studerus, X. 1982. *Sekundärströmung im Offenen Gerinne über Rauhen Längsstreifen.* Unpub. PhD Thesis, Institut für Hydromechanik und Wasserwirtschaft, ETH Zürich, 185 pp.

Tait, S.J., Willetts, B.B. & Maizels, J.K. 1992. 'Laboratory observations of bed armouring and changes in bedload composition', in Billi, P., Hey, R.D., Thorne, C.R. & Tacconi, P. (Eds) *Dynamics of Gravel-bed Rivers*, John Wiley & Sons Ltd., Chichester, 205–225.

Tait, S.J. & Willetts, B.B. 1992. 'Characteristics of Armoured Bed Surfaces', in *Proc. Int. Grain Sorting Seminar*, Mitteilugen **117**, ETH Zürich, 207–225.

Wang, J., Dong, Z., Chen, C. & Xia, Z. 1993. 'The effects of bed roughness on the distribution of turbulent intensities in open channel flow', *J. Hydraul. Res.*, **31**, 89–98.

Wang, Z., Larsen, P. & Xiang, W. 1994. 'Rheological properties of sediment suspensions and their implications', *J. Hydraul. Res.*, **32**, 495–516.

Whiting, P.J., Dietrich, W.E., Leopold, L.B., Drake, T.G. & Shreve, R.L. 1988. 'Bedload sheets in heterogeneous sediments', *Geology*, **16**, 105–108.

NOTATION

D_{50} D_{84}	represents grain diameter with the subscript indicating the percentage of material finer than that grain diameter
K_3	the maximum vertical difference between three adjacent bed elevations

32 On the Origin and Effects of Large-scale Longitudinal Flow Structures in the Outer Humber Estuary

JACK HARDISTY, HELEN L. ROUSE & NICK E. HUGHES
School of Geography and Earth Resources, University of Hull, Hull, HU6 7RX, UK

ABSTRACT

Preliminary evidence for recurring, large-scale, longitudinal flow structures at the mouth of the Humber Estuary on the North Sea coast of central England was provided by remotely sensed satellite and aircraft images showing the distribution of suspended sediments during the early stages of the flood tidal cycle. The features were confirmed by high resolution field measurements of the near-surface suspended sediment concentrations obtained during National Rivers Authority Sea Vigil cruises on the Humber Flux Curtain as part of the Natural Environment Research Council's Land-Ocean Interaction Study (LOIS) programme. The structures consist of narrow (c. 10^2 m) and elongate (c. 10^3 m) streaks of less saline water within which suspended sediment concentrations are three or four times greater than in the surrounding water. These observations are discussed together with theoretical concepts and it is suggested that the features represent the surface expression of what is termed Estuary Mouth Secondary Flow (EMSF). EMSF is composed of flow-parallel vortices associated with three-dimensional secondary flow generated by tidal and topographic controls at the estuary mouth. Characteristic signatures of the near-bed flow are defined for EMSF circulations. Predicted EMSF properties concur with measurements collected using an instrumented tripod deployed at the sea-bed. Finally, preliminary calculations are described which suggest that the features may be responsible for the transport of more than 35% of the 100 000 tonnes of suspended sediment which is carried into and out of the estuary during each tidal cycle.

INTRODUCTION

River drainage systems discharge freshwater and sediment into the marine waters of the World Ocean through any one of a wide range of geomorphological features (Viles & Spencer, 1995). An estuary may be defined as "a semi-enclosed coastal body of water which has a free connection with the open sea and within which sea water is measurably diluted with freshwater derived from land drainage" (Pritchard, 1952). The Humber Estuary has a mean freshwater input of 250 $m^3 s^{-1}$ and a

Coherent Flow Structures in Open Channels.
Edited by P.J. Ashworth, S.J. Bennett, J.L. Best and S.J. McLelland.

Spring tidal range in excess of 6 m. Many estuaries are considered to be gradually infilling (e.g. Jago & Hardisty, 1984), and the estimation and control of suspended sediment flux at estuary mouths is currently topical within the pollution pathway debate.

One, fairly obvious, feature of estuaries is that they are usually aligned normal to the coast, whilst the flood-ebb tidal currents in the open sea are usually aligned parallel to the coast. In addition, estuary mouths often consist of steep-sided, deep channels running seawards between shallow sub-tidal banks outside the estuary mouth. Hence, the water must not only change direction by about 90° but it must also spill over and down the step-like bathymetric features during the flood. The potential therefore exists for estuary mouths to generate large-scale flow structures which may influence sediment flux.

THE HUMBER ESTUARY

The Humber Estuary drains a catchment of around 25 000 km^2 which is more than one quarter of the land area of England. The tidal reaches of the Humber extend 120 km inland on the Trent and 110 km inland on the Ouse and the plan shape of the estuary flares from about 1 km wide at the Ouse-Trent confluence to over 8 km wide at Spurn Head (Figure 32.1). Approximately 120 000 years BP the mouth of the estuary was close to the Humber Gap which is the site of the modern Humber Bridge (IECS, 1987). During the last glaciation, waters from the Humber Basin

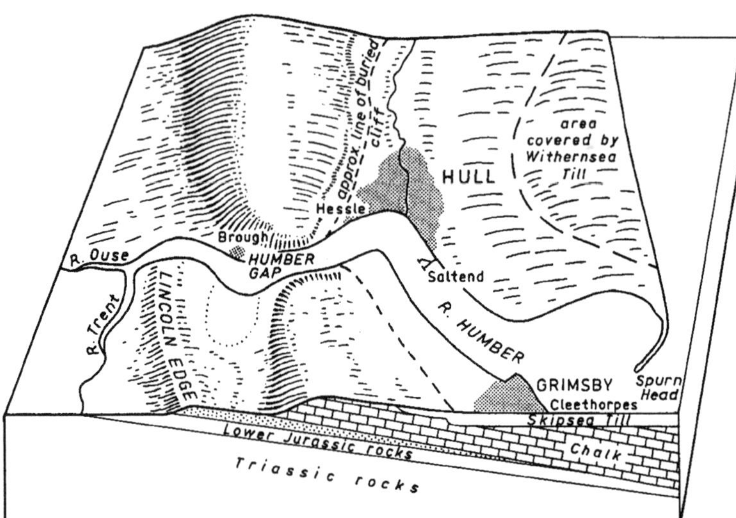

Figure 32.1. Location map of the Humber Estuary indicating the geological and geomorphological history of the region

(the present day Vale of York) were diverted through the Humber Gap and then south leading to the distinct change in the orientation of the estuary to the east of Hessle. Rising sea-levels flooded the river valley about 6000 years BP and gave rise to the present-day estuary shape. Natural and artificial accretion of peripheral sediments and erosion of the Holderness coastline has occurred since the flooding so that the estuary mouth, which was probably once 25 km wide about 20 km seawards of the present-day mouth, is now only 8 km wide between Spurn and Cleethorpes.

ENVIRONMENTAL CONDITIONS

The most frequently occurring and strongest onshore winds at the mouth of the estuary are from the north and east (Shellard, 1976; Palutikof *et al.*, 1985; Ward, 1987) whilst the central and upper estuary are dominated by prevailing south-westerly winds. The mean annual days of gale (DOG: wind speed in excess of 34 knots, 17 m s^{-1} or Beaufort Force 8 for several minutes) increase from 2.5 at Hull to 23.7 at Spurn Head. In general, the estuary experiences a climate which is an amalgam between the dry east-coast lowland weather and the maritime influence of the North Sea. Rainfall in the area is low, rarely rising above 650 mm $year^{-1}$ and dropping to less than 600 mm $year^{-1}$ close to the coast. Other forms of precipitation are relatively rare with snow being recorded on only 5 days per year and hail on only 4 days per year during the period 1960–1970. The mean annual days of fog (DOF: visibility less than 1 km in the direction of worst visibility) decrease slightly from 17 at Spurn Head to 12 at Hull. Temperatures in the Humber area show a typical winter-summer contrast with average values of 3.5° and 15.2° C respectively.

WAVE CONDITIONS

There are two wave regimes which affect the Humber: those generated in the North Sea; and those generated within the estuary itself. North Sea waves are larger and have been recorded at the Dowsing Light, approximately 25 km off the estuary mouth, by the Institute of Oceanographic Sciences (Fortrum, 1980; Bacon, 1989). Analysis of data give 50 year return values of the significant wave height (H_s) as 8.23 m (using a Fisher-Tippett Type I (FT-1) distribution) with this winter maximum reducing to 7.41, 6.91 and 5.35 m for autumn, spring and summer respectively. More recent results from deployments of Wave Rider buoys off Holderness are reported by Bacon & Carter (1988) and suggest an FT-1 50 year return H_s of 5.99 m for all data. North Sea waves approaching the estuary mouth are refracted as they enter the shallow water of the estuary and, for the north-easterly storm conditions, incoming wave energy is focused on the south-bank around Cleethorpes. Waves within the estuary are locally generated and substantially smaller. IECS (1987) suggests that a westerly wind blowing along the estuary may generate waves of 1–2 m.

TIDAL CONDITIONS

The Humber tides are dominated principally by the M_2 amphidromic point in the eastern-central North Sea (Doodson & Warburg, 1941). In general, the tidal wave progresses in a southerly direction down the coasts of Holderness and Lincolnshire (Hardisty, 1990) and enters the mouth of the estuary some 6 hours before High Water (HW) at Dover. The tide takes about 3 hours to progress up the estuary from Spurn Head to the Ouse-Trent confluence and higher order harmonics are generated during this transition. Spring tidal range in the Humber is 6.5 m at the mouth of the estuary rising to a maximum of 7.2 m at Saltend and then decreasing progressively upstream. The pattern is semi-diurnal with a pronounced neap-spring inequality (IECS, 1987); for example at Saltend the tidal range varies from 3.9 to 7.2 m over a neap-spring cycle. The tide at Spurn Head is sinusoidal and becomes increasingly asymmetric upstream. Spring tide current velocities can exceed 3 m s^{-1} at the mouth of the estuary. The tidal asymmetry causes increased velocities on the flood tide because of the shorter flood tide period and the velocity also varies with depth, decreasing rapidly towards the bed.

REMOTELY SENSED AND TRANSECT DATA

Preliminary evidence for the existence of large-scale, coherent structures in the mouth of the Humber Estuary during the flood tide was identified in remotely-sensed satellite and aircraft images and by high resolution suspended particulate matter measurements obtained during routine estuary mouth transects. These data are presented in order to develop a qualitative model of the flow features.

SATELLITE IMAGERY

Early indications that sediment transfers between fluvial and marine systems could operate through streaky structures was apparent in the Landsat images of the 1970s, 1980s and early 1990s (Figure 32.2). However, the pixel resolutions were generally too low and it was unclear how suspended sediment concentrations could best be derived from particular bandwidths and with various atmospheric correction algorithms. The possibility that such features represented oceanographic fronts was considered to be unlikely since such convergent fronts usually lie normal or oblique to dominant flow directions in estuaries. An alternative, in which the features represent vertical shear planes remained a possibility, but was later discounted when field measurements demonstrated that the water on either side of the feature retained very similar properties.

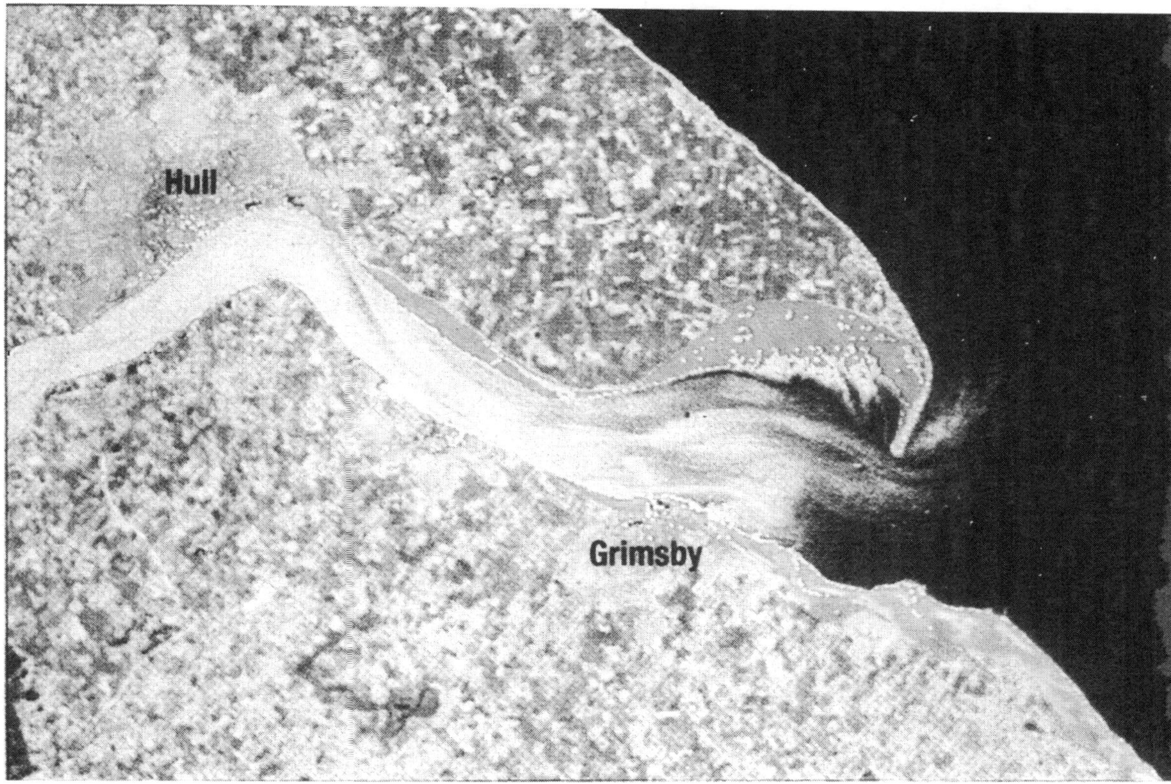

Figure 32.2. Landsat image of the mouth of the Humber Estuary showing streaky structure with a general flow-parallel orientation

SEA SURFACE MEASUREMENTS

Detailed observations at the mouth of the Humber Estuary were undertaken as part of the Flux Curtain experiment in the Natural Environment Research Council's Land-Ocean Interaction Study (NERC-LOIS; Hardisty & Rouse, 1995). In particular, the streaky structures were detected during instrument transects along the flux curtain, which runs from Spurn to the south-bank near Cleethorpes, in May 1994 (Figure 32.3). A Partech IR40C transmissometer was towed at a depth of approximately 1 m astern of the National River Authority (NRA) Survey Ship *Sea Vigil* and measurements logged at approximately 12 m intervals across the estuary mouth. It is apparent that narrow bands of very high suspended particulate matter were encountered some 1–2 km south of Spurn Head between Low Water +2 hours and LW+3 hours (during the flood tide). These results were taken to correlate with the estuary mouth streaks revealed from the satellite images.

CASI IMAGERY

The mouth of the Humber estuary has been studied in more detail as part of the NERC-LOIS programme using the Compact Airborne Spectrographic Imager (CASI). CASI overflights obtained on 18 October 1994 provide the first high resolution images of the spatial extent of the structures and are shown in Figure 32.4. The five images are in sequence taken at 5 minute intervals from 1400 GMT, and their axes lie along the line of the flux curtain. Low Water (Immingham) on 18 October was 10:48 GMT (1.4 m CD) followed by High Water at 17:01 (6.7 m CD) with a tidal range of 5.3 m, a reasonable spring tide as detailed above. The images are therefore taken at approximately LW+3 hours and are interpreted in the following section.

QUALITATIVE DESCRIPTION

It is evident from the CASI images and the transect data that during the flood tide a streak like structure has developed by the time of the first image, and that during the following twenty minutes the structure advects upstream and becomes more diffuse. Comparison of the CASI images with Admiralty Chart 109 *River Humber and Rivers Ouse and Trent* (1990) suggests that the structure's head (i.e. its easterly surface expression) lies in about 20 m of water at LW+3 hours 10 minutes in Figure 32.4a, drifting south from a position 1.5 km along the flux curtain from the Spurn to 2 km along the curtain from the Spurn at LW+3 hours 30 minutes in Figure 32.4d. Although the bed of the estuary is mobile, it is considered that this depth in the vicinity of the deep channel is accurate enough for the purposes of the present study. The lateral drift rate is therefore approximately 0.3 m s^{-1}. Tidal diamonds F and H on the same chart suggest that the surface current at such a time is about 3.3 knots (1.65 m s^{-1}) with a set of 310° magnetic (i.e. normal to the flux curtain). The eastern end of the feature advects some 2.1 km between

(a)

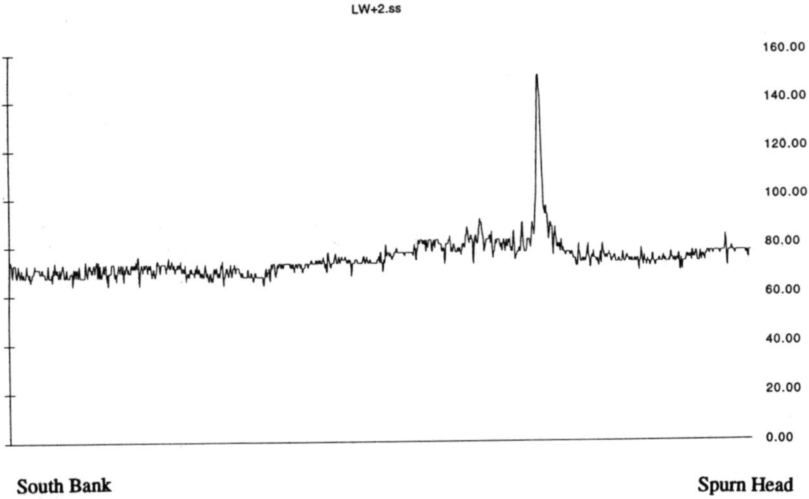

(b)

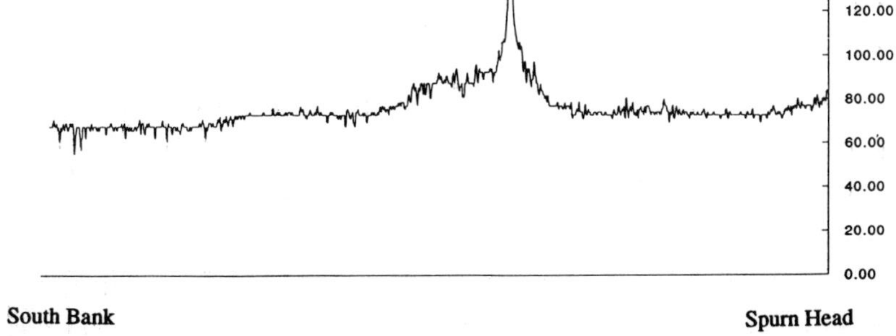

Figure 32.3. High resolution suspended particulate matter concentration data obtained from Partech Transmissometer measurements during NRA *Sea Vigil* transects across the Flux Curtain, May 1994 for (a) Low Water +2 hours and (b) Low Water +3 hours showing near-surface, narrow, high concentration features interpreted as being the streaks revealed in the remote sensed images

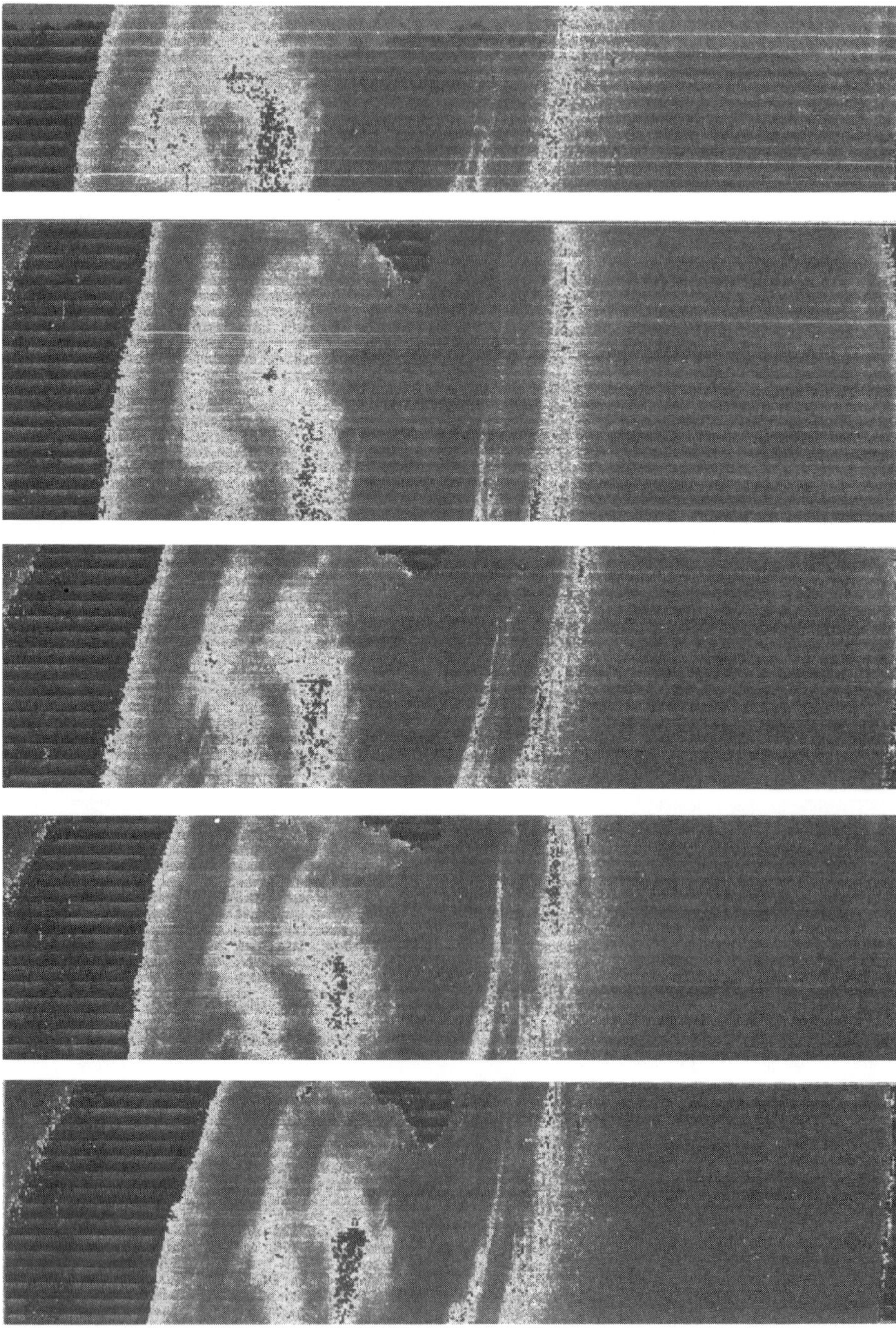

Figure 32.4. Sequence of CASI images along the Flux Curtain (left to right), 5 minutes apart starting from just after LW+3 hours. North is at the top of the image. Each image has a lateral width of approximately 1 km (not geometrically corrected)

LW+3 hours 10 minutes and LW+3 hours 30 minutes suggesting a similar velocity of about 1.75 m s^{-1}. The structure clearly becomes more diffuse so that along the flux curtain (which is more or less the axis of the CASI images) it appears to increase in width from about 100 m at LW+3 hours 10 minutes to about 500 m at LW+3 hours 30 minutes and, possibly, separates into two distinct streamwise streaks each approximately 150 m wide separated by a similar width of clearer water. The following sections discuss the possible origin of the structure observed in the CASI images and describe a set of field experiments designed to test some of the theoretical considerations.

THEORETICAL CONSIDERATIONS

It may be tentatively suggested from the satellite and CASI images that the near-surface suspended sediment streaks are a result of streamwise secondary flow structures. Furthermore, it can be postulated that a pair of counter-rotating helical cells are generated during the early flood tide at the estuary mouth with convergent flow at the bed raising water and near-bed sediment up towards, and through, a surface divergence. Here, we consider the literature on such features and attempt to illustrate near-bed flow and sediment concentration data. Such coherent secondary flow structures have been recognized in natural fluid flows for some time and have been associated with the transport of sediment and with various bedforms and bed structures (Kenyon, 1970; Werner & Newton, 1975; Swift & Ludwick, 1976; Karl, 1980; Pantin *et al.*, 1981).

There are, perhaps, six candidates: Prandtl (1952) mentions three well-known types of secondary flow (i) curved channel secondary flow (CCSF); (ii) non-circular channel secondary flow (NCSF); and secondary flow around an oscillating body (which is not discussed further here). Pantin *et al.* (1981) include (iii) differential roughness secondary flow (DRSF); and (iv) Langmuir circulation secondary flow (LCSF). Later Pantin *et al.*, 1987 added (v) boundary layer secondary flow (BLSF). Finally, the literature on bedform development has numerous examples of secondary flow generated within the lee of ripples, sand dunes or sand waves. This phenomenon is referred to as (vi) step separation secondary flow (SSSF) even though the rotating flows are parallel to the primary flow, and non-parallel flow is only induced if the primary flow direction is oblique, or becomes oblique, to the step. Each of these is shown schematically in Figure 32.5.

(i) *Curved Channel Secondary Flow (CCSF).* The flow of water in a channel bend gives rise to an excess of fluid pressure on the outer bank and a corresponding deficit on the inner bank. Throughout the main body of the flow (Allen, 1970, p. 131) the radial pressure gradient thus set up is balanced by the gradient of the centripetal force. At the bed, however, frictional losses unbalance the gradient and water moves inward at the bed, down the radial gradient, and a compensating outwards flow at the surface is established (Figure 32.5i).

(ii) *Non-circular Channel Secondary Flow (NCSF).* This phenomenon is also widespread (Pantin *et al.*, 1987) and also does not depend upon the presence of sediment in the fluid. Perhaps the best known example of NCSF is the flow into the corners, and out from the sides, of a polygonal channel as described by Prandtl (1952, pp. 148–149). Townsend (1961) has explained this in terms of the accumulation of turbulent Reynolds pressure in the corners. The experiments of Karcz & Kersey (1980) demonstrate the influence of this process on sediment distribution; for example NCSF would self-reinforce flow parallel bedforms sweeping sediments from the troughs to the crests (Figure 32.5ii).

(iii) *Differential Roughness Secondary Flow (DRSF).* McLean (1985) describes DRSF as secondary circulation driven by Reynolds stresses resulting from lateral variations in bed roughness. DRSF can be self-reinforcing if a rough substrate is overlain by less rough banks of sediment (e.g. sand ribbons over a gravel surface; Pantin *et al.*, 1987; Figure 32.5iii).

(iv) *Langmuir Circulation Secondary Flow (LCSF).* Langmuir circulations consist of alternate left and right-handed roll vortices generated by wind in the surface layers of lakes or the sea (Pantin *et al.*, 1981). Their axes run more or less parallel to the wind, and the circulations probably depend upon the interaction of waves with near-surface currents (Garrett, 1976; Pollard, 1977; Figure 32.5iv).

(v) *Boundary Layer Secondary Flow (BLSF).* Pantin *et al.* (1987) state that BLSF is now widely recognized, and occurs most characteristically on hydraulically smooth boundaries. Instability in the viscous sublayer results in longitudinal vortices (e.g. Leeder, 1982; Best, 1993), with on average equal numbers of left and right-handed vortices. Such vortices are observed in the every day environment (e.g. sand blowing on a beach) and provide a convincing explanation of primary current lineations (Hesse & Chough, 1980). These vortices usually form in patches which grow as they move downstream, and the spacing of these vortices varies with thickness of the viscous sublayer and flow speed (Figure 32.5v).

(vi) *Step Separation Secondary Flow (SSSF).* Onslow *et al.* (1993) describe the long recognized flow separation over dune or ripple crests by generalizing the case to flow over a backward-facing step. They note that Driver *et al.* (1987) and Simpson (1989) describe flow detachment causing the formation of a free shear layer which rolls up into discrete vortices due to Kelvin-Helmholtz instabilities (Müller & Gyr, 1986). A pair of counter-rotating vortices can develop with flow convergence in the recirculation zone at the bed (Figure 32.5vi).

It is, we believe, not unreasonable to discount the fourth and fifth of the mechanisms when seeking an explanation for the streaky features at the mouth of the Humber Estuary. There is no apparent correlation with wave activity and

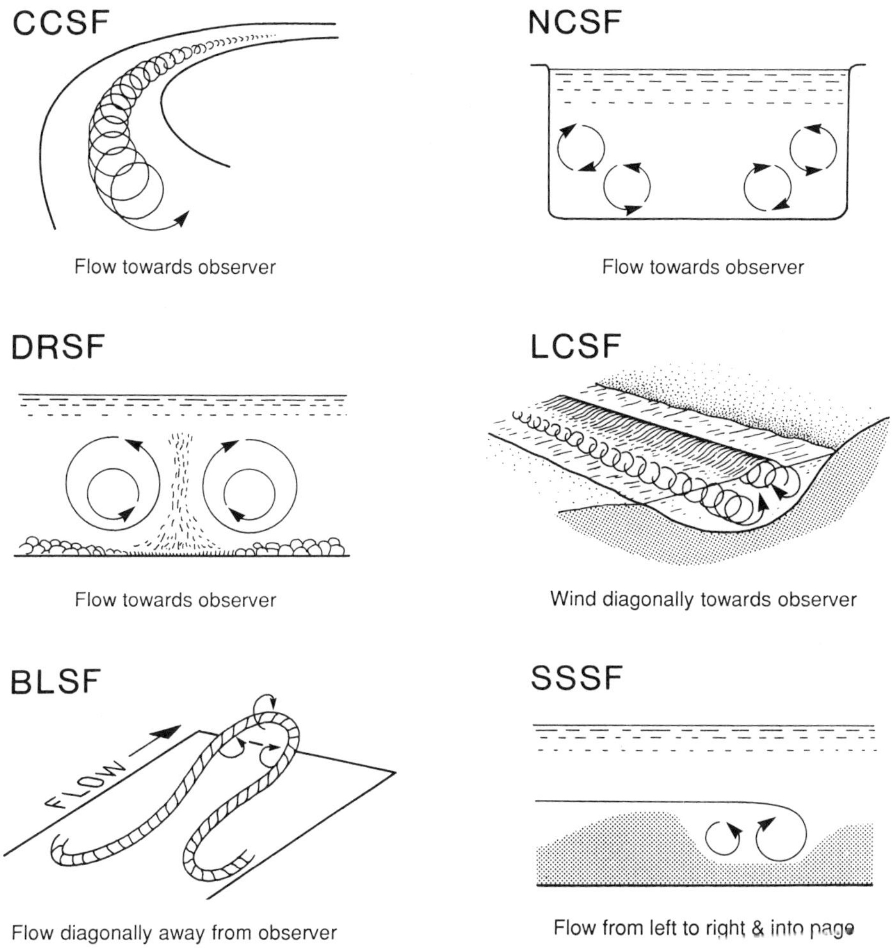

Figure 32.5. Schematic representation of (i) curved channel secondary flow (CCSF); (ii) non-circular channel secondary flow (NCSF); (iii) differential roughness secondary flow (DRSF); (iv) Langmuir circulation secondary flow (LCSF); (v) boundary layer secondary flow; and (vi) step separation secondary flow (SSSF) over an oblique step

therefore LCSF can be discounted and the stability of the feature in 20 m of water seems to preclude any sub-layer mechanism and thus we discount BLSF. With less justification we shall proceed by tentatively discounting DRSF since there does not appear to be any particular change in sea-bed characteristics shown on any of the detailed field survey charts. Neither the Admiralty nor the Associated British Ports bathymetric charts (which contain some annotations of bed characteristic) indicate the required flow parallel boundary between different roughness elements. Again, the British Geological Survey sediment maps show only sand with discrete and irregular gravel patches further offshore. NCSF may also be discounted since the channels are not particularly angular or rectangular. Therefore, the discussion will consider the mechanisms of Curved Channel Secondary Flow and Step Separation Secondary Flow in more detail and introduce a derivative which is termed Estuary Mouth Secondary Flow (EMSF). In particular, it is suggested that EMSF is a common feature which results in part from the rapidly changing direction of the flow at the (right angle) junction between shore-parallel tidal streams at the coast and the shore-normal orientation of the estuary. In addition, most estuary mouths contain a deep channel and the flood tide will therefore flow over a backwards-facing step. The proposed mechanism has four distinct phases as shown in Figure 32.6:

I From low water the flood tide (running parallel to the Holderness Coast in a southerly direction) begins to flow in over the shallow water of (in the case of the Humber Estuary) the Binks, Outer Binks and Chequer Shoals.

II This phase appears to operate from about LW+1 hour to LW+3 hours. Published work on shear, mixing and boundary layers (particularly the extensive literature on flow over sand ripples, waves and dunes e.g. Onslow *et al.,* 1993) is relevant here and, in particular, flow over a backward-facing step (Simpson, 1989; McLean *et al.*, this volume; Keogh & Addison, this volume) provides an analogy to that over the shallows and into the deep channel at the estuary mouth. It is suggested that flood flow detachment at the northern edge of the deep channel causes the formation of a free shear layer. The vortex grows by entrainment and generates a counter-rotating partner with flow convergence at the bed and divergence at the surface. Onslow *et al.,* (1993) note that the eventual impingement of the structure on the bed and the free surface are the factors limiting further growth of the features in purely unidirectional flows.

III The bed convergence and surface directed flow which results from the vortex pair will transport more concentrated suspended sediment towards the surface.

IV Between about LW+2 hours and the end of the flood tide, currents will advect the structure up estuary and, if the feature becomes less well defined in the shadow of the coast (Spurn Head in the case of the Humber Estuary), then it is plausible that the sediment may well continue to roll over the vortices and clear water develop separating two streaks.

The near-bed signature of EMSF will therefore be as depicted by Figure 32.7. In the case of a stratified or partially mixed estuary the converging and then upwelling flow may lift sediment above the bed, there may be evidence of mean near-bed flow direction being deflected north and then south within the vortices, and there may be a slight increase in the flow velocity due to the secondary flow itself. It is also thought that salinity may be lowered due to mixing of water bodies.

(I)

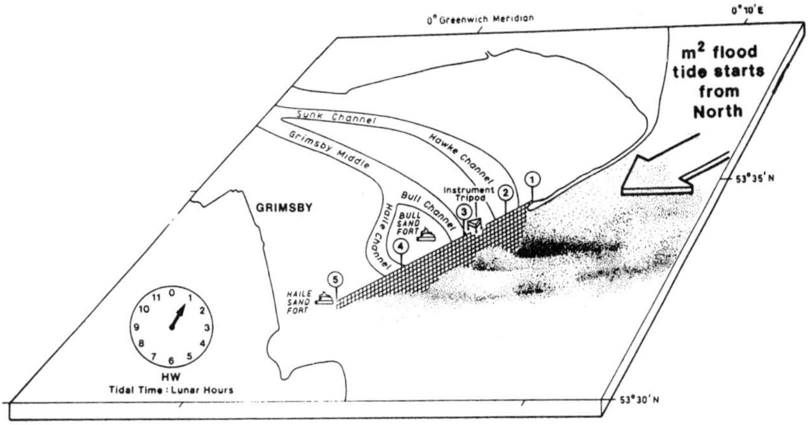

(II)

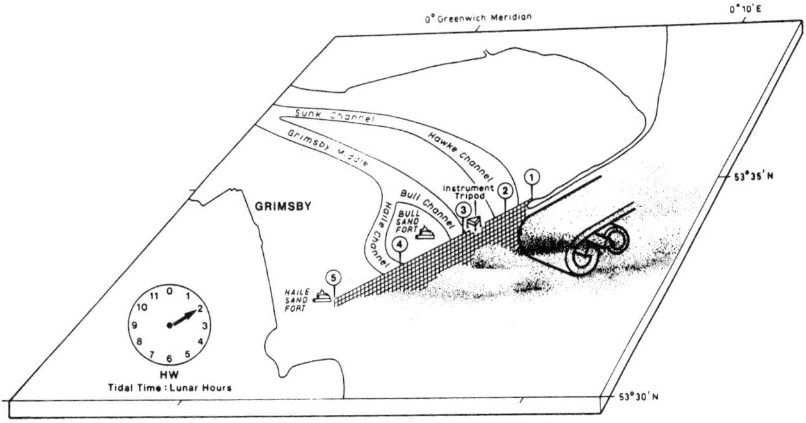

(III)

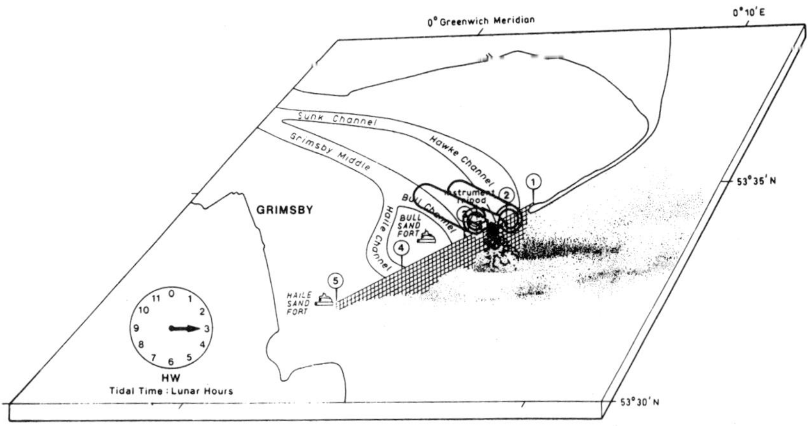

Figure 32.6. Schematic diagram of Estuary Mouth Secondary Flow (EMSF), showing the first three stages (I–III) outlined in the text. *Continued on next page*

(IV)

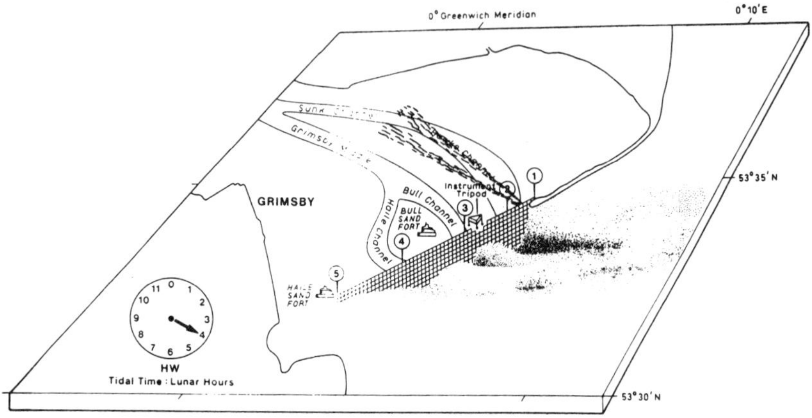

Figure 32.6. *continued* Schematic diagram of Estuary Mouth Secondary Flow (EMSF), showing stage (IV) as outlined in the text. See previous page for stages (I–III)

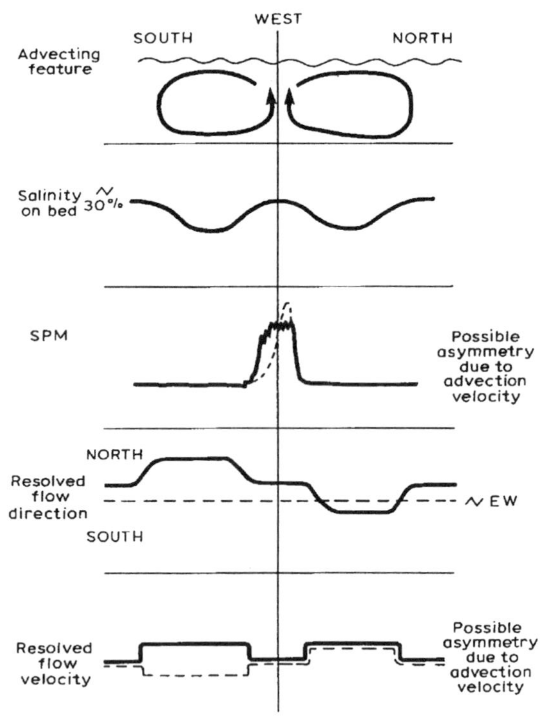

Figure 32.7. Schematic diagram of the near-bed signatures of EMSF at a fixed point on the bed over which the feature rolls for an estuary (like the Humber) in which the streamwise direction is approximately E–W with the flood current from the east and the feature approaches from the northwards

SEA-BED MEASUREMENTS

INSTRUMENT SYSTEMS

In order to obtain additional information on the streak-like features and to test the proposed mode of origin, two separate sets of field measurements were undertaken from the NRA *Sea Vigil* in March 1994. In each experiment the tripod shown in Figure 32.8 was lowered to the sea-bed and various data collected. The equipment consists of a WS Ocean Systems SAL-1 conductivity-depth-salinity-temperature probe mounted with a Partech IR40C transmissometer and an Alec Electronics ACM-200PC electromagnetic current meter on a small open frame. The instruments were lowered to the sea-bed and data transferred by cable to a National Instruments analogue-to-digital board mounted within an Apple Macintosh IIcx microcomputer. Data were logged on all channels at 2 Hz using specially written programmes within the *LabView* language running on the microcomputer. The software samples 2048 measurements on all channels in each record so that about three records are obtained in each hourly interval.

SITE J RESULTS

The tripod was deployed from about 14:00 GMT on 2 March 1995 on which day the predicted Low Water Spurn was 12:07 GMT with a predicted range of 6.2 m. The instrument tripod was deployed at Site J on the Humber Flux Curtain some 1050 m to the East of the Bull Sand Fort. Seven data records were obtained of which the most interesting was the fifth, SK020395.05. This record begins at 15:07:08 GMT (LW+3 hours) and runs for 1024 s (i.e. slightly more than 17 minutes). Data from this record are shown in Figure 32.9 where two features, in particular, are noted.

Firstly, there is a small surge in the flow velocity at approximately 7 minutes 30 s into the record which corresponds with a swing to the North from about 260° to 300° magnetic and an increase in the suspended sediment concentration by about 60 mg dm^{-3} from 300 mg dm^{-3}.

Secondly, there is a much more significant event evident at about 14 minutes after the start of the record. The current velocity increases and the flow direction swings from about 250° up to about 290° magnetic and then down to about 220° magnetic. Again the suspended sediment concentration shows the most significant changes moving from a value of 300 mg dm^{-3} up to a peak of about 800 mg dm^{-3}. Salinity shows a distinct decrease from 30‰ to 15‰.

SITE M RESULTS

The tripod was redeployed from about 10:00 GMT on Monday 13 March 1995 on which day the predicted Low Water Spurn was 08:40 GMT with a predicted range of 3.1 m. The instrument tripod was deployed at Site M on the Humber Flux Curtain some 425 m to the East of the Hawke Light Float. Seven data records were obtained of which the most interesting was the fourth, SK130395.04. This record begins at 10:57 GMT (LW+2 hours) and runs for 1024 s (i.e. slightly more than 17 minutes). Data from this record are shown in Figure 32.10.

A small increase in suspended sediment is noted at about 6 minutes into the record, corresponding to a swing south from 250° to 210° magnetic and a slight increase in flow speed. However at about 14 minutes into the record, there is a dramatic decrease in salinity from 30 to about 8 ppt and a simultaneous rapid increase in SPM from 100 to 1200 mg dm^{-3}. However, there does not appear to be a corresponding change in signature for either current speed or direction.

If the feature is as described above, with a lateral drift of some 0.3 m s^{-1} and approximately 100 m wide, the near-bed signature should be seen to pass through the instruments in about 3–5 minutes. This is roughly the time-scale identified in the results from site M.

DISCUSSION

In this paper we have described a coherent secondary flow structure, and after considering various theories we have postulated a new and possibly common secondary flow feature termed Estuary Mouth Secondary Flow. Certain aspects of this hypothesized EMSF can be identified from various data sources: the CASI images, *Sea Vigil* transects and the specially deployed sea-bed rig. Further detailed field experiments are obviously required to confirm the origins of this coherent flow structure.

Another possibility may be that this feature may be an elongate body of upper estuary water (with lower salinity and higher suspended sediment concentrations) that was moved downstream on the preceding ebb and has since been slower to move back up-estuary.

In order to further investigate this feature, we have attempted to simulate the occurrence of the surface sediment streaks using the commercially available computational fluid dynamics modelling package, *FLUENT* (Version 4.26), operating on a 100 MHz, 486DX4 microcomputer. The entrance to the Humber estuary was implemented as a flood channel (depth z_a) entering a deep channel (step depth z_b , therefore total depth is the sum of z_a and z_b) with a flood-parallel, horizontal extent of x_c, a step length of x_d and a horizontal flood-normal model length of x_e. Flood velocities, U_f, were delivered over the step and advected at velocity U_a into the estuary (Figure 32.11). A range of model inputs were examined (Table 32.1) and five distinct flow types have been identified: (i) bed flow when near-bed water was not lifted vertically; (ii) weak wall rise when near-bed water was lifted a few metres up the step and promontory wall; (iii) strong wall rise when near-bed water was lifted all the way up the step and promontory wall; (iv)

Figure 32.8. Upper: The Sea Vigil survey boat Lower: Sea-bed tripod showing the electromagnetic current meter, salinity-temperature-depth probe and transmissometer

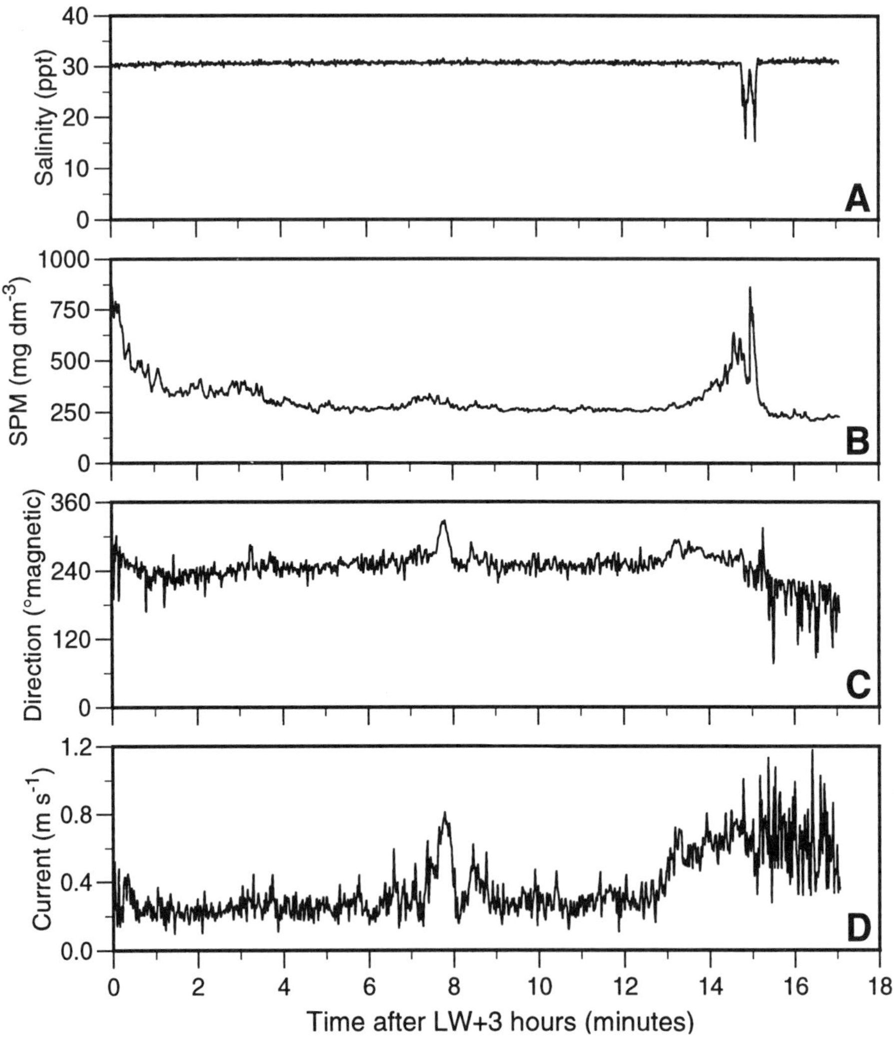

Figure 32.9. Data from SK020395.05 showing: (s) salinity; (b) SPM (taken to be indicative of suspended sediments); (c) flow direction; and (d) flow speed

bed flush when flow is moved out into the channel at the bed; and (v) streak flow when vortices developed which raised near-bed water to the surface and moved it away from the wall into the main channel. In an attempt to summarize the numerical results we have plotted a non-dimensional step depth ($d_s = z_a/z_b$) against a step Reynolds number ($Re_s = U_f z_b/\nu$) as shown in Figure 32.12 for 15 simulations with $z_a = 2$–8 m, $z_b = 15$ m, $U_f = 0.2$–1.25 m s^{-1}, $U_a = 0.25$ m s^{-1}, $x_c = 750$ m, $x_d = 100$ m, $x_e = 1500$ m and horizontal and

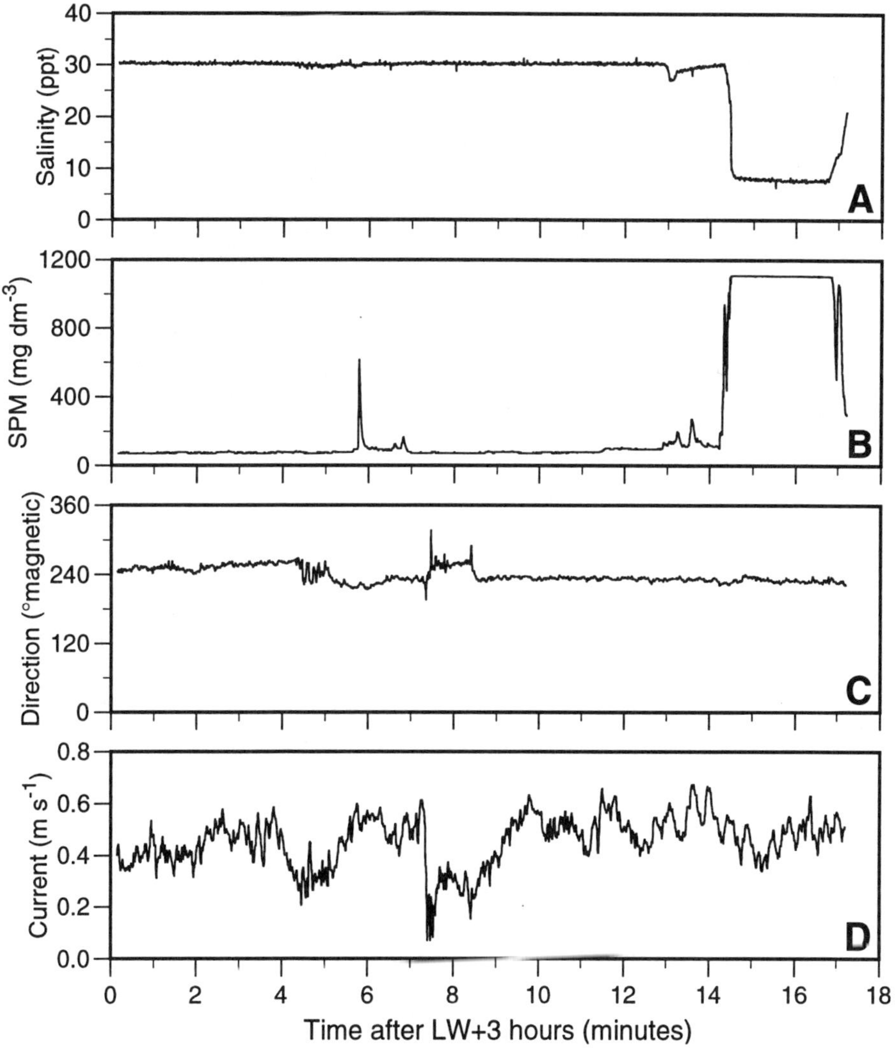

Figure 32.10. Data from SK130395.04 showing: (s) salinity; (b) SPM (taken to be indicative of suspended sediments); (c) flow direction; and (d) flow speed

vertical resolutions of 50 m and 1 m respectively. These values correspond reasonably-well to actual values at the mouth of the Humber Estuary. The results show that increasing step Reynolds Number increases the development of EMSF but such development is suppressed by increasing step number.

Whatever this feature proves to be, it clearly possesses a very high flux potential. For example, much work has recently been undertaken as part of the

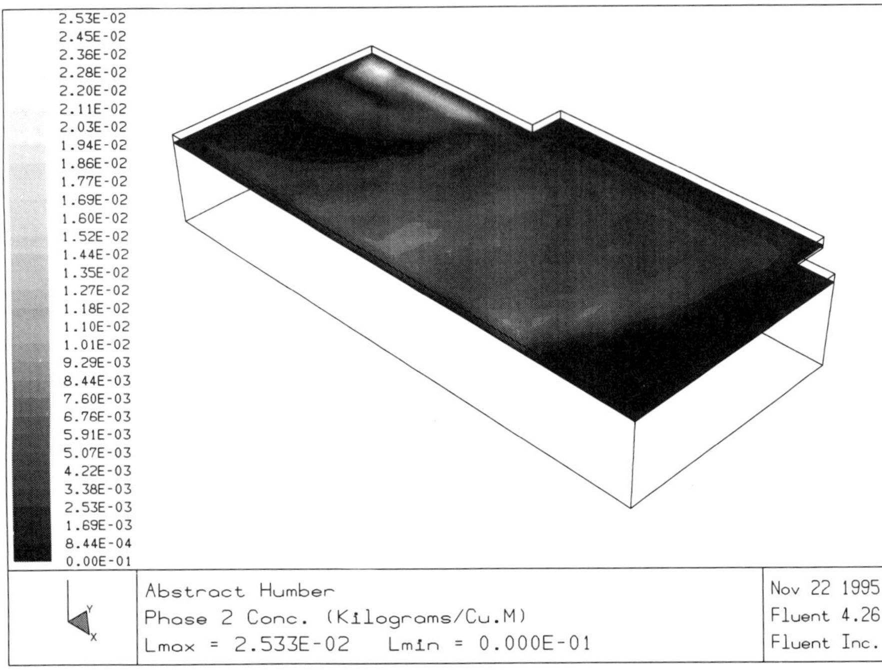

Figure 32.11. *FLUENT* output. See text for further details

Table 32.1. Results from *FLUENT* tests for different input channel depths (z_a) and flood velocities (U_f)

Flood velocity U_f	Input channel depth (z_a) 2 m	5 m	8 m
0.25	very slight wall rise	bed flow	no vertical flow
0.50	slight wall rise	weak wall rise	very slight wall rise
0.75	streak (100 m)	strong wall rise	slight wall rise
1.00	streak (200 m)	streak (150 m)	wall rise, bed flush
1.25	wide streak (500 m)	wall rise, bed flush	slight wall rise and flush

Humber Flux Curtain research project on the analysis of tidal and net suspended particulate matter fluxes through the mouth of the estuary. In general, we find that approximately 0.7 km^3 of water enters the estuary during a neap tidal cycle and that the volume approximately doubles during a spring tide. Taking, as an average, a mean volume of 1 km^3 and a mean suspended particulate matter (SPM) concentration of 100 mg dm^{-3} it is apparent that some 100 000 tonnes of solid enters the estuary during the flood. The field programme carried out as part of the

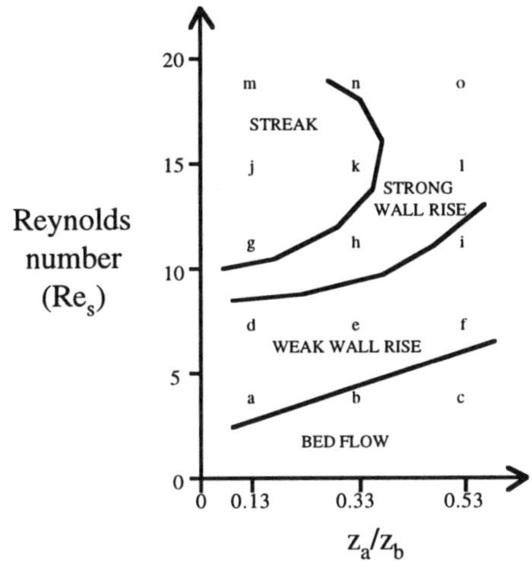

Figure 32.12. Phase space representation of existence fields for non-linear counter rotating vortices in terms of the step Reynolds number and the dimensionless step height

flux curtain experiments has demonstrated that there is not a great deal of horizontal or vertical variation in SPM. The very high concentrations associated with the EMSF feature may, however, represent a significant proportion of this total load. The weight of material which enters the estuary in the streak can be estimated from:

$$Load = width \times depth \times advection\ \ velocity \times duration \times concentration$$

where it may be estimated that: width is 100 m, depth is 20 m, advection velocity is 1.5 m s^{-1}, duration is from LW+2 hours to LW+5 hours (equivalent to 10 800 s) and the sediment concentration is 1000 mg dm^{-3} (equivalent to 0.001 tonnes m^{-3}). Thus, the load per flood tide is 32 400 tonnes. It is apparent, therefore, that between 30 and 40% of the suspended particulate matter which enters the estuary may be transported by Estuary Mouth Secondary Flow related processes as described in this paper.

ACKNOWLEDGEMENTS

This work was funded as a Special Topic by the Natural Environment Research Council's Land-Ocean Interaction Study (LOIS) initiative. This publication is LOIS publication number 58.

REFERENCES

Allen, J.R.L. 1970. *Physical Processes of Sedimentation: an Introduction*, George, Allen & Unwin, London, 248 pp.

Bacon, S. 1989. *Waves Recorded at Dowsing Light Vessel 1970-1985*, Inst. Oceanographic Sciences Deacon Laboratory, Unpub. Report No. **262**, 60 pp.

Bacon, S. & Carter, D.J.T. 1988. *Waves and Wave Spectra Recorded at Two Sites off the Holderness Coast*, Inst. Oceanographic Sciences Deacon Laboratory, Unpub. Report No. **260**, 88 pp.

Best, J.L. 1993. 'On the interactions between turbulent flow structure, sediment transport and bedform development: some considerations from recent experimental research', in Clifford, N.J., French, J.R. & Hardisty, J. (Eds) *Turbulence: Perspectives on Flow and Sediment Transport*, John Wiley & Son Ltd., Chichester, 61–92.

Driver, D.M., Seegmiller, H.L. & Marvin, J.G. 1987. 'Time dependent behaviour of a reattaching shear layer', *Am. Inst. Aeronautics and Astronautics J.,* **25**, 914–919.

Doodson, A.T. & Warburg, H.D. 1941. *Admiralty Manual of Tides,* HMSO, London, 270 pp.

Fortrum, B.C.H. 1980. *Analysis of Waves Recorded at the Dowsing Light Vessel Between 1970 and 1979*, Inst. Oceanographic Sciences, Unpub. Report **126**.

Garrett, C.J.R. 1976. 'Generation of Langmuir circulations by surface waves - a feedback mechanism', *J. Marine Res.,* **34**, 117–130.

Hardisty, J. 1990. *The British Seas: an Introduction to the Oceanography and Resources of the North-west European Continental Shelf*, Routledge, London, 272 pp.

Hardisty, J. & Rouse, H.L. 1995. 'The Humber Observatory: monitoring, modelling and management for the regional environment', *Environ. Int.,* **21**, 143–149.

Hesse, R. & Chough, S.K. 1980. 'The northwest Atlantic mid-ocean channel of the Labrador Sea II: Deposition of parallel laminated levee-muds from the viscous sublayer of low density currents', *Sedimentology,* **27**, 697–711.

IECS, 1987. *The Humber Estuary Environmental Background*, Inst. Estuarine and Coastal Studies, Univ. of Hull for Shell UK, 59 pp.

Jago, C.F. & Hardisty, J. 1984. 'Sedimentology and morphodynamics of a macrotidal beach, Pendine Sands, SW Wales', *Marine Geol.,* **60**, 123–154.

Karcz, I. & Kersey, D. 1980. 'Experimental study of free-surface flow instability and bedforms in shallow flows', *Sediment. Geol.,* **27**, 263–300.

Karl, H.A. 1980. 'Speculations on processes responsible for mesoscale current lineations on the continental shelf, southern California', *Marine Geol.,* **34**, M9–M18.

Kenyon, N.H. 1970. 'Sand ribbons of European tidal seas', *Marine Geol.,* **9**, 25–39.

Leeder, M.R. 1982. *Sedimentology: Process and Product,* Allen & Unwin, London, 344 pp.

McLean, S.R. 1985. 'Theoretical modelling of deep ocean sediment transport', *Marine Geol.,* **66**, 243–265.

Motyka, J.M. 1986. *A Macro Review of the Coastline of England and Wales. Volume 2. The East Coast. The Tees to the Wash*, Report **SR107**, Hydraulics Research Station, Wallingford, 112 pp.

Müller, A. & Gyr, A. 1986. 'On vortex formation in the mixing layer behind dunes', *J. Hydraul. Res.,* **24**, 359–375.

Onslow, R.J., Thomas, N.H. & Whitehouse, R.J.S. 1993. 'Vorticity and sandwaves: the dynamics of ripples and dunes', in Clifford, N.J., French, J.R. & Hardisty, J. (Eds) *Turbulence: Perspectives on Flow and Sediment Transport,* John Wiley & Son Ltd., Chichester, 279–293.

Palutikof, J.P., Kelly, P.M. & Davies, T.D. 1985. 'An analysis of the spatial and temporal distribution of the wind field over the British Isles', in Palz, W. (Ed) *Proceeding of the European Wind Energy Conference*, Stephens & Associates, Bradford, England, 65–70.

Pantin, H.M., Hamilton, D. & Evans, C.D.R. 1981. 'Secondary flow caused by differential roughness, Langmuir circulations, and their effect on the development of sand ribbons', *Geo-Mar. Lett.*, 255–260.

Pantin, H.M., Hamilton, D. & Evans, C.D.R. 1987. 'Theoretical modelling of deep ocean sediment transport – Comment', *Marine Geol.,* **76**, 163–167.

Pethick, J.S. 1994. *Humber Estuary: Coastal Processes and Conservation*, Unpub. Report to English Nature by Inst. Estuarine and Coastal Studies, Univ. of Hull, 56 pp.

Pollard, R.T. 1977. 'Observations and theories of Langmuir circulations and their role in near surface mixing', in Angel, M.V. (Ed) *A Voyage of Discovery: George Deacon 70th Anniversary Volume*, Pergamon Press, Oxford, 235–251.

Prandtl, L. 1952. *Essentials of fluid dynamics,* Blackie, London, 452 pp.

Shellard, H.C. 1976. 'Wind', in Chandler, J.T. & Gregory, S. (Eds) *The Climate of the British Isles*, Longman, 39–73.

Simpson, R.L. 1989. 'Turbulent boundary layer separation', *Ann. Rev. Fluid Mechanics,* **21**, 205–234.

Swift, D.J.P. & Ludwick, J.C. 1976. 'Substrate response to hydraulic process: grain-size frequency distributions and bedforms', in Stanley, D.J. & Swift, D.J.P. (Eds) *Marine Sediment Transport and Environmental Management*, John Wiley & Sons, New York. 159–196.

Townsend, A.A. 1961. 'Turbulence', in Streeter, V.L. (Ed) *Handbook of Fluid Dynamics*, McGraw-Hill, New York, 10/1–10/33.

Ward, R.C. 1987. 'Meteorology of the Humber Estuary', in *The Humber Estuary: Environmental Background*, Inst. Estuarine and Coastal Studies, Univ. of Hull, 21–30.

Viles, H. & Spencer, T. 1995. *Coastal Problems*, Edward Arnold, London, 350 pp.

Werner, F. & Newton, R.S. 1975. 'The pattern of large-scale bedforms in the Langeland Belt (Baltic Sea)', *Marine Geol.,* **19**, 29–59.

NOTATION

d_s	non-dimensional step depth
Re_s	step Reynolds number
U_a	flood wave advection velocity
U_f	flood velocities
x_c	horizontal flood-parallel channel length
x_d	step length
x_e	horizontal flood-normal channel length
z_a	flood channel depth
z_b	step depth

33 Incoherent Structure: Turbulence as a Metaphor for Stream Braiding

CHRIS PAOLA
Department of Geology & Geophysics and St. Anthony Falls Laboratory, University of Minnesota, Minneapolis, MN 55455, USA

ABSTRACT

The coherent structures of a braided stream system are obvious: the major ones, including bars, confluences and diffluences, are the focus of a number of papers in this volume. However, it can also be useful to think of a braided channel network in its entirety as a stochastic system, with some features in common with turbulence. This paper demonstrates a set of spatially averaged equations by using a procedure analogous to Reynolds averaging. The resulting set of equations is severely under-constrained. Using a set of heuristic first approximations, the free parameters in the averaged equations can be reduced to a non-dimensional measure of stress variance relative to the mean and a velocity-depth correlation coefficient. For data from the Ohau River in New Zealand, these variables are fairly constant over a four-fold range of discharge. After scaling the non-dimensional stress variance, the model can be used to predict spatially averaged variables such as total flow width and mean sediment flux. The model is tested using laboratory data which shows the model slightly over-predicts sediment discharge and under-predicts system width, but is much closer to the observations than the same theory with shear-stress variability removed. Probably the most important result of the work is to underline the need for system-scale data sets on the stochastic structure of flow and sediment flux from a variety of braided rivers.

INTRODUCTION

One of the most obvious properties of turbulence is its apparently stochastic nature, and so it is not hard to see why turbulence studies were dominated by probabilistic approaches for many years. Indeed, as is discussed in several of the other papers in this volume (e.g. Smith, this volume; Grass & Mansour Tehrani, this volume), the idea that buried within the apparent randomness of turbulence is a whole menagerie of coherent structures emerged only after decades of research on turbulence. The papers in this volume collectively attest to the effort required to extract the traces of these structures from the 'noise' in which they are embedded.

The theme of this paper is almost the exact inverse of that of the rest of the works collected in this volume. A braided stream, viewed as a whole, may be considered as a system in which the coherent structures are obvious. This paper

Coherent Flow Structures in Open Channels.
Edited by P.J. Ashworth, S.J. Bennett, J.L. Best and S.J. McLelland.

investigates what happens if some of the methods developed in the early years of turbulence research are applied to estimating the average behaviour of a braided river system. In view of the effort that has gone into extracting coherent structures from turbulence, smearing the beautiful structures of a braided river into a set of averages might well seem like a step backwards. But on a practical level, solution of problems with long time-scales, like the filling of a sedimentary basin by rivers, requires a clear understanding of the time-averaged behaviour of stream systems. There are also cases, such as the tailings fan discussed by Parker (this volume), in which deposition is so rapid that time-averaged dynamics are important even on a scale of a few years. On a more philosophical level, an understanding of the time-averaged behaviour of a system helps provide a context for evaluating the significance of coherent structures. Thus, for instance, we measure the importance of sweep and ejection events (both coherent structures) by computing their relative contributions to the Reynolds stress (a time-averaged quantity; Smith, this volume). One might analogously ask what the averaged contribution of confluences, is to the overall sediment transport rate of a braided stream system. At present no framework exists within which to pose this question. The present rather speculative paper (i) shows how one can construct spatially averaged models of braiding that use insight from more detailed studies of specific features (like those in this volume) to construct integrated views of braided stream systems; and (ii) shows that even a very simplified spatially averaged model, based on a form of Reynolds averaging over the width of a braid plain, suggests that fluctuations in stress away from the mean are extremely important in determining the mean properties of the system.

BRAIDING AND TURBULENCE

To begin with, it is important to stress that in this paper turbulence is considered as a metaphor for braiding *as a whole*; this view point has no connection with whether the flow in individual braid channels is turbulent or not. In this context, the coherent structures of a braided stream are obvious: they are the bars, channels, confluences and diffluences that give the stream its distinctive interwoven pattern. The stochastic nature of the system becomes clear upon observation over time-scales long enough to allow significant channel migration. The precise nature of this apparently unpredictable behaviour is not crucial here; it is sufficient to note that, although the basic equations governing braided streams are reasonably well known, the detailed configuration of an active stream cannot be predicted very far into the future. This unpredictability is intrinsic to the system and does not require external forcing. Nonetheless, it has been suggested that braided streams are, in a statistical sense, the most stable of river types. Recently Murray & Paola (1994) have shown that even a severely simplified representation of the governing equations can give rise to apparently realistic braided patterns that exhibit the same two properties: statistical stability and continuous, unpredictable reconfiguration of the detailed plan pattern.

Pursuing the analogy more closely, we find that both stream braiding and turbulence are characterized by a hierarchy of scales (see Sidorchuk, this volume). In turbulence the objects forming these scales are generically referred to as 'eddies'; for braiding, we will use the term 'bars' analogously to refer to all forms of sediment islands, including the whole zoology of bar types described by previous workers (Smith, 1974). Turbulence and braiding both exhibit fractal behaviour (Bergé *et al.*, 1984; Peinke *et al.*, 1993) within this hierarchy of scales. Also, in both types of system, interactions between structures give rise to short-lived events (ejections and sweeps in turbulence; confluences in braiding) that contribute disproportionately to the overall net transport (of momentum in turbulence; of sediment in braiding).

Like all metaphors, this one eventually breaks down. In turbulence, the scale hierarchy is associated with the well known energy cascade from large to small scales. The small-scale structures play an essentially different role (dissipation of kinetic energy to heat) than the large ones (extraction of kinetic energy from the mean flow). Does an analogous cascade exist in braiding? None has yet been proposed, nor has there been any suggestion of a qualitative difference between the dynamics of small and large-scale bars and channels.

There are two essential differences between stream braiding and turbulence that would lead one to expect that the cascade process would not occur in braiding. One is that turbulence is essentially three-dimensional; the vortex-stretching mechanism by which energy is passed down the cascade cannot exist in two-dimensional flow. Braiding, on the other hand, while involving depth variation, is basically a two-dimensional, planform instability (although the flow is strongly three-dimensional in places). The other difference is in the nature of the parameters governing stability in the two systems. For turbulence, the stability parameter is the Reynolds number (*Re*). The only mechanism available in a flowing fluid to dissipate energy to heat is viscous dissipation; this occurs at the classical Kolmogorov fine-scales, in which the length and velocity scales of the dissipative motion have an *Re* of order one. The fact that *Re* includes an external constant (the viscosity) provides a basis for setting absolute (dimensional) values for fine structure of turbulence.

The stability conditions governing braiding are more complex. In general, the main parameter governing stability of a straight channel is the channel aspect (width-to-depth) ratio A (Figure 33.1; note that the weaker effect of relative roughness is ignored). The initial instability that develops as A exceeds its critical value is alternate bars. From a theoretical point of view, it is generally assumed that the alternate-bar instability implies a meandering planform and in many natural rivers it probably does. Nonetheless, it has been pointed out that this is strongly mediated by bank stability (Ferguson, 1987); in the absence of some form of bank cohesion, a channel with alternate bars will eventually break up into a weakly but unmistakably braided pattern. This has been observed experimentally (Schumm *et al.*, 1987), as well as in our model runs (Murray & Paola, 1994). Thus, here it is

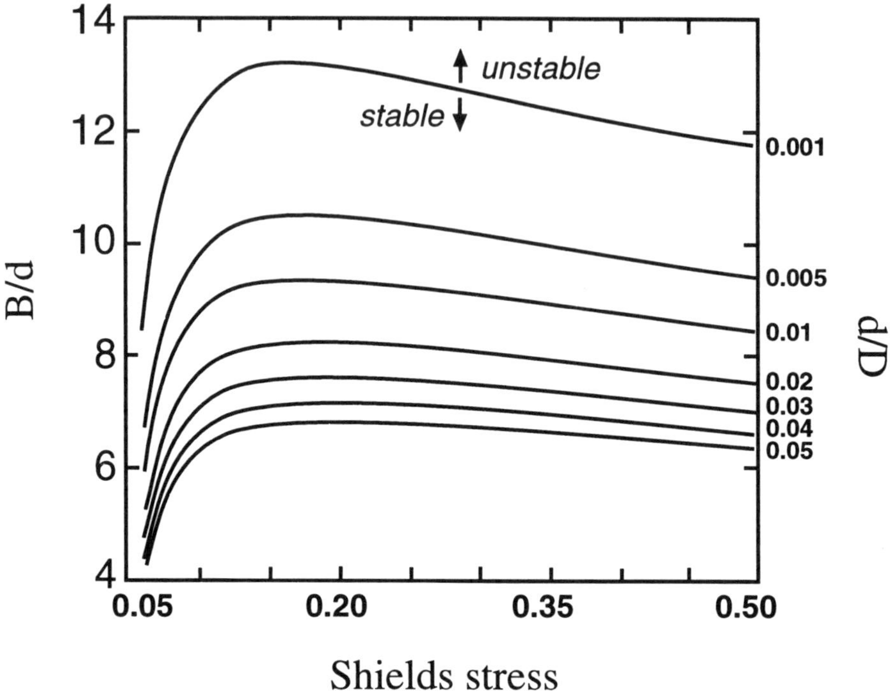

Figure 33.1. Stability criteria for formation of alternate bars. *B* is channel half-width, *d* is centreline depth and *D* is grain size. After Seminara & Tubino (1989)

assumed as a first approximation that *any* channel instability in a cohesionless sediment will lead to braiding and so the marginal stability criterion is that shown for alternate bars in Figure 33.1, (note that the reasoning would not be substantially altered by using a larger critical value for A). Unlike the Reynolds number, the aspect ratio A is purely geometric, therefore, it is possible for a channel of *any* width to satisfy the marginal stability condition. Hence, one of the basic causes for the scale hierarchy in turbulence is eliminated – whereas only eddies with small length and velocity scales can satisfy the marginal stability condition $Re \sim 1$, channels of any scale can satisfy the condition of marginal stability. Observationally, there does seem to be an identifiable fine-scale in braided streams, but at this point it is not clear what defines this scale.

A SPATIALLY-AVERAGED MODEL FOR BRAIDED STREAMS

Despite its limitations, the comparison between braiding and turbulence outlined above suggests that one might gain insight into the overall dynamics of braided

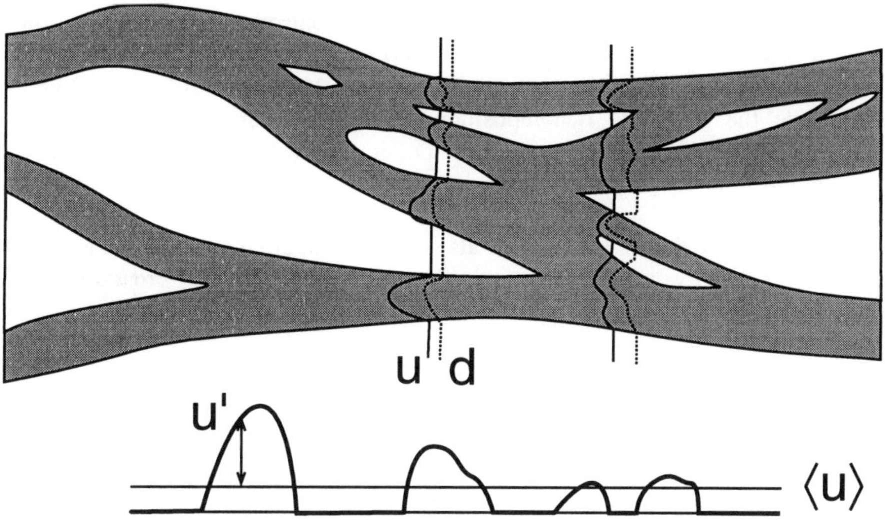

Figure 33.2. Definition sketch a braided river section showing mean (angle bracketed) and fluctuating (primed) quantities. Averaging includes only wetted areas, shown in grey

streams by applying some of the techniques developed in turbulence theory to braided streams as a whole. Consider a section of a braided stream as shown in Figure 33.2. As in classical Reynolds averaging, our first step is to define an averaging area that is large enough to provide meaningful average values but is still small compared to the overall length of the system. Assuming that lateral variations within the braid plain are not of interest, we will take the width of the plain as our averaging scale. Our final equations will thus be one-dimensional in x (where x is the distance downstream). We will also neglect the effects of temporal variation in the analysis. Temporal variation arises both from the intrinsic dynamics of the system (e.g. shifting of flow from one channel to another) and from external causes (e.g. changes in discharge). In a flow that is macroscopically steady and statistically uniform (stationary), spatial averaging should be equivalent to averaging the intrinsic fluctuations (as is the case for turbulence). We do not attempt to deal here with externally imposed time variations.

Unlike a turbulent flow, in a braided river the flow over which we are trying to form averages occupies only part of the averaging domain. Defining A_w as the wetted part of the averaging area A_{avg} (shown in grey in Figure 33.2), we define an averaged quantity $\langle f \rangle$ as:

$$\langle f \rangle = \frac{1}{A_w} \int_{A_w} f(x, y) \mathrm{d}A \qquad (33.1)$$

where y is the distance normal to the bed.

It is important to note that because A_w is in part controlled by the distribution of flow, it is possible that, if A_{avg} is chosen too small, A_w could vary with x. This issue does not arise in classical Reynolds averaging. Variation in A_w would greatly complicate the Reynolds decomposition of the equations of motion because averaging and differentiation could no longer be freely interchanged. We will assume that the length scale L_{avg} defining A_{avg} is large enough that A_w is independent of x.

For the sake of simplicity, we will also assume that at length scales greater than L_{avg}, the sinuosity is small enough that we can ignore the cross-stream (z) momentum balance. The x component of the steady St. Venant equation is:

$$du\frac{\partial u}{\partial x} + dw\frac{\partial u}{\partial y} = -gd\frac{\partial(\eta + d)}{\partial x} - \frac{\tau_0}{\rho} \tag{33.2}$$

where u and w are local velocities in the x and z directions respectively, d is depth, g is the gravitational acceleration, η is the bed elevation, and τ_0 is the boundary shear stress.

Conservation of mass for the fluid (continuity) is expressed by

$$\frac{\partial(ud)}{\partial x} + \frac{\partial(wd)}{\partial z} = 0 \tag{33.3}$$

The standard Reynolds decomposition is to split each fluctuating quantity into its mean and fluctuating part, i.e. $f = \langle F\rangle + f'$ (Figure 33.2). With the aid of Equation 33.3, we can apply this decomposition to Equation 33.2 and then average the entire equation to obtain:

$$\frac{\partial(R_u + gR_d)}{\partial x} = g\left\langle \eta' \frac{\partial d'}{\partial x}\right\rangle + gS\langle d\rangle - \frac{\langle \tau_0\rangle}{\rho} \tag{33.4}$$

where

$$R_u = \langle u\rangle^2\langle d\rangle + 2\langle u\rangle\langle u' d'\rangle + \langle u'\rangle^2\langle d\rangle + \langle u'^2 d\rangle$$

$$R_d = \frac{1}{2}(\langle d\rangle^2 + \langle d'^2\rangle + \langle d' \eta'\rangle)$$

S is the spatially averaged slope $-\partial\langle\eta\rangle/\partial x$ and ρ is the fluid density. Note that we have assumed that no water flows into or out of the system from the sides.

Although a braided stream is only likely to be locally uniform for very short distances, it is reasonable to assume a state of statistical uniformity (stationarity) on length scales larger than the averaging length. In this case the term on the left disappears. However, the term $\langle\eta'\partial d'/\partial x\rangle$ is a new correlation term that remains in the overall momentum balance even for the case of statistically uniform flow. It represents a correlation between rate of change of local flow depth and local topographic height (Figure 33.3). In the form $\rho g\langle\eta'\partial d'/\partial x\rangle$, it has the dimensions of a stress and thus could be thought of as analogous to a Reynolds stress. However,

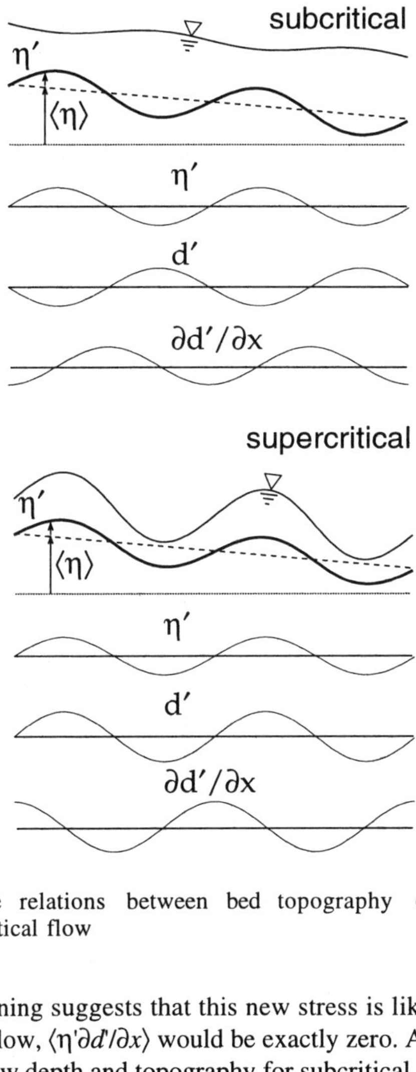

Figure 33.3. Phase relations between bed topography (η) and depth (*d*) for subcritical and supercritical flow

simple physical reasoning suggests that this new stress is likely to be small. For a rigid-lid free surface flow, $\langle \eta' \partial d'/\partial x \rangle$ would be exactly zero. At first order, the phase difference between flow depth and topography for subcritical flow is 180°, while for supercritical flow the phase difference is zero. For either of these cases the correlation $\langle \eta' \partial d'/\partial x \rangle$ is also zero (Figure 33.3). Hence, non-zero values of this correlation term would arise from non-linear effects that would cause the mean phase between the local flow depth and topography to be other than 0° or 180°. For present purposes, we will assume that first-order effects dominate and therefore that $\langle \eta' \partial d'/\partial x \rangle \cong 0$. However, this issue is certainly worth investigating further.

With this assumption we recover, in spatially averaged form, the standard long-wavelength momentum balance for a statistically stationary open-channel flow:

$$\langle \tau_0 \rangle = \rho g S \langle d \rangle \tag{33.5}$$

Again assuming that no water is added to the system from the sides, Equation 33.3 can be averaged as

$$Q = W\langle q\rangle = W\langle ud\rangle = W(\langle u\rangle\langle d\rangle + \langle u'\,d'\rangle) \tag{33.6}$$

where Q is the total water discharge, q is unit water discharge and W is the wetted width of the cross-section, averaged over A_{avg}. W is to be determined as part of the solution. Note that we are assuming that the averaging area is large enough to encompass the intrinsic spatial variability in W that is a natural part of braided stream dynamics (Murray & Paola, 1996).

Now we turn to the sediment dynamics. Ignoring the effects of local variations in slope, we assume that the sediment transport is related to the shear stress via the Meyer-Peter & Müller relation:

$$q_s = \frac{8}{g(s-1)}\left[\frac{\tau_0 - \tau_{0c}}{\rho}\right]^{3/2} \tag{33.7}$$

where q_s is the unit sediment flux by volume, s the sediment specific gravity, τ_0 the boundary shear stress, τ_{0c} the critical stress for initial motion and ρ the fluid density. From the point of view of sediment transport, the most important effect of the channel dynamics of the braided system is to produce fluctuations in the shear stress. In the absence of any detailed understanding of how to model the stress distribution, we will see how far we can get by assuming a simple form. The two-parameter gamma probability density function f_γ is given by:

$$f_\gamma(\tau_0) = \frac{\alpha^\alpha \tau_*^{\alpha-1} e^{-\alpha\tau_*}}{\langle\tau_0\rangle\Gamma(\alpha)} \tag{33.8}$$

where $\tau_* = \tau_0/\langle\tau_0\rangle$, α is a non-dimensional measure of the width of the distribution, equal to the inverse square of the coefficient of variation (standard deviation/mean) and $\Gamma(\alpha)$ represents the standard gamma function. In the absence of any direct measurements of the overall distribution of shear stress in a braided river, we can estimate the distribution approximately using the invaluable data set assembled by Mosley (1982) for the braided Ohau River, New Zealand. The river discharge was set to a series of increasing values, each of which was maintained for a week while a research team made exhaustive measurements of depth and velocity through the system. These measurements allow us to estimate the complete joint distribution of depth and velocity over a series of discharges. Here we focus attention on the discharge Q (240 $m^3\ s^{-1}$) nearest the mean annual flood (280 $m^3\ s^{-1}$) and the two nearest values (104.5 and 506.5 $m^3\ s^{-1}$). We can estimate the shape of the stress distribution from the velocity distribution assuming that the stress locally satisfies a standard drag law

$$\tau_0 = \rho c_f u^2 \tag{33.9}$$

where c_f is the drag coefficient and is taken to be constant. Note that the assumption of constant c_f, made because we have no other way of constraining the stress distribution, should have the effect of causing our estimate of the variance in stress to be slightly high, since deeper flows, which generally produce the highest stresses, would be expected to have relatively lower relative roughnesses and hence lower drag coefficients. Plots of estimated probability density from the Ohau data are shown in Figure 33.4 along with a normalized gamma distribution with a value of $\alpha = 1.14$, the mean of the values estimated for the three discharges (Figure 33.5). There is a weak tendency for α to increase with discharge: from 0.99 at $Q = 104.5\ m^3\ s^{-1}$ to 1.25 at $Q = 506.5\ m^3\ s^{-1}$.

The parameter α sets the variance in the distribution relative to the mean; we will constrain the mean using physical arguments based on the analysis of Parker (1978). Sediment particles on a surface sloping laterally with respect to the local

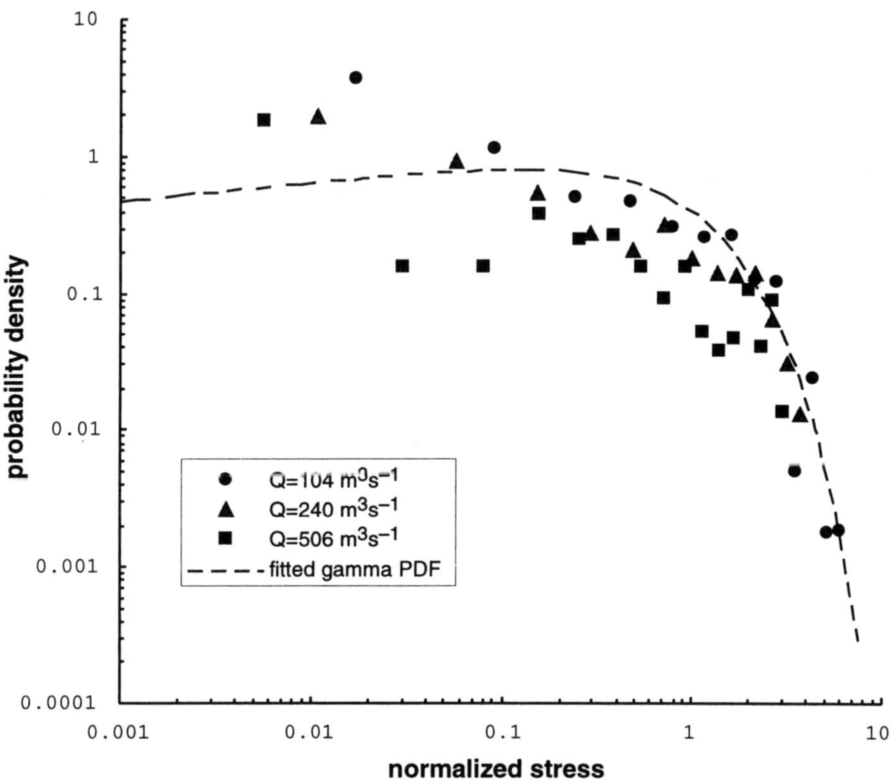

Figure 33.4. Estimated probability-density functions (PDF) for normalized stress $\tau_0/\langle\tau_0\rangle$ for three discharges Q from the Ohau River data of Mosley (1982), along with a two-parameter gamma PDF with variance parameter $\alpha = 1.14$

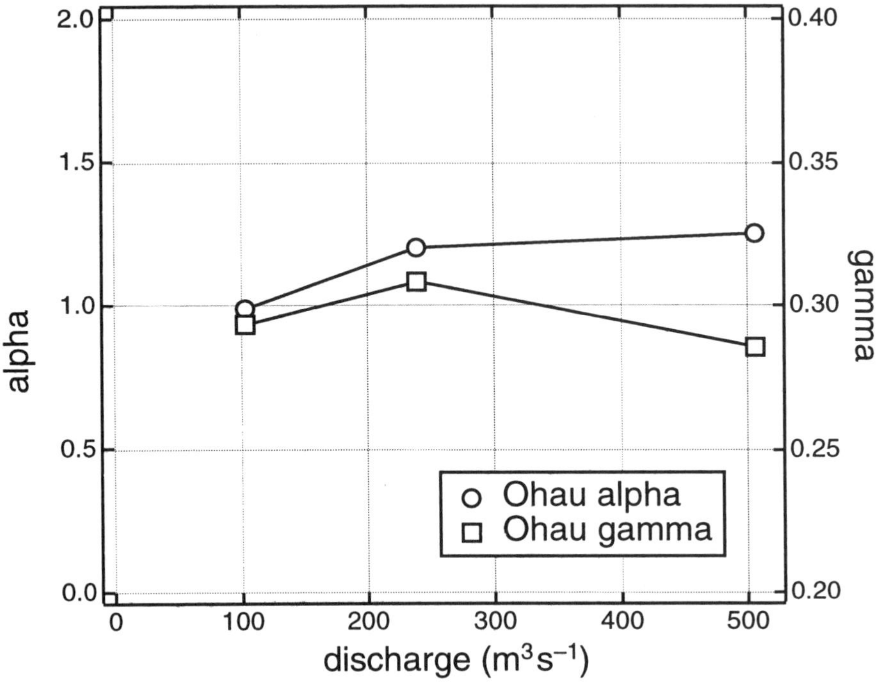

Figure 33.5. Estimated values of stress-variance parameter α and velocity-depth correlation parameter γ for three values of discharge. Estimated from Ohau River data of Mosley (1982)

flow direction cannot be transported downstream without developing a component of lateral transport as well. Thus, any transport occurring on a stream bank leads to removal of material and hence bank erosion. Parker (1978) suggested that, for bedload-dominated non-cohesive channels, the equilibrium channel width is set by the condition that lateral drag from the banks be just sufficient to bring the stress on the sloping bank region of the channel below the critical value for initial motion τ_{0c}. Parker showed that this condition implies that the shear stress in the channel centre is limited to

$$\tau_0 = (1 + \varepsilon)\tau_{0c} \tag{33.10}$$

where ε has a theoretical value of about 0.2. If the stress does rise above this value, the higher stress induces bank erosion that widens the flow, reduces flow depth, and thus reduces the stress. In this sense, an equilibrium channel is self-regulating.

In natural rivers, conditions of bedload-dominated transport and non-cohesive banks would be expected to lead to braiding. The observation that braided systems with active sediment transport continuously evolve demonstrates that Parker's equilibrium-channel hypothesis could not apply locally in every channel.

Nonetheless, it is worth asking whether the average conditions over an ensemble of channels might not approach the equilibrium condition, at least in the limit of low sediment supply. In that case, the mean shear stress is set by

$$\langle\tau_0\rangle = (1+\varepsilon)\tau_{0c} \qquad (33.11)$$

where the critical shear stress τ_{0c} is based on the reach-averaged median grain size. A simple test of this hypothesis is shown in Figure 33.6, which was produced with the invaluable data compilation of Church & Rood (1983). It shows surface slope predicted from Equations 33.5 and 33.10 compared with measured slope. The compilation includes, along with grain size, depth (in many cases for a range of discharges) and slope, information on planform and behavioural trends. To obtain the most data for the comparison whilst still satisfying the basic assumptions, the rivers shown are all those in the compilation with non-cohesive bed and banks that either (i) have plan forms described as anastomosed to braided, or (ii) have 'split-channel' planforms but are described as being laterally unstable, aggrading, distributary or torrents. The geometric mean ratio of predicted to measured slope in Figure 33.6 is 0.85; the best-fit value of ε in Equation 33.10 is 0.4. This is remarkably close to the theoretical value, especially since we have not included

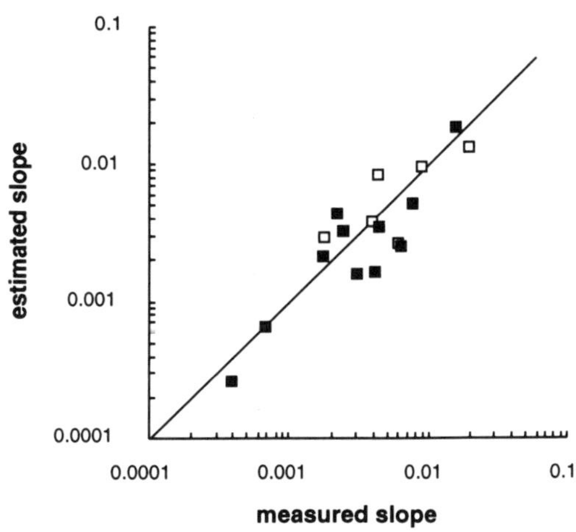

Figure 33.6. Slopes estimated from Equations 33.4 and 33.11 using a critical Shields stress of 0.045 and $\varepsilon = 0.2$. Data from the compilation of Church & Rood (1983) were selected as follows: only gravel-bed rivers with non-cohesive bed and banks were considered. Where available two-year flood data were used; otherwise bankfull flow referred to active floodplain. Filled points are for planform types 'anastomosed' and higher (i.e. more braided), including stable channels; open points are for planform types 'split-channel' and higher but restricted to channel types 'laterally unstable', 'aggrading', 'distributary' and 'torrent'. Duplicate points between the two data sets have been removed

form drag on bars in using Equation 33.10 to estimate the mean stress which could easily account for most of this discrepancy (Parker & Peterson, 1980).

At this point the shear stress distribution is completely constrained, although we have no information on the extent to which α can be expected to remain constant for different braided rivers. The spatially averaged sediment flux is then given by

$$\langle q_s \rangle = \int_{\tau_{0c}}^{\infty} \frac{8(\tau_0 - \tau_{0c})^{3/2} f_\gamma(\tau_0)}{g(s-1)} \mathrm{d}\tau_0 \qquad (33.12)$$

and the spatially averaged equation of conservation of mass is

$$\frac{\partial \langle \eta \rangle}{\partial t} = \frac{-1}{C_0} \frac{\partial (I \langle q_s \rangle)}{\partial x} \qquad (33.13)$$

where C_0 is the volume concentration of sediment in the bed and I is the wetted fraction of the alluvial plain.

The main remaining step is to constrain the fluctuating part of the equation of fluid continuity Equation 33.6. In turbulence one generally does not expect the correlation terms to be scaled by their mean-flow counterparts, since many of them (the Reynolds shear stress, for instance) can be large even when one of the corresponding components of the mean flow is zero. However, in the present case this is not true. Hence, we examine the hypothesis that

$$\langle u' d' \rangle = \gamma \langle u \rangle \langle d \rangle \qquad (33.14)$$

where γ is a non-dimensional correlation coefficient. Values of γ can be determined from the Ohau-River data obtained by Mosley and used in the previous section. The joint probability density function for velocity and depth for the intermediate discharge $Q = 240\ \mathrm{m^3\ s^{-1}}$ is shown in Figure 33.7; as would be expected velocity and depth are strongly correlated. This gives rise to non-zero values of γ. For the same three discharges used above, we obtain values of γ of 0.29–0.31 (Figure 33.5). There is no indication of systematic variation of γ with Q, therefore, γ is taken to be constant at 0.3.

We now have all the pieces we need to put together a complete first-pass theory for the averaged dynamics of a braided river system. The spatially averaged sediment flux becomes

$$\langle q_s \rangle = \frac{8\tau_{0c}^{3/2} K_1}{g(s-1)} \qquad (33.15a)$$

where K_1 is an integral function of α and $\tau_{c*} = \tau_c / \langle \tau \rangle$:

$$K_1 = \frac{\alpha^\alpha}{\Gamma(\alpha)} \int_{\tau_{c*}}^{\infty} \left[\frac{\tau_*}{\tau_{c*}} - 1 \right]^{3/2} \tau_*^{\alpha-1} e^{-\alpha\tau_*} \mathrm{d}\tau_* \qquad (33.15b)$$

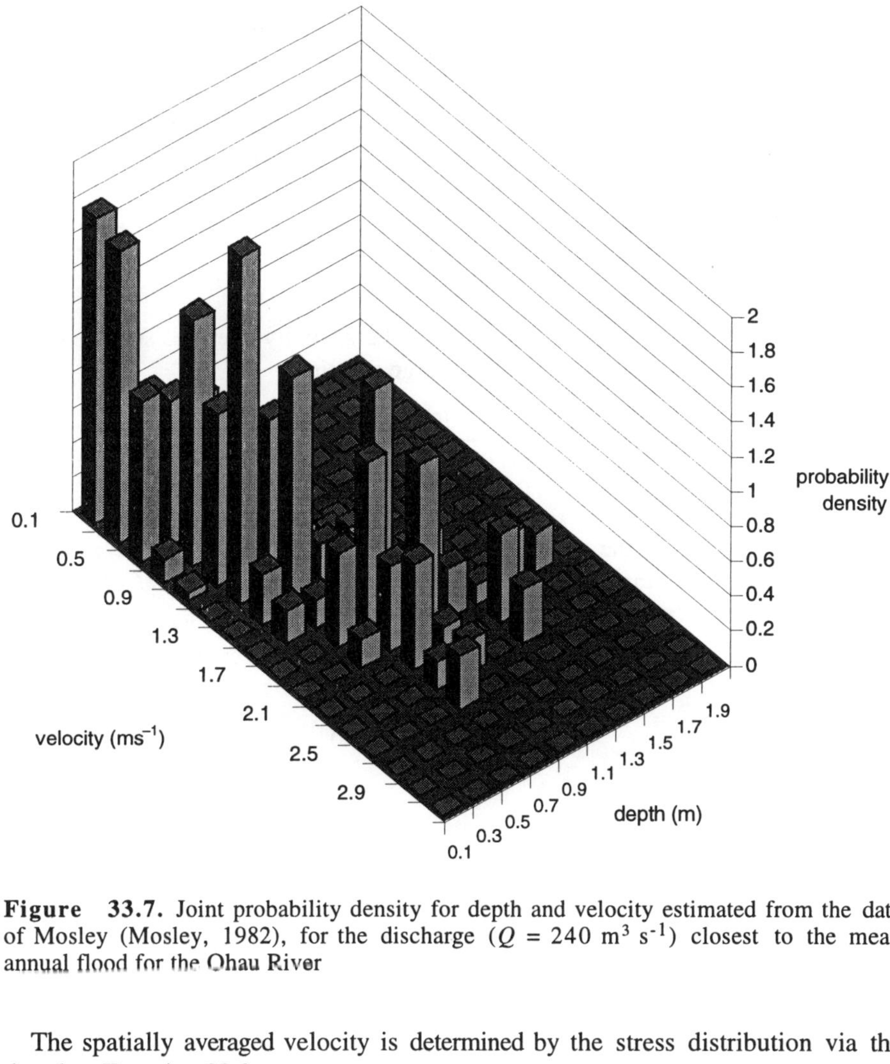

Figure 33.7. Joint probability density for depth and velocity estimated from the data of Mosley (Mosley, 1982), for the discharge (Q = 240 m^3 s^{-1}) closest to the mean annual flood for the Ohau River

The spatially averaged velocity is determined by the stress distribution via the drag law Equation 33.9:

$$\langle u \rangle = K_2 \sqrt{\langle \tau_0 \rangle / c_f} \tag{33.16a}$$

where

$$K_2 = \frac{\alpha^\alpha}{\Gamma(\alpha)} \int_0^\infty \tau_*^{(\alpha-0.5)} e^{-\alpha \tau_*} d\tau_* \tag{33.16b}$$

This equation can be viewed as defining an overall 'effective' drag coefficient for the flow given by c_f / K_2^2. For the value of α estimated above for the Ohau River

(1.14), $K_2 = 0.90$, so the effective drag coefficient is about 1.24 times the local value. This is due solely to averaging over the non-linear drag relation Equation 33.9 and does not include any additional effects of form drag around bars.

Combining Equation 33.16 with the overall momentum balance Equation 33.5, conservation of mass for water Equation 33.6 and our mean-stress hypothesis (Equation 33.11) we find a relation for the fractional width I:

$$I = \frac{\langle q \rangle g \sqrt{c_f}}{K_2(1+\gamma)[(1+\varepsilon)\tau_{0c}]^{3/2}} \frac{\partial \langle \eta \rangle}{\partial x} \tag{33.17}$$

Finally, this relation allows us to solve for the spatially averaged sediment flux $\langle q_s \rangle$ in terms of the spatially averaged discharge and slope:

$$\langle q_s \rangle = \frac{-8K_1\sqrt{c_f}\langle q \rangle}{K_2(s-1)(1+\gamma)(1+\varepsilon)^{3/2}} \frac{\partial \langle \eta \rangle}{\partial x} \tag{33.18}$$

Note that combination of Equation 33.18 with the sediment mass balance Equation 33.13 leads to the diffusion equation for the evolution of the sediment surface η. This derivation is analogous to that presented in Paola *et al.* (1992), except that their derivation did not include any of the effects of fluctuations discussed here. The effective diffusion coefficient ν is given by

$$\nu = \frac{8K_1\sqrt{c_f}\langle q \rangle}{K_2(s-1)(1+\gamma)(1+\varepsilon)^{3/2}} \tag{33.19}$$

TEST OF THE MODEL

There are two relatively simple tests we can apply to the model to see if it can predict the spatially averaged properties of braided stream systems: Equation 33.18 allows us to find the spatially averaged sediment flux in terms of slope and water discharge (or, equivalently, to find the slope needed to transport a given imposed sediment flux and water discharge), and Equation 33.17 allows us to find the total wetted width of the stream system in terms of the slope and water discharge. These tests require as parameters α, γ, and the stress ratio τ_{c*}. A data set providing all the necessary observations is that of Ashmore (1985). These data were measured in a laboratory scale model braided river (with a median grain diameter (D_{50}) of 1.16 mm and a slope range 0.01–0.015).

Seal & Paola (1995) found that application of the Ohau River stress distribution to a river with a much wider range of grain sizes led to unreasonably low probabilities for moving sizes in the coarse tail such as D_{90}, and hence argued that the best way to scale the stress distribution was to require that the probability of moving sizes coarser than a reference size such as D_{90} remain the same from system to system. This means that α is determined by the ratio D_{90}/D_{50}. For the

Ohau, $D_{90}/D_{50} = 4.5$ (bulk size distribution; Mosley, 1982) while for Ashmore's experiments, $D_{90}/D_{50} = 2.4$. Applying the same reasoning here, the scaled value of α that produces the same relative probability of moving sizes greater than D_{90} in Ashmore's experiments as in the Ohau is 4.0 (i.e. the stress PDF is about half as wide in Ashmore's experiments as in the Ohau River).

Sediment discharges for Ashmore's experiments predicted from Equation 33.18 with $\alpha = 4.0$, $\gamma = 0.3$, and $\varepsilon = 0.2$ are shown in Figure 33.8. Also shown for comparison are discharges predicted using the theory above with no stress variations (i.e. for limit $\alpha \rightarrow \infty$), to give some idea of the relative importance of stress variance in determining the overall sediment flux. To the extent that the tail of

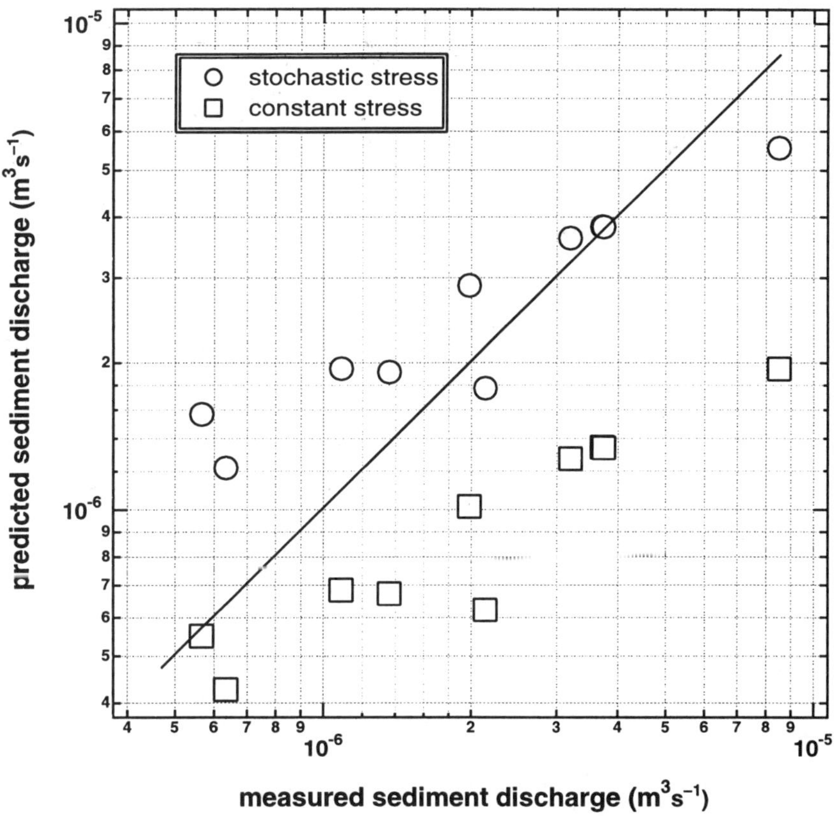

Figure 33.8. Comparison of predicted and measured sediment discharge for the experimental data of Ashmore (1985). Stochastic stress estimates were derived from Equation 33.18 using parameter values $\alpha = 1.14$, $\gamma = 0.3$, and $\tau_{c*} = 1/1.2$; the constant stress values were estimated using the same theory for zero stress variance ($\alpha \rightarrow \infty$, $\gamma = 0$)

high-stress events is produced mainly by confluences, this comparison can be viewed as indicating the net effect of confluences on the overall sediment flux. Evidently it is fairly important since the theory using a realistic stress variance generally provides substantially better predictions, although it tends to over-predict somewhat. Without the variance the predicted transport rates are consistently too low. These trends are clearest for high sediment fluxes which makes sense if the production rate of confluences increases with higher rates of sediment transport and channel activity.

A more sensitive test is obtained using Equations 33.17 and 33.18 to predict the total width needed to pass the measured sediment flux through the system. These results are shown in Figure 33.9. Consistent with the results in Figure 33.8, the theory including variance gives much better agreement with the observations than

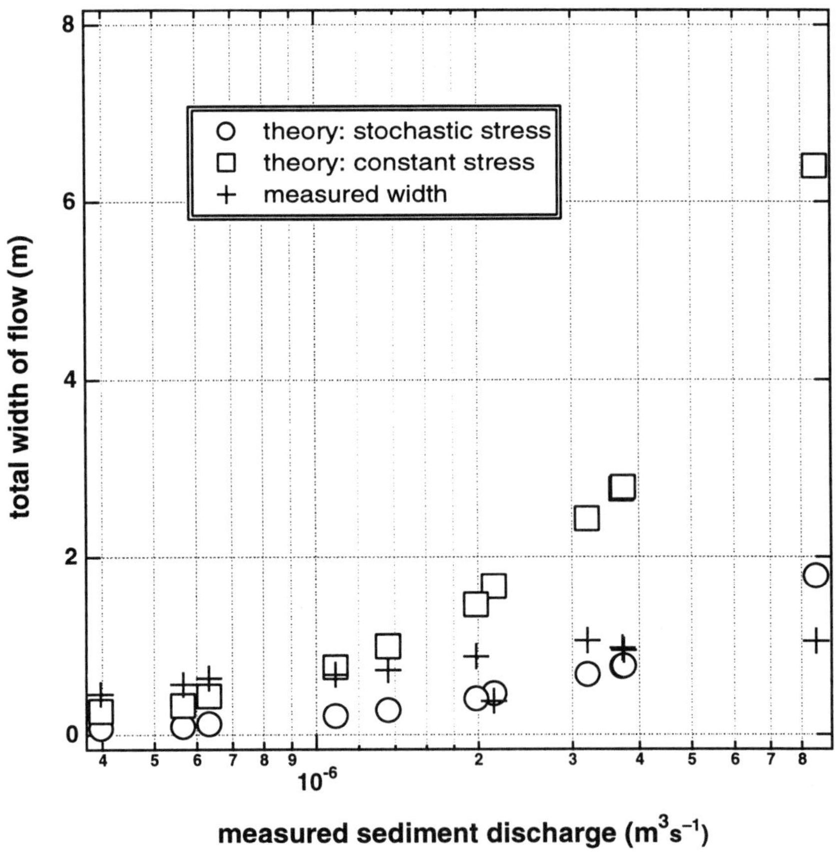

Figure 33.9. Comparison of predicted and measured total flow width for the experimental data of Ashmore (Ashmore, 1985), using Equation 33.15. Parameter values as for Figure 33.8

the zero-variance case, particularly for the higher sediment fluxes. These two results suggest that the total stress variance in Ashmore's model experiments is indeed smaller than in the Ohau River. Use of the unscaled stress variance parameter α leads to over-prediction of the sediment flux and under-prediction of the total width.

CONCLUDING COMMENTS

The results presented here suggest that the understanding of braided rivers has advanced to the point where it is possible to construct a complete, although crude, theory of their overall (spatially averaged) behaviour. This paper examines the fundamental problem of setting the flow width by imposing a shear stress distribution. The crucial question which can be answered only by developing a better understanding of channel-network dynamics, is how this stress distribution behaves in general, especially for systems driven far from equilibrium by, for instance, high imposed sediment loads. However, there is much to do even for the relatively simple near-equilibrium case. In particular, this paper demonstrates that that it would be impossible even to begin to constrain the parameters needed to describe a braided stream system in stochastic terms without the data sets like that measured by Mosley (1982) and his co-workers in New Zealand. If nothing else, it is hoped that the analysis presented here makes clear the fundamental utility of system-scale data sets for braided river networks. More such data are needed, both from laboratory and field studies. Among other things more system-scale data and improved averaged theories would help provide a context for integrating much of the excellent work that has appeared in recent years on flow and sediment dynamics of specific structures (e.g. bars and confluences) within braided river systems.

ACKNOWLEDGEMENTS

I would like to thank Gary Parker, David Mohrig and Kelin Whipple for many helpful and stimulating discussions of the ideas discussed in this paper, and Rob Ferguson and John Bridge for useful reviews. Financial support was provided by the US National Science Foundation, grant NSF/DMS 93-16078.

REFERENCES

Ashmore, P.E. 1985. *Process and Form in Gravel Braided Streams: Laboratory Modelling and Field Observations*, Unpub. PhD Thesis, University of Alberta, 414 pp.

Bergé, P., Pomeau, Y. & Vidal, C. 1984. *Order Within Chaos: Towards a Deterministic Approach to Turbulence*, John Wiley & Sons, New York, 326 pp.

Church, M. & Rood, K. 1983. *Catalogue of Alluvial River Channel Regime Data*, Department of Geography, University British Columbia, Vancouver, 99 pp.

Ferguson, R. 1987. 'Hydraulic and sedimentary controls of channel pattern', in Richards K.S. (Ed) *River Channels: Environment and Process*, The Institute of British Geographers, Sp. Pub. Series, **18**, 129–158.

Mosley, M.P. 1982. 'Analysis of the effect of changing discharge on channel morphology and instream uses in a braided river, Ohau River, New Zealand', *Water Resour. Res.*, **18**, 800–812.

Murray, A.B. & Paola, C. 1994. 'A cellular model of braided rivers', *Nature*, **371**, 54–57.

Murray, A.B. & Paola, C. 1996. 'A new quantitative test of geomorphic models, applied to a model of braided streams', *Water Resour. Res.*, in press.

Paola, C., Heller, P.L. & Angevine, C.L. 1992. 'The large-scale dynamics of grain-size variation in alluvial basins, 1: Theory', *Basin Res.*, **4**, 73–90.

Parker, G. 1978. 'Self-formed straight rivers with equilibrium banks and mobile bed. Part 2. The gravel river', *J. Fluid Mech.*, **89**, 127–146.

Parker, G. & Peterson, A.W. 1980. 'Bar resistance of gravel-bed streams', *J. Hydraul. Div., ASCE*, **106**, 1559–1575.

Peinke, J., Klein, M., Kittel, A., Okninsky, A., Parisi, J. & Roessler, O.E. 1993. 'On Chaos, Fractals and Turbulence', *Physica Scripta*, **T49B**, 672–676.

Schumm, S.A., Mosley, M.P. & Weaver, W.E. 1987. *Experimental Fluvial Geomorphology*, John Wiley & Sons, New York, 413 pp.

Seal, R. & Paola, C. 1995. 'Observations of downstream fining on the North Fork Toutle River, Mt. St. Helens, Washington', *Water Resour. Res.*, **31**, 1409–1419.

Seminara, G. & Tubino, M. 1989. 'Alternate bars and meandering: free, forced and mixed interactions', in Ikeda, S. & Parker, G. (Eds) *River Meandering, Water Resources Monograph*, **12**, American Geophysical Union, 267–320.

Smith, N. D. 1974. 'Sedimentology and bar formation in the upper Kicking Horse river, a braided outwash stream', *J. Geology*, **82**, 205–223.

NOTATION

A	width-to-depth ratio
A_{avg}	averaging area
A_w	wetted part of the averaging area
B	channel half-width
c_f	drag coefficient
C_0	volume concentration of sediment in the bed
d	depth
D_{50} D_{90}	50th and 90th percentile of the grain-size distribution
f	averaged quantity
f_γ	two-parameter gamma probability density function
g	gravitational acceleration
I	wetted fraction of the alluvial plain
K_1	integral function of α and $\tau_{c*} = \tau_c/\langle\tau\rangle$
K_2	integral correction to spatially averaged drag coefficient
L_{avg}	length scale
q	unit water discharge by volume
q_s	unit sediment flux by volume

R_d	momentum correlation term
R_u	depth-topography correlation term
Re	Reynolds number
Q	total water discharge
s	sediment specific gravity
S	spatially-averaged slope
W	wetted width of the cross-section, averaged over A_{avg}.
x	distance downstream
y	distance normal to channel bed
z	distance cross-stream
α	non-dimensional measure of the width of the distribution
γ	non-dimensional correlation coefficient
Γ	standard gamma function
ε	non-dimensional excess shear stress
η	bed elevation
ν	effective diffusion coefficient
ρ	fluid density
τ_0	boundary shear stress
τ_{0c}	critical stress for initial motion
τ_*	$\tau_0/\langle\tau_0\rangle$
'	indicates instantaneous quantities
$\langle\ \rangle$	spatially averaged quantities

Index

Index compiled by Geoffrey Jones